Origin of Group Identity

Luis P. Villarreal

Origin of Group Identity

Viruses, Addiction and Cooperation

Luis P. Villarreal
University of California, Irvine
School of Biological Sciences
Viral Vector Facility for Gene Therapy
3221 McGaugh Hall
Irvine, CA 92697
USA
lpvillar@uci.edu

ISBN: 978-0-387-77997-3 e-ISBN: 978-0-387-77998-0
DOI: 10.1007/978-0-387-77998-0

Library of Congress Control Number: 2008934794

Printed on acid-free paper

springer.com

Preface

A sense of belonging is basic to the human experience. But in this, humans are not unique. Essentially all life, from bacteria to humans, have ways by which it determines which members belong and which do not. This is a basic cooperative nature of life I call group membership which is examined in this book. However, cooperation of living things is not easily accounted for by current theory of evolutionary biology and yet even viruses display group membership. That viruses have this feature would likely seem coincidental or irrelevant to most scientist as having any possible relationship to human group identity. Surely such simple molecular-based relationships between viruses are unrelated to the complex cognitive and emotional nature of human group membership. Yet viruses clearly affect bacterial group membership, which are the most diverse and abundant cellular life form on Earth and from which all life has evolved. Viruses are the most ancient, numerous and adaptable biological entities we know. And we have long recognized them for the harm and disease they can cause, and they have been responsible for the greatest numbers of human deaths. However, with the sequencing of entire genomes and more recently with the shotgun sequencings of habitats, we have come to realize viruses are the black hole of biology; a giant force that has until recently been largely unseen and historically ignored by evolutionary biology. Viruses not only can cause acute disease, but also persist as stable unseen agents in their host. In this, they attain stability in evolution. It is from such a persisting relationship that viruses can inform us regarding the strategies and mechanisms of group membership. In order to persist in their host, a virus must be able to resist both themselves and all other competitors. In this persistence, viruses were the first genetic elements to show a strategy called an addiction module. In addition, persistence often involves mixtures of viruses that can even include nonviable, defective, even 'dead' members which work cooperatively to attain fitness. With these characteristics, viruses can introduce into their host new genetic identities that create group identities and group immunity, including altruistic-like individual self-destruction used for the protection of the group. From this relationship, we can trace how genetic parasites and the strategy of addiction modules have contributed to the evolution of group identity, a pathway that leads us directly to humans. Addiction modules can compel cooperation and symbiosis between individuals. They are composed of an essential combination of harmful and protective components that must act together to specify and stabilize group identity. From this perspective of genetic parasites and addiction modules, we can identify events leading to the evolution of group identity (immunity) and group cooperation in all life. We see that the molecular basis of olfaction of small molecules (pheromones) may have stemmed from bacterial toxins–antitoxins but has long been adapted by the earliest animals to identify and learn group identity. Both invertebrate and vertebrate animals have mostly retained this method for specifying group membership. In mammals, the basic maternal–offspring social bond has also maintained an olfactory component. However, during their evolution, the African primates experienced a great endogenous virus invasion (ERV colonization) that led to the disruption of much of this olfaction-based group identification. This loss promoted the emergence of a highly visual-based (facial) social bonding and identity system seen only in the African primates amongst mammals. This also required specific brain adaptations to support such enhanced social (and color) vision. This ERV colonization also altered the relationship between African primates and their viruses such that colonization by exogenous retroviruses has continued to promote hominid and human evolution. However, only human hominids evolved an additional system of group identity based on sound that utilized the enlarged primate social brain in order to learn song and language. Language thus originated as a human-specific system of group identity. The emergence of language-based group identity placed further demands on the evolution of an even larger social brain of humans. However, for language to be used as a system of group identity, its learning must be stable, linked to human development and also linked to the relevant addiction modules which are mostly based on emotion. Thus the emergence of language-based group identity also required the emergence and evolution of stable belief states along with the neurological substrates for belief attribution that link audio learning to emotion for social purposes. This development essentially liberated the evolution of human group identity from dependence on genetic parasites, promoting cultural evolution. However, the fundamental strategy of information-based identity and emotional addiction have prevailed. With the emergence of language and the large social brain of humans, we also see the emergence of a new dynamic mental state that we call the mind which is mediated by internal (language based) dialogue. In modern humans, the mind, not the genome, became the substrate for learned (acquired) group identity, aided by the development of reading. It is with our minds, not our noses, that we learn to belong.

Irvine, CA, USA Luis P. Villarreal

Acknowledgments

This book is the product of decades of thought and investigation, but this was a line of thought that was seeded in the 1970s by my graduate mentor John J. Holland and later encouraged by my late friend and fellow virologist Edward K. Wagner. Without the early insight and unending encouragement of John Holland, I would have been unable to appreciate and pursue the fundamental significance of persisting viruses. Thus, this early interest in persistence has itself persisted during my career, in spite of the fact that its value to our understanding biology has long been ignored dismissed due mainly to our intense, although necessary, focus on virus disease. Edward Wagner himself had also long been interested in viral persistence with the human herpesviruses. This viral lineage can trace its roots to the most numerous phage of bacteria; so it is indeed ancient. Edward Wagner was one of the few brave individuals willing to read and seriously critique the early (and often highly redundant) drafts of my writing. Through his early comments and suggestions, I grew more willing to follow the chain of concepts and evidence that took many years to complete and led me into fields that would normally be considered to far out for the normally over-focused area of expertise typical of most current scientist. The years of finding and evaluating the relevant primary scientific observations that led to the thesis of this book, however, became self-catalyzed and sustained my effort.

The last chapter of this book concerns issues related to human learning. This chapter has also benefited from decades of my development and experimentation in science education. For this I must thank the many students I have taught over the years, both those that benefited from my efforts and those that did not, as I learn as much from failure as from success. Like most scientist, I long, but incorrectly, believed that science education should be a relatively rational process requiring mainly clear and organized explanations. However, my observations of students did not support this belief well, as I experienced a large variation in both successes and failures. Unlike my experience with science, experimentation with science education has been a more isolated activity. There were many more set beliefs to contend with but many fewer scientific resources (studies and data) to guide development. For this I thank NIH and NSF for grants supporting my extensive (but now limited) involvement in minority science education in the 1990s. Ironically, I came to realize that science education is not often done by a very scientific process. My own development as a graduate student under mentoring by John Holland, however, was a purposely unstructured but heuristic experience intended to develop independent (asocial) scientific thinking. This type of development seemed starkly at odds with the well-structured passive lecture-based process I had been adapted to and believed in as an undergraduate. When I became responsible for mentoring others, I started a long-term process of experimentation with student development. From this I concluded that problems in observation, in objective description and in the separation of these activities from thinking, interpretation and belief were prevalent. The language and the structure of thought in science was (and remains) a problem for most students. Later, I realized these were essentially the same problems that had historically inhibited the evolution of science itself. There exists a natural tendency to rationalize explanations and defend beliefs that reflects an inherent biological legacy of our social mind. Science thinking which is individual and critical is essentially an unnatural (asocial) skill.

As mentioned, some of my efforts in science education were supported by research, training and Presidential award grants (from NIH and NSF). I thank both these federal agencies for their support; I should also thank them for their lack of support on some issues. This will seem odd to most scientist. Grants, in general, are the coin of the realm for supporting science in the last century. But grants also have an undesirable feature regarding the development of new science. They inherently operate by criteria based on a strong social consensus, especially in times of limited funding, and not by some self-selecting feature that would be more likely to inform our social mind of new (non-consensus) thinking. In this, grants inherently tend to oppose new ideas and beliefs so we write them conservatively, tightly focused and with partially completed results. Indeed, we have no effective mechanism to review (or select) new ideas outside of this consensus process. I came to see the entire science enterprise as operating by the principles of an addiction module as outlined in this book. Our consensus-based funding compels stability of thought and maintains the direction or focus of research. Thus, in spite of my long-term interest and activity in the study of virus persistence, I have never had a grant on this topic, which meant I was free to follow the relevant evidence regardless of funding concerns. Getting grants to study virus disease is not too hard. However, if successful, many people (students, technicians, follows) come to personally depend on the ability of the PI in maintaining grants for their personal welfare. I realized years ago I would need to extricate myself from such a dependent state. Addiction modules are indeed effective strategies at maintaining the stability of thought, but science should seek alternative strategies that are less dependent on consensus to promote the development new effective ideas.

I also wish to acknowledge the excellent assistance in writing, compiling and organizing this book provided by Allison Kanas, the administrator for the Center for Virus Research at UCI. Her effort turned an otherwise unmanageable task into a simply arduous and laborious chore. Others have also provided valuable feedback on my writing, including Frank Ryan, Paulo Casali, George Striedter, David Fruman, Barbra Baker, Nancy Reich, Rowland Davis, Massimo Palmarini, Raphael Sagarin and Günther Witzany. I also wish to thank Dave Krueger for his excellent assistance in rendering numerous illustrations as well as the cover artwork.

Contents

Chapter 1
An Overview: Identity from Bacteria to Belief

Cooperation and group behavior is a key feature of human social interactions. It seems our large social brain is especially adapted to develop and support such social behaviors. When people are introduced to each other, they are typically interested to know various things, such as national origin (and language), cultural background, education and beliefs. These are all aspects of social identity that we have learned during our upbringing and socialization which are central to our lives. These are also basic elements of human social group identities. Such questions would typically be addressed during initial casual conversations between people who do not yet know each other. In the last chapter of this book, I will indeed consider and evaluate the issues of human social membership in some depth and how it relates to the biology of our social brain. However, human social groups are most puzzling from an evolutionary perspective. They are known for various features, such as cooperation and altruism which can be difficult to explain. Indeed, many have argued that the prevalence of such a human social feature presents a dilemma to Darwinian theory. This is because Darwinian theory emphasizes competition and selection of the individual as a main guiding force in evolution (not cooperation or group membership). And in order to account for altruistic or cooperative group behaviors, some indirect versions of competition and selection of an individual have been proposed (such as cost–benefit, game theory and kin selection) which provide a rationale for how a selfish individual can indirectly promote group cooperation and altruism. However, most human social structures are far too diluted for game theory to work well in such situations. Thus, in general, game theory and other related concepts work poorly when applied to extended but cooperative human group behavior. Yet various forms of group behavior can clearly be demonstrated in essentially all life forms. In contrast to humans, when dogs, our favorite companion, meet each other, they are especially interested to know the olfactory signatures of each other and this will be the subject of intense sniffing and interest. Such olfactory identities are also learned by most other animals (including most vertebrates, insects and nematodes). So the importance of identifying identity is crucial to all species. And when that identity is social (generally the case), dilemma posed to Darwinian theory appears to be broad, applicable to many life forms and not human specific.

L.P. Villarreal, *Origin of Group Identity*, DOI: 10.1007/978-0-387-77998-0_1,

In this book, I will focus on the issue of group identity and its evolution, which is presented as a fundamental topic relevant to all life. In essence, this book develops the theory behind the origin of group identity. It will also examine sexual identity and sexual behavior, when it exists in a particular species, as an element of group identity. Group identity will allow us to understand group cooperation and conflict. However, the molecular foundations that are applied for this development will be far from intuitive for most readers, and the evidence to be presented will likely be perceived to be dense for many readers. This is unavoidable due to the heavy dependence on supporting experimental evidence that has a highly specific character. In addition, how the evidence is organized may also add to its apparent density. In all cases, I will first examine the role of genetic parasites, such as viruses, in the evolution of group identity. This is an entirely new 'virus first' perspective and will provide the overall organization for all the chapters. The choice of this perspective stems directly from my prior book, 'Viruses and the Evolution of Life'. There, evidence was presented that supported the view that viruses and other genetic parasites are central participants in the evolution of all life and that they have a role in the biogenesis of life, not simply by providing negative selection. That book also introduced another concept that will be central for the evolution of group identity; that is the nature and role of addiction modules. Addiction modules were defined by persisting viruses of bacteria. In my prior book, I also generalized this concept to include a state of 'virus addiction' in which viruses or their defective elements reside in host and provide protection from related lytic viruses. This topic will also be further developed in the next chapter of this book and provides a central framework for understanding group identity. Addiction is also a process that favors symbiosis. Another fundamental lesson that viruses can provide us concerns the importance and molecular nature of cooperation. Viruses can evolve by mechanisms that involve quasi-species, large populations of genetically variable genomes in which there may be no 'fittest type'. In a quasi-species, the consortium is often the purveyor of fitness, and cooperation between otherwise unfit genomes has been established. This is a major distinction with Darwinian evolution that applies directly to the evolution of group identity. Thus, in this book, I present a genetic perspective and develop a mechanistic approach to understand the origin of group identity. Our recent dramatic advances in comparative genomics now provide vast amounts of information that allows us to re-examine how complexity has evolved and how it relates to group identity from the perspective of genetic parasites. Much genomic evidence will thus be considered. However, historically parasitic DNA has been thoroughly dismissed by evolutionary biologist as selfish DNA of no substantial significance to the evolution of complexity (i.e., genetic junk). Nor has this material thought to be of relevance to the evolution of group identity. However, in the context of genetic addiction modules, a different role for this DNA can now be proposed. All living things have ways to respond positively and negatively to both similar living things and other forms of life that are dissimilar. The pervasive recognition of biological similarity provides us with a definition of group identity. In some cases, the existence of group identity appears well established and unquestionable. For example, our bodies have adaptive immune systems that can clearly recognize self from non-self. This develops following a process we call 'immunological education'; hence the concept of 'education' as acquired group identity applies to cells. Our cells recognize groups of related but distinct cells that are part of our bodies (except during autoimmune disease). These are clearly cellular systems of group identity. Most clonal multicellular metazoan organisms must also have an ability to recognize self from non-self, although distinct mechanisms are often used. This capacity is often studied under the heading of immunity or cell biology. However, even prokaryotes, such as bacteria, have cellular mechanisms for recognizing themselves and other groups, not thought of as immunity. The increasing interest in bacterial communities (such as mat formation and quorum sensing) now provides many examples of the ability of bacteria to recognize related groups, and a resulting bacterial group behavior (such as motility) is clearly involved. However, even groups of individual metazoan organisms can and do exist in very large and extended group structures, such as algal blooms, shoals of vertebrate fish, flocks of birds or troupes of chimpanzees. Such groups are generally considered to have social or behavioral links that bind them together and provide their group identity. However, biochemical (pheromone) communication also applies to such extended groups. The concept of group identity thus defines a very broad issue relevant to all life and provides the central focus of this book. The concept of self and non-self identity and how individual identity relates to group identity is inherent to this issue. However, it will surprise most readers to learn that even viruses can and do have the capacity to recognize self from non-self and that selection for such capacity has been well maintained in evolution. Indeed, viruses appear to inform us of the most basal molecular strategies that define, promote and enforce group identity. It was not an accident of nature that viruses defined the addiction module. Nor is it an accident that our human adaptive immune system evolves mostly by the action of viruses. Viruses can depend on cooperative group phenomena for their highly efficient evolution. This establishes the rationale as to why it is crucial to consider first the viral examples. From this perspective, we can see the acquisition (colonization) of new genetic identity during evolution along with the frequent displacement of prior identity systems. In this, we see a strong inherent tendency to increase genetic complexity associated with the accumulation of displaced genetic parasites as the host evolves.

Evolutionary biologist have for the most part rejected the concept of group selection as lacking experimental support. And the reader may be starting to wonder if the concept of group identity, as I have outlined it, is equivalent to the concept of group selection. Groups do not appear to be under direct selection as we can find no particular host genes that show a group-related phenotype. Group behavior (which can readily be observed) has thus been explained by various indirect forms of 'selfish' behaviors that might generate group-like interactions. Yet 'group identity', as I define it, clearly exists and numerous examples will be presented in considerable detail. Viruses provide particularly strong examples of agents that can differentiate otherwise similar groups. But these differences may be epigenetic and may not relate to specific or variable host genes. Indeed the two concepts, group identity and group selection, are related but not equivalent. One major distinction is that group identity can be externally imposed, such as by persisting genetic parasites that need not colonize the genome. Such states can be experimentally established and are not simply theoretical. However, the resulting groups that are persistently colonized can become very harmful to competitors with different parasite identities. Such a state, however, is often extragenomic and frequently missed or ignored by field studies of species survival. Such studies seldom consider the possible involvement of viruses (genetic parasites) in group identity. But data that clearly support this idea can be found when sought. This book therefore develops the theme that extragenomic and genomic genetic parasites will often be involved in group identity and group behavior, and this analysis will rely heavily on molecular genetic evidence.

The overall pattern in this book will be to follow the evolution of group identity in all life that lie in the pathway from bacteria to humans (see Fig. 1.1). Group identity is predicted to have emerged along with the very earliest prebiotic replicators. Thus, I will trace the origin and evolution from such early replicators to virus and to bacteria then to populations of bacteria in the first three chapters. With virus, we can more clearly see the emergence of population-based group behavior and identity. We can now strongly support the assertion that a virus population (quasi-species) can be experimentally demonstrated to display

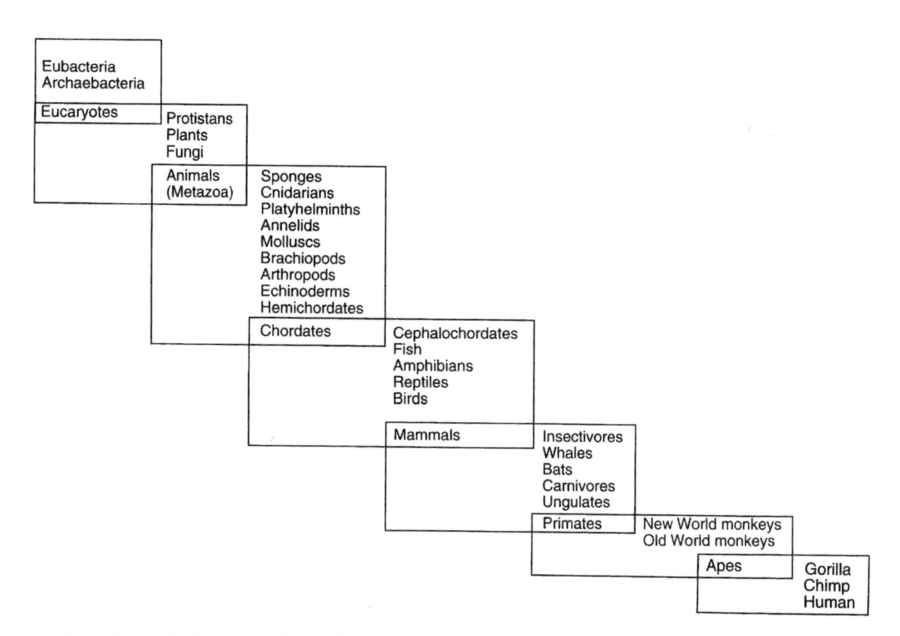

Fig. 1.1 The evolutionary pathway from bacteria (prokaryotes) leading to humans (reprinted with permission from: The Crucible of Creation: The Burgess Shale and the Rise of Animals, 1998)

cooperative and group selective properties (i.e., survival of a population, not the fittest type). This assertion will seem incorrect to most evolutionary biologists who have long thought of viruses as purely selfish, non-cooperating individual agents. But the concept of persistence by consortia of genetic parasites can oppose the view that evolution is strictly aimed at benefiting the individual. Persistence as a central concept provides us with a major clarification in our thinking. The early chapters of this book will therefore develop and present this central concept or theme in molecular detail. From this, the role of addiction modules in persistence and group identity will be made clear. Addiction modules promote both group immunity and identity. The initial concept of an addiction module was first proposed in the 1980s to explain human drug addiction, but was also adapted in 1991 to explain the stability of persisting extragenomic genetic parasites (P1 bacteriovirus) in *Escherichia coli*. An addiction module is a two-component molecular strategy that enforces the stability of the parasite by killing those host that lose the parasite, but preserving those that retain the parasite (see Chapter 2). However, as will be developed, an addiction module also inherently and simultaneously creates the circumstances both of immunity and of group identity by being active against other competing parasites. Thus, genetic parasites that have persistently colonized their host will often be involved in the origin and evolution of addiction modules and group identity. Addiction modules also explain an inherent tendency for group identities to depend on and use the destruction of an individual cell (apoptosis/altruism) in order to enforce or create extended cellular group identity. It is by considering the role of addiction modules that we can then trace the evolution of group identity in all life, including the highly social animals. This pathway eventually leads to human social group behavior (Chapter 9), although humans no longer require genetic addiction modules but instead depend on learned cognitive and emotionally mediated addiction modules for their group identity.

After presenting prokaryotes, the origin of eukaryotes, the evolution of metazoan, aqueous vertebrates, insects and terrestrial mammals group identity will be presented in subsequent chapters. In these metazoans, the emergence of the nervous system and its ability to control group and sexual behavior is initially presented. A special focus will be to trace the evolution of the sensory (mostly olfaction) detection systems, which provides the nervous system with the external sensory information needed for group behavior and identity. The molecular components of the olfactory system were first introduced in the bacterial quorum-sensing systems, which control bacterial motility behavior. Similar sensory mechanisms are adapted to odor and visual sensory detection in metazoans and later used as sensors by the central nervous systems to control group behavior. The emergence of a central nervous system with a capacity for sensory memory allowed the memory of sensory input to become stable and affect stereotypical emotional behaviors (and emotional memories). Thus, memory and emotions could now participate heavily in and compel group (and reproductive) identity. The stereotypical emotional behavior of fear, aggression, association and protection was thus developed with the origin of the central nervous system in sets that constituted addiction modules. Accordingly, with the origin of CNS, we also see the use of memory and behavior-based addiction modules that must affect emotions and provide group identity. Following the origin of the nervous system, I will then present the emergence of adaptive immune systems in vertebrates as a cellular-based system of group identity. Genetic parasites (especially endogenous retroviruses) were highly involved in the origin and emergence of this elaborate system.

Later in the book, mammals are presented which evolved a placenta and from this new organ developed a hormonal system that promoted extended social bonding between mother and offspring. This basic social bonding initially used immune olfactory markers of the adaptive immune system to help specify identity, in addition to the placental hormones. This maternal social bond was to provide the basal system and mechanism that was to evolve into the much more elaborate human social bonds. In mammals, there is strong and prevalent evidence of a role for genetic parasites in the evolution of placenta and social regulation (especially endogenous retroviruses). Indeed, the frequent involvement of ERVs and other genetic parasites in all systems of group identity will be closely evaluated in all chapters. The emergence of new systems of group identity and social bonding also defines key innovations for new orders of life. Thus, I will seek to understand how new molecular genetic identities are forced onto a host and how this process affects group identity. Another theme that will emerge is that of diversity. The diversity of identity systems is peculiar for and specific to the order of life being considered. Recently emerged or currently active identity systems will generally show great diversity due to intense competition and displacement of the related or similar strategies. Indeed, the existence of molecular diversity will be a diagnostic indicator of the currently active systems of group identity for any particular species. Hence, genomic data will provide excellent clues regarding the extant group identity systems that are currently active in a species.

Following the above themes leads us to some surprising topics at the end of this book. Genetic parasites (ERVs, LINEs) were especially active in the origin and evolution of primates. One unexpected consequence of such broad genomic activity was the genetic inactivation of much of olfactory (and immune) based group recognition, including its involvement in the basal maternal–offspring bond as seen in all other mammals. This olfactory-based system was ancient and had been preserved in aquatic vertebrates and all other terrestrial vertebrates. Its loss in African primates led to a compensatory and large expansion of brain-based visual capacity (including color) that was used for group recognition (especially face). This required special sensory, memory and emotional neurological adaptations (including mirror neurons). In the great apes, the maternal–offspring bond was extended and stabilized by the abilities of this enlarged social brain in which vision now played a central role. Humans underwent a further expansion in their social brain and evolved audio systems (language) that allowed a further extension in the duration of the maternal bond. This large human brain also allowed for other extended social bonds via both vision and voice (song and language). Accordingly, humans became much more dependent on cognitive and emotional memory systems for their group identity and social bonding. As such, identity depends on stable memory (language and belief); this also required specific brain adaptations and structures, but still retains the essential features of addiction modules. Our large brain not only initially evolved for social purposes, but also produced a social mind (a dynamic conscious state). The human mind thus became a substrate for group identity, colonized by learning (belief) and competing language instead of genetic parasites. But learning and language also considerably increased human cognitive capacity and promoted the emergence of a modern individual critical mind. This modern mind has frequently found itself in conflict over the beliefs and resistance to learning of our more ancient social mind.

Various themes presented in this book will seem counterintuitive to most readers. This is especially likely to apply for the role that viruses and other genetic parasites are proposed to play. Genetic parasites are universally thought of as bad things, either destructive or at best selfish genetic entities with little capacity to create complexity. Indeed, there is much evidence in human history that supports such negative views. That they can also promote the evolution of group identity and complexity will thus seem counterintuitive to most. Yet there is also much evidence that also supports this view. The most significant reason for this misunderstanding, in my view, has been the historic and continuing failure to acknowledge the basic role of persistence in virus and host evolution. Persistence needs to be considered as a major concept, one that provides a significant force regarding fitness in the evolution of not only virus, but also host. Its importance in this book, however, will be crucial, as will be the other related concepts of addiction modules as first presented in my prior book. As noted above, this book will delve into the considerable (and often obscure) details that relate to each relevant topic. These details not only provide essential evidence but also depend on highly specific and dense scientific terminology for accuracy. Since this may overwhelm some readers less familiar with scientific terminology who have more general backgrounds, each individual chapter has thus been structured to attempt to provide overviews at the beginning as well as summaries at the end in less dense language and less evidence. In this way, it is hoped that the chain of reasoning remains correct and clear to the reader.

Specific Concepts to Keep in Mind

Viruses are now established to be the most abundant and adaptive genetic entities on Earth. They are both found as free virions in all habitats and incorporated into the genomes of all life. In this, they can be considered as the 'dark matter' of the biological universe. Thus, all life has had to deal with this viral world in order to survive.

New host immunity can derive from persistent virus colonization. For example, all of us are persistently infected with various human-specific viruses. This is also true for most other species, including bacteria. These viruses affect the host susceptibility to related viruses. In doing so, they can also provide systems of immunity and promote host survival.

Viruses can evolve by cooperative population-based process. This is known as the quasi-species theory. In such populations, there is no fittest type. Instead, the population provides various interactions (including cooperative, competitive, defective and even lethal types) that together promote fitness.

Viruses make genes and gene sets in large numbers, including some core genes found in all life. They are genetically creative and not simply thieves that take genes from host genomes. By far the most productive sources of viral genes are those large DNA viruses that infect prokaryotes and unicellular eukaryotes. Here, novel viral genes are the products of high levels of recombination, mostly between viruses. The entire pool of such dsDNA viruses may constitute one immense, dynamic and most ancient of gene pools.

Persisting extrachromosomal viruses of bacteria led to the molecular definition of an addiction module. The concept of an addiction module provides the bases by which symbiosis and cooperation can be promoted during evolution. It also applies to social symbiosis or cooperation. Addiction modules also promote systems that use programmed cell death (apoptosis, altruistic death of individuals) for the purposes of group identity.

Persisting viruses compete fiercely with one another to occupy a host. They use addiction and immunity systems for this purpose and also preclude competing lytic viruses. The addiction modules kill group members that have lost or disrupted their group identity. Such persisting viral genomes are the most adaptable and dynamic part of the host genome.

Host populations that are colonized by aggressive persisting viruses have also acquired a system of group identity and immunity. Genetically identical host populations that are not so colonized can be killed following contact with a persistently infected host population. This defines survival of the persistently infected group, which provided group selection via virus. This situation also tends to genetically isolate populations and promotes speciation.

The colonization of host genomes by sequential waves of genetic parasites tends to increase the genetic complexity and increase systems of group identity during evolution. Thus, identity tends to be cumulative and increases complexity, but not necessarily more efficient. Displacement of prior identity systems is also common. However, some host lineages are colonized by a highly efficient genetic parasite that imposes immunity systems able to preclude most subsequent competitors, hence generate fit host lineages that henceforth evolve slowly.

Persisting genetic parasites are often composed of populations that include defectives and hyperparasites (parasites of virus and other parasites). This situation closely resembles the concept of selfish DNA. Such parasitic elements, however, were essential participants of a population that allowed the needed competition and preclusion during persistence. They are also often elements of addiction modules.

The relationship between group identity, self-identity and mate identity is inherent to the addiction modules being used. Mate selection will often be a specific application of group identity. The acquisition of group identity occurs during limited developmental windows.

Group behavior (cooperative, aggressive, even lethal) is a product of behavioral addiction modules which operate via the central nervous system. This requires emotional systems to set general behavior patterns. Olfaction was the important early sensory system to regulate such behavior, but vision and sound (language) later became more important.

Group identity and group competition is most intense within similar systems. Species will tend to have the most intense competition within their own species or similar species. This is mediated by the use of highly related addiction modules that directly interact with each other. This also promotes the diversity of such addiction modules. The destructive potential of an addiction module will often be most effective against species that are in related lineages.

Group identity will even allow competition with self. Identity is set or transferred during imprinting periods (developmental windows), but once set can oppose even the same types of cells or individuals. This is a ubiquitous process that drives the continued evolution of identity.

Extragenetic systems for group identification. Identity systems are not restricted to genetic similarity. Because an 'imprinting' process can be sensitive to extragenetic alterations, genetically identical systems can still acquire distinct group identities. A spectrum of mechanisms are able to do this, including cell surface modification and recognition, DNA marking by methylation, altered transcriptional control, small or antisense RNA expression, education of adaptive immune (or apoptosis system to kill cells), regulation of nerve cell function, such as receptors and ion channels. Persistent virus colonization also remains as an extragenetic system for group identity. In primates, learning and psychological alterations can also set extragenetic group identity.

An ultimate objective of this book is to explain the origin and evolution of the vertebrate nervous system from the perspective of the evolution of group identification systems. Memory and behavior become key. Although odor detection and its memory may have been basal in most animals, in primates this system was displaced by visual and audio sensory memories which came to predominate. Thus, learning (imprinting) evolved to become a main process to set great ape group identity. Here, the role of emotion and addiction modules in group behavior becomes central. Thus, primates evolved to depend heavily on visually based group identification. Humans extended this social dependence to include audio learning (language), which required major brain adaptations. Our large social brain became the mediator of all social bonds.

The product of the evolution of group identity was the emergence of human culture, a product of the human mind. The human mind (a conscious brain state) itself is the product of a large social brain and learning. The mind became a human specific vehicle for group membership (a social mind) which used cognitive information and learning processes for imprinting and used emotional memory to establish social bonds. Belief is a learned but stable cognitive content, and belief states were necessary for a mind to participate in such social group membership. Thus, a biological selection required brain systems dedicated to belief attribution and emotional memory. But to provide group identity, a belief state must also necessarily resist subsequent learning and displacement by competing identities. Belief system thus characterizes all human cultures and many human social bonds.

Chapter-Specific Issues

Many of the above basic principles of group identity can be demonstrated in prokaryotes. Viruses and other genetic parasites provide the exemplars of how most such prokaryotic systems operate. Indeed, the molecular details of addiction modules are only really understood to any detail in prokaryotic viral systems. The topics of individual bacterial identity systems and bacterial communities are presented in Chapters 2 and 3, respectively. The prokaryotes are the most ancient, diverse and adaptable of all cellular life forms on Earth. But even more abundant and diverse and adaptable are the viruses of prokaryotes, now known to be the most numerous biological entities on Earth in all habitats. During the initial but long period of unicellular prokaryote evolution on early Earth, viruses and other genetic parasites played (and continue to play) the major role in the origin and evolution of group identity systems. These identity systems include restriction/modification systems, immunity, bacteriocins and phage typing genes. Lysogeny (and the cellular immunity it provides) is also considered as a major process of group identity and will be presented as such. In Chapter 2, the origins of the group identity processes as it applies to large populations of individual cells are introduced in considerable detail.

Most of the mechanisms and strategies used by prokaryotes for group identification can also be found in at least some simpler oceanic eukaryotes. The solutions that were adapted by these representative early eukaryotes for group identity were often preserved for use by many metazoan oceanic microorganisms as well. The considerable, possibly obscure, detail in the first two chapters may be overly challenging for the more general readers. As these chapters will provide the molecular foundations for all the following chapters, such detail is justified and necessary. The microbiological world is also a highly dynamic genetic realm, and this dynamism is mostly mediated by viral entities and other genetic parasites. Viruses, cryptic and defective virus, plasmids and other parasitic genetic elements will be presented to constitute one seamless domain of interacting genetic parasites (and hyperparasites). Viruses are defined as host-dependent genetic parasites so these other genetic parasites share this definition. These elements have come to have many names, such as plasmids, transposons, IS elements, cryptic phage, homing introns, intenes, and are often studied as autonomous elements. However, the assertion will be presented that such elements frequently (generally) work in concert with (i.e., are mobilized by or preclude) viruses and together can form stable associations which determine host group identity. This concerted (cooperative) feature of hyperparasites and their role in group identity will be a common theme frequently maintained in the genomes of eukaryotes and presented as such in all subsequent chapters. That prokaryotic genomes are indeed highly dynamic is now an accepted view, especially due to comparative genomic analysis. The concept of horizontal transfer has also become an accepted term to characterize the fluid appearance of novel genes and gene sets into distinct bacteria. However, the role of virus in this 'horizontal transfer' concept has simply been that of a vehicle, or a genetic truck that transports genes to and fro. I will take issue with this concept in several respects. The persistence of a genetic parasite is a transformative event for host group identity. The presence of genetic parasites in host genomes determines the group identity of the host. In addition, the horizontal transfer concept fails to acknowledge the established and vast potential viruses have created new gene combinations and gene sets by recombination mostly between genetic parasites. Thus, these genetic solutions originate in the genetic parasites, not the host. Genetic parasites are creators of genes and host group identity, not simply genetic trucks. In this way, the concept of 'horizontal transfer' obscures this role. These interdependent genetic parasites need to be understood much as we now understand quasi-species, as a cooperative set that uses seemingly defective elements to persist and compete. The confusion that has been promoted by thinking of them as autonomous elements needs to be discarded. The real world of genomes involves networks of hyperparasites that function in concert. Thus, all bacteria (and even eukaryotes) can be specifically identified based on patterns of colonization by virus-like parasites. Phage typing, phage conversion and phage (and plasmid) exclusion will all be presented as major group identifiers for bacteria. The immense diversity of restriction/modification systems, identity and addiction systems will also be presented. In addition, the role of parasite-mediated addiction strategies in maintenance of these identities will be presented. Addiction modules can induce 'altruistic'-like behavior (programmed cell death), even in unicellular bacteria; thus, addiction can explain death-mediated community maintenance. Viruses and other related genetic parasites can themselves act as addiction systems, and the footprints of their past colonization will often be diagnostic of evolving prokaryotic group identity.

However, we have come to learn that the microbiological world is also a highly social world and bacterial group behavior has become a well-studied topic. Bacteria can clearly be social and their communities are seen as blooms, mats, stromatolites and films and have been preserved as fossils in some ancient and large communities (cyanobacterial stromatolites) which were crucial for the evolution of photosynthesis and O_2-dependent higher life forms on Earth. The overall events of the geological and biological history of the Earth are shown in

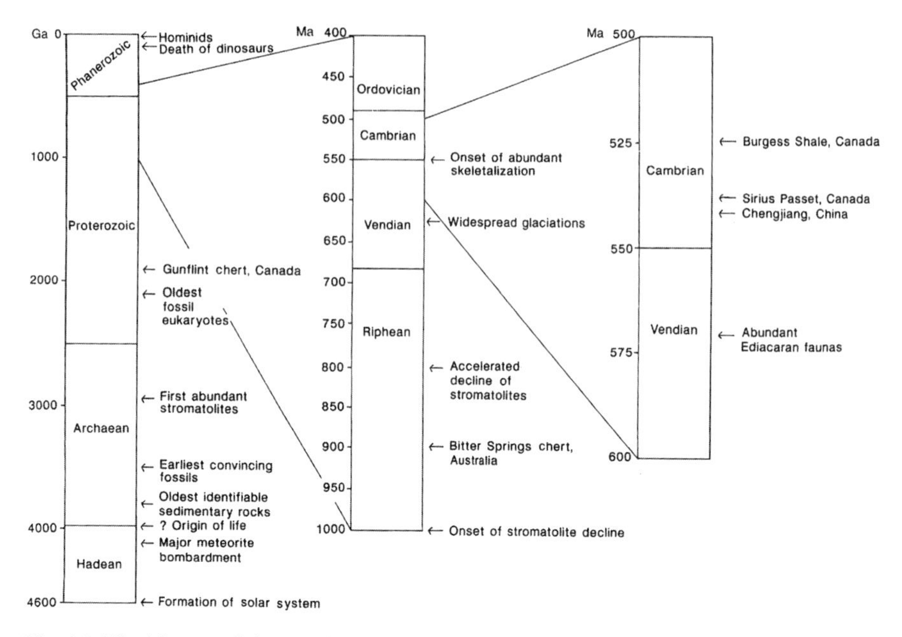

Fig. 1.2 The history of the Earth showing geological and biological events (reprinted with permission from: The Crucible of Creation: The Burgess Shale and the Rise of Animals, 1998)

Fig 1.2. Extended bacterial communities are an early event and an example of evolved group behavior that we now know using quorum-sensing systems. Quorum sensing is known to regulate biofilm formation and host colonization (for pathogens). It is also of special interest in the context of cyanobacteria group behavior (i.e., blooms and mats). Cyanobacteria use both light production/detection and small diffusible molecule production/detection in their quorum-sensing response. Thus, cyanobacteria can provide models for early sensory systems that affect behavior. Here we can see the origin of small diffusible molecules (olfaction) involved in sensing self-like systems that can regulate transcription. We see too the use of sensory receptors and signal transduction systems that control behavior (motility) via the use of ion channel proteins. In the cyanobacteria, we also see distinct relationships to their viruses, including highly complex viruses that can provide core photosynthetic activity to cyanobacteria. Some viruses are clearly lytic and can terminate large blooms of cyanobacteria. Cyanobacterial evolution (including its photosynthetic apparatus) appears to be substantially mediated by colonization with virus. The relationship of group survival to phage genomic colonization, lytic phage sensitivity and group sensing systems is presented in this context. The concepts of a bacterial 'species', sex and the role of genetic parasites in defining populations will also be presented. These persisting parasitic systems also often appear to be involved in symbiotic (cooperative) relationships with various host. Since cyanobacteria are thought to have provided the symbiotic origin of chloroplast, the origin of plastids (chloroplast and mitochondria) is also examined from the perspective of the likely role of viruses and genetic parasites. Other stories involve bacterial symbiosis, such as that of Wolbachia involved in insect sexual compatibility. Here I also briefly present the perspective of potential role of phage in group identity. Thus, in these bacteria we see the origin of communication systems. But we see the origin of social communication that becomes linked to cellular differentiation systems. Differentiating social bacteria (dictyostelium) are then examined and show terminal differentiation in the formation of spores and fruiting bodies. Dictyostelium is the most social soil bacteria and displays an altruistic-like feature of cell differentiation. Parasitic DNA also appears to be involved in this differentiation processes. Thus, from ancient mats of cyanobacteria, to oceanic blooms, to biofilms to spore forming soil bacteria, the molecular (often parasitic) mechanisms of sensory systems, social behavior and group identity of bacteria are presented as the foundation for the evolution of metazoan group identity.

In Chapter 3, the simplest multicellular oceanic eukaryotes are examined for their group identity. These simple aquatic organisms are considered as examples of the basic systems of group behavior that evolved into more complex life forms. Red algae are the most basal of eukaryote on our lineage of interest. Red algae present some interesting issues regarding their self-identification due especially to the existence of transmissive-infectious but simple nuclei in some species. The genetic identity and defense mechanisms of this sessile multinuclear network or mat organism are thus considered. Another topic raised by the unicellular or simple multicellular oceanic life is the abundance of diverse toxin production (also a characteristic of cyanobacteria). The toxins are especially active against ion channel protein associated with motility and nerve cell function. This situation is examined from the perspective of the role of toxins in group identification and whether they are constituents of toxin/antitox (T/A) in systems. Filamentous fungi are then considered and are of special interest for several reasons. Like red algae, they are a sessile multinuclear cellular network (many nuclei in a common cytoplasm) but one that exchanges genetic information by hyphal fusion. Viruses (dsDNA/dsRNA) are found in all such fungi and appear highly relevant to the mating systems that allow or prevent such sexual exchange. In addition, due to their basal relationship to all forms of higher metazoans, filamentous fungi represent a genetic foundation that led to the evolution of great genetic complexity of these higher organisms. Although filamentous fungi are static, they can live very long and move by hyphal growth. A static organism is physically constrained and must withstand the onslaught of all potential genetic parasites they will encounter in their long life. Thus, the mechanisms of self-definition and identity in a colonial organism are especially crucial. Some filamentous fungi use RNA and DNA parasites to distinguish colonial identity via killer or addiction systems. These parasites are also important for and involved in hyphal fusion. In terms of genome colonization by parasitic DNA, fungi are highly significant regarding retroparasites (retroposons, endogenous retroviruses) and show a bifurcation in their relationship to these elements. Such parasites were rare in prokaryotes. In the simple unicellular eukaryotes, we start to see the occurrence of chromovirus (a retrovirus with a LINE-like RT) in numerous genomes. But in filamentous fungi, specific species can be seen to efficiently preclude or to require retroparasites for their development. As will be presented, neurospora has a highly efficient RIP system that will genetically destroy retroparasites (LINEs, ERVs) during sexual reproduction; hence their genomes are essentially free of such elements. Other fungi, such as pathogenic fungi (*Magnaporthe*), use the presence of retroparasitic DNA to promote genetic adaptability and acquire pathogenicity. The more recently evolved fungal yeast species appear to use retroviral genomic parasite (Gypsy viruses) in their mechanism of mating control. Thus, fungi represent a crucial point of divergence in the evolution of genomes and their tolerance for retroposons, including the elements we call LINEs.

With the oceanic microorganisms we also turn focus on the role of light in group identity. The light produced by cyanobacteria and dinoflagellates is thus re-considered from the perspective of a system of group behavior. When cyanobacteria are examined in this way, we also see much evidence for an involvement of cyanobacteria viruses in their evolution. These phage show an intimate involvement with the light physiology and mechanisms of their host. These organisms also present models for the origin 7 TM receptor molecule used for light detection. This receptor is the basal sensory receptor, whose structure is applicable to all other sensory receptors. It thus represents the origin of sensory detection systems used for group identification, which we can now trace to all sensory evolution, including recent primate evolution. Eukaryotic algae (red and green) are also examined for their light detection and response and association with group identity. The eyespot of algae represents a particularly rich system for molecular mechanisms and models of more complex (but non-nervous) sensory function. Here, we can link also the receptors to the behavior of the organism and in so doing we see a role of ion channels and transduction, the origin of G-coupled transducers. We also see the role for receptor-mediated kinases. Thus, the ocean, in particular, has an overabundance of light emitting and detecting organisms that affect their group behavior. In oceanic multicellular organisms, light is also used in various modes, such as the patterns and spectrum of flashes in group identification. In addition, coincidence detectors and temporal detection are seen which introduce the importance of timing for light-based group detection, especially regarding sexual reproduction in which light and taxis are frequently linked. Swarms of individual organisms often move, differentiate and sexually reproduce in response to light. As noted above, many of these same organisms (i.e., dinoflagellates and cyanobacteria) were also notable for the diversity of toxins they produce, suggesting links between light and toxin-based group identity.

In Chapter 4, I extend the analysis of sensory-driven motility as a basis of group behavior that was presented above in oceanic microorganisms. Chapter 4 examines metazoan animals that transition to evolve nerve cells and brains which became a principle system for the control group behavior. Sponges are basal oceanic animal organisms, lacking nerve cells but with very large genomes that are highly colonized by genetic parasites. They can sexually reproduce and their sexual cells are chemotactic (like oceanic microbes). They do have various cellular identity systems, including highly diverse Ig-like cell surface markers that can recognize (and kill) non-group members, but lack any nerve cells or their role group behavior. With hydra and jellies, however, we see the first examples of animals with nerve cells, but they are not organized into central nervous systems. Interestingly, these nerves express prolactin-like neuropeptides which will be of considerable importance for the group behavior of higher organisms. Here we also find the box jellyfish, which has an advanced eye but without brains. Such 'brainless' vision is used by box jellyfish for group detection, group movement and sex. However, it is in the worms (*Caenorhabditis elegans*) that we examine in detail the origin of a central nervous system and the function of neurons in group behavior. Here too, we retain a linkage between behavior and odor receptors. But in *C. elegans*, the learning of an odor (pheromone) becomes a central process for group behavior. This happens via 7 TM receptors which link olfaction to neuron function and to epigenetic events that set learned states of group behavior. Here we can explore the genetics of odor receptors and how they affect the behavior of chemotaxis, especially as it relates to sex. *C. elegans* will forage for hermaphrodites by using diffusible pheromone cues. Clearly, sexual identity and sex and behavior are learned (imprinted) states; thus with the emergence of a CNS, learning becomes a central process for specifying group identity. Since genomic analysis is advanced in *C. elegans*, we can understand the genetic and evolutionary events that underlie the ability of nerve cells to differentiate and create memory. Terminal neuron differentiation (apoptosis) is essential. And for this, RNAi molecules play a crucial role. RNAi transcripts are themselves mostly produced from parasitic retroposon DNA that has colonized the *C. elegans* genome. Thus, genetic parasites and their products played a basal role in the evolution of the central nervous system. Mutants in the RNAi production will reactivate genomic retroviruses and retroposons, indicating RNAi has a continuing role in persistence of these genetic parasites. This same RNAi system also confers onto *C. elegans* efficient resistance to all known viruses and its loss makes them susceptible. Thus, it has the features of a combined identity/immunity system. In *C. elegans*, we see the first clear example in which retrovirus and retroposon colonization, RNAi production, neuronal apoptosis and defense against other genetic parasites are all linked to the evolution of a central nervous system.

Other, more developed oceanic animals with central nervous systems still appear to uses toxins as systems of group identity. Cone snails, for example, produce a very high diversity of highly active and often neuron-specific channel toxins. These snails also evolved a system to generate toxin diversity via combinatorial peptides. The suspicion is that such toxin diversity, which is aimed at neuronal ion channels, is a form of T/A module involved in group identity. With *Apalsia*, olfaction memory remains important for sexual identity. Light also remains as a common mediator of group and sexual identity, as seen with various invertebrates, such as glow worms, in which light provides mating information. And in various other oceanic invertebrates, such as snapping shrimp, we see the emergence of fast neurons, neuron sheaths and highly advanced visual systems capable of seeing extended frequencies of light as well as the polarity of light. Thus, it seems clear that such highly developed visual capacity is used for group identification in these species. As shrimp-like organisms are thought to be ancestral to terrestrial animals (insects and vertebrates), their sensory systems and CNS are relevant.

In Chapter 5, we briefly consider the first terrestrial animal that emerged onto the land with regard to group membership, the insects. Insects are the most successful of all animals and as many are also social, they have many lessons to teach us. Although the evolution of insects stems from invertebrate ancestors that resemble a fairy shrimp, there are no oceanic insect species. The most numerically successful of all insect species are the parasitoid wasp, which are also the ancestors to the most numerous social insects (bees, wasps and ants). Odor and pheromone detection are crucial to insect group behavior. In parasitoid wasps, we see that sexual behavior (oviposition) can be altered by complex learning. In many parasitoid wasps, we also see that reproductive biology is highly dependent on and manipulated by persisting and genomic virus. Some of these viruses can alter both odor detection and oviposition behavior. Most insects undergo metamorphosis during development. It is interesting that learning that occurs in larval stages can persist into some adult forms, in spite of the generally severe restructuring (apoptosis) of the larval brain. Learning of olfactory cues by social insects is the chief mechanism by which social (hive, nest) mates are identified. Thus, we see a primary role of odor detection in the discrimination of self, from non-self often via cuticle hydrocarbons. Olfactory learning is also the process by which social insects resolve internal colony identity, such as hierarchy and social status within the social order. A bee, for example, smells like a queen or drone. Odor detection is highly developed and sensitive in insects and is used for mate, food and toxin detection as well as for predator defense (via alarm pheromones). The aggressive group reaction (aggression) that alarm pheromones can induce indicates that systems of stereotypical group behavior have developed in a social context. These states closely resemble emotional states in animals, such as anger. There also appears to be a role for endorphins, and neuropeptides in insect social bonding and group behavior (i.e., addiction module). Behaviors that elicit both social benefit and social harm define the two essential components of group identity as determined by a behavioral addiction module and can induce both supportive and aggressive group behavior. Moths are especially known for their exquisite sensitivity to sex pheromones, but also learn to identify their plant host and nesting site. There seems to exist a basal odor (sex pheromone) based link between flowering plant and insect, as seen in wasps that implant eggs and larvae into plant flowers (i.e., figs). There is also evidence of much activity of what appears to have been insect-derived retrovirus and other genetic parasites during the evolution of flowering plants. However, although odor-based group identification is prevalent in insects, many insects also used advanced visual and audio sensory detection for group and sexual behavior. For example, the songs of crickets used for sexual identification are characteristic of each species. Thus, in insects we see the emergence of audio-based group identification in species that are otherwise mostly using sex pheromones for this purpose. This required the development of receptors, neurons and memory that were able to identify and learn specific audio and temporal patterns. Thus, the learning of sound and rhythm became linked to sexual identity in crickets.

The detailed molecular mechanisms that insects use for learning (especially olfaction) is best understood with Drosophila and will be presented. The brain of a bee is composed of about 340,000–400,000 neurons. This is tiny compared to the brains of vertebrates but large compared to *C. elegans*. Bee genomes appear to have less immune-related genes compared to Drosophila, for unknown reasons. They do have, however, a large number of genetic parasites and odor receptors. In the Drosophila genome, 60 OR (odor receptors) genes have been identified as well as numerous related pseudogenes. These ORs are 7 TM receptors, similar to those of *C. elegans*. Also similar is the odor receptor linkage to neuron function as well as a role for signal transduction, G-coupled proteins and ion channels. Although courtship behavior in drosophila has an odor basis, it also uses visual discrimination. Drosophila, like most insects, have trichromic color vision that uses 7 TM receptors. Many insects use color vision for group and sex identification. With Drosophila, some learning mutants are known. Chief among these is the Dunce mutant, which affects learning associated with sexual behavior. With this mutant, a role cAMP as well as *roo-copia* LTR (retroposon) activation is seen. The maintenance of genetic parasites in Drosophila and their role in OR receptor evolution are discussed.

Social insects present a very interesting situation with regard to their expected interactions with viruses. Unlike social vertebrates, social insects are clonal, hence genetically homogeneous. This along with their colonial or high-density living patterns, their corresponding high rates of contact with their environment and with each other would be expected to make then exquisitely sensitive to infection by acute viral agents (similar to domesticated animal species). Bees, due to their commercial importance, are the one social insect which has been examined from the perspective of virus interactions. Indeed persistent infections of bee hives with various types of picornaviruses (such as black queen syndrome virus) are highly prevalent. These viruses show complex interactions that may well affect the group or hive identity of bees. There is also evidence that viruses can affect the aggression of their host. Other insect species show sexual behaviors that appear to be modulated by symbiotic bacteria, such as Buchnera, implicated in insect mating and speciation. Viruses of this bacteria are also examined for their role in bacterial-host symbiosis.

Chapter 6 presents the evolution of group identity in the aquatic and terrestrial vertebrates. Since this path leads to human evolution, it provides more applicable example of systems for group identity. Many of the basic processes and systems used for group identification, communication and behavior were developed in the early vertebrates of this habitat. The oceans are also the place that supports the greatest number of vertebrate species. The most successful of these species are social fish (teleost fish) that live in very large schools. For example, Pacific herring have been estimated to form extended schools of 30 billion individuals in off Alaskan waters. It is interesting that like other large oceanic populations (cyanobacteria, dinoflagellates, coral, shrimp) oceanic viruses play a big role in teleost populations, and large crashes in herring populations, for example, have been observed (mainly due to rhabdoviruses). Water is an excellent habitat for the diffusion of virus as well as an excellent habitat for water-soluble molecules and pheromones for olfactory detection.

A major development in biological identity of bony fish is the emergence of the adaptive immune system. Ironically, bony fish also appear to support an enhanced array of viruses. A component of this adaptive system, the MHC locus, was also adapted for group and sexual identification via olfaction. These vertebrates have Class I OR receptors (7 TM proteins linked to neuronal TSR ion channel) which are much used for group and sexual behavior. These neuron-specific receptors are expressed in a highly regulated and individual cell-specific way. Immunity and group identity become linked as fish also have early elements of vomeronasal (VNO) neuronal tissue that can detect MHC-derived peptides. Like the origin of the nervous system, the adaptive immune system also depends much on programmed cell death (apoptosis) to 'educate' and set cellular group identity (immunity). There is much evidence that retroviruses and retroposons were fundamentally involved in the origin of this system which has continued to evolve by their action. In general, there was a major expansion of such ERV elements in fish genomes at the origin of vertebrates, but only a small number have been fully conserved. Fish genomes vary considerably in the content of their parasitic elements. The puffer fish (fugu), for example, has 1/8 the genome of a physiologically similar zebrafish. Fish also show a bifurcation in the evolution of genetically determined sex. Many species show a socially determined sex in which females can differentiate into males based on visual sensory and social input. This involves an epigenetic process that results in relatively stable (seemingly compulsive) sexual behavior. Other species show environmentally determined sex (i.e., temperature), which was conserved within and common to amphibians and reptiles. Most interesting and relative to mammals are also some fish species that show genetically determined sex involving sex chromosomes. Here too there is evidence of involvement of retrovirus and retroposon colonization in the origin of this process.

The linkage of MHC to olfaction and group identification was maintained through the evolution of amphibians and most mammals (i.e., mice) and remains used for group and sexual identification in most mammals. In fish, both olfaction and vision are involved in social behavior, such as shoal membership. In terms of sexual behavior and reproduction (spawning), olfaction, however, appears to be more generally important. Although fish can exist in large shoals, they are not usually considered very social species as these behaviors seem to have physiological determinants and they do not often show individual specific social bonds. However, a few species do show social bonding such as in biparental care as well as polygamy and monogamy. This social bond is obviously a learned state, but it is not a well-studied system. In terms of learning, fish clearly learn odors via OR and VNO receptors, and fish brains show a considerable expansion in their olfactory bulb. Fish also developed REM sleep, thought necessary to consolidated (non-cognitive) memory. These learned odors can induce and contribute to compulsive (even 100% lethal) reproductive behaviors. Fish clearly can learn emotions and can be taught to elicit a fear response to specific stimuli. In goldfish, for example, avoidance learning can be taught and appears to consist of emotional learning and special learning. This learned fear response is also socially contagious which can also be mediated by pheromones. The speed by which fish recognize visual and odor patterns and respond emotionally indicate that it is likely an early evolved state, but one that is 'subconscious' and does not involve a frontal cortex. The CNS of fish does use endorphins and opioids regarding their behavior, but in ways that are not well understood.

With the terrestrial vertebrates, we see the development of reptilian minds that have a significant emotional characteristic, but tend not to be very social minds and have no frontal cortex. Olfaction remains a major sensory system for social detection in reptiles. The VNO system has been preserved or expanded and is used mainly for sexual activity (mate detection). Its use in social behavior is much more limited, with a few interesting exceptions that are discussed. In general, there is little pair bonding or maternal bonding in reptiles. Thus, aside from some large gatherings for mating, group social behavior is limited in reptiles and there is usually no care of the young. Sex determination is mainly environmental (i.e., temperature dependent) process and there are no X or Y chromosome-determined sexes. Snakes are an interesting example that appears to identify similar species that are both oviparous and viviparous. It seems that placental biology has evolved several times independently in snakes. Endogenous retroviruses also appear to be associated with reproductive tissue and CNS disease in snakes.

Avians are not on a direct path that led to human evolution, so their social evolution is not examined in great detail. However, avian species do have some distinct and strong social tendencies that are worth considering, as well as considerable general intelligence. A major shift in social identification and sexual genetics happened in birds. And in some bird species, there is also an expansion of the cortex. Like the great apes (see below), bird species also lost most all of their VNO genes by pseudogene formation. Thus, birds do not depend on odor markings for social identification. In this, birds seem to mostly use visual information for identification. Bird genomes show a distinctly different pattern of colonization by genetic parasites, relative to mammals. Besides highly reduced ERVs, their genomes mostly lack the alu, LINEs and SINEs elements of mammals. Birds, however, do tend to form mating pair bonds and provide maternal and sometimes paternal care to their young. Thus, strong social bonds are clear. They also have complex mating behaviors and genetically determined sex in which ZZ chromosomes are male and ZW are female. In contrast to mammals, the Z chromosome is large and gene rich (X-like), whereas the W is small and heterochromatic (Y-like). There is a preponderance of social monogamy in birds. Some bird species show species-specific audio (vocal) learning and social imprinting. Birds can also recognize kin and live in large gregarious populations. Thus, they are much more social than reptiles. Some birds are also efficient general learners (blackbirds, crows). Of special interest are some social hunting bird species that show paternal or other male care of young. In these species, there is good evidence that prolactin production is associated with social bonding of males to young.

Terrestrial mammals have all preserved their VNO system for sexual and other social (group) identification and use it for maternal bonding. Unlike the reptiles, the placental mammals are monophyletic in origin. In addition, the mammary gland (lactation) of placental is also monophyletic. In both these systems, there is evidence for the involvement of ERVs colonization at their origin. The maternal bond of all mammals provides the basal social link that was absent from reptiles. This social bond is the model that will later be adapted to evolve more extended social bonds of humans. In the case of rodents, the VNO genes are genetically linked to the MHC I locus and together are involved in maternal recognition of young. Mice have 150 VNO receptor genes, some of which are able to affect activation of stereotypical social behaviors. The activation of olfactory receptors (OR) shows a single neuron pattern of expression. These genes expanded significantly in the mouse and represent 2% of all mouse genes. Mice have three types of OR genes and many OR pseudogenes. Thus, OR gene inactivation (via pseudogenes) has been frequent event and such inactivation is lineage specific (comparing rat to mouse). Curiously, testis express 50 OR genes, but these have no pseudogenes. In terms of maternal bonding, prostaglandins are involved in birth and subsequent maternal behavior. Care of young requires obsessive (addictive) behaviors, and it appears clear that prolactin, oxytocin, and AVP are all involved. Emotional behaviors, including protection and aggression mediate bonding. Opioids, endorphins and addiction modules also seem likely to be involved. Thus, VNO-mediated social recognition, lactation, (maternal) imprinting, extended care of young, REM sleep, social–emotional memory, and an expanded olfactory bulb in the brain all characterize mammals. Visual color resolution is reduced in mammals relative to other animals, as they have dichromatic color vision (M/LWS opsins). In most mammals, there was a large expansion OR and VNO receptors (7 TM proteins), also seen in the prosimians (which retain dichromatic visual receptors). With respect to chromosomes, sex is now determined by XY chromosomes and sex determination is via SRY. There has also been a major expansion of mammalian ERVs, LINEs and SINEs. The Y chromosome, in particular, is a host for large amounts of genetic parasites and these are species specific. The rodent lineages underwent radiation to become the most numerically successful of all mammals. In contrast to the avians, in mammalian social behavior monogamy is uncommon. One model of monogamy versus polygamy (voles) indicates a central role for VNO and AVP (V1aR) receptors in species-specific associative mating behaviors. There is a compelling role for genetic parasites (microsatellites) in vole social evolution. Social pleasure and pain as well as protection and aggression all seem involved in pair bonding. Olfaction is also used for kin avoidance in breeding (in mice urine order and MHC makeup inbred mice avoid maternal mates). Some voles have also evolved audio (vocal) signaling to identify mates. This rodent model defines the basic character by which socially mediated group identity functions and evolves in mammals.

Chapter 8 focuses on the African great apes and presents an evolutionary path from shrews to humans. In the prosimian ancestors, both dichromic color vision and the VNO organ have been preserved but all were lost in the African apes. Thus, olfaction lost its central and ancient social role in the African apes. African apes underwent a great genetic colonization involving HERV Ks, LTRs and their SINEs in large numbers relative to other mammals. In contrast, New World primates did not undergo this expansive genetic colonization and did not all lose their VNO systems. The inactivation of the entire VNO receptor system which had been preserved in all terrestrial vertebrates is thus a striking evolutionary development raising the question of how such an ancient system of social identification and bonding could have been lost in a mammal which must maintain its maternal bonds. In addition, the African apes show the emergence of a trichromatic color visual system that is X chromosome associated and aids social identification. In contrast, some New World monkey species show distinct, female-specific patterns of color vision. In spite of the lost olfactory systems for social bonds, the African apes extended the duration of their maternal social bonds. For this, the apes adapted and expanded their visual and brain capacity which was applied for the purposes of social behavior. Thus, this large expanded primate brain (not the VNO) became a main social organ. Visual-mediated social links were thus established by neurological adaptations in these primate brains (including visual cortex, face recognition and mirror neurons). This also required the establishment of links between vision (face, gesture) and emotional memory (via the amygdala). This resulted in an expanded maternal bond which no longer required signals from the placental of immune molecules (MHC peptides) to set social identity. Liberated from female or placental biology, this form of visual social bonding was also readily adapted to male bonding and the bonding of more extended social groups. The expanded visual and memory capacity of this enlarged social brain also enhanced the cognitive capacities of apes. Learning, in particular, benefited from these social developments. Thus, movement imitation and visual learning became a by-product of this visual social brain. The resulting expansion of social bonding and visual learning also allowed chimpanzees to develop social hunting in which they were able to 'envision' the prey, before it had acted. This cognitive capacity was the product of chimp social communication, but also relied on the inherent tendency for group identity to also provide group aggression. Thus chimpanzee group identity is mainly mediated by a visual social brains, but is also mainly directed at other chimpanzee regarding social groups. The variation of ape social structures is thus the product of learning, not MCH or pheromone biochemistry.

For stable social bonding to develop that was less biochemically determined, the ancient neurological systems of mammalian maternal bonding, involving oxytocin and AVP receptors, which control emotional addiction modules, needed to become linked to visual input and memory. Group identity and behavior thus became a product of visual and emotional memory in which addiction modules provide supportive and negative behavorial elements working together. We can thus understand the preponderance of associative interactions in primate social behavior and the failures of selfish gene concepts involving aggression, competition and selfish individuals and why they have failed to explain ape sociology. In social group identity, social pleasure and social pain must work in concert (T/A modules) to maintain group stability. Social aggression especially applies to conspecifics that are non-group members. Thus, internal competition is not the primary force for social stability.

Chapter 9 presents the human-specific developments regarding group identity. Here, we are especially interested to consider the developments that led to the much larger social brain of humans and how social behavior is selected and generated. Genetic colonization has continued to be a main creative and disruptive force in the evolution of the human genome, as human-specific HERV K and LINE acquisitions are clear. The human and chimpanzee genome differ by 68,000 indels (insertions and deletions), which are mostly the product of alu, LINE and LTR parasitic element activity. Most of these indels average about 300 bp and show a clear relationship to ERVs. A preponderance of such parasitic DNA is found on the Y chromosome, which is the most distinct between human and chimp chromosomes. All this is consistent for a continued role of parasitic DNA in recent human evolution. In terms of genes, odor receptor genes account for the largest and most recent differences between hominid species. Humans have 1,000 OR genes (half of mouse), but half of these loci are pseudogenes, inactivated by indels. The major loss in OR genes (and total loss of VNO genes) is consistent with a reduced role of olfaction in human group identity. The biggest difference between humans and chimpanzees is the substantially enlarged human brain, especially the hypertrophic and more invasive neocortex. As in the great apes, it is here proposed that this enlarged brain was mainly evolved for social purposes and promoted by genetic parasites.

The basal social bond in humans continues to be the maternal bond. However, its characteristics as emotional addiction modules are now fully mediated by our social brain. And its emotional T/A elements are clearly apparent following the break of such bonds (i.e., grief following death of offspring). Humans have extended the duration of the maternal bond as well as required a much more supportive social relationship as human infants initially have significantly less inherent behavioral capacity than any other mammal (including chimpanzees) and required much more care. However, humans have added an additional process of social bonding and identity, that of voice and language which was absent in other hominids. Because such extended social bonding is mediated by learning, it was able to be adapted to male bonding and other more extended forms of social bonding. Language is learned and processed via specific neurological adaptations which provided group identity. This required a stability of the learned state of language. Learned stable cognitive information can also be defined as a belief state. Thus, belief acquisition and retention also required brain-specific adaptations (including lateralization). These adaptations considerably enhanced the cognitive capacity of the human mind (a product of learning of a social brain). Social learning for the purpose of social group identity includes the colonization of brains by language. Other social functions (theory of mind, human-specific social learning) were also the products following the selection for a large social brain. With language, a more capable and conscious social mind also became possible. And with this adaptation, the mind itself (a conscious brain state of imprinted or learned identity) became the vehicle for creating group identity and extended human social structures.

The human mind evolved from a dynamic (consciousness) processing of sensory, memory and emotional information in which emotions mediate social bonds. A social mind was thus selected which was dependent on language, memory, emotions as well as belief states (accepted sensory data or identity) to set group identity. The development of childhood mind appears to recapitulate the evolution of many of the features of this social mind. A social mind links to the minds of others via vision and voice (e.g., fear contagion, empathy, aggression), but it is a fragile dynamic state that can be manipulated by drugs, disease, sensory physiology and external language (i.e., hypnosis). However, such mind-based group identity still requires addiction modules to set stable identity.

The lateralization of the human brain was a neurological adaptation that provided the biological substrate that allowed language and belief to serve the purpose of group identity. The two brain halves involved are mostly associated with two distinct sensory modes of social learning, the ancestral primate visual (facial) system and the new human voice and language-based system. The left brain was more dedicated to 'verbal' control, such as language, names, math, the retention of details, facts and patterns. The left brain also was to provide the basis of rational thought. Left brain seizures, however, can result in language deficits and in the case of some savants with such damage, the right brain compensates (sometimes resulting in enhanced and detailed perception and cognitive performance abilities, but reduced empathic/social abilities). The right brain was dedicated to belief and meaning, mostly with respect to language and social identity. It is more responsible for 'non-verbal' control, such as special perception (music, tonal stimuli) and its emotion, explaining symbols, imagination and generating the big picture. These functions serve to promote various social and cognitive group identities in which social identity is stably learned as a belief state (i.e., culture, religion). Brain studies using fMRI now identify specific regions and patterns associated with belief attribution. Various forms of brain damage also indicate some of the neurological basis of belief attribution. Strong social bonds, such as intense romantic love, can also elicit clear fMRI patterns, suggesting involvement of addiction systems.

The abstract and recursive nature of human language required major adaptations in the capacity of our brain to learn essentially infinitely complex pattern recognition. Language memory and its reproduction led to the ability to be able to 'hear' an inner voice associated with modern consciousness. This ability to synthesize non-sensory sources of voice also provided the origin of cognitive creativity. Language acquisition has structural effects on brain development. Schizophrenia as a disease associated with the development of language, brain lateralization, associative social learning and the inner voice (auditory hallucinations or active mirror circuits). Schizophrenia is also a disease associated with higher cognition and imagination. There is also a clear association between endogenous retroviruses, the human neocortex and schizophrenia that is not at all understood.

Our large social brains thus serve as host for epigenetic group identity which is based on learned cognitive content, involving language and beliefs, and provides our social identity. The resulting social (emotional and addictive) bonds were initially evolved to extend the maternal–infant bond. Due to their non-maternal (non-biological/placental) and learned nature, such social bonds were able to adopt to other social identities involving males (e.g., mates, nuclear families, tribes, nations, language, cultures and religion) whose strength can equal or surpass the ancestral maternal bond but still uses similar emotional mechanisms. Like all group identity systems, such 'congnitive' (learned) identity must also resist subsequent identity displacement. This defines a central role for belief in extended social identity and also indicates the requirement that learning attain closed or resistive states in which new social learning is resisted (as cognitive immunity). The combined mental state that results is defined as a social mind, which links similar individuals and resists competition and displacement by other information sources. This social mind resembles an early idea of a bicameral mind (J. Janes; associated with the brain lateralization and development of reading), but is distinct in several aspects. However, since this social mind is mediated by addiction modules, it has retained the strong capacity to elicit aggression to non-members.

The views in this book reasserts the importance of a biological tradition to understanding human behavior. It defines some inescapable biology legacies at the base of our social behavior and intelligence from the context of group identity. In contrast to commonly accepted views, belief-based cognition is proposed to have precisely such a biological foundation. Our capacity to hear an inner voice, susceptibility to schizophrenia and the character of early human cultures and religion were all affected by this biological legacy. Social learning and social group identity are inherently linked to belief states. Social pleasures and social discomforts, such as feelings of guilt, shame, embarrassment, pride, fame, glory and general social awareness, are seen as recently emerged social (emotional) addiction systems that provide group coherence. The infectious transmission of social states, such as laughing, panic, mob anger and fear, also supports social cohesion and identity. Thus, the human tendency to place others within social groups (the various -isms; sex, race, national, language, religious) that persist till today is an inherent feature of our social mind. And language was the most basal and biologically crucial social identifier that colonizes and alters the brain and mind. The innate constraints on the biology of language (i.e., universal grammars, language competition and exclusion) resulted from biological selection for group membership. Like language, the extended groups that have resulted still depend on belief states and all early cultures held beliefs in leaders, including various forms of kings or family-based leaders. Such beliefs, however, have acquired great social power and often transcend any sensory reality becoming imagined (abstract) beliefs in their leaders as gods. This also evolved to beliefs that were fully abstract, beliefs in gods that control social cohesion, including their leaders. Such beliefs transcend death and grief and can provide the strongest of all social bonds that now include dead family members and leaders. This social power can resemble that seen in social insects in its ability to transcending the will and survival of individuals. Like most addiction modules, such behavioral cohesion and power can promote death of individuals in defense of group identity (appearance of altruism). This was the character of the evolved and native social mind.

The emergence of the modern (asocial) critical mind is the last topic to be presented in this book. Our social brains were biologically adapted to learn (and host) a spoken language, for social purposes. However, with the origin of written phonetic language about 5,000 ybp (such as Sumerian or Greek language), a human mind emerged with significantly enhanced symbolic and cognitive capacity. Writing helped to stabilize, propagate and elaborate (rationalize) belief systems, such as texts for religion which still dominate writing regarding quantity of publication. Thus, writing served a major social purpose, was often taught for such purposes and greatly expanded and extended social structures. However, the arduous process of learning to read promoted the cognitive integration of both the visual and audio capacities for abstraction, thinking and self-reflection. This combination of these two parallel systems of social intelligence with some inherent symbiotic capacity led to the emergence of a significantly enhanced cognition (and altered brain connectivity). Such learning (silent reading) also promoted the development of the inner voice that could be applied to self-evaluation and self-development. An unintended consequence of the invention of writing was the origin of modern (albeit fragile and partial) individual consciousness. This emerged individual mind was more able to react after reflection (thinking), and not fully compelled by rapid social or emotional reasoning. Thus, along with the development of writing was the development of a process of learning that used observation to set beliefs. Such an approach initiated the development of the modern mind (critical and scientific). Direct visual and experimental observation, however, could confront social beliefs. Science slowly developed reproducible criteria by which it accepts or 'believes' facts and ideas. Thus, this emerged modern mind can sometimes confront our ancient social mind regarding the role and stability of beliefs. Since belief states remain a basal system of human group identity, science is often viewed as an alternative version of 'belief-based' group identity. Science, however, is not a group identity mediated by any belief. Yet scientists themselves retain these ancient biological characteristics of a social mind, and can be prone to belief-based group identity. Belief states when confronted are typically defended by rationalization and emotionally charged language and such defenses are often employed when new paradigms challenge the beliefs of scientists. A curious nested version of a 'belief-state' situation is that is seen in scientists is that they too tend to 'believe' in rational thinking and rationale education as a much more prevalent human process for accepting new information than is supported by evidence. Secular law has historically recognized the role belief plays in human social identity, and the tolerance of religious beliefs in law has been a heuristic solution to the endless belief-based group conflict. However, such practical solutions should not be confused by placing religious beliefs on the same footing as science regarding the reality of our world. Our biological legacy has left us with a strong tendency for belief-based reasoning (including superstitious and irrational thought). Our very language has evolved mostly for the purpose of group identity and is thus rich in vocabulary used to defend group identity (i.e., prevalence of metaphor and emotionally charged rhetoric in social discourse). Thus, there is an evolutionary basis for why so many people can resist the ideas of science (including evolutionary thought). Science needs to assume a much stronger role in education and culture. It should seek a better understanding that can help us confront the essential but negative elements of our biological legacy. Group identity can be a very positive and symbiotic force in society that promotes social cooperation and mutual support. These are inherent features of human social structures. However, group identity can also resist new information and promote the denial of empathy to non-group members and allow group conflict. Science must seek to educate, develop and promote the positive social aspects of our modern social mind. Rational but empathic-based reasoning can form the foundation for an expanded transnational group identity in which decision making and ethics stem from scientific evidence, not social beliefs.

Recommended Reading

1. Margulis, L., and Punset, E. (2007). "Mind, life, and universe: conversations with great scientists of our time." Chelsea Green Pub., White River Junction, Vt.
2. Ryan, F. (2002). "Darwin's blind spot: evolution beyond natural selection." Houghton Mifflin Company, Boston.
3. Villarreal, L. P. (2005). "Viruses and the evolution of life." ASM Press, Washington, D.C.
4. Villarreal, L. P. (2004). Can viruses make us human? *Proc Am Philos Soc* **148**(3), 296–323.
5. Villarreal, L. P. (2008). From Bacteria to Belief: Immunity and Security. *In* "Natural Security: a Darwinian approach to a dangerous world" (S. R. D., and T. Taylor, Eds.), pp. 25–41. University of California Press, Berkeley, California.

Chapter 2
The Prokaryotes: Virus, Hyperparasites and the Origin of Group Identity

Goals of this Chapter

This chapter will focus on the microbiological prokaryotic world to consider the origins and evolution of the molecular systems that control group identity and behavior. I will examine the role that genetic parasites have played in providing systems of group identity. Group behavior is often considered as a characteristic of complex multicellular eukaryotic life forms, including humans. It has always presented a conundrum for evolutionary biology as the self-sacrifice (altruism) of genetically distant group members, in particular, is theoretically problematic. Microbes have traditionally been thought of as free-living individual (selfish) organisms that display little group identity and group behavior. Viruses of microbes are even less thought of in the context of a role in group identity. However, we have recently come to realize that our world is predominantly prokaryotic, such as seen with the most visible example of life from space: blooms of cyanobacteria (see Fig. 2.1). We have also come to realize that this prokaryotic world is itself often communal (living in blooms, mats, biofilms) and that group behaviors are also prevalent. Prokaryotes thus provide the beginnings of molecular systems that regulate group identity. Below, I present the evidence and arguments that such systems often originate from persisting viral agents. Persisting genetic parasites (such as P1) have helped define a most fundamental strategy of group identity: addiction modules, which will provide a central thesis for this book. The concept of an addiction module can be generalized to also include the action of lytic viruses and other genetic parasites (hyperparasites) in defining group identity. Group identity systems also provide the foundation for group (multicellular) immunity. Viruses also help us define the basal sensory systems that are used to communicate group identity (via pore proteins). The role of addiction modules (toxin/antitoxin gene sets), sensory surface receptors, small molecule pheromones and programmed cell death will all be developed from the context of group identity. Most of these systems are also important for sexual identity, a version of group identity. These basal sensory systems have been preserved and further developed in metazoan evolution. Thus, the strategic and molecular foundations presented in this chapter will be directly traced to the evolution of complex group behavior, including that of humans.

L.P. Villarreal, *Origin of Group Identity*, DOI: 10.1007/978-0-387-77998-0_2,

Fig. 2.1 A planktonic bloom (cyanobacteria) in the ocean as seen from space (*See* Color Insert, reprinted with permission from: Wikipedia)

Communities

The existence of communities of organisms is well known for animals. In many cases involving communities, however, the community can for the sake of community survival subjugate the welfare of the individual. This is particularly evident in social animals, such as insects and vertebrates in which some members do not breed or contribute directly to the offspring. In extreme cases, individuals will die for the good of the group. Such altruistic or cooperative behavior has always presented a theoretical problem for Darwinian evolution, which emphasizes the survival of the individual. However, the problem of group cooperation is not generally thought to apply to prokaryotes (bacteria and archaea), which often exist as populations of individuals and where selection on individual behavior is most apparent. Yet here too, group behavior and group identity can be clearly demonstrated. For example, the very large blooms of cyanobacteria, as shown in Fig. 2.1, can demonstrate circadian or light-based group behavior. The formation of microbiological communities (colonies, mats) has become a well-studied field in the last decade and is presented in Chapters 3 and 4. We now know that such organisms sense and identify each other by a process called quorum sensing, and in so doing, will respond by affecting group behavior. In this chapter, I will present the molecular details and possible origins of systems that define group membership. Many of these molecular circuits have been maintained in the evolution of sensory systems of higher organisms. However, in this prokaryotic chapter, I will focus on the role viruses and genetic parasites have played in the origin of programmed cell death (PCD). The case will be presented that PCD can be considered as a suicidal (altruistic) response by an individual to an altered group identity which disrupts an addiction module (group identity) and induces death.

Microbiological communities were of crucial importance for the origin of complex life on Earth, and much of this early history takes place in the oceans. For the first billion years of life on Earth, unicellular prokaryotes ruled the roost and represented the most advanced life forms.

Paleobiology: Stromatolites and Evidence of Early Cyanobacterial Communities

The oldest fossils of life on Earth correspond to stromatolites and cherts that were the result of colonial growth by mat-forming cyanobacteria. These ancient bacterial communities underwent fossilization and can be seen in the sometimes dramatically large stromatolite fossils from the Proterozoic period (2,500–2,590 million years ago). Extant living examples of stromatolites can still be found in shallow warm ocean waters (see Fig. 2.2). These fossils provide direct evidence for the early existence of very large microbiological communities and often show a rather homogeneous cellular morphology, suggesting that these ancient cellular communities may have had a clonal nature. The living extant, although much smaller, stromatolites that can still be found in Sharks Bay, Australia, appear to have mixed communities of bacteria, although cyanobacteria are dominant. The initial age of large stromatolites on Earth was followed by an era in which they underwent a dramatic decline in quantity and scale. Although some stromatolite reemergence occurred later, at no time did they again attain levels equivalent to this early period. Their large-scale deposition was indicative of massive growth by photosynthetic cyanobacteria that had organized into huge communities. The biology of these communities is presented in detail in Chapter 4. It is currently felt

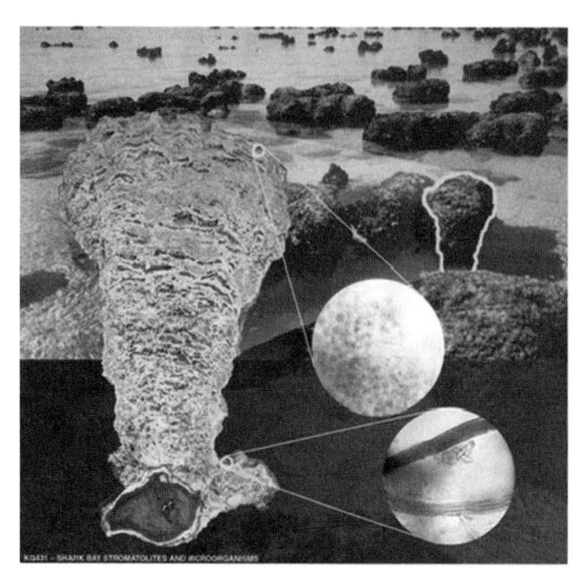

LEFT: Living stromatolites at Shark Bay, Western Australia, showing the laminated structure of a single column and the microscopic cyanobacteria that construct them. Each cyanobacterial filament is about 100th of a millimetre in width. These organisms are rarely preserved in fossil stromatolites, although some have been discovered in chert (silica-rich rock) at other localities in the Pilbara that are about the same age as the Trendall locality.

The bacteria precipitate or trap and bind layers of sediment to make accretionary structures, which can be stratiform, domical, conical or complexly branching. They can range in size from smaller than a little finger to larger than a house. Some branching stromatolites resemble modern corals.

Fig. 2.2 A current stromatolite, similar to those 2,500–2,590 million ybp (*See* Color Insert) (Source: http://images.google.com/imgres?imgurl = http://www.doir.wa.gov.au/Images/GSWA/gsdImg_strom_sharkbay2.jpg&imgrefurl = http://www.doir.wa.gov.au/GSWA/D8826E646FC746C7BCE99DB5CF5F7C01.asp&h = 371&w = 400&sz = 187&hl = en&start = 3&tbnid = aD1tGAEmfNF70M:&tbnh = 115&tbnw = 124&prev = /images%3Fq%3Dstromatolites%26gbv%3D2%26ndsp%3D20%26svnum%3D10%26hl% 3Den%26sa%3DN.)

that this large-scale expansion of cyanobacterial growth contributed to decreased atmospheric CO_2 and increased O_2 that transformed Earth to an oxidizing atmosphere and allowed more energetically active eukaryotes to evolve. Thus, bacterial communities are felt to have participated in major alterations of the Earth's atmosphere. In modern stromatolites, cyanobacteria contribute to the associated mats, which must tolerate intense light, salinity and desiccation. These conditions are not well tolerated by many diverse bacteria, and it has been suggested that this may select for a more homogenous cellular makeup. Recent PCR-based evaluations suggest that one strain of cyanobacteria may predominate in each individual colony. If so, this suggests how a clonal character to the cyanobacteria of some of these communities can be maintained.

Genomes and Parasites of Communal and Free-Living Cyanobacteria

It is not now possible to compare the genomes of extant cyanobacteria in stromatolites to those of the large ancient fossils. However, since most extant prokaryotic genomes remain rather similar to each other in overall characteristics, it seems likely these ancient genomes would also have been similar in basic features. The sequenced cyanobacterial genomes will be presented in detail in Chapter 3. With respect to patterns of genetic parasites, mat-forming *Synechoccus* genomes harbor many genes derived from P2-like phage tails and base plates which closely resemble bacteriocidins (discussed below). Clearly, phage colonization and phage-derived toxins have been a significant factor for the evolution of these mat-forming genomes; and in this, they differ significantly from the non-mat-forming cyanobacterial genome which lack these P2-related genes. It is therefore likely that the identity of mat-forming cyanobacteria has been phage modified. Why we might expect or predict this situation will be developed below.

How do bacterial communities or mats interact with viruses? Currently, we know that oceanic phage of cyanobacteria are abundant and thus can assume that phage must have also been abundant during the early era of stromatolites. A cyanobacterial mat community that is clonal and physically fixed would appear to present a very interesting host ecology with respect to virus interactions. As will be presented below with the P1 exemplar, a bacterial colony persistently infected with a temperate phage is capable of using resident phage genes as immunity/persistence functions to affect the community survival. For example, a lytic T4 like virus could quickly destroy the entire colony without the existence of physical or molecular genetic barriers to phage infection and transmission. And, lytic cyanophage have indeed been directly implicated in the termination of blooms of various photosynthetic microorganisms, including cyanobacteria. Thus, the mat communities would need to initially evolve some defense against lytic phage before they could develop and persist as a dense community. Physical barriers to phage transmission may well be part of the mat-forming cyanobacterial life strategy such that these bacteria may effectively limit external access of virus to individual cells. The mats of cyanobacterial communities (such as *Synechoccus*, Chapter 4) are the result of the secretion of a slime filament from pores that will form a tube that encloses individual cells, as well as propels cell movement. Cells appear to move within these filaments by a flagella-free gliding process, which most likely results from propulsion due to a polyelectrolyte ejected from limited numbers of terminal pores that then undergo cationic hydration to provide the motive force. The involvement of a pore structure in such movement is interesting from the perspective of possible viral origins, as will be discussed further below. Also, cellular replication is coordinated with cellular movement in these mats, which is clearly a crucial issue for virus replication. It is well established that the surface chemistry of a cell has strong affects on its permissiveness to phage attachment and penetration. It therefore seems rather likely that the physical barrier produced by the surrounding slime would effectively limit phage access to the underlying gram-negative cell wall, and also likely affect access of phage-related particulate toxins (i.e., P2-like bacteriocidins). Some phage of other hosts, such as SH1 halophage, indeed appear to be inhibited by thick, walled host cell variants so the accessibility of a virus to the cell exterior can clearly affect phage infectivity. However, this concept has not been well evaluated experimentally in the context of slime or mat production.

Since the filaments that make up stromatolites (slime filaments) are fossilized, the origin of slime production also presents an interesting topic in regard to ancient bacterial group identity. Slime has a big affect on the bacterial surface chemistry, and phage must also interact with this surface. Resistance to phage colonization would likely be much affected by slime production. Slime production is also directly involved in formation colonies by numerous other bacterial species (discussed below in the context of quorum sensing). This issue is of further relevance to community formation under the topic of 'virulence factors'. With mat-forming cyanobacteria, coordinated movement within these slime filaments occurs in relationship to light flux (toward or away). Coordinated movement and group identity is a general issue in bacteria, and their group identity and movement are frequently linked topics. The free-swimming cyanobacteria that make up blooms also move (up and down water columns) in coordination with light flux. Both the free-living and mat-forming cyanobacteria also divide cells in relationship to light flux, which is directly linked to virus production. Many cyanobacteria also produce light, in a community (quorum sensing)-sensitive way. With respect to bacterial movement and light, it is interesting to note that most oceanic bacteria (other than cyanobacteria) are motile and also produce light, but many of these bacteria are non-photosynthetic. However, even these non-photosynthetic bacteria tend to have sensory photodetectors and move in coordination to light. Below, I will develop the argument that photodetection is a common group-identification system of oceanic microorganisms.

Cyanobacteria are photosynthetic and, as such, tend to reside in shallow waters. It is interesting to note that the first photosynthetic bacteria to evolve, such as purple bacteria, would not likely have resided near the surface of the ocean due to intense and lethal levels of UV light radiation present in an early reducing atmosphere. Sensitivity to UV-induced inactivation remains a major factor to the survival of cyanobacteria and their viruses and accounts for much of their turnover in the oceans. It is not clear how UV sensitivity might have affected the evolution of bacterial group interactions. However, it is likely that due to surface UV intensity, any such early group interactions would have to have occurred at greater depths of water. The origin of the ozone layer, which allowed for surface growth, is associated with the emergence and proliferation of stromatolites and their associated photosynthetic production of oxygen along with the generation of the Earth's oxidizing atmosphere. Thus, the resulting ozone itself would have favored even more microbiological growth near the upper photic layer by limiting UV damage that would otherwise have significantly limited cellular growth. Early cellular life had to contend with many harsh environments including intense and toxic levels of light. Viruses (phage) can significantly affect this toxicity, and many have life strategies that are linked to UV host damage and can even replicate in an UV-killed host. In the last two decades, we have come to appreciate just how rugged and tenacious these prokaryotic life forms are, as they are able to survive and thrive in what was previously considered inhospitable environments of extreme temperature, pH, salinity and radiation. Following this age of stromatolites, the Earth underwent a period of significant decline and few stromatolites were present in the fossil record. Although this fossil decline was later followed by a period of some resurgence in fossilized stromatolites, these early huge bacterial colonies would never again attain the prevalence and stature they once held. Giant bacterial communities thus appear to have been central in the transformation of Earth into a habitat that allowed for the evolution of more complex eukaryotic organisms that respire by O_2 oxidation. Let us now develop the molecular strategies that support bacterial group identity.

'Black Holes' of Biology: Prokaryotes and Their Viruses

The stromatolites and cherts are fossils that emphasize the basal role that prokaryotes had in the early evolution of life. However, even today, the world remains mostly prokaryotic in nature if we judge this by biomass, quantity of species or consequence to the Earth's climate. Prokaryotes predominate in all habitats, even in inhospitably hot, salty, frozen waters, or in deep rocks. Since the times of the stromatolites, bacteria have retained their dominant role in the biology of Earth. It has been suggested by E. O. Wilson that prokaryotic organisms represent the 'black hole' of our current living planet. These mostly unseen and uncharacterized individual organisms remain essentially unmeasured in most habitats and are likely to exist in numbers almost beyond comprehension. Thus, our world initially was and remains mostly prokaryotic. But Professor Wilson's thoughtful insight does not raise the specter of a second, even larger and less visible black hole, or of the existence of other ancient and vast biological community. In the last several years, systematic evaluation of the prokaryotic makeup of the oceans has been undertaken by shotgun sequencing methods. From this we have started to get a picture of what that vast bacterial community looks like. As shown in the dendogram in Fig. 2.3, we have roughly defined the evolutionary relationship between many of the prokaryotes that live in the oceans. Although these prokaryotes exist in vast communities, these communities are known to harbor their own, even more massive, but specific populations of autonomous and endogenous (temperate) viruses. The existence of this massive virus population identifies a second even larger 'black hole' which represents a mind-boggling level of genetic diversity. This issue is presented in detail below. As will be presented, viruses and other related genetic parasites of prokaryotes are the masters of genetic novelty and are the chief determinants of prokaryotic evolution. They will now be examined for their possible role in the evolution of host group identity (including cyanobacterial mats) and behavior. In this chapter, it will be shown that persisting genetic parasites are indeed capable of superimposing group identification systems onto their host. Throughout this book, essentially all observations concerning

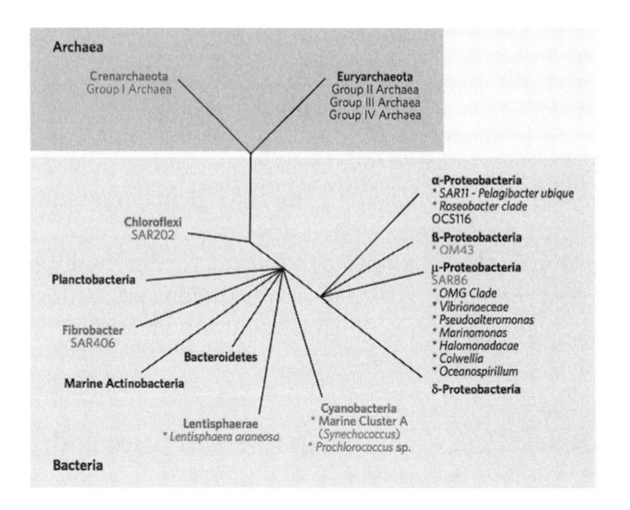

Fig. 2.3 Evolutionary relationships of oceanic prokaryotes (reprinted with permission from: Giovannoni, Stingl (2005), Nature, Vol. 437, Pg 344)

group identification will also be reexamined for the role of various genetic parasites. For example, one very early observation along these lines was by M. Bijernick. In 1892, he was first to isolate cyanobacteria, already known for their blooming and mat-forming group behavior. He reported that a fraction of isolated colonies would spontaneously stop producing light or 'go dark', which was also correlated with altered colony surface morphology. Subsequently, it was also seen that phage sensitivity was simultaneously altered in dark variants. Curiously, all three of these phenotypes could together revert back to the initial state. Why these features were all linked has remained a long and mostly forgotten mystery. As I will present at the end of this chapter, phage sensitivity and group behavior of social bacteria are frequently associated. Cyanobacteria have recently been shown to evolve via genomic phage colonization events (see Chapter 3). In this chapter, I will present the basic case for why we should expect viral involvement in the evolution of most complex host characteristics. Although we have long known that bacteria represent the most ancient and adaptable of all cellular life forms, even more ancient and adaptable are the viruses that infect and colonize them. And, it is the viruses that provide the answers as to how to evolve group identity and behavior.

Viral Origins of Prokaryotic Complexity

Our world is mostly microbiological/viral. Thus, the bulk of life on our world is made up of haploid 'viral' and prokaryotic organisms. Such a haploid world provides a vast and dynamic genetic foundation, but one that tends not to conform to accepted views of evolution, due to high rates of genetic colonization from distinct lineages. The iconic tree-based dendogram commonly used by evolution biologists to represent evolution fails when applied to this prokaryote-viral situation. For example, a dendogram-based analysis fails to identify a particular lineage as the 'Last Universal Common Ancestor' (LUCA) to the three domains of cellular life, due to the high levels of 'horizontal' transfer that prevailed. Instead, a diffuse unresolved 'cloud' defines the ancestral core of such dendograms. Prokaryotic evolution more closely resembles a network of reticulated genetic invasions by viral-like elements derived from host-disassociated trunks. The thesis of this chapter is that most answers concerning origins of systems of biological group identification can be found to have originated within oceanic microorganisms, especially in their viruses. The concept of 'horizontal transfer,' although accurate in a strict sense, has unfortunately become an accepted concept in which viruses are bit players, accessory elements that simply move genes from one host to another. As noted in my prior book, 'Viruses and the Evolution of Life,' this accepted view fails to acknowledge the much more fundamental relationship between viruses and the evolution of prokaryotes. It especially ignores the well-established and vast creative capacity of these genetic parasites to evolve complex solutions. Viruses (and other persisting genetic parasites) are the ultimate providers of complex multigene solutions to issues of host identity. Thus, the creators of this new host identity are the genetic parasites that can prevail on their host, not other hosts. But the parasites themselves are numbingly complex, highly interactive, and encompass many distinct strategies for competition, cooperation and persistence. In this chapter, I delve into the often unintelligible nomenclature and classification of these genetic parasites and attempt to understand how they function collectively and how they have affected evolution of host group identity. Especially complex are the interactions between hyperparasites and viruses as outlined below. The passive reader will likely get lost in this nomenclature, but it is hoped that the chosen exemplars will be clear enough to communicate the general principles involved.

Prokaryotic Evolution: Parasite Persistence as Fundamental

Clearly, much biological evolution must have been occurring during the first billion years of life on Earth when prokaryotes represented the most complex life form. However, with scant fossilized evidence that represents this period, we are hard-pressed to evaluate in any detail what developments had taken place. Yet we do know much about how prokaryotes evolved due to substantial progress in comparative genomics. I previously presented evidence that most of the genetic novelty associated with prokaryotic evolution occurs by a process that involves stable genetic colonization by parasitic, mobile genetic elements (i.e., the so-called horizontal transmission). Acquisition of various exogenous gene sets, at specific integration points, is the common difference between two closely related prokaryotic genomes. Viruses (called phage in bacteria) and their defective and parasitic relatives are the main movers of gene sets between species and are directly involved in integration/colonization and evolution. Historically, viruses have not been considered as members of the tree of life; thus, any such genetic role in host evolution was dismissed. I have previously written several essays in which I argue that they are indeed major architects of life and must now be included in the tree of life (see Recommended Reading). Much support now exists for this view. With the advancements of comparative genomics, we can now begin to clearly see the footprints left by viral activity in the genomes of all life forms, especially all prokaryotes. Yet, although evidence of viral involvement is clear, the question has remained as to how a genetic parasite can contribute to the evolution of their host or generation of host complexity. Why would a lytic virus (i.e., a runaway replicon) evolve such genes? It would not. For this, we need to consider viral persistence as the defining, fundamental and constructive life strategy. Persisting viruses must become one with the host, establishing a stable colonization that then defines a new viral/host persisting identity. This is a key point: persisting viruses can create and superimpose new molecular genetic identity into their host. Once this has occurred, the evolutionary trajectory followed by the host/virus is strongly affected. For such a process to be successful in an evolutionary timescale, the parasite must have some mechanism that compels the long-term (temporal) virus–host stability. Time and longevity matters for fitness and temporal stability (not simply rate of offspring generation) are of central importance. Thus, a key objective of a new genetic colonizer is to somehow 'force' this stable association. It is from this very process that not only will a new individual genetic identity be established for specific host, but will also a new 'group identity' (population of host harboring the same genetic parasite) can be established. The establishment and evolution of such group identity is the main theme of this book. Since this process can be seen to occur in prokaryotes, and since prokaryotes and their viruses were the progenitors of all complex life, this topic will here be examined in considerable detail in order to properly evaluate existing and relevant information and to establish the authority of this conceptual foundation. However, such molecular detail will likely tax readers that are less expert on all the relevant terminology and experimental detail that will need to be presented. In an effort to make this section more intelligible to a wider audience, I have attempted to summarize in more general language how the cited evidence supports my thesis. These summaries will precede the presentation of relevant results, allowing readers to understand the general arguments without having to know the experimental details.

As the vast and ancient viral-microbiological world was and remains mainly unseen, it has not had the appropriate historical impact on the development of our thinking of evolutionary biology. Darwin could not have understood a virus. Similarly, modern rationalizations that dismiss viral-like elements as junk or selfish DNA do not acknowledge or understand the viral persistence. Thus the prevailing view is that things are different down there with the prokaryotes. Evolution tends to occur by strange 'horizontal' process that we don't see much in eukaryotes. Sex is also different, resulting from a mainly infectious and transmissive process that resembles virus transfer. We have come to accept that although prokaryotes are the most highly adaptable and prevalent form of all cellular life, they are oddly 'outside' of the normal evolutionary process due to their established ability to acquire gene sets by 'horizontal' (colonizing) genetic events. Viruses, even less visible and more highly adaptable than their bacteria host, have been precluded from serious consideration by neo-classical evolutionary biology.

Addiction as a Central Theme

Comparative genomics now establishes that genetic parasites (including prophage and plasmids) are most common mediators of complex (multigene) and rapid prokaryotic evolution. Such elements represent the main differences that best characterize an otherwise most similar bacterial genomes and account for recent genetic difference between bacterial species. As noted above, temporal stability seems key to the successful host colonization by genetic parasites. In my prior book, I presented the argument that the persistence of a genetic parasite needs a mechanism to create stability, which I have called an 'addiction module'. The concept of an addiction module was originally coined to explain the stability of P1 phage (plasmid) in its *E. coli* host and the fact that a host cured of the virus would die (see Recommended Reading). The module consists of two components, a stable and potentially harmful component counteracted by an unstable but protecting or beneficial component. Both of these components are encoded by the parasite and can be contributed as genes, toxic molecules, or even virus infection itself. Loss of the genetic parasite initially results in loss of the unstable protective (immunizing) component of the addiction module, leading to activation of the stable harmful component and cell destruction. Thus, colonized cells are 'addicted' and must stably maintain the protective immunity function of the parasite for their own survival (see schematic in Fig. 2.4). In this chapter, we will extend this concept to not only include the gene products produced by viruses and other genetic parasites, but also include the overall immunity and host lysis that persisting genetic parasites can induce in their host. This latter feature has an infectious or transmissive character that will affect neighboring cells, thus creating the foundation of population-based group identity. Thus, the basal role of persistence via addiction for the origin and evolution of host group behavior will be maintained as a central thesis throughout this book. The molecular process by which genetic parasites attain stability will be presented as of major importance for group behavior. A basic linkage between group identity and individual or group immunity will also be developed. In fact, the identity of an organism cannot be separated from the process of immunity (e.g., recognition of foreign cells), and it will always be necessary to consider both identity and immunity when examining group behaviors indicated in Fig. 2.4.

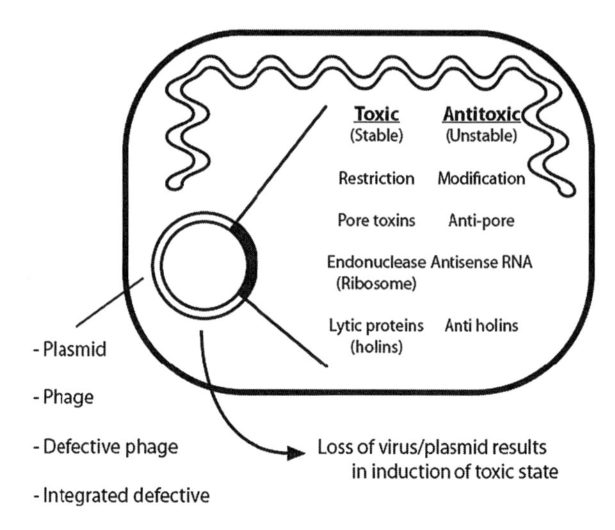

Fig. 2.4 Schematic of addiction module (with self-killing)

Group Lessons from Bugs

The unicellular world has much to tell us concerning the mechanisms that underlie group behavior. The basic molecular systems that are used to control group behavior were initially invented in the prokaryotic world and have mostly been retained and further elaborated in the eukaryotic world. These systems include the major sensory systems, visual, odorant and mechanical (sound) detection systems. A component of bacterial odor detection is the detection of pheromones, associated with self or group identity. Bacteria have also evolved several systems of motility also associated with sensory-mediated group behavior. Both pore-based and flagella-based motility will be discussed. Along with sensory detection and motility, bacteria have also evolved various systems (e.g., pore toxins) that can attack and kill non-group members. Prokaryotes have thus solved most of the molecular problems associated with group identity, and similar (or same) types of molecular solutions were used later in evolution of unicellular eukaryotic life forms, such as the dinoflagellates, algae, fungi, and protozoa (see Chapter 4). As these unicellular eukaryotes have further developed most of the basic systems used for group identification and behavior, their application to the evolution of multicellular, metazoan eukaryotes will be presented in subsequent chapters. Sensory detection, in particular, will be emphasized as a theme throughout this book as this topic provides a pathway leading directly to human cognitive evolution (see Chapter 9). However, regarding molecular origins in prokaryotes, cyanobacteria are especially useful. These photosynthetic oceanic bacteria are known to be able to respond in groups to light. Light sensation and group response systems in particular will be considered in detail and shown to provide a crucial chain of clues to origins of group control (see Chapters 3 and 4). This conceptual thread, light and odor detection (including pheromone) leads directly to visual, odor and sound detection of all animal life and relates to the evolution of their group behavior.

Complexity, Entanglement of Group Identity, Immunity and Genetic Hyperparasites

I have asserted that genetic parasites will play a central role in the evolution of group identity. But persisting genetic parasites also involve their own versions of parasites that I call 'epiparasites' or hyperparasites (parasites of parasites). A simple version of a hyperparasite is the prevalent defective elements of bacterial virus (cryptic phage). More distant hyperparasites are also included (satellites viruses, plasmids, homing introns, intenes, etc.). These more defective elements depend on (require) interactions with other elements for new host colonization. They often participate in mixtures during host colonization and resulting parasite stability. In this book, I will make no effort to present these additional elements as distinct or autonomous elements as is typical in the literature. Instead, I will consider them as interacting networks of linked parasites. This network approach can result in seemingly unnecessary system complexity in apparent violation of the maxim of Occam's razor. William Ockham was a fourteenth-century logician who presented the maxim that 'entities should not be multiplied unnecessarily'. The simplest or most parsimonious explanation is to be favored. However, in the context of evolution of group identity, this maxim does not usually apply. Group identity is frequently a mixed affair involving hyperparasites. Group identity from mixtures is an inherent feature of virus-related (quasispecies) evolution. The main difficulty stems from the problem that group immunity, group identity and epiparasite persistence are inseparable concepts and processes. Instead of each of these characteristics having distinct potentially simple origins and solutions, these three characteristics are always affected simultaneously and also result from more complex interactions. They cannot be further separated or reduced to autonomous constituent elements and phenotypes, hence their origin and evolution and function will be tangled. A schematic of some of the hypothetical interactions of hyperparasites is presented in Fig. 2.5. A more detailed example of this situation will be presented later, in Chapter 4 using the molecular biology of sex determination in *Caenorhabditis elegans*. The result of such hyperparasite-based evolution is the emergence of seemingly, overly – complex solutions to simple problems. For example, the molecular basis of the signal transduction system appears to resemble a Rube Goldberg-like design in being overly complex for the task at hand. This complexity is a necessary product of epiparasite colonization and displacement of existing identity systems, not efficiency. Efficiency is the mainstay of theories of evolutionary biology; and by linking efficiency to survival, it can predict a system that might appear very similar, but it should be much less complex. But, seemingly unnecessary complexity is

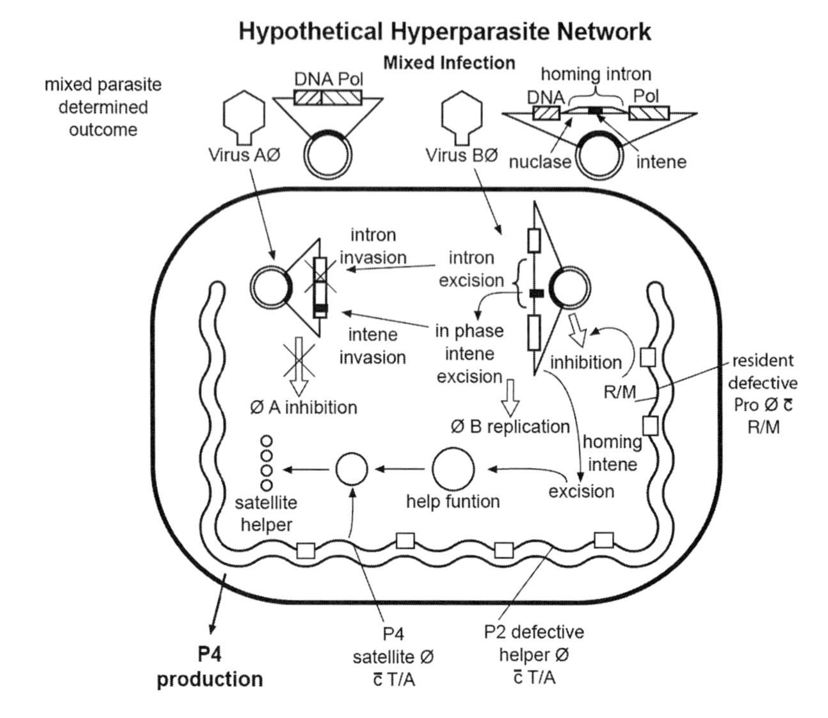

Fig. 2.5 Schematic of hyperparasites (plasmid–phage–intron networks)

common in genomes of higher organisms, particularly due to parasitic elements. Evolutionary biology has tended to explain this apparent and unnecessary complexity by applying the concept of 'selfish DNA' to account for the parasitic elements. Parasitic elements are not linked to group identity. In a section below, I will explain why the concept of selfish DNA fails to account for the evolution and accumulation of group identity.

Further Entanglement: Molecular Diversity and Identity

I will now provide what appears to be a further assault on Occam's maxim. Another entangled theme that is directly related to issues of group identity and the persistence of addiction modules is the relationship between identity and molecular diversity. Although identity and molecular diversity may seem to be distinct topics that can be defined separately, they are also not separable. Identity and diversity are two necessary states of the same issue, and that issue also includes immunity. Like the particle-wave duality of light, these three issues together behave like quantum entanglements: no matter how we try, we cannot separate them from each other. The concept is that competition between related groups (usually within the same species) will drive the evolution and diversification of its corresponding recognition system. That group recognition system will need to persist to define group identity. The common molecular strategy that will allow persistence and link these two issues is the use of addiction modules. But the addiction module itself will provide the basis of the group immunity system, protecting identity and harming non-identity. For example, restriction/modification enzymes (responsible for modifying DNA and cleaving unmodified DNA) are amongst the most diverse of all bacterial proteins. Restriction/modification enzymes are also the most prevalent form of genetic immunity systems found in prokaryotes and are closely associated with group immunity. Further, persisting genetic parasites can encode restriction modification enzymes and use them as addiction modules that insure maintenance and stability in their host (see Fig. 2.4). Thus, a link between molecular diversity, immunity, group identity and molecular addiction can usually be seen. As will be developed below, the existence of molecular diversity itself (visible by genomic analysis) will often be a strong clue concerning the mechanisms of group identity used by any particular organism. Special attention will thus be paid to the systems in any organisms or lineage of organisms that show a particular pattern of high molecular diversity. Along these lines, the diversity of genetic parasites peculiar to any and all organisms will also be examined to evaluate their potential role in group identity and group behavior.

Renaissance in the Study of Bacterial Communities

Another development in our understanding of bacteria that has principally occurred since the 1990s is the realization that many prokaryotes exist in microbiological communities, such as biofilms or colonies. Contrary to the prevailing view of bacteria as isolated individuals, bacteria often operate in social groups that coordinate their colonization, movement, feeding, differentiation and mating in groups. Thus, we now know that such social or group behavior is surprisingly common for many microbes. Bacteria are clearly capable of social movement, moving toward or away from each other or from other signals, and this is especially apparent in the oceans in association with light and small-molecule detection. Bacteria are also capable of both harmful (toxic) and beneficial effects on each other and have surprisingly elaborate communication systems to affect such behavior, especially colonization. The term 'quorum sensing' (QS) has become accepted, and it is often applied to the group detection systems of bacteria. However, some bacterial species maintain multiple parallel QS systems that affect related genes (see Chapter 3 for details). Such multiple QS systems raise questions concerning the purpose of such systems. It will be asserted here that such systems are actually better understood as group identity systems and that the molecular diversity of the chemical identifiers of systems are characteristic of and peculiar to the specific species. In fact, I will posit that group identification is a general and basic issue, important for, and applicable to, understanding the fitness of all life forms. And as a corollary, I am also asserting, as noted previously, that such identity systems tend to be especially directed at related species and as a consequence of interspecies competition. Identity molecules tend to be the most diverse molecules for a specific species. Thus, for example, the role of diverse toxins in colony or group maintenance can be considered from this perspective.

Awakening to the Viral Ocean

As mentioned above, since the early 1990 s, there has been another radical shift in our perception of the microbiological ecology of the oceans. It had long been thought that viruses were present in the oceans in insignificant numbers and that they had little consequence to the ecology. With the development of molecular screening and proteome technology, we now know that this early view was incorrect. Viruses clearly predominate in the oceans, a seeming black hole of biology and have major consequences on ocean ecology. Current estimates based on electron microscopy of viral particles in ocean water suggest that there are about 10^{31} viral particles in all the oceans (see Recommended Reading). How might these ubiquitous genetic parasites affect bacterial group identity and/or group behavior? Most people understand viruses as simply killers that consume their host, like a predator kills its prey. Such destructive parasites would seem incapable of having any positive consequence with respect to host group identity. But, as I have asserted, viruses often silently colonize and persist their host for its lifetime and sometime even the lifetime of their offspring. It is such persisting infections that directly affect bacterial group fitness and identity, returning us to the trinity of concepts: identity, immunity and addiction. As I will show below, such stably persisting viruses provide molecular genetic identity to their host. They superimpose these new molecular genetic group identification systems and can permanently alter the group survival and often the genetic composition of their host. A most surprising realization that stems directly from this colonization is that the very same system that promotes persistent colonization also compels altruistic group behavior (programmed cell death) onto its host. Such colonization events not only alter group identity but also can initiate an entirely new evolutionary trajectory for the lineage. It is the selective pressure created by such colonization that can also result in new species with distinct sexual and group identities. Thus, this vast number of viruses found in the oceans has an enormous collective potential to affect bacterial group identity and behavior and provide the genetic powerhouse that drives the evolution of complexity in prokaryotes.

Early Observations, Group Immunity from Virus: Mixing Bugs that Kill Each Other

Most readers are likely to have heard the term lysogenic. Modern definitions indicate that this term represents a relationship between a bacterial virus (phage) and its host in that the virus is residing as an essentially silent genetic element within the chromosome of the host, but can be induced to produce more virus and lyse its host (often following damage to the host with UV light or drugs, i.e., mitomycin C). However, the origin of this term describes a somewhat different situation. In the 1920 s, many researchers were excited by phage research because it appeared to offer a way in which bacteria could be specifically killed, hence this research could potentially provide possible cures for many bacterial diseases (1915 F. W. Twort 'glassy transformation,' F. D. Herelle and bacteriophage). However, it was sometimes observed that when two related strains of bacteria were grown together, one of them would lyse the other, hence the term a lysogenic strain. Later, it became clear that the lysogenic strains of bacteria contained a silent and persisting virus (prophage) that would infect and destroy the other non-lysogenic bacterial strain. The lysogenic strain itself was not destroyed in these mixed cultures. In addition, it was subsequently discovered that the lysed strain of bacteria could sometimes be turned into the lysogenic strain of bacteria if it was exposed to the virus, but then survivors of infection were selected. In fact, it was determined that the only difference between the lysed bacterial strain and the lysogenic bacterial strain could be the presence of the lysogenic phage. The host need not change any genes, the virus had become part of the host. This early observation had defined a mode of group identity that led to decades of subsequent research into trying to understand the details of how lysogenic phage functions. Such studies eventually provided the very foundations of molecular biology. However, some confusion and disagreement developed early on when other respected researchers also described phage (such as the T-even phage), which always killed their host (now called lytic phage). We now know that these lytic phage have a clearly different relationship with their host, consuming them in the more accepted way, like a predator consumes it prey. But let us further reconsider the implications of the lysogenic host relationship. What results is one group of bacteria will kill a very closely related second group of bacteria due to the virus that it silently carries. This clearly seems to be some type of bacterial group recognition. The virus in the lysogenic bacteria does not normally (or frequently) kill its own host because in this situation since the virus expresses stable 'immunity genes' that prevents virus production. The non-lysogenic bacterial strain, lacking these same viral immunity genes, is thus killed following infection. On the face of it, this might seem to simply represent the actions of a clever and selfish genetic parasite that has forced itself onto its host. However, there are significant fitness consequences to the now colonized host, besides the fact that they have acquired a whole set of new viral genes. Principle amongst the fitness consequences is that the colonized host, by expressing the viral immunity genes, has modified it own molecular genetic identity and can respond to a new and distinct set of genetic invaders. The viral immune genes can now recognize related (and sometimes unrelated) viruses, bestowing onto its host a broader immunity to other lysogenic and lytic phage. This has highly significant consequences to the survival of this host in virus-infested oceans as it has acquired immunity and group identity. If related lysogenic or lytic viruses are in the habitat of this colonized host bacteria, its likelihood of surviving this viral exposure is completely transformed. Furthermore, to acquire this new 'viral identity', the host need not be colonized by a fully infectious virus, but it can also be colonized by a partial (defective) virus that still retains the immunity function or genes, but might lack many or all of the genes needed for the production of infectious virus. These relationships and outcomes are outlined in Figs. 2.4 and 2.5. This last observation makes a crucial point: seemingly defective genetic parasites can nevertheless fully transform their host such that the colonized host has acquired a new molecular genetic identity and viral immunity. In some cases, no actual viral genes might be involved as the presence of viral regulatory DNA itself can affect the outcome.

Virus as Toxins

There is much more to this parasite–host relationship than this simple outline has indicated. For example, it seems that essentially all prokaryotes (Archaea and Bacteria) express toxins principally aimed at killing similar species of bacteria (described below). These toxins can define group membership, but as will be presented, also appear to have evolved from defectives of early viral parasites. Toxin and antitoxin production appears to be an especially effective means by which colonizing genetic parasites and hosts can attain a high degree of temporal stability.

What It's Not, Selfish DNA: Persisting Parasites Affect Fitness

It is worth emphasizing a key point at this time. A 'defective' virus (or hyperparasite) as presented above is not simply a selfish DNA element. Historically, the frequent presence of defective and nondefective elements of genetic parasites in genomes has been explained by another theory, called selfish DNA. Dawkins book of 1976 initially developed the concept of the selfish gene and selfish DNA in the context of evolution. But in 1980, Orgel and Crick as well as Doolittle and Sapienza proposed the concept of selfish DNA to explain the large numbers of such noncoding defective parasitic elements in eukaryotic genomes. Since it had long been observed that the genomes, especially of more complex eukaryotes, have large amounts of parasitic DNA, it was clear this would need to be accounted for. This explanation, which has become widely accepted, was that this DNA is present simply because it has the capacity to be copied (in a selfish way) and the host lacks the molecular capacity to efficiently eliminate it. Thus, it persists in the genome simply because it can. It otherwise has no fitness consequence to its host and only a small price of efficiency is paid for its presence (i.e., junk). In contrast, I argue that the colonization (and expansion) by essentially all such parasitic DNA is associated with parasite competition and preclusion. Numerous examples are presented throughout this book that support this assertion. These elements once promoted the superimposition of new molecular genetic identities that occurred at the time of parasite colonization (usually during a selective sweep by infectious agents). During this period, they were highly selected to provide persistence or preclusion of the same or related and competing parasites. In so doing, the group identity of the host is usually altered. Furthermore, these new colonizers will frequently interrupt extant molecular identity/immunity systems that opposed colonization. Thus, there will generally be a basic connection between such genetic parasites and the extant host group identification and immunity systems. However, in contrast to the selfish DNA hypothesis, these genetic colonizers have had major consequences to host fitness, especially when fitness is measured against related genetic parasites. Another difference between this view and past views concerning genetic parasites concerns the definition of fitness. Persistence of parasites requires a fitness that is not defined solely by increasing host reproduction. Rather it can prevent host destruction of those that are colonized. These parasites will especially affect host longevity by providing immunity against prevalent genetic parasites at the time of colonization. Hence, the fitness effects of these elements can be major, but have often been transient, dependent on previously prevalent genetic parasites that may no longer prevail. Due to this, such elements may no longer appear to provide a selective advantage, thus they adhere to the tenants of selfish DNA.

Dynamics of Parasite Colonization

One implication of the above assertions is that waves of colonization can be expected with increasing host group identity. This is because a successful colonization, in order to persist, must preclude infection with the same genetic parasites. However, once this happens, the colonized host is immune and may no longer be under selection by that same exogenous parasite. This tendency to extinguish exogenous parasite susceptibility can lead to a host genome in which waves of colonization and subsequent exoparasite extinction have occurred. Furthermore, the prevalent genetic parasites need not be constituted by a single genetic entity. Frequently, mixed parasites (epiparasites) and hyperparasites (parasites of parasites) will be involved together to form stable host colonization. In fact, such mixed parasite situations appear to be much more common than is generally appreciated. For example, some of the most-studied mammalian viruses we know (SV40, Polyoma, Adenovirus, MuLV, KSV) were initially isolated from mixed persistent infections. As will be presented below, this is also true for the iconic phage of *E. coli.* The nature of these parasite–parasite interactions span a large range of positive, negative, dependent and independent relationships and is considered in greater detail below. The main point to emphasize here is that it is often incorrect to consider the action of a single genetic parasite as an autonomous agent. This is seldom the situation in natural habitats, as parasites appear to have established networks of interaction that can together build complex host identities. Here again we seem to violate the dictum of Occam's razor. The simplest solution may not be the best. The concept of epiparasites as a source of host identity is frequently confusing to some readers. To better understand this idea, some background will now be presented on how genetic parasites can be expected to add to host complexity and identity.

The First Persisting Virus/Replicator

The concept that genetic parasites can be expected to contribute to evolution of host complexity and host identity will be counterintuitive to most readers. This issue was developed in greater detail in my prior book, 'Viruses and the Evolution of Life,' and readers wishing to evaluate the details of this proposal are advised to read the early relevant chapters of that book. Here, I will simply provide an outline of the ideas in order to provide a better context for the remaining chapter. When discussing the ultimate origins of viruses or all genetic parasites, it has often been noted to me by colleagues that all viruses need cells; therefore, cells must have evolved first. This conundrum can also be considered as the chicken and egg argument in terms of which came first. Well, this analogy is not exactly correct. From the perspective of evolution, even the chicken and egg argument has a clear solution. An egg, being a single cell must represent an earlier life form than the much more complex multicellular chicken. Single cell organisms clearly evolved before multicellular organisms. In terms of the first virus, we are relatively certain that fully functional cells were not the first life forms. Acellular, proto-life forms must have predated extant DNA-based cells. We can replicate virus in acellular (in vitro) situations, such as those done by Eckhard Wimmer and colleagues, who were able to synthesize poliovirus in vitro and replicate virus in a test tube (sans cells). However, even putting aside this observation, since it might still be argued that the complexity of this mixture can only be found in a cell, the basic requirement still seems to call for evolving cells first in order to evolve virus second. But this is a false analogy. In the last two decades, it has been generally accepted that earlier life forms based on RNA genomes must have predated DNA-based life (known as the RNA world). How these initial RNA-based cells evolved also presents a problem, which will not be addressed here. However, prior to this RNA-based cell, there must have also existed a pre-cellular world in which chemical replicators were the prevalent proto-life forms. Possibly, this replicator world was restricted to chemical surfaces, such as clay-like interfaces or membranes. In its simplest form, a chemical replicator catalyzes its own synthesis from available substrates. For this to lead to a biological replicator, it requires two functions: it must encode the information or instructions for itself, as well as encoding the catalytic function for its own synthesis. The linkage of these two functions (catalytic phenotype to genetic genotype) has been proposed by Manfred Eigen and colleagues to involve a system called hypercycles. These hypercycles provide a potential solution to the problem of maintaining and building genetic complexity from a chemical replicator that has high error rates. However, even at such very early periods in the evolution of proto-life, genetic parasites are expected to have been prevalent. Any replicator can become parasitized since a parasitic replicator is usually a simplified version of the replicator that needs only encode its instructions, and it can dispense with the catalytic synthesis since it has parasitized the replicator for that function. In fact, there is reason to think that any system of information can be susceptible to parasitic information systems that copy themselves at the expense of the whole system (such as computer viruses). The implications of this simple situation are profound. It is likely that there was never a circumstance in which cells, or protocellular replicators, were free of genetic parasites. As soon as the first replicator evolved, parasitic replicators could also have evolved. Thus, the first genetic parasite (virus) would not have required a fully functioning cell; viruses were present all along. However, such a situation might make it difficult to envision how this proto-life managed to evolve greater complexity, as this would result in a collapse of the hypercycles, as noted above. The entire system would seem prone to rapid degradation due to parasite loads that selfishly replicate themselves without contributing to replicating the entire system. However, as presented in my earlier book, the evolution of *persisting* genetic parasites has a very different outcome from this scenario. Since in order to persist, such stable parasites must preclude other competing parasites, including themselves, they impart onto their host a new replicator identity system that allows for differentiation of disallowed replicators. Thus, they create a more specific identity and lead to the cumulative building via colonization of increasing host complexity and host identity. Such a role would represent a most fundamental character for the evolution of host.

Is there any evidence that can support the fundamental role of persisting replicators in the evolution of life? As outlined in my previous book, such very early events in the evolution of life are necessarily highly conjectural at this time since no remnants from that age have survived. Very little, if any, direct evidence thus exists concerning such early events in the evolution of life. However, it is possible to make computer models that evaluate some of the ideas noted above; and such models lend support for the role of genetic parasites for prebiotic or informational replicators.

The First DNA Virus/Cell

Unlike prebiotic replicators, the evolution of DNA-based cellular life forms represents an event for which at least some clear phylogenetic information remains, and this information can be used to evaluate various ideas concerning the origin and evolution of cells. All three domains of cellular life (Bacteria, Archaea, Eukarya) have dsDNA genomes, but differ from each other in many fundamental ways, including the specific proteins involved in DNA replication, the chemical nature of their membranes and the specifics of the translational system used. Yet all three domains use the same genetic DNA code, which has been used to argue for a common origin as a DNA-based cell. Sequence analysis, however, is only able to identify about 325 genes that are in common to all three domains of life. These genes are thought to represent a gene set that was contained by the Last Universal Common Ancestor (LUCA) to all cellular life. Thus, it is exceedingly curious that the genes involved in DNA synthesis and replication are not amongst this conserved gene set. It has thus been enigmatic to explain how these three distinct DNA-based cells originated, given their significant differences, which make a common ancestor to the three DNA-based systems seem most unlikely. However, it is equally unlikely that three independent origins of DNA-based cells might explain the current domains given the common genetic code. Recently, Patrick Forterre (Pasture Institute) has used phylogenetic evidence to propose that the three distinct sets of DNA replication systems originated from viral sources. He suggests that persisting retroviruses and DNA viruses were the likely progenitors that provided the specific machinery of all three domains of DNA-based cells, but that distinct sets of persisting viruses were associated with each cellular domain. The hosts for these viral colonizations are proposed to have been three distinct lineages of cells based on RNA genomes, but with distinct membranes and translational systems. Thus, a most fundamental and complex system of cellular biology has been proposed to originate from persisting viruses. If persisting viruses might provide such basic cellular machinery, it also seems likely that viruses could also contribute to other complex systems of the host.

The Viral Nucleus

Another equally profound conundrum of evolutionary biology is the eukaryotic nucleus. The nucleus has many features that are not found in prokaryotes. These include the separation of DNA replication and transcription from translation via a double nuclear membrane, the existence of pore complexes on this membrane, the processing, capping, polyadenylation splicing and transport of nuclear RNA, the structure and nature and segregation of chromosomes, the function and complexity of the cell cycle plus other features, essentially none of which are found in prokaryotes. There appears to be no candidate prokaryote that could have been the ancestor to the eukaryotic nucleus. In recent years, various researchers have suggested a possible viral solution to the origin of the nucleus. As outlined in my prior book, a strong argument can be made that a persisting cytoplasmic large DNA virus, similar in many ways to poxviruses, is the likely candidate to have provided essentially all of the features needed for nuclear function. Thus, there are several arguments based on phylogenetic evidence and analysis in support of the idea that persisting viruses were involved in major host adaptations during evolution. In this light, their possible contribution to the evolution of host identity systems would seem not to pose a major problem.

Colonization Versus Horizontal Transfer: Gene Trucks or Gene Factories?

Comparative genomic analysis has established that prokaryotes are subjected to the frequent invasion of exogenous DNA sequences. Many such sequences are not found in any likely ancestral cellular organisms but instead seem to have originated outside of the species lineage. For example, the biggest differences between the two, first sequenced related bacterial species (i.e., *E. coli* and *B. subtilis*) are to be found in limited numbers (several hundred) of multigene chromosomal domains that differ between the species. Since most of these differences are also adjacent to tRNA sequences, it seems clear that such regions were acquired from non-ancestral sources via DNA integration. Given that phage (not hosts) are known to utilize tRNA genes for integration, it would appear that colonization by phage carrying various sets of genes, including plasmid genes, is frequently involved. In Chapter 3, we will see that even very similar species of cyanobacteria differ from each other in precisely such a way. However, this process has generally been referred to as horizontal transfer, indicating that such genes originate in one bacterial species but are thought to be transferred, via the action of viruses and other genetic parasites, into the chromosome of another species of bacteria (aka lateral transfer). In fact, it has been proposed that the early common ancestor to all three domains of cellular life must have existed in essentially a 'horizontal' cloud involving rampant horizontal transfer which later 'condensed' into the three domains of extant cellular life. The concept of 'horizontal' or 'lateral' transfer is thus widely accepted with such apparently strong support. However, in my prior book, I took issue with this prevalent view. The problem, as I presented it, is that this concept assumes that the role of the genetic parasite is simply to act as a vehicle to move genes from one lineage to the other. Thus, viruses are simply cargo carriers that play no role in originating any of the transferred genes. As I have argued, this limited view ignores the vast evolutionary and creative power of viruses and is not consistent with many observations concerning genes that are known to be unique to viruses and genetic parasites. Phylogenic analysis clearly shows that viral lineages, which can involve hundreds of genes, mostly have genes that are unique to the viral lineage and are not host derived. In addition, comparative viral genomics also shows that viruses often create and acquire new genes at very high rates, usually by recombination between various other viral lineages. In some viral lineages, essentially no viral genes are host derived. In fact, viral genetic variation will by far exceed that of any host, indicating very high rates of viral gene evolution. Clearly then, viruses can be gene factories, inventing new genes in high numbers. Viruses can also readily colonize hosts in highly stable relationships, thus potentially adding numerous genes in a single event that can alter host identity systems. Taken together, these observations support a different concept for the role of viruses and other genetic parasites in the evolution of hosts. Genetic parasites have vastly greater genetic plasticity than do hosts. This plasticity stems from very high copy error rates, very high recombination rates, the existence of viral quasispecies populations that have higher fitness, and the ability to complement (or tolerate) very large numbers of defective variants within the viral population. Viruses (such as HIV) have been measured to evolve at rates up to one million fold greater than their host. This enormous capacity for evolution suggests that viruses are the creators of many new interacting gene sets, not simply trucks transporting genes between hosts. And, some of these viral gene sets can clearly be used to colonize hosts in persistent ways, since immunity genes are often the most variable in some viral lineages. However, if different host lineages were to be colonized by related viral lineages, they would subsequently appear as if one host lineage had transferred gene sets to another host lineage, using viral trucks. Thus, the prevalent view that viruses simply transfer genes between host lineages developed. A careful re-examination of phylogenetic data, however, will almost always support the concept that such transferred genes originated outside of any host cell and must be derived from the genetic parasite itself. I therefore suggest that the term ‘horizontal’ or ‘lateral’ transfer is generally misleading and should be supplanted by the term ‘colonization’. This issue will come up repeatedly in the rest of this book, especially in the context of the evolution of mammalian genomes and the differences between closely related species, such as chimpanzee and human genomes. Genetic colonization involves the superimposition of new genetic identity, whereas horizontal transfer, or selfish or junk DNA, has no identity consequence. Colonization allows us to understand why there exists a seemingly unending process of acquisition and extinction of genetic parasites, and why this would be associated with greater and greater host identity and complexity. Genetic parasites are creators of new identity, not simply vectors of a few genes. Group identity and selection, not simply individual fitness, thus results from stable colonization. With these concepts, we are now able to approach the evolution of communities and how community behavior is linked to group identity. The rest of this chapter will therefore focus on the types of genetic parasites that colonize microbes, the various mechanisms and strategies they use to impose stable new identity and the consequences these parasites have on bacterial group identity.

The Categorization of Genetic Parasites: Babble of Reductionism

In this book, I define a virus as originally proposed by Salvador E. Luria. A virus is a molecular genetic parasite that depends on its host for replication and/or maintenance. In its simplest form, a virus is a genetic string that can parasitize its host cell and be maintained or reproduced. This is a broad and encompassing definition, but one that can include the full range of genetic products made during the colonization and replication of most viruses. With this definition, there is no necessary linkage between virus, host lysis or disease. It can happen, but it is not necessary. Nor does this definition differentiate between viruses that exist outside of their host cells as physical viral particles (virions) and those that do not (temperate viruses, endogenous viruses, intracellular viruses, other genetic parasites). Nor does it differentiate between viruses that encode the genes for their own replication from those that do not (satellite viruses, defective viruses, hyperparasites). This definition is inclusive of the full spectrum of genetic parasites and can include many agents that some scientist would probably define as nonviral or 'transposable elements'. However, historically, the scientific literature has differentiated almost all of the above elements and situations and given them many different and specific names and roles. This has created a very confusing (babble-like) nomenclature. Within even one field, such as endogenous retroviruses for example, the nomenclature is daunting even to the cognoscenti. Here, I consider them all as genetic parasites (hence, essentially viral). They are all dependent on, or highly influenced by, viral colonization events. Yet it is important to understand these more specific names and their associated characteristics to understand how they can interact. Thus, some babble needs to be introduced. Entire books have been written in which these individual elements are examined in great detail as essentially autonomous (selfish) genetic entities. I will avoid this reductionist approach and will attempt to keep track of the interrelationships between all the various linked parasitic agents. Below, I provide many of the accepted definitions for these genetic parasites. However, in an effort to provide some integration, I will consider their origin from the perspective how they have come to colonize their host and how this has affected host interaction with other genetic parasites, as well as with host identity. I will pay special attention to interactions that I consider being epiparasitic or hyperparasitic. That is, many of these agents are parasites of the parasites, and as such, can function to either assist or oppose parasite colonization. When looked at this way, many networks are clearly visible. The common names of these agents include plasmids, transposons, insertion elements, cryptic phage, homing introns, and intenes. As noted, the schematic in Fig. 2.5 outlines the hypothetical role of several of these agents in the context of *E. coli*. Within the accepted definition of transposons, there exists a whole slew of additional classifications: such as type I, type II DNA transposons, LTR, non-LTR retrotransposons, SINES, LINES, MIRS, Alus, etc. These specific elements will be better defined as they are introduced later. I feel all these specific names, and their being considered as essentially autonomous agents (especially as being non-viral or dissociated from virus), has created a confusion that has inhibited a more general understanding of interrelationships. Specifically, the existence of networks of parasites and the relationship of such networks to the evolution of host identity have been especially obscured. Since this view of hyperparasite networks will likely violate the beliefs (and career commitments) of some scientists, it is likely that some will claim that their favorite genetic parasite does not conform to my generalizations. But my main assertion will remain. Like viral quasispecies itself, the fitness of such genetic parasites resides in the survival of the network, not in individual elements. The confusion caused by reductionist thinking as autonomous elements has prevented such a consideration. The real biological world is very much like the ancient or current ocean: a vast pool or network of interconnected parasitic agents. All characterized natural habitats have this same viral characteristic. Thus, the natural fitness of any organism cannot be understood in the absence of the network of genetic parasites it faces. For example, if we were to delete a region of parasitic DNA from an *E. coli* or a mouse and see no effect on survival in the lab, we must then ask how this organism survives the feral world in which genetic parasites predominate. How would such a mouse fare following exposure to mouse hepatitis virus, which is essentially inevitable, for example. Reductionism prevents such

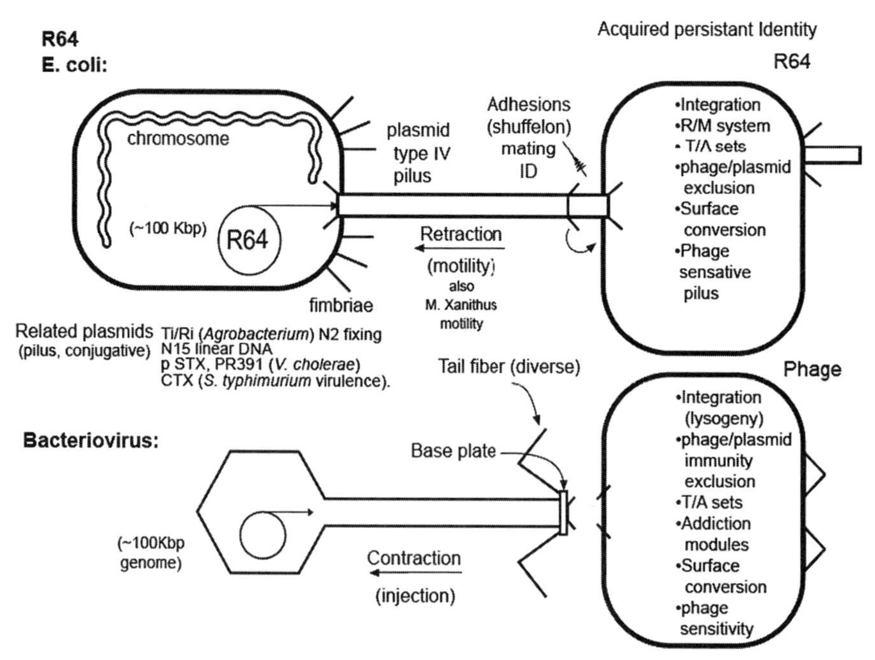

Fig. 2.6 Schematic of the R64/lambda exemplar

realistic assessments. Just as we cannot understand how a television functions by studying the autonomous function of a single type of transistor, we cannot understand the function or fitness of a genome by studying the functions of an autonomous ERV, LINE SINE or Alu element, for example. We must ask how the network works. Therefore, in this book, I will first consider the basal role of genetic parasites as we seek to understand the evolution of group identity. We start from the schematic in Fig. 2.6 and will then consider how persisting genetic parasites deal with such agents.

The R64 Exemplar (How Plasmids and Viruses Are Equivalent)

The assertion above, that viruses and various 'transposible' elements are equivalent to and originate from each other, and thus should be defined and considered together as 'persisting viruses', will likely be met with considerable objection. Although this issue will be further developed in various specific situations later, at this time it would be worthwhile to consider an important exemplar of this assertion. The real difference between these various types of genetic parasites is twofold: one is the importance of persistence and exchange, and the other is the importance of the extracellular packets, called virions. Only viruses have the latter. In contrast, conjugative plasmids, responsible for much prokaryotic sexual exchange, are generally thought to be distinct entities from viruses in that they make no virions. Let us contrast the lifestyle of a phage and the R64 conjugative plasmid as our exemplar (Inc II class, similar to F-plasmid). This comparison is outlined in Fig. 2.6. In a phage particle, the parasitic DNA is encapsidated into a virion (such as T4 or lambda). T4 after lytic release must then find and attach to a new host using two types of viral encoded attachment protein structures: the thin legs and the central thick tail/base plate. The thin legs make the first low affinity, multiple contacts to the host surface, allowing the virus then to be drawn in for the baseplate/tail fiber to attach to and penetrate the host cell membrane. The virus moves toward its host via these structures. Viral DNA is then injected, via the tail, into the host. If the virus is temperate (e.g., lambda), viral integrase will allow DNA integration into specific host regions of the chromosome, which then expresses various viral genes, including some for phage surface conversion. In contrast, the R64 plasmid encodes two types of surface pili, a thin polar structure expressed as a type IV pilus and a thick rigid sex pilus, through which plasmid DNA must replicate and pass during DNA transfer. The R64 type IV pilus requires 14 genes and is involved in initial surface mate recognition (which in some cases can also be involved in reeling cells together). The thick sex pili then penetrate (or creates a channel) in the second cell, allowing the R64 DNA to transfer into the host cell. R64-expressed integrase then allows integration into specific regions of the host chromosome, which then expresses surface pili. Both the viral and R64 systems then use toxin/antitoxin modules to insure parasite stability and will kill any host that might lose the parasite following cell division (post-segregational killing via R64 *pnd* gene). The T/A modules and post-segregational killing are generalized elements that will be further developed below; it was presented in Fig. 2.4. Both these genetic parasites have attained genetic transfer to a previously uncolonized host using very similar molecular strategies. The main difference is that the virus exited the initial host as a virion in order to find the new host, whereas the plasmid expressed the equivalent function as surface proteins on its host and used the host cell itself to find another new host. As will be discussed below, there are several examples of conjugative plasmids that are hybrids with phage DNA (PSSVx). In this case, there is little to distinguish these two types of genetic parasites; thus they fit well into one common definition. Some might argue with this assertion, noting that these plasmid-encoded genes are really host-derived and do not originate from the genetic parasite. For example, most prokaryotes appear to express surface fimbrae similar to type IV pilus. The tips of these fimbrae are thought to provide general function as adhesions between cells. Surely, this can't be of viral origin? However, given the impressive system for the generation of pilus diversity present in R64, I would argue the converse situation is more likely true: host pili are most likely of viral origin. The R64 plasmid appears to be under intense selective pressure to express diverse forms of pili, needed to recognize appropriate mates for sexual DNA transfer (genetic colonization). Thus R64-like parasites are more likely to have originated from diverse types of IV pilins. R64 has what has been called a 'shufflon system' that uses site-specific recombination, acting on a 19 bp repeat, that can invert one of four DNA segments, allowing the selection of one of seven pilV genes with constant N-termini but distinct C-regions. As discussed below, the N-terminal pil genes have many additional roles in viral and host and eukaryotic RNA identity. This highly elaborate plasmid system allows R64 to create and express a set of pili genes, well in excess of any host capacity for generating novelty in this same pili system. In addition to closely resembling persisting viruses in lifestyle, the conjugative sex plasmids are themselves subjected to viral parasitization. For example, various phage (such as PO4) use the type IV pilus as receptors for viral entry, and other phage adsorb to the sex pilus. Thus, only those cells harboring conjugative plasmids become hosts for specific types of virus. In addition, some conjugative plasmids appear partially defective and actually depend on such viral superinfection for their own mobilization. Finally, sequence analysis of integrase genes of F, P1 and P4 plasmids compared to those of lambdoid phage show that the integrase found in F1-like linear plasmids (like N15) links the evolution of plasmid and viral clades. Thus, there is clearly a seamless continuum of situations in which plasmids and viruses are linked hyperparasites of host. In addition, there are several genera of prokaryotes that do not appear to support any known plasmids (myxobacterium, euryarchaea), yet these too appear to evolve mainly by colonization of genetic elements, indicating a basal role of viruses. It is thus asserted that this represents a general situation: host and their genetic parasites (including parasites that appear to be non-viral or defective) together constitute a network of parasitic identity systems that can recognize (or oppose) other host cells and other genetic parasites.

Archaeal Parasites, Toxin Specificity and Identity

The archaea are composed of crenarchaeote (inclusive of thermophiles) and the euryarchaeota (inclusive of halophiles), which are able to grow in or tolerate extreme habitats with respect to temperature, salinity, pH and desiccation. In addition, they are metabolically diverse, able to derive energy from various chemicals, organic and light-based resources. This capacity makes them attractive candidates to have been the first free-living cells on Earth, predating the cyanobacteria, although phylogenetic evaluation is not able to unambiguously assign them such a basal role. Although, I will examine them first (due to this likely early development in the evolution of life on Earth), it should be noted that this domain of life is much less well understood than the bacteria and their viruses. Therefore, in depth development of several specific issues with respect to group identity systems will be presented below using bacteria, and especially their well-studied viruses like T4. Archaea do appear to have some group behavioral aspects, however. Some are known to grow in colonial mats; and others, such as halobacteria, are known to undergo population blooms in specific habitats (such as the Dead Sea) that prevent the growth of competing bacteria and algae. However, details of such group behavior are essentially unknown. As far as we can tell, it appears that the archaea have their own peculiar and very interesting relationships with viruses and genetic parasites relative to bacteria. Like the bacteria, archaeal genomes also show strong evidence that their DNA has been molded by viral interactions. For example, all free-living archaeal genomes avoid palindrome sequence word bias associated with restriction enzyme sites, but encode restriction modification systems. The extremophiles living at high temperatures (crenarchaeote) have some very interesting viruses. PSV (Pyrobacterium spherical virus), a 28 kbp DNA virus, is spherical with an external lipid envelope, but it is chronically produced and does not lyse its host. Similarly, SSV a spindle-shaped temperate virus that integrates at host tRNA genes is also non-lytic. In contrast, in bacteria, chronic production by dsDNA viruses is rare and spindle-shaped viruses are unknown. The ability of these crenarchaeal cells to chronically produce virus is probably related to an absent peptidoglycan cell wall and the outer S-layer nature of most archaeal species' cell surfaces. Non-lytic chronic virus production also questions the use of the word phage in this context, since this term is defined in relationship to host lysis. So far, all of these viruses are DNA viruses, the majority being dsDNA viruses (no RNA viruses have yet been observed). However, unlike the common morphological-tailed phage types seen in the oceans, a large array of

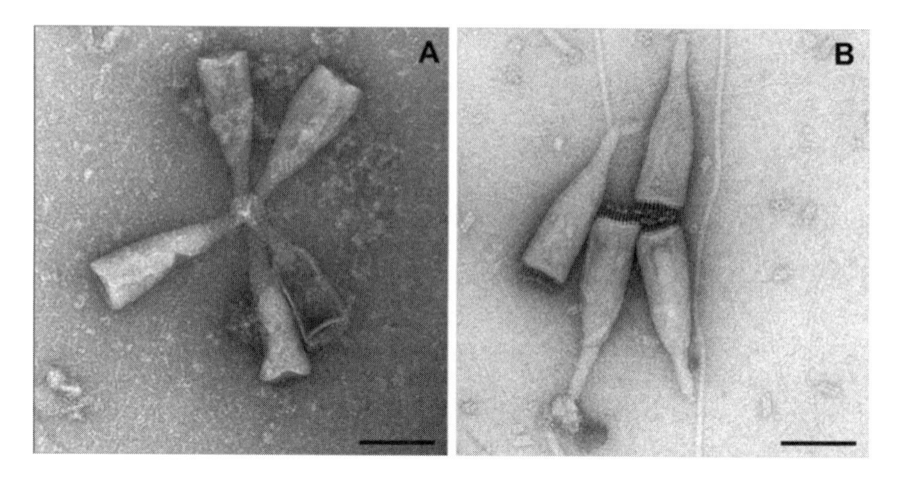

Fig. 2.7 Archaea viruses of unusual shape as imaged by electron microscopy (reprinted with permission from: Haring, Rachel, Peng, et al. (2005), Journal of Virology, Vol. 79, No. 15)

previously unknown morphological types of virus appears to be supported by these archaeal prokaryotes. These viruses include bizarre shapes that were not previously suspected to have existed, such as lemon drops, self-assembling rods with clamps on the ends and bottle forms (see Fig. 2.7). Many of these viruses are known to persist as episomal DNA elements, frequently in complicated sets of mixed viruses. In terms of conjugative plasmids, the distribution is rather specific and unlike bacteria. It is most interesting that all known self-transmissible plasmids so far found in archaea are of the genus *Sulfolobus*. One of these plasmids, PSSV+, which depends on a helper virus for spreading, is a hybrid with the SSV viral genome. PSSV+ is a member of the pRN plasmid family (RCR replicon), which encodes their own replication and origin recognition proteins. Other plasmids, such as pRN1 and pIPN2, are always found together even though they can be propagated separately in culture. Clearly, there is a highly complex and poorly understood relationship between these crenarchaeotes and their mixtures of genetic parasites; but viruses are common in these species and their genomes show strong evidence of viral effects.

The virus–host relationship in euryarchaeotal species (such as halobacterium) is distinct from that of crenarchaeotal species. Although the numbers of characterized viral species are much fewer with these organisms, many of their viruses seem more similar to bacterial viruses. For example, ϕH is a temperate P1-like phage of haloarchaeal species. Other viruses, such as SH1, are lytic and resemble PRD-1 and have external membranes. As noted below, this host, *Haloarcula hispanica*, can differentiate into thick-walled multicellular cell clusters that are resistant to SH1 infection, implying a major selective role for viruses in host cell identity. Spindle-shaped viruses are also known. In fact, in some extreme environments such as the Dead Sea, which has 10× the dissolved salt concentration of ocean water, only halobacterial species thrive (no bacteria or bacterial predators are found). Here large concentrations of spindle-shaped haloviruses (7×10^9 VLP/ml) can be found, associated with termination of blooms of halophilic archaea.

These hearty organisms thrive under habitats that are extreme in pH, temperature, and salinity; yet here too, we find that viruses dominate. The extreme pH 10 of Mono Lake, the extreme salinity of the Dead Sea, the extreme temperature of Yellowstone National Park Hot Springs or the extreme pressure and temperature of hydrothermal vents of the mid Atlantic ridge, all support archaeal viruses in very large numbers (all about 10^9 VLP/ml). In addition, serial isolations from the same hot spring, for example, show that these DNA viruses can display an unending pattern of genetic change for reasons that are not understood. Clearly, highly prevalent and dynamic viral genomes are seen in natural archaeal settings. Both chronic and temperate viral infections are common, so merging with host genomes is frequent. Therefore we can expect archaeal host identity to be modified by the immunity systems of these genetic colonizers. However, few experiments have examined this issue.

Archaeal Toxins Don't Poison Us

It is clear that essentially all prokaryotes, including all archaea, are able to produce toxins and antitoxins. Toxin production is often discovered by the ability of one strain to inhibit the growth of a related strain (via lawn spot halo assays, see Fig. 2.8). Clearly, toxins can differentiate groups of archaea from one another. As will be presented below, toxin/antitoxin gene pairs are used by persisting viruses and plasmids to enforce the maintenance (temporal persistence) of the genetic parasite by post-segregation killing of daughter cells that have lost the parasite. With respect to specific archaea, all examined *Sulfolobus* species produce proteinaceous toxins that kill other strains of *Sulfolobus*, but not the producer strain. Such toxins are made after cells enter the stationary phase, indicating an essential role in persisting, non-growing host populations. These sulfolobicins are often about 20 kdal proteins, related to the leader peptide of

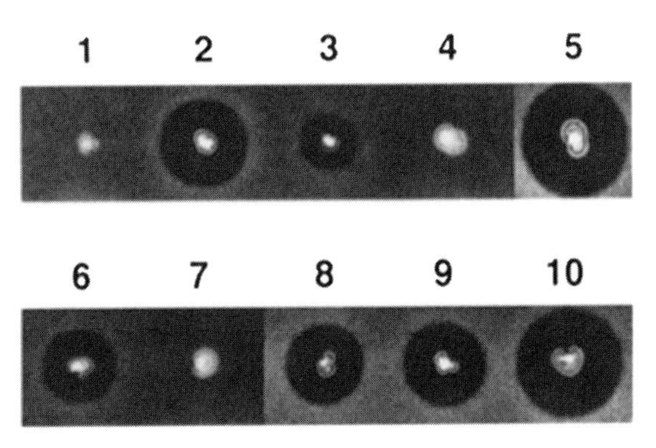

Fig. 2.8 Bacteriocin halo assay (reprinted with permission from: Haruvoshi Tomita, et al. (1997), Journal of Bacteriology, Vol. 179, No. 24)

type IV prepilins (discussed further below). These are not soluble proteins and thus seem to require close cell contact and can be produced in either cell-associated form or as S-coated associated vesicles, but their mechanism of toxicity has not been elucidated. However, the situation with toxins of halobacterium is different. Halocins (e.g., H4, H6, S8, R1, A4) are mostly secreted soluble proteins and more closely resemble bacteriocins. For example, H4 shows single hit killing kinetics similar to the particulate bacteriocins described below. H6 halocin disrupts the $Na+/H+$ antiporter. In addition, microhalocins (small peptide toxins of about 36 amino acids) are also made, but their mechanism of action is unknown. These small peptide toxins are reminiscent of the peptide pore toxins often made by blooming cyanobacteria (described below). Curiously, however, in spite of the similarity of some of these archaeal toxins to those found in bacteria, they do not affect either bacteria or eukarya. Archaeal toxins are specific to archaeal species, usually closely related species. This domain-restricted toxin specificity may help explain why archaeal species are not pathogens in eukaryotes. Most mammalian pathogens produce toxins that affect the host. Thus, all such pathogens known are either from the bacteria or from the eukarya domain, none are archaeal. Yet all archaea (like all bacteria) produce toxins (such as RelE-related genes) and antitoxins (either as RNA or protein). It therefore seems clear that the main purpose of these toxins cannot be to poison competing but phylogenetically distant species. They are mainly intended to limit competition by related species. Put another way, we can think of these toxin/antitoxin modules as identity modules that differentiate related species. The reason archaeal toxins do not affect eukaryotes is that eukaryotes and bacteria share similar identity modules, but do not share them with archaea.

Bacteria and Their Viruses: Foundations of Identity

Bacteria have many well-characterized genetic parasites. And in many cases, the mechanisms and interactions employed by these genetic agents are exceedingly well understood, often the best understood in all of molecular biology. This includes the iconic T4 phage that I will now consider in detail to understand what bacteria can tell us regarding the mechanisms of group identification and the role that genetic parasites may have played in origin of these processes. According to earlier arguments, it should be kept in mind that the existence of molecular diversity in any specific species will be looked on as a possible clue and as a potential source of an identity system. Furthermore, special attention will also be paid to stably colonizing parasites for their possible involvement as providers of new host identity systems. In terms of bacterial identity, it has long been known that phage susceptibility and bacterial type identity are closely linked. Thus, phage typing has long been used, and continues to be used, to identify bacterial strains, especially pathogenic bacteria of humans and domestic animals. However, there are additional effects of phage on host identity that will also be examined. Phage have long been known to induce phage conversion of their host, in which the surface molecules are altered. Phage are also known to often exclude one another, indicating that phage identity systems can be intended to affect other genetic competitors, not simply their host. Furthermore, these various host effects need not involve fully infectious phage. Frequently, cryptic or defective phage can induce stable alterations in host identity and immunity. Since cryptic phage can closely resemble plasmids or other mobile genetic elements, as the R64 exemplar above asserted, there exists a continuum of genetic parasites, from infectious virus to defective to plasmid, all able to persist in their host and affect its immunity. These will also be examined together and not considered as autonomous agents. However, both the phage typing of bacteria and the R64 plasmid add an additional concept to host–parasite identity. The precise chemical makeup of the surface of bacterial cells provides recognition specificity; this specificity can strongly affect binding and entry by parasite receptors. Phage conversion can alter this chemical surface, such as via glycoprotein expression, resulting in the loss of sensitivity to phage infection. As mentioned, the R64 plasmid also depends on type IV pili interaction with specific surface receptors and uses a recombination-based shufflon system to vary pili expression and alter receptors. In these cases, identity is external. This surface chemistry is one of the most variable elements of a bacterial cell.

Bacterial Gene Diversity: Internal Identity Systems

Although bacterial surfaces are highly variable, what are the most diverse genes of bacteria and do they have any role in immunity and/or parasite colonization? Comparative genomics allows us now to address this question. Amongst the most diverse genes found in all bacterial genomes are the restriction/modification enzymes. Since restriction and modification (R/M) enzymes act only on DNA within cells, the R/M system can be considered an internal identity system. As noted above, analysis of palindrome nucleotide bias sequence supports the idea that the genomes of all prokaryotes have been highly selected against the presence of recognition sequences of these enzymes. Restriction/modification enzymes thus seem to represent the most prevalent molecular immune system of prokaryotes, clearly aimed at preventing the susceptibility to genetic parasites. Yet, as we will see below, the very origin of restriction modification systems themselves appear to be viral (horizontally acquired) and can also be considered to have originated as part of an addiction module. Addiction strategies also seem to be crucial for the origin of group identities (see P1 exemplar below).

A Brief History of Early Phage Studies, Beliefs and Concepts

The study of phage of *E. coli* is filled with accounts in which the genetic parasites present in the bacteria were observed to significantly alter the phage and group biology of the bacteria. Most often, early *E. coli* isolates had mixtures of genetic parasites. For example, the highly studied model phage T7 was originally isolated from an *E. coli* that also maintained a conjugative Fertility (F) factor. We now know that there are important interactions between F-factor-like plasmids and phage that also affect interactions with other genetic parasites, hence the survival of the cell or its interactions with other cells (such as mating) will be substantially affected by what parasites colonize it and what parasites attempt to colonize it. Other early studies also saw evidence of alterations in group behavior as a consequence of genetic parasites. One of the most obvious of these was the near simultaneous group emergence of phage from an entire *E. coli* population following UV light exposure. The genetic capacity for a population of *E. coli* to make virus was noted by Bordet and Ciuca, who reported early on that such a capacity for phage production was genetically stable, in that these populations could not be readily freed from phage. In addition, there appeared to be some special forms of bacteria in which phage is not pathological, but was still released and able to infect other bacteria. It was also observed early on that one phage can affect the immunity to a second phage, as noted by Bail and Der Colistamm (1925). For many years, there followed a period of sharp disagreement over the definition of a phage as to whether it was strictly lytic or a conditionally produced host genetic element (see Chapter 9 for the relationship between science beliefs and human group cognition). However, in the early 1950s, these beliefs had to be abandoned due to compelling experimental observations. This was especially the case with regards to bacterial identity alterations due to prophage. In a 1949, in a paper published by J. Lederberg (PNAS 35 178 (1949)), it was observed that F-mediated mating between an *E. coli* strain harboring a lambda prophage and an *E. coli* recipient lacking the lambda prophage [cross of K12 (lambda) to F- (ly-)] resulted in segmental elimination of heterozygote diploids. The prophage, when transferred as DNA to a non-immune recipient, would induce virus production and kill the uncolonized zygote. The lack of lambda immunity genes destroyed the zygote. Thus, the prophage is not simply an inactive 'seed,' but is a genetic colonization that has significantly altered its host, creating a situation of sexual isolation between otherwise identical *E. coli* strains. Other phage similar to lambda (e.g., ph 434) showed similar mating behavior. However these phage differ from each other significantly in their corresponding immunity regions – along with phage base plates, an area of phage DNA that generally shows the greatest genetic diversity. Here, then, is our first specific example in which genetic diversity, immunity and group identity are all tightly linked by the same genetic colonizer. In addition, the highly diverse surface of bacteria mentioned above (originally shown by surface-mediated compliment fixation) is also related to phage colonization and group identity.

Viral Identity Can Be Affected by Host

The above example indicated that host identity could be altered by a colonizing phage. The converse situation can also occur to some degree. This process is known as host controlled phage variation. A single passage of phage in specific host can yield altered phage features that affect its host range. Thus the last viral host can determine, to some degree, the host range of the progeny. An example of this is seen with P22 grown in salmonella host. The host-specific restriction/modification enzymes will modify and/or digest the P22 DNA at the time of phage DNA replication. Infection of subsequent host will thus depend on the occurrence of matching restriction/modification enzymes. Other mechanisms of host-dependent variation are also known, such as membrane composition and protein modifications that are host specific. This type of situation indicates that the history of a parasite and its host can also affect the subsequent host specificity of the virus.

Mechanism of Group Fitness and Hyperparasites

The concept that persisting genetic colonizers are very much involved in the generation and evolution of new systems of bacterial host identity needs a mechanism. It is not simply that a stable host–parasite interaction determines the outcome, but also how other prevalent genetic parasites interact with this colonized host that determines the actual fitness. How can this happen? The colonized host must withstand the prevailing competition by other genetic parasites. It must become the 'king of the hill', remain functional and retain its identity after all prevalent genetic onslaughts have been faced. This success is not simply measured by offspring numbers, but also by duration of life and by encounters with competing parasites. Such interactions can be fierce, exclusive of each other, and highly competitive. However, some other parasite interactions are clearly complimenting, interdependent and cooperating. How do we understand these opposing situations? How does new identity get transferred into new host? In some cases, the defective parasites are the most stable, but the non-defective ones are the most mobile. The net result of such opposing situations is that networks of positive and negative interacting epiparasites are frequently found in natural populations. Can such networks also affect group behavior? A common mechanism for attaining stable host colonization by genetic parasites is the use of addiction modules (as shown in Fig. 2.4). As previously proposed, these provide a fundamental genetic strategy for persistence. Examples in which two populations of bacteria can kill each other depending on what genetic parasite they harbor are well known. Such colonized bacteria can also differ as to how susceptible they are to being killed by other genetic parasites. The molecular details of this killing process also define the inner workings of group behavior and how addiction compels group behavior or even cause 'altruistic,' or suicidal, behavior. The exemplar below thus presents the details of the role of addiction modules in group identity. The very molecules that stabilize persisting genetic parasites also maintain identity and group behavior of a cellular community.

The P1 Exemplar: Altruistic Group Behavior from a Simple Phage

P1 phage was particularly important for identifying and defining concepts of how a persisting genetic parasite can attain temporal stability (Fig. 2.9). In some publications, P1 is often referred to as a plasmid, not a virus. Yet it is clearly a virus, as can be seen by its genome content, but it is a virus that persists as an episome, hence the reference as being a plasmid. P1 is unusual. Most stable temperate phage integrate into the host chromosome. But P1 persists an extremely stable episome. Because of this, it provides much insight into the general (non-covalent) strategy of persistence and was used to coin the concept of 'addiction module'. In addition, this seemingly simple and unapparent genetic parasite also informs us about host-group identity. The genetic stability of P1 is known to involve the use several addiction modules (including a restriction/

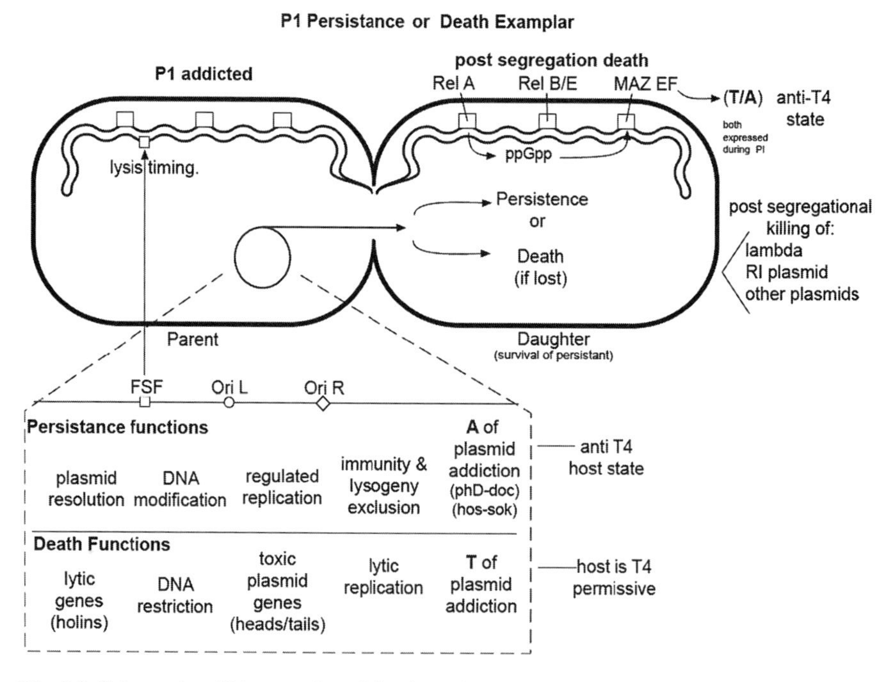

Fig. 2.9 Schematic of P1 exemplar of death on loss

modification gene pair). As previously outlined in Fig. 2.4, an addiction module is a pair of genes or functions, one of which is stable but toxic, the other which is unstable but is antitoxic – often by being a protein that binds to the toxin (aka T/A module). The loss or disruption of the parasite that harbors the T/A module during reproduction results in the degradation of the unstable antitoxin, which leads to the death of the cell that has lost the genetic parasite. P1 plasmid stability was initially characterized with the analysis of Phd-doc, a T/A gene pair. These are two P1 proteins that are physically complexed to each other as a toxin/antitoxin pair (doc is a metalloenzyme). They are thus named for the post-segregation killing that occurs when P1 is lost (via death of plasmid-free cells, similar to the lambda story above). The precise mechanism of cell killing is not fully worked out, but it does seem to involve a resident *E. coli* T/A module, the mazEF genes which are analogues to Phd-doc. Disruption of this mazEF T/A module allows the maze gene to function as a ribonuclease that will cleave ribosome RNA, terminate translation and kill cells. P1 has a second T/A pair known as *hos-SOK* genes. However, in this case, the antitoxin is regulated by an antisense RNA molecule. In addition, P1 has an ISI element (involved in C-inversion), which is also involved in lysis timing, a function similar to holins described below. P1 itself thus resembles a hyperparasite network! If the ISI element is lost, P1 lysis is delayed and plaques are minute. P1 also has a *sim* gene that prevents P1 superinfection, thus immunity against itself is also operating,

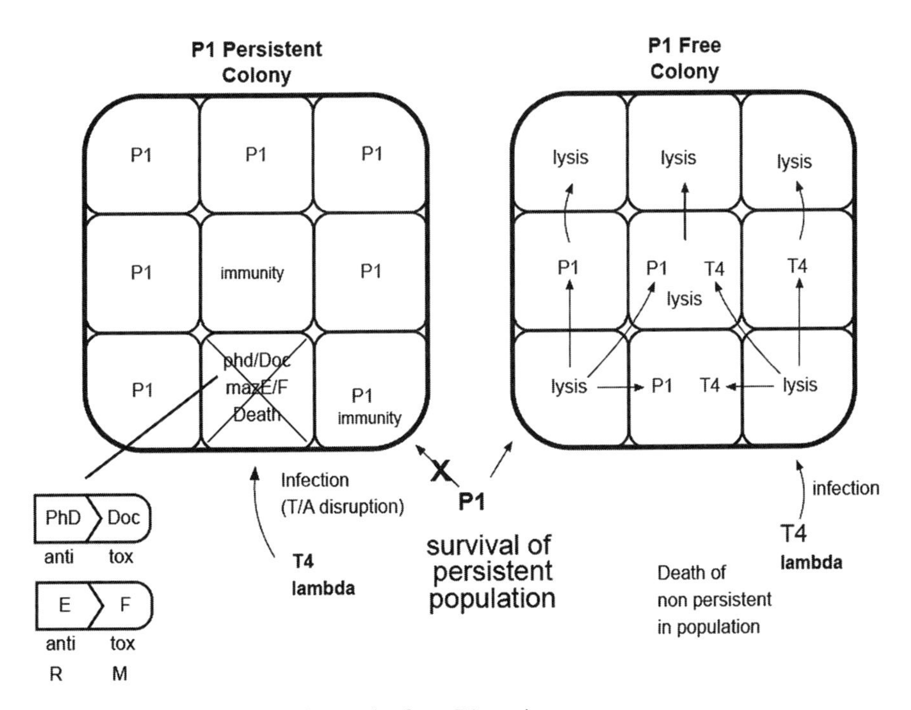

Fig. 2.10 Schematic of group immunity from P1 persistence

similar to most persisting genetic parasites. The net result is when *E. coli* is colonized with P1; it not only expresses immunity functions, but also expresses the P1 Phd-doc and cellular mazEF addiction modules that kill cells that lose P1. However, it is the expression of these two addiction modules that also compels group identity behavior. When P1-colonized *E. coli* are infected with other phage (T4 for example), the antitoxin gene is destabilized and the liberated toxin gene will then kill individual infected cells of the P1 colonized population, preventing T4 transmission to uninfected P1 harboring neighbors (see Fig. 2.10). Thus an individual P1 colonized cell will induce a suicide response that will prevent T4 replication and transmission thereby protecting the P1 harboring colony. The very process that insures plasmid persistence also induces death of cells that have been colonized with other foreign genetic parasites. This is clearly a molecular form of altruistic behavior in what is considered to be a free-living bacterium. This viral-mediated altruism is proposed to identify a common strategy for exclusion of competing genetic parasites (lytic phage).

The Lambda Exemplar, Toxic Pores

Such suicidal killing of cells infected by other agents also closely resembles *Rex*-mediated cell killing by lambda. The lambda *Rex A* gene was one of the very first genes to be studied in molecular biology but has long been elusive in terms of its function for lambda. It appears to have no role in the lambda life cycle itself. Its only known function appears to be to kill T4 or other phage-infected cells, preventing transmission to *E. coli* persistently colonized by lambda. *Rex* has the potential to be both protective of, or to kill, cells, clearly resembling a T/A module. Interestingly, *Rex A* may act by activating *Rex B*, a membrane channel protein allowing exit of monovalent ions that lead to rapid cell death. This observation introduces another potentially generalized idea; viral-encoded pore proteins are fast acting and effective toxins that can be associated with group identity and group behavior. This issue is discussed below in the context of holins and bacteriocins, both large families of phage pore proteins. In addition, lambda-colonized populations appear to rccognize other aspects of foreign viral infections. Such cells respond to T4 rII early gene expression as a group, resulting in coordinated control of lambda prophage expression in a population. Group behavior, mediated by persisting genetic parasites in response to infection with competing genetic parasites, appears to be common situation, and given that non-essential but highly conserved genes such as *Rex* are involved, we can conclude that such functions are highly selected in nature, even when they have limited function in laboratory experiments. If we can generalize this situation, we can propose that the very genes that allow persistence by genetic parasites (i.e., addiction modules) also bestow upon their host the capacity for 'altruistic' (suicidal), self-destruction of individuals that have lost their group identity due to infection with other genetic parasites. Thus, such altruism does not result from any 'kin selection' or 'game theory' or other such reasoning, but results directly from the use of an addiction module to establish molecular genetic identity and promote persistence of a genetic parasite.

A Generalized Relationship: Persistence Compels Group Identity Via Addiction

In the section below, we will examine numerous examples and details of systems that affect bacterial identity. For the most part, these are systems that allow one bacterial strain to recognize and often harm a related bacterial strain. These systems, therefore, constitute group identity systems of bacteria. I will also present the assertion that these bacterial group identity systems have generalized features that also appear to apply to higher organisms. I have already introduced the concept that genetic parasites are frequently involved in bacterial identity. Historically however, numerous other identity systems have been reported that did not appear to be the result of genetic parasites. These 'non-viral' identity systems will include colicinogenic or bacteriocinogenic factors and other toxins active in strains of *E. coli*, often against related strains. However, as we examine these systems, evidence that many (if not all) of them also originated from genetic parasites (generally viral) will be presented. In addition, the relationship of genetic parasites to mating behavior will be further examined (as introduced above for lambda and P1). One example is the strain-specific role of F-sex factors and their relationships with other genetic parasites. The evaluation of this F factor is of historical significance since many early mating observations involved *E. coli* K12 (which has a lambda prophage) and *E. coli* K12 plus P1. The P1 persisting plasmid is highly stable due to addiction modules and other strategies, and this provides T4 resistance. Thus P1 affects host fitness in the context of other genetic parasites, and not simply as determined by host-specific genes or chromosomal makeup. Figure 2.10 outlines how a persisting parasite like P1 can now establish group identity and immunity onto a colony of *E. coli* in which PCD will be induced with infection by T4 (or lambda) phage. Here, we can see that the P1-persistent colony is protected, whereas the identical colony not P1 infected is killed by T4 (or P1). As evolutionary biologists generally consider fitness only from the context of cellular genes, the major fitness consequences of prevailing genetic parasites and the interactions of such parasite network have been completely overlooked. And even the presence of a 'defective' (selfish) genetic parasite can have major consequences to host fitness in the context of parasite networks, since defectives frequently affect parasite self-immunity, competition and persistence with other parasites. Thus, a trinity of fundamental but linked concepts is now apparent. Temporal stability (via persistence or addiction modules), immunity systems (harmful response to altered identity) and group identity (individual harm for group good) are all inextricably linked at the molecular level. This trinity appears to define a generalized relationship that applies to most organisms (including eukaryotes) and relates to how they establish group identity. This trinity, however, has several common molecular futures that deserve emphasis at this time. The harmful component of the immunity or addiction module is frequently a pore molecule that can rapidly depolarize target cells. In bacteria, most of these pore proteins are evolutionarily related to phage holin, tail and base plate proteins. Such a stable toxic molecule will also require either an unstable antitoxin protein to bind the toxin or other preventive strategy (unfolded or unprocessed proteins) that prevent self-destruction but respond to foreign identity. Another common molecular feature mediated by phage is the surface receptor molecule. In terms of bacteria and their phage, surface receptors affect phage susceptibility, but the phage itself will often encode genes that alter host surface expression (phage conversion). Bacterial identity is thus highly associated with surface expression and corresponding phage colonization and/or susceptibility.

Sensory Receptors, the Story Begins

Although phage are clearly involved in much of bacterial host surface receptor expression, surface receptors themselves, especially sensory receptors, have taken on a major role in group identification and have a deep evolutionary history. These bacterial sensory proteins can now be used to trace a path from bacterial sensory use for group identity to that of all higher organisms including human. The rest of this book will therefore attempt to follow this evolutionary path through sensory receptor. During this trace, I will frequently return to consider the relationship to the trinity: stability, immunity and group identity. This will be done from the perspective of possible relationships to genetic parasites and genetic diversity. It is this pathway that leads to a cumulative and more complex group identity.

Self-Killing, Identity Imprinting and T/A Modules

As we will see, effective group behavior frequently involves an altruistic outcome, in which individuals (cells or organisms) will die as directed by addiction modules that insure persistence of group identity. Group identity is typically a consequence of stable colonization and expression of corresponding immunity functions. This state, although symbiotic in the simplest sense of the term, is not necessarily benign since it can kill colonized host. However, this concept differs considerably from the prevailing view of kin selection and how it would apply to group behavior. According to the view I have just outlined, identity system did not evolve because colonized cells should be 'good' citizens for overall kin survival, giving up their lives for the good of the genetically related community, but in so doing promoting the survival of close genetic relatives (i.e., kin selection). Rather altruism (programmed cell death, PCD) is a direct consequence of a disturbed addiction module. This was the very same addiction/immunity system that insures the stable colonization by new genetic parasites. This new identity and stability precludes alternative molecular identities by PCD. It is an essential and central feature of the fitness of a persisting genetic parasite and thus will normally originate via the successful colonization by a genetic parasite of its host. This superimposition of new identity, however, normally has several temporal (imprinting) features that are required in order to attain or transmit the identity while still precluding competing identities. A good example of this 'temporal imprinting' characteristic of identity transfer is found with the bacterial restriction/modification identity system as used by P1. Restriction/modification (R/M) systems provide a clear example of why an 'imprinting' process is often needed to transfer group identity. Identity transfer generally occurs only during a critical developmental time frame (window) by the action of some transient marking process. Thereafter, all unmarked copies of even the same genetic element are attacked as foreign. In this case, the unstable DNA methylation enzyme (activity) must be present during DNA replication (cell division) in order to methylate the newly replicated DNA and imprint both daughter cells with the appropriate epigenetic identity tag (via methyl group transfer from SAM). If the P1 plasmid were to be destroyed or otherwise incapacitated by epigenetic means (such as alternate DNA modification, or intron, transposon or other hyperparasite invasion), the stable but toxic endonuclease gene would cleave any newly synthesized DNA and kill the resulting cells. T/A gene pairs will generally have such a temporal characteristic since the toxic component is usually stable.

Temporal Fitness and Serial Group Identity

New R/M identity is both genetic (requiring the two R/M addiction module genes) and epigenetic (requiring the marked/methylated newly replicated DNA), but the epigenetic component has a clear temporal window during which it can transfer identity. Stability is attained once the T/A system has been set. This temporal stability makes possibly the most important general point of this book, for it raises the issue of the relationship between the duration of life and the definition of fitness. The fitness of a persisting genetic parasite depends very much on its temporal stability and its corresponding ability to withstand the onslaught of all the competing identities by T/A action. Fitness is therefore not simply defined by relative replication or over replication of the genetic parasite. I have argued that the attainment of this temporal stability requires either a complex phenotype (i.e., multigene addiction module) or a clear addiction strategy (such as defective-mediated viral immunity).

These are complex solutions requiring the coordination of the participating elements. Thus T/A colonization frequently favors, or enhances, the stability of the host while promoting parasite stability. However, any host successfully colonized will also have acquired the new complex parasite identification/addiction system. Because of this new system, the colonized host can now recognize and respond to an expanded set of genetic parasites that at the least must include the new parasite itself, but often will also recognize other genetic parasites. Colonization, new immunity and identity are all necessarily acquired together. Because stability is attained by self-destruction when identity is modified or lost, we can also see how the molecular seed of 'altruistic' behavior (PCD) has been acquired and why this is under constant positive selection.

Serial Acquisition of Multiple Identities

The process as outlined above leads us to see why identity acquisition has a serial, cumulative and episodic nature to it. Waves of colonization can result in greater and greater redundancy or overlay of identity. Thus a general (but not essential) evolutionary trend will be to evolve toward greater complexity (but not necessarily greater efficiency). Yet the very act of a successful genetic colonization can also oppose further colonization by subsequent parasites. A truly efficient identity module could actually extinguish further colonizations by competitors, thus resulting in an evolutionary lineage that no longer undergoes diversification at high rates. Thus a 'near-perfect' genetic immune system should be possible, but once it is attained, the host would become much less genetically adaptable. Chapter 5 presents the case that *C. elegans* may represent such a situation in their efficient resistance to virus. Successful colonization, however, may also require that the new parasite circumvent or incapacitate existing host identity systems that oppose it and superimpose a new identity. Here we can see the considerable significance of epiparasites or hyperparasites. A parasitic agent (virus) that either carries, cooperates with, or complements a hyperparasite (intron, intene, transposon, retroposon) has as part of its genetic armamentarium a potentially mobile and amplifying agent that can be deployed to inactivate other and competing genetic identity systems (see Fig. 2.5). This is like having genetic 'smart bombs' that can amplify the reach and extend the effects of a genetic colonizer. The deployment of such agents could result in the large-scale inactivation of resident colonizers that are distributed. Thus, there would be a general tendency for serial displacement of the most recently acquired genetic parasites by invasion of such epiparasites (as seen in mammalian LINE evolution). These inactivated parasites would then remain as a historical archive of prior identities that were subsequently displaced.

From Multiple to Group Identities

We can now outline the principles by which group membership and group behavior originated and evolved from parasite identities. Successful colonization by new parasite T/A identity systems results in a population of organisms that have a common group identity, and will recognize each other but will kill off individual members that lose or temporally modify that identity. We can euphemistically call this process 'altruism'. I use the term euphemistic because it can also be considered the ultimate selfish strategy (not selfless). Selfish in that it is better to induce self-destruction than allow identity displacement by a competitor. In a sense it is like saying, 'If I can't have this host, then nobody can', not a charitable strategy. Yet, ironically, it is this very same strategy that we can now use to trace the evolution of group behavior, including cooperative behavior, in more complex organisms. So why should such a process tend to originate from virus and other genetic (and information) parasites? Why not simply evolve this from host-maintained group identity systems? The problem is the basic need to create novel but interacting gene sets (T/A gene assemblies) that define identity. Simple Darwinian point changes or gene shuffling via recombination does not readily or simultaneously create matched gene sets. However, since viruses evolve in mixed, interacting and even defective laden populations, they can assemble and evaluate gene functions in large (nonfunctional) sets that are not necessarily needed for the normal lytic life style. Yet evolutionary pressure does select for a virus able to establish stable host colonization and competitive exclusion. Thus, in one colonization event, these preselected gene sets can be selected and then be acquired by the host, which then becomes available for Darwinian evolution. Since the novelty of viral protein configurations far exceeds that of the host, viral–host colonization allows the hyperexploration of multigene sequence space to occur outside of the host genome. The vast and ancient viral content of our oceans provides this multidimensional sequence space. This evolutionary scenario is consistent with the comparative analysis of the genome and the patterns of genetic parasites observed.

Bugs in Milk Vats: Evidence for Viral-Derived Gene Sets

The logic or appeal of the above scenario for T/A and group evolution does not necessarily make it so. Can this proposal be experimentally evaluated, specifically in bacteria? Is there evidence that bacteria actually evolve consistent with the above proposal? Perhaps the best-studied genetic system in this regard is with lactic acid bacteria since these bacteria have been grown in huge commercial quantities in the milk industry for many decades. It has long been known that lytic and temperate phage are a frequent occurrence in these cultures, which can strongly affect the yield of fermented milk products. Thus, phage diversity and evolution have been very well studied here (see Recommended Reading).

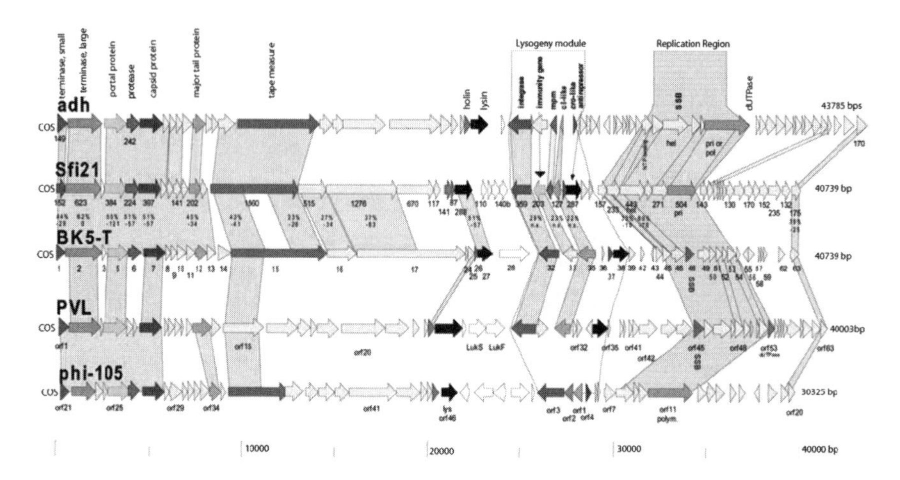

Fig. 2.11 Patterns of lactic acid phage evolution (reprinted with permission from: Frank Desiere (2004) Antonie van Leeuwenhoek, Vol. 82, No. 1)

It has been shown, for example, that new strains of virulent phage can evolve from temperate phage that lose immunity, such as via an acquired intron in a lysis gene (similar to Fig. 2.5). In terms of prophage evolution, as with most bacteria, it now appears that prophage behave like swarms of related prophage, allowing recombination of common domains, which results in the majority of the strain-specific DNA of their host genome. With the phage of lactobacteria, it seems clear that phage are a major source of genetic novelty and that phage-derived protein domains are most often used to generate these novel proteins (see Fig. 2.11, from dairy phage evolution). However, it also seems clear that issues of host genetic identity are similar in lactic acid bacteria. For example, lactococci bacteriophage and/or plasmids, such as pAH90 (a resistant plasmid with a restriction/modification system that also has a group II self-splicing intron) appear to also superimpose new molecular identity following successful host colonization. Thus practical experience studying large bacterial populations is consistent with the above proposal.

Host T/A Modules in Chromosomes

The existence of T/A modules is not restricted to viruses. With the sequencing of what is now over 100 prokaryotic genomes, it has become clear that T/A modules (related to MazEF or relBE) are highly prevalent, found in most prokaryotes. The interesting exception to this is bacteria that are parasites to other, mostly eukaryotic host (such as mycoplasma). Consistent with this, no archaea appear to be parasitic to eukaryotes, nor are any of their sequenced genomes free of T/A modules. The consensus has been that these T/A modules are associated with stress responses and that because parasitic bacteria live essentially stress-free lives, they no longer require such modules. I find this idea unconvincing. Somehow the idea that living within a lung macrophage, with their enormous potential for generating ROS and degrading microbes, doesn't strike me as stress free. Yet that idea 'has legs' and is often repeated in the literature. Conversely, slow-growing cells such as mycobacterium tuberculosis are thought to have a large number of T/A modules in order to deal with their stressful habitats (somehow the lung macrophage that mycobacterium colonize is now stressful). Given the important role of T/A modules outlined above for persisting genetic parasites, I suggest that a rather different function for chromosomal T/A modules can be proposed. They are identity modules able to counteract genetic colonizers and are often associated with longevity. As mentioned previously, the R/M systems of prokaryotes can also be considered as T/A modules, but here the function of R/M systems is clearly aimed at genetic parasites. However, as previously mentioned, all prokaryotic genomes have a word sequence bias against the occurrence of palindrome words, the recognition sequence for restriction sites (except those of bacteria that are obligate parasites). Since this R/M and word bias pattern overlays exactly with T/A module occurrence, these two observations support the role of T/A modules being aimed at genetic parasites. The absence of T/A modules from parasitic bacteria results from the fact that such internalized parasites are also protected from genetic parasites by residing within their eukaryotic host (hence also no R/M systems). The apparent T/A relationship to stress might simply result from any condition that perturbs the unstable antitoxin from protecting cells, resulting in a general tendency for stress to induce cell death. Slow growing, persisting cells (*M. tuberculosis*) would therefore be more likely to encounter large numbers of genetic colonizing events during their extended life span, hence need an extended set of T/A identity modules.

Evidence that *E. coli* T/A Modules Oppose Virus

If the above proposal were correct, such chromosomal T/A modules would seem unnecessary when deleted and not challenged by infectious agents. There is some good evidence for this idea. The *prr* locus of *E. coli* is known to contain several T/A gene sets, including an anticodon nuclease (ACNase) that cleaves specific tRNA molecules and inactivates translation. This is a nonessential locus. Its only known function is to counteract infection by T4 and lambda. In fact, studies of T4 led to the initial discovery of the *prr* locus. T4 has two genes (polynucleotide kinase, RNA ligase) whose only known function is to counteract the *prr* ACNase. When *E. coli* is not T4 infected, the nuclease is kept in an inactive form by complex to a type 1C R/M protein, also encoded by the *prr* locus. This is possible because the ACNase protein is actually fused onto the same peptide as the restriction enzyme. This is the only known example of a type 1C restriction/modification that is not encoded by a conjugative plasmid directly, implying that *prr* originated from some type of genetic parasite. This very same R/M system will also restrict infection by lambda. In addition, *S. mutans* also has an *hsd-prr* locus that encodes a suicide system. Here, the rel BE genes are within the prr locus. There is a very interesting link between the rel ABE genes and the mazEF genes noted above in P1-mediated killing (see Fig. 2.9). Rel BE is a T/A pair that encodes a ribonuclease that can cut mRNA in the ribosomal A site (the general importance of rel BE genes is described below). Rel A is a regulator of rel BE. Rel A encodes ppGpp production, which is considered an 'alarmone', ppGpp also inhibits the mazEF T/A module. All these T/A systems, R/M systems and their associated regulatory genes appear to constitute internal identity modules aimed principally at recognizing and inactivating competing genetic parasites.

Genetic Networks: Parasites, Epiparasites, Hyperparasites as Fitness Nets. Opposing Reductionism in Defining Fitness

Phage Typing, Tropism and Receptor Variation

The occurrence of prophage or their remnants in bacterial genomes is nearly universal. All 115 currently sequenced bacterial genomes harbor phage or phage remnant sequences. Within this 115 genomes, 190 clearly identified prophage sequences were observed. Such prophage sequences are especially prevalent in human bacterial pathogens, accounting for much of the strain-specific sequence variation found in those strains. These strain-specific DNA sequences are also mostly integrated next to tRNA genes, indicating they likely originated from integrations that used phage mechanisms. Most of these prophage retain some genetic activity and express immunity and/or lysis conversion genes. The presence of such phage immunity genes would clearly affect the ability of bacteria to host other phage. As previously mentioned, phage typing is a very practical process that allowed the early identification of pathogenic strains or variants of bacteria, responsible for most human infections. Such practical use of phage typing continues to this day. A current example applies to *Salmonella enteritidis* strains. A recent clinical study in Poland showed the existence of 41 different phage susceptibility types that could be classified into five different kinds of phage of *Salmonella enteritidis*. Interestingly, this study also observed four distinct plasmid profiles involving both large and small DNA plasmids and DNA particles in these same isolates. This is clear evidence for the normal existence of diverse phage/plasmid networks on specific but diverse host strains in a natural setting. Phage typing of a host bacterial strain is really a way to measure phage tropism. Phage tropism can be mediated by the binding of the phage base plates and attachment proteins to the surface of the bacteria. Given the enormous pool of phage that exist natural habitats, phage base plates and attachment proteins represent the most diverse of all genes on Earth. These proteins are basically receptors and pore-forming proteins that allow the recognition and entry of viral genetic information into their host. Although some of this protein diversity can result simply from the inherently high rates of phage genetic variation and recombination, it seems clear some phage also have specific and ingenious mechanisms to create even greater diversity of these receptor genes. For example, inversion elements are often used to switch expression cassettes and provide alternative sets of attachment/receptor genes to effect host tropism. One of the most intriguing and intricate examples along these lines is seen with a phage of Bordetella. This phage resembles a hybrid of the well-studied P22 and T7 phage. This Bordetella phage uses a phage-encoded reverse transcriptase system to mediate tropism switching. Two variable regions (VR1 and VR2) are used in combination to create receptor amino acid sequence variation. VR1 is modified via a single amino acid substitution at a defined location of the receptor gene. VR2, however, is highly dynamic via a variable number of 19 bp repeats separated by one of three 5 bp repeats. This gene encodes an adenine methylase, so it is likely part of a phage identity system. Thus, we see a bacterial system that uses of reverse transcriptase (RT) to create a variable gene product involved in host and self-recognition. Like the R64 plasmid above, we see genetic parasites can have complex, non-host systems specifically for the purpose of creating variable molecular receptors. This system of Bordetella phage clearly resembles the vertebrate adaptive immune system in how it creates variability in the heavy chain immunoglobulin surface molecule.

Epiparasites: Interactions Between Parasites

Lysogeny and Exclusion

Lambda is the best-studied lysogenic virus. However, as presented above, two camps of thought that developed early held essentially black and white views and dominated the history of the study of phage–host interaction. Phage were either strictly lytic agents or they were endogenous genetic agents that were sometimes capable of cell lysis. Lysogeny was a state that appeared to produce a lytic agent that was essentially of spontaneous intracellular origin. To some researchers, this situation seemed too much like spontaneous biogenesis and was looked on with great suspicion or simply dismissed. A hereditary capacity to make virus in every cell seemed to some to violate the very definition of what was a virus. Later, with the development of molecular biology, when hereditary viruses were firmly established, it also became clear that lysogenic phage could not only control themselves but also the closely related phage. The lytic and lysogenic world of phage are linked as indicated with lactic acid bacteria. We can now see how the presence of lysogenic phage provides a major source of host diversity, identity and immunity. Consistent with this, the currently characterized strains of the ECOR (*E. coli)* database vary mostly due to the presence of various lysogenic phage and plasmids, all of which appear to have originated outside of the host chromosome. This strongly suggests a fundamental role of integrating phage and other genetic parasites in large-scale host speciation. Lambdoid phage can clearly interact amongst themselves. They are classified into about 20 immunity groups based on exclusion phenomena. This exclusion process involves various mechanisms, but outer membrane protein, such as FhuA, are commonly affected, which also results in a stable host cell that has converted its surface-recognition properties. Such surface or membrane interactions also apply to persistent-lytic phage interactions. For example, T7 (lytic phage of *E. coli)* will damage membranes of F-plasmid harboring *E. coli.* This damage affects the transmembrane potential and also allows nucleotides to leak out of affected cells. It seems clear that some type of molecular 'hole' is involved in this PifA-mediated leakiness. However, other viruses, such as T3, can clearly counter PifA-mediated membrane leakage as T3 encodes gene 1.2 that prevents this T7 damage. This T3 gene appears to be a 'counter pore' that plugs the T7-mediated leak in F-plasmid harboring cells. Clearly this is another example of a hyperparasite situation in which important viral genes are directed at other genetic parasites, not simply host genes. So it must be that the selection for, or fitness of, these viral genes is determined by the prevailing genetic parasites that colonize the host. Surface receptors and molecular pores seem to be commonly involved in such interactions. Here we can also see how viruses might mediate the evolution of T/A modules by competing with one another for host, resulting in matched-pore and antipore proteins. As alluded to above and presented in detail below, surface identity via molecular receptors and pores represents a major process by which host identity is attained, modified or rejected. Such structures also provide the foundation from which sensory systems of higher organisms have evolved. Thus, as outlined in Fig. 2.9, hyperparasites and epiparasites are prevalent and participate directly in parasitic networks, which have major consequences on host survival.

Essential Cost of Epigenetic Methylation

Besides surface receptors and pores, persisting genetic parasites can alter host identity by internal mechanisms already mentioned, such as the R/M and other T/A modules discussed above. Lysogenic viruses and plasmids can also use systems that involve post-replicative (epigenetic) modifications to the genome of the virus and host to mark identity. The term epigenetic generally means that DNA sequences are not altered, but some process has stably modified the genome to affect recognition and/or gene activity (R/M enzymes being the most common). A majority of R/M systems utilize site-specific methylated DNA to protect against site-specific endonucleases. Such DNA methylation, however, is rather costly in terms of bioenergetics, requiring a methyl group to be donated during DNA replication to the appropriate nucleotide from S-adenosyl methionine (SAM). Lytic phage are known to often counter such epigenetic marking by other broad epigenetic DNA modifications that preclude restriction/modification. For example, lytic T4 encodes an HMC modification that results in the methylation of all cytosines and thus precludes sensitivity to many other restriction enzymes, but also protects T4 DNA against T4 encoded endonucleases, which will cut non-HMC DNA. This HMC-modified DNA can also counter recognition systems of other persisting genetic parasites. For example, HMC DNA is not sensitive to *mcr* restriction. This is a host-strain-specific restriction that is encoded by the *mcr* e14 prophage gene (which cuts non-HMC modified T4 DNA). It is interesting that T4 can be genetically modified so that both the HMC genes and the corresponding T4 endonuclease genes have been deleted. This modified T4 will grow to wild type levels in non-mcr strains of *E. coli*, producing unmodified T4 DNA. Clearly, neither HMC modification nor the endonucleases are essential for the phage life cycle in this limited lab situation. Yet HMC modification is highly conserved in phage field isolates and is used as a phenotypic marker to identify T4-related phage. The conclusion is that T4 opposing mcr-like lysogenic phage are so common in the field that T4 is compelled to maintain this HMC identity system, and that such maintenance is stable on an evolutionary timescale. Clearly, prophage matter to both host identity and lytic phage sensitivity in natural settings, directly supporting the concept of hyperparasite networks. The long-term fitness of a host is entangled with both lysogenic and lytic viruses. Thus the needed internal identity systems (such as SAM-dependent methylation) although energetically costly are highly conserved in a world dominated by genetic parasites.

Phage Group Inclusion and Exclusion (Networks)

What is the point of lysogenic conversion? Why do prophage frequently modify the surface expression or phage susceptibility of their host? Lysogenic conversion is not pertinent to the process of lysogeny for most phage, as genetic studies have often indicated, so why do it? It appears that lysogenic conversion is most frequently aimed at competing phage, in order to exclude superinfection of genetic competitors. However, not all phage and not all other genetic parasites are excluded. Some are allowed and may even be necessary, whereas others are not. Why is this? Is there an 'allowed' confederation (network) of phage/plasmid, as well as a precluded confederation of genetic parasites? If so, does this identify the existence of a process of network complementation or group selection amongst sets of cooperating and competing genetic parasites? There are clearly both positive and negative interactions between genetic parasites and their host. For example, phage not grown in *E. coli* harboring persisting

P1 won't acquire P1 'molecular markings' (methylation) and will not be able to infect other P1-harboring bacteria. In a sense, P1 is allowing, or not allowing, other phage to grow in P1-harboring *E. coli*. Such interactions have long been established in the bacterial literature. Some interactions can be even more complex. For example, P2 was originally isolated from P1- and P3-harboring *E. coli*, but this mixed colonization still allows the replication of P4 (a satellite phage of P2, see below). P1 and P2 will receive special attention in subsequent sections due to their ability to significantly alter host identity and evolution, as well as how the host interacts with other genetic parasites. P1, besides providing us with the crucial concept of a genetic addiction model, has additional mechanisms for precluding other plasmids and phage. For example, P1 (which is related to P7) can also preclude other plasmids via interactions of the origin of DNA replication with DNA replication proteins (representing a whole additional internal process of identity control). This and other interactions can also result in a partition function very similar to F-plasmid (sexual isolation by post-segregational host killing). Although P1 is normally maintained as an extra-chromosomal episome, it also carries a hyperparasite, an IS1 insertional element, that can be used for host chromosome integration (see Fig. 2.9). When P1 becomes integrated into the host chromosome, it can suppress host replication function and allow the bacterial chromosome to replicate from prophage origin. Thus P1 has the capacity to even displace the molecular identity of the host *E. coli* replicon system. Since replicon identity is fundamental to genetic identity that a P1-like genetic parasite can superimpose, even this most fundamental identity shows that some genetic parasites can have the ultimate capacity to redefine their host. P2 is a contractile tailed 34 kb dsDNA phage that normally integrates and is discussed below due to its relationship with the satellite P4 phage. P2 is of special interest for two reasons. One, it uses a small RNA as a regulator of expression of immunity factor (a strategy much used in eukaryotes as discussed later), and two, its receptor also resembles glucocorticoid/estrogen receptor, which is also important in eukaryotes. As we will see, P1- and P2-like phage have had big impacts on evolution of bacterial host communities (see cyanobacteria, Chapters 3 and 4). Thus again we see that the consequence of phage colonization to their host, and how their host interact with other cells and other genetic parasites, is crucial to the survival of the host; but when it is a multiagent situation as just described, it can be exceedingly complex to envision and this situation appears in direct opposition to Occam's maxim. Trying to get the big picture of how these parasites, hyperparasites and their inter-relationships affect host and viral fitness is likely to leave one in a dizzy or frustrated state, especially a casual reader. They will struggle in a seemingly futile effort to retain the various names and positive or negative interactions of phage, plasmids, prophage, competitors and host genes involved in an attempt to resolve a linear logic. Surely, I am not explaining things clearly, they are likely to think. However, if, as I have suggested, networks of hyper- and epiparasites are essential in understanding host fitness (via fitness nets), this presents a situation that resembles a multidimensional or hyperspace problem in terms of trying to comprehend the system as a whole. Any particular perspective that is used to evaluate the fitness landscape will give a partial and ultimately misleading view. For example, if we seek to understand the effects of lytic phage or fertility plasmids on host fitness, in reality this issue intersects with other entire genetic planes, such as defective prophage or satellite phage, whose presence completely transforms the fitness landscape. Yet we purposely create laboratory models that eliminate these other planes. We simplify *ad absurdum*. We seem to lack the mental facility for such real-world network comprehension and analysis. Little wonder we have failed to appreciate the deeper significance of so many well-established observations, such as the mixed natural circumstances from which nearly all these phage and plasmids were initially observed. In addition, the fitness consequence of parasitic agents appears to be mostly conditional to wild settings. When evaluated in our simplified laboratory models, free of competing or epi-genetic parasites, we comfort ourselves with the observation that neither the parasitic agents nor their mixtures are important for the fitness of the host (at least according to how we define fitness). All these agents seem unnecessary until we wade into parasite-infested natural settings. The autonomous and linear view of say, selfish DNA, presents a simpler, much more comfortable and less demanding vision of genetic parasites, consistent with our belief in basic concept of simplicity (i.e., Occam's razor). This has imprinted us with beliefs in which genetic parasites have no fitness consequence to the host and has allowed an entire generation of evolutionary biologists to ignore a vast wealth of observational evidence to the contrary. Genetic parasites matter in a big way to all life.

Phage Surface Conversion, Slime and Bacterial Colonies

As we just presented, essentially all bacteria express surface markers specific to the strain, which is often altered by phage conversion. Phage surface conversion can also be involved in bacterial community (colony) formation in that some phage conversions result in bacteria that produce a surface slime coating. As a specific example, we can consider phage D3, which lysogenized *Pseudomonas aeruginosa*. Following colonization, this phage employs three phage genes that convert the bacterial serotype from O5 to O16. A result of this conversion is that a glycosylated surface protein is now highly expressed. Similar phage-mediated surface conversion also appears to be frequently involved in the origin of bacterial surface slime, such as originally observed with pathogenic variants of pneumococcal. Slime production is typically a central component of bacterial colony formation and habitat colonization that has major implications for the evolution of group behavior and multicellularity (Chapter 4). In addition to surface conversion, phage colonization is also frequently associated with toxin production. An example of this would be phage-mediated virulence seen with streptococcus, in which both slime and pore toxin production are involved (see below).

Phage Compatibility and Gene Pools: Origin and Dynamics of Genetic Pools

I have presented several examples above in which phage are interacting and excluding each other in complex ways. The topic of phage compatibility is a historic topic with many early observations. Such phage–phage interactions include essentially all the well-studied phage models, such as Lambda, T2, P1, and P2 PRD1. I have already mentioned how some phage (P1) will compete with other phage for replication origin DNA complex. Others will compete for replication complex on the plasma membrane, such as MB78 of *S. typhimurium*, which doesn't allow P22 replication. P22 itself has the sieA gene that excludes superinfection of other related phage. It is clear that many mechanisms are involved in such phage interactions, but they basically all recognize molecular genetic identity. The issue of phage compatibility also leads to the issue of extended phage-based gene pools. Do phage exchange genes in a defined or restricted manner associated with such compatibility? If so, does this define a type of group (pool) selection or clade? Although some extended groups of phage clearly appear prone to recombining with related phage (such as rampant recombination between the T-even phage), this is also clearly not always the case. For example, the D3112 and B3 phage of *P. aeruginosa* recombine at very low frequency, indicating some capacity to preclude certain types of genetic recombination. Lysogens can also raise the issue of gene pools via compatibility. We know that large numbers of bacterial species are lysogenic, and some can simultaneously harbor up to eight known lysogens. Yet, due to the specificity of immunity regions, incompatibility of lysogens must also exist. What then specifically allows some epiparasite interactions but precludes others? How can one bacteria establish multiple and complex colonization and does this produce a host that precludes all such colonizers? As we will see, the genomes of social bacteria and eukaryotes do indeed have specific, generally well-defined patterns of colonization and preclusion of genetic parasites. These patterns are observed at even order level of organisms. For example, some prokaryotic orders, such as crenarchaea, are prone to multiple chronic colonization by mixtures of viruses. As we will see in Chapter 4, fungi also show peculiar and highly prevalent patterns of virus colonization (dsRNA viruses). In bacteria, we do know colonization of some specific phage can preclude others, such as Shigella colonized with P2 doesn't support T2, T4, T5 or T6 phage. Of course the historic *E. coli* K12 (with lambda) precludes T4. It seems likely that such interactions must define gene creation, recombination and gene flow patterns to some significant degree. Gene transfer through mating is also affected by phage. For example, mating of *Bacillus* can induce plasmid integrative prophage (J7W-1) in which phage invades the plasmid. If Bacillus harbors defective prophage (PBSX), mating between closely related species can be antagonistic during co-cultivation. Here, we might think that mating and competition between host is mediated by their resident phage. This link between mating competition and colonization by genetic parasites is often observed. Thus both the gene pool and its dynamics are strongly affected by the specific mix of genetic parasites.

Satellites – Effective Hyperparasites

The common existence of satellite phage makes the clear point that a defective 'hyperasitic' phage can make a 'good living' in the real world (Fig. 2.5). The best-studied satellite phage system is the P4, which is a satellite of P2. Satellites generally lack some or all structural genes and therefore are dependent on a helper phage for propagation as a virus. Thus the 11.6 kbp DNA P4 satellite lacks the capsid genes provided by the larger 38.8 kbp DNA of the P2 helper and can be defined as a 'hyperparasite'. In several respects, satellites resemble defective virus, but differ from defectives in that they are usually a distinct replicon from that of their helper and generally express some helper-specific replication proteins. Satellite viruses can be prevalent in some natural host populations, so they are clearly fit in many circumstances. It is interesting that the P4 satellite is also lysogenic and following integration will express immunity genes, similar to P1 immunity genes. P2 helper virus also expresses immunity functions, usually a transcriptional repressor of lytic functions, but may also express a 'second layer' antirepressor to this repressor, which itself will require a second repressor to establish immunity. With phage 186 (a close relative of P2) this 'second-layer' anti-immunity repressor is also expressed, but in this case it is from the lexA gene product. Thus both satellite and helper viruses encode immunity or identity systems that can affect immunity to other phage. Because P4 can also establish a non-integrated multicopy plasmid state, a satellite phage and a plasmid are not always clearly distinguished from one another, since the defectives of a phage can be defined as a plasmid. Both of these can show strong interactions with phage and could clearly contribute to a less apparent element of a parasitic network.

Phage–Intron Interactions: Stable, Homing Epiparasites

Another epiparasite of phage (as well as other genetic parasites) are introns. Introns are RNA sequences that can be removed by an RNA splicing process, but can be encoded as DNA. Mobile introns may not depend on a specific parasite for their replication, as do satellite and defective viruses, thus they can be considered as 'epiparasites'. However, the specific relationship between an intron and its host genetic parasite (phage) can be both highly specific and highly conserved. In such cases, they appear to provide strong evidence for the existence of stable parasitic networks. It has been recognized that different mechanisms of RNA splicing can be classified into three general classification groups of introns: group I introns, whose RNA splicing is via a guanosine initiated pathway, and groups II/III, as found in eukaryotic tRNA/archael intron class, respectively, and which use a lariat-mediated RNA splicing. Here, the nuclear RNA of eukaryotes is processed in a complex spliceosome structure, suggesting the simpler group I and group III introns were first to evolve. Introns were first discovered in eukaryotes as a processing event of nuclear RNA of some DNA viruses. Some introns (type I) were later shown to be undergo RNA-based autocatalytic excision of the RNA. These catalytic introns are thought to be most ancient and to possibly represent RNA molecular parasites that date back to the RNA world. Soon after their discovery in eukaryotes, it was discovered that phage T4 also had a group I intron within the thymidylate synthase (TS) gene. This is a good example of a stable intron–phage interaction. Curiously, related introns are not found in the *E. coli* host chromosomal TS gene, but were found in various genes of other phage, such as SPO1 and SP82 DNA pol genes of *Bacillus subtillis*. These group I introns have an open reading frame within them which encodes an endonuclease. This endonuclease appears to cleave sequences with some homology to the exons that flank the introns, so they are called homing endonucleases and appear able to promote the transposition of the intron. In this regard, these homing introns clearly appear to be parasitic genetic elements able to invade related sequences. With respect to phage, however, homing introns behave like epiparasite ‘smart bombs’, in that they can incapacitate competing phage or plasmids. For example, T2 phage lacks the intron endonucleases of T4. During mixed T2–T4 infections however, T4 production will predominate. This exclusion is partially due to the T4 segA gene (similar to endonuclease of group I intron), which preferentially cleaves T2 DNA. In a related observation, the corresponding homing endonucleases of SPO1 and SP82 show preference for the sequences of each other’s DNA. During mixed infections between these two phage, SP82 intron endonuclease cleaves DNA of SPO1 at the related DNA pol sequence, promoting SP82 propagation. These results strongly support the idea that homing introns are highly adapted epiparasites of their corresponding phage and can function as part of an extended identity system that reaches out to incapacitate related phage or plasmids. These epiparasites can thus arm a genetic invader with targeted molecular machinery needed to incapacitate already established (lysogenic) genetic competitors or identity systems. Some (but not most) host bacterial lineages have clearly acquired such mobile intron systems. They are found in alpha proteobacteria (at tRNA genes) and also in cyanobacteria. In terms of diversity, viruses seem to provide the lion’s share. Some DNA viruses have very high numbers of homing endonucleases. For example, a field study of 62 thermophilus phage isolated from ecological settings showed that 1/2 had group I introns (conserved in one site). The best example of homing endonuclease diversity, however, is PBCV-1, a large DNA virus of unicellular green algae (also unique for encoding eukaryotic restriction/ modification enzymes). PBCV-1 has amazing 36 such homing endonuclease genes, although the function of so many endonucleases is yet to be evaluated.

Homing endonucleases also have the clear potential to be sequence-specific toxins. As described below, some classes of holins are homing endonucleases that are strain-specific bacterial toxins found in gut bacteria that can kill related strains of bacteria. As described elsewhere, T/A (toxin/antitoxin) gene pairs that are thought to encode RNA endonucleases along with antitoxins to these endonucleases are found in the genomes of most prokaryotes.

Type I introns also can harbor intenes. Intenes are in-frame peptides that self-splice the excision of an amino acid chain from within a precursor protein, in this case the homing endonuclease protein (HEGs). Intene positions can be highly conserved in introns. HEGs are observed to occur in two independent families. Family I has 130 members, found in archael type 1 introns. HEGs are also found to be involved in massive invasion of higher plant mitochondrial DNA (via the coxI gene), possibly from fungal origin. Intenes thus appear to be parasites of epiparasites (homing introns) of parasites (phage) that can be directed at yet other parasites (phage and plasmids). Talk about a dizzying parasite network! Clearly these hyperparasites can have a group identity role given their clear pattern of distribution and activity. Along these lines, it is interesting to note a role for intene–homing endonucleases in terms of sexual identity. The yeast *Saccharomyces cerevesae* utilizes the HO endonuclease to control mating type interconversion switching between sex types. This system, which recognizes and cleaves specific DNA sequences, appears to have evolved from a degraded intene and homing endonucleases. Thus the concept is that there exist networks of interdependent genetic hyperparasites that are involved in host identity that may extend into eukaryotes.

Phage–Phage Interactions: T2 Exclusion of T4 as a Model of Gene Flow

In the section above, I considered the possibility that phage gene pools might be directly associated with phage compatibility. I have presented several examples in which phage can actively preclude other phage from mixed or superinfection situations. Conversely, I have presented examples with plasmids and satellite phage which show highly conserved positive interactions between parasites. Since host mating behavior can be strongly affected by colonizing phage (which also represent the greatest species-specific genetic novelty), and as mating interactions can define host species and gene flow, I will now revisit the issue of the role genetic parasites play in the identification of common gene pools by examining some specific, well-studied examples, T-even phage. By now, it should be clear that phage interference with other phage seems to be rather common situation. The T-even phage are all known as lytic phage that often appear able to recombine genetic information amongst themselves. In fact, some researchers have suggested that the T-even phage can be considered not so much as one species of phage, but one large shared pool of phage genes, implying a common genetic identity and gene flow. Yet some T-even phage interactions are tightly restricted and such restrictions can be pronounced and sometimes systemic. For example, all the T2 alleles are strongly excluded in crosses with T4 phage, especially the allele for gene 56 of T2. The exclusion is at the level of DNA sequence identity and involves the use of the phage-specific recombination system. Recombination occurs as a consequence of recombination-dependent DNA replication, a replication process that is unique to some phage. By using a 'join-copy' or 'join-cut-copy' mechanism, related DNA sequences can form a heteroduplex that are susceptible to ssDNA invasion (via strand displacement and heteroduplex formation) and results in a specific region of DNA that can be cleaved by nucleases, leading to the destruction of the non-matching sequence. In this way, the heteroduplexed sequences for T2 gene 56 (a dCTPase) are preferentially cleaved, but the T4 sequence is not. The process appears to involve T4 Seg F site-specific endonuclease, which as mentioned above is very similar to GIY-YIG group I homing introns that are able to recognize regions of exon homology. Interestingly, T4 (like PBCV-1) has many optional homing endonucleases like these, but these endonucleases are rare in other T-even phage. This process suggests the origin of a 'speciating' system within the T-even phage, based on recombination control, leading to the isolation of otherwise common viral DNA lineages from each other. In a further hyperparasite tit for tat, T4 itself can also be excluded by a 'defective' *E. coli* prophage e14 (via Lit genes) which appears to operate by different mechanisms (involving inhibition of late T4 gene expression). Besides this rather amazing process of systemic DNA recognition, T-even phage can block superinfection by a variety of other phage systems. One system of special note prevents DNA transfer across plasma membrane. As we will see below, genes affecting transfer of molecules across the plasma membrane have a major role in host identity. Such systems are not unique gram-negative bacteria. Some lactococcal phage can also exclude other phage at the point of DNA injection through plasma membrane via membrane proteins.

Phage–Plasmid Interactions: Exclusion and Compatibility

DNA entry exclusion systems as noted above are not only common in phage but also found in parasitic plasmids. For example, plasmid R1 (famous for its early role in the development of recombinant DNA cloning) has a *hok* (host killing) *sok* (suppressor of killing) gene pair (addiction module) needed to maintain low copy of R1. These same genes also block T4 superinfection by preventing DNA entry, and (reminiscent of the P1 story above) will also induce self-destruction (PCD) in individual R1 colonized cells that become T4 infected, thus protecting the larger R1 colonized population in an altruistic-like process. There appear to be many examples of related interactions between plasmids and phage. It has in fact been proposed that the dynamic selection pressures between plasmids and phage may define the main selective pressure for the maintenance of various plasmids, as well as lysogenic phage or phage defectives in bacteria. The early observation of F-plasmid exclusion of T7 phage has already been mentioned. This exclusion is now known to operate via the plasma membrane protein (PifA). It is possible to select for T7 capsid mutants that will grow in such F^+ *E. coli* cells, indicating the role of specific receptor–capsid interactions for such exclusion. However, T7 is not defenseless against F-plasmids harboring *E. coli*. T7 also makes two proteins (gp 1.2 and gp10) that are lethal to (but can be regulated in) F-plasmid containing *E. coli*. This function involves host transmembrane protein FxgA, which is a four-domain transmembrane (TM) protein that interacts with PifA noted above. F exclusion also operates by not allowing late T7 proteins to be made in *E. coli*, although late T7 mRNA is made indicating posttranscriptional control. In addition, this F exclusion of T7 can be alleviated by several antibiotic resistance genes, such as rifampin resistance (rspL) gene, which probably relates to the involvement of a post-transcriptional block in T7 exclusion. However, since these drug resistance genes are themselves mostly found on plasmids, this implies that another intricate (dizzying) and dynamic selection is operating within networks of plasmids, phage and their *E. coli* host. It is noteworthy that in *E. coli* (and other gram-negative bacteria), T7 is a strictly lytic phage. However, as will be presented in Chapters 3 and 4, T7-like phage are common in cyanobacteria (cyanophage), but these cyanophage are temperate. It will also be noted that T7-like DNA polymerases are basal to all the DNA polymerases of eukaryotic chloroplasts.

Plasmids and Host Identity

Phage typing is a common process to identify bacterial strains. And as discussed, both F-plasmid and phage susceptibility are also highly dependent on the presence of plasmids and/or defective phage elements. Bacterial identity is therefore largely dependent on phage and plasmid colonization. Whether the plasmids remain as episomal or become integrated, their affect on host identity is similar. When integrated, such sequences have sometimes been called fitness islands. Like a prophage colonization event, acquisition of a fitness island results in the rapid host adaptation to a complex habitat and is associated with acquisition of gene sets, such as multidrug resistance, that affect competition with other genetic parasites. I suggest that such fitness islands can also be considered as identity modules, which impart systems for self-recognition and frequently use T/A modules to insure stability. Like their phage counterparts, many plasmids will also inhibit redundant or related plasmids via affects on plasmid transfer systems (ICE). In some cases, genes involved (such as traG) also encode inner membrane proteins that are expressed in both donor and recipient cells. Interactions of these proteins can provide exquisite specificity to the transfer process. This is clearly a component of a bacterial self-recognition system.

Plasmid Compatibility Groups

That plasmids can be compatible or incompatible with one another has long been recognized and is also used as criteria to classify them. In some cases, this compatibility results from competition for common replication proteins, and incompatibility results from one plasmid being a more efficient replicon that displaces the competing plasmid. Some of these plasmids have broad host features in common. For example, plasmids of incompatible group P (Inc P) are conjugative and can transfer DNA to almost all gram-negative bacteria. It is also very interesting that Inc P also codes for the PRD1 receptor, a lytic phage able to infect a broad spectrum of gram-negative bacteria harboring P-, N- or W-type conjugative plasmids. PRD1 was first discovered in connection to antibiotic resistance plasmids, and is unique phage in *E. coli* in that it has a protein-primed DNA replication system as well as an internal lipid membrane. PRD1 is also of special interest due to its clear evolutionary relationship to various eukaryotic DNA viruses. Plasmid incompatibility is currently classified into numerous groups, such as N, P, W, F11, A/C, Q, X, etc. Some (P, N, W) can affect plasmid fertility in both positive and negative ways. Other plasmids have broad but complex affects. For example, the 54 kbp plasmid p06670 of *Salmonella enterica* is an X-group conjugative plasmid and encodes antibiotic resistance. When this plasmid is transferred into new host, it results in phage type conversion (affecting 8/10 phage types), but both losses and gains of phage type sensitivity occur. These plasmids are found in nonrandom distributions of natural populations of their host, but their distribution is complex and often dynamic. Plasmid incompatibility can also be due to interactions of addiction modules described above for persisting phage. Some of these addiction modules encode toxins, often referred to as virulence factors that can be involved in group identity and behavior. For example, *Enterococcus faecalis* can harbor plasmid pCF10, which results in cellular aggregation in response to pheromones by the recipient (see Chapter 3 for pheromone discussion). This plasmid also modifies the cell wall in a potentially toxic way but also encodes suppressors of this modification (characteristics of an addiction module). Another toxin encoding plasmid is pPD1, which is also related to the antibiotic AS-48 of pMB2 plasmid. pPD1 is of special interest in that it encodes bacteriocin 21 (described below) but is also pheromone-responsive plasmid. Since both bacteriocins and pheromone responsiveness are common elements of bacterial group communication and group behavior systems, this plasmid suggests interesting possibilities regarding the parasitic origins of such genes. Bacteriocins are well-known toxins used by most bacteria to kill related strains of competing bacteria (see Fig. 2.8 for halo killing assay). Bacterial pheromones are small molecule diffusion-based signals commonly used for group identification (quorum sensing) and group control. Detection of such soluble molecules may represent early versions of odor detection for the purpose of group identification.

The P2 Exemplar: Success of Defectives

P2 was discussed above in the context of a helper phage for the satellite P4 phage. However, P2 alone is also a well-studied phage system, with relatives infecting many different host, that provides many important generalizations concerning 'defective' genetic colonizers, host identity, addiction modules and exclusion of competitors. P2-like phage are common in clinical *E. coli* strains (approximately 30% of the ECOR strains harbor a P2-related genome with relatively limited variation). P2 becomes 'defective' by integrational recombination, which severs its transcription unit to allow quiescence. That means, as a consequence of integration, the P2 genetic map is interrupted in a way that disrupts its transcription and otherwise prevents the integrated copy from making phage. P2 has three known lysogenic conversion functions – *old, tin, fun* – which constitute 12% of genome. As this region is of high AT base composition and the rest of the genome is not, this suggests that these immunity regions are derived from a non-P2 origin. The *Z/fun* gene is of special interest with respect to host identity. This is a multivariable locus composed of 10 completely different areas, suggesting high diversity, hence the suspicion of involvement in viral/host identity. In fact, P2 also appears to maintain an additional genetic (epi) parasite IR element that may be directly involved in generating and maintaining diversity. This system operates on all the *Z/fun* gene regions in that they are surrounded by inverted repeats (IR) that can result in gene mobilization. The P2 Ch1 capsid gene is also flanked by IR sequences that can be inverted by flip expression. Immunity is established using several systems, including the *old* gene, which can exclude many related phage such as lambda. The P2 *tin* gene provides resistance to T-even phage. The P2 prophage, when encountered by sequencing projects, usually resembles a defective or leftover genome from a 'failed' P2 infection since it is interrupted. Furthermore, actual defectives of P2 (not simply interrupted but missing regions of DNA) are also known to be common in field isolates. However, all the defective P2 harboring strains still maintain the P2 R/M system, suggesting that the R/M addiction module is under positive selection even in defectives. Since it is common in natural populations, these P2 R/M modules must frequently alter host genetic identity. Besides being common in natural environments, P2 is often found in mixed infection with other phage and plasmids. It was responsible for the original restriction of lambda infection in *E. coli* reported by J. Leaderberg in 1957. P2 is member of a large phage family which including PhiR67 and 86, a defective prophage that contains a retron (reverse transcriptase encoding element). All these P2-related phage help P4, suggesting that this P4 cooperation must also be under some type of positive selection. Phi CTX of *P. aeruginosa* is a P2-related phage with a 36 kbp genome. CTX, however, is a cytotoxin converting phage (see Fig. 2.12). P2 and CTX both use a Ca^{++}-dependent receptor for binding for entry. P2 also encodes gene *Y*, strikingly similar to holins (described below), which are

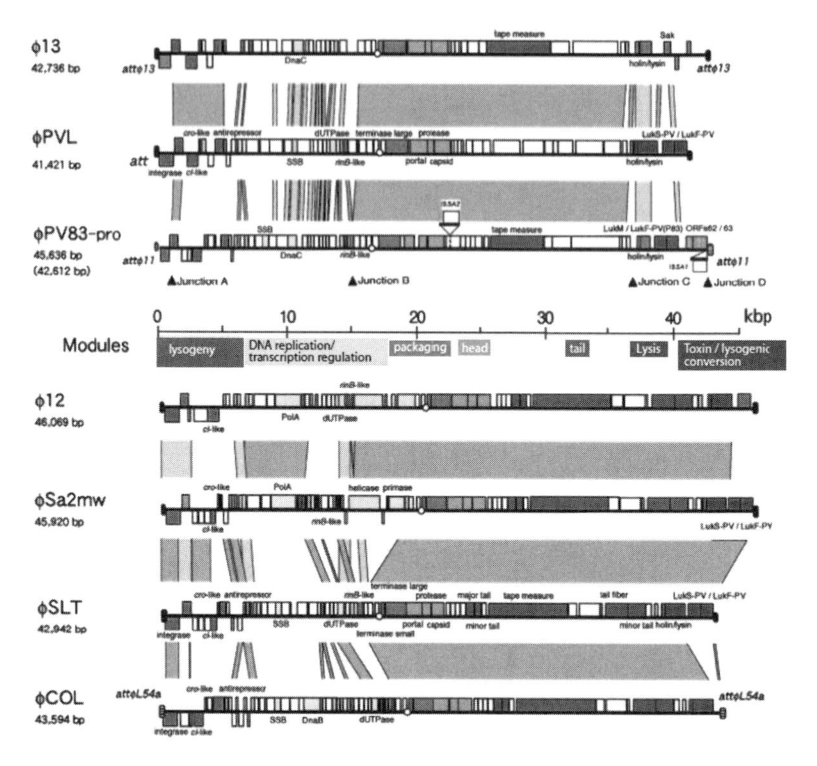

Fig. 2.12 CTX (toxin) relationship to P2 and other phage toxins (reprinted with permission from: Frank Desiere (2004) Antonie Van Leeuwenhoek, Vol. 82, No. 1)

often used as toxic agents against competing bacteria. The P2 tail fiber is similar to those of many other phage (Mu, P1, T2, lambda), all of which appears to have experienced a lot of recombination. As will be presented below, some additional types of bacteriocins are also closely related to P2 tail fibers. *Vibrio cholerae* also has a P2-related phage, K139. Like P2, this cyanophage also encodes a Dam methyltransferase and shows exclusion of other phage via orf2, which prevents superinfection by affecting the g10 gene found on the periplasmic membrane. These latter two genes are also associated with bacterial colonization and virulence in mice (as described further in Chapters 3 and 4). As will be presented, communal mat-forming cyanobacteria have been colonized by sets of tail fiber genes that closely resemble P2 and it is likely that these phage genes contribute to host group identity (Chapter 4). Thus, cells harboring P2-like phage and their defectives are common in natural settings and occurrence of these phage-derived R/M systems, holins and tail proteins is ubiquitous. These P2-related phage genes appear to have made important contributions to the evolution of bacterial communities. P2 is thus an excellent example of a successful 'defective' parasite that affects host interaction with other genetic parasites and creates new host identity.

Phage Remnants (Tail Fibers, Base Plates, Defective Prophage) and the Origin of Host Identity (Restriction Modification and Holins)

With the development of genomic sequencing along with phylogenetic analysis, we can now compare total genomes and infer the likely origins of many chromosomal genes with some confidence. Thus, bacterial genomes contain many genes that were likely of viral origin. However, as most of these genes were described before we understood their likely origin, their corresponding names and functions often fail to connote any relationship to phage. This has certainly been the case for many bacterial toxin genes, such as the bacteriocins, pyocins and holins. There are highly diverse set of proteins associated with the killing of related bacterial species. It now seems clear that the majority of such toxin genes appear to have originated from genetic parasites (via common parlance of horizontal transfer).

For example, let us consider the R/M system (restriction modification enzymes), as the best-studied bacterial immunity system. R/M enzymes are both highly prevalent and highly polymorphic and are found in essentially all non-symbiotic prokaryotes. For example, one strain of *Helicobacteria* has 52 homologues of type II restriction enzymes, possibly representing the high mark for R/M content in the genetic database. These endonuclease enzymes are potentially toxic, cleaving all unprotected DNA. Therefore cells with sets of restriction enzymes must also have protection against the entire set, mostly via DNA modification enzymes that methylate cytosine and adenine at specific sites. I have already noted that DNA methylation is via *S*-adenosylmethionine (SAM), and is an energetically costly molecule that consumes ATP during its synthesis. Thus, to maintain DNA modification and prevent DNA degradation, enough SAM must be around as substrate to provide methyl groups for all the required DNA sites just after DNA synthesis (the temporal window allowing epigenetic identity transfer). Therefore cells should not commit to DNA synthesis until their energy stores are adequate for the required SAM production. As will be presented with the cyanobacteria, it appears that this dependence on SAM might have also led to new group identity (quorum- and light-producing circuits in many bacteria). Therefore, high level of DNA methylation is not without considerable biochemical cost to the cell, and it must be selectively essential for some organisms to maintain so many restriction enzymes. It is likely that phage are providing this selective pressure. However, this situation presents a conundrum when trying to understand the origins of diverse cellular R/M systems, especially species with large R/M sets. The R/M systems function as a matched pair of genes, which requires that the two genes evolve together. For a genome to have acquired 52 restriction enzymes, it must also simultaneously acquire 52 modification enzymes (unless uniform DNA methylation occurs, such HMC, which is not the case here). This presents a daunting evolutionary problem that requires the acquisition of 52 ‘complex phenotypes’ (multiple interacting genes). A Darwinian process involving natural variation, recombination and selection has difficulty explaining this state. Where then did all these matched R/M systems originate and how did they become host genes? I have noted that both persisting episomal phage (P1) and plasmids are known to use highly diverse R/M systems (type II) as addiction modules to enforce parasite persistence and restrict mating outcomes. Since these gene sets provide molecular identity for these genetic parasites, they are also often involved in resistance of, or displacement by, other genetic parasites, phage and plasmids, including post-segregation host killing. Therefore, a more likely origin of the diverse R/M systems is that they are mostly acquired from genetic parasites via stable and cumulative colonization, and not from the ancestor bacteria. This view is consistent with current phylogenetic analysis R/M systems, which concludes that they are mainly acquired by 'horizontal' gene transfer. Thus, the most basic and diverse host identity systems of prokaryotes mainly result from sequential but stable colonization by genetic parasites. This is a general concept that will apply to the identity systems of eukaryotes as well (including the origin of the adaptive immune system, see Chapter 7). As presented below, other bacterial identity systems, such as colicins (like R/M), can also be genotoxic and can fragment DNA as well as confer limited resistance against lytic phage (i.e., colicin 1b inhibits T7 phage). Thus DNA fragmentation and protection from degradation is the most prevalent prokaryotic T/A system for cellular identity. However, DNA fragmentation itself is interesting to consider from an evolutionary perspective. Some sporulating social bacteria and unicellular eukaryotes, such as dinoflagellates, use DNA fragmentation as a normal component of cellular differentiation (Chapter 4). As will also be presented in Chapter 5, DNA fragmentation and PCD were to be maintained as major systems of cellular identity in metazoans. Here, DNA fragmentation has also been adapted to control cell fates and identity, especially in the nervous system (see Chapter 4) and adaptive immune system (Chapter 6).

Toxins: Central Role in Group Identity

Phage-Derived Toxins: Endolysins and Holins

Toxins are a topic of broad and deep relevance to the issues of immunity and group identity. As will be presented throughout this book, the many natural pore-like toxins that can rapidly disrupt the physical or conductive integrity of the plasma membrane are often used to specify cellular identity. Such eukaryotic pore proteins are central to the function of the nervous system and the origin of apoptosis. Some of the bacterial pore toxins are clearly derived from phage base plates or phage proteins associated with viral exit from host (including holins). Other bacterial pore toxins are clearly the harmful half of an addiction module, and are used by genetic parasites. Another class of phage-derived bacterial toxin genes are the endolysins. These are a diverse set of phage enzymes that digest the cell wall (peptidoglycan), allowing release of progeny phage. All dsDNA phage use endolysins. Since bacterial cells are under about 4 atm of internal osmotic pressure, degradation of the cell wall (i.e., peptidoglycan 'cage') leads to osmotic rupture of the plasma membrane. Phage endolysins are tightly regulated by phage holins, which are pore-forming proteins that allow the endolysin enzymes to exit past the membrane in order to digest the cell wall. Since holins are pore-forming proteins, they too can induce membrane damage and must be tightly regulated, often by physical association to antiholin immunity proteins. Both holins and endolysins are often used by plasmid and prophage as virulence factors in pathogenic bacterial strains. Since enterobacterial toxins are often active against eukaryotes, they present a particularly well-studied system. Some virulence plasmids clearly resemble prophage and will integrate into host chromosomes using phage-like tRNA sites for integration (such as pSG1, pSAM2). Some phage (T5, T-even, Cholera Phi 149) carry their own tRNA genes that are involved in homologous integration. Thus holins, antiholins and endolysins are all highly prevalent powerful phage toxins that appear to have several viral origins.

Bacteriocin/Colicin

Bacteriocins are toxic proteins made by one bacterial strain that are active against related bacteria. With *E. coli* strains, it has long been known that some strains can kill other nearly identical strains. The ability of one population of organisms to kill another related population is a basic character of group identity and group behavior that will be maintained throughout evolution. I have already mentioned the early observations of population-based killing due to lysogenic phage. However, frequently, one strain can kill another strain by processes that are not infectious. This type of interstrain killing is common and is often mediated by genes known as bacteriocins. Current estimates are that about 50% of human isolates of *E. coli* harbor a killing function, mostly specific to other *E. coli* strains. Other bacteria, such as field isolates of *Klebsiella*, also harbor similar killing function at significant rates. Such antimicrobiological activity is therefore prevalent in natural populations. These killing functions are often associated with K^+ efflux, affecting membrane polarization and frequently involve some type of pore protein. Usually, host harboring such antimicrobiological proteins also harbor a second gene that provides immunity to toxic function (clearly T/A pairs). Some of these bacterial toxins are produced as 'particles' composed of large protein complexes which have very high toxicity to other bacteria. Some of these particulate toxins are clearly phage-like particles, such as colicin 15 killer particles which clearly resemble a 'head of phage' with or without corresponding tails, but lack DNA. Such particulate toxins are thought to act very similar to a phage in binding to the surface and producing a hole into the target cell. Other toxins are much smaller molecules, but still resemble phage conversion and phage addiction modules, such as Colicin Kp O, a somatic antigen involved in cell wall lysogenic conversion. In fact, colicins were the very first sequenced proteins capable of forming voltage-dependent pore channels, able to vanquish their intended target cells by membrane depolarization. In spite of this early attention, these channel proteins are still poorly understood. Structurally, it is known that pore-forming colicins are in extended H_2O-soluble conformations soon after synthesis, but following membrane interactions, the proteins will fold themselves into

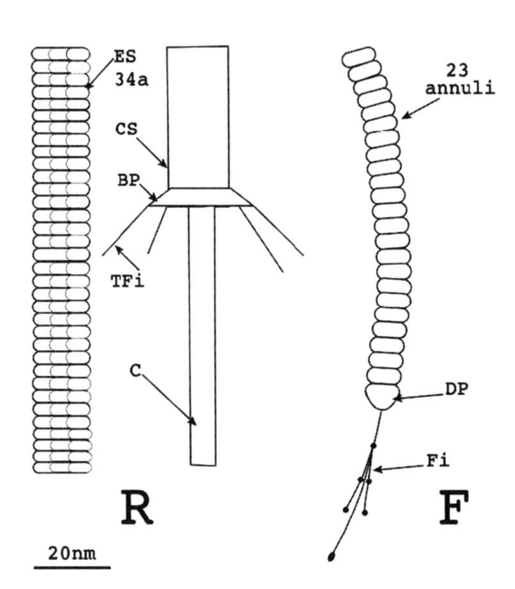

Fig. 2.13 Bacteriocins resemblance to phage tail fibers (reprinted with permission from: Y. Michel-Briand, C. Bausse (2002), Biochimie, Vol. 84)

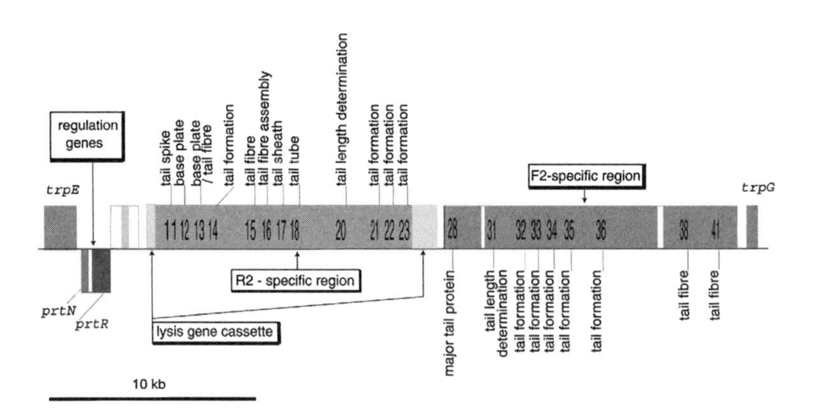

Fig. 2.14 Pyocins toxin genes' similarity to phage genes (reprinted with permission from: Y. Michel-Briand, C. Bausse (2002), Biochimie, Vol. 84)

membrane proteins, after which they become channel proteins with inscrutable properties. These pore proteins represent one of three families (R, F, S) of colicins, all which clearly have phage origins (see pyocins below and Figs. 2.13 and 2.14).

Colicin-Like Pore Toxins

One family of colicin-like proteins is the S-type (soluble, non-particulate). ColE1 is one example which is a secreted protein that induces entry exclusion of other plasmids. An immunity protein protects against the toxic action in ColE1-harboring cells by binding to ColE1 toxin. However, ColE1 is not very common in human clinical isolates, compared to the occurrence of lysogenic phage. ColE1 is famous for its role in the development of genetic engineering and its ubiquitous use as a cloning vector. These colicin plasmids resemble phage in several ways. Some are induced by mitomycin (ColE7, K317). Also, such induced colicins also provide immunity functions that can protect against various phage, such as phage M13K07 and lambda. The colicin plasmids are also often mobilized in the presence of conjugative plasmids. Some, like Col V, encode virulence plasmids of type 2 pilins. A distinct subclass of the colicin plasmids encode pore-forming toxins which kill susceptible cells by expression of a voltage-dependent ion channel. Five such subclasses are known: 10, E1, Ia, Ib and K. These pore toxins tend to be about 600 aa. The corresponding antitoxin immunity proteins are often small (about 100 aa) and are usually integral membrane proteins. Some colicins can be homing endonucleases (see the pyocins below, E2-like DNAse). Clearly bacteriocins are derived from and/or have many phage-like features. However, the original name of bacteriocins or colicins generally fails to communicate any phage relationships.

Holins: Group Sensory Pores for Timing Phage Lysis

Sixty years ago, Hershy (1946) observed that T-even phage can delay lysis when homologous phage particles are present (altering burst size from 100 to 1,000). T-even phage thus seem to sense the presence of similar phage and adjust their molecular program accordingly. In addition, the T-even phage can also block superinfection via membrane DNA transfer mechanism (see phage exclusion above). Both these functions suggest that the T4 phage senses and temporally responds to the presence of other phage of the same type. This sensory process can also be considered as a 'quorum-sensing' system for phage. As presented in the next chapter, quorum sensing is a major topic for the evolution of group identity. In 1981, Streisinger et al. reported on the genetics of this lysis sensing (lysis inhibition system, LIN). Since then, the existence of holin proteins

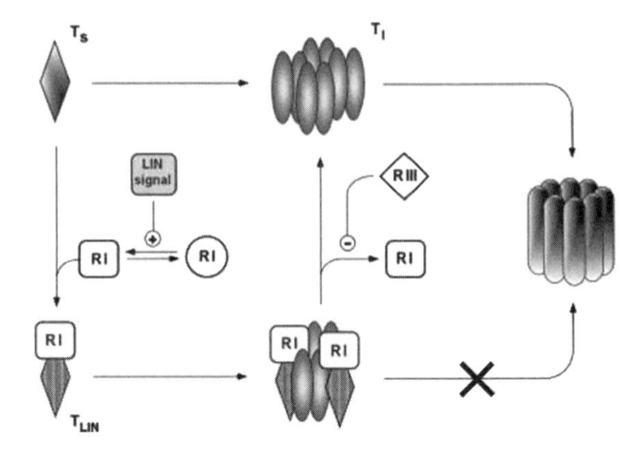

Fig. 2.15 Phage T4 quorum sensing by holin lysis delay (reprinted with permission from: E. Ramanculov, R. Young (2001), Molecular Genetics and Genomics, Vol. 265, Pg. 345–353)

involved in LIN has been well established (see Fig. 2.15). Work with lambda has also indicated that it too has a holin system. The lambda system works via *rI* gene (RI protein) and *t* gene (**T** protein, the holin) which encodes 3 TM domains, periplasmic protein. **T** protein allows the phage lysozyme (endolysin) to lyse/kill the host. Holins are known to exist in two major classes. Class I has 3 TM domain proteins of which the lambdoid-like holins are the most common. These function by depolarizing membrane. Class II has 2 TM domains. Ironically, T4 holin is unusual with only TM domain and doesn't fit these two classes. Class II holins are very diverse, containing over 100 members in 30 orthologous groups, making it (like restriction enzymes) one of the most diverse of all bacterial proteins. These holins will form a channel that allows passage of pre-folded 15 kD endolysin (via 5 μm dia hole). Some phage, such as P1, use host *sec* protein transport systems for this endolysin transport. Holins, like all known cellular and ion proton channels, are found in the cytoplasmic membrane and are in helical bundles. Holins function as part of the phage clock system aimed at the induction of timed host lysis. The clock seems to be attained by the rate of proton loss leading to a de-energized membrane (Fig. 2.15). A de-energized membrane may itself trigger holing function (oligomerization and assembly of pore), which would allow lysis at optimal time for the phage with respect to the spent cell physiology when the cell is unable to maintain the proton gradient. Given that bacterial cell walls are external to the plasma membrane, all lytic phage would require some process to breach this peptidoglycan cell wall and release the assembled progeny virions. However, filamentous phage extrude in a chronic manner from their host and use a pilus-like system, not holins, for virion extrusion. Holins are highly damaging, toxic proteins when expressed alone. The T-even *rI* gene, that makes RI antiholin binding protein, initially accumulates in inactive form. Clearly, holin–antiholin protein sets have the characteristics of an addiction module as well as an identity system. That they are also used in a sensory and temporal capacity suggests how pore toxins can be used for several crucial aspects of group identity. Holin functions seem to have some general characteristics. For example, the Lambda LIN gene S is an antiholin, but this lambda antiholin can substitute for a T-even one. Common structural features are maintained between the holins/antiholins of these two normally competing phage systems. Both P1 and P2 also make holins and presumably antiholins. Thus the holin antiholin system provides both temporal and group sensory functions for phage.

Pyocins and Pseudomonas

Pyocins are toxic proteins of *P. aeruginosa* that are clearly related to the bacteriocin system described above. *P. aeruginosa* is a common gram-negative bacteria that serves as a model system for the study of soil and oceanic bacteria. Due to lung infections, *P. aeruginosa* is also known for its ability to induce lung disease following lung colonization, such as in patients with cystic fibrosis. Pyocins are molecules that kill related species of *Pseudomonas*. In 1954, F. Jacob named these toxins pyocins, after he observed their induction following UV-light irradiation or treatment with mitomycin C and noted that P (pseudomonas)-like relation to *E. colicins*. Like colicins, these molecules are also most active against the same species, but are also self-toxic. Pyocins also have an immunity gene that binds to and inactivates the toxin, protecting the expressing cell as an addiction module. In terms of natural populations, greater than 90% of field isolates of *P. aeruginosa* carry pyocins as chromosomal, and sometimes extrachromosomal, elements. Some of these elements are IR flanked, such as with TnAP41- and Tn3-like elements, so pyocin mobility and variation via epiparasites seems important. Common lab strain of *P. aeruginosa* also carry the AP41 plasmid that encodes a DNAse domain (large killing component) and an immunity subunit (small component). This plasmid is flanked by a transposon (Tn3-like) and is similar to E2-colicin group, but lacks a transposition gene, apparently depending on an exogenous gene source for its motility. Because pyocins are mitomycin C-inducible and can kill related species via particulate material, it initially appeared that pyocins might be proviral products; they are not. Like colicins, there are also three types of pyocins. Some strains of *P. aeruginosa* can produce all three families of colicins, so they have been particularly well studied. All three of these colicin families also bear some resemblance to phage proteins (see Figs. 2.13 and 2.14). The R-type colicin family closely resembles contractile, nonflexible phage tail structures and assembles into the surface of target species as a pore structure that depolarizes membrane of target cell, resulting in a rapid loss of 90 mV potential, killing cells in 20 min. Phylogenetic analysis suggests such colicins are related to P2 phage ancestors. F-type proteins are flexible but non-contractile phage tails. These appear phylogenetically similar to phage lambda. Both R- and F-type colicins are therefore clearly derived from structural protein sets of defective prophage. As we will see in Chapters 3 and 4, the acquisition of such phage-related genes is of particular interest with respect to the genomic distinctions between mat-forming colonial cyanobacteria and bloom-forming cyanobacteria.

The R-type pyocins are surprisingly effective toxins and display a single hit kinetics such that one molecule inserted into a target cell membrane is sufficient to kill that target. The corresponding immunity protein is small (87–153 aa) and acts by binding to the C-terminus of toxin. These pyocins are organized into gene cassettes. Phylogenetic analysis clearly indicates that not only these genes but their corresponding gene order (16 ORFs) are closely related to P2 family of phage. This similarity also includes phage CTX of cyanobacteria (see Fig. 2.12), since antibodies of CTX phage tail proteins also neutralize pyocins. However, only the phage-like genes for tails structures, not phage replication genes, are present in these genomes. F-type pyocins (rod-like flexible, non-contractile phage tails) are significantly less active in killing host cells, requiring about 10-fold greater protein levels relative to R-type. The third type of pyocins is the S-type (soluble), and it also resembles pore-forming colicins that was discussed above. However, included in the S-type are colicins that are a DNAse (homing endonuclease) class. Both of these S-type pyocins are also associated with their corresponding immunity proteins. Phylogenetic analysis indicates that these colicin-like pore proteins also appear to have been derived from P2-like phage holins.

With the colicins and pyocins, we can see strong evidence that pore and antipore protein sets are common and diverse systems by which bacteria recognize and kill related competing bacterial species. These toxin/antitoxin protein sets also fit well into the definition of T/A-based addiction systems that insure group identity and temporal stability. They appear to provide highly effective systems for inducing rapid harm in competing species, especially in their capacity to function as ion channel or pore proteins. The depolarization of membranes appears to be a highly effective toxic process that we will see repeatedly used in the context of group identity. In addition, as seen with T4, membrane depolarization can also be used to generate a quorum-based temporal or lysis timing system. In this, pores provide the basic mechanism of a time-dependent sensor that can kill cells. This basic capacity will now be traced to the evolution of human sensory function. Also, the role of pore proteins and ion channels as virulence factors and bacterial colonizing factors will be developed in greater detail below. In addition, the following chapters will show that similar pore-based addiction systems are present in eukaryotes, especially the filamentous fungi in which dsRNA viral persistence and killer factors are common. Later we will also see that ion channel proteins themselves are the frequent target of small molecule toxins found in oceanic microorganisms, such as cyanobacteria and dinoflagellates, and that such toxins are also used as group identity markers whose toxic activity is also likely to be directed against related and competing species. And in Chapters 5 and 6, I will introduce the central role of pore proteins as mediators of apoptosis and neuron development and in the development of the adaptive immune system. However, it needs to be re-emphasized that the origins of all these identity systems appear to be essentially viral in nature. The restriction–modification systems, the holins, the colicins and the pyocins are all undoubtedly derived from persisting viral genetic parasites, especially from phage related to P2 and lambda. Thus, a central role of persisting genetic parasites in the creation of these host identity systems in prokaryotes appears to be clear. The significance of this context of the evolution of cyanobacterial communities can now be presented in the next two chapters.

General Lessons from Bacteria: Genetic Parasites, Mechanism of Group Identity and Immunity and the Basis of Sensory Detection

This chapter has sought to establish the molecular foundations of group identification in prokaryotes. It is asserted that the basic molecular strategies of group identification first evolved in the prokaryotic world, and that these systems can then be traced to recapitulate the evolution of group identification in higher organisms. Bacteria do indeed often recognize each other in groups and respond with often lethal systems. It has been especially evident that viruses and other genetic parasites can provide and/or affect such group recognition and response. Comparative bacterial genomic analysis confirms that most species-specific genomic differences are mediated by genetic invasions by viruses and related genetic parasites. These genetic parasites often operate together (as epiparasites and hyperparasites) to create complex, network-like situations in which the surviving host cell must be fit with respect to prevailing genetic parasites' networks. Resulting host and parasite fitness often has a temporal or persistent character to it. The parasite–host relationship must be stable to both the division and growth of the host cell, but also stable with respect to potential genetic competitors that would seek to displace a parasite from its host. It is precisely the processes that control this stability that also provides the underlying mechanisms of group identity. This stability generally employs an addiction module, in which a stable toxic (T) and an unstable antitoxic (A) gene function together (as a T/A set) to compel the maintenance of the parasite. The T/A gene concept was made clear from the P1 virus exemplar, which provides the molecular details of recognition and destruction, and it mediates the maintenance of group identity and immunity. The perturbation of this T/A set, such as invasion by another parasite, also disrupts the unstable antitoxin and causes the stable toxic function to destroy the host via programmed cell death. The most common version of this system in bacteria is the MazE/F gene system of *E. coli* responsible for programmed cell death (see Fig. 2.16). Thus altruistic-like self-destruction is an inherent response to lost molecular identity which will be seen to prevail in eukaryotes. However, the toxic component of the T/A set can also be exported and used to destroy nearby non-group member cells. Toxins are thus common elements of group identification and the main mechanism by which bacteria recognize and destroy other bacterial groups. These toxins are mostly evolved from genes of

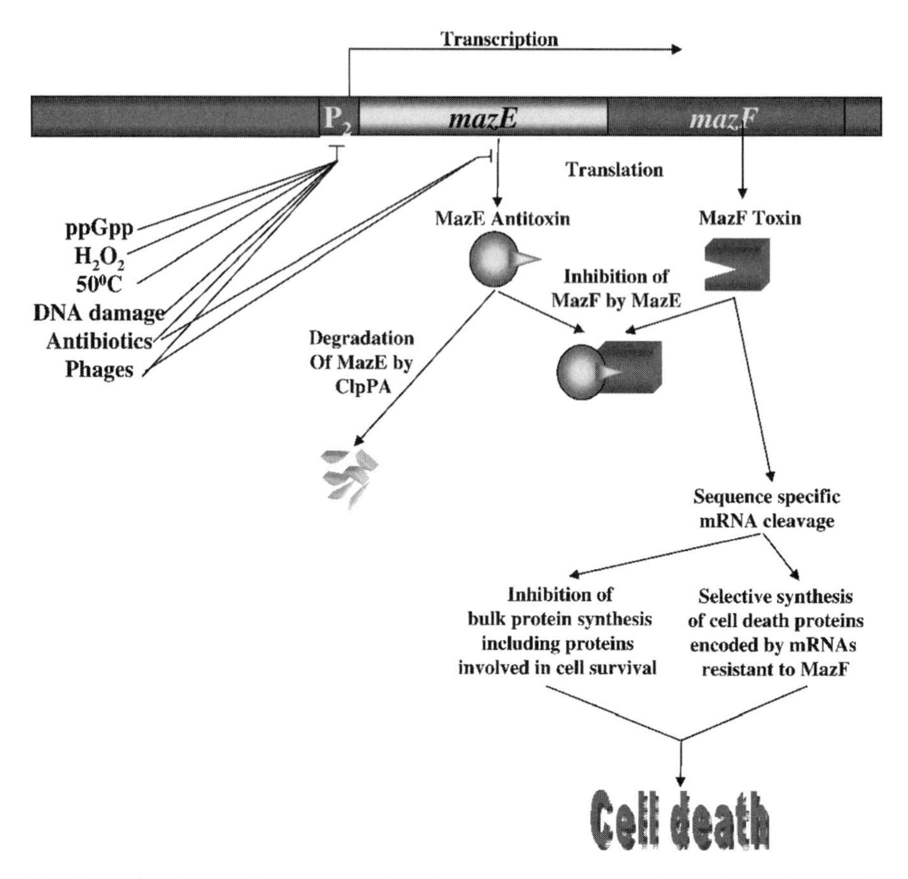

Fig. 2.16 The Maz E/F toxin/antitoxin addiction module and cell death (reprinted with permission from: Journal of Cell Science, Published by The Company of Biologists (2005), Vol. 118, Pg. 4327–4332)

genetic parasites and often are pore or pore-affecting molecules that disrupt membrane function and rapidly kill cells lacking the antitoxin.

Recommended Reading

Addiction Modules, Toxin/Antitoxin Gene Pairs

1. Engelberg-Kulka, H., and Glaser, G. (1999). Addiction modules and programmed cell death and antideath in bacterial cultures. *Annu Rev Microbiol* **53**, 43–70.
2. Gerdes, K., Christensen, S. K., and Lobner-Olesen, A. (2005). Prokaryotic toxin-antitoxin stress response loci. *Nat Rev Microbiol* **3**(5), 371–382.
3. Hayes, F. (2003). Toxins-antitoxins: plasmid maintenance, programmed cell death, and cell cycle arrest. *Science* **301**(5639), 1496–1499.
4. Jensen, R. B., and Gerdes, K. (1995). Programmed cell death in bacteria: proteic plasmid stabilization systems. *Mol Microbiol* **17**(2), 205–210.

5. Kamada, K., Hanaoka, F., and Burley, S. K. (2003). Crystal structure of the MazE/MazF complex: molecular bases of antidote-toxin recognition. *Mol Cell* **11**(4), 875–884.
6. Osborn, A. M., and Boltner, D. (2002). When phage, plasmids, and transposons collide: genomic islands, and conjugative- and mobilizable-transposons as a mosaic continuum. *Plasmid* **48**(3), 202–212.
7. Yarmolinsky, M. B. (1995). Programmed cell death in bacterial populations. *Science* **267**(5199), 836–837.
8. Ziedaite, G., Daugelavicius, R., Bamford, J. K., and Bamford, D. H. (2005). The Holin protein of bacteriophage PRD1 forms a pore for small-molecule and endolysin translocation. *J Bacteriol* **187**(15), 5397–5405.

The R64 Exemplar and Relationship to Virus

1. Boltner, D., MacMahon, C., Pembroke, J. T., Strike, P., and Osborn, A. M. (2002). R391: a conjugative integrating mosaic comprised of phage, plasmid, and transposon elements. *J Bacteriol* **184**(18), 5158–5169.
2. Dziewit, L., Jazurek, M., Drewniak, L., Baj, J., and Bartosik, D. (2007). The SXT conjugative element and linear prophage N15 encode toxin-antitoxin-stabilizing systems homologous to the tad-ata module of the *Paracoccus aminophilus* plasmid pAMI2. *J Bacteriol* **189**(5), 1983–1997.
3. Hartskeerl, R., Zuidweg, E., van Geffen, M., and Hoekstra, W. (1985). The IncI plasmids R144, R64 and ColIb belong to one exclusion group. *J Gen Microbiol* **131**(6), 1305–1311.
4. Ishiwa, A., and Komano, T. (2003). Thin pilus PilV adhesins of plasmid R64 recognize specific structures of the lipopolysaccharide molecules of recipient cells. *J Bacteriol* **185**(17), 5192–5199.
5. Oshima, K., Kakizawa, S., Nishigawa, H., Kuboyama, T., Miyata, S., Ugaki, M., and Namba, S. (2001). A plasmid of phytoplasma encodes a unique replication protein having both plasmid- and virus-like domains: clue to viral ancestry or result of virus/plasmid recombination? *Virology* **285**(2), 270–277.

The P1 Exemplar for Stable Persistence by Addiction

1. Engelberg-Kulka, H., Reches, M., Narasimhan, S., Schoulaker-Schwarz, R., Klemes, Y., Aizenman, E., and Glaser, G. (1998). rexB of bacteriophage lambda is an anti-cell death gene. *Proc Natl Acad Sci USA* **95**(26), 15481–15486.
2. Gazit, E., and Sauer, R. T. (1999). The Doc toxin and Phd antidote proteins of the bacteriophage P1 plasmid addiction system form a heterotrimeric complex. *J Biol Chem* **274**(24), 16813–16818.
3. Hazan, R., Sat, B., Reches, M., and Engelberg-Kulka, H. (2001). Postsegregational killing mediated by the P1 phage “addiction module” phd-doc requires the Escherichia coli programmed cell death system mazEF. *J Bacteriol* **183**(6), 2046–2050.
4. Klassen, R., Tontsidou, L., Larsen, M., and Meinhardt, F. (2001). Genome organization of the linear cytoplasmic element pPE1B from Pichia etchellsii. *Yeast* **18**(10), 953–961.
5. Lehnherr, H., Maguin, E., Jafri, S., and Yarmolinsky, M. B. (1993). Plasmid addiction of bacteriophage P1: doc, which causes cell death on curing of prophage, and phd, which prevents host death when prophage is retained. *J Mol Biol* **233**(3), 414–428.
6. Lobocka, M. B., Rose, D. J., Plunkett, G., 3rd, Rusin, M., Samojedny, A., Lehnherr, H., Yarmolinsky, M. B., and Blattner, F. R. (2004). Genome of bacteriophage P1. *J Bacteriol* **186**(21), 7032–7068.

7. Magnuson, R., Lehnherr, H., Mukhopadhyay, G., and Yarmolinsky, M. B. (1996). Autoregulation of the plasmid addiction operon of bacteriophage P1. *J Biol Chem* **271**(31), 18705–18710.
8. Yarmolinsky, M. B. (2000). A pot-pourri of plasmid paradoxes: effects of a second copy. *Mol Microbiol* **38**(1), 1–7.

Viruses of Archaea

1. Bettstetter, M., Peng, X., Garrett, R. A., and Prangishvili, D. (2003). AFV1, a novel virus infecting hyperthermophilic archaea of the genus acidianus. *Virology* **315**(1), 68–79.
2. Koonin, E. V., Senkevich, T. G., and Dolja, V. V. (2006). The ancient Virus World and evolution of cells. *Biol Direct* **1**, 29.
3. Lipps, G. (2006). Plasmids and viruses of the thermoacidophilic crenarchaeote Sulfolobus. *Extremophiles* **10**(1), 17–28.
4. Ortmann, A. C., Wiedenheft, B., Douglas, T., and Young, M. (2006). Hot crenarchaeal viruses reveal deep evolutionary connections. *Nat Rev Microbiol* **4**(7), 520–528.
5. Prangishvili, D., Forterre, P., and Garrett, R. A. (2006). Viruses of the Archaea: a unifying view. *Nat Rev Microbiol* **4**(11), 837–848.

The Virology of the Oceans and World

1. Angly, F. E., Felts, B., Breitbart, M., Salamon, P., Edwards, R. A., Carlson, C., Chan, A. M., Haynes, M., Kelley, S., Liu, H., Mahaffy, J. M., Mueller, J. E., Nulton, J., Olson, R., Parsons, R., Rayhawk, S., Suttle, C. A., and Rohwer, F. (2006). The marine viromes of four oceanic regions. *PLoS Biol* **4**(11), e368.
2. Comeau, A. M., and Krisch, H. M. (2005). War is peace–dispatches from the bacterial and phage killing fields. *Curr Opin Microbiol* **8**(4), 488–494.
3. Filee, J., Tetart, F., Suttle, C. A., and Krisch, H. M. (2005). Marine T4-type bacteriophages, a ubiquitous component of the dark matter of the biosphere. *Proc Natl Acad Sci USA* **102**(35), 12471–12476.
4. Hambly, E., and Suttle, C. A. (2005). The viriosphere, diversity, and genetic exchange within phage communities. *Curr Opin Microbiol* **8**(4), 444–450.
5. Hendrix, R. W. (2003). Bacteriophage genomics. *Curr Opin Microbiol* **6**(5), 506–511.
6. Lepage, E., Marguet, E., Geslin, C., Matte-Tailliez, O., Zillig, W., Forterre, P., and Tailliez, P. (2004). Molecular diversity of new Thermococcales isolates from a single area of hydrothermal deep-sea vents as revealed by randomly amplified polymorphic DNA fingerprinting and 16S rRNA gene sequence analysis. *Appl Environ Microbiol* **70**(3), 1277–1286.
7. Paul, J. H., Sullivan, M. B., Segall, A. M., and Rohwer, F. (2002). Marine phage genomics. *Comp Biochem Physiol B Biochem Mol Biol* **133**(4), 463–476.
8. Rohwer, F. (2003). Global phage diversity. *Cell* **113**(2), 141.
9. Sano, Y., Matsui, H., Kobayashi, M., and Kageyama, M. (1993). Molecular structures and functions of pyocins S1 and S2 in *Pseudomonas aeruginosa*. *J Bacteriol* **175**(10), 2907–2916.
10. Short, S. M., and Suttle, C. A. (2002). Sequence analysis of marine virus communities reveals that groups of related algal viruses are widely distributed in nature. *Appl Environ Microbiol* **68**(3), 1290–1296.
11. Suttle, C. A., and Chen, F. (1992). Mechanisms and rates of decay of marine viruses in seawater. *Appl Environ Microbiol* **58**(11), 3721–3729.

The Evolution and Adaptation of Lactic Acid Bacterial Phage

1. Brussow, H., Bruttin, A., Desiere, F., Lucchini, S., and Foley, S. (1998). Molecular ecology and evolution of Streptococcus thermophilus bacteriophages – a review. *Virus Genes* **16**(1), 95–109.
2. Brussow, H. (2001). Phages of dairy bacteria. *Annu Rev Microbiol* **55**, 283–303.
3. Desiere, F., Lucchini, S., Canchaya, C., Ventura, M., and Brussow, H. (2002). Comparative genomics of phages and prophages in lactic acid bacteria. *Antonie Van Leeuwenhoek* **82**(1–4), 73–91.
4. Piknova, M., Filova, M., Javorsky, P., and Pristas, P. (2004). Different restriction and modification phenotypes in ruminal lactate-utilizing bacteria. *FEMS Microbiol Lett* **236**(1), 91–95.
5. Silander, O. K., Weinreich, D. M., Wright, K. M., O'Keefe, K. J., Rang, C. U., Turner, P. E., and Chao, L. (2005). Widespread genetic exchange among terrestrial bacteriophages. *Proc Natl Acad Sci USA* **102**(52), 19009–190014.

Selfish DNA

1. Dawkins, R. (1976). "The selfish gene." Oxford University Press, New York.
2. Dawkins, R. (2006). "The selfish gene." 30th anniversary ed. Oxford University Press, Oxford, New York.
3. Doolittle, W. F., and Sapienza, C. (1980). Selfish genes, the phenotype paradigm and genome evolution. *Nature* **284**(5757), 601–603.
4. Orgel, L. E., and Crick, F. H. (1980). Selfish DNA: the ultimate parasite. *Nature* **284**(5757), 604–607.

Phage Self Recognition and Lysis Inhibition

1. Abedon, S. T. (1992). Lysis of lysis-inhibited bacteriophage T4-infected cells. *J Bacteriol* **174**(24), 8073–8080.
2. Ramanculov, E., and Young, R. (2001a). An ancient player unmasked: T4 rI encodes at-specific antiholin. *Mol Microbiol* **41**(3), 575–583.
3. Ramanculov, E., and Young, R. (2001b). Genetic analysis of the T4 holin: timing and topology. *Gene* **265**(1–2), 25–36.
4. Slavcev, R. A., and Hayes, S. (2003). Blocking the T4 lysis inhibition phenotype. *Gene* **321**, 163–171.
5. Tran, T. A. T., Struck, D. K., and Young, R. (2005). Periplasmic domains define holin-antiholin interactions in T4 lysis inhibition. *J Bacteriol* **187**(19), 6631–6640.

Prokaryotic Genome Evolution

1. Canchaya, C., Fournous, G., Chibani-Chennoufi, S., Dillmann, M. L., and Brussow, H. (2003). Phage as agents of lateral gene transfer. *Curr Opin Microbiol* **6**(4), 417–424.
2. Canchaya, C., Fournous, G., and Brussow, H. (2004). The impact of prophages on bacterial chromosomes. *Mol Microbiol* **53**(1), 9–18.
3. Cheetham, B. F., and Katz, M. E. (1995). A role for bacteriophages in the evolution and transfer of bacterial virulence determinants. *Mol Microbiol* **18**(2), 201–208.

4. Hacker, J., and Kaper, J. B. (2000). Pathogenicity islands and the evolution of microbes. *Annu Rev Microbiol* **54**, 641–679.
5. Hurtado, A., and Rodriguez-Valera, F. (1999). Accessory DNA in the genomes of representatives of the Escherichia coli reference collection. *J Bacteriol* **181**(8), 2548–2554.
6. Jain, R., Rivera, M. C., Moore, J. E., and Lake, J. A. (2002). Horizontal gene transfer in microbial genome evolution. *Theor Popul Biol* **61**(4), 489–495.
7. Riley, M., and Serres, M. H. (2000). Interim report on genomics of Escherichia coli. *Annu Rev Microbiol* **54**, 341–411.
8. Smeltzer, M. S., Hart, M. E., and Iandolo, J. J. (1994). The effect of lysogeny on the genomic organization of Staphylococcus aureus. *Gene* **138**(1–2), 51–57.

Bacteriocins/Pyocins/Holins and Relationships to Phage

1. Bull, J. J., and Regoes, R. R. (2006). Pharmacodynamics of non-replicating viruses, bacteriocins and lysins. *Proc Biol Sci* **273**(1602), 2703–2712.
2. Chavan, M., Rafi, H., Wertz, J., Goldstone, C., and Riley, M. A. (2005). Phage associated bacteriocins reveal a novel mechanism for bacteriocin diversification in Klebsiella. *J Mol Evol* **60**(4), 546–556.
3. Heo, Y. J., Chung, I. Y., Choi, K. B., and Cho, Y. H. (2007). R-Type pyocin is required for competitive growth advantage between *Pseudomonas aeruginosa* strains. *J Microbiol Biotech* **17**(1), 180–185.
4. Nakayama, K., Takashima, K., Ishihara, H., Shinomiya, T., Kageyama, M., Kanaya, S., Ohnishi, M., Murata, T., Mori, H., and Hayashi, T. (2000). The R-type pyocin of *Pseudomonas aeruginosa* is related to P2 phage, and the F-type is related to lambda phage. *Mol Microbiol* **38**(2), 213–231.
5. Riley, M. A. (1998). Molecular mechanisms of bacteriocin evolution. *Annu Rev Genet* **32**, 255–278.
6. Sano, Y., Matsui, H., Kobayashi, M., and Kageyama, M. (1993). Molecular structures and functions of pyocins S1 and S2 in *Pseudomonas aeruginosa*. *J Bacteriol* **175**(10), 2907–2916.
7. Tikhonenko, A. S., Belyaeva, N. N., and Ivanovics, G. (1975). Electron microscopy of phages liberated by megacin A producing lysogenic Bacillus megaterium strains. *Acta Microbiol Acad Sci Hung* **22**(1), 58–59.
8. Young, R. (2002). Bacteriophage holins: deadly diversity. *J Mol Microbiol Biotechnol* **4**(1), 21–36.

Chapter 3
Sensory Systems (Light, Odor, Pheromones) in Communities of Oceanic Microbes

Microbiology and Sensory Light Detection of the Oceans

Cyanobacteria are ubiquitous photosynthetic bacteria found in the oceans. They are also photosynthetic. In addition, the phage that infect them are also abundant. These phototactic and phototactic bacteria are only part of a much larger aquatic micro-ecosystem that also responds to light. Recent large-scale shotgun sequencing projects have started to more quantitatively estimate the microbiological makeup of the oceans (such as the Sargasso Sea). From these studies, a clear picture emerges. Various members of the proteobacteria dominate this ecology ($\alpha > \delta > \beta$ proteobacteria). This bulk of proteobacteria is followed by cyanobacteria in prevalence. Thus, gram-negative bacteria (such as *Vibrio* species) are present in the greatest numbers. As previously mentioned, the photosynthetic component of this microbiological biomass is dominated by *Prochlorococcus* and *Synechococcus* cyanobacteria. However, the distribution of genetic parasites in this photosynthetic ecology is much less well characterized. It is known that bacteria from marine coastal sediments, composed mainly of alpha and gamma proteobacteria, harbor lots of diverse plasmids, so hyperparasite networks seem possible. However, since most of these plasmids lack sequence homology to known plasmids, they have been difficult to characterize. Since many of these plasmids do show similar replication processes to known plasmids, they are likely to be distant members of known plasmid classes. The ecology of marine viruses is also poorly understood, although their abundance is well established. In terms of group identification and group behavior, cyanobacteria were crucial in the identification and study of a system known as 'quorum sensing' (QS), now known to be present in many gram-negative bacteria. QS systems allow bacteria to detect the nearby presence of conspecific bacteria and alter their behavior (genetic programs) accordingly. Some bacteria harbor multiple QS 'systems', suggesting that these are actually multiple 'identity' circuits, not simply systems to detect a conspecific neighbor. These systems were initially discovered due to their ability to control the production of light as well as toxins in cyanobacteria, than in other gram-negative bacteria. Yet, although able to sense and produce light, most oceanic gram-negative bacteria are not photosynthetic; therefore, they do not need to detect light for metabolic

L.P. Villarreal, *Origin of Group Identity*, DOI: 10.1007/978-0-387-77998-0_3,

reasons. The reasons for their QS/light link have not been established, but I suggest that group behavior may be the common thread that maintains a link between QS and light production in these diverse bacteria. This frequent link of quorum-sensing systems to photosensory systems (and motility) will be of special interest regarding origin and evolution of sensory systems and group behavior. In terms of cyanobacterial mat communities, the topic of light-based motility and group behavior will also be considered in Chapter 4. The sensory protein most involved in light detection is rhodopsin (described in detail below). It was, however, most unexpected that the Sargasso Sea shotgun sequencing project would discover that the greatest number and most diverse genes characterized were rhodopsin-like sensory proteins. Thus, sensory rhodopsins are the most ubiquitous bacterial gene family yet isolated. The large majority of these sensors appear to be ion channels. The biological reason for this remains mysterious since, as noted, relatively few bacteria are photosynthetic. Fossil evidence makes clear that cyanobacterial oceanic communities were crucial to the evolution of all higher life forms. This development is the focus of the next chapter. In this chapter, I develop the assertion that light detection played a crucial role in the evolution of bacterial group behavior and that for the most part, the issue group identity and group behavior has not been the focus of evolutionary biology, especially regarding bacteria. Here, I will present this as a foundation for the evolution of the basic mechanisms of community formation.

An Oceanic Community of Light

Many organisms of greater complexity than bacteria that dwell in the oceans appear to have a very broad and ancient relationship to both the detection and emission of light. It is currently estimated that there are 30 different and independent origins of bioluminescence. For example, organs for light production in fish is common (i.e., 109 species are known) especially the reef-associated (Pelagic) fish. We have already noted the very large numbers of sensory rhodopsin genes isolated from the Sargasso Sea, so clearly light detection in microbes is highly prevalent. The relevant mechanisms in the context of community formation have been most studied in cyanobacteria and *Vibrio* species. However, many other oceanic microbes also sense and make light, including dinoflagellates, red algae, coral, invertebrates and, as noted, vertebrate fish. In many of the microbiological species, light production is also associated with blooms and toxin production. In some established cases, the source of this light production is essential for bacterial colonization of the host, such as the squid light-producing organ. Fish have light organs of various designs and biochemistries, including some that are not dependent on microbiological symbionts. Light production can be found in 30 genera of teleost fish. Here, it is mainly (and clearly) used for intraspecies communication. In some cases, such as glowworms, light production has a well-established sexual function as well. In this case, light is undoubtedly used for group identification purposes. The *Vibrio* bacterial species that produces light for the octopus light organ is also very similar to *Vibrio* and cyanobacterial species that colonize the guts of many vertebrate hosts and also produce light. Like these light organs, fish guts are also colonized by dense bacterial communities having 10^6–10^7 bacteria per ml of fecal material. These communities can make visible light. This ability of gut-dwelling bacteria to make light has always been a curious observation that not only applies to vertebrate fish, but can also be found in some insects such as mole crickets, mayflies, ants, millipedes and even in guts of nematodes (although here the numbers are small). Below, the theme that light production provides a group identity function in all these diverse oceanic species will be developed.

Sensory rhodopsin shows a clear relationship between that found in bacteria and the eye pigment of eukaryotes, suggesting that a direct evolutionary relationship exists between these bacterial sensory receptors and those of eukaryotes. With this in mind, it is interesting to consider various arguments concerning the importance of the evolution of the eye for the Cambrian radiation that would take place. In his book, 'In the Blink of an Eye,' Gehring makes the assertion that the evolution of eyes was the main trigger associated with the Cambrian explosion, although others question this view. In this regard, thus the eye cells of the larvae planula (with single-cell photoreceptors) might be good models for such a putative transition to complex eye structures. There are various differences between bacterial and eukaryotic light production. For one, bacterial light production is continuous and depends on autoinducers (described below) to keep production going. In contrast, eukaryotes will often control the timing of light production by various mechanisms involving different biochemistry and shutter systems. These shutter systems are often used to control light production when bacterial symbionts are the source of eukaryotic light. Thus, eukaryotes are using temporal patterns of light production for group identification. As we will see in the following chapter, glowworms and fireflies also use temporal light patterns for sexual identification.

High Light Sensitivity in Bacterial Group Identity

Gram-negative bacteria thus have highly diverse sensory photoreceptors, well beyond that found in eukaryotes. But in the eukaryotes, we see the evolution of advanced photosensory and behavioral systems. In bacteria, it appears that the sensitivity of photodetection is under strong selective pressure that also originates receptor diversity and maintains receptor function. Why should so many non-photosynthetic bacteria need photosensory receptors? The usual (rationalized) answer has been that bacteria must sense and avoid light due to potential damaging effects of intense light, such as cell killing that is sensitive to UVB wavelength (280–720 nm). Ironically, the cyanobacteria, which depend on solar energy, are very sensitive to this UV wavelength, more so than many other bacteria. The problem with this explanation, that bacteria sense light to avoid UV killing, is mainly due to the sensitivity of detection. Many bacteria can sense and respond to very low flux levels of light, such as moonlight. This appears to be much too sensitive for the purpose of UV avoidance due to bright sunlight near surface waters. Yet, UV light accounts for much bacterial mortality in the oceans. Interestingly, bacterial UV light sensitivity can also be highly affected by phage infection, which encodes phage-specific DNA repair enzymes. Also, UV light can induce cyanobacterial light production for unknown reasons. This induction would not seem to serve a role regarding UV avoidance and may simply reflect a DNA damage response, as it does not appear to involve quorum or group sensing systems (described below). What then is the point of such light sensitivity or production? High sensitivity could be used for group detection, as it is in oceanic metazoan organisms. For this, light would need to provide a physical system with sufficient range and resolution to communicate with and differentiate between competing groups. The spectral resolution of light production and receptor sensitivity would seem insufficient for such use. Perhaps the very diverse rhodopsin receptors provide some resolution capacity. Light has several attractive physical characteristics that could be used for group communication and identity. In transparent water, it can travel a large distance between free-growing cell populations (such as blooms), to maintain a community or population structure. For this to provide group-specific information, either light production and/or light detection would need to be highly tuned to a particular wavelength (which is not observed) or it must be linked to an additional signal that provides the needed resolution for identification. I now consider the genomes of representative cyanobacteria to evaluate how specific quorum-sensing systems might work along with light to provide the needed resolution for group identity purposes.

Let us reconsider the potential role of light in gram-negative bacteria found in the guts of several animal species that are known to produce light. In the large majority of gut-colonizing bacteria, photosynthesis is clearly not a factor. Clearly, light produced in guts of organisms cannot be related to ambient light states or a need to avoid UV-mediated light killing or photosynthesis. Instead it must serve some purpose applicable to its gut habitat, such as group or colony behavior. These cyanobacteria are initially free-living (non-symbiotic) bacteria found in the ocean. Only following gut colonization do they form dense communities and produce light. How then might light production relate to such microbiological group behavior? The most likely answer is via sensory light detection. Light production is also often linked to quorum sensing (QS) and also motility in many *Vibrio*-related bacterial species. These systems might need to work together, hence be under common regulatory control, if both are needed for control of group behavior. In this case, the adaptive function of light would serve to communicate group identity or quorum status. These dispersed bacteria must somehow find and colonize a common habitat, form a colony that precludes competition and provides group recognition in fish guts.

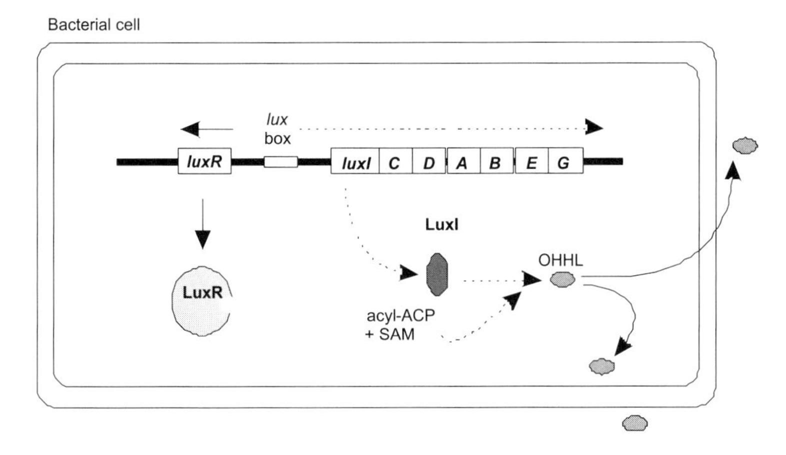

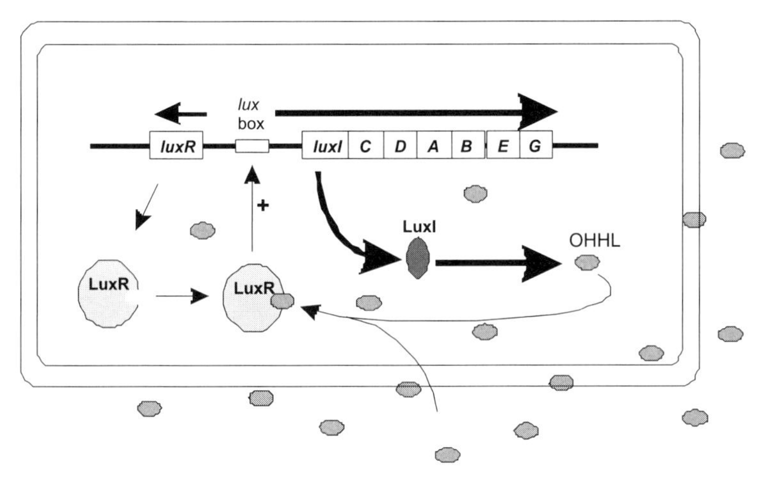

Fig. 3.1 Lux operon (reprinted with permission from: Whitehead, Barnard, et al. (2001), FEMS Microbiology Reviews, Vol. 25)

A similar scenario would also apply to blooming bacteria that may also use light production and detection to coordinate the coherence and movement of large bacterial populations. The main circuit that links light production to QS in cyanobacteria is shown with the Lux operon as shown in Fig. 3.1. In this circuit, the detection of a small diffusible molecule (QS pheromone) is also essential for light production. This provides the molecular origin of an olfactory-like system for group detection. This system can also affect bacterial motility in some cases

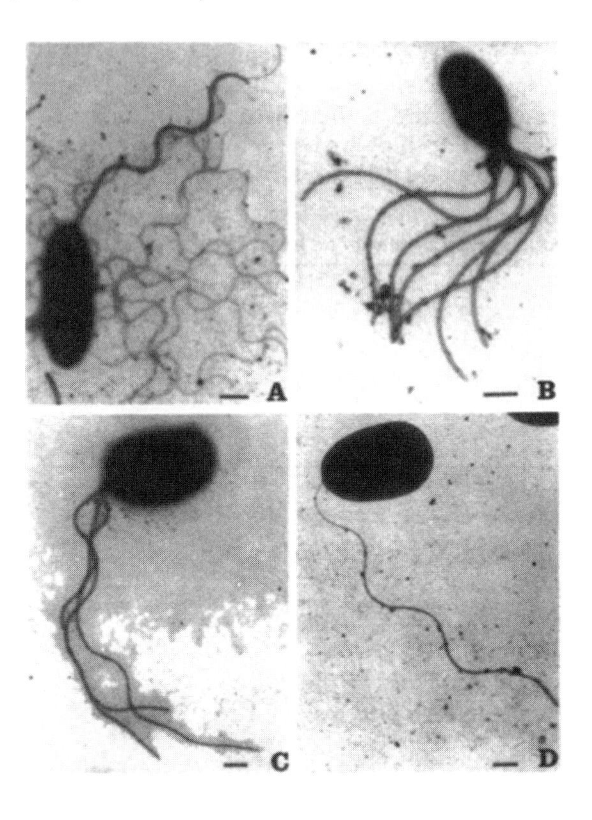

Fig. 3.2 Light-producing cyanobacteria (reprinted with permission from: Reichelt, Baumann (1973), Arch Mikrobiol, Vol. 94, pg. 283–330)

(presented below). Thus, small molecule (pheromone) detection, light production, light detection and motility can all be functionally linked in cyanobacteria. In eukaryotic terms, these equivalent functions would link olfaction and vision to group behavior (a theme presented in Chapter 5). An example of cyanobacteria that link all these functions is found in the squid light-producing organs as shown in Fig. 3.2. Of particular interest are the complexity of flagella and motility associated with these cyanobacteria. These flagella provide both sensory functions and motility. As will be developed in subsequent chapters, flagella-related structures were retained and used for the evolution of sensory organelles (olfaction, visual, etc.) in eukaryotes. It is interesting to note that besides light production and blooming, most of these cyanobacterial species are also well known for their ability to produce toxins and they display an extraordinary chemical diversity of such molecules. These toxins are especially active against the ion channel proteins of related bacteria and eukaryotes, but not archaea. This is discussed in Chapter 4, which considers the possible role of toxins as T/A modules in group identity. Curiously, it was most surprising that the gut-isolated bacteria found in the Tyrolean iceman were also mostly cyanobacteria able to produce light, suggesting humans once harbored a fish-related bacterial fauna.

The Biochemistry of Light Production

The chemical origin of biological light production presents an interesting dilemma regarding oxidation states (see Fig. 3.3). The luciferase-catalyzed production of light requires an oxidized state with the availability of molecular O_2. The luciferase enzyme is like a mixed function oxidase and operates via the oxidation of flavin mononucleotides by O_2, which will oxidize aliphatic long-chain aldehydes. This light chemistry is unique to bacteria. Luciferase is one of the slowest enzymes known, producing 10 quanta per second per enzyme at room temperature. This O_2 requirement strongly suggests that such light production would not have been possible in the Earth's early reducing atmosphere. It would seem that photosynthesis and O_2 generation would need to predate (or co-evolve with) the wide-scale evolution of light production. Curiously, some light-producing bacteria (*V. fischeri*) do not make light near surface water, due apparently to high O_2 levels. Thus, there is a general inhibition of luciferase with high O_2 levels. One of the interesting possibilities regarding the origin of light production is that it initially served as an internal system that would detoxify the production of damaging levels of O_2 within cells, generated during photosynthesis (see Recommended Reading). If so, we might imagine that photosynthesis and luciferase could have constituted a highly elaborate toxin/antitoxin functional pair that required their co-evolution. Such an evolutionary history could also explain the general inhibition at high O_2 levels, which would be extracellular and provide a feedback loop to the T/A function. If light production and photosynthesis indeed had such a T/A relationship early in its evolution, we might also understand why light production would continue to be associated with group density and remain a process under QS control.

Slow Light for Group Identification?

Thus, the slow timing of luciferase photon release could have been used by the cell for group identity purposes. Given the low of light production, Hastings and Nealson (see Recommended Reading) proposed that fish guts provide the most numerous and concentrated occurrence of cyanobacteria in the ocean and that light production in these gut colonies could be used for bacterial identification and colonization purposes, required in order to colonize fish. This was based on a calculation that bright, fully induced (10^6–10^7/ml@hi quorum) bacteria can emit 10^3–10^4 quanta/cell/second. Since the bacterial sensory photoreceptors can detect and behaviorally respond to a light flux of 5–10 quanta, one could expect colony behavior in this circumstance. I suggest that this reasoning could be extended to include large colonies of blooming and mat-forming cyanobacteria, which might also use light for communication and group behavioral purposes. In this regard, it is interesting to recall the original observations of M. Beijerinck. At elevated temperatures, cyanobacteria can transform into dark forms (K variants) that now make very low levels of light. Along with this transformation,

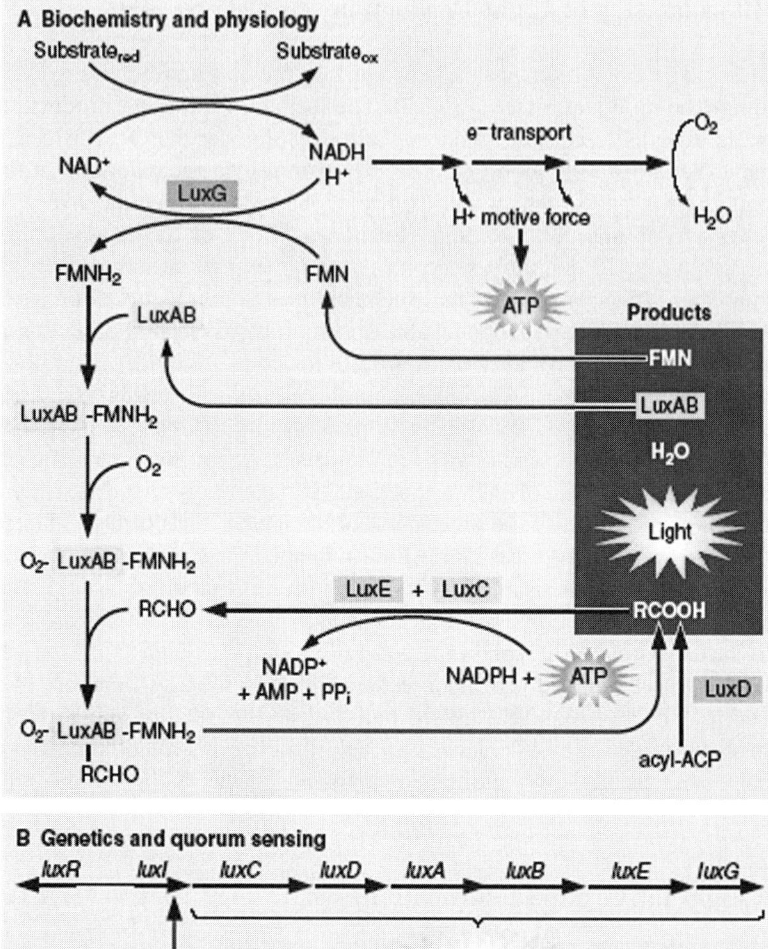

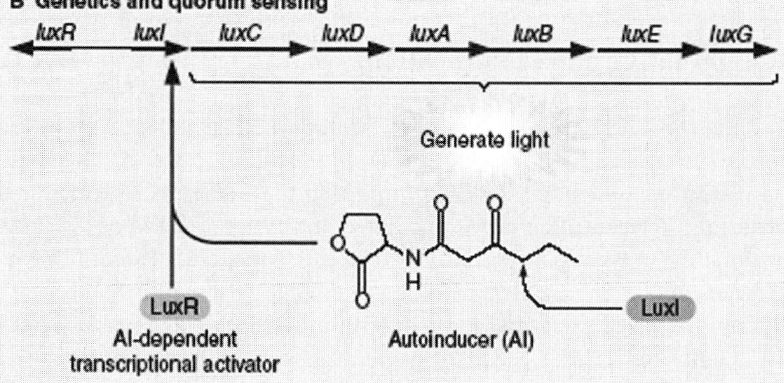

V. fischeri bioluminescence. (A) Biochemistry and physiology. (B) Genetics and quorum sensing.

Fig. 3.3 The biochemistry of light production (reprinted with permission from: Eric Staab (2005), ASM News, Vol. 5)

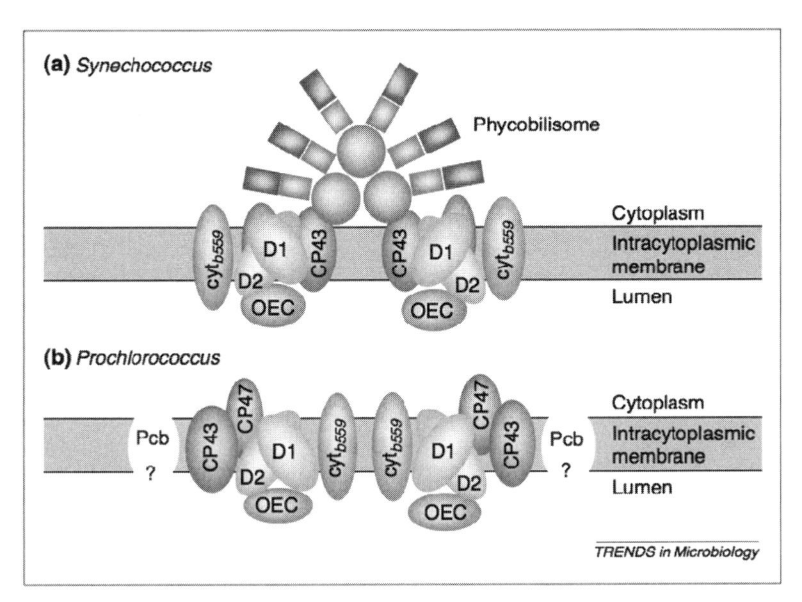

Fig. 3.4 Cyanobacterial photosynthesis (*See* Color Insert, reprinted with permission from: Ting, Rocao, King, Chiholm (2002), Trends in Microbiology, Vol. 10, No. 3)

colony morphology is also altered (losing slimy mucoid appearance) as is phage sensitivity. The reason for the linkage of these three features remains unknown. However, if these features are all important for colonization, as seen in fish guts or the squid light organ (below), we might predict that these K variants would be inefficient colonizers of fish guts. It is known that all three of these phenotypes can also spontaneously revert to light production, mucoid colonies and restored phage sensitivity. If indeed a main purpose of light production in aquatic microorganisms is for group identification and behavior, along with the corresponding small-molecule autoinducers, we might be able to understand this linkage, as light would constitute one of the major identity systems in the oceans.

Cyanobacteria are also photosynthetic (see Fig. 3.4). This highly complex system of interacting genes was to have profound impact on the evolution of life. As these bacteria utilize the type II of photosynthesis, which is related to that used by eukaryotic plants, they are considered as the source of this system. (Chapter 4 will consider the role of horizontal gene transfer and virus-mediated evolution in the evolution of Algae.)

Sensory Rhodopsin, Mechanism of Light Detection and Origin of G-Transduction

Rhodopsin, a Sensor and Transducer of Light Detection

There are two distinct types of rhodopsin molecules, based on their sequence similarities and the type of chromophore in which light induces either a *cis* or

Current Biology
R589

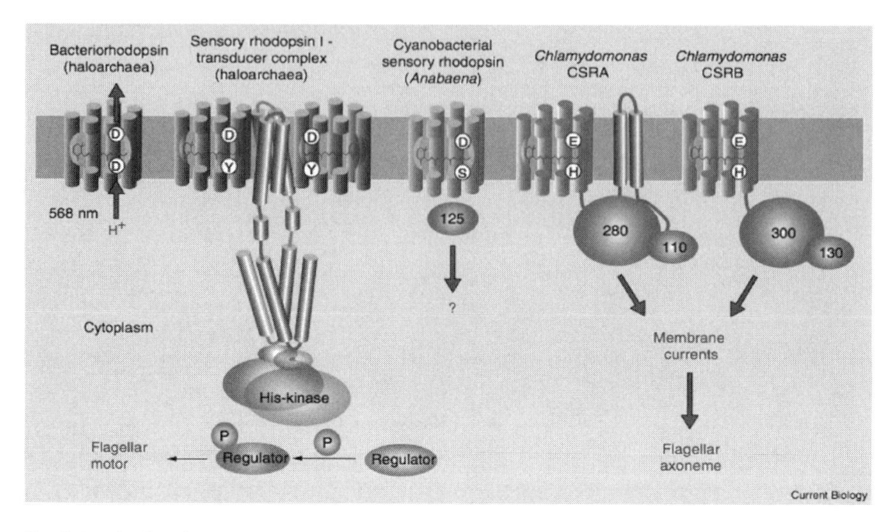

Fig. 3.5 Rhodopsin sensory proteins (reprinted with permission from: Kevin D. Ridge (2002), Current Biology, Vol. 12)

trans shift in conformation, respectively (see Fig. 3.5). All bacterial rhodopsins use trans-retinal as a light-harvesting pink pigment. Type I rhodopsin is found in archaea and bacteria, type II includes the sensory visual receptors of higher animals. These two rhodopsin types have no sequence similarity between them, but have retained clear structural and functional similarity. They are all transmembrane proteins with seven transmembrane (TM) domains, which form tetramers that make a central channel that encloses the retinal. This 7 TM structure is highly conserved in sensory and other transducing proteins as used by eukaryotes. A very similar 7 TM structure is used by the beta-adrenergic and the muscarinic acetylcholine receptors (see Chapter 5 in origin of the nervous system). All these sensory receptors function via either a light-gated ion channel or a signal transducer or energy capture in the case of photosynthetic bacterial rhodopsin. In the case of type II rhodopsin, signal transduction is via a G-coupled protein, an amplifying membrane transducer that interconverts between bound GDP and GTP resulting in cGMP hydrolysis, cGMP directly binds and opens cation channels. Receptor control via G-coupled proteins is a major control system for all eukaryotes. This point deserves some emphasis. As we will see in all the subsequent chapters, membrane receptors linked to G-coupled protein transducers are of enormous importance, especially in terms of sensory and communication functions for higher organisms. With the rhodopsins, we can see situations in which the G-protein transducer is both present and absent, suggesting that the origin of light detection predates the evolution of G-coupled protein transduction. The simplest sensory rhodopsins are proton pumps with no apparent cytoplasmic proton donor. However, unlike the passive ion channels for the various toxins, these rhodopsins pump protons against the cellular gradient using light as an energy source. Eukaryotic rhodopsins are all integral membrane proteins, synthesized by ribosomes attached to the endoplasmic reticulum. Some prokaryotic rhodopsins, however, are of a simpler structural class, which also lack ER signal sequence, allowing direct insertion into membranes. Rhodopsins of halobacteria (archaea) are able to transport chloride ions, which have considerable interest with respect to the origin of chloride-based signaling, as in neurons. The sensory rhodopsin of *Anabaena* (heterocyst, blooming cyanobacteria) resembles those of archaea and is a 14 kDa 7 TM tetramer. This photoreceptor appears to be able to sense two distinct wavelengths of light, for unknown reasons. Recent structural studies with this receptor (ASR) and its associated 125 aa transducer protein (ASRT) suggest that the cytoplasmic transducer has clear structural similarity to G-beta transducer and further supports the evolutionary link between sensory systems G-coupled-like transducers. Cyanobacteria, unlike most bacteria, have eukaryote-like protein histidine kinase of the Pkn2 family that is used as a transducer to control motility.

Light and Group Movement

Rhodopsin-based light detection in photosynthetic organisms has a clear role in host survival. It links light detection to motility and allows bacteria to move toward light for photosynthetic purposes. Here, it is clearly important in order to move toward light and capture energy. However, the majority of sensory rhodopsin genes are not associated with photosynthetic bacteria. Figure 3.6 shows the light antennae of green algae which also use rhodopsin-based receptors. Here, we see that the flagella are integrated into the sensory apparatus for light detection. This molecular integration of flagella and sensory systems was to be preserved in the sensory organelles of many eukaryotes. In the motile cyanobacteria, the link between light detection and motility has been best

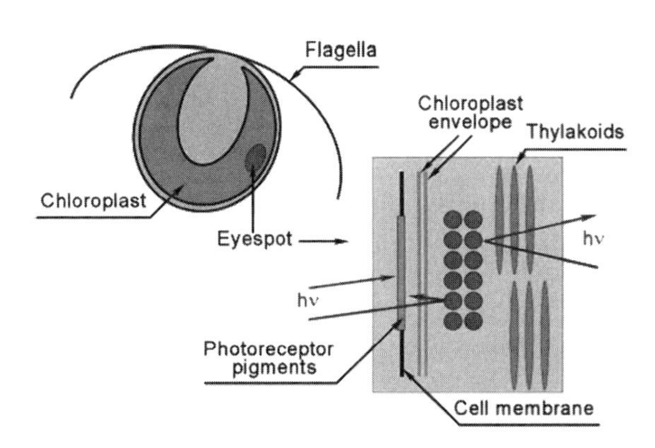

Fig. 3.6 The light antennae and flagella of green algae (*See* Color Insert, reprinted with permission from: Thomas Ebrev (2002), PNAS, Vol. 99, No. 13)

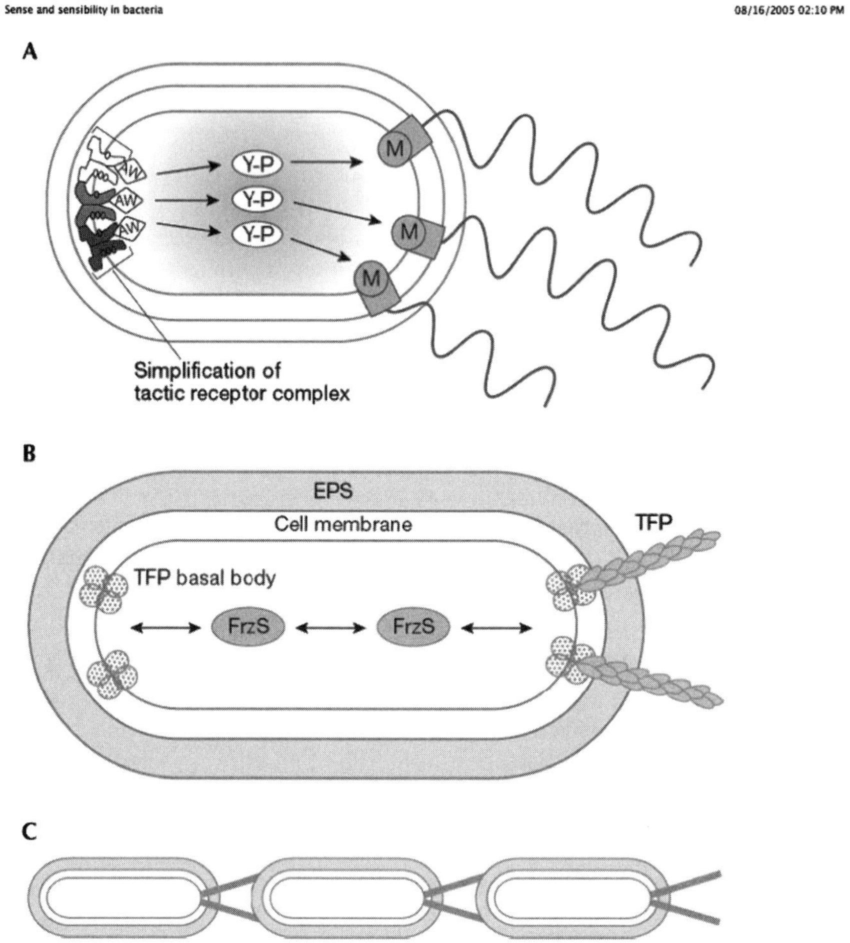

Fig. 3.7 Light transduction to flagella (reprinted with permission from: EMBO Reports 6,7: online, 6/2005)

studied. As shown in Fig. 3.7, sensory detection is transduced which affects the action of flagella and induce motility. Some motile cyanobacteria, however, do not have flagella and use pore extrusion-based mechanisms to propel bacterial movement. This process will be of special interest regarding mat-forming cyanobacteria and social bacteria as described in Chapter 4. A linkage between rhodopsin-based light detection, photosynthesis and motility has also been maintained in photosynthetic unicellular eukaryotes.

The mechanics of signal transduction to affect cellular movement in motile species involves proton gradients and ion channels that affect the control of the flagellated motor. Here the rhodopsin photosensor is a light-driven H^+ pump and ion channel that appears to alter Ca^{++} levels in which an altered electrical potential leads to the breakdown of proton gradient. The altered proton gradient will open Ca^{++}-specific channels, allowing increase of Ca^{++} influx. This bears some similarity to cGMP action noted above. Thus, we can see an early example of a voltage-gated channel for signal transduction. This process occurs in light to dark transition. Most motile cyanobacterial species form long nonpolar filaments with flagella at one end and the light antenna at the other end. Motility also involves type IV pili (see *V. cholera* below). However, as mentioned not all light tactic cyanobacteria and *Vibrio* bacteria have flagella (see Chapter 4). Some blooming species move coordinately by gas vacuole formation to affect buoyancy in response to light. In other cases, cyanobacteria movement is enabled by a gliding mechanism. This is especially true for the benthic communal mat-forming species. In fact, as will be discussed in Chapter 4, other highly social bacteria below also use both gliding and type IV pili-mediated movement, but flagella do not, suggesting a common type of surface interaction for social bacteria. Bacterial gliding appears to work by formation of slime sheath around individual cells. Gliding results from specific pore structures (nozzle) at one end of the cell walls that extrudes slime and propels movement (see Fig. 3.6). Movement is directional and can be synchronous, as seen in the entire mat. Cell division is also synchronous and light-associated. However, all three forms of movement (flagella, buoyant, gliding) are in response to light detection via rhodopsin receptors. Therefore, the more basic consequence of light detection in these systems is to affect group behavior and movement.

Communicating Group Identity: Light, Pheromones and Quorum Sensing

The light organ of the squid was originally used to discover not only the luciferase light-producing system, but also quorum sensing. *Vibrio fischeri* are the symbiotic bacteria responsible for the formation of a slime colony in mucosa below the squid eye that provides the light organ. It was discovered that light production was also dependent on a high density of organisms, initially establishing that a form of social control operates over the production of light. Quorum sensing is essential for *Vibrio fischeri,* not only to produce light in the squid light organ, but also for the initial colonization. Thus, community formation and slime production are both involved. Free-swimming *V. fischeri* is common in the ocean but is not luminescent and only becomes luminescent following colonization of its squid host. Newly hatched squid of *Euprymna scolopes* have no light organ and are not colonized with the symbiotic *V. fischeri.* Some hours after hatching, low numbers of free-swimming *V. fischeri* attach to and colonize the ciliated epithelia. A pilin A-like bacterial protein is important for this initial colonization. The local epithelial cells respond by producing squid mucus, which is high in *N*-acetylneuraminic acid. *V. fischeri* chemosenses this and migrates using flagella toward this mucus and, using a quorum-sensing response, will concentrate

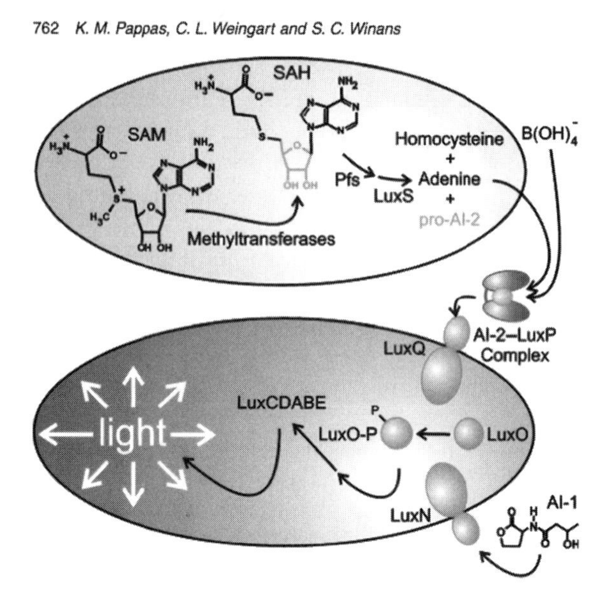

Fig. 3.8 Multiple control of ALH (*See* Color Insert, reprinted with permission from: Pappas, Weingart, Winas (2004), Molecular, Microbiology, Vol. 53, No. 3)

and aggregate into mucus-based colonies. These cells migrate into ducts and produce acyl-homoserine lactone (AHL), which in large amounts functions as an autoinducer (see Fig. 3.8). The cells then form enclosed crypts via LPS-based induction of epithelial apoptosis, resulting in highly concentrated colonies that constitute the squid light organ. Resulting light production is continuous. In a sense, this process clearly resembles a living bacterial cyst, enclosed by host cellular response. As mentioned above, cyanobacteria slime patches also colonize fish guts, so the mucosal eye epithelia bears some similarity to this. Such mucosal colonies probably represent the numerically dominant version of these bacterial

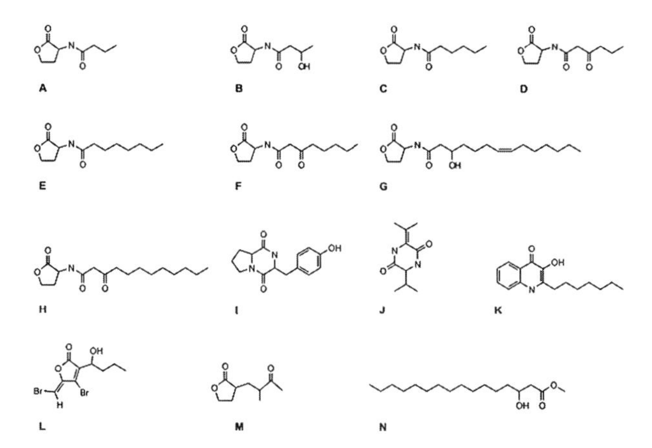

Fig. 3.9 Variation in autoinducers (reprinted with permission from: Whitehead, Barnard, Slater, Simpson, Salmond (2001), FEMS Microbiology Reviews, Quorum-Sensing Gram Negative Bacteria)

populations. The luminescence of the *V. fischeri* is needed for the persistent colonization of the epithelia, since mutants that affect this also do not colonize, so light appears to be important group identity signal, similar to the proposal above for fish gut colonization by *Vibrio* species. These symbiotic cyanobacteria most closely resemble the free-living *V. harveyi*, which also has a QS Lux system. Other free-living photosynthetic prokaryotes, such as *Rhodobacter sphaeroides*, a purple bacteria, also have QS system, but use novel autoinducers, in this case 7,8 *cis N* (tetradecenoyl) homoserine lactone. Autoinducers (AIs) are now known to exist in many chemical variations (see Fig. 3.9). The existence of many variations of AI chemistry also suggest that AIs are used to

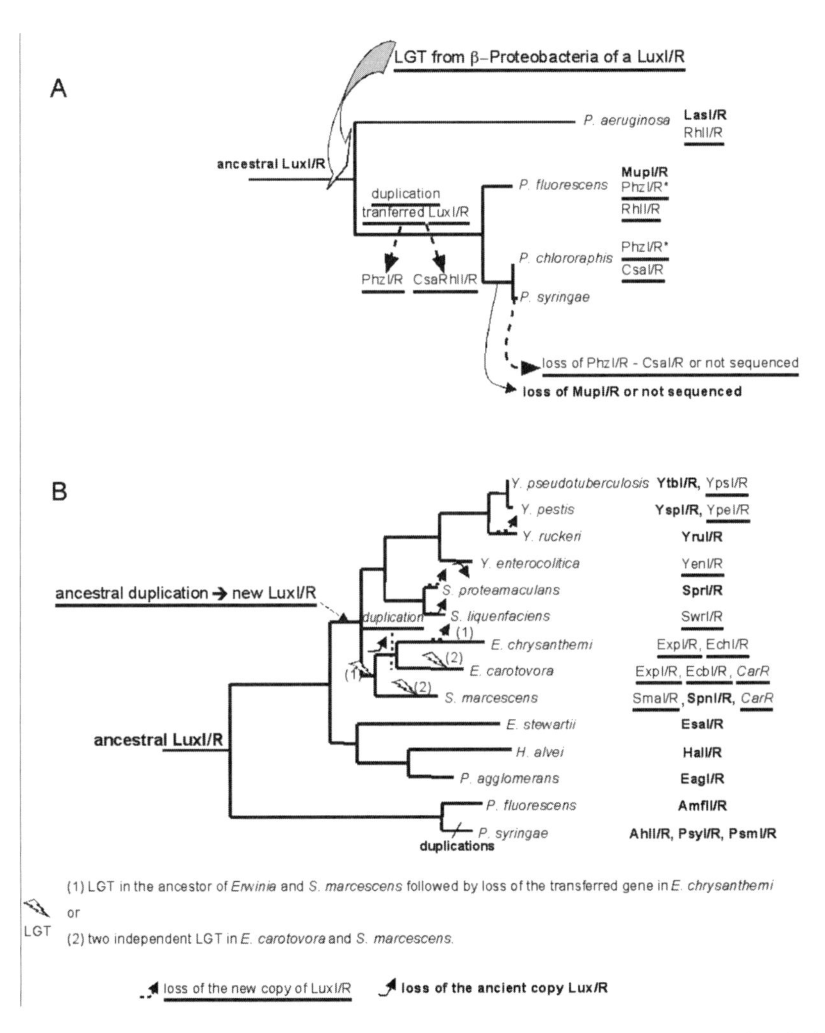

Fig. 3.10 Phylogenetics of LUX evolution in cyanobacteria (*See* Color Insert, reprinted with permission from: Lerat, Moran (2004), Molecular Biology and Evolution, Vol. 21, No. 5)

designate group identity and as discussed below, also suggest the origin of systems that evolved in cyanobacteria to differentiate pheromone olfaction of related compounds. It is worth noting that for some years following the discovery of the Lux I/R system in squid light organ, it was thought to be a rather unique system. However, it is now known to be an ancient and broad system that can be found in all bacteria. Figure 3.10 shows a phylogenetic analysis of the evolution of this broadly distributed I/R system, via homology to Lux R. Note that the 'I' and the 'R' components of an I/R system must be acquired together, as a set, to function properly. In this dendogram, there is much evidence for the acquisition of I/R genes from exogenous sources. The authors of this figure suggest the frequent horizontal transfer of I/R along with the decay of some earlier versions of I/Rs. This suggestion is in good keeping with the idea of sequential colonization and displacement by I/R-based group identity systems.

Autoinducers, Multiple Identities and the Origin of Pheromones

Quorum sensing depends on the production of diffusible small molecules that function as autoinducers (AI) and control their own production. Although the term 'quorum sensing' has become the accepted general concept of this function, this sensing system is not strictly a quorum-sensing system; thus, it is a bit of a misnomer. This is because multiple QS systems exist for the same conspecific detection process. Several types of bacteria are now established to have multiple sensing systems all of which can affect light production and movement. In each of these systems, a small molecule (autoinducer, AI) will induce the expression of the enzyme (synthase) that synthesizes the autoinducer. Thus, their function cannot simply be to evaluate the concentration of nearby cells, but must be able to differentiate the various specific autoinducer signals produced by those nearby cells. It has been suggested that such multiple detection systems are identity detectors and not simply quorum sensors. For example, *V. harveyi* has three known quorum-sensing systems that control not only bioluminescence, but type III protein secretion and metalloprotease production. AI 1, 2 and 3 are all related lactone molecules (such as HSL). What is the point of such multiple detection systems, all of which can affect movement behavior in similar ways? With *V. harveyi*, the AI-2 (Lux PQ) detection may be a negative regulator that also needs AI-1 (Lux N) to overcome regulatory control. Increasing AI-1 levels have been reported to overcome the inhibitory effects of AI-2. AI-2 induction by itself (sans AI-1) requires 25–45 fold greater concentration (as measured in *V. fischeri*). This mixed positive and negative induction system could constitute a two event or coincidence detector and would also introduce a temporal element to sensory detection. Thus, the two QS systems working together may constitute a coincidence detector that can provide a temporal control mechanism for movement. If so, this might constitute the earliest beginnings of a pattern detection system able to affect group behavior. One target of these multiple detectors is flagella motor, whose activation is via histidine kinase as an internal signal transduction system. As asserted, these multiple pheromone detectors can also be considered as the earliest examples of odor detectors, since they both must recognize multiple small diffusible molecules that identify self and self-like groups. The theme of odor and identity detection is a major theme that will extend all the way to primate evolution and will be frequently considered in the subsequent chapters. Multiple AI system can also be considered as a multiple level of identity detection. For example, AI-2 is thought to be able to communicate (via *N*-octanoyl-L-HSL) between divergent bacterial species, and thus it might provide a system of extended identity detection, whereas AI-1 is more specific. Another example of an alternative autoinducer signal with a distinct transduction is seen with *R. solanacearum*. Here the autoinducer is an acyl-HSL variant, palmitoyl-ACP (3 OH, PAME), which binds a periplasmid receptor (Phcs), which phosphorylates PhcR, leading to negative regulation of transcription (see Fig. 3.9). The Phcs gene is an example of secondary identification systems that occur via surface binding. All these AI molecules are small volatile molecules (in *V. fischeri*), able to diffuse past membranes and directly bind internally to regulate allosteric transcriptional factors. However, as noted previously, some AIs are highly hydrophobic and appear to form micelles that display much more limited diffusion. These would not be prone to diffuse past the external membrane. If we consider these external AI systems to be early examples of pheromone detection, then we also see the beginnings of membrane-based sensory detection, which was also initially linked to rapid control of transcription.

In the context of cyanobacterial aquatic colonies and biofilms (including submerged stones – stromatolites), ALH has been detected, indicating that QS is likely operating as an element of group detection in these colonies. However, any possible QS affect on group behavior is unknown is such settings. Clearly, light is known to affect cellular group behavior in these settings (see Chapter 4). However, it is not known if QS is able to affect light production or light detection via sensory rhodopsin in mat colonies.

Did Phage Resistance Contribute to the Origin of QS?

A theme of this book is that genetic parasites, like phage, will often be involved in the origins of host group detection and identity systems. Can such a theme also apply to the likely origin of QS systems? This would probably strike most readers as unlikely. What might be the purpose for the original QS of bacteria? As comparative genomics indicates, the Lux QS system is deeply conserved in bacterial species (Fig. 3.10). Yet most of these species are not known to live in extended communities. Such sequence analysis also indicates that these conserved Lux QS systems are not usually generated by either gene duplication or retrotransposition events (a process that is common in eukaryotes). Rather it appears that QS evolution is occurring by a process colonization from external genetic sources (i.e., lateral transfer). The conundrum is that although the Lux systems appear to evolve by a lateral colonization process, these genes are mostly congruent with host cell phylogenetics. Thus it appears that once a QS gene set has been acquired, they tend to stay together as a set during subsequent evolution, but the acquisition of additional new sets can still occur. Since QS acquisition needs a matched gene set in order to provide both the AI autoinducer and its matched synthase together, this process resembles colonization by a complex addiction module. It is also interesting to note that *Vibrio* species appear to use small RNA to regulate quorum-sensing systems. As noted previously, regulation by small antisense RNA is a feature common to some addiction modules. *Vibrio cholerae* uses a similar system small regulatory RNA for Lux mRNA regulation. However, none of these observations directly supports a role for genetic parasites in QS. So how might this be rationalized? A curious common feature of all the QS AI molecules is that they require SAM for their synthesis. Since the quantities of AI made are substantial, the corresponding level of SAM needed is also substantial. As noted previously, SAM production is energetically costly. In fact, SAM recycling appears essential to all bacterial life in that all 138 (non-parasitic) prokaryotic genomes had conserved SAM recycling systems. Proteobacteria tend to use Pfs-Lux system to recycle SAM, whereas archaea, eukarya and cyanobacteria tend to use SahH (hydrolase) pathway. Although such phylogenetic arguments might be interesting, they are far from definitive. Can we experimentally connect SAM and AI production to bacterial phage biology? The answer is yes. Consider *V. fischeri* AI-2, which is derived from SAH recycling (S-adenosylhomocysteine). Depletion of SAM occurs when HML is made. By using the *V. fischeri* regulon to code for autoinducer synthase in a recombinant *E. coli*, the effect of autoinduction on lambda infection was evaluated. AI-2 induction alleviated restriction of lambda replication 10–20 fold by depleting SAM. Did this observation have any evolutionary implications? Although we have no direct evidence on this, given the crucial role of virus in bacterial evolution, AI would seem able to alter the bacterial phage evolution.

Plasmids and the Origin of Quorum Sensing and Virulence

Lux I- and Lux R-related genes can also be found in plasmids. A very interesting version of Lux I- and Lux R-like sequences are found in Rhizobium species (i.e., *R. agrobacterium*). These are components of the N_2-fixing megaplasmids described below. What is most striking from a phylogenetic perspective, is that these Rhizobium versions of Lux I/R are the most phylogenetically basal of all I/R genes.

N_2-Fixing Plasmids and I/R

Rhizobium species have multiple versions of these I/R systems, but only two of them are most host congruent. It is most interesting that these two genes are found on the symbiotic plasmids needed for N_2 fixation, not on the corresponding host chromosome. Specifically, *Rhizobium leguminosarum* has four Lux I-type systems and six Lux R-type systems. These systems are found both on the chromosome and the N_2-fixing megaplasmid. These Lux systems function to regulate biofilm formation as well as N_2-fixing symbiosis. This implies that these plasmids may identify the most basal version of the Lux-inducer system also associated with community formation. Thus, this plasmid-based and colonizing capable version of Lux sequence is indeed representative of the ancestral version as found in *Agrobacterium tumefaciens*. This bacteria, along with a large resident plasmid, is also responsible for crown gall tumors of plants. It has TraI/TraR system that is homologous to Lux R and uses opines made in plant tumors as an autoinducer in detecting nearby *A. tumefaciens*. This system controls vegetative growth of cells. In order to establish a plant tumor, the bacteria must do two things. First, it must colonize the plant, and then it must transfer its large 200 kbp plasmid (Ti) DNA into nuclei of plant cells in order to express plasmid genes and exert growth control over the plant. As with other Lux-like sensors, colonization is regulated by this system. In addition, the transfer of Ti DNA is also under the control of Lux-like identification system. Although most of the attention on this system has focused on the ability to transform plant cells, it is likely the entire system originally evolved as a system of infectious plasmid transfer between *A. tumefaciens* cells. This is because Ti DNA retains the ability to be transferred between bacterial cells and this system is also under Lux control. As with most plasmid transfer systems, described above, it is likely that Ti DNA system must also have some regulation (immunity) over allowed transfers. As the Rhizobium and Agrobacterial Lux I/R proteins represent the most phylogenetically basal set of all bacteria, taken together, these observations are consistent with the idea that the Lux identification system originally evolved as part of system involved in transfer and colonization of parasitic plasmid DNA, but then became essential to host identification. Such a scenario would explain the overall evolutionary pattern of Lux-like systems in which specific matched pairs of I and R genes have evolved together as a matched set (resembling a T/A addiction module). Subsequent colonization events have then added (or deleted) additional matched I/R systems.

Other N_2-Fixing Plasmids

Another case in point along these lines is found with *Rhizobium etli* (CFN42), which constitutes a significant soil species. Like many other bacteria, *R. etli* can be typed by specific phage sensitivities, which are affected by surface lipopolysaccharide composition. Genomic analysis has demonstrated that *R. etli* can contain up to six plasmids. One of these, p42a, is a high-frequency self-conjugative plasmid that is essential for the mobilization of the N_2-fixing symbiotic plasmid (pSym), which encodes the type III secretion system. P42a is dependent on a QS system for the transfer of pSym (which itself is also QS regulated). In fact, *R. etli* has three Lux R/I-like genes. In addition, the *R. etli* plasmid transfer genes are highly related to those of *Agrobacterium tumefaciens*. Therefore, like *Agrobacterium*, here too we see that the QS system is controlling plasmid mobilization and transfer. In addition, some of the Rhizobium species are also known (*Sinorhizobium meliloti*) to harbor restriction/modification systems (likely an addiction module). With *R. meliloti*, two symbiotic plasmids are present. pSymA is a 1.35 Mbp plasmid with low G/C content that encodes import and export genes and is an efficiently mobile genetic element. pSymB is a 1.7 Mbp plasmid with high G/C content and is less mobile. The G/C content of these two plasmids is evolutionarily interesting in that all known *nod* genes that are unique to Rhizobium have low G/C content, but chromosomal genes have high G/C content. Since G/C content is though to be a reliable indicator of ancestry, this suggests that *nod* genes were derived from a non-Rhizobium genetic source. In addition, these same species generally harbor co-evolving temperate phage which are common in field isolates, and which encode complex immunity systems unlike those seen in the phage of most proteobacteria. They are also known to support numerous lytic phage (which have highly methylated DNA), indicating a likely prevalent dynamic between persisting and acute genetic parasites and the need for genomic identity systems to deal with them (SAM). The genomes of *R. meliloti* are also interesting with regard to reiterated DNA sequences, an uncommon situation in proteobacteria (although common in many eukaryotes), but similar to *Anabaena* heterocyst cyanobacteria (Chapter 4). These genomes are known to contain reiterated multigene families (such as those of N_2 fixation), in which sequence identity appears to be maintained by frequent recombination events involving large plasmids. This recombinational-identity process is reminiscent of some T-even phage described above. Other N_2-fixing bacteria are also known to frequently harbor cryptic prophage and plasmids. For example, field studies of N_2-fixing salt marsh bacteria identified 521 field isolates and found 134 plasmids (ranging from 2 to 100 kbp). These bacteria tended to harbor one plasmid, which resembled cryptic phage of 71 different types. The picture we are left with is genetically complex. These N_2-fixing bacterial species exist with a nested complex and interacting set of internal and external genetic parasites that are controlled via QS systems (Lux I/R). It might be best to think of these as networks of persisting genetic elements that must also transmit DNA and resist pressure from competing genetic replicators. Thus Lux I/R evolution is here associated not only with network of genetic parasites, but also with the acquisition of high host complexity. This system appears to have evolved by non-Darwinian character and has both a highly horizontal method of colonizing character and retaining more ancient ancestral Lux I/R versions.

AI Interference

The small diffusible AI molecules associated with QS systems may provide an additional process by which organisms can establish, communicate and enforce group identities. If such AI molecules could interfere with each other, they would also provide a system for competition. One example of this might be the HtdeDKL autoinducer, which is a fatty acyl variation of the more common AHL. HtdeDKL has a 14-carbon acyl side chain added to the lactone. Interestingly, when this autoinducer was first discovered, it was thought to be a type of bacteriocin since it was able to inhibit the growth of other strains. Subsequent characterization established that it was an AHL chemical variant, not a bacteriocin, but one that is inhibitory to related bacterial strains. This did demonstrate a clear capacity to affect related competing species. Other observations also support the idea that species competition can occur via autoinducers. For example, bacillus species can code for the enzyme AiiA that will hydrolyze the AHL lactone ring. This probably serves to allow bacillus to better compete with organisms such as *Erwinia carotovora* as a plant pathogen. In the ocean setting, other non-bacterial species may also compete via autoinducers. The red algae, *Delisea pulchra*, releases halogenated furanones that likely block AHL by increasing proteolysis. Thus, the ability of some autoinducers to be toxic to other species and their destruction by competing species do suggest a likely role for these small molecules in species identification and competition.

Some bacteria have also used this same Lux I/R-like systems to develop more complex social behavior. One example of this is with *Serratia liquefaciens*. Some of these strains are known to differentiate into swarmer cells at the edge of colonies that can rapidly swarm out and colonize an extended area. These swarmer cells are long multinucleated, multiflagellated cells. This system is under a Lux I/R-like system control. As we will see below, swarming behavior is also seen in more socialized bacteria and algae.

Virulence via Slime-Based Group Identity

Surface adherent biofilms appear to represent a most common process by which bacteria can recognize themselves to form multicellular and often biologically complex colonies. The self-adherent surface is the product of slime production, which is usually involved in bacterial community formation. This slime-based colony formation can often be under the control of the Lux I/R quorum-sensing system. Thus group detection, via pheromones, can be associated with surface slime expression that can in turn affect social affinity and behavior in bacterial populations and appears to provide a common theme that can now link previously disparate bacterial relationships. Many bacteria sense each other and induce a social behavior following this sensory event. However, our focus on microbiology of disease has overemphasized the view that such characteristics should be called virulence factors. This has led to names of these systems that tend to obscure their more basic role in group biology. For example, most often the biofilm-associated bacterial genes have been called virulence factors or pathogenic elements, even when isolated from some non-human habitat, such as oceanic *Vibrio* species. Yet, as noted, avirulent cyanobacteria also use these same highly conserved quorum-sensing systems for the control of their group behavior. Thus, pheromone production, group sensing, slime production and colony formation are all linked to community behavior in many non-pathogenic species and should not be thought of strictly from the perspective of pathogenesis. In addition, many of these oceanic bacteria also retain their ability to produce light in response to quorum-sensing autoinducers (such as *V. salmonicida*), even if they are not photosynthetic or symbiotic in metazoan light organs. Thus, light production in oceanic bacteria also appears to serve a community function. We can now seek to understand why a relative of *V. fischeri, V. cholera*, employs the same quorum system for the purposes of colonizing gut epithelia of mammals. Like *V. fischeri, V. cholera* mutants in the Lux system also fail to colonize and form biofilms in the small intestines of mice. This QS-controlled colonizing feature also appears to be characteristic of many other pathogenic gram-negative bacteria. One example that might be considered due to 'non-viral' element would be enterohemorrhagic *E. coli* 0157: H7. *E. coli* 0157: H7 harbors a 'pathogenic' (fitness) island responsible for the highly harmful attaching and effacing lesions. However, *E. coli* 057: H7 is also known to harbor a Mu-like prophage and interacts with many other phage in peculiar ways. Of 26 screened phage, 9 were able to lyse these bacteria, but the others were able to convert the host. The pathogenic island (genetic element) responsible for this complex phenotype is under the control of AI-2, a Lux I homologue responsive to homoserine lactone autoinducer. Interestingly, this *E. coli* is also able to use the eukaryotic epinephrine molecule for QS. It also encodes a type III secretion system that provides a colonizing factor that is similar to that in *Vibrio* species. Similar pathogenic islands with colonization-associated AI under Lux I-like control are found in *Salmonella typhimurium* and *Enterococcus faecalis*. Here, pheromone-responsive pPO1 conjugative plasmid is involved in cell aggregation and is known to mediate killing of many other gram-negative bacteria. Additional plasmids (associated with high-frequency transfer and mating) pAD1, pMB2, pYI17 are all pheromone responsive and can also encode various 'virulence' factors, including hemolysins. Thus, bacterial biofilm formation and group behavior appear to be common themes for bacterial group success and are typically mediated by genetic parasites.

Other, non-human pathogens also appear to employ very similar group behavior and colonization strategies. *Pseudomonas aeruginosa* and plant-associated *Agrobacterium tumefaciens* (discussed above) both utilize virulence factors that are involved with host colonization and are under Lux I/R like control, responsive to HSL autoinducers. In the case of *P. aeruginosa*, biofilm formation uses type IV pilin gene expression that is needed for inducing lung disease of cystic fibrosis patients. This bacterium uses a modified version of AI, PAI or 3-oxo-*N* (tetrahydro-2-oxo-furany) dodecanamide. In addition, a second AI 2-heptyl-3-hydroxy-4-quinolone is also used. Here too, virulence involves exoenzyme transport to the surface of bacteria via two distinct pathways. It is worth noting that the type III secretion system (TTSS) that is so often used for toxin secretion is specific to bacteria, but is composed of several proteins closely related to flagellar export proteins and of likely evolutionary relationship to flagella.

Phage, Virulence and Toxic Channel Pores

The above section emphasized the role of 'pathogenic island' genetic elements formation of host virulence factors involved in group (colony) formation. Another very important function associated with virulence is the production of toxins, especially pore toxins. Here, I will consider the role of phage in toxic 'channel-affecting' pore proteins. These toxic proteins are often under a QS-controlled process, thus are community controlled. The first channel-forming protein discovered was produced by virulent *Staphylococcus aureus* and was associated with the lysis of human and/or rabbit red blood cells. Since its discovery and to this date, the use of blood agar plates is principally aimed at identifying bacterial colonies that produce such ion channel toxins. These toxins are two component pores, which were known only for Staphylococci. We now know these pore-based toxins are produced by extragenetic elements, plasmids and phage that colonize susceptible bacteria. The alpha-toxins proteins will oligomerize into a membrane-spanning pore. The three classes of pore-forming toxins are gamma hemolysin (Hlg), leukocidin (Luk) and Penton–Valentine leukocidin (PVL). These are secreted as monomers (about 34 kD), bind to a membrane (via cholesterol-binding domains) and assemble into pores in a process reminiscent of holins described above. The resulting ion flux will provoke K^+ efflux and cellular toxicity. These pores bear no homology to other molecules. Since related toxins are also often found in bacteria that are not pathogenic in known eukaryotic host (such as gut epithelia), the original purpose and target of such toxins is not clear. One proposal suggests that their main environmental targets are various grazers in oceans that feed on bacteria (such as nematodes). As these grazers consume bacteria by phagocytosis (resembling macrophage in this function), it seems possible such animal (predator) toxicity may be selected for bacterial toxin production. Consistent with this, the occurrence of such toxin-encoding elements is common in bacterial field isolates including genes clearly related to Luk/Hlg as seen to occur in other non-pathogenic oceanic bacteria. Alternatively, eukaryotic organisms may share identity systems (T/A pairs) with these oceanic bacteria and hence tend to be susceptible to identity toxins made by bacteria due to common evolutionary histories. Such bacterial toxins are also frequently virus encoded. This observation suggests that toxin genes must confer a significant selective advantage to the viral life cycle. What might this advantage be? To evaluate this, let us consider that toxin-related genes as found on prophage, such as phage PV 83-pro (40 kb DNA), which converts host to produce PVL-like pore toxin. The host colonized with this seemingly defective phage will induce high-level PVL toxin production following induction with mitomycin C (a T/A-like response). The phage toxin production is thus responding to cellular damage. This is a frequent virus–host state. A non-defective version of related phage, SLT, is also known which is able to convert *S. aureus* to toxin production. In fact, 30 phage versions of PLV-like genes have been observed. These toxin genes all occur within the phage lysis cassette, reminiscent of late gene holing function. One of these, phage 187, is also known to encode an endolysin holin gene, clearly associated with timed phage lysis. What function might all of these phage toxin genes provide regarding phage fitness? It should be recalled that Staphylococci are diverse bacteria which are typically typed by phage sensitivity and also have complex phage ecology. Such phage toxin diversity would be consistent with a role in phage–phage competition or host colonization and exclusion. Thus, these phage toxins, along with their corresponding antitoxins, are likely to provide identity T/A markers that can be induced to destroy colonized host that become infected by competing phage. Phage-encoded toxins would thus be common systems involved in phage colonization and competition. *V. cholera* also has similar but ‘bi-component’ toxins, some of which are interesting in that they can allow Na^+ and K^+ efflux, but not the larger Ca^{++} ion. As will be discussed later, this K^+ efflux is also of interest with respect to the origin of apoptosis, since the mitochondrial membrane is the target membrane permeable initiator of such a process.

General Lesson from Oceanic Bacteria

The bacteria of the oceans conserve and display many community features. One of such feature to first be discovered and evaluated was light production and detection. Such production is often linked to quorum-sensing (QS) systems that also require the detection of small diffusible molecules. Light production may be a principal process by which marine bacteria control the group behavior of large populations. Light production can thus serve a communal function. Light production, QS signal production and detection are all linked to motility. Thus, in these oceanic bacteria we see the origin of the basic mechanisms to control group behavior involving a sensory system, a group response signal (light) and altered group behavior (motility). The molecule most responsible for light detection is sensory rhodopsin, one of the most diverse genes known. This sensory process now provides an exemplar of basic sensory and transduction system that links sensory detection to group movement and involves G-coupled proteins and the flagella. This system also provides both chemical and visual (light) based communication. The small

QS molecules involved (AIs) may sometimes act as identity molecules. Involvement of phage and other genetic parasites in such sensory systems seems likely. Plasmids associated with host survival (fitness islands) also employ similar QS-based communication. These genetic elements strongly affect group living (behavior) and are known as colonizing or virulence elements. New I/R genes (QS systems) appear to evolve by genetic colonization. Mostly, these elements show a clear evolutionary relationship to phage genes, especially phage-encoded toxin/antitoxin genes. Other plasmids that have strong effects on host fitness, such as N_2-fixing plasmids, have similar colonizing and QS systems. The toxins produced by these various genetic elements are likely mediators of group identity, aimed mostly at competitor between genetic parasites. A very similar sensory system is used by many oceanic bacteria for light detection and production. This response is also under pheromone social control and frequently affects motility. Bacteria, which have evolved small molecules (pheromones/olfaction) that are used to detect the presence of group members as well as receptors for these small molecules, introduces an evolutionary pathway that can now be traced to eukaryotic evolution. The ability to detect these molecules by surface receptors that transduce signals to the flagella in order to control motility identifies the most basic sensory circuit, which has been conserved into the sensory systems in higher eukaryotes.

Recommended Reading

Lux Operon and Quorum Sensing

1. Anand, S. K., and Griffiths, M. W. (2003). Quorum sensing and expression of virulence in Escherichia coli O157 : H7. *Int J Food Microbiol* **85**(1–2), 1–9.
2. Dunlap, P. V. (1999). Quorum regulation of luminescence in Vibrio fischeri. *J Mol Microbiol Biotechnol* **1**(1), 5–12.
3. Dunn, A. K., and Stabb, E. V. (2007). Beyond quorum sensing: the complexities of prokaryotic parliamentary procedures. *Anal Bioanal Chem* **387**(2), 391–398.
4. Meyer-Rochow, V. B. (2001). Light of my life – messages in the dark. *Biologist (London)* **48**(4), 163.
5. Mok, K. C., Wingreen, N. S., and Bassler, B. L. (2003). Vibrio harveyi quorum sensing: a coincidence detector for two autoinducers controls gene expression. *Embo J* **22**(4), 870–881.
6. Montgomery, B. L. (2007). Sensing the light: photoreceptive systems and signal transduction in cyanobacteria. *Mol Microbiol* **64**(1), 16–27.
7. Parsek, M. R., and Greenberg, E. P. (2005). Sociomicrobiology: the connections between quorum sensing and biofilms. *Trends Microbiol* **13**(1), 27–33.
8. Schauder, S., Shokat, K., Surette, M. G., and Bassler, B. L. (2001). The LuxS family of bacterial autoinducers: biosynthesis of a novel quorum-sensing signal molecule. *Mol Microbiol* **41**(2), 463–476.
9. Tiruppathi, C., Freichel, M., Vogel, S. M., Paria, B. C., Mehta, D., Flockerzi, V., and Malik, A. B. (2002). Impairment of store-operated Ca2 + entry in TRPC4(-/-) mice interferes with increase in lung microvascular permeability. *Circ Res* **91**(1), 70–76.

10. Visick, K. L., and Ruby, E. G. (2006). Vibrio fischeri and its host: it takes two to tango. *Curr Opin Microbiol* **9**(6), 632–638.
11. Wisniewski-Dye, F., and Downie, J. A. (2002). Quorum-sensing in Rhizobium. *Antonie Van Leeuwenhoek* **81**(1–4), 397–407.
12. Ziedaite, G., Daugelavicius, R., Bamford, J. K., and Bamford, D. H. (2005). The Holin protein of bacteriophage PRD1 forms a pore for small-molecule and endolysin translocation. *J Bacteriol* **187**(15), 5397–5405.

Photosynthesis/Light Production

1. Hastings, J. W., and Nealson, K. H. (1977). Bacterial bioluminescence. *Annu Rev Microbiol* **31**, 549–595.
2. Hess, W. R., Rocap, G., Ting, C. S., Larimer, F., Stilwagen, S., Lamerdin, J., and Chisholm, S. W. (2001). The photosynthetic apparatus of Prochlorococcus: insights through comparative genomics. *Photosynth Res* **70**(1), 53–71.
3. Massey, V. (2000). The chemical and biological versatility of riboflavin. *Biochem Soc Trans* **28**(4), 283–296.
4. Nelson, N., and Yocum, C. F. (2006). Structure and function of photosystems I and II. *Annu Rev Plant Biol* **57**, 521–565.
5. Rees, J. F., de Wergifosse, B., Noiset, O., Dubuisson, M., Janssens, B., and Thompson, E. M. (1998). The origins of marine bioluminescence: turning oxygen defence mechanisms into deep-sea communication tools. *J Exp Biol* **201**(Pt 8), 1211–1221.
6. Sproviero, E. M., Gascon, J. A., McEvoy, J. P., Brudvig, G. W., and Batista, V. S. (2007). Quantum mechanics/molecular mechanics structural models of the oxygen-evolving complex of photosystem II. *Curr Opin Struct Biol* **17**(2), 173–180.
7. Timmins, G. S., Jackson, S. K., and Swartz, H. M. (2001). The evolution of bioluminescent oxygen consumption as an ancient oxygen detoxification mechanism. *J Mol Evol* **52**(4), 321–332.
8. van der Staay, G. W., Moon-van der Staay, S. Y., Garczarek, L., and Partensky, F. (2000). Rapid evolutionary divergence of Photosystem I core subunits PsaA and PsaB in the marine prokaryote Prochlorococcus. *Photosynth Res* **65**(2), 131–139.
9. Wilson, T., and Hastings, J. W. (1998). Bioluminescence. *Annu Rev Cell Dev Biol* **14**, 197–230.

Pathogenic Islands/Nitrogen Fixation/Plasmids

1. Carniel, E. (1999). The Yersinia high-pathogenicity island. *Int Microbiol* **2**(3), 161–167.
2. Finlay, B. B., and Falkow, S. (1997). Common themes in microbial pathogenicity revisited. *Microbiol Mol Biol Rev* **61**(2), 136–169.
3. Hacker, J., and Kaper, J. B. (2000). Pathogenicity islands and the evolution of microbes. *Annu Rev Microbiol* **54**, 641–679.
4. Lau, R. H., Sapienza, C., and Doolittle, W. F. (1980). Cyanobacterial plasmids: their widespread occurrence, and the existence of regions of homology between plasmids in the same and different species. *Mol Gen Genet* **178**(1), 203–211.
5. MacLean, A. M., Finan, T. M., and Sadowsky, M. J. (2007). Genomes of the symbiotic nitrogen-fixing bacteria of legumes. *Plant Physiol* **144**(2), 615–622.
6. Muniesa, M., and Jofre, J. (2004). Abundance in sewage of bacteriophages infecting Escherichia coli O157:H7. *Methods Mol Biol* **268**, 79–88.

7. Osborn, A. M., and Boltner, D. (2002). When phage, plasmids, and transposons collide: genomic islands, and conjugative- and mobilizable-transposons as a mosaic continuum. *Plasmid* **48**(3), 202–212.
8. Perret, X., Kobayashi, H., and Collado-Vides, J. (2003). Regulation of expression of symbiotic genes in Rhizobium sp. NGR234. *Indian J Exp Biol* **41**(10), 1101–1113.

Phage Toxins

1. Betley, M. J., and Mekalanos, J. J. (1985). Staphylococcal enterotoxin A is encoded by phage. *Science* **229**(4709), 185–187.
2. Casas, V., Miyake, J., Balsley, H., Roark, J., Telles, S., Leeds, S., Zurita, I., Breitbart, M., Bartlett, D., Azam, F., and Rohwer, F. (2006). Widespread occurrence of phage-encoded exotoxin genes in terrestrial and aquatic environments in Southern California. *FEMS Microbiol Lett* **261**(1), 141–149.
3. Chavan, M., Rafi, H., Wertz, J., Goldstone, C., and Riley, M. A. (2005). Phage associated bacteriocins reveal a novel mechanism for bacteriocin diversification in Klebsiella. *J Mol Evol* **60**(4), 546–556.
4. Cheetham, B. F., and Katz, M. E. (1995). A role for bacteriophages in the evolution and transfer of bacterial virulence determinants. *Mol Microbiol* **18**(2), 201–208.
5. Coleman, D., Knights, J., Russell, R., Shanley, D., Birkbeck, T. H., Dougan, G., and Charles, I. (1991). Insertional inactivation of the Staphylococcus aureus beta-toxin by bacteriophage phi 13 occurs by site- and orientation-specific integration of the phi 13 genome. *Mol Microbiol* **5**(4), 933–939.
6. Coleman, D. C., Sullivan, D. J., Russell, R. J., Arbuthnott, J. P., Carey, B. F., and Pomeroy, H. M. (1989). Staphylococcus aureus bacteriophages mediating the simultaneous lysogenic conversion of beta-lysin, staphylokinase and enterotoxin A: molecular mechanism of triple conversion. *J Gen Microbiol* **135**(6), 1679–1697.
7. Dziewit, L., Jazurek, M., Drewniak, L., Baj, J., and Bartosik, D. (2007). The SXT conjugative element and linear prophage N15 encode toxin-antitoxin-stabilizing systems homologous to the tad-ata module of the Paracoccus aminophilus plasmid pAMI2. *J Bacteriol* **189**(5), 1983–1997.
8. Goshorn, S. C., and Schlievert, P. M. (1989). Bacteriophage association of streptococcal pyrogenic exotoxin type C. *J Bacteriol* **171**(6), 3068–3073.
9. Hambly, E., and Suttle, C. A. (2005). The viriosphere, diversity, and genetic exchange within phage communities. *Curr Opin Microbiol* **8**(4), 444–450.
10. Johnson, L. P., Tomai, M. A., and Schlievert, P. M. (1986). Bacteriophage involvement in group A streptococcal pyrogenic exotoxin A production. *J Bacteriol* **166**(2), 623–627.
11. Kimura, K., Fujii, N., Tsuzuki, K., Murakami, T., Indoh, T., Yokosawa, N., Takeshi, K., Syuto, B., and Oguma, K. (1990). The complete nucleotide sequence of the gene coding for botulinum type C1 toxin in the C-ST phage genome. *Biochem Biophys Res Commun* **171**(3), 1304–1311.
12. Pullinger, G. D., Bevir, T., and Lax, A. J. (2004). The Pasteurella multocida toxin is encoded within a lysogenic bacteriophage. *Mol Microbiol* **51**(1), 255–269.
13. Riley, M. A. (1998). Molecular mechanisms of bacteriocin evolution. *Annu Rev Genet* **32**, 255–278.
14. Weeks, C. R., and Ferretti, J. J. (1984). The gene for type A streptococcal exotoxin (erythrogenic toxin) is located in bacteriophage T12. *Infect Immun* **46**(2), 531–536.
15. Zielenkiewicz, U., and Ceglowski, P. (2005). The toxin-antitoxin system of the streptococcal plasmid pSM19035. *J Bacteriol* **187**(17), 6094–6105.

Chapter 4
Subjugation of the Individual; Prokaryotic Group Living – Blooms, Slime and Mats

Goals of this Chapter

The last two chapters traced the molecular origins of prokaryotic group identification and sensory systems. Genetic parasites, toxins and antitoxins were presented and stressed as relevant and basic elements. The role of pheromones, flagella motility, light detection and slime production are all part of the basic systems used by bacteria to control group identity and behavior. In this chapter, I will extend these issues to consider the evolution of large bacterial communities as initially seen in stromatolite fossils. The focus will be on filamentous and mat-forming oceanic cyanobacteria. As before, the role of genetic parasites (phage) as well as the presence of addiction (T/A) modules will be evaluated as possible sources of genetic novelty and group identity. Thus, these ancient and large communities of cyanobacteria provide the foundation for the evolution of eukaryotes. How these specific bacteria developed systems of group identification and maintain colony coherence will be of central importance, but they will also be compared to other more social bacteria, such as *Myxobacteria xanthus,* and how programmed cell death relates to group identity. Finally, the blooming and light-producing unicellular eukaryotes, the dinoflagellates, will be presented as examples of eukaryotic microbial communities. This chapter thus develops the base from which the evolution of multicellular eukaryotes will be presented in Chapter 5.

Major Earth Transition: Iceball Earth and the Role of Bacterial Communities

Geologists and paleontologists have recently come to accept the view that about 2.3 billion years ago, Earth underwent the first of several planet-wide freezings in which the oceans froze over. This 'iceball' Earth remained frozen for about 100 million years. Even the oceans at the equator froze and remained so for a depth of up to 1 mile thick. This planet-wide freezing also marked a major punctuation and transition in the evolution of life on Earth. In fact, it now

L.P. Villarreal, *Origin of Group Identity*, DOI: 10.1007/978-0-387-77998-0_4,

appears that the evolution of new life strategies was to blame for the freeze. The cause of this global calamity appears to have been a first ever, but massive build-up of atmospheric O_2. This oxidized and depleted the extant reducing atmosphere that had, up until then, kept the world warm by a greenhouse effect. The prior atmosphere had been high in methane, a greenhouse gas that has 22 times the insulating capacity of CO_2. O_2 would react with methane to generate CO_2. O_2 also provided the ozone layer, which blocked much of the UV light reaching the surface that limited surface life. Although this early Earth was warmed by a weaker sun (84% current radiation), the high methane atmosphere allowed Earth to be temperate. Thus, the evolution of O_2 production due to the emergence of photosynthesis was catastrophic for the Earth's climate. Eventually, CO_2 outgassing by volcanoes restored some greenhouse gas to reestablish a temperate Earth, but this took about 100 million years. Most readers interested in this issue are not likely to consider any possible role for viruses in this development. However, as will be developed below, even here a strong argument can be made for a possible viral involvement in this catastrophe.

The large-scale production of O_2 was due to the evolution of giant cyanobacterial communities' ability to oxidize H_2O as an electron acceptor during photosynthesis. For each O_2 generated, four electrons are transferred to two H_2O molecules. These ancient, giant communities of cyanobacteria are now visible as fossilized stromatolites (shown in Fig. 2.2). Before 2.3 billion years ago, the prevailing metabolism of life on the Earth was anaerobic. Prokaryotic life had up until then been essentially anaerobic and bottom-dwelling (with corresponding low UV intensity). Although oxidation had been used by some early cells as a source of energy, here electrons were transferred between H_2, H_2S and CH_4 in reducing environments, resulting in less energetically efficient metabolism. Thus, the presence of O_2 allowed the development of much more efficient oxidation, generating four times the energy for glucose metabolism, for example. Although photosynthesis had likely evolved during this early period, it mostly used less energetic infrared photons to transfer electrons to CO_2 or SO_4. The energy to oxidize water into O_2 required 1.2 V per electron and only visible light had sufficient energy for this reaction. Thus, the invention of photosystem II, which is able to use visible light as an electron source for an H_2O acceptor, provided a more efficient metabolic engine for most life on the planet. By making water the acceptor, a vast increase in bacterial communities that increased atmospheric O_2 to very high levels became possible.

Is Photosynthesis a T/A Module?

Thus, two events seem crucial and linked: the origin of photosystem II and the creation of huge cyanobacterial communities. With respect to photosystem II, a conundrum is immediately apparent. Early life evolved in the presence of endogenous reductants, hydrogen sulfide and alkylthiols, which are free-radical scavengers. O_2 is a highly reactive and potentially toxic substance, reacting with many basic components of the cell (such as membranes and DNA). But, O_2 is the necessary product of photosystem II. This toxic O_2 has the clear capacity to provide a potent T of a T/A module. O_2 can readily lose an electron to generate peroxides (H_2O_2), superoxides (O_2^-), hydroxyl ions (OH^-) and hydroxyl radicals (•OH), known together as reactive oxygen species (ROS). These are among the most potent general toxins known. It seems clear that for prevailing cells to survive O_2 production, it is essential that they also provide protection against O_2 and ROS. Could this simultaneous acquisition have been brought about by T/A modules? It is currently proposed that photosystem II was likely adapted from some of the earlier photosynthetic systems. Lateral transfer of gene sets from some earlier bacteria has been proposed, most likely originating from the green sulfur bacteria photoreaction center to purple photosynthetic bacterium. However, the photosystem II is quite distinct from these earlier systems, yet it appears to have emerged from a massive lateral transfer event. What then was the source of this complex photosynthetic gene system? The photosystem II evolution has not remained static, as genomic data from *Prochlorococcus* cyanobacteria strongly indicate ongoing evolution by lateral transfers (discussed below). If this was transferred as a T/A module, a potent O_2 scavenging system would also be needed to accompany these proposed lateral gene transfer events. One interesting and common way to consume O_2 is via luciferase since all known luciferases utilize molecular O_2 to generate light. Thus, it has been proposed that the photosystem II/luciferase paired functions might well constitute T/A module, which could detoxify O_2 production and have also established a new identity into the host cell. As we will see in the next chapter, oxidation and reactive oxygen species are indeed frequently used as toxins against pathogens by eukaryotes, and thus are associated with cellular identity systems. Photosynthetic O_2 thus has toxic features that could provide the T of a T/A module. And, since photosynthetic organisms and plastids are often in symbiotic associations with other cells, this would also imply that photosynthesis and some O_2-consuming function could help stabilize symbiosis. This idea is also interesting with respect to a possible relationship to the origin of mitochondria. Photosynthesis in bacteria appears to have clearly evolved prior to the evolution of eukaryotes. However, the mitochondria that are a characteristic of eukaryotes can also be thought of as a more efficient plastid that consumes O_2 and detoxifies the cell. Therefore, it is interesting that when O_2 is limited, mitochondria will generate superoxide anions and hydroxyl ions. Thus, in an early eukaryote, such as algae, photosynthesis generates O_2 whereas their mitochondria consume O_2. The resemblance to a T/A module and the possible involvement of mitochondria in enforcing symbiosis seems clear. As to the possible origin of 'lateral' photosystem II, it is worth mentioning now that the two major classes of cyanovirus both have their own photosystem II core enzymes (discussed below), which have greater resistance to photobleaching than that of cyanobacteria. The likely role of viruses in cyanobacterial evolution is thus examined below.

Community, Fossils, N_2 Fixation and Terminal Cells

As mentioned, the earliest fossils (stromatolites) of any community life form on Earth appear to correspond to lithified remains of ancient cyanobacterial mat communities once found in shallow ocean waters. The cyanobacterial morphologies (i.e., filaments) that produced this radiation as found in these Precambrian microfossils can still be seen in the orders of extant cyanobacteria (see Fig. 2.2). The classification of extant cyanobacteria is mainly based on morphological criteria, although toxin production is also used as a classification criterion. Some extant cyanobacteria can also form communities in the form of chains or mats (see Fig. 4.1 for filamentous forms). Mat formation is the product of filamentous bacteria, which move by gliding. When they glide, they do so by propulsion in which they extrude the extracellular filament. It is this filament that becomes thickened by repeated gliding and eventually becomes fossilized in stromatolites. Since cyanobacteria are photosynthetic and since photosynthesis was so important in the oxidative

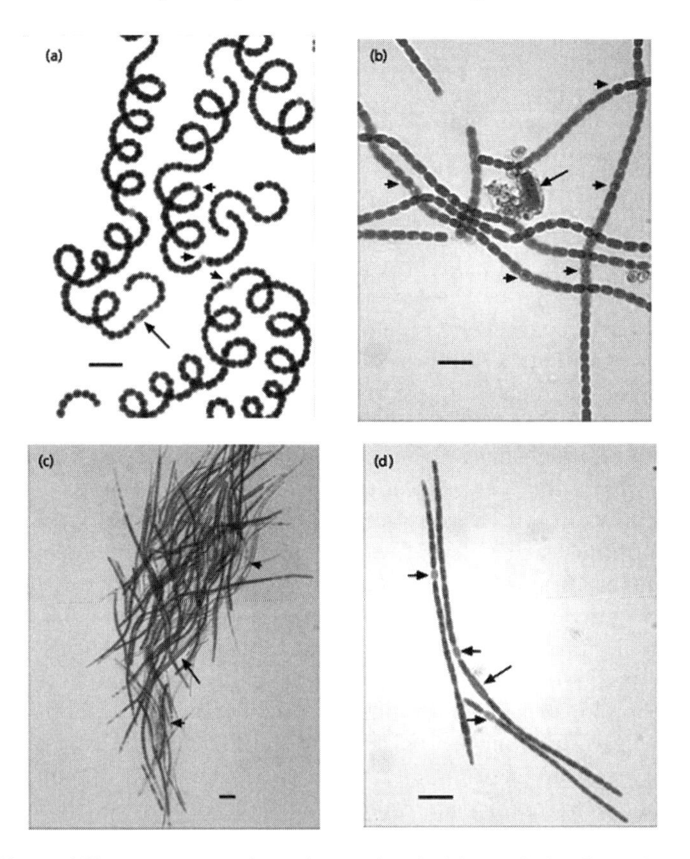

Fig. 4.1 Photo of filamentous cyanobacteria (reprinted with permission from: Gugger, Lyra, et al. (2002), International Journal of Schematic and Evolutionary Microbiology, Vol. 52, pg. 1867–1880)

transformation of the Earth's early reducing atmosphere, most attention has been focused on the photosynthetic capacity of cyanobacteria. However, cyanobacterial participants of extant stromatolites also efficiently fix nitrogen. But O_2 as a product of photosynthesis is chemically incompatible with the reducing environment needed for N_2 fixation. Thus some cyanobacteria (non-heterocyst formers) fix N_2 mostly at night, separating daylight O_2 production from nighttime N_2 fixation. However, other cyanobacteria can also fix N_2 during daylight; these are filamentous heterocyst-forming cyanobacteria that are also known participants of stromatolite communities. Heterocyst-forming cyanobacteria will terminally differentiate an N_2 fixing cell (heterocyst) that is physically separated from a photosynthetic (O_2-producing) germ line. These differentiating cyanobacteria introduce a point of major significance. These cyanobacteria have evolved cells that are 'altruistic-like' or terminal/soma cells that cannot produce more bacteria. This is the first example of terminal differentiation we have considered and also the first example of the separation of germ line from soma. The reason this is of intense interest is that with this terminal differentiation, a clear example of social cell biology in seen in which some individual members of a community of cells can no longer continue their lineage, for the benefit of the larger community. This is a subjugation of the individual by the community. Such a feature is a common characteristic of multicellular eukaryotes, not prokaryotes. However, neither the capacity to photosynthesize nor that to fix nitrogen, as important as they are for cyanobacteria, appears able to offer an explanation as to how these communities evolved from individual cells to form differentiated and organized groups of cells. What features identify, support or protect large and stable bacterial communities? What selective forces promote such large and stable communities and was the invention of terminal differentiation somehow important for this? Were genetic parasites and their T/A modules involved? Extended community behavior of cells clearly suggests the existence of systems that maintain and establish group identity, immunity and behavior. What specifies community membership? Are such communities genetically identical (clonal) or are they more complex mixtures of unrelated cells, as seen in extant stromatolites? Extant stromatolites clearly display more complex communities, although a considerably clonal makeup has also been reported in some communities. Some algal mats are clearly complex communities, containing *Nostoc*, *Nodularia* and *Anabaena* species. Some reports have observed stratified communities of cyanobacteria within mats, whereas other reports do not see community stratification. However, some stromatolites, resembling structures that are lithified or partially desiccated, are found in low (or hot) water communities and may have a different, often simpler species makeup. In such structures, *Synechococcus* species seem to predominate. How such community members sense and respond to each other thus becomes a crucial issue to understand. As presented in the last chapter, many bacteria recognize and respond to related groups via temperate phage, bacteriocins and toxins. Thus we will seek to understand if cyanobacterial temperate phage, toxins, light detection, light production and motility were involved in group recognition as they are in other bacteria. Since photosensory detection is much more common in oceanic bacteria than is photosynthesis, I will focus on the possibility that cyanobacterial photosensing is a group behavioral characteristic. We have already observed that photosensing, quorum sensing and motility can all be linked systems in cyanobacteria (via Lux operon, Chapter 3). I will now examine how these systems relate to the filamentous mat-forming cyanobacteria and later consider the group behavioral role for photosensing in oceanic eukaryotic microorganisms as well.

Mat and Bloom Behavior and Photosensing

In terms of current stromatolites and fungal mats, a community may present a complex situation with respect to light response. Benthic cyanobacteria of the *Nostoc* order (heterocysts, filamentous) appear to contribute to structures very similar to those found in ancient fossils. Such filamentous cyanobacteria respond to light by altering their growth and gliding movement. Each newly divided cell will surround itself with the filamentous slime sheath as mentioned above, which will eventually harden into a filament and it is with such filaments that the bacteria glide to move. Gliding will normally be towards appropriate light flux, so growth is normally away from shade, clearly indicating they have light sensors. However, although these filaments contribute to the basic matrix for stromatolites, they are not likely the only cyanobacteria present. Most of these cyanobacterial species (both blooming and mat-forming) are also capable of light sensation and light production. A most intriguing demonstration of this highly refined light sensing and motility capacity in cyanobacteria is shown in Fig. 4.2. The image in this figure resembles a German cathedral. In fact, it is an image of a cyanobacterial lawn that has been exposed to the visual image of the cathedral and resulted in bacterial motility that generates the image. In stromatolites, small diffusible lactone molecules (QS pheromones, see Chapter 3) associated with both the induction of light production and quorum-sensing have also been directly detected. Therefore it seems likely that group detection and light detection are both involved in mat communities, but few studies have evaluated any specific relationship.

Cyanobacteria Classification, Genome Evolution and Light

Bacteria dominate the current ocean's ecology in terms of both biomass and diversity. Of the bacteria, cyanobacteria have a crucial role with respect to photosynthesis and have been noted in the early evolution of O_2-based life. Cyanobacteria are classified into several distinct types, whose evolutionary

Fig. 4.2 Photo of cyanobacterial plate forming image of cathedral (reprinted with permission from: Donat Hadar (1984), Biologie in unserer Zeit, pg. 79–83)

relationship is presented as a dendrogram (see Fig. 4.3B). Two dominant genera of cyanobacteria are the *Synechococcus* and the *Prochlorococcus*. These genera differ from each other especially in their light-harvesting and antennae systems (phycobilisome versus chlorophyll a_2/b_2, respectively), although the rest of their genomes are surprisingly similar. Each of these genera is further classified into clades that differ with respect to motility, pigments and, in the case of *Prochlorococcus*, high and low light-harvesting capacity. In terms of cellular abundance, *Synechococcus* are found in about 10-fold greater numbers, undergo blooms, but are more restricted to temperate waters. *Synechococcus* are well studied with respect to phage interactions, supporting a wide variety of lytic podoviruses (T7-like, short tails) and myoviruses (T4-like, long contractile tails). This T7-like and T4-like phage biology will be presented in detail below. In contrast to mat-forming cyanobacteria, blooming cyanobacteria have a distinct population organization and structure. A dendrogram of various cyanobacteria based on 16S rDNA sequence analysis is shown in Fig. 4.3. This analysis indicates the evolutionary relationship between *Prochlorococcus* and *Synechococcus* species. Here we see that *Synechococcus* species appear more basal to the *Prochlorococcus* species and that *Prochlorococcus* continue to show recent evidence of divergence with respect to high-light and low-light adaptations. This *Prochlorococcus* pattern of evolution is of special interest regarding the evolution of eukaryotes as they are thought to best represent the origin of the type II photosynthetic system found in all plants. Evidence will be presented below that the photosynthetic genes in *Prochlorococcus* have undergone wide-scale lateral gene transfer.

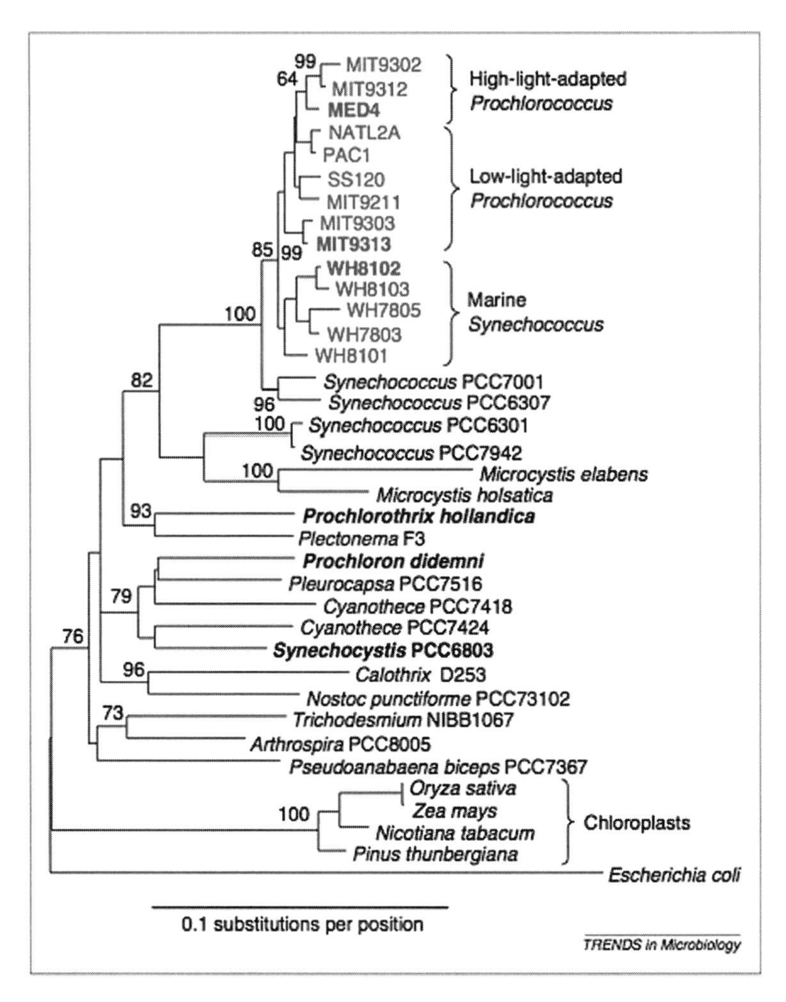

Fig. 4.3 Dendogram of *Prochlorococcus* evolution (reprinted with permission from: Rudi, Sckulberg, Jakobsen (1998) Journal of Bacteriology, Vol. 180, No. 13)

Viral Threats

In the context of a bacterial mat, it would seem crucial for large sessile and genetically similar community that will be continuously exposed to the high viral loads within the oceans to have evolved systems that will protect them from destruction by virulent genetic parasites. This is especially an issue for a cellular network situation, in which one virulent virus would seem able to threaten the entire colony. Systems of virus immunity would thus seem to be an essential prerequisite for the survival of a sessile cellular community. In addition, the steady exposure to high UV flux in shallow waters would be expected to damage and destroy large cellular DNA genomes. Curiously, some cyanobacteria, such as *Prochlorococcus*, lack several important DNA repair enzymes, whereas their phage have many such DNA repair enzymes. Clearly, light and phage would seem to present serious hurdles to stationary bacterial colonies in shallow waters.

Lights On, Lights Out and Light Movement

The filamentous cyanobacteria not only move and divide relative to light flux but do so in a synchronous, communal fashion. This light phasing of cell division would also be expected to have big effects on virus–host interactions. Interestingly, the ability of cyanobacteria to sense, move to and to make light in populations was one of the earliest observations in the bacteriological literature. In 1893, M. Beijernick (prior to his seminal discovery of virus) reported on the isolation and study of bloom-forming cyanobacteria. He also went on to observe the ability of these bacteria to produce light and showed that some isolates of these bacteria would spontaneously ‘go dark’ and stop producing light, yet these dark bacteria would grow normally. Thus with this first report, the question regarding the purpose of light production was raised. To this day, this question remains essentially unanswered. It is usually rationalized that light sensing is associated with photosynthesis and that it allows the bacteria to move towards the desired light energy source. However, as indicated above, light sensing is also common in non-photosynthetic bacteria, so clearly it does not fulfill such a role in those more numerous bacteria. But what about light production? Why would both photosynthetic and non-photosynthetic bacteria produce light, and why might this be switchable state as seen by Beijernick? Later studies went on to establish that the cyanobacterial isolates that had ‘gone dark’ (K variants) could also be back-selected to again produce light. Curiously, the light ‘on’ or ‘off’ state was also linked to phage sensitivity and colony morphology. The basis of this linkage was not determined and remains a mystery after 113 years. The sensitivity of cyanobacteria to light can be exquisite. The motile cyanobacteria move in relationship to specific light flux (towards and away) to maintain a specific range of light intensity. It is this feature as shown in Fig. 4.2 that allows photographic-like reproduction.

Why so Sensitive?

The sensitivity of cyanobacteria to light is surprisingly high and even light scatter can attract bacterial movement. Such high sensitivity is well beyond that which would be expected simply for daytime photosynthesis. However, such high sensitivity would be appropriate if it were used to detect nearby populations of fellow light-producing cyanobacteria, which would produce much less light. The movement demonstrated in Fig. 4.2 was due to a motile, flagellar cyanobacteria. In natural settings, similar motile cyanobacteria live in lakes and will move as a population from one shore to the next in relationship to light, but remain as a group offshore, not simply follow light sources to the shore. In such population-based movement, what keeps the bacteria together? Non-motile cyanobacteria can also move up and down a water column in response to even very low light flux of moonlight. Clearly, this moonlight reaction is a sensory, not photosynthetic response. Such upwelling movement is of central importance for the behavior of water blooms. Starting rather deep, where light levels are very low, and sometimes at night, blooms can travel up the water column by the coordinated production of gas vesicles that will affect buoyancy. The scale of these bloom communities can be massive. As previously shown in the NASA picture in Fig. 2.1, cyanobacterial water blooms are the most visible life form from space and can constitute some of the largest biomasses on the planet. As presented in Chapter 2, in some cases (fish guts, light organs), it seems clear that light production and light detection are used to sense and control group behavior in some free-living cyanobacteria. Thus it also seems possible that mat-forming and blooming cyanobacteria could also use light for group behavior purposes although this issue has not been evaluated.

Cyanobacterial Genomes and Parasites, Mats/Slime

Synechococcus elongates is known to participate in algal mats and in some extant stromatolites. Since its genome has been sequenced, we can now examine it for colonization events by genetic parasites that might have affected bacterial group identity. Doing so, we indeed see lots of evidence of viral influence. Specifically, *S. elongates* contains no known plasmids but does have 15 genes that are closely related to phage P2-like tail and base plate proteins. These also closely resemble the pyocins R-proteins in organization (see Chapter 3). It is therefore predicted that this group of proteins is likely to function as the pyocins (cyanocins?), producing toxins against related species of cyanobacteria. If so, this would represent a group identity system aimed mainly at competing species. However, there is currently no data that relate to this prediction so it remains a speculation. In addition to these P2-like structural proteins, *S. elongates* also codes for six ORFs that are putative restriction enzymes, three ORFs that are SAM-dependent methyltransferases, six ORFs that are known toxin genes and three transposases, some belonging to families with known antiviral activity. All these ORFs suggest that the *S. elongates* does indeed have systems designed to resist colonization by genetic parasites. Interestingly, and in stark contrast to other cyanobacteria, this genome contains no retrons (RT-encoding introns). Also curious, given the presence of P2-like late structural genes, few identifiable prophage elements or phage immunity regions seem present in the genome, suggesting that *S. elongates* is indeed able to resist the type of frequent phage colonization common to so many gram-negative bacteria. However, the natural phage biology of mat-forming cyanobacteria is not well studied, although a few studies have failed to find any inducible prophage. Still, it appears that these cyanobacteria effectively resist new genetic colonizers. Some reports using PCR-based evaluation suggest that cyanobacteria in stromatolites are genetically rather homogeneous, almost clonal in nature. If this proves to be correct, then it would be very interesting from the perspective of group identity in a non-motile, static bacterial community, indicating the existence of highly effective, essentially clonal, group immunity systems. How might bacterial mats resist so wide a diversity of virus? A mechanical solution might apply. With the filamentous cyanobacteria, the production of a slime sheath, for example, would seem capable of providing an effective physical barrier to the adsorption and entry of exogenous phage. Indeed slime-producing pathogenic bacteria have also been shown to resist phage. Although non-sheath producers can also participate in these mat communities, it is these sheaths that provide the structural filaments that determine physical integrity of the community. These filaments also provide an external habitat for the unicellular cyanobacteria. In this setting, a pyocin-like identity system could help maintain the 'extra-sheath' system of colonial homogeneity and stratification of the unicellular cyanobacteria by exclusion and toxic death of non-members. These issues have not yet been evaluated.

Blooms and Genomic Parasites

In contrast to mat-forming cyanobacteria, blooming cyanobacteria have a distinct population organization and structure. They also have distinct phage relationships. *Trichodesmium erythraeum* (a filamentous, blooming, N_2-fixing non-heterocyst-forming cyanobacteria) genome has been sequenced and has some very interesting differences relative to *S. elongates*. Most notable and numerous are the 229 transposases, like ORFs, encoded in its genome (of which probably only a fraction are expressed). This number seems wildly excessive and leads to the question as to why so many transposases might be selectively maintained in this genome. Since transposases are often used to interrupt genetic colonizers, this might represent a high flux of genetic parasites and be a cellular strategy to invade and inactivate such genetic colonizers. However, little experimental information is available regarding this possibility. Another significant difference with the genome of *S. elongates* relative to the *T. erythraeum* is that the latter genome also has 28 retrons but lacks any identifiable phage-tail or base plate proteins. However, it does have four phage-like lysozymes and four phage-like heat shock proteins. In addition, it has 17 restriction enzymes and 42 toxin genes, as well as perhaps the largest intene sequence yet observed (1650 aa in the helicase B gene). The picture that emerges is quite different from our mat-former exemplar above. These blooming cyanobacteria appear to posses many genes that are set to contend with lots of invasion by phage and other genetic parasites. The movement of bloom populations through vertical water columns tends to be seasonal and affected by both high light and warm water temperatures. The periodic occurrence of such large populations of moving bacteria may subject them to infection with acute viral agents and the termination of these blooms has in fact been proposed to occur via lytic phage. If so, such blooming populations would be subjected to unending invasion by acute viral parasites.

Heterocyst Cyanobacteria

A third sequenced cyanobacterial genome is from terrestrial (non-marine) *Nostoc* species (also *Anabaena*; filamentous, heterocyst forming, N_2 fixing). This bacteria harbors six plasmids which encode many ORFs, the bulk of which are of unknown function with little similarity to Genbank sequences. Its genome has no phage proteins and no retrons, but does encode 78 transposase ORFs. Only three restriction enzymes are apparent, and one gene which appears to be a toxin to ABC transporter. A most interesting distinction with the other cyanobacteria and most prokaryotes, however, is the occurrence of repeat elements. These repeat elements are a eukaryotic-like DNA feature, but here they are involved in terminal heterocyst formation. DNA rearrangements (deletion within the nitrogen-fixing *nifD* and *fdxN* gene) are needed for heterocyst terminal differentiation and such deletions involve viral-like transposase. *NifD* is a N_2 reductase that cannot work in O_2 environment of photosynthesis. Another very interesting eukaryotic-like sequence feature is the presence of several long tandemly repeated repetitive (LTRR) sequence elements (17 copies of inverted 37 bp). This non-coding element is flanked by two genes, one which is clearly homologous to T4 gene 15, directly implying a viral involvement in the origin of this repeat. This same element is conserved in other *Anabaena* and, most curiously, is also found in mitochondria of *Vicia fabia*. Repeated elements (of completely different sequence) are also known for other unicellular cyanobacteria (such as *Microcystis aeruginosa*, but not *Synechocystis*), suggesting a rather basic function of the repeats. In the case of these heterocyst-forming cyanobacteria, I will note that it is the germ line that maintains the 'parasitic' (junk DNA) genetic repeat, whereas the soma heterocyst is terminal (dead) and deletes the parasite. This represents a shift in cellular identity associated with a shift in its genomic parasite. A similar germ/soma theme with respect to genetic parasites will also be seen in some eukaryotes, such as hagfish (see Chapter 5).

Cyanobacteria, Lytic Phage and Photosynthesis

Given the discussion in Chapters 2 and 3 on the role genetic parasites can have on bacterial systems of group identification, what is known regarding the lytic viruses of cyanobacteria? Is there any reason to think they might have contributed to the evolution of colonial cell populations as found in stromatolites? Most evidence relevant to this issue comes from studies of free-living oceanic cyanobacteria. In the last 10 years it has become clear that the ocean can be considered as a viral soup which presents an ideal habitat for viruses to persist, diffuse and find new host. The world's oceans have recently been estimated to contain about 10^{31} viral particles *in toto*. The large majority of these particles resemble large dsDNA phage of bacteria. Actual population measurements of this viral soup, however, are incomplete, so the complexity and relative makeup of this vast population can only be guessed at for now. For example, we are unable to say much about the likely life strategy of the viral particles that have been observed, although it is likely to contain large numbers of both lytic and persisting (lysogenic) viruses. Clearly, however, lytic cyanophage are common and viral-induced lyses of cyanobacteria also seem common. Some have estimated that up to 51% of oceanic cyanobacteria can be destroyed by phage every day. *Synechococcus* are better studied with respect to phage interactions and support a wide variety of lytic podoviruses (T7-like, short tails) and myoviruses (T4-like, long contractile tails). Some specific cyanophage, like N-1, can infect blooming cyanobacteria and have been proposed to contribute to the termination of blooms (although this claim is contested). N-1 infects filamentous N_2-fixing *Nostoc muscorum*. An interesting side note is that *Phaeocystis globosa* dying blooms in the English channel also appear to be mediated by lytic virus infection. These dying blooms release large quantities of dimethylsulfoniopropionate and dimethyl sulfate which act in the atmosphere as a major source of cloud seeding and resulting in rain overland. Both the cyanobacteria and their viruses also have strong associations with light. For example, the cyanobacteria chlorophyll and the photosynthetic O_2 release occur in flattened sac structures called thylakoids. Some phage are known to replicate in and displace these photosynthetic lamellae. Other phage, such as AS-1 (*Anacystitis nidulans*), show replication that is tightly linked to photosynthesis which results in low yields in the dark. Some phage are photosynthetically obligate and need light for reproduction and will not replicate in the dark. Until recently, phage of *Prochlorococcus* were much less studied (see below). However, genomic analysis has made it apparent that the *Prochlorococcus* lineages show strong evidence of relatively recent evolution of diverse and plant-like chlorophyll systems. Cyanobacteria themselves appear to have acquired photosynthetic capacity via horizontal transfer of DNA (virus/plasmid) as noted above. However, in *Prochlorococcus*, the specific variation of these sequences of photosynthetic genes also suggests an ongoing role for horizontal gene transfer in this evolution. Thus *Prochlorococcus* offers a system that can be used to understand the evolution of photosynthesis and has received much more attention as such a model. A strong case can be made that viruses were involved in this evolution. As will be presented below, *Prochlorococcus* phage, like P-PSSP7, actually encodes two photosynthetic core proteins that have higher photobleaching resistance than that of the host cell and are expressed at high levels during the late virus replication cycle. Most phage also have group ID systems that preclude other (often similar) phage. For example, AS-1, mentioned above, codes an endonuclease that cleaves the host but not viral DNA via 5-methyl cytosine, which also protects against various other restriction enzymes. Clearly, these lytic cyanophage must have large effects on host populations. Natural population studies of the ocean show cyanophage to be numerous (titers ranging from 10^4 during summer to 10^2 per ml during winter) and diverse (genome sizes ranging from 22 to 300 kbp). In one coastal habitat, 20 different cyanophage types were identified. The cyanobacteria can frequently show seasonal shift from predominantly filamentous to predominantly coccoid bacteria populations. Such shifts in natural bacterial populations can also be emulated in laboratory settings. Still, other natural bacterial populations seem much more resistant to phage effects and such populations appear stable. It has been proposed that the lytic phage eliminate those host cells that correspond to the winners, that is, eliminate those cells that have numerically prevailed. Ecologists that study the ocean tend to only focus measurements on this type of virus–host relationship. However, as we saw in Chapter 2, the cells that survive assaults by lytic phage often harbor lysogenic or persistent phage that provides the needed identity and immunity functions.

Cyanobacterial Lysogens

A stable population of cyanobacteria would raise the possibility that such cyanobacteria harbor persistent, defective or lysogenic phage that confer immunity to lytic phage. In such a persistent state, cells would be expected to produce T/A products that allow stable persistence and aid in the competition with other parasites and uncolonized but related bacteria. Recent phage and metagenomic screens of natural ocean populations, such as the Sargasso sea project, are starting to provide some information concerning natural occurrence of viral genes. These studies are consistent with prior estimates, showing that a large diversity of morphology of phage types exist, although some types (podoviruses) clearly predominate. However, lysogenic phage in natural populations can be difficult to induce and are not well studied. Some lysogens, however, have been observed. Two *Synechococcus* phage especially well studied in this host are P60 and S-PM2 (see below).

Host/Phage Phylogenetics

Prochlorococcus is the numerically dominant prototroph in tropical and subtropical waters, and its phage are also expected to dominate in these oceans. *Prochlorococcus* are classified into low light and high light adapted species, and these species appear to have their corresponding characteristic phage (see Fig. 4.3). In high light adapted species, the corresponding phage appears to be exclusively of the podovirus family and these phage tend to be highly host specific (some appear to be unique to their host). In low light adapted species, phage tend to be of the myoviridae family and show a much broader host range. The marine podoviruses (T7-like) encode DNA polymerases that can be used to evaluate relationships to other viruses and host (see Fig. 4.4). As will be presented below, this T7-like DNA pol is also related to those found in chloroplast and mitochondria. From a broad perspective of phage–host dynamics, it would be very interesting to evaluate virus–host interactions of cyanobacteria that are from colonial mats relative to those cyanobacteria that undergo phased population blooms. In this context, it would be important to understand if phage, prophage-like elements and plasmids are contributing to the distinct host life strategy or community dynamics. As mentioned, cyanobacteria are classified based mainly on morphological criteria and toxin production. However, recent phylogenetic analysis now indicates that some of this morphologically based classification is clearly not correct. The three major morphological classes are (1) simple filamentous forms (Oscillatoriales order; *Plectonema*); (2) filamentous forms that also form differentiated cells, including N_2-fixing heterocysts formers (Nostocales order; *Anabaena*, *Nostoc*, *Nodularia*, Stigonematales order; thermophilic *Chlorogleopsis*); (3) unicellular cyanobacteria (Chroococcales order; *Microcystis*, *Synechococcus*, *Anacystis* and Pleurocapsales order). The heterocyst-forming filamentous cyanobacteria appear to be mostly monophyletic (as supported by phylogenetic analysis if 16S rDNA) and include the much studied *Anabaena* and Phormidium species. These heterocyst

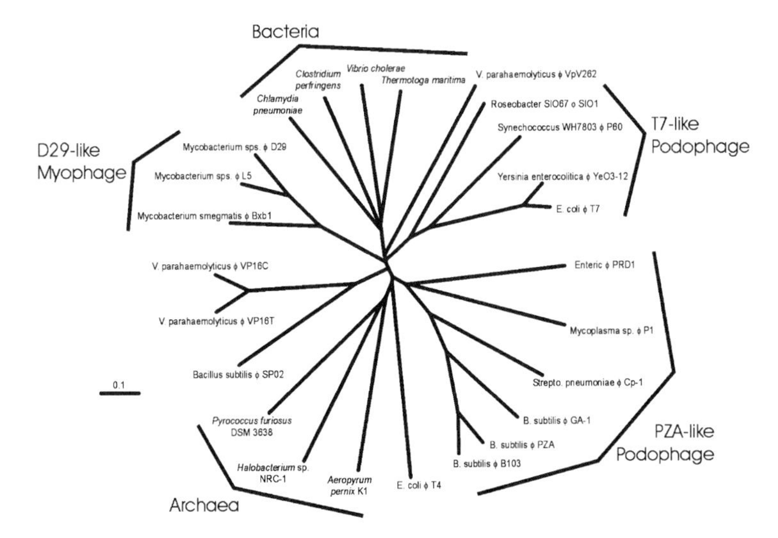

Fig. 4.4 Dendogram of phage DNA pol (reprinted with permission from: Sullivan, Segall, Rohwer (2002) Comparative Biochemistry and Physiology Part B: Biochemistry and Molecular Biology, Vol. 133)

species have no flagella, make gelatinous cell walls that cover an underlying gram-negative cell wall and can be associated with dense water blooms and toxin production. The heterocyst formers include the order Nostoc and many members of this order can make gas vesicles (to rise in the water column by controlled buoyancy) as well as make many toxins (neurotoxins and hepatotoxins). The non-heterocyst cyanobacteria are more dispersed on phylogenetic trees according to 16R rDNA sequence analysis. Phage infect both unicellular and filamentous forms, but have not been reported for all cyanobacteria orders (e.g., Pleurocapsales, Stigonematales). It should be emphasized, however, that although T7-like phage are common in cyanobacteria, unlike lytic-only T7 of proteobacteria, some of these cyanophage will integrate to establish a lysogenic life cycle. Thus a significant shift in virus–host relationship has occurred between T7-like phage and their cyanobacterial host. According to the concepts presented in this book, the persistence of genetic parasites is expected to be highly host specific and have significant effects on host evolution. Thus this relationship, I argue, represents a significant evolutionary shift. One such T7-like lysogenic phage is the LPP-1 phage of non-blooming cyanobacteria (considered below). In addition, cellular differentiation can also be associated with phage biology. Some lytic cyanophage will lyse vegetative cells but will not infect N_2-fixing heterocyst. However, few cyanophage are well studied in natural settings or with respect to such issues.

Toxins and Phage: Vibrio cholerae

Cyanobacteria are well established to often produce toxins and can even be classified according to toxin production. However, possible phage relationships to the evolution of toxin production is not well studied or understood in cyanobacteria. Yet *V. cholera* (a gram-negative oceanic bacteria and distant relative of planktonic cyanobacteria) has been much better studied along these lines due to its medical importance. *V. cholera* can be infected by virus CTXϕ, which is a persisting filamentous phage that produces toxin. The toxin is secreted by the same system that transports the CTX filamentous phage. CTXϕ can persist either as an extrachromosomal plasmid or an integrated prophage. It was because of this characteristic that CTX was originally described not as a phage, but as a genetic element encoding several toxin genes found in the chromosomes of some pathogenic *V. cholera* strains. The integration of this element was mediated by a repetitive RS element also encoded by CTXϕ. However, sequence analysis now make it clear that CTX is a defective element of CTXϕ, a virus clearly related to ssDNA M13 phage. Thus this toxin production is a phage conversion event. Furthermore, toxin production is co-regulated with phage receptor expression (via a second TCP phage). CTXϕ infection uses type IV pilus as a bacterial surface receptor. However, this type IV pilus itself appears to be encoded by a second cryptic prophage – TCP. Since it is chronic,

CTXϕ needs to establish itself as either a persisting plasmid or an integrated state. The persisting plasmid can also be considered as a 'fitness island' since it is a gene set that allows *V. cholera* to express a toxin which functions to allow the colonization of the gut of its animal host. In fact, the colonizing factor (GbpA) of *V. cholera* in human small bowl gut infection is also the same as the factor for attachment to the chitinous exoskeleton of zooplankton, and thus may well have evolved from a marine environment. A general role of phage-acquired fitness islands in cyanobacteria evolution and adaptation is described below. CTXϕ can also be considered as a parasite not only to the *Vibrio* host, but of host specifically colonized with TCP. As we will see, the type IV pilus is frequently involved in bacterial colony formation, so phage-mediated group identity can transform its host and host relationship to other parasites. Thus the CTXϕ, TCP and RS parasitic elements can be considered together as a hyperparasite–epiparasite network that can be crucial for host identity and fitness and do not simply kill their host. This situation is clearly reminiscent of other hyperparasite networks described in Chapter 2 for *Escherichia coli* (see Fig. 2.5). Thus viruses can be intimately involved in many of the basic characteristics we associate with cyanobacteria. This includes photosynthesis, N_2 fixation, differentiation, bloom formation and toxin production. As argued below, these parasite associations are proposed to represent the evolutionary forces that promote the creation of such host complexity.

A Viral Role in Light and Toxin?

I previously presented the link between phage of *V. cholera*, the quorum-sensing system, light tactic motility and light production (Chapter 3). Similar interactions could be expected to be virally associated. For example, the large episomal prophage (pCTX) described above also uses a photosensory related quorum-sensing system to control photoreceptor expression. In addition, the receptors for this virus (pilis IV) are themselves prophage-encoded. So a viral role in this specific light sensory process also seems clear in this case. However, the specific host–virus relationships regarding the light response as seen in our *V. cholera* are not likely to be the same as those of cyanobacteria. As I have stressed, persisting virus–host relationships tend to be highly specific to the host species and might not be conserved.

T7-Like Exemplars

In terms of cyanobacterial ocean ecology, *Prochlorococcus* and their corresponding phage numerically dominate. These phage are mainly related to T7-like podovirus and also myovirus and have been reported to moderate host populations. Of the T7-like phage, one (P-SSP7) has been more extensively studied and has shown some remarkable features (such as encoding photosystem II core genes). Unlike the strictly lytic T7-related phage of proteobacteria, these cyanophage also encode an integrase and accordingly, the phage DNA can be found integrated into some host chromosomes as prophage. In addition, there are other families of T7-like marine phage, such as that of *V. parahaemolyticus* (VpV262,), which are clearly similar to T7 phage, but show intriguing differences, such as the apparent lack of a T7-like RNA polymerase. Furthermore, this phage uses distinct promoters from those of T7. These observations suggest a very different relationship between these specific T7-like marine phage and their host compared to that seen in T7 of gram-negative bacteria. It is interesting to note that cyanophage DNA is often resistant to cleavage by numerous restriction enzymes, indicating the common occurrence of DNA modification. Clearly molecular identity systems (R/M) are abundant in marine bacteria and their viruses. There are also T7-like phage that infect *Synechococcus*, such as P60. Curiously, P60, unlike T7, lacks any identifiable holin or endolysin genes. Ma-LBP is a T7-like phage of *Microcystitis aeruginosa* (freshwater cyanobacteria). Ma-LBP is clearly a T7-like phage in that it has a T7-like DNA pol and gene order. Ma-LBP is able to lyse 95% of host populations in some lab settings, so it is clearly mainly a lytic virus. But unlike related phage of *E. coli*, the replication cycle cyanobacterial phage is slow, requiring about 11 hours. Also in contrast to T7 of *E. coli*, Ma-LBP has a small burst size of only 28 PFU. Thus small burst sizes and slow replication cycles appear to be a characteristic of many lytic cyanophage.

T7-Like Phage and Plastid Evolution

Recently P. Forterre has suggested a very interesting proposal concerning the possible role T7-like cyanophage may have played in the symbiotic evolution of the chloroplast. It is now well accepted that chloroplasts evolved from a cyanobacteria and established a permanent symbiotic relationship with some eukaryotes. These plastids (chloroplast) all encode their own (non-eukaryotic) version of DNA polymerase. However, it has become clear that chloroplast (and mitochondrial) plastids encode a DNA polymerase that is not related to that found in bacterial cells, but is instead clearly similar to that of a T7-like virus. In fact, as previously shown in Fig. 4.4, cyanophage DNA pol sequence appears basal phylogenetically to even that of T7 with the *V. parahaemolyticus* version being the most basal (VpV282). These cyanophage encode lots of genes that modify photosynthesis and appear to also depend on photosynthesis. Such observations have led P. Forterre to propose that all chloroplasts and mitochondria plastids evolved soon after the displacement of the bacterial DNA replication enzyme with a T7-like prophage. It therefore seems T7-related viruses had very much to do with the bacterial and symbiotic origin and evolution of both mitochondria and chloroplasts of eukaryotes.

T4-Like Phage and Photosynthesis

In addition to T7-like phage, the T4-like phage (myovirus) of *Prochlorococcus* cyanobacteria are also common. Recall that T7 and T4 (even) phage of bacteria occupy distinct phylogenetic clades and classifications. Of the T4-like cyanophage, the best characterized are the P-SSM2 and P-SSM4 (which have 42 of the 75 T4-like core genes, respectively). Remarkably, these otherwise distinct phage also encode several photosystem II core genes (psbA, hliP), just like the T7-like cyanophage above. These viral genes are more resistant to photodynamic inactivation compared to host genes and restore photosynthetic activity to infected but damaged host photocenters. However, these viral photosynthetic genes show no homology to the host genes, thus they appear to be viral creations. There are many other observations that relate cyanophage biology to photosynthesis (light-dependent viral replication, dark-dependent immunity, interaction with photosynthetic thylakoids). These viruses also all encode other genes found specifically in cyanobacteria. Given that two distinct families of phage encode photosynthetic and other host-related genes, it is clear that these phage have intimate and highly host-specific interactions. S-PM2 is a T4-like cyanophage that infects *Synechococcus*, but it is distinctive in several regards. It was the first lipid-containing phage ever isolated, a feature not found in any T-even phage (but seen in P2). S-PM2 is also a very large phage, with a genome of 196 kbp. Such genome complexity for a phage was initially surprising. Although S-PM2 encodes many genes that modify photosynthesis, it also encodes many tRNAs, including self-splicing introns. Clearly, it has considerable capacity to modify host photosynthesis and genome architecture. Interestingly, S-PM2 encodes many cell envelope-associated proteins, including holins, endolysins and S-layer proteins, a unique surface feature of some cyanobacteria. S-PM2 also encodes some extremely large proteins (3800 aa) of unknown function. S-PM2 appears more related to the giant K2 phage of *Pseudomonas aeruginosa* (a 280 kbp DNA with 306 ORFs, but lacking DNA polymerase).

It is difficult to determine how such giant phage have affected host evolution. The potential for major consequences, especially for host responses to other phage and light, seems clear. Whether colonization, persistence or addiction modules are involved remains to be determined. However, as will be argued below, addiction modules, toxin production and group identification systems are frequently one in the same gene set. Cyanobacteria (especially *Prochlorococcus*) appear to have repeatedly acquired various versions of photosystem II via some type of 'lateral' transfer process. Similarly, toxin production, a rather complex characteristic of many cyanobacterial orders, has also been the product of frequent 'lateral' acquisition. Many of these gene-set transfer events also appear to have been rather recent in cyanobacterial evolution (e.g., the chlorophyll systems and toxin systems are often not phylogenetically congruent with 16S rDNA dendograms). Clearly, large DNA viruses have the now-established capacity to originate and transfer such complex multigene systems. Interestingly, cyanobacteria are also polyphyletic, with respect to the occurrence of group 1 introns (within tRNA leu). This too could be explained by stable virus colonization. The likely viral involvement in basic host processes makes it also likely that viruses were involved in evolution of other complex host characteristics.

Recent evidence adds considerable weight to the idea that cyanobacteria (*Prochlorococcus*) are mostly evolving by phage-mediated colonization events. *Prochlorococcus* exist in six known ecotypes that differ with respect to their habitat (temperature, depth, light intensity). Based on their 16rRNA sequences, these ecotypes differ by less than 1% in sequence. Since *Prochlorococcus* have compact genomes (1.7–2.4 Mbp, only 17,000 genes), it was possible to evaluate all the genetic differences between these ecotypes. The variation was found to predominantly occur in a limited number of gene islands (i.e., five). These islands clearly appeared to have been acquired by a phage-mediated process (due to various telltale signs, such as tRNA integration and viral-specific repeat sequences). Thus they closely resemble the pathogenic islands described in Chapter 2. Furthermore, within these islands, up to 80% of the genes were clearly phage-like (including integrases, DNA methylases, endonucleases, surface receptors). Many of these phage-like genes were transcriptionally responsive to light and nutritional states. Thus it seems clear that the process by which *Prochlorococcus* is adapting to habitat changes is mainly mediated by persisting virus colonization. Furthermore, this colonization is a dynamic, ongoing and likely competing (group identifying) process, since multiple ecotypes of *Prochlorococcus* exist within ocean gradients of light and temperature. From this it is concluded that virus colonization is the major engine of adaptive evolution in cyanobacteria (and indeed all bacteria). It therefore seems reasonable that the origin of the major adaptations of *Prochlorococcus* (photosynthesis and N_2 fixation) would also likely have been mediated by viral colonization events.

See the Light: Eyes and the Cambrian Radiation

Light detection, motility and light production seem to have a special significance for oceanic biology. Many diverse organisms appear to use light for social purposes. As previously mentioned, *V. fischeri* is symbiotic in squid (*Euprymna scolopes*) and provides a light organ. Here we can see several parallels with pathogenic *V. cholera* or *V. parahaemolyticus* in control systems. For example, all make group III catalyase (katA) upon addition of some hydrogen peroxide. Without this very gene, *V. fischeri* will not colonize the squid light organ. Also, the squid light organ makes lots of peroxidase, which requires halide ions for activity, and these are provided by *V. fischeri*. The halide is typically an antimicrobiological agent. The combination thus looks very much like T/A addiction module that compels cooperation between the bacteria and the squid. The next chapter will introduce the basal importance of ROS as a component of

T/A modules and self-identification (see red algae innate immunity). As mentioned, most microorganisms and many multicellular organisms in the ocean make luciferase (which consumes O_2). But the bioluminescent systems are not evolutionarily conserved. Luciferase proteins are not homologous to one another and luciferins are also very different from each other, falling into many unrelated chemical classes. Therefore there is tremendous molecular diversity in light production and detection in the ocean. The diversity of light-producing organisms includes non-photosynthetic bacteria, cyanobacteria, dinoflagellates, diatoms, algae, fungi, worms, cnidarians and fish. All emit and detect light for social purposes. In some cases, filters have been evolved allowing spectral resolution of the light, as commonly seen in fish and cnidarians. In these cases, a role for light in group/sex identification is well established. The production of light for social purposes, however, is much less characteristic of terrestrial organisms, including microorganisms. There are, however, some interesting exceptions, such as glow worms and fire flies. It is interesting that the most spectrally advanced visual detection system found in biology is in a shrimp species (see mantis shrimp, Chapter 5). Why is light in the ocean so often used for social detection? Was this simply the byproduct of too much photosynthesis in which the resulting O_2 needed to be eliminated by light pumps? We already noted the high levels of photosensory rhodopsin gene present in environmental clones of Sargasso sea. Light clearly has a special role in ocean biology, from bacteria to fish.

A Sea of Channel Toxins: Competing Self-Identification?

Besides light production, oceanic blooms of cyanobacteria, red algae and dinoflagellates are all associated with the production of numerous toxins, the majority of which are active against various ion channels. Why is this also such a prevalent situation for microorganisms in the oceans? Recall that the Archaea have no such propensity for making such toxins. As noted above, there has been a strong bias in the medical literature to consider toxins strictly from the perspective of human disease or virulence factors. Often such pathogenic genes are also under Lux-like QS group-sensing control. However, toxin production is a surprisingly broad occurrence in non-pathogenic cyanobacterial and *Vibrio* species as I noted. We know that some pore toxins (e.g., holins) and other pore-forming molecules (bacteriocins) can be the product of a genetic parasite that constitutes part of an addiction module that can kill related, uncolonized species and also colonized host infected with competing viruses and plasmids. As we will see in the following chapter, fungi are also well known for their ability to produce pore-based toxins encoded by genetic parasites which can similarly kill related species. Thus we can pose the following question: do ion channel proteins and their toxins constitute a prevalent system for identity modules in oceanic organisms derived from parasites? If so, are they are involved in competition or group survival? From the perspective of human disease, one of the best studied toxins is that produced by *V. cholera*. And as discussed above, this toxin production is due to the presence of a defective CTX phage. *Vibrio* species are common in the oceans so this involvement of genetic parasites in toxin production might represent a general situation. However, pore toxin-producing phage are not known for most of these oceanic species. Yet, we can recall that the *Synechococcus* genomic DNA does encode P2-like tail proteins. And we can also recall (Chapter 2) that P2-like tail proteins are used by other bacteria as bacteriocin pore protein. Currently, we do not understand the significance of these P2-like genes.

Small Toxins

It seems clear that bacteriocin-like toxins are not generally known for cyanobacteria. Instead, there is a very large diversity in the production of small-molecule toxins. The most recognized of these toxins is microcystin, which is also a dominant type of neurotoxin as well as toxins that inhibit eukaryotic protein phosphatase. Freshwater cyanobacterial species are most associated with a diversity of such toxins (e.g., microcystins) and are made by *Microcystis*, *Anabaena*, *Oscillatoria* and *Nostoc* species. Over 100 cyanotoxins and over 60 isoforms are known. These cyclic heptapeptide neurotoxins are not made on ribosomes. In addition, many cyanobacteria can make other types of toxins, such as depsipeptides, cyanopeptolin, microginins, aeruginosins and tricyclic microviridins. Together, these toxins comprise an extraordinarily large family of peptides. Although many are ion channel neurotoxins, many toxins also inhibit eukaryotic protein phosphatase 1 and 2A. Some toxins, such as cyanopeptolin, inhibit serine protease or trypsin-like enzymes but these are not toxic to eukaryotes. There is little evidence that these toxins are derived from genetic parasites. Nor are the cellular targets of all these toxins defined. Some have proposed that micrograzers, such as *Daphnia galeata*, might be the targets as they can be poisoned by some toxins. The synthesis of these peptide toxins occurs in a large non-ribosomal synthase complex. In the filamentous cyanobacteria *Planktothrix agardhii*, this synthase complex corresponds to a 56 kbp microcystin synthase gene cluster encoding peptide synthase and polyketide synthase. Unlike the photosynthesis genes noted above, genetic analysis suggests this toxin gene cluster has not been horizontally transmitted and is thus phylogenetically congruent with 16s rRNA. Thus, these toxins thus show a tight association with species identity. Generally, toxins are made in early log growth and are produced and exported (via putative ABC transporter) in blooms and also in association with light exposure (although intense light inhibits toxin production). The role of light in toxin production is intriguing and raises unanswered questions regarding possible roles. Some have proposed that ongoing photosynthesis might be needed to provide substrates or energy required for large-scale toxin production. However, energy-demanding N_2 fixation clearly occurs in the dark with many cyanobacteria, so this explanation seems problematic.

The toxin–light association, however, does suggest a possible role in the communal biology of cyanobacteria. Although the high toxicity to eukaryotes is clear for some cyanotoxins, not all cyanotoxins are toxic to eukaryotes. This suggests that the broader or main biological role for toxins does not include eukaryotes. As I have asserted, the existence of molecular diversity in any cellular systems suggests a role in group identity. In some specific cases of toxin production, it has been shown that toxin production by one cyanobacteria species is inhibitory to competing species (as an antibiotic). Although this observation remains limited, it clearly suggests that these toxins could be systems of group identity that act as growth suppressors of related species. If so, toxins would constitute a common element of group identification of cyanobacteria. Thus, cyanobacteria that produce specific toxins would also be expected to produce products that prevent self-toxicity. Clear evidence supports this idea. A screen of 65 field isolates of filamentous cyanobacteria, found that 7 of the *Nostoc* isolates produced antibiotics that were very active against other cyanobacteria. These same isolates failed to find any that could induce temperate phage. In addition, these seven isolates also made antiantibiotic proteins that provide immunity. Thus the diverse toxin production seen in cyanobacteria might best be explained as an identity system that allows competition via toxin–antitoxin modules. It is worth noting that terrestrial cyanobacterial *Nostoc* species are also often symbiotically associated with lichen, which are composed of a mixture of eukaryotic and prokaryotic cells. These lichen cyanobacterial species also make microcystins, six of which were specific to symbiotic species, including three that were new variants not seen in other species. Clearly this mixed-cell, symbiotic system must also be protected against any toxic actions of these microcystins. It therefore seems likely that such toxins–antitoxins themselves might act as addiction modules that stabilize the participation of all the cellular participants in this symbiotic state.

Gram-Positive Bacteria

Most of the oceanic bacteria considered above are gram-negative proteobacteria. Gram-positive bacteria also represent another major prokaryotic lineage found in many natural environments. Although this issue will not be developed in any detail, here too there is strong evidence of an important role of quorum-sensing operons that control group identity and the protein secretion system. However, QS system control in most gram-positive bacteria does not appear to use lactone-related autoinducing signals. Instead, gram-positive bacteria tend to use peptides for such signaling. The physiological results, however, are similar to the induction of cytolysins in the enterococci. For example, the dental bacteria *Streptococcus mutans* encodes genomic toxin–antitoxin modules similar to relBE and mazEF modules of *E. coli*. In *S. mutans*, the relBE genes are within the hsd-prr region, which contains an anti-T4 addiction module in *E. coli*. *S. mutans* expressing this addiction module make biofilms that are much more resistant to stressful environmental conditions. Also, a common low copy plasmid found in pathogenic antibiotic-resistant *S. pyogenes* (pSM19035) encodes an addiction toxin–antitoxin module, showing strong association with plasmid maintenance. Thus, although much less evaluated, gram-positive bacteria appear to also use some of the same general strategies of toxin–antitoxin addiction modules for their success.

Social Bacteria – Myxobacteria xanthus

The existence of some species of social bacteria, such as *M. xanthus*, presents us with an opportunity to examine in considerable detail the dilemma of social control of cell fates, division, differentiation and death in the context of a prokaryote and to also envision how this situation led to the evolution of metazoan eukaryotes. The main Darwinian dilemma is that individual cells are under social control and often exhibit altruistic-like behaviors in which 'non-selfish' behavior, including death of cells, is needed to attain the biological goal of a social or colonial organism. In this chapter, I have argued that group identification (addiction) systems are essential for any such group behavior. Furthermore, I have argued that persisting genetic parasites can often impose onto their host systems of persistence that involve toxin–antitoxin addiction modules which will also provide systems of group identity. This trinity of characteristics is normally inseparable. We will now consider if such a process can help us understand the origin and evolution of social bacteria. *M. xanthus* is a social soil bacteria that is capable of coordinated movement during nutritional stress that results in the coalescence of cells into colonial structures that will then form differentiated fruiting bodies and produce bacterial spores. *M. xanthus* is a non-flagellated, rod-shaped, gram-negative bacteria. During sporulation, most of the cells within the structure will subsequently die by a cytolytic process that resembles programmed cell death. Thus it is clear that much of the colonial population of cells is behaving in an altruistic way, similar to most multicellular organisms. In Chapter 2, I discussed how the presence of an addiction module can result in programmed cell death, as in the case of T4 and lambda. I noted that a colonized cell will self-destruct when perturbed by either another competing genetic colonizer or by a physiological situation that upsets the metastable addiction module (i.e., mitomycin C). Thus the induction of cell death can be an inherent part of a group identity system. Besides the formation of fruiting bodies and death of cells, *M. xanthus* is also capable of individual cell movement as it 'hunts' for food (bacteria). Cells leave the colony as individuals or swarms of individuals. *Myxococcus xanthus* does not cooperate with other species or genotypes of same species to form fruiting bodies. Also, *Myxococcus xanthus* has a large number of distinct social types. In one study that evaluated possible cooperation, nine strains were compared and paired with one competitor during starvation. It was observed that strong mutual antagonism seems to prevail in this situation that can reduce spore production by 90%, leading to extinction of some populations. The mechanism of this antagonism is unclear. Since Myxobacteria are lacking any known plasmids or conjugation systems, bacteriocins seem unlikely to be involved in such mutual antagonism. However, *xanthus* species do support temperate phage, although this phage DNA is unusual and circularly permuted.

Motility, Pilins and T/A Modules

The movement of individual *Myxococcus* cells during scavaging occurs by a process called A-motility (for adventurous) and uses a gliding mechanism very similar to the gliding system of cyanobacteria. Propulsion is attained by the expulsion and cationic hydration of a polyelectrolyte via a pore 'nozzle' at one end of the bacteria. The other mechanism of movement used during fruiting body formation is the S-motility (for social motility). Here, the cells need to be within 2 μm of each other, which is about the length of a thin unipolar type IV pili. S-movement occurs when the pili attaches to an adjacent cell and is then retracted, pulling the cell in a twitching motion towards its neighbor. Retraction occurs via pili disassembly following contact and signal transduction. The pili are able to extend from one end of the rod up to fivefold their length to bind to the surface matrix at the end of an adjacent cell, eliciting a C-signal sensory chemical excitation. This signal transduction system is homologous to the chemosensory system of *E. coli*. *E. coli* also has five families of MLP-family chemoreceptors (ligand-specific proteins on periplasma membrane). Interestingly, these receptors are also found on the end of the cell and appear to drive signal transduction to a flagellar motor, affecting swimming behavior. Although these receptors bind various sugars (food, taste), they also bind serine and respond to Tsr at high concentration, and therefore are not simply food receptors. *M. xanthus* uses a series of five chemical sensor systems (autoinducer-like) to initiate movement and control genetic programs for fruiting body formation in a series of 5-timed signal systems (involving amino acids). This signal transduction involves MglA, a Ras/Rab/Rho family protein which is a membrane-spanning tyrosine kinase. A schematic of this signal transduction process and its relationship to programmed cell death is shown in Fig. 4.5. With this process, we see a much more 'eukaryotic-like' transduction system which is used to initiate social control of this bacteria. The resulting spores are non-motile, which is an interesting contrast to motile unicellular zygotes of algae described later. Thus, one result of close contact in the colony is the initiation of programmed cell death in most of the cells. As previously mentioned, there

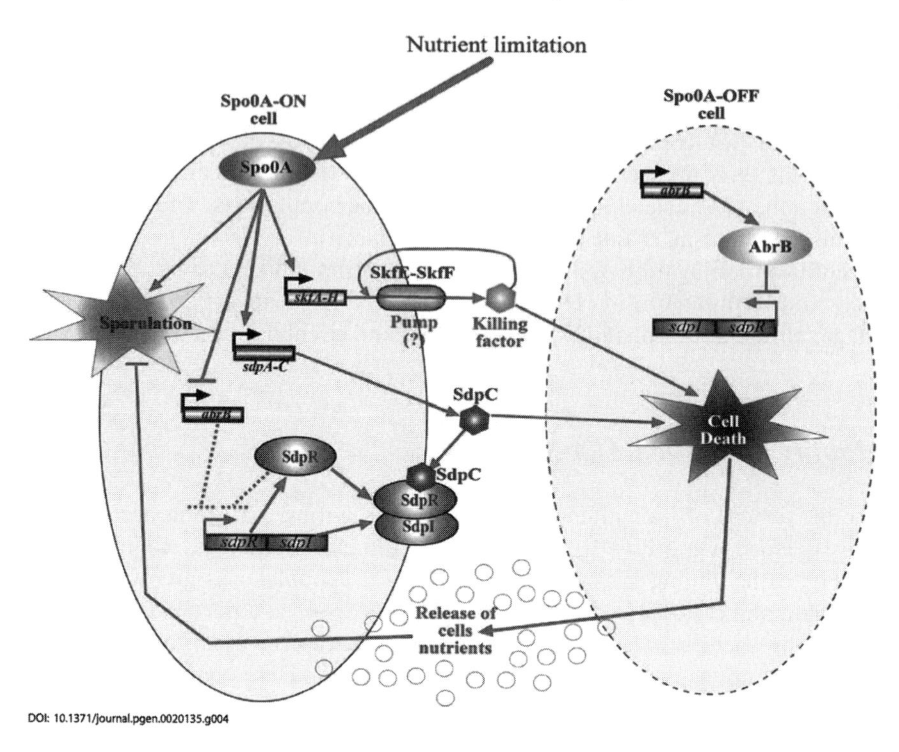

Fig. 4.5 Schematic of myxo cell death (reprinted with permission from : Plos Genetics (2006), Vol. 2, No. 10, e135)

is a fascinating link between type IV pili and mating with conjugative plasmid R64. Recall the R64 encodes seven different type IV pilin genes, used as adhesion surface molecules to identify mating specificity. These pilins are made from a precursor protein in which a highly conserved leader peptide (of about 20 aa) is removed via a leader peptidase. The N-terminal 140 aa of type IV pilin genes, however, is also recognized and conserved in another situation. Within this region is a heximeric ATPase domain that has been called the PIN domain (for Pil-N terminus). PIN domains are now thought to be T/A modules that can be found in all prokaryotic genomes. As mentioned previously, they are especially prevalent in the genomes of bacteria with long life spans (*M. tuberculosis*). These T/A domains, such as Maz EF, have been proposed to be induced during programmed cell death. A specific example of a PIN would be the RelE-like proteins. RelBE is a toxin–antitoxin gene pair that is also a transcription factor. However, RelE-like proteins are known for their ability to cleave RNA (mRNA in the ribosomal A site). More recently, it has become clear that genes related to RelE are abundant in eukaryotes and associated with RNA decay of various types (such as RNAi and dicer), a clear form of molecular recognition. Although fascinating, experimental analysis has not yet examined if these pili or T/A associations apply to *M. xanthus* and its social phenotype. *M. xanthus* has no know conjugative plasmids, such as R64. Curiously, plasmids are essentially unknown for any Myxobacterium species. However, *M. xanthus* is known to support various temperate phage. Interestingly, phage Mx8 has a recombination system very similar to that of R64, used to generate pili diversity. Mx8-like phage usually produce turbid plaques in their host as they lysogenize, but it is possible to isolate clear plaque mutants that no longer lysogenize. Interestingly, if mutants of *M. xanthus* are selected to be resistant to this lytic phage variant, these bacteria are also no longer social, motile or able to form fruiting bodies, but otherwise normal. The reasons for linkage between phage and social phenotype are not known, but it does establish that virus replication can have strong links to the host social phenotype.

The Dinoflagellates: The Guiding Light

Dinoflagellates are eukaryotes and accordingly have nuclei, chloroplasts and mitochondria. They are clearly distinct from the prokaryotes we have considered above. This group of oceanic organisms represents another simple light-producing, blooming, motile unicellular eukaryote, also know for their diverse and high level of toxin production. The toxic red tides (not to be confused with red algae) responsible for destruction of many fish and marine bivalve stocks is a consequence of this light-dependent toxin production by such blooms. The majority of dinoflagellate species are photosynthetic and free living and the remainder are symbiotic or parasitic species (distant relatives to the malaria organism). Yet dinoflagellates retain many life-style characteristics of ocean-dwelling prokaryotes. Also, dinoflagellates have several features which clearly distinguish them from all other eukaryotes, and are thus monophyletic. For one, they lack histone proteins, a basic characteristic of all other eukaryotic chromatin. Another rather amazing distinction is that they have very large genomes whose DNA contains large amounts of non-coding sequences. This DNA is highly methylated, or modified, and is permanently condensed and enclosed within a nuclear envelope. Nuclear division is thus 'closed,' in that the nuclear membrane does not dissolve during S-phase. The DNA is organized into a densely packed 'liquid crystal' state with filaments of actively transcribed DNA projecting out of the nucleus into the cytoplasm. Such nuclear organization and metabolism is unique amongst all eukaryotes and is probably also of direct relevance to the distinct relationship dinoflagellates have with virus. Although dinoflagellates are mostly unicellular marine organisms, they can aggregate in blooms and thus have some clear group behavior.

Dinoflagellate Viruses

Currently, only large DNA viruses that replicate and mature within the cytoplasm have been reported in dinoflagellates (i.e., HaV01, HcV, GM6, GM7).

Such viruses are common in coastal marine habitats. These viruses are generally lytic for growing but not static dinoflagellate host cells. Since dinoflagellates undergo most cell division at the end of the dark cycle, viral replication would also be linked to host light cycles. Such a circadian process also involves the photooxidation of melatonin, a process conserved in most eukaryotes. It also appears that lytic viruses of dinoflagellates are often involved in the termination of blooms. In natural settings, dinoflagellates appear highly susceptible to these lytic viruses and only 5% of isolates were seen to be resistant to the viral lysis. Persisting viruses of dinoflagellates, if they exist, have not yet been described. Thus a broad relationship to large, dsDNA cytoplasmic and lytic viruses applies to the dinoflagellates. In addition to large DNA viruses, *Heterosigma akashiwo* can be lytically infected by new family of picornavirus.

Dinoflagellate Virus-Like Plastids

Another major distinction of dinoflagellates is the structure and genetics of their plastids. For one, they are surrounded by three membranes and composed of limited sets of single minicircular DNAs corresponding to genes (including rDNA and photosystem II genes). These single-gene minicircular DNA seems a very odd genome for an early representative eukaryote. Nor do they resemble the genomes of any prokaryote. Many have proposed that these plastids have evolved from the reduction of an ancestral cyanobacterial genome. However, these minicircular plastids also have highly recombinogenic inverted terminal repeats, features which are not characteristic of cyanobacterial genomes but are clearly found in various viral genomes. Thus it is likely that they have evolved on multiple independent occasions from an ancestor that used a viral-like recombination-based DNA replication process, similar to that of some T-even phage.

Sex Cells and Eyespots

Since dinoflagellates are sexually reproducing and some can make cysts, they have evolved systems to differentiate the germ line from soma. Thus they produce differentiated sex cells. In addition, sexual reproduction also requires that they have evolved mechanisms to find and identify appropriate sexual partners. In this, it would appear that cellular motility is involved. An inherent feature of sexual reproduction is the need for a partner identity system. This identifies a major development in the evolution of cellular group identity, a sexual partner. In dinoflagellate, gametes indeed are motile and move towards and find each other, then bind to each other using a fusion pore that connects the two sex cells. This process must employ some form of cellular identity. In the context of dinoflagellates' oceanic biology, they show several interesting characteristics associated with light and sex (group) behavior, reminiscent of cyanobacteria. Thus, sensory light detection, sex and group behavior are all associated in dinoflagellates. They have a more complicated eyespot organelle than bacteria which bears a clear resemblance to the eyes of higher eukaryotes. These eyespots are directional (often referred to as light antennas) and display diverse morphology. Eyespots are usually associated with the flagella-based motility system and are often found near the base of the longitudinal flagellum. The evolution of their directional light antenna is a clear step towards evolution of a more advanced vision-based sensory system as seen in animals. An array of such antennae with different directionalities would constitute an early vision system, although dinoflagellates lack nervous cells needed to process any such multichannel information. Thus light detection and motility are physically linked in dinoflagellates. The eyespot photoreceptors are of the rhodopsin type, similar to that of cyanobacteria, but are also clearly more related (via 6-TMs) to those of higher eukaryotes. Some species, such as *reinhardtii*, also have archaeal-like (type I) rhodopsins (such as CSRA and CSRB, both light-gated proton channels), so dinoflagellates are the only eukaryotes to also have such archaeal and bacterial-like photoreceptors. In addition, it seems clear that distinct single transduction and regulators are used in photo motility. Here, it is interesting that catecholamines (DOPA/dopamine) are known to decrease dinoflagellate light sensitivity, suggesting a regulatory system that uses signal molecules that are much more like those of nervous cells of multicellular animals.

Blue Light Sex

Light is involved in the sexual behavior of dinoflagellate as well as in their sexual life cycle. *Chlamydomonas reinhardtii*, for example, uses blue light as a cue to control its sexual life cycle program. *Pelagia noctiluca* also undergoes light-dependent sexual differentiation, using a fusion pore to connect the two gametes. The sexual life cycle of other dinoflagellates, however, can be somewhat complex. *Nocticula scintillans* is an interesting luminescent marine dinoflagellate that can (for unknown reasons) also emit light flashes in a nearly simultaneous way from many cells following mechanical stimulation. This organism has two sexual stages, vegetative and swarmer. The swarmer stage follows a strict calcium-dependent cell division program that can result in synchronous sporulation of these sperm-like cells, which then sense and swim to multinucleate vegetative cells. Clearly such gamete motility resembles that of sperm, and like sperm, pheromone chemosensory cues are likely involved. However, here the pheromone-like sexual attractor remains undefined. In terms of the evolution of group identity, dinoflagellates have two clear patterns of group behavior: blooming behavior and sexual behavior. Both of these group behaviors involve motility and light. Although dinoflagellates are also renowned for toxin production, the possible involvement of toxin production in group behavior (either sex or bloom) has not been reported. Toxin is produced during blooming and also in some cells that are symbiotic with corals. One blooming dinoflagellate (*Pyrodinium*) has been reported to make up to five distinct toxins, most of them neurotoxins. Although such molecular diversity implies a role in group identity, this has not been experimentally established. However, in the context of cell communities, dinoflagellates appear to live essentially as communities of single cells and do not form more intimate social cell structures. Such social cellular associations are characteristics of metazoans presented in the next chapter.

Mechanical-Induced Light Production

As mentioned, some dinoflagellates (*Gonyaulax polyhedra*) can make light in response to mechanical stimulation (such as the wake of a boat). This light production is itself mediated by light detection as light is known to inhibit this mechanical induction. The inhibitory process involves light-gated proton channels. It is worth noting that this mechanical light response can be considered to represent the first known sensory system for detecting mechanical wave energy, related to evolution of sound detection. Since motility and light detection are linked, it is interesting to consider the group behavioral implications of such a sensory system. Clearly light production and detection have the capacity to control group motility. Interestingly, this same organism also shows some clear resemblance to the light-dependent regulatory systems as seen in multicellular animals. These dinoflagellates, like animals, make high levels of melatonin in a circadian manner, maximizing production at night and undergoing photooxidation during the day. Here it appears that melatonin may play a role as a component of a photosynthetic O_2 detoxifying system. Thus in the dinoflagellates we see for the first time the full trinity of sensory detection: light, smell and mechanical senses are all present in this oceanic microbe.

Diatoms and Toxic Tides

As mentioned, the terms red algae and red tide have historically been rather confusing, similar to the historically confused use of the term algae itself, which was previously inclusive of both photosynthetic prokaryotes and eukaryotes. Red tide designates the color of the toxic water produced by blooms of organisms like dinoflagellates. Red algae (seaweeds) are discussed in the next chapter. Red tide blooms are known for their poisonous affect on marine animals, hence there has been much interest in them from the aquaculture community, continuing the trend that blooming photosynthetic oceanic organisms tend to produce toxins. There is also the organism known by the common term of brown algae, or diatoms, involved in toxic tides. Diatoms are ecologically very important and highly species-diverse (estimated in the tens of thousands). These unicellular organisms are mostly sexually reproducing and belong to the class

Bacillariophycae. Diatoms are distinct for their ability to manipulate walls of silica into elaborate patterns, forming plates that surround the cell. Some are bilaterally symmetrical and can move by actin-mediated gliding over attached surfaces. Toxin produced, especially domoic acid and its chemical derivatives, is generally neurotoxic. Also, polyunsaturated aldehydes are also known, but these toxins, although pervasive, are much less diverse than that seen in the cyanobacteria. The relationship, if any, of toxins to diatom community structures is not clear. Diatoms encode both blue- and red-sensitive photoreceptors, but the relevance of these receptors to group behavior is also not known.

Diatoms are also evolutionary distinct from other algae, but with some clear relationship to the red algae and also *C. reinhardtii.* Like the red algae, they appear to have relatively small genomes (e.g., *Thalassiosira pseudonana* is 34 Mbp) and therefore they seem able to significantly limit the content of parasitic or repeat DNA elements. Unlike dinoflagellate presented above, their nuclei appear more typical of eukaryotic nuclei and undergo dissolution of the nuclear membrane, chromosome segregation and mitosis. Although diatoms are blooming and photosynthetic eukaryotes, there has been little study of mechanisms of immunity or group identity in these organisms. Since most diatoms are sexually reproducing and can make flagellated sperm cells, some form of gamete recognition must exist and the Sig1 gene is known to be involved. Eggs cells are encased in a silicate frustule which presents a physical barrier. Lytic DNA viruses that infect marine diatoms have recently been well established in the genus *Chaetoceros* and such infections may limit bloom formation by these diatoms. This virus (CsNIV) is a small ssDNA circo-like virus (6,000 bp) that replicates within the nucleus. This nuclear replication of a small DNA virus is in sharp contrast to the large cytoplasmic DNA viruses of dinoflagellates described above. How this small virus breaches or evades the silicate cell wall of its diatom host has not been reported. Nor is it known if the host has immune systems against these or other agents or if host sexual reproduction relates to virus susceptibility or if there are species-specific versions of virus. Interestingly, some dinoflagellates are parasitic to diatoms, but molecular details of this interaction are unknown. Plasmids have been reported in 5 of 18 diatom species surveyed. Curiously, these plasmids show some similarity to chloroplasts but are clearly distinct from them. For example, the DNA can have an ORF related to Tn3 resolvase, but unlike dinoflagellate, no inverted repeats are present in the sequence.

In summary, we see that the unicellular organisms of the oceans have formed extended community structures that have had tremendous impact on the Earth and its life. These communities are able to photosynthesize, produce O_2 and fix N_2, which transformed the world. But in so doing, these organisms were directly aided by the colonizing actions of viruses. O_2 and its ROS products were perhaps the ultimate toxins that contributed T/A-based addiction and identity modules, leading to the more complex symbiotic relationships. However, a basal role of ROS in cellular self-destruction was set from which other more complex cellular and organismal identities could evolve. These microorganisms developed all the basic molecular machinery for sensory detection, transduction and action (motility via flagella). Light detection and production in particular were used for social control purposes, but odor detection (pheromones) was a common media for group and sexual identification. Also developed was the use of self-destruction (programmed cell death) as a way to eliminate cells that lost their T/A-based identity. This very process led to the establishment of terminal differentiation in some bacteria. We have now set the stage with all the needed components for the evolution of multicellular eukaryotes to be presented in Chapter 5, which will thus extend these same issues and systems to consider the evolution of multicellular eukaryotes. Starting with red algae, to green algae, filamentous green algae, brown algae, fungi, sponges and early animals (nematodes), the evolution of their group identity will be presented. The main focus will remain on examining mechanisms of group identity, immunity from the context of relationships to genetic parasites.

Recommended Readings

Stromatolites

1. Hofmann, H. J. (1975). Australian Stromatolites. *Geol Mag* **112**(1), 97–100.
2. Hofmann, H. J., Grey, K., Hickman, A. H., and Thorpe, R. I. (1999). Origin of 3.45 Ga coniform stromatolites in Warrawoona Group, Western Australia. *Geol Soc Am Bull* **111**(8), 1256–1262.

Marine Photosynthesis, photosensory

1. Hess, W. R. (2004). Genome analysis of marine photosynthetic microbes and their global role. *Curr Opin Biotechnol* **15**(3), 191–198.
2. Hess, W. R., Rocap, G., Ting, C. S., Larimer, F., Stilwagen, S., Lamerdin, J., and Chisholm, S. W. (2001). The photosynthetic apparatus of *Prochlorococcus*: insights through comparative genomics. *Photosynth Res* **70**(1), 53–71.
3. Raymond, J., and Blankenship, R. E. (2004). The evolutionary development of the protein complement of photosystem 2. *Biochim Biophys Acta* **1655**(1–3), 133–139.
4. Sproviero, E. M., Gascon, J. A., McEvoy, J. P., Brudvig, G. W., and Batista, V. S. (2007). Quantum mechanics/molecular mechanics structural models of the oxygen-evolving complex of photosystem II. *Curr Opin Struct Biol* **17**(2), 173–180.
5. van der Staay, G. W., Moon-van der Staay, S. Y., Garczarek, L., and Partensky, F. (2000). Rapid evolutionary divergence of Photosystem I core subunits PsaA and PsaB in the marine prokaryote *Prochlorococcus*. *Photosynth Res* **65**(2), 131 139.
6. Montgomery, B. L. (2007). Sensing the light: photoreceptive systems and signal transduction in cyanobacteria. *Mol Microbiol* **64**(1), 16–27.
7. Nelson, N., and Yocum, C. F. (2006). Structure and function of photosystems I and II. *Annu Rev Plant Biol* **57**, 521–565.

Cyanobacteria

1. Dodds, W. K., Gudder, D. A., and Mollenhauer, D. (1995). The Ecology of Nostoc. *J Phycol* **31**(1), 2–18.

2. Montgomery, B. L. (2007). Sensing the light: photoreceptive systems and signal transduction in cyanobacteria. *Mol Microbiol* **64**(1), 16–27.
3. Muhling, M., Fuller, N. J., Millard, A., Somerfield, P. J., Marie, D., Wilson, W. H., Scanlan, D. J., Post, A. F., Joint, I., and Mann, N. H. (2005). Genetic diversity of marine *Synechococcus* and co-occurring cyanophage communities: evidence for viral control of phytoplankton. *Environ Microbiol* **7**(4), 499–508.
4. Vaulot, D., Marie, D., Olson, R. J., and Chisholm, S. W. (1995). GROWTH of *Prochlorococcus*, a photosynthetic prokaryote, in the equatorial Pacific-Ocean. *Science* **268**(5216), 1480–1482.
5. Zhang, C. C., Laurent, S., Sakr, S., Peng, L., and Bedu, S. (2006). Heterocyst differentiation and pattern formation in cyanobacteria: a chorus of signals. *Mol Microbiol* **59**(2), 367–375.

Cyanophage

1. Angly, F. E., Felts, B., Breitbart, M., Salamon, P., Edwards, R. A., Carlson, C., Chan, A. M., Haynes, M., Kelley, S., Liu, H., Mahaffy, J. M., Mueller, J. E., Nulton, J., Olson, R., Parsons, R., Rayhawk, S., Suttle, C. A., and Rohwer, F. (2006). The marine viromes of four oceanic regions. *PLoS Biol* **4**(11), e368.
2. Bailey, S., Clokie, M. R., Millard, A., and Mann, N. H. (2004). Cyanophage infection and photoinhibition in marine cyanobacteria. *Res Microbiol* **155**(9), 720–725.
3. Chen, F., and Lu, J. (2002). Genomic sequence and evolution of marine cyanophage P60: a new insight on lytic and lysogenic phages. *Appl Environ Microbiol* **68**(5), 2589–2594.
4. Douglas, A. E., and Raven, J. A. (2003). Genomes at the interface between bacteria and organelles. *Philos Trans R Soc Lond B Biol Sci* **358**(1429), 5–17; discussion 517–518.
5. Hambly, E., Tetart, F., Desplats, C., Wilson, W. H., Krisch, H. M., and Mann, N. H. (2001). A conserved genetic module that encodes the major virion components in both the coliphage T4 and the marine cyanophage S-PM2. *Proc Natl Acad Sci USA* **98**(20), 11411–11416.
6. Hambly, E., and Suttle, C. A. (2005). The viriosphere, diversity, and genetic exchange within phage communities. *Curr Opin Microbiol* **8**(4), 444–450.
7. Lindell, D., Sullivan, M. B., Johnson, Z. I., Tolonen, A. C., Rohwer, F., and Chisholm, S. W. (2004). Transfer of photosynthesis genes to and from *Prochlorococcus* viruses. *Proc Natl Acad Sci USA* **101**(30), 11013–11018.
8. Lindell, D., Jaffe, J. D., Johnson, Z. I., Church, G. M., and Chisholm, S. W. (2005). Photosynthesis genes in marine viruses yield proteins during host infection. *Nature* **438**(7064), 86–89.
9. Mann, N. H., Cook, A., Millard, A., Bailey, S., and Clokie, M. (2003). Marine ecosystems: bacterial photosynthesis genes in a virus. *Nature* **424**(6950), 741.
10. Mann, N. H., Clokie, M. R., Millard, A., Cook, A., Wilson, W. H., Wheatley, P. J., Letarov, A., and Krisch, H. M. (2005). The genome of S-PM2, a "photosynthetic" T4-type bacteriophage that infects marine *Synechococcus* strains. *J Bacteriol* **187**(9), 3188–3200.
11. Muhling, M., Fuller, N. J., Millard, A., Somerfield, P. J., Marie, D., Wilson, W. H., Scanlan, D. J., Post, A. F., Joint, I., and Mann, N. H. (2005). Genetic diversity of marine *Synechococcus* and co-occurring cyanophage communities: evidence for viral control of phytoplankton. *Environ Microbiol* **7**(4), 499–508.
12. Pope, W. H., Weigele, P. R., Chang, J., Pedulla, M. L., Ford, M. E., Houtz, J. M., Jiang, W., Chiu, W., Hatfull, G. F., Hendrix, R. W., and King, J. (2007). Genome sequence, structural proteins, and capsid organization of the cyanophage syn5: a "horned" bacteriophage of marine *Synechococcus*. *J Mol Biol* **368**(4), 966–981.

13. Sullivan, M. B., Coleman, M. L., Weigele, P., Rohwer, F., and Chisholm, S. W. (2005). Three *Prochlorococcus* cyanophage genomes: signature features and ecological interpretations. *PLoS Biol* **3**(5), e144.
14. Sullivan, M. B., Lindell, D., Lee, J. A., Thompson, L. R., Bielawski, J. P., and Chisholm, S. W. (2006). Prevalence and evolution of core photosystem II genes in marine cyanobacterial viruses and their hosts. *PLoS Biol* **4**(8).
15. Yoshida, T., Takashima, Y., Tomaru, Y., Shirai, Y., Takao, Y., Hiroishi, S., and Nagasaki, K. (2006). Isolation and characterization of a cyanophage infecting the toxic cyanobacterium Microcystis aeruginosa. *Appl Environ Microbiol* **72**(2), 1239–1247.

Bioluminescence

1. Hastings, J. W., and Nealson, K. H. (1977). Bacterial bioluminescence. *Annu Rev Microbiol* **31**, 549–595.
2. Meyer-Rochow, V. B. (2001). Light of my life–messages in the dark. *Biologist (London)* **48**(4), 163.
3. Nealson, K. H., and Hastings, J. W. (1979). Bacterial bioluminescence: its control and ecological significance. *Microbiol Rev* **43**(4), 496–518.
4. Nieto, C., Pellicer, T., Balsa, D., Christensen, S. K., Gerdes, K., and Espinosa, M. (2006). The chromosomal relBE2 toxin-antitoxin locus of *Streptococcus pneumoniae*: characterization and use of a bioluminescence resonance energy transfer assay to detect toxin-antitoxin interaction. *Mol Microbiol* **59**(4), 1280–1296.
5. O'Kane, D. J., and Prasher, D. C. (1992). Evolutionary origins of bacterial bioluminescence. *Mol Microbiol* **6**(4), 443–449.
6. Rees, J. F., de Wergifosse, B., Noiset, O., Dubuisson, M., Janssens, B., and Thompson, E. M. (1998). The origins of marine bioluminescence: turning oxygen defence mechanisms into deep-sea communication tools. *J Exp Biol* **201**(Pt 8), 1211–1221.
7. Salmond, G. P., Bycroft, B. W., Stewart, G. S., and Williams, P. (1995). The bacterial 'enigma': cracking the code of cell-cell communication. *Mol Microbiol* **16**(4), 615–624.
8. Schauder, S., Shokat, K., Surette, M. G., and Bassler, B. L. (2001). The LuxS family of bacterial autoinducers: biosynthesis of a novel quorum-sensing signal molecule. *Molecular Microbiology* **41**(2), 463–476.
9. Shimomura, O. (1985). Bioluminescence in the sea: photoprotein systems. *Symp Soc Exp Biol* **39**, 351–372.
10. Timmins, G. S., Jackson, S. K., and Swartz, H. M. (2001). The evolution of bioluminescent oxygen consumption as an ancient oxygen detoxification mechanism. *J Mol Evol* **52**(4), 321–332.
11. Visick, K. L., and Ruby, E. G. (2006). Vibrio fischeri and its host: it takes two to tango. *Curr Opin Microbiol* **9**(6), 632–638.
12. Warrant, E. J., and Locket, N. A. (2004). Vision in the deep sea. *Biol Rev Camb Philos Soc* **79**(3), 671–712.
13. Wilson, T., and Hastings, J. W. (1998). Bioluminescence. *Annu Rev Cell Dev Biol* **14**, 197–230.

Dinoflagellates

1. Nagasaki, K., Tomaru, Y., Tarutani, K., Katanozaka, N., Yamanaka, S., Tanabe, H., and Yamaguchi, M. (2003). Growth characteristics and intraspecies host specificity of a large

virus infecting the dinoflagellate Heterocapsa circularisquama. *Appl Environ Microbiol* **69**(5), 2580–2586.
2. Onji, M., Nakano, S., and Suzuki, S. (2003). Virus-like particles suppress growth of the red-tide-forming marine dinoflagellate Gymnodinium mikimotoi. *Mar Biotechnol (NY)* **5**(5), 435–442.
3. Regel, R. H., Brookes, J. D., and Ganf, G. G. (2004). Vertical migration, entrainment and photosynthesis of the freshwater dinoflagellate Peridinium cinctum in a shallow urban lake. *J Plankton Res* **26**(2), 143–157.
4. Soyer, M. O. (1978). Virus like particles and trichocystoid filaments in dinoflagellates. *Protistologica* **14**(1), 53–58.
5. Wilson, W. H., Francis, I., Ryan, K., and Davy, S. K. (2001). Temperature induction of viruses in symbiotic dinoflagellates. *Aquat Microb Ecol* **25**(1), 99–102.

Social Myxobacteria

1. Berleman, J. E., and Kirby, J. R. (2007). multicellular development in *Myxococcus xanthus* is stimulated by predator-prey interactions. *J Bacteriol* **189**(15), 5675–5682.
2. Brown, N. L., Burchard, R. P., Morris, D. W., Parish, J. H., Stow, N. D., and Tsopanakis, C. (1976). Phage and defective phage of strains of *Myxococcus*. *Arch Microbiol* **108**(3), 271–279.
3. Fink, J. M., and Zissler, J. F. (1989). Defects in motility and development of *Myxococcus xanthus* lipopolysaccharide mutants. *J Bacteriol* **171**(4), 2042–2048.
4. Goldman, B. S., Nierman, W. C., Kaiser, D., Slater, S. C., Durkin, A. S., Eisen, J. A., Ronning, C. M., Barbazuk, W. B., Blanchard, M., Field, C., Halling, C., Hinkle, G., Iartchuk, O., Kim, H. S., Mackenzie, C., Madupu, R., Miller, N., Shvartsbeyn, A., Sullivan, S. A., Vaudin, M., Wiegand, R., and Kaplan, H. B. (2006). Evolution of sensory complexity recorded in a myxobacterial genome. *Proc Natl Acad Sci USA* **103**(41), 15200–15225.
5. Gyohda, A., Furuya, N., Ishiwa, A., Zhu, S., and Komano, T. (2004). Structure and function of the shufflon in plasmid R64. *Adv Biophys* **38**, 183–213.
6. Kaiser, D. (2006). A microbial genetic journey. *Annu Rev Microbiol* **60**, 1–25.
7. Magrini, V., Salmi, D., Thomas, D., Herbert, S. K., Hartzell, P. L., and Youderian, P. (1997). Temperate *Myxococcus xanthus* phage Mx8 encodes a DNA adenine methylase, Mox. *J Bacteriol* **179**(13), 4254–4263.
8. Mignot, T., Merlie, J. P., Jr., and Zusman, D. R. (2007). Two localization motifs mediate polar residence of FrzS during cell movement and reversals of *Myxococcus xanthus*. *Mol Microbiol* **65**(2), 363–372.
9. Muller, S., Shen, H., Hofmann, D., Schairer, H. U., and Kirby, J. R. (2006). Integration into the phage attachment site, attB, impairs multicellular differentiation in Stigmatella aurantiaca. *J Bacteriol* **188**(5), 1701–1709.
10. Ruiz-Vazquez, R., and Murillo, F. J. (1984). Abnormal motility and fruiting behavior of *Myxococcus xanthus* bacteriophage-resistant strains induced by a clear-plaque mutant of bacteriophage Mx8. *J Bacteriol* **160**(2), 818–821.
11. Sun, H., Zusman, D. R., and Shi, W. (2000). Type IV pilus of *Myxococcus xanthus* is a motility apparatus controlled by the frz chemosensory system. *Curr Biol* **10**(18), 1143–1146.
12. Zusman, D. R., Scott, A. E., Yang, Z., and Kirby, J. R. (2007). Chemosensory pathways, motility and development in *Myxococcus xanthus*. *Nat Rev Microbiol* **5**(11), 862–872.
13. McDaniel, L., Houchin, L. A., Williamson, S. J., and Paul, J. H. (2002). Lysogeny in marine *Synechococcus*. *Nature* **415**(6871), 496.

Chapter 5
Animal Group Identity: From Slime to Worms, Emergence of the Brain

Goals of This Chapter, from Slime to Worms

The prior two chapters have presented many details concerning mechanisms used by unicellular (mainly oceanic) organisms to attain group identification. Quorum-sensing systems, T/A addiction modules (often redox based), ion channel toxins and rhodopsin sensory receptors were all evaluated, especially regarding how such systems can affect single or group cellular motility of flagellate or gliding organisms. Genomic and epigenomic parasites were often involved in these systems. Also, all of these molecular systems are often found to be associated with flagella as a focus of sensory receptors and motility effectors. In this chapter, I will examine how these same processes have been involved in, been modified by or contributed to the evolution of motile multicellular organisms. Since it is our final aim to understand human systems of group identification, our pathway will stay focused mainly on organisms that have clear relationship to human ancestors, initially centered on early multicellular eukaryotes. I will, however, examine well-studied model systems in greater detail for it is only in those systems that we will often have sufficient knowledge to make a specific or rational evaluation. In the context of a motile multicellular metazoan, this chapter will start its journey with red algae but aim to understand the origin of the group and social behavior of the nematode worm, *Caenorhabditis elegans*. The main innovation in *C. elegans* that will be considered will be the nervous system. Along the way, I will examine the various known mechanisms of cellular identity and immunity. Special attention will be paid to the role of 7 TM membrane receptors in sexual and group behavior. And as always I will consider in detail the possible and fundamental role of genetic parasites in the origin of these processes.

The evolutionary transitions that we will examine will often seemingly appear to represent rapid, non-Darwinian acquisition of phenotypic complexity (such as neurons). In addition, these systems will often also involve seemingly harmful events, such as the example of the acquisition of self-destructive capacity of cellular apoptosis. These complex acquisitions will be presented from the perspective of a genetic colonization. I will also evaluate the possible role of T/A addiction modules in the identity and stability of the cells involved.

L.P. Villarreal, *Origin of Group Identity*, DOI: 10.1007/978-0-387-77998-0_5,

As we presented in the prior chapter, the microbiological world has many examples of social and group behavior. Often light and small-molecule membrane receptors are directly involved, which were part of a quorum-sensing group behavior system. Thus the origin of light or small diffusible molecules and self-sensing systems, signal transduction, transcriptional regulators, the role of ion channels and flagella are all connected with group behavior and traceable to prokaryotes. The relationship of these same systems to bacterial colony (group, biofilm) formation, host colonization and host virulence has already been presented. These same molecular issues will continue to be examined as we now trace the origin of motile metazoans. As in the prokaryotes, the relationship of group identification, group sensing systems and sex will remain a central focus. In fact, it appears that sexual identification frequently provides the earliest examples of group identification systems that later become adapted and are used in tissues and for other purposes. This chapter thus seeks to provide the mechanistic foundation by which metazoans originated their various systems of group identification and provide a trail head for understanding the origin of systems used by humans.

In the last chapter we did consider some simple eukaryotes, especially dinoflagellates, from the perspective of group identification. Normally, eukaryotes are not considered along with prokaryotes in most evolutionary analysis. The eukaryotic nucleus is considered to represent a big difference between prokaryotes and eukaryotes, thus regulatory systems will often be distinct. Dinoflagellates, red algae and hypotrichs, however, all appear to represent separate lineages of an independent early eukaryotic nucleus. However, these lineages were not equally associated with the evolution of multicellularity. See the phylogenetic dendrogram tracing animal evolution based on SSU rRNA in Fig. 5.1. In this dendrogram, dinoflagellates, brown algae and red algae (Rhodophyta) are independent and predate green algae, plants, fungi and animals (our focus of interest). However, according to fossil evidence, only red algae are well represented as an early eukaryote since coralline red algae deposit calcite in their cell walls resulting in extensive fossilization. Thus, these three primitive orders of eukaryotes may represent early but independent origins, not the derived relationship as shown in the dendrogram. In the case of dinoflagellates, besides their distinct nuclear structures (no nucleosomes), many bacteria-like group characteristics seem to apply. For example, flagellated motility, light production and light taxis, bloom formation and the production of many types of toxins are all characteristics we saw in many bacteria. Dinoflagellates, however, like many cyanobacteria, do participate in various symbiotic relationships and symbiosis also raises the issue of how organisms identify themselves and their partners. The general problem is how to make cells recognize and cooperate with each other. I have presented the idea that T/A, or addiction modules, can provide the systems that compel organisms into mutual living states and kill off the 'uncolonized' individual. In a sense, sex also presents a similar 'symbiosis' problem, in that two haploid gamete cells must cooperate to make a fused diploid progeny. It is worth recalling that bacteria can and do participate in sexual exchange, but that this is

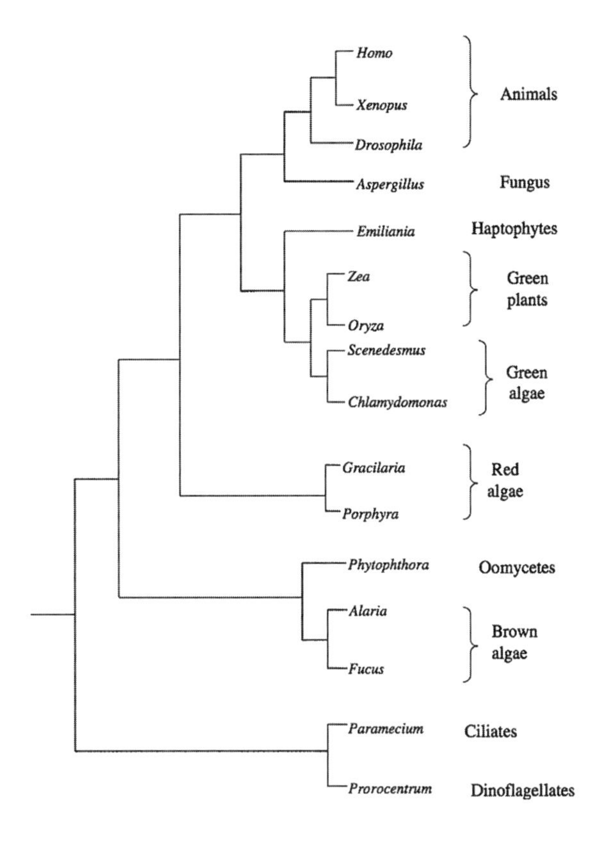

Fig. 5.1 Dendrogram of the evolution of animals (reprinted with permission from: Donald Kapraun (1995), Annals of Botany, Vol. 95, pg. 7–44)

generally mediated by a persisting genetic parasite and also frequently involves addiction modules as a system of identity. A perturbation of this T/A module by an attempted colonization of a 'non-identity' sex partner results in the self-destruction of the cell. This, it was proposed, was the underlying mechanism by which molecular altruism can be imposed onto an individual organism. Such a strategy should also apply to multicellular eukaryotes and allow the social control over fates of individual cells of a multicellular organism. However, if an addiction module can operate at a distance such as via light or by diffusible toxins, it can coordinate a loose association of individual (i.e., blooming) microorganisms. But, if addiction modules operate by contact, via cell adherence, communication or cell to cell destruction, they could then provide the basis of contact determined by social addiction or multicellularity. The issues of self-immunity and self-harm are thus central to the possibility of T/A-mediated social control.

At this point, a redefinition of group identity that is now inclusive of symbiosis can be considered. Group identity can include stable associations of cells and genetic entities that have distinct evolutionary histories. This includes symbiosis. Recall that I had proposed a trinity of concepts stemming from the P1 persistent phage exemplar. A genetic parasite represents a new non-ancestral genetic identity that must be superimposed onto its host. This new identity is stabilized (persistent) with the aid of an addiction module, involving a parasite-derived harmful but stable component along with a parasite-derived protective but unstable component. Thus, stability along with new identity and immunity are all acquired from the colonizing agent. To apply this concept to symbiosis, the colonizing agent becomes a cell, not simply a genetic parasite (similar to the p64 exemplar presented earlier). In this case, the colonizing cell produces toxins and the host can produce antitoxins. Thus the toxin and antitoxin components can be provided by the symbiotic cell pairs. In the case of a photosynthetic symbiont, the toxin would be the product of photosynthesis, the highly oxidizing O_2 (from a cyanobacterial genome), and the antitoxin could be reductive systems from the nuclear genome (or in the case of mitochondria, another symbiotic cell that consumes O_2). The early evolution of dinoflagellates also suggests that they were present in an early Earth, where the habitat was O_2 poor. Thus, photosynthetic O_2 production would have been a generally toxic product at this time. The high tendency for dinoflagellates to form symbiotic association with various other cell types (and also produce light) may thus stem from their early biological strategy involving photosynthetic- and O_2-based addiction modules.

Early Ingredients for Multicellularity

As we contemplate the origin of motile metazoans, let us now tabulate identification systems and mechanisms that would likely have been present prior to the evolution of worms and evaluate if any of these systems might have contributed. At such an ancient time (circa 1.5 billion ybp), all life appears to have been oceanic. The paths of the life forms we will be tracing became distinct by about 600 million years ago. In the existing habitat, we know the living world is predominantly bacterial. There was also a large quantity and diversity of virus, prophage, defectives and epiparasites, mostly of the large dsDNA type. These systems impose a large and diverse variety of identity (T/A) modules onto their hosts, but it is also likely that a large array of QS systems, most using diffusible small molecules (some being toxins), are part of the extant social microbiology. Social structures are represented via colony formation, biofilms (slime) and some differentiation (spore-forming fruiting bodies) systems. Although programmed cell death has already been developed (as in prophage addiction modules), bacterial cells for the most part have not employed this process to generate multicellular groups and its use is primarily to destroy cells acquiring foreign genetic identities (immunity). Light production and light-directed motility are also generally under QS control and the target of this control is either flagella or gliding motility systems. Photosynthetic O_2 production is abundant (with H_2O acceptor) and it is now possible to employ the highly oxidizing nature of O_2 (and ROS) as a T/A module. The highly destructive and non-specific character of this oxidation might also require the evolution of terminal differentiation as a way to isolate and use non-specific toxins for group identity (immunity). Programmed cell death can thus now be promoted by ROS. However, in extant prokaryotes, ROS use is not common in PCD. Instead, the molecular ID systems employ immunity systems that are composed of other modules, such as R/M and other prevalent T/A systems (i.e., RelB RNAse), as well as more specific toxins (holins, pores proteins and antiholins, antipores). Protean secretion is used to transmit pore toxins between nearby cells. Rhodopsin-like sensory (7 TM membrane) receptors are highly abundant and capable of transducing signals via changes in Ca++ levels and can affect transcription factor activity and expression programs. There is also a large abundance of toxins specifically targeting ion pore proteins. A relationship between sensory detection (light, pheromone) and sexual behavior is prevalent. Most of these prokaryotic ID systems operate on a single-cell basis. This ancient oceanic bacterial habitat is rich in large dsDNA parasites, but poor in other types of genetic parasites such as retroviruses (or pararetroviruses) and also dsRNA viruses or any negative strand RNA viruses. The extant bacterial genomes are gene-rich although most are also colonized by parasites (proviruses, defectives proviruses, satellite viruses and parasites of genetic parasites). Prokaryotic genomes do not harbor the *large* numbers of retroparasites (products of RT) as will become common in many eukaryotes (although some retrointrons are found, e.g., cyanobacteria). The existing bacterial communities do not have common cytoplasmic networks suitable for the transmission of genetic parasites, thus there are no examples of strictly intracellular viruses that transmit via common cell network connections. However, bacterial sex pilis might function in parasite transmission.

As we will see, not all of these extant immunity systems are to be employed by motile metazoans. R/M systems, for example, so highly prevalent in the prokaryotic world, are not used in any metazoans, except for those found in DNA viruses of unicellular green algae. However, the RelB system is retained in eukaryotes. Oxidation of PUFA (polyunsaturated fatty acid) becomes a common system for early metazoan innate immunity. However, with the symbiotic acquisition of the mitochondria, multicellular eukaryotes now employ pore toxins for the disruption of mitochondrial oxidation systems and evolve a prevalent system of ROS production and apoptosis able to control cell identities and fates. With the evolution of the nucleus, DNA prophage, so common to all prokaryotes, virtually disappears from the genomes of eukaryotes (with the possible exception of filamentous brown algae). Also, DNA modification (methylation), so prevalent as a genome identifying system for the DNA viruses of prokaryotes, becomes adapted in eukaryotes to silence a new type of genetic parasites (endogenous retroviruses and retrotransposons) and control the epigenetic fate of cells by silencing transcription. Endogenous retroviruses and retroposons and their silencing control systems (RNAi) also become available to control viral immunity and multicellular cell fates. This same RNAi system (along with apoptosis) is adapted in worms to evolve nerve cell terminal differentiation. This RNAi/apoptosis system becomes linked to sensory detectors (such as olfaction) which can alter cell expression programs and cell fates, providing a cellular and molecular basis for learning and memory. All of the identity systems retain the essential strategy of working as elements of an addiction (T/A) system to define group identity. These processes thus become adapted in eukaryotes to provide complex systems promoting multicellular life strategies. Thus we may now start to understand why all metazoan genomes are colonized by their own peculiar versions of genetic parasites. They are not simply genetic residue of selfish replicators, but the sources of new host identities. I am proposing that the origin or acquisition of new genetic identity and group identity is fundamentally intertwined. Below I present a chain of evidence and reasoning that leads to this generalization.

Early Clonal Multicellular Eukaryotes

The Red Algae

The red algae (Rhodophyta) will be examined first since the fossil record indicates that they are likely the oldest of multicellular eukaryotes in paleontology. It is not likely that red algae are well known to most readers as the majority of extant seaweed is brown algae. The one exception to this is the commercially produced red algae used in sushi (*Porphyra*). Based on rRNA and other sequence analysis, red algae can be clearly differentiated from green and brown algae and dinoflagellates (sometimes called golden algae, see Fig. 5.1). Rhodophyta are basal to the animal and plant eukaryotes and a sister group to green algae, brown algae and dinoflagellates. Multicellular Rhodophyta appear to represent the first marine 'plant' life form. The origin of red algae appears to be independent of that of animals or green plants and is related to the basal lineage that evolved into the green algae and fungi. Red algae represent 700 genera and 6,000 species. The Bangiophycidae subclass or red algae are generally unicellular and considered basal to the Florideophycidae which are mostly multinucleate and more species rich. Unicellular red algae generally have small genomes (16 Mbp) and small flat nuclei. A multinucleate red algae is shown in Fig. 5.2. In some species, nuclei are organized into hexagonal arrays in a common cytoplasm, so they differ from dinoflagellates significantly in this. Red algae, like dinoflagellates, also possess light-harvesting plastids but these have distinct chromophores (phycoerythrin). Red algae members range from unicellular red algae to rather large seaweeds (such as commercially grown *Porphyra*). Red algae are unique in several other features. Although they are sexually reproducing, cycling through haploid and diploid stages, they lack flagellated gametes or other motile cells (this is in contrast to both dinoflagellates and green algae). Therefore, they are sessile organisms that must stay put and successfully oppose the genetic parasites and predators that are so prevalent in the oceans. As will be discussed below, in this sessile lifestyle they resemble filamentous fungi (and sponges). For this they

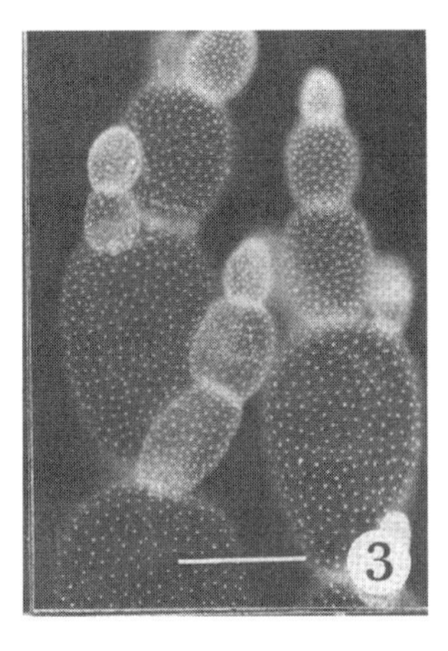

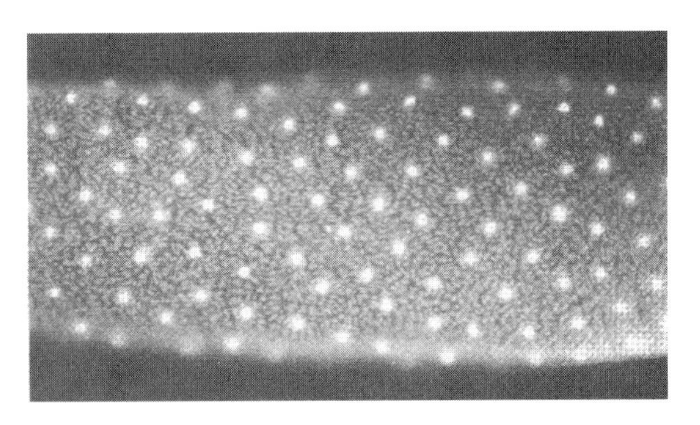

Fig. 5.2 Multinucleate cells of red algae. www.jochemnet.de/fiu/bot4404/BOT4404_23.html (reprinted with permission from: Biology of Red Algae (1990))

appear to have evolved an effective innate immune system, described below, which may provide a basal strategy for immunity of all multicellular organisms. Many red algae also have the curious feature of harboring transmissive (parasitic) nuclei, that is, nuclei that can move from one cell to another and reproduce in the new 'host' cell often via a 'pit-plug' connecting structure (see Fig. 5.2). The ability of red algae to transmit nuclei is shown schematically in Fig. 5.3. This nuclear transmission is used in both sexual process and cellular differentiation, and as a system to parasitize other red algal networks (mycels). A cell parasitized by small nuclei is shown schematically in Fig. 5.3 and a cell with a mixture of large and small parasitic nuclei is shown in Fig. 5.4. In sexual reproduction, nuclear fusion (transmission) occurs post-fertilization. Since red algae represent such an early eukaryote, they can also be used to consider the origin of the nucleus itself. In fact, it is felt that the complex regulatory control that a eukaryotic nucleus allows was essential for the evolution of multicellular organisms. The origin of the nucleus represents the greatest conundrum for evolutionary biology. A eukaryotic nucleus has many distinct and complex characteristics not found in any prokaryote that might have been a direct ancestor (membrane-bound chromosomes, chromatin-bound linear chromosomes, cell cycle control, pore complexes, distinct system of DNA initiation control and chromosome segregation, complex RNA processing, etc.). Therefore, explaining the origin of all these complex and coordinated systems has posed a serious problem. In my previous book, I outlined the arguments that support the idea that a persisting cytoplasmic large DNA virus could have provided essentially all of the complex characteristics needed to have originated the nucleus. If viruses can indeed represent the origin of the nucleus, then organisms like red algae, which represent some of the earliest eukaryotes, should also provide evidence that supports the 'viral origin' hypothesis. Indeed, they do

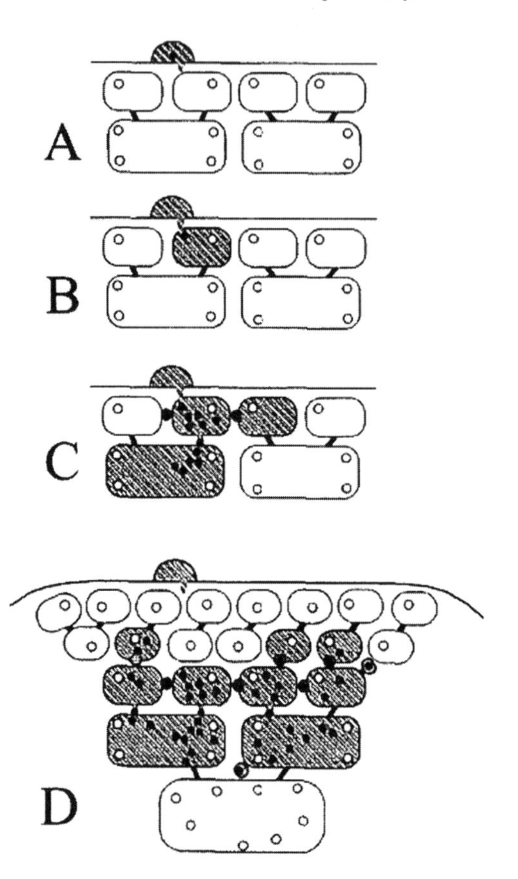

Fig. 5.3 Red algae as nuclear parasites (reprinted with permission from: Goffa, Coleman (1995), The Plant Cell/Plant Physiology, Vol. 7)

provide some support for such a view. As mentioned, the red algae are distributed into two major subclasses, Bangiophycidae (3–4 orders, mainly unicellular) and Florideophycidae (14 orders, mainly multinucleate). The Florideophycidae are monophyletic but have two distinct pit-plug features or groups that differ in the plug cap layer versus cap membrane. Thus it is the system that is directly involved in nuclear transmission that best defines the host classification and complexity. Since transmitted nuclei can also be parasitic and are able to degrade the resident nucleus over replication of the newly transmitted nucleus, these features (which are found in no other eukaryote) are clearly consistent with an infectious and possibly viral origin. In addition, red algae show other features that suggest strong viral involvement. For example, a search of the *Porphyra* EST database turns up a surprisingly large number of viral-like expressed genes. These include an array of large DNA viruses of Halobacteria, *Chlorella*, *Ectocarpus*, insect nuclear polyhedrosis virus, amphibian granuloviruses and mammalian herpes viruses. The algae-like genes include viral elongation factors, viral regulatory proteins, capsid proteins and DNA polymerases. Some viral-like proteins from RNA viruses are also seen including proteases and RNA-dependent RNA

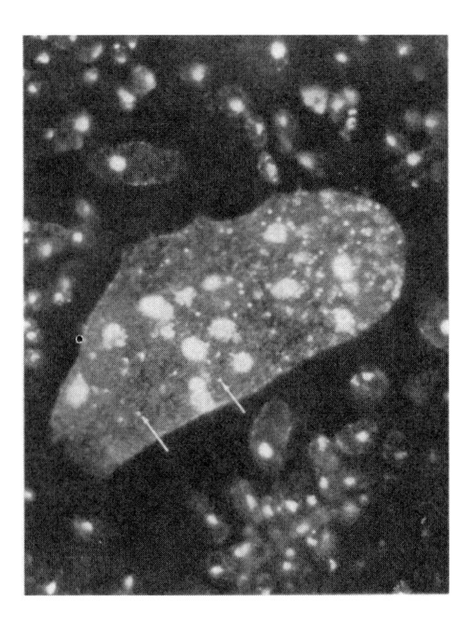

Fig. 5.4 Red algae pictured with multiple large and small nuclei (reprinted with permission from: Goff, Coleman (1984), Proc. National Academy of Sciences, Vol. 8)

polymerases from picornaviruses and receptors from various mammalian retroviruses. Some red algae are also known to make viral-like particles (Fig. 5.5) as discussed further below. Some investigators are likely to question the significance of such similarities, regardless of the surprisingly large numbers involved. The likely consensus is that such similarities most probably represent examples of viruses having 'lifted' host genes for their own purposes, not the converse. Two points strongly counter such a belief. In many cases, viral genes (not other cellular genes) are the most similar to these viral-like host genes, yet some of these viruses infect hosts that are only distantly related to red algae. More importantly, these viral genes include some clear examples of 'core' viral proteins, such as capsids, polymerases and basal regulatory proteins. These genes are clearly of a viral origin and they are the very genes used to assign and trace viral lineages. Thus the occurrence of so many viral-like genes supports a strong viral role in the origin of the red algae nucleus.

In general, nuclei of red algae have small chromosomes. Recently, a very large DNA virus infecting *Paramecium* has been described, the 1.2 Mbp mimivirus. This virus has a genome that is in excess of the size of the genome of some red algae and encodes more proteins, including genes previously thought to be only found in cells (tRNA, translation and metabolic proteins). Another curious and possibly viral-like nuclear feature is the existence of what has been called 'enslaved' nuclei by other eukaryotes. The 'enslaved' cryptomonad red algae nucleus is very small (only 551 kbp of DNA), thus unlike the dinoflagellates, its genome harbors much less non-coding DNA. Given the viral-like nature of red algal nuclei described above, however, the concept 'enslaved' might need to be replaced by the concept of 'stable nuclear

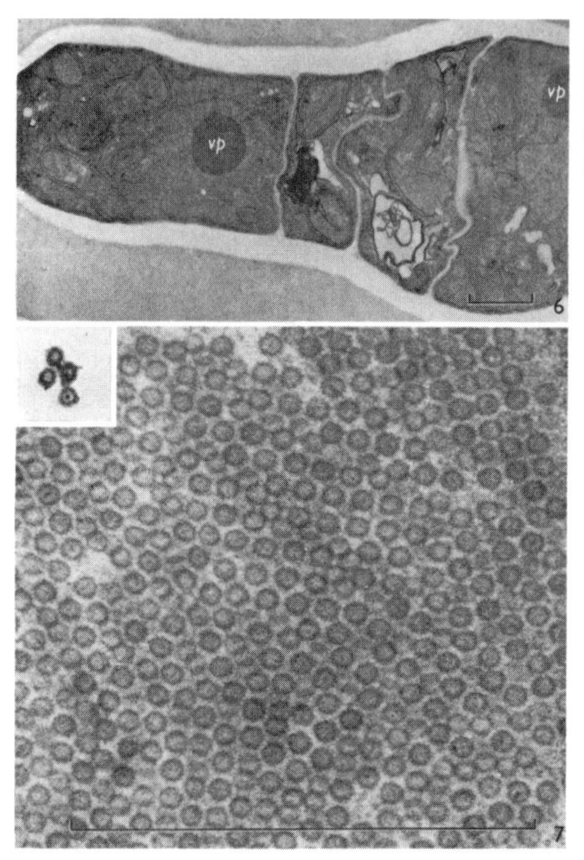

Fig. 5.5 Virus-like particles seen in red algae (reprinted with permission from: RE Lee (1971), Journal of Cell Science, Vol. 8, pg. 623–631)

colonization'. This red algal genome is actually significantly smaller than that of the largest DNA virus (mimivirus of protozoa). *Cyanidioschyzon merolae* is a unicellular red algae that is found in hot acidic waters. The symbiotic cryptomonad nucleus is present within chlorarachnean host algae. Given the transmissive nature of these nuclei, it seems more likely that this symbiotic red algal nucleus resulted from the stable transmission from an ancestral red algae into its green algae host. It would thus be very interesting to understand the relationship between red algae and their viral and genomic genetic parasites (see below).

Red Algae Group Identity

In considering the lifestyle of red algae, much attention has been focused on both the free-living unicellular algae and also the symbiotic algae. For example, sequencing the genome of *C. merolae* has been the major focus of study. The main interest of this chapter, however, is to consider the systems of group identity and immunity. These are essentially the same mechanisms that allow a multicellular organism to function as a group. For this, *Porphyra* and other multicellular red algae are therefore of more interest as these are multinucleate mycelar organisms. *Porphyra* has two morphologically distinct life phases: a free-living gametophyte and filamentous sporophyte. Formation of the multicellular structure must involve some type of self-recognition systems. Nuclear transmission is used during both sexual reproduction and parasitic colonization of other red algal networks. Individuals can live as extended filamentous structures, or mycels, and can contain hundreds of nuclei, acquired and replicated from neighboring red algae. Since the resident or acquired nuclei can clearly be distinguished and are often degraded, this raises an interesting question of how red algae identify their own nuclei and how nuclear colonization is regulated. Unfortunately, little is known concerning this issue. No relevant molecular studies have been reported. In terms of group identity, however, some information has recently come to light. The degradation of unallowed, unrecognized or displaced host nuclei happens frequently between encounters of red algae, and fusion post-fertilization appears related to this issue. Clearly the identity of nuclei is important and recognized.

Innate Immune System, Lipid Toxins and ROS as a Foundation

Sessile multicellular red algae must clearly prevent predation and colonization by genetic parasites, and thus they should have some type of effective innate immune system. As noted, unlike prokaryotes, restriction/modification systems appear to be absent, so the question of how they oppose genetic parasites is not resolved. The nucleus itself might be involved somehow, by providing an internal membrane/pore-bound barrier to parasite access to the genome, requiring nuclear transport through pore complexes. Thus, a potential red algae prophage cannot simply inject its DNA and expect that it will attain access to the host genome. In addition, unlike the cyanobacteria and dinoflagellates, red algae do not appear to code for the production of a diverse set of known toxins and antitoxins (such as Uma2 T/A modules of cyanobacteria) which could also provide identity modules. However, there are some characteristics of red algae that do suggest the existence of a toxic group-specific identity system. For one, the multicellular *Porphyra* EST database (but not the *C. merolae* genome) shows a surprisingly large number of genes (29) related to anthrax toxin precursor and receptor. However, since no reports so far suggest the presence of anthrax-like toxins in red algae, the function of these genes is unknown. Although humans can consume some types of red algae (i.e., *Porphyra yezoensis*), other types are highly toxic. However, these are not the usual endotoxins and for the most part, this toxicity results from the oxidation of polyunsaturated fatty acids (oxylipins, oxygenated fatty acids) present in these specific red algae. These oxylipins appear to elicit B1 and B2 prostaglandins, resulting in the high-level generation of free radicals and cellular destruction. It is therefore very interesting that red algae (much more than green algae) often have broad antibiotic activity against many multiantibiotic-resistant bacteria. The general solvent-soluble nature of this antibiotic activity suggests the involvement of oxylipins. It is interesting that some cyanobacteria can also employ very similar fatty acids as potent antibacterial toxins. Filamentous *Spirulina* species can produce jasmonic acid (a PUFA) which has high toxicity (equal to cyanotoxins) against other species of cyanobacteria. To prevent self-toxicity, these species of cyanobacteria also produce potent antioxidants, such as C-phycocyanin (an apparent T/A module). This observation supports the possibility that a major addiction module based on oxidation/antioxidation may exist and be linked to photosynthesis. Since photosynthesis must produce molecular O_2, which presents a highly oxidizing molecule to cyanobacteria, the cells must also produce an antioxidation system to prevent self-destruction and consume O_2. In this case, polyunsaturated fatty acids (PUFA) would provide a diffusible toxic oxidizer and be the T component of a new diffusible T/A module. Possibly relevant to this idea, *Porphyra* species show a remarkable ability to uptake O_2 when linoleic acid is presented as a substrate. This is due to efficient fatty acid oxidation involving a 13 kDa cytosolic protein that has a heme-like chromophore. The X-ray crystal structure of this protein indicates it is most similar to cytochrome *c*-like proteins from cyanobacteria, and not like cytochrome *c* proteins found in green algae. Further evidence for the role of oxidation systems in the identity systems of red algae has recently come to light following studies of the innate immune system. *Chondrus crispus* has a diploid/haploid life cycle in which the two states show relatively small differences in cellular morphology. *Chondrus crispus*, however, can be parasitized by a filamentous algal endophyte (*Acrochaete operculata*), but only the diploid cells are susceptible as the haploids resist infection. This is because only the haploid gametophytes make C20 and C18 oxylipins that are metabolized into hypoperoxides and cyclopentenones (prostaglandins) which produce an oxidative burst that kills the infecting cell. These haploids also produce dehydrogenase. *C. crispus* induces a specific form of NADPH oxidase during the oxidative burst, similar to one found in unicellular red algae. Thus red algae employ an oxylipin-mediated oxidative burst to kill foreign colonized cells in a sex-associated manner. Since red algae represent an early multicellular organism, we might expect this PUFA-based identity system to have had important contributions to the evolution of other multicellular lineages. This clearly seems to be the case. Oxylipins have long been established to be crucial signal molecules (developmental, hormonal, defense) in both animals and terrestrial plants. In animals, C20 eicosanoids (prostaglandins, leukotrines) are involved in cell differentiation, homeostasis and immune reactions. In terrestrial plants, C18 octadecanoid and C16 fatty acids (jasmonic acid) are important developmental and defense hormones. The red algal fatty acid system includes components of both the plant and the animal lineage (C20, C18, C16), thus it appears basal and ancestral to both of them. Consistent with this idea, animals and fungi have type I fatty acid synthase, whereas bacteria have type II fatty acid synthase. The flip side of this oxidation-based system is the need for potent reducing capacity. Along these lines, red algae have long been recognized as a most excellent source of antimutagenic (antioxidizing) substances.

Red algae also make chlorinated C12 versions of fatty acids. In fact, organohalogens are common and diverse products made by many oceanic organisms; 3,800 such compounds are known in which chlorine, bromine, iodine and fluorine are incorporated. Producers include seaweeds, sponges, corals and tunicates, suggesting a common selective pressure for their presence in an ocean habitat. These compounds can be antibacterial in function, although their major purpose is not generally understood. Given this diversity, it seems possible they could be involved in toxic or identity systems. An interesting footnote is that such compounds made by red algae have been used commercially as flame retardants.

Genome and ID Genes

C. merolae is a free-living unicellular red algae that can be found in harsh environments, such as acidic hot springs. It has 20 chromosomes with a combined 16.5 Mbp genome that encodes 5,331 genes. Most of these genes are related to genes found in other eukaryotes. Curiously, only 26 genes have introns, but the reason for this paucity is unknown. Only a few histone genes are present. It does have blue light photoreceptors and its cell division is light synchronized, so rhodopsin-like receptors seem present. No phytochrome gene, however, is present. Most significantly, there is no evidence for trimeric G-protein or cAMP signaling. This most major metazoan sensory pathway therefore seems absent early in multicellular eukaryotes. In general, the transduction systems of red algae seem simple. Also of significance with respect to our interest in motility, neither myosin genes nor autophagy genes are found. But in keeping with a prominent role of oxidative bursts and ROS, this algae does have well-developed peroxisomes. *Porphyra yezoensis*, the multicellular red algae, also has a small genome of 20 Mbp. Interestingly, it only encodes two known membrane-spanning ion channels. It does encode an apparent killer toxin resistance-like gene (Hkr1p), but of unknown function. Of special interest to metazoan immune and ID systems, no dicer genes and no RNAi-related genes have been found. However, a scan of the EST database does indicate the presence of many apoptosis-related genes. Only two genes that seem similar to autophagy proteins are found in the *Porphyra yezoensis* EST database. In summary, it seems red algae genes are clearly like other eukaryotes in many ways. However, the main genes associated with immunity and self-identity are those involved in oxidative burst and a subset of apoptosis-related genes. But the multinucleated (network) nature of red algae suggests that oxidation and apoptosis systems would likely be harmful to the entire organism if this was to be induced in the adult plant. That may explain why ROS bursts are restricted to haploid germ cells, but also suggests that this system may have originally evolved as a sex cell identification system. RNAi and G-proteins are absent.

Genetic Parasites

Interestingly, red algae have been an excellent source of antiviral compounds, especially against herpes and HIV (anti RT). The significance of such compounds is unknown. Red algal chromosomes are known to harbor some copies of retrovirus-related sequences (called chromoviruses or metaviruses), and thus unlike prokaryotes, they are colonized by retroviruses of some type. A genetic map of a chromovirus is shown in Fig. 5.6 and outlines the essential features of an endogenous retrovirus (i.e., LTRs, GAG, PR, RT, RH and IN genes). The numbers of such elements are small and most copies are incomplete in red algae. However, since the gag gene is found in the *Porphyra yezoensis* EST database, this indicates the activity of a virus, not simply retroposon. Still, it is important to note that although not highly expanded, chromoviruses or their related elements are found in the genomes of red algae. The importance of retrovirus colonization of the genome is discussed in detail below.

Plasmids

Seldom considered are the genetic parasites of red algae. DNA plasmids do indeed occur throughout Rhodophyta. Circular dsDNA at high-copy levels are maintained with laboratory passage of these red algae. The size of such plasmids is specific to regional isolates of the same red algae species. No known function or phenotype has been associated with these plasmids. However, a few species of red algae seem not to harbor any detectable plasmids. How such high-copy plasmids attain stable persistence is unknown. Since they

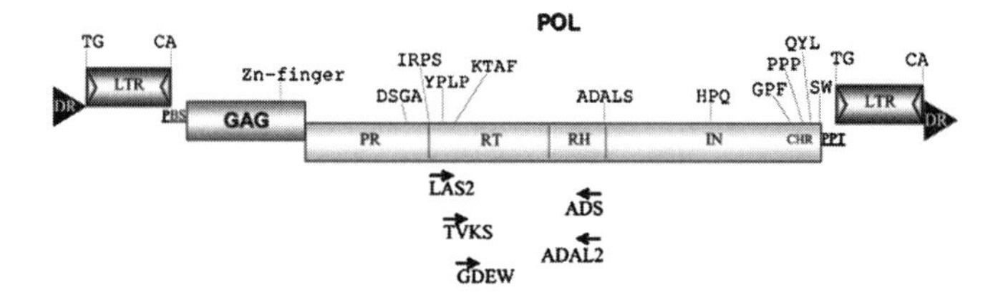

Fig. 5.6 Chromovirus genetic map (reprinted with permission from: Garinsek, Gubensek, Kordis (2004) Molecular Biology and Evolution, Vol. 21, No. 5)

are stably maintained during cell passage and their ratios are also stable, the plasmid ORFs and regulatory DNA must provide some form of stability functions. Given arguments already presented, and since these plasmids are persisting efficiently, it seems likely they are imparting addiction modules and identity systems onto their host. There have been no laboratory or field tests of this possibility. However, in one field survey, 5 of 21 red algal genera were found to contain circular dsDNA plasmids, usually two or more per species. These plasmids do not cross-hybridize to any other DNA source, so they have unique sequences. *Porphyra pulchra* has been best characterized to harbor five plasmids (6.8, 6.4, 1.9 and 2×2.1 kbp). The three smaller plasmids have related sequences, including inverted repeats (290 bp) and long direct tandem repeats (4×216 bp and 5×21 bp) of unknown function. It is interesting that the mtDNA of *Porphyra* also has two 291 bp inverted repeats, a very unusual feature for mtDNA. Although such elements are often recombinogenic in genomes, red algal mtDNAs are unusually uniform in size and gene order (in contrast to fungi described below). The two larger plasmids have distinct sequences with no prominent inverted repeats. These can potentially code for five ORFs with no known similarity to any other genes. Many questions are unanswered concerning these plasmids. For example: How are they transmitted? Do they affect nuclear recognition and nuclear transmission? Do they have compatibility, toxin or antitoxin systems? Are they associated with, or do they affect, sexual reproduction? It seems likely they have strong effects on host recognition systems, but nothing is know about this.

Other Viruses?

The possibility that other virus systems occur in red algae has also not been well evaluated. EM observations, however, often suggest that additional agents may exist. *Gracilaria verrucosa*, which can either be free of or develop simple galla, was examined by EM. Galls showed unusual structural features, rows of fusiform bodies, connected to the ER in the cytoplasm located near plastids. These bodies clearly resemble rhabdovirus-like structures, but further evaluation has not been reported. Another example of virus-like structures from red algae was reported with a multicellular algae (freshwater *Sirodotia tenuissima*). Here, EM evaluations of the filamentous stage observed that all apical cells of the thalus had prominent, central, and uniform inclusions in their cytoplasm that upon closer examination corresponded to paracrystalline arrays of viral-like particles, clearly resembling the ER-associated viral factories characteristic of polygonal RNA virions (see Fig. 5.6). This material was inherited, not transmitted and not found in other species, but was found in filamentous Chantransia, clearly resembling a persistent vertically transmitted virus. Unfortunately, this early report was never further investigated. Both of these reports relate persisting virus-like structures to sex cell differentiation. Since characterization is still lacking, little can be firmly concluded other than to note the strong possibility that species-specific persisting viruses may be associated with differentiation in red algae.

Virus-Like Genes

In the above section, I discussed the occurrence of virus-like genes in the *Porphyra yezoensis* in the EST database that support the possible viral origin of the red algal nucleus. In addition to those examples, which included an array of PBCV-1 and CSV-1 related genes, other virus-like genes can also be observed. These include ORFs with similarity to Pseudorabies-like nuclear antigens, HSV4-like EBNA-2, and nuclear polyhedrosis virus capsid genes. As many of these virus-like genes are proline rich (inducers of internal bends in proteins), it is possible that these apparent similarities are a simple artifact of these proline-rich regions and do not imply any ancestral relationship. However, other examples, such as an ORF similar to TTV1 (rod-shaped temperate *Thermoproteus tenax* virus 1, extreme thermophile), do not correspond to proline-rich regions but no evaluations of these similarities have been reported.

Retroviral Genetic Parasites and Metazoans

We have seen shifting patterns of host lineages with respect to their genetic parasites. For example, in the social bacteria (i.e., Xan.) no DNA plasmids are known, but many examples of temperate phages occur. LTR retroposons are not common in bacteria and true retroviruses are unknown. With the origin of the nucleus, we see a major shift in virus–host relationships. Aside from filamentous brown algae, no eukaryotic DNA virus persists as an integrated temperate DNA which was so common to prokaryotes. It appears that a nucleus precludes some but allows new types of genetic parasites to persist, in isolation from cytoplasmic systems that might otherwise recognize them. Elements that can persist as DNA, but replicate and express as RNA (retrotransposons), thus become common in most (but not all) eukaryotic lineages. At this juncture, red algae represent a base of ancestors leading to all metazoans, and it is important to consider in detail how genetic parasites, especially retroviruses, may affect group identity since this will affect all subsequent metazoan descendants. It is estimated that red algae originated as long as 1.5 billion years ago. However, other nucleated eukaryotes are also rather basal and probably as old. Dinoflagellates, as mentioned in the last chapter, have a clearly distinct nuclear structure and also distinct relationship with large cytoplasmic DNA viruses. In addition, diatoms, diplomonads, euglenozoa and alveolata all represent early eukaryotes with clearly distinct nuclear structures and replication strategies and which are not basal ancestors to the metazoans.

Recently, however, it has become clear that all these lineages can also be distinguished by the type of genetic parasite they harbor. Specifically, red algae, green algae, fungi, slime molds and early animals (worms), which are all related, are also colonized by a family of retrovirus (see Table 5.1), now called chromovirus, but the above other early eukaryotes are not colonized by chromoviruses. (See Fig. 5.7 for eukaryotes that have chromoviruses.) What might be the significance of this ERV colonization?

The chromoviruses are an endogenous retrovirus genome that can easily be distinguished by sequence analysis of its genes (see Fig. 5.6). They consist of an LTR (long terminal repeat) regulatory sequence followed by a structural group-specific antigen (GAG), protease (PR), reverse transcriptase (RT), RNase H (RH) and integrase (IN) and another LTR. This viral family is very similar to the retroviruses of vertebrates. Missing, however, is the envelope gene of exogenous retroviruses which allows the virus to exit infected cells and enter a new uninfected host. In the case of the red algae, the existence of pit-plugs and multicellular mycels makes it likely that any possible chromovirus could be transmitted without the need to make an extracellular virion. A similar situation (with no external virion) appears to apply to the viruses of filamentous fungi, another sessile cellular networked organism (see below). True retroviruses, as opposed to retrotransposons, are absent from prokaryotes, since no LTR- or GAG-containing viruses are found in prokaryotes. However, it does appear that the retroviral-related reverse transcriptase and the integrase can be found in genetic parasites of prokaryotes. For example, the IN gene bears clear similarity to that highly efficient integrase necessary for phage Mu replication. In addition, as LTR-like elements were seen in cyst-forming cyanobacteria (see prior chapter), this too could have prokaryotic origins. It therefore seems likely that retroviruses (chromovirus) evolved from a combination of genetic elements present in prokaryotes but now define a new viral order not present in prokaryotes. In the case of *Porphyra*, it is clear that the chromovirus IN sequence is well expressed, indicating that this chromovirus is transcriptionally active. As we will see, this virus family in particular gives us a very interesting marker to trace the evolution of metazoan organisms (see Fig. 5.7 and Table 5.1). It has undergone lineage-specific expansions and eliminations, especially later in metazoan evolution (i.e., absent from avians and mammals). Its role in the biology of red algae, unfortunately, has not been evaluated. That it is absent from the other early unicellular eukaryotes, however, suggests it has a basal role in metazoans. This element is highly prevalent in plants, and there its expression can be stress-induced, suggesting a possible T/A module connection.

Complex and Simple Repeat DNA

The occurrence of chromoviral LTRs in some host genomes, however, raises another topic that will persist for the rest of this book. Repeat sequences, like

Table 5.1 Distribution of chromovirus (reprinted with permission from: Gorinsek, Gubensek, Kordis (2004) Molecular Biology and Evolution, Vol.21, No. 5)

Phylum	Class	Order	Suborder	Superfamily	Family	Species	Number of Sequences
Chordata	Reptilia	Squamata	Scleroglossa	Colubroidea	Viperidae	*Vipera ammodytes*	3
						Vipera palaestinae	2
						Echis coloratus	3
						Crotalus horridus	3
						Bothrops alternatus	3
					Elapidae	*Walterinnesia aegyptia*	2
						Notechis scutatus	3
					Colubridae	*Natrix tessellata*	3
						Clelia rustica	3
				Henophidia	Boidae	*Boa constrictor*	3
					Pythonidae	*Python molurus*	1
				Teiioidea	Teiidae	*Tupinambis teguixin*	3
					Gekkonidae	*Hemidactylus turcicus*	3
		Testudines	Cryptodira	Testudinoidea	Emydidae	*Trachemys scripta elegans*	–
		Crocodylia	Eusuchia		Crocodylidae	*Caimon latirostris*	2
	Amphibia	Anura	Mesobatrachia	Pipoidea	Pipidae	*Xenopus laevis*	2
		Caudata		Salamandraidea	Salamandridae	*Salamandra salamondra*	3
	Aves	Galliformes			Phasianidae	*Gallus gallus*	–
	Actinopterygii	Perciformes	Blennioidei	Blenniidae		*Blennius sp.*	3

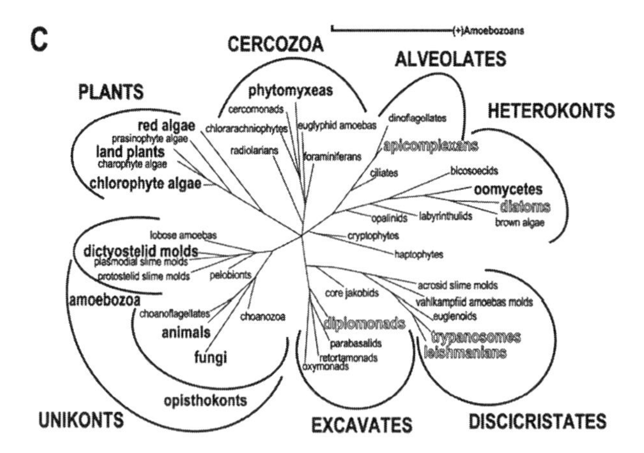

Fig. 5.7 Chromovirus dendogram of its evolution in eukaryotes (reprinted with permission from: Gorinsek, Gubensak, Kordis (2004), Molecular Biology and Evolution, Vol. 21, No. 5)

these simple LTRs, will be seen to often represent major components of eukaryotic genomes, although they are only a small fraction of red algae genomes. As we saw in the heterocyst-forming cyanobacteria, LTR-related sequences (and their loss) were associated with prokaryotic terminal differentiation. They are also found within the inverted terminal repeats of various plasmid DNAs and are often associated with recombination (see the R64 plasmid). LTRs, however, are not tolerated in all metazoan genomes (see *Neurospora* below). In other cases, such repeats can be transcribed and affect host biology; for example, VL30 transcripts affect infection with mouse retrovirus (MuLV). LTRs represent a more complex version of all repeat DNA. Some of this repeated DNA is called 'simple repeat', as it represents reiterated di-, tri- and tetra-nucleotides found in all eukaryotes. Such repeats tend to be highly variable and are generally thought to result from the slippage of DNA polymerase on its template. However, these simple repeats also show a highly lineage-specific pattern of occurrence, and thus they are excellent markers of a host or group lineage. Such a lineage specificity would not be predicted if it resulted only from template slippage-based origin. Thus, the conservation of simple repeat DNA has not been adequately explained. Since some of these repeat sequences can occur in coding regions, they can affect protein function (see below).

In summary, red algae represent a basal eukaryote that led to the origin of algae, fungi and metazoans. They have unusually small nuclei that can be transmitted via pit-plug structures in sexual and parasitic relationships, in which displaced nuclei are degraded. They are also multinucleated and sessile organisms. They use polyunsaturated fatty acids (oxylipins) as toxins that resemble cyanobacterial toxins. They also have an identity system in the form of a sex-linked innate immune response that operates via oxidative burst and a high level of antioxidation activity. This system seems basal to related systems found in plant and animal innate immune systems. They lack RNAi and autophagy systems but have some components of apoptosis. They also have a unique collection of genetic parasites. Many are colonized by persisting small circular episomal DNA plasmids of unknown function. Their genomes represent the first eukaryotic example of colonization by chromovirus, an endogenous retrovirus lacking env genes. Their genomes also have lots of other viral-like genes, but are low in toxins, ion channel and G-proteins.

Did Viruses Invent Red Algae? Origin of Identity

As discussed above, red algae have numerous chromosomal genes that closely resemble those found in large DNA viruses and transmit nuclei through pit-plugs in a process reminiscent of the transmission of cytoplasmic DNA viruses. I have presented the idea that a large DNA virus might have provided the origin of the nucleus. Let us now consider what is known about viruses present in unicellular oceanic organisms to evaluate if these viruses could have also contributed to the origin of host identity systems. Recently, it has become clear that large cytoplasmic and nuclear DNA viruses of unicellular eukaryotes are much more prevalent than previously appreciated, but these diverse viruses are also surprisingly interrelated. Blooming dinoflagellates (golden brown algae) can be lytically infected with HcV01 (heterosigma akashivo virus), as presented in the previous chapter. Phylogenetic analysis of HcV01 DNA polymerase genes indicates that this virus exists as species-specific isolates, but the entire group is a sister group to viruses of blooming unicellular eukaryotic green algae, the much more studied phycodnaviruses, even though green algae are phylogenetically distant from dinoflagellates. Other blooming algal species, such as *Emiliania huxleyi*, are hosts for EhV-86, another viral sister group to the phycodnaviruses, which lytically infects the blooming calcareous nanoplankton. Other phycodnaviruses lytically infect the harmful *Phaeocystis globosa* and contribute to bloom termination as observed in the English channel. Especially intriguing has been the discovery of the very large mimiviruses of amoeba (*Acanthamoeba polyphaga*). Mimivirus is currently the largest known DNA virus, with a genome of 1.2 million bp and a gene repertoire of 911 protein-coding regions, including many genes (such as core translational proteins), not previously thought to have been virally encoded, as well as many genes of no similarity to other genes in the database. In contrast to phycodnaviruses, this virus lacks R/M enzymes, but does encode a lot of proteases. Furthermore, recent comparison of the mimivirus genome to sequences cloned from environmental isolates of the Sargasso Sea clearly indicates that unknown viruses related to mimivirus exist in high abundance in the Sargasso Sea. The world may harbor a very large number of such large DNA viruses able to infect unicellular eukaryotes. Ironically, this virus, the largest and most complex of all eukaryotic DNA viruses, infects the simplest of unicellular eukaryotic host. Why is that? This oceanic habitat is filled with an abundance of large, complex and poorly understood DNA viruses of prokaryotes and eukaryotes. In addition, there are both surprising links and differences among these large DNA viruses. One unexpected link is that HaV, phycodnavirus and mimivirus all harbor highly related intein elements within the active site of their corresponding DNA polymerase genes. Clearly, there seems to be significant genetic flow between these virus groups infecting diverse hosts. The mechanism that maintains this intein similarity is unknown, but it should be recalled that inteins in the DNA pol genes of bacillus phage contributed to phage–phage competition as hyperparasites (Chapter 2). The phycodnaviruses, mimiviruses, poxviruses, iridoviruses and herpes virus all share a similar and distinct type of primase genes involved in DNA replication, consistent with some common ancestral links. However, a surprising difference among these large DNA viruses is the presence or absence of genes that code for several subunits of DNA-dependent RNA polymerase. Such genes are absent from the phycodnaviruses (and phaeoviruses below), which are both nuclear viruses, but present in the coccolithovirus (EhV-86) and in mimivirus. Since the mimiviral and EhV-86 sequences are phylogenetically basal to those of the phycodnaviruses, the implication is that the ancestral eukaryotic DNA virus to all of these extant families was most likely a very large cytoplasmic DNA virus that coded for multisubunit DNA-dependent RNA polymerase. Was this the proto-nucleus virus? Another intriguing difference is that the DNA polymerase of mimivirus is of a distinct type than that found in the phycodnaviruses. Whereas the phycodnaviral DNA polymerase is most related to the DNA pol in herpes virus (an alpha polymerase related to host extension polymerase), the mimiviral DNA pol (B family) and topoisomerase are more similar to the corresponding DNA pol and topoisomerase of poxviruses, large cytoplasmic DNA viruses of animals. It seems that significant divergences in DNA and RNA polymerase strategy occurred in these viruses.

It is difficult to evaluate how these diverse DNA viruses have affected host immunity, group identity and evolution, since genetic studies on natural host populations are mostly absent. Most are lytic viruses of large blooming host populations, and thus they seem to frequently provide negative selection on host survival. There is evidence, however, that these viruses can compete with each other and also modify host-identifying characteristics. For example, various isolates of chlorovirus, PBCV-1, have been characterized to encode for the synthesis of polysaccharide fibers on infected host cell walls. Two viral enzymes are known: hyaluronan synthase (HAS) and chitin synthase (CHS). Interestingly, both hyaluron and chitin are surface molecules highly characteristic of multicellular animals and insects, not unicellular organisms. More interesting yet is that the PBCV-1 HAS is phylogenetically basal to the three versions of this gene characteristic of all metazoans. It is therefore curious that PBCV-1 appears to have two genetic types that have changed from HAS virus to CHS virus by exchanging these genes. The implication is that one of the two surface expression states is favored by natural selection. Is this an identity module? Another possible example of PBCV-1 modifying host identity is with respect to viral K+ ion channel proteins. It was surprising to learn that one of the most variable genes found in natural PBCV-1 genomes was the K+ ion channel proteins they encode (the smallest known ion channels, 124 aa). This gene is also found in the phaeoviruses (EsV, see below). Since natural variant of K+ ion channels are common, it was possible to compare the biological consequence of such variation. Recent results suggest that these proteins are involved in competitive exclusion of a second virus via membrane depolarization (a situation reminiscent of bacterial holins, see Chapter 2). Ion channels appear to provide viral identity systems. Along these lines, viral modification of host identity may also work via signaling alterations. For example, it is very interesting that EhV-86 codes for four proteins that synthesize sphingolipids, absent from all other phycodnaviruses. These are regulatory signal transduction proteins that result in the synthesis of ceramide, a signal of apoptosis in metazoans. In terms of total gene numbers, the PBCV-1 homing endonucleases are clearly the most diverse in their genome and although their function has not been evaluated, arguments presented in Chapter 2 suggest they are likely involved in exclusion of other endogenous and exogenous genetic competitors. The existence of many viral signal transduction proteins as well as ion channels implies that viruses could well regulate these same processes in their host. And although some results support a role for viral ion channels in identity and viral competition, effects on hosts are not studied. One thing is known, however. Infected cells undergo highly modified surface expression, resulting in the appearance of numerous and long fibrils, taking on a furry appearance. The purpose of this modification is not yet clear. One idea is that such surface modification might allow cellular aggregation and promote the transmission of virus between cells. However, since these phycodnaviruses are uniformly lytic, such alterations would seem to be transient. All of these examples suggest that these large DNA viruses are modifying host identity and competing with one another in very diverse and active ways. Thus, they are well equipped with the genes needed to have provided major new identity systems for their unicellular host as well.

The Green Algae Host

Chlorophyta contains both unicellular and multicellular green algae and is inclusive of 13 species of flagellated freshwater unicellular glaucophytes. Glaucophytes have primitive 'cyanobacterial-like' phycobilin containing chloroplasts (also found in red algae) and thus may represent the earliest and simplest green algae. Current estimates based on the geological record indicate that the red algae evolved prior to green algae, with green algae diverging from the red algae about 1.4 billion ybp. This would place the origin of this green algal lineage after the first proposed period of iceball earth. Green plants are thought to have been descendents of green algae that went on to generate over 30,000 species, most of which are multicellular organisms. Since it is the aim of this book to trace the evolution of group identity to the human lineage, we will not consider in any depth the issue of green plant evolution. However, early green algae are also thought to be ancestral to fungi, so we are interested in examining what is known concerning the group identification systems of simple green algae. Glaucophytes are unicellular, mostly haploid and asexual organisms. Unlike the dinoflagellates, however, they have more typical nuclei and undergo open mitosis, requiring the dissolution and reformation of the nuclear envelope. Both flagellated motile and non-motile species are known, but it appears that the flagellated species are basal to the non-motile species. Some of these organisms have large genomes, with substantial quantities of non-coding DNA, so in this they differ from the small genomes of red algae or the diploid and sexual life cycle of brown algae and green plants. Another major type of eukaryotic unicellular algae that are thought to have descended from red algae is the diatoms (heterokonts). Although most are unicellular they can undergo blooms, form colonies and other communities (sometimes in association with filamentous fungi). Although we do not understand the mechanisms by which they form and maintain their community structures, it is very interesting that they show an amazing degree of diversity in their silicate shells (see Fig. 5.8). Since diversity and group identity are often linked this would be an intriguing topic for further study.

The Origin of Sensory Cilia from Flagella

Many unicellular chlorophyta species can grow as blooms of individual cells. They are clearly phototactic, but how this relates to group behavior is not well studied. Like the dinoflagellates, phototaxis (positive and negative) occurs via links between sensory rhodopsin receptors in the eyespot and the flagella. This linkage involves Ca^{++} and a large array of signal transduction proteins, many of which are physically part of the flagella complex. In fact, it appears that the algal flagella, besides providing motility, are also highly involved in sensory function. Moreover, this involvement in both motility and sensory function appears to provide a surprisingly strong evolutionary link between unicellular algae and humans. Flagella are highly similar to cilia found on the cell surfaces of many types of metazoans. Many cilia, like those in mucosal epithelia, are motile in that they move in coordinated patterns to provide fluid movement over cells. However, other cilia are clearly sensory organelles, such as vertebrate odor receptors, which are modified cilium that function as chemical and mechanical antennae for many cell types. Along these lines, all vertebrate cells express primary cilium, which is a non-motile organelle that appears to have sensory function (with the very interesting exception of vertebrate myeloid or lymphoid cells). Some primary cilia display polycystin-2, a Ca^{++} channel protein involved in kidney differentiation and apoptosis. Primary cilia can also express SST_3, the somatostatin receptor found in specific regions of brain

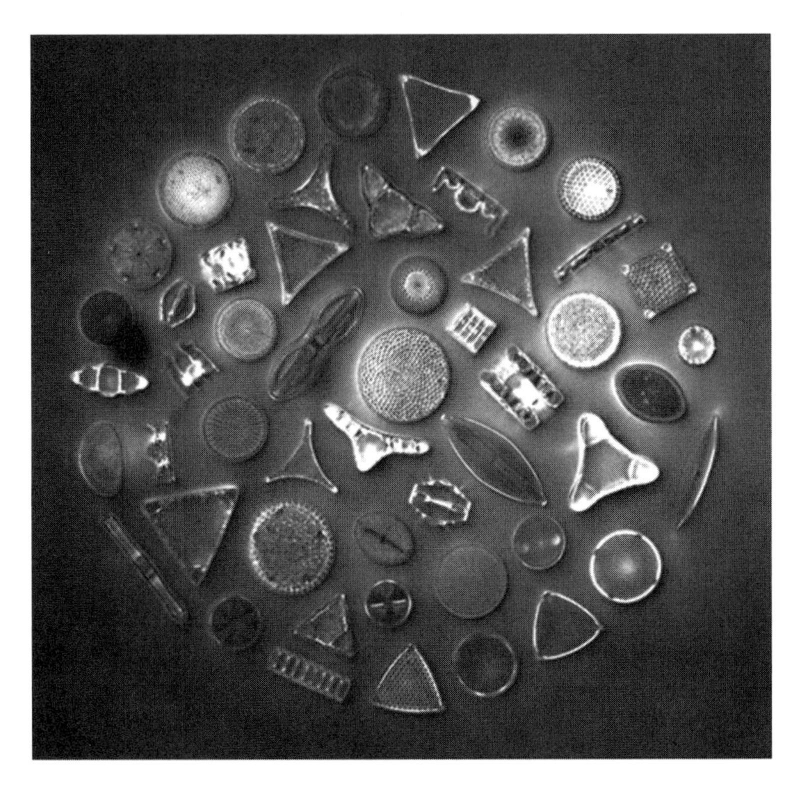

Fig. 5.8 Physical diversity of diatom shells. courses.bio.psu.edu/fall2005/biol110/tutorials/tutorial30.htm (reprinted with permission from: http://images.nbii.gov/details. phy?id = 65320 & cat = Animal%20 products Public Domain)

neurons as well as 5-HT serotonin receptor. Both the SST_3 and the 5-HT receptors are G-coupled 7 TM proteins associated with apoptosis. Given the broad importance of G-coupled proteins for light, odor and mechanical sensory detection in higher organisms, this association with cilium is very intriguing. Interestingly, non-vertebrate animals, such as insects, do not express primary cilia in non-neuronal cells. Recent analysis of the flagellar proteome of the unicellular biflagellate *Chlamydomonas reinhardtii* (C.r.) adds much strength to the idea that algae and humans show strong conservation. C.r. is a haploid unicellular algae which is mostly asexual and employs light-dependent cell division. The flagellar proteome of C.r. has been fully measured by mass spectroscopy and is composed of about 336 proteins, most of which have matching sequences in the human genome. In fact, 132 of these human-related genes are not found in the genomes of lower eukaryotes, establishing a high level of conservation between this algal organelle and human proteins. And although green algae are also the ancestors of higher plants, plants lack flagella. A clear bifurcation in the plant and animal sensory and motility systems must have occurred in evolution. Within this set of C.r. flagellar proteins, an amazing 90 correspond to various types of signal transduction proteins, 6 of which are ion channel pumps of various types. This includes the blue light receptor for phototropin. It also includes polycystin-2, the primary cilia-associated Ca ++ channel protein noted above. The implication of these similarities is that motile unicellular algae have many of the same motility and sensory transduction proteins as those found in humans. We can essentially say that humans experience their environment through cilia (light, odor, sound) which we can closely trace to flagella of algae. This theme, tracing sensory and identity systems, will be closely followed for the rest of this book. It is interesting to contrast the situation with the multicellular, spherical green algae volvox. The organism is organized into a sphere in which most cells are haploid and flagellated, but moving in unison to propel the whole organism. How volvox forms a multicellular organism is not well understood. Volvox, however, has separated somatic cells from germ cells, but in a motility pattern converse to many eukaryotes. Motile cells are somatic and cannot divide whereas gonadia are totipotent, immortal, but not motile. This is converse to the situation with the brown algae, described below, in which the germ cells are non-motile and the somatic cells are motile.

Small-Molecule Identifiers?

Algae are known to produce a large array of secreted small molecules which might function in group identification. Algal halogenated furones are known to be specific and potent inhibitors of the QS surface moieties of *Serratia liquefaciens*. The same molecules are active against *Escherichia coli*, suggesting generalized activity against QS. Oxylipins may also serve related QS functions as they do in red algae. However, few studies have evaluated this issue.

The Filamentous Sexual Brown Algae (Phaeophyta) and Their Viruses

The filamentous brown algae represent some significant transitions in the evolution of multicellular organisms. Here we see the emergence of non-motile adults, but the retention of motile chemotactic gametes. The adult is also diploid and the product of sexual reproduction. Brown algae also show a distinct relationship with their DNA viruses, the phaeoviruses. As noted above, the viruses of green algae (such as PBCV-1, a chlorella virus) showed many prokaryotic or phage-like features (R/M enzymes, homing endonucleases, external virion injecting the DNA into the cell). Like all other eukaryotic DNA viruses, the phaeoviruses have lost these phage-like characteristics. Along with this viral change we also see significant and seemingly basic changes in the virus–host relationship: a shift from acute (lytic) to persistent (lysogenic) lifestyle. The lytic viruses of green algae infect motile unicellular haploid hosts, some of which can escape infection by becoming symbiotic and immotile within other eukaryotic hosts, such as *Paramecium*. In contrast, essentially every species of filamentous brown algae so far examined is infected by species-specific DNA proviruses, with clear similarity to the chlorella viruses. Unlike chlorella viruses, phaeoviruses persist as proviruses integrated into host genomes and are maintained in a Mendelian fashion. This point also requires some emphasizing. Although chromosomal integration was a common feature of many DNA viruses of prokaryotes, it is a much less common feature of the DNA viruses of eukaryotes: in fact it is unknown for the normal biology of any large eukaryotic DNA viruses. There is one possible exception to the statement that no other eukaryotic DNA virus uses proviral integration. The endogenous, or genomic, polydnaviruses of parasitoid wasps appear to be germ line-integrated elements and will be described in the next chapter. Why was this phage-like integration capacity gained in the phaeoviruses, but is essentially unique for any large DNA viruses of eukaryotes? What selective forces might favor this situation? The most apparent distinction between chlorella algae and the brown algae is the sexual reproduction of the brown algae. Phaeovirus reproduction is tightly linked to host sexual reproduction, so this seems like a significant link between virus and host. The brown algae reproduce sexually, making flagellated or motile gametes that combine to make diploids, and it is the infection of these cells by phaeoviruses that characterizes viral host biology. Phaeovirus, such as *Feldmania* species virus (FSV and FirrV-1) or *Ectocarpus silicosis* virus (EsV-1), only infect the wall-less free-swimming spores or gametes of brown algae. The viral genomes integrate into the resulting germ cells and are subsequently transmitted into the genomes of all the cells of the host where they remain latent until sexual reproduction. Thus, there is a very intimate relationship between host sexual biology and virus biology. However, we are currently unable to articulate how this situation could provide a specific selection for a proviral lifestyle. Possibly episomes are excluded from gametes. Although viral latency or persistence implies the existence of some type of viral immunity or T/A module, existence of such systems or how they might affect host identity has yet to be documented. It would be expected that phaeoviruses must have more sensory transduction circuits to link virus replication to host reproduction. Comparison between the phaeovirus and the chlorella virus genomes, as well as comparisons within the phaeoviruses, does suggest some interesting differences regarding transducing proteins. The genome of EsV-1 is 335 kbp, FsV is 160–180 kbp and FirrV-1 is 180 kbp. These EsV-1 genomes are significantly smaller than that of the lytic chlorella virus (or mimiviruses) and share only 28 genes (mainly core genes) with them. Amongst the phaeoviruses, however, many genes are conserved and 93 ORFs are similar between FsV and FirrV-1. However, in contrast to most other DNA and RNA viruses, the gene order between FsV and FirrV-1 is not conserved. Comparative phylogenomics suggest all the phycodnaviruses are likely derived from a common ancestor (the proto-nucleus?) that had a more complex genome and that the individual lineages were mostly the result of gene losses. However, phaeoviruses have gained gene sets that were absent from the lytic chlorella viruses. These gene sets are involved in sensory processes or signal transduction, especially hybrid histidine kinases pathways, genes characteristic of bacteria, algae and fungi, but absent in animals. These transducing proteins are the most variable set of phaeovirus genes. It is interesting to note that in bacteria these proteins are pathogenic genes. In the FirrV-1 genome, one of these histidine kinases is a phytochrome chromophore-binding protein, which is interesting with respect to the strong linkage to light-dependent biology of sex cells release and virus release. Phaeoviruses also have genes for threonine/serine protein kinases. Phaeoviruses, however, are not simply eukaryotic versions of bacterial prophage, as they differ significantly in the process of integration. Recent evidence indicates that the phaeovirus integration is highly unusual in that the viral DNA becomes dispersed into many subgenomic regions, integrated into various chromosomal locations. This unusual process might also be relevant to the observation that the FirrV-1 virion packages contain a very large quantity of subgenomic viral DNA, a feature reminiscent of the polydnaviruses (discussed below) but absent from other phycodnaviruses. How the DNA gets dispersed or reassembled to form a complete viral genome is not understood. It is very interesting that in many of these characteristics (high species specificity, high prevalence, dispersed genomic integration, packaging of subgenomic DNA, reproduction in reproductive cells) the phaeoviruses resemble the genomic polydnaviruses of parasitoid insects. Thus in the brown algae, we see significant but linked transitions in lifestyle for both the virus and its host. The resulting relationship is that of a highly successful and prevalent persistence infection that has apparently evolved from a lytic virus ancestor. Viral and host identity appear highly linked and it seems that sensory (transduction) gene sets were acquired by the virus to allow this lifestyle. The consequences of these viruses to host biology and evolution, although poorly understood, suggest that affects regarding the sexual identity of the host would seem likely.

Pheromones, Sex and Viruses

Phaeovirus release is triggered by the same environmental cues that trigger release of motile gametes or spores. Prominent amongst these cues is exposure to blue light followed by a period of darkness, a situation that might be associated with a burst of photosynthesis and increased oxidation. As this release is synchronous and essentially systemic for many gametes, it seems clear that both virus and host must have, or use and coordinate, sensory systems that allow such synchrony and increase the likelihood that sex cells will find each other, and virus cells will also encounter gametes prior to diffusion. Since EsV-1 is prevalent in natural brown algae populations of coastal waters (100% worldwide in *Ectocarpus* populations), clearly this virus has a highly successful biological strategy. Also, since EsV-1 infection is species specific, it is interesting that infection of other species can result in reproductive pathology. Thus EsV-1 would likely need sensory functions. It does encode for the production of K+ ion channel proteins, which are likely to provide competitive viral exclusion function as seen with PBCV-1 channel proteins. All of these circumstances are consistent with a major role for virus selection in host reproductive biology. In a sense, the very high levels of virus replication in reproductive cells can be considered as a substitution of virus reproduction for host reproduction, since these viral-producing gametes are no longer capable of host reproduction. How then can such a relationship be sustained? Given the high prevalence of EsV-1, how can colonized hosts sustain their own reproductive inefficiency? Clearly, some type of balance between virus and host reproductions has been achieved and not all gametes are producing virus. This close association (exclusion) between virus and host reproduction and the existence of motile sex cells is of considerable interest, from the perspective of group identity.

Algae are considered the progenitors of both plants and animals but animals, not plants, use motile sex cells. How then do motile sex cells identify each other in a diffuse water environment, and how does the virus affect or use this process? Although there is little information on the latter point, much is known of the former issue. Female gametes (motile eggs) of over 100 species of brown algae are able to release chemical signals that attract their conspecific motile male cells (spermatozoid). In many cases (45 species), these pheromones are oxylipin fatty acids (sometimes halogenated), suggesting a possible linkage to the major signal of the innate immune system discussed above for red algae. Following reproduction, female cells will swim for some distance then settle on a surface. At this point they will start to produce unsaturated acyclic and/or alicyclic C11 hydrocarbons (11 distinct chemical backbones are known, plus 50 stereoisomers, such as alkylated cycloheptadiene in *Ectocarpus*). Brown algae also make an array of large quantities of other halogenated hydrocarbons (such as brominated phenols), but their biological function is not clear. These C11 hydrocarbons are chemo-attractant pheromones for the flagellated motile male cells. Presumably, like phototaxis, pheromone surface adsorption induces signal transduction to the flagella that directs swimming direction. Thus female cells lure male cells using these volatile olefinic hydrocarbons, a process of odor-based mate identification that has been conserved in most metazoans. Specific receptors for these molecules are present in the male cells, some of which are found in the biflagella structure. Thus a link between flagella and sensory detection is basic in this first example of a sexually reproducing eukaryote. As mentioned, both gamete and virus release are synchronized by following light exposure. In addition to the specific pheromone, a second process of identification occurs upon contact between the flagella. Surface glycoproteins and lectins are also involved in fusion of sex cells and fusion is species specific. All of these C11-related pheromones are biologically degraded by oxidative pathways of various types, some of which involve light. Thus, there seems to be a clear similarity between the chemistry of these pheromones and the oxylipin pheromones described above for the red algae innate immune system. Red algae tend to have an abundance of C20 PUFAs whereas brown algae tend to have C18 PUFAs. It is also interesting to note that diatom blooms have also been observed to produce oxylipins and C11 hydrocarbons. However, knowledge concerning the details of innate immunity of brown algae is limited. Massive amounts of oxylipins can be accumulated in male cells of brown algae. The fate of this material following fusion is not clear, but its potential for an oxidative burst would certainly be a possibility. In some species, high levels of pheromone production are not simply sex-related or sex-phased. In these situations, pheromone production may be part of a defense strategy so that these continually released pheromones can interfere with the sex cell motility of competing species of brown algae (i.e., mating disruptants). Thus we see a strong link between potentially toxic oxylipin pheromones, sexual identification and sex-based group competition.

Although the attached large brown algae species reproduce primarily by sex, not all brown algae are attached or sexually reproducing. *Fucus spiralis* is a hermaphroditic species and reproduces primarily through selfing, which can result in clonal natural populations. The brown algae thus represent a bifurcation in the sexual life strategy. Little is known concerning the viral ecology of this asexual species.

RNA Viruses and Algae

As a footnote, it is worth noting that although most algae have not been clearly established to harbor RNA viruses, some are known. Curiously, there are dsRNA viruses found in mitochondria of some green algae (agents prevalent in fungi, see below). Bryopsis and Codium species can be usually found to harbor complex species-specific mixture of dsRNA within their mitochondria. Some of these dsRNA viruses are visible as viral-like particles. However, mostly they appear to lack capsid genes needed for extracellular transmission. Analysis of their RdRp ORF indicates that these viruses belong to the partiviridae family, which have two RNAs encoding two ORFs. How these agents may have affected the evolution of mitochondrial function is unknown. However, as will be presented below, the partiviridae are exceedingly common parasites persisting in fungi.

Other Algal Models

One of the additional well-studied unicellular models of algae is *Chlamydomonas r*. This is a unicellular, flagellated algal model that reproduces by both asexual budding and mating-type cell fusion, a process very similar to many fungi yeast species. Nitrogen starvation induces gamete formation, resulting in the production of mt + and mt– cell types. It seems likely that this starvation response will involve T/A modules of some type, but molecular characterization is lacking. In the case of yeast species, the production of mating-type cells and their fusion appears to promote the transmission of endogenous retroviruses (such as Ty elements described below). In the case of *Chlamydomonas*, possible transmission of endogenous retroviruses, such as chromovirus, has not been investigated. However, in contrast to the filamentous brown algae, there do not appear to be any exogenous viruses associated with mating. Similar to the brown algae, *Chlamydomonas* also uses a pheromone sex attractant (but not oxylipin), and gametes swim towards each other. Initial contact between gametes is through the flagella (via surface glycoproteins), which allow matched cells to agglutinate. This can be considered a self-identity system which excludes the agglutination of cells of the same mating type. Following a signal transduction process, agglutinated cells will fuse to form the zygote. This is clearly a distinct process from that used by the brown algae or chlorella. *Chlamydomonas* also displays clear light-associated swimming behavior in that blue light exposure induces oxidation that helps in swimming. Light exposure is also associated with cell division. *Chlamydomonas* has rhodopsin-mediated photoreceptor currents, which likely operate via Ca^{++} levels and can also induce sexual differentiation to make gametes. In terms of ion-gated ion channels, *Chlamydomonas* has two major types, CSRA and CSRB, which are involved in membrane current regulation. CSRB is also related to neuron-specific synapsins. Another significant adaptation seen in *Chlamydomonas reinhardtii*, but not in most other algae, is the ability to silence genes post-transcriptionally via dsRNA. C.r. has at least two genes associated with the RNAi response (discussed in detail below); a *dicer*-like and an *argonaute*-like gene. It does not appear to have an RdRp gene associated with the amplification of the RNAi response. Closely associated with these two RNAi genes is a gene that codes for reverse transcriptase. One implication of this association is that these two RNAi genes may have originated following the colonization by a retrovirus-like element. In addition, it appears that retrotransposons are often a target of silencing, via degradation of aberrant RNAs they can produce. Thus C.r. has acquired a new gene-silencing system, seemingly from a retrovirus-like agent that also silences other retroviral-like agents. A similar association of RNAi to endogenous retroviruses, gene silencing and antiviral immunity is discussed in fungi and *C. elegans* sections below.

Other Simple Plants: Volvox

Volvox (multicellular algae) physically resembles the early embryo (blastocyst) of animals and were initially studied as possible models leading to the evolution of early multicellular animals. However, genomic data clearly show volvox is not representative of, or basal to, early animals. It has only two cell types, a flagellated somatic cell and an immotile reproductive cell. This situation is the converse of what was seen in brown algae. Sex production occurs in response to O_2 levels and is also stress induced. Volvox has a visual rhodopsin that controls phototactic behavior and uses an eye structure to detect light and induce motility. Some gonidia are eyeless, yet express this photoreceptor, so it must serve non-motile functions. As our main interest is to trace a path to the evolution of animals, volvox represents an interesting segue to plants. But it is interesting to note that like all plants, volvox gametes are non-motile.

Light and Social Life

In the algae we see frequent associations between light and social (often sexual) behavior. In terms of light production (generally via luciferase), it seems that most of the unicellular oceanic microbes produce light in a rather steady on/off manner. Light organs of animals that use these unicellular sources often overlay a shutter to control light production in a temporal (flashing) pattern. Interestingly, the flagellate protozoa, such as *Pyrocystis* species, can also produce light in a flashing pattern with characteristic temporal cycling and such light production also tends to follow circadian patterns. The purpose of this flashing capacity is not clear, but it is suspected that it must be involved in group or sexual behavior, as it is in various marine animals. In some cases, the metabolic investment in maintaining circadian light production is considerable, such as *Gonyaulax polyhedra*, which curiously destroys its luciferase proteins daily and must remake them every new day to initiate another photosynthetic cycle. Such apparent high metabolic inefficiency would seem illogical and might instead imply the existence of a T/A module based on photosynthesis or the O_2 produced. Light and sex seem frequently associated.

Algae, General Conclusions

With the red, green and brown algae, we see distinct systems of group identity in unicellular and multicellular eukaryotes. We also see distinct relationships with their viruses. Large DNA viruses are highly prevalent. As we will see below, related large DNA viruses are also prevalent in protozoa, marine crustaceans and marine vertebrate animals, but are absent from fungi, early animals (nematodes) and all plants. In the filamentous brown algae, we see the only eukaryotic example of an integrating DNA provirus, with an unusual dispersed genome whose reproduction is sex-linked. Genomic retroviruses (*env* defective chromoviruses) make a more substantial evolutionary appearance in algae, and, although in low numbers and of unknown consequence, they are also expressed. We also see the use of flagella as a light and pheromone sensory organelle linked to sex cell motility. This organelle has conserved a surprisingly large number of protein functions from algae to human, as can be seen in the sensory cilia of metazoans. This includes the participation of G-proteins in sensory transduction, a theme we will further evaluate in detail. This molecular sensory system provides a foundation that will now be traced to human evolution for it will be seen to often contribute to group identity and behavior mechanisms. With the red algae, we also see a new system of innate immunity based on oxidative burst and oxylipins. Similar oxylipins function as pheromones for the motility of brown algae gametes, linking immunity to sexual identity systems (reminiscent of conjugative parasites and phage of bacteria). No complete RNAi system is present in most algae, but C.r. clearly has a simplified (non-amplifying) version of an RNAi system. This RNAi system may have originated from a retrovirus-like colonization and appears able to silence retroviral expression, suggesting involvement in immunity. Although some components of the apoptosis system are also found, somatic tissue differentiation does not appear to be controlled by apoptosis as it is in animals.

Fungi: A Path to the Animal Kingdom

Fungi are currently designated to be composed of six phyla. They are estimated to contain about 100,000 known species and up to 1,000,000 possible species yet to be identified. Thus fungi represent one of the most diverse domains of all life, but also represent the ancestor base for all metazoan organisms. Fungi also represent a major species expansion relative to the brown and red algae. A main role of fungi in the ecosystem is that of decomposers of other dead organisms. In addition, fungi are known to include plant parasites which often also have genes and metabolites associated with pathogenesis. One fifth of all fungal species are obligate symbionts with green algae and/or cyanobacteria (as photobionts). Lichens are symbiotic organisms in which 15 of the 18 fungal orders are found to participate. Of these, the Ascomycota constitute 60% of total Ascomycota members, but 95% of lichenized species are within this fungal group. Current views are that lichenized species evolved early and the non-lichenized species followed. Therefore, parasitic/symbiotic lifestyle seems basal to fungi and as we will see, there is clear evidence for a role of genetic parasites on this relationship. Like red algae, fungi are also sessile, mostly multinucleate hyphal species, some of which are clonal mats with extended life spans of up to 2,000 years. It now seems clear that essentially all filamentous fungi are also infected with various types of species-specific viruses, especially dsRNA viruses, although DNA viruses of mitochondria are also known. If these infections extend to the vast array of yet uncharacterized species of fungi, we can extrapolate that the numbers of fungal RNA viruses are staggering. Mostly, these RNA viruses are intracellular and not made into virions that function for external transmission. Rather, it appears that virus transmission is typically cytoplasmic via hyphal fusion. This hyphal transmission represents a major life strategy of fungal viruses, which includes highly defective RNA, DNA and retroviral parasites of mitochondria. Some virus transmission also seems to occur via host sexual reproduction, but few fungal viruses are vertically transmitted, which is in contrast to the phaeoviruses of the brown algae. Thus, the fungi represent a major shift in virus–host lifestyle from that of algae. Included in this lifestyle shift is the occurrence of non-motile gametes in a non-motile organism. Fungi are both clonal and sexually reproducing diploid/haploid organisms. Hyphal growth is used as a primary mechanism to explore and change the habitat of the organism, but also to detect pheromones, which can both attract and repel hyphal growth that will allow sex or nuclear exchange between compatible partners, resulting in hyphal fusion and nuclear (and viral) transmission, similar to red algae. The algal systems of group identity (immunity) seem to operate between organisms at the level of hyphal fusion. Allowed hyphal fusion results in the formation of sycitium and transmission of nuclei. The resulting mixed nuclei in one cytoplasm is a heterokaryon. This involves a clear identification system known as vegetative incompatibility. If the participating fungi are not compatible, nuclear destruction of the transmitted nuclei will occur. This process has clear resemblance to the nuclear transmission seen in the red algae.

Primacy of Genetic and Genomic Parasites

With the fungi and their viruses, we see a major shift from both algae and the dinoflagellates. No large DNA viruses are known for fungi, which were so common to these other hosts. Instead, all fungi seem prone to colonization by persisting dsRNA viruses (many of which make VLPs). These viruses are found in all major fungal lineages and are mostly co-evolved with their host. These viruses include many examples of defective RNA and satellite viruses of the cytoplasm and mitochondria, as well as mitochondrial DNA viruses that closely resemble defective adenovirus. In the unicellular yeast species, dsRNA viruses can confer the production of killer toxins (described below). Virus and toxin production can also be linked to host sexual reproduction. With the fungi, not only are extragenomic viruses distinct, but we also see a significant shift in the pattern of genomic parasites. In particular, the expansion of the chromoviruses (endogenous retroviruses) and Ty1/3/5 elements in some fungal lineages is observed (as well as related LINEs). In fact, fungi appear to represent a bifurcation point in the evolution of host genomes with regard to genomic colonization by endogenous retroviruses, their defectives and LINE elements. Currently, it appears that the more basal species of fungi have relatively larger genomes, but the nature of extra DNA (likely repeat) has not been characterized. This suggests that some type of major genetic colonization by unknown agents occurred at the base of all fungal lineages, but other lineages of higher fungi have compacted their genome and eliminated much DNA. As described below, some well-studied fungal lineages do have small genomes and systems that strictly prevent such colonization, (via RIP, MIP) whereas others allow it. But as will be presented, the colonization by these endogenous retroviruses shows a strong relationship to the origin and function of siRNA and the link to host sexual reproduction (haploid/diploid/meiosis).

Fungi Evolution

A main distinction between lower and higher fungi is the complexity of their hyphae, which in lower fungi are structurally less complex. It now seems likely that the early fungi were symbionts (or parasites) of other organisms. This suggests the employment of various types of T/A addiction modules to compel stable colonization (likely involving ROS). The phylogenetic basal fungal group is now considered to be the Zygomycota. One of these is *Rhizopus oryzae*, a marine fungus parasitic to Oomycetes. The unicellular zoospore of *Rhizopus* shows an interesting resemblance to those of green algae. Zygomycota, however, appear to not harbor chromovirus in their genome. However, *Rhizopus* are known to produce VLPs as paracrystalline arrays of 40 nm VLPs are seen in their mitochondria of all field isolates so far examined. These VLPs are induced by heat shock, are highly stable and lab strains apparently cannot be cured of such virus production. This suggests vertical transmission is involved, but such viral transmission does not appear to have been maintained in other fungal lineages. However, it is very interesting that induction of *Rhizopus* VLPs prevents this fungus from colonizing its Oomycetes host. This suggests that the reactivating persisting virus has some affect on (interruption) an addiction (T/A) relationship between *Rhizopus* and its host, which would be consistent with the VLP itself using an addiction module for its own persistent stability. As noted, this virus is highly species specific and highly conserved. If such virus–host relationships are common and crucial for the ability of the fungal host to colonize other organisms (as with *Rhizopus*), it would therefore be expected that similar persisting viruses (VLPs) would also be common in other fungi and this relationship could be conserved in the evolution of higher fungi. If the conserved virus–host relationship is predominantly persistent, it will also generally be without pathology to the specific host, but retain viral superimposed identification and immunity systems that could be harmful to competing fungi. Thus, a direct viral role in the evolution of these fungal lineages could be expected and the very common occurrence of such viruses in all fungi would support this proposal. Most fungal viruses are transmitted to individuals within the same or closely related vegetative compatibility group via hyphal anastomosis and heterokaryon formation. Less common is virus transmission via host spores. Fungal spores can be made by sexual and asexual processes. In general, sexually produced spores tend to eliminate viruses, but such elimination can be both virus and host specific and is not always seen. For example, mushroom basidiospores efficiently transmit ss- and dsRNA of mushroom viruses. The current scenario suggests that early fungi were mainly symbiotic/parasitic to other organisms. Higher fungi developed more complex hyphae as well as more independent life strategy. The latest evolutionary development of the fungi appears to have been the evolution of yeast species, which followed the evolution of high-carbohydrate fruit in plants. But as we will see below, even yeast species appear to be prone to specific relationship to RNA and retroviruses that affect host identity.

Origin and Diversity of Fungal RNA Viruses

What might be the origin of the ubiquitous persisting dsRNA viruses of fungi? As mentioned, dinoflagellates are known to harbor ssRNA viruses, such as HaRNAV, a virus with clear resemblance to picornavirus (narnavirideae **arena-viridae?**). RNA viruses of red algae have been observed, but not well evaluated. Some green algae, such as *Bryopsis* species, usually contain defined levels of 4.5 kbp dsRNA in mitochondria, so mitochondrial RNA parasites seem present in related ancestral hosts. It therefore seems that various types of ssRNA and dsRNA viruses were present early in eukaryotic evolution. In fungi, we can find five major groups of RNA viruses, suggesting a radiation of viral lineages. Partiviridae is the largest and is a family of isometric viruses that usually have two RNA segments and which are also found in higher plants. The fungal versions of these viruses are monophyletic, suggesting a basal relationship to the other members. Narnaviridae are mostly found in fungi and correspond to about a 3 kbp RNA genome that encodes a phage-like ss RdRp. Totiviridae are found in both fungi and protozoa and consist of one 5–7 kbp RNA genome with 2 or 3 ORFs, one of which codes for the capsid fused to the RdRp gene. The class of fungal viruses with the biggest genomes is the Hypoviridae/Endoviridae, also found in higher plants. These have ss + RNA genomes of >10 kbp. They are associated with the production of cytoplasmic vesicles and show a less common, more specific host distribution. Clearly, not all of these viruses were maintained in metazoans, as many became restricted to higher plants. Ironically, in fungi we see the evolution of molecular immune systems able to counter RNA viruses, such as the siRNA response which uses a dsRNA as a common trigger to initiate recognition. Yet dsRNA viruses are prevalent in all fungi.

Genomes of Ascomycetes and Basidiomycetes: Chromovirus, LTRs and GPCRs

Most fungal genomes are relatively small for a eukaryote, ranging from 10 to 30 Mbp. This suggests that fungal genomes are not highly populated by parasitic DNA sequences. However, it is striking that a family of endogenous retroviruses, the chromoviruses, makes their appearance as an expanded full-length agent in fungal lineages. Numerous full-length chromoviruses are found in fungi, although many copies have become inactivated in specific lineages. As mentioned above, chromoviruses are related to Ty3/Gypsy retroviral elements as well as the metavirus of plants. One clear relative is the Skipper element of Dyctyostelium discoideum. DIRS, also mentioned previously, is also distantly related. It is worth trying to understand the significance of this colonization event with respect to host evolution as it provides an important genetic marker. These chromovirus sequences show patterns of clade distribution that match their corresponding fungal host clades (including the large ascomycetes and basidiomycetes groups). Distinct chromoviral clades are found in ascomycetes and basidiomycetes, but most of the elements have been inactivated (RIPed, see below) and are not transcribed, suggesting waves of lineage-specific colonizations followed by inactivation. Later, we will see evidence that such waves of viral colonization resulted in the displacement of previously integrated viruses. Still, it is most interesting that the smaller numbers of intact viral copies are highly conserved and expressed. Thus chromoviruses are tightly associated with the evolution of their fungal host (a pattern also seen in mammals). But chromoviruses are not found in all eukaryotic lineages (as shown in Fig. 5.7). Although they are retained in plants and teleost fish, some fish, such as the small puffer fish genome, has only one copy of chromovirus. As mentioned, red algae (*Porphyra*) have a chromovirus integrase-like sequence (found in the EST database), but no full genome has been seen. Other lower eukaryotes also have related elements but lack full chromovirus sequences, including plasmodiophorids and bryophata. And still other simple eukaryotes, such as diplomonads, euglenozoa, aveolata and diatoms have no detectible chromovirus-like sequences. Green algae genomes do have some chromoviruses, but do not have distinct chromoviral clades as do fungi. Chromovirus colonization thus seems to provide a marker that was acquired in fungi, but maintained in key vertebrate and amniota lineages (including reptiles), so it is directly on our path towards human evolution. Curiously, as we will see later, chromoviruses were lost in the genomes of mammals and avians. In addition, they are absent from the genomes of most invertebrates. They are not found in protostoma or most basal deuterostoma (echinoderms or urochordata). Recently, the sequence of *Aspergillus oryzae* has been determined. This fungi is used for sake fermentation and represents a non-lichenized fungal order. This order has lots of human and plant pathogens, suggesting ID or T/A systems similar to those maintained in plants and animals. It was surprising that *Aspergillus oryzae* was found to have a very large number of ORF (33,500), considering the relative simplicity of this organism. It has also conserved 24 chromoviruses (most of which are lineage specific). However, for such a large number of ORFs, it has a compact genome of only 95 Mbp. We can compare this to the human genome of 30,000 ORFs but 3,000 Mbp. In general, fungal genomes are much less colonized with parasitic DNA than mammals (especially retroviral-derived). Why would a structurally simple hyphal fungus need so many genes? What do they all code for? One answer appears to be that this fungus makes a lot of secondary metabolites. Many of these metabolites, such as aflatoxin, are toxic to other organisms, but require complex biosynthetic pathways. Also, this fungus has many genes for the peculiar metabolism related to fatty acid oxidation. Since gene order in these accessory genes is conserved, it seems clear that these genes did not evolve from a major gene duplication event. It may be, therefore, that fungi have complex metabolite-based identification systems.

Viruses of Filamentous Fungi

Viruses infecting ascomycetes filamentous fungi do show some common patterns. Most of them are efficiently eliminated during spore formation of the host. This is in contrast to various yeast species in which both ssRNA and dsRNA viruses are effectively transmitted via ascospores (see below). As in the algae, in fungi we also see a close association between viral and host sexual biology.

Ascomycetes

The history of viruses of ascomycetes is related to the history of interferon. In the 1950s, the Lilly labs were commercializing and improving penicillin production with *Penicillium stoloniferous*. It was observed that a generalized antiviral activity could be isolated from this fungus, which eventually came to be known as interferon. Later, it was shown that this antiviral activity was due to persistent production of dsRNA containing a particle that was shown to be penicillium stoloniferous virus (PsV-S), and that this particle was inducing an antiviral state in animal cell cultures. This was the first description of a mycovirus. The virus had two RNA components: S1, which encoded RdRp of the partiviridae type, and S2, which encoded another gene. Virus fungal infection was inapparent, and all industrial isolates were subsequently shown to be persistently infected. Some early reports suggested that this virus could lyse related penicillium strains (P. variable). PsV-S also appeared able to affect early sporulation. Virus infection was very stable in its host. In fact, after decades of passage for industrial usage, there are essentially no changes in nucleotide sequence of this virus. Recall that this is an RNA virus with an RdRp that has been measured to have very high error rates. Such genetic and biological stability strongly suggests the use of a persistence module by the virus, but this has not been described. The S2 RNA segment of PsV-S also appears to be related to dsRNA (of RdRp) of chestnut blight virus, cryphonectria parasitica (see below). This RdRp has a mitochondrial codon bias, a curious feature of many fungal viruses. Since then, numerous other fungal partiviruses have been described, some of which show sequence relationships to other dsRNA viruses. Of some interest is P. chrysogen virus (Pcv), which has the distinction of four dsRNA segments (with individual ORFs), one of which encodes an RdRp. The chrysovirus virion is thus related to partivirus, but with more segments. However, in contrast to most other dsRNA viruses, this virus has unusual capsid structure (T = 1, not T = 2). Other virus-harboring species of ascomycetes include *Coccidioides*, *Histoplasma*, *Pneumocystis*, and *Aspergillus*. The highly aerobic character of *Aspergillus* is interesting from the context of ROS, evolution and T/A modules. These fungi can tolerate highly oxidizing environments and are known for the production of lots of secondary metabolites. However, unlike the basidiomycetes described below, these ascomycetes metabolites are less prone to be highly toxic to humans. Thus, for example, *oryzae*, an ascomycetes species, is used for sake fermentation. Edible species (mushrooms) are also mostly ascomycetes. These edible species includes *Penicillium*, which makes blue cheese. In contrast, human toxicity due to basidiomycetes includes the highly poisonous toad stools. As basidiomycetes also has many species of plant pathogens, such *Ustilago hordei* (smut) and *U. maydis* (smut fungus), it seems their tendency to produce toxins also applies to plants. This general difference in animal and plant toxicity between ascomycetes and basidiomycetes reminds us of the differences between bacteria and Archaea, in that Archaea tend not to produce human toxins, and suggests some interesting implications regarding the conservation of fungal T/A or identity modules and the evolution of animals and plants.

Neurospora intermedia mtDNA Parasites

As *Neurospora* is a highly studied model of ascomycetes species, it has been the source of much molecular information. *N. intermedia* can be colonized by Kalilo 'linear DNA plasmids' that are associated with host life span. These elements are able to induce senescence in their host via integration into mitochondrial DNA and interruption of the *cox* gene. These are also able to persist as extragenomic elements. Unlike similar elements found in some yeast species, these do not appear to encode 'killer' functions. L-DNA of this 'linear' plasmid has an attached 5′ terminal protein and also encodes corresponding viral-like DNA polymerase. These two features are basic molecular characteristics of the adenovirus family. Thus, this seems to represent an early (persisting) lineage of adenoviruses. Similar Ad-like DNA parasites are prevalent in some yeast species, such as kluyveromyces. as described below. Field isolates of *Neurospora* show that L-DNA-like elements are common, but the corresponding life span of the colonized host does not usually show alteration. It is interesting to contrast this with other known genetic parasites of *Neurospora crassa*, such as TAD. Field isolates rarely (1:100) harbor this element. The much referenced TAD retroposon element is a favorite of the LINE community that likes to refer to its occurrence in fungi as evidence that LINEs predated the evolution of retroviruses. This field clearly indicates that TAD is a recent and rare colonizer of *Neurospora* (which harbors chromovirus sequences) and thus makes TAD of questionable general biological significance regarding *Neurospora*.

Podospora mtDNA Parasites

Although *Podospora* has a similar life strategy to *Neurospora*, a sessile hyphal organism with a potentially long life span, it appears to differ significantly in how it is affected by known persisting genetic parasites. *Podospora* can be parasitized by an element associated with longevity, known as PAL2-1. This is a linear DNA parasite that can also integrate into mitochondrial DNA or remain as an extragenomic element, much like L-DNA above. However, in this integration, PAL2 renders the mitochondrial DNA more stable from normal degradation in an age-dependent way. This is thought to extend the life span of the host. Numerous forms (78) of PAL2 have been observed from natural isolates. Also like L-DNA, they have a 5′ terminal protein. The DNAs encode two ORFs which include a DNA and RNA polymerase, both related to the types of polymerases specific to adenovirus. DNA replication is also similar to that of adenovirus (protein-primed DNA extension). It is therefore very curious that rather than inducing senescence as in *Neurospora*, PAL2 induces longevity. In this case, integrated copies of PAL2 are suppressive of mitochondrial genomes, suggesting some form of identity module is at work. Curiously, in both *Neurospora* and *Podospora* fungi, these linear DNA parasites are organelle-associated, whereas in animals, adenoviruses are non-integrating nuclear parasites.

Mitochondrial dsRNA Parasites

The fungal agent of Dutch elm disease is *Ophiostoma novo-ulmi*. However, the ability of the fungus to be pathogenic is associated with the presence of a mycovirus, a dsRNA genetic parasite. The mitochondria of these fungi are colonized by dsRNA, which encodes an RdRp that has a mitochondrial codon bias. The viral sequence shows phylogenetic variation that is congruent with that of its basidiomycote fungus host, indicating that the evolutionary split in the host is also reflected by the related genetic parasites. In addition, this RdRp is also related to the 20S and 23S yeast plasmids. Chestnut blight (*Cryphonectria parasitica*, a filamentous ascomycete) pathogenicity is also negatively mediated by a mitochondrial parasite, a small dsRNA that also encodes an RdRp. This agent has received much attention due to the highly destructive nature of this disease. The chestnut blight hypovirus reduces the virulence of the disease and has been highly studied as a possible biological control agent. Hypovirus biological control has been effective in Europe, but not America. Hypovirus is made into VLPs that lead to poor asexual sporulation (conidiation) and result in female infertility, which in turn results in a lowered pathogenicity. It is very interesting that the virus encodes genes that are able to suppress the fungal host mating-type response, a mating system which is similar to *S. pombe* mating factor. The virus codes for Mat-1 and Mat-2 which produce two pheromone precursors. The pheromone signals are transmitted in the host fungi via an alpha subunit of membrane-bound G-protein. Hypoviruses often affect host growth via G-proteins and cellular signal transduction. This example links several crucial issues. A persistent genetic parasite uses and affects host sexual identity and communication via a signal transduction system (G-proteins). As will be presented, this process is an example of a theme that is of central importance to the control and evolution of group identification in metazoans. Hypovirus transmission in natural settings has also been studied. In some cases, such as in Southern Switzerland, vegetative incompatibility is not a barrier to hypovirus transmission as might be expected. In other cases, in the USA, especially Michigan, natural transmission was more problematic. Here, distinct hypoviruses were hypovirulent by inducing mitochondrial dysfunction which was not well transmitted. Hypoviruses are not transmitted during mating of ascospores, indicating a further link to host reproductive strategy. The link between mycovirus and virulence is also seen in several other fungal examples, such as Nectria radicicola, responsible for ginseng root rot. This fungus harbors four dsRNAs (seen in about 30% of field isolates) that are associated with increased virulence and sporulation. In this case, however, the viral RdRp is directly involved in virulence. This RdRp is related to that of plant cryptovirus. Virulence would seem an odd phenotype for a viral RNA polymerase and suggest a much more intimate, but unknown role for RdRp in host biology. As will be presented below, host RdRp plays major role in host innate immune systems (siRNA). In Nectria, viral virulence is regulated via signal transduction involving cAMP-dependent protein kinase pathway. Thus the themes of persisting genetic parasites, sexual identity, signal transduction, mitochondrial dysfunction (toxin production) were all present and interrelated in filamentous fungi.

It is curious to consider why these persisting RNA and DNA fungal parasites were found in mitochondria, as this is uncommon in higher organisms. One implication is that there may be a link between the destructive oxidative potential of mitochondrial respiration and fungal host identity modules. For example, mitochondria could provide T/A (ROS) identity systems that the viruses would need to interrupt. A relationship of mitochondrial parasites to the life span of its fungal host would be consistent with such an idea. It is also interesting that some fungal mitochondria appear to be persistently infected with an apparent zoo of dsRNA elements. Such complex mixtures imply the existence of hyperparasitic relationships amongst these elements since many dsRNA sets are commonly observed together. One example is Gremmeniella abietina, in which five distinct dsRNA can be commonly seen. Within these five elements, representatives of a mitovirus, a partivirus and also a totivirus-like RdRp are all present. There is also a general split between ascomycetes and basidiomycetes with regard to mitoviruses. Helicobasidium mompa tanaka (violet root rot) hosts a dsRNA in mitochondria (HmVI-18) which is related to several mitoviral RdRp of ascomycetes. However, this was the first example of 'mitovirus' of ascomycetes, as most are from basidiomycetes. Viruses of basidiomycetes mitochondria are significantly more numerous as indicated below. Other mitochondrial viruses are found in Rhizoctonia solani. Fusarium-pose is a pathogen of wheat in which host damage is via mycotoxins. Here a graminearum hypovirus is involved and VPLs are made. This virus has a partivirus-type RdRp that is highly conserved in its host and similar to RdRp found in protozoa. Divergence of these fungal and protozoan RdRps matches divergence of their host, implying a very deep phylogenetic relationship between viruses and host. In the case of this hypovirus, vegetative compatibility is directly associated with the specific hypovirus found. Field studies showed that hosts of the same vegetative compatibility group have the same virus, whereas hosts of different compatibility groups have different viruses. This observation supports the idea that a main role of vegetative incompatibility (group identity) is to control mycovirus transmission.

Although there is a general tendency for fungal viruses to be asymptomatic and not be vertically (spore) transmitted, this is not always the case. Agaricus bisporus (white mushroom) is susceptible to La France disease virus due to infection with an isometric virus (LIV). Unlike most mycoviruses, however, this virus is pathogenic especially in fruiting bodies (mushrooms). The isometric LVPs contain dsRNA. In fact, up to nine dsRNAs can be observed in diseased fruit, but some of these RNAs are also found in healthy fruit, suggesting asymptomatic persistence for these elements. The disease situation may be more complex, however, as an ss+RNA bacillus particle can also be seen in disease, suggesting that it may result from a mixed viral colonization that could perturb resident persistence by virus.

Yeast Species of Fungi: Retroviral Colonization and Sex (Pheromones)

A large majority of fungal species are filamentous. Some, however, have mixed lifestyles and undergo morphological transition between filamentous and unicellular yeast forms. Relatively few fungi are strictly yeast species. Phylogenetic analysis now supports the idea that yeast species evolved significantly later than the filamentous forms. It has been proposed that not until the evolution of high sugar fruit of plants did the alcohol-fermenting yeast species develop. However, alcohol fermentation is clearly energetically inefficient so its production cannot be rationalized as the result of improved yeast metabolism. But alcohol is toxic to most fungi and thus it may represent a toxin part of T/A strategy that kills competing organisms unable to detoxify alcohol. By creating a habitat on fruit with high levels of alcohol, yeast may preclude colonization by non-fermenting fungi. Yeast species have relatively small genomes and have been well evaluated from the perspective of endogenous retroviral-like elements and association with host biology. No RIP system is present in yeast and this seems to allow some retrovirus colonization. The first element to be characterized was the Ty1 element, which should now be properly called a retrovirus. Ty1 has received much attention as an example of a transposon representing the apparent success of selfish DNA. Ty1 encodes both a functioning RT and GAG and was shown to be highly active, leading to the production of lots of RNA, but with very low rates of transposition. It now appears that the Ty1 copy number is inhibited by the RNAi system. Recently, about 40 host co-factor genes have been characterized to restrict Ty1. The common yeast lab strain has about 32 Ty1 elements that produce substantial quantities of cytoplasmic RNA, assemble them into VLPs, and express sufficient RT activity to make cDNA copies. Since yeasts have a closed mitosis, this cDNA must have systems for import into the nucleus. Currently, it appears that Ty1 virus can be transmitted during mating, indicating that it may not be trapped within one cell as is often proposed. Ty1/3/5 all appear to insert into silent regions of chromatin, which should result in silencing of the virus, a beneficial situation for viral persistence. Most interestingly, a natural field isolate of yeast (*Saccharomyces paradoxus*) was observed to be missing Ty1, although many solo LTRs were present. Clearly Ty1 maintenance is not essential in all *Saccharomyces* species. However, when this strain was repopulated with Ty1 and mated into a heterozygote diploid strain, the Ty1 RNA expression was suppressed and heterozygotes tended to lose Ty1 element. Thus, as with the filamentous fungi, with yeast species we again see a link between viral colonization and host mating and sexual reproduction. A brief outline of yeast mating system as it relates to pheromone production is presented below.

Killers and Group Immunity in Yeast Species

Yeast species can also be colonized by dsRNA and dsDNA parasites that are very similar to those described above for the filamentous fungi. However, in the case of some yeast species, these agents are clearly viral systems that use addiction modules and impose a group identity onto their host, similar to the bacterial story presented in the last chapter. These are the killer viruses found in yeast and smuts. Such genetic elements are common in field settings. In one field study, 600/1,800 yeast strains evaluated harbored such cytoplasmic elements. One well-studied DNA element is found in the Kluyveromyces lactis killer strains. This strain is colonized by pGKL1, which is able to kill related yeast species. Like the DNA elements noted above, the pGKL1 element encoded an Ad-like DNA polymerase as well as an Ad-like terminal replication protein. This also resembles Phi-29 phage of *B. subtilis*. In addition, three subunit killer toxin genes and immunity to toxins are also encoded. The toxins are membrane-bound proteins that are antiproliferative via zymocin production. PGKL2 is another related element but in addition to the Ad-like DNA pol, it also encodes an Ad-like RNA pol gene. PGKL2 is needed for pGKL1 maintenance and immunity. Pichia farinose encodes the killer toxin, SM KT. A killer element found in Pichia is similar to pGKL2. However, this element is not species specific and can be transmitted to other species, such as lab yeast strains (*S. cerevisiae*). But the resulting colonized yeast only maintains the plasmid in a haploid host. Other examples, such as pichia inositovora, also have three cryptic plasmids. Thus stable plasmid persistence is often linked to the host reproductive state in a species-specific way. Along these lines, it is very interesting that asexual Fusarium oxysporum (yeast-like ascomycetes) are 100% colonized by related plasmids that are transmitted via asexual spores. Although the molecular details are often lacking, all major fungal groups appear able to encode killer toxins that are often membrane proteins. This includes Pichia, Zygosaccharomyces, Hansenula, Kluyveromyces, Peninillium and Gibberella species. *U. maydis* has three killer toxins: KP4, KP6 and zygocin, and KP1 and KP4 are known to block Ca++ channel function in mammals. This point will be emphasized in the next section dealing with the evolution of ion channels in animals as seen in *C. elegans* and their involvement in sensory function.

Yeast also harbors killer dsRNA viruses that can be vertically transmitted. In 1963, a strain of *S. cerevisiae* was reported to harbor dsRNA killer viruses that were later shown to be composed of an L1 dsRNA 4.7 kbp (capsid encoding), an M1 dsRNA 1.9 kbp (encoding dimeric exocellular membrane toxin) and specific immunity to that toxin. Some strains also harbor M2, which codes for a second killer toxin. These strains maintain the dsRNA at several thousand copies per cell so the viral load and viral gene products are substantial. The virus is maintained by vertical transmission and defective interfering copies of dsRNAs are abundant. These viral genes are also involved in incompatibility with other viruses and genetic parasites. Clearly, they fit the definition of addiction modules that also provide group identification and immunity to colonized yeast. By the late 1960s, the K27 toxin-producing strain of yeast was isolated. K27 toxin production was shown to be a widespread phenomenon and due to a related dsRNA encapsulated mycovirus. It is now known these are three classes of the K1, K2 and K28 viruses, which are composed of a mix of helper virus and toxin-encoding satellite virus. Immunity to toxin is virus-encoded and the helper virus alone seems to have no phenotype. Many yeast genera have killer systems which are mostly asymptomatic infections for the colonized host but lethal to similar, uncolonized host species. All are transmitted horizontally via mating (or heterokaryon formation for hyphal-forming species). Interestingly, only in *Candida* species (which has no known sexual cycle) and *Podospora* species have no dsRNA viruses been found. Clearly, dsRNA and DNA viruses have had profound effects on fungal sexual biology and group identity and directly contribute to the main systems of group or self-identity (vegetative compatibility, RIP, siRNA and killer toxins).

Other Linear DNA Yeast 'Plasmids'

Over 20 linear dsDNA plasmids of fungi are known, but not all of them code for killer functions. They are, however all stable persistent parasites that do tend to have the common characteristic of coding for a terminal Ad-like protein as well as an Ad-like DNAdep pol. However, the phenotypic or evolutionary consequence to their host or to other competing genetic parasites is mostly unknown.

Yeast Mating, Pheromones and Innate Immunity

Possibly, the most basal form of group identification is that used to identify (and allow) mating partners. In yeast species, this identification involves the fusion of haploids of opposite mating types. The best studied fungal mating system is that of *S. cerevisiae*. It has two mating types, a and alpha, that affect the activity of a single MAT locus. Peptide pheromones are directly involved and these have chemically attached farnesyl group, which limits diffusion and promotes formation of chemical gradients. As mentioned above, the secreted alpha factor affects cell division and creates a clear zone of inhibited cell growth via G1 cell cycle arrest. This clearly resembles a toxic agent. Haploid cell fusion is then followed by nuclear fusion. Although diploids can be stably passed, if they are starved for nitrogen (disrupting T/A modules), meiosis is induced. The process of induction involved membrane receptor G-protein signal transduction that targets protein kinase (STE20) that will then affect transcription factor of MAT locus. The link of this pheromone signaling process to virus transmission and innate immunity is therefore clear. For example, treatment with alpha factor pheromone plus exposure to H_2O_2 will induce apoptotic markers, such as DNA fragmentation, chromatin condensation and accumulation of reactive oxygen species (ROS). Thus innate immune responses are involved in haploid generation, reminiscent of red algae sexual reproduction. With respect to the above killer virus section, it is interesting that K28 killer toxin will also induce apoptosis in which it is clear that ROS is the basal trigger. High toxin doses will prevent this induction but may induce apoptotic independent killing. As discussed below, the induction and use of apoptosis in a multinucleate hyphal organism to control cell fates (other than gametes) presents problems. Virus, sex and immunity systems thus seem to be intermingled systems in yeast species.

Programmed Cell Death, Mating and Pheromones in Multinucleate Cells

As mentioned above, in some filamentous fungi, a relationship between hypovirus pathogenicity and host mating type was mediated by a pheromone produced by viral genes. In this case, the virulence affecting gene was the virally produced pheromone. Mating compatibility and virus transmission appear to be inextricably linked in filamentous fungi. All known species of Ascomycota have HMG genes of alpha mating type, even asexual species. Mating-type recognition is thus more conserved than is sexual reproduction. In a hyphal species, filamentous fungi like ascomycetes can fuse two organisms via anastomosis of the growing hyphal tips. This results in a common cytoplasmic bridge that provides a nuclear/viral transfer system and abrogates the need for an extracellular virus. Since fungi are sessile, the tips will grow towards or away from the appropriate mate for fusion. The pheromone produced by one organism is an inhibitor for the growth of that same species. These pheromones are clearly a component of a system of self-identification that prevents self-'invasion' and promotes the 'colonization' of nuclei/virus into genetically different neighbor organisms, but also allows mixing of mitochondria along with resident mitochondrial viruses. If the hyphal tip encounters a genetically different fungus, it can form a heterokaryon (a parasexual mating). This will be dependent on the *het* loci. Various ID systems seem to operate, resulting in vegetative incompatibility. If compatibility is correct, nuclear division results. If compatibility is not correct, nuclear decay or programmed cell death (PCD) will result. Thus, this version of programmed cell death originally evolved in fungi to control nuclear transfer and appears to be an identity T/A module. Sexual mating also occurs and involves fusion of haploid cells. Haploids and sexual reproduction are often induced as a consequence of nitrogen starvation. Here too, mating factors are also involved in the induction. Haploid mating can involve various identities and occur between two sexual types or involve multiple mating types. Some fungi can fuse with no specific mating factors. In other fungi, homeodomain proteins are involved and appear to provide an internal identity system. Fungi such as *U. maydis* also use internal transcription factors along with external mating factors to regulate this fusion. Thus the early use of these homeodomain proteins and the induced transcription factors appears to have been for the purpose of internal mating recognition and compatibility. Often, meiosis is initiated by a peptide hormone, such as lipopeptides with attached isoprenoid groups. The alpha factor of yeast MAT noted above is this type of pheromone. The chemistry of these pheromones bears some resemblance to a mixture of red algae and gram-positive bacteria pheromones. The level of production of these factors can be substantial. For example, in *N. crassa*, mta-1 pheromone is the most abundant transcript in nitrogen-starved cells. Since these pheromones can be toxic, such high-level pheromone production could be part of a defense system as was seen in algae. Mutants in mta-1 do not produce ascospores. Higher fungi like *N. crassa* produce specialized sex organs and the organs undergo fusion in response to pheromone production. Sexual cell fusion between unlike mating-type individuals generally triggers cell death, which is a version of genetic incompatibility. In *Podospora anserina*, the genes involved in this PCD process are homologous to those of autophagy (PrB protease). However, the use of nuclear decay (PCD) as a system of genetic identification poses some interesting dilemmas in a multinucleate hyphal organism. In a clonal multinucleate cell, essentially the entire organism would be susceptible to destruction if PCD were induced against a resident nucleus. To avoid this, the new genetically distinct individual nucleus (heterokaryon product of either sexual fusion or vegetative hyphal transfer) must be identified for safe PDC destruction without also risking total self-destruction of resident nuclei. This targeted destruction seems to be the main outcome of vegetative incompatibility. However, it is currently thought that vegetative incompatibility exists primarily to limit the transmission of all the cytoplasmic viruses found in filamentous fungi so its activity against nuclei would seem to be a byproduct of this antiviral activity. But, as noted above with hypoviruses (chestnut blight disease) and others, transmission in the field across such incompatibility barriers has been observed so it seems this is not a stringent system of virus control.

What is the relationship between pheromone production, innate immunity and sexual versus asexual reproduction? *Aspergillus nidulans* can secrete hormone-like oxylipin (used by red algae for gamete-specific innate immunity). In *Aspergillus*, these oxylipins mediate the balance between asexual and sexual spore ratio. Oxylipin production is dependent on three conserved fatty acid oxygenase, PpoA/B/C (present in both ascomycetes and basidiomycetes). Mutations in these genes increase asexual spore production and fail to produce the mycotoxin sterigmatocystin (and also overproduce penicillin). Thus in *Aspergillus*, we again see a basic link between pheromones, toxin production and sexual reproduction. Oxylipins regulate toxins and other secondary metabolites via surface interaction with a G-protein receptor (FadA allele) that transduces signals at transcriptional level to positively and negatively control metabolite expression. Similar regulation is seen in Fusarium sporotrichioides' production of trichthecene mycotoxin. Recall that large numbers of secondary metabolite genes present in fungal genomes are related to fatty acid oxidation. Oxylipin production was also important for host colonization of pathogenic and parasitic fungi, similar to the situation in red algae.

RNA Silencing, dsRNA and Retroviral Immunity

Another significant innovation in genetic identification seen in many fungi is the presence of the siRNA system of post-transcriptional silencing. This represents a new identification system that is able to recognize RNA and repeat DNA via dsRNA regions and results in the silencing of RNA expression. SiRNA homology-dependent silencing is found in most, but not all, fungi. It is not present in budding yeast, but is present in basal zygomycetes. Therefore siRNA appears to have been a basal acquisition, but was lost in some subsequent fungal lineages. The main purpose of siRNA appears to be to silence expression or eliminate the sequence of genomic retroviruses and retroposons. The MIP and RIP silencing systems of *Neurospora* appear to be systems that recognize repeat elements and induce DNA methylation of the template. The RIP system of *Neurospora* renders this fungus intolerant of genomic retroviruses and retroposons present in more than one copy. Recall that TAD is a LINE-like element found in lab strains of neurospora, but which is absent from the bulk of field isolates. However, *Neurospora* appears to harbor lots of remnants of TAD-like retrons that have been RIPed. Thus the *Neurospora* genome has very few repetitive loci, outside of rRNA (apparently under nucleolus protection). Homology-dependent gene silencing also operates during the process known as quelling, which functions in vegetative haploid cells. Quelling uses dsRNA (RdRp – qdc-1) to induce specific mRNA degradation and operates via the action of dicer (RNaseIII) to generate short interfering siRNA resulting in post-transcriptional gene silencing (PTGS) via DNA methylation. It seems likely that such a process may contribute to the fact that most fungi do not transmit fungal dsRNA viruses during ascospore formation. However, the main target of the siRNA machinery also appears to be suppression of endogenous retroviruses (chromovirus) and retrotransposons. For example, repression of LINE-1-like TAD element requires qde-2 plus dicer (not qde-1, qde-3). This indicates existence of distinct pathways for silencing repeat elements. Repression targets can be evaluated with mutants, such as lys 9H3 methylation loci. Such mutants show that 90% of silenced loci map to sequences that are relics of RIPed transposons, thus clearly identifying them as a major target of RIP. RIP is a premeiotic process that results in 5-methyl cytosine, which deaminates to yield a T resulting in a C to T point mutation. Meiotic silencing of unpaired genes occurs during meiosis (at the diploid stage), but requires gene transcription to silence. This operates via semidominant sad-1, a mutant which suppresses sexual phenotypes. The sad-1 + wild type allele is an RdRp and it is noteworthy that RdRp activities are core RNA viral functions that also suggest a likely viral origin of this entire system. The point mutating capacity of RIP was not retained in higher eukaryotes, however. Since multiple chromovirus colonization and subsequent inactivation is highly associated and congruent with the evolution of fungal lineages, post-transcriptional silencing seems to operate in concert with viral colonization to mold fungal genomes and identity systems. However, given the presence of siRNA system, it is highly curious that fungi are so prone to persistent infection with dsRNA viruses. This suggests that siRNA systems are not really designed as adaptive immune systems against extragenomic dsRNA virus, but rather siRNA is needed to silence resident retroviruses into persistence and eliminate genetic competitors (retroviral and dsRNA) during sexual reproduction.

Neurospora *and Rice Blast: LTRs and the Origin of Genetic Novelty*

GPCRs are one of the largest protein families in human and animal genomes and are involved in much sensory and group detection. As mentioned above, in fungi, some G-proteins have a sex-associated function for group identification.

Also, in yeast, the membrane receptor for alpha mating factor is a G-protein with clear similarity to both rhodopsin and odorant receptors. G-proteins are also involved in H_2O_2-induced apoptosis. However, fungi have a paucity of G-proteins and only 10 typical GPCR genes have been identified in sequenced fungal genomes. This contrasts with the large number of typical G-proteins as will be presented in *C. elegans* below. However, with the sequencing of the *Magnaporthe grisea* genome, it was possible to probe for more distant GPCR family members in this fungal lineage. Using both the cAMP receptor and steroid receptor (mPR) from Dyctyostelium as phylogenetic probes, a novel class of receptors, called PTH111, was identified. These genes were related to pathogenic genes present in rice blast; 61 PTH111 related proteins with homology in 7 TM domains and EGF-like extracellular domains were seen. Related proteins were found in other Ascomycota (pezizomycotina) but none were found in other fungal lineages such as Basidiomycota, yeast or in other eukaryotes. Instead, the other G-protein receptors were conserved across the Ascomycota and Basidiomycota genera. It should be recalled that oxylipin control of asexual to sexual spore ratios and toxin/antibiotic production was mediated via one of these heterotrimeric G-protein receptors. Why would rice blast fungi have acquired a unique set of G-proteins not present in other fungi? That many of these genes are involved in virulence suggests they have a role in induction of toxic situations. Since the G-protein role in MAT pheromone transduction is more broadly conserved, this would appear to have been the ancestral role of such G-proteins. The toxic (growth-suppressing) potential of pheromones could also have been relevant. The large diversity of new PTH111 genes can also infer their role in host identity or immunity. Another intriguing difference between rice blast and *Neurospora* is that during evolution, rice blast has rapidly acquired sets of genes associated with virulence. These new gene sets are associated with LTR elements, thus indicating that retrotransposition has contributed to the virulence adaptability of rice blast and involves repeat LTR elements. For such a process to be allowed, however, rice blast must not have a RIP system like that present in *Neurospora* (expressed during mitosis). This, in fact, seems to be the case as neither meiosis nor RIP occurs in rice blast. Rice blast therefore appears to present an example of a major bifurcation in fungal evolution regarding the role that retroviral-derived LTRs have played in host genome; *Neurospora* prevents LTR repeats, but rice blast uses them. Most fungi are similar to *Neurospora* and limit repeat DNA elements in their genomes (hence their generally small genomes). With rice blast, the loss of the RIP/MIP system that prevents the colonization by retroparasites seems to have allowed retrotransposition to now function as a main process for the acquisition of genetic novelty associated with virulence. This characteristic tolerance for repeat (LTR) DNA has been preserved in plant and animal genomes as well.

What can we say about the likely source of novel GPCR proteins? Where did they come from and why are they lineage specific, especially the PTH111 receptors of rice blast? This question is of great importance since this family of proteins has evolved to become primary sensory detectors of group identity in all metazoans, including recent human evolution. We do not know the origin of the PTH111 genes. However, later I will consider the role of viruses, like specific lineages of large DNA viruses (e.g., herpes gamma) as possible originators of GPCR genes.

General Lessons from Fungal Group Identity as Basal Eukaryote

The concept that a trinity of linked functions, introduced in the first chapter (immunity, identity and persistence), were involved in host group identity also appears to apply to sessile, multinucleate fungal organisms. Persisting genetic parasites can clearly affect fungal group identification systems, such as vegetative compatibility. Several significant biological differences, however, are seen between the fungi and the filamentous algae. For one, fungi use non-motile sex cells, which no longer require a chemotactic pheromone response to find or identify their mates. Instead, they use directed growth of hyphal tips in response to pheromones to find mates. Fungi also show major differences from algae in their relationship to viruses. The absence in fungi of large lytic or lysogenic DNA viruses, common to algae, dinoflagellates and all prokaryotes, is particularly noteworthy, since this represents a significant transition of a most ancient and long-lived virus–host relationship. It is especially noteworthy that we now see in the fungi the prevalence of persistent infections by asymptomatic dsRNA and small linear dsDNA viruses, many of which are found in mitochondrial agents not common in algae. Many of these mitochondrial agents can also encode killer/immunity functions that provide group identification to colonized host. Fungal genomes have also been colonized by waves of lineage-specific endogenous retroviruses called chromoviruses. In this, the fungi are distinct from other early eukaryotes as these viruses were not found in their genomes. Most of these chromoviruses, however, have been inactivated by mutation and RIP/MIP systems. This observation also identifies another significant innovation in fungi. Fungi have evolved siRNA (dsRNA) based genetic identification systems (including RIP/MIP) that suppress or silence endogenous retroviruses and their derivatives. These gene-silencing systems limit the colonization of fungal genomes by retroviral agents in most fungi. However, some fungi, such as rice blast, have lost their RIP systems, but utilize LTR-mediated transposition to rapidly evolve and acquire gene sets (G-like proteins) involved in virulence. Fungi have also evolved vegetative compatibility systems that target and affect transmission by dsRNA viruses. These dsRNA viruses are highly common, but their transmission is mainly horizontal via mating or heterokaryon formation, a frequent occurrence in nature. Vegetative incompatibility is thus a novel antiviral genetic identity system that can also destroy nuclei transferred as parasexual heterokaryons that are incompatible. This nuclear destruction occurs by a process of programmed cell death that has many (but not all) of the components of mammalian apoptosis. Fungi are also noteworthy for the large number of secondary metabolites (toxins, antibiotics) that are produced by complex synthetic pathways. Finally, the most crucial generalization is that fungi represent the basal eukaryote that went on to evolve into all metazoan eukaryotes. Thus these characteristics should be considered as providing molecular foundations for that evolution.

Sponges (Porifera)

Sponges represent an important transition in evolution as the crown group for all of the metazoan kingdom. This is the lowest of the metazoan phyla, ancestral to all animals, and appears to have evolved about 800–650 million years ago. It now consists of about 5,000 species, so it is much less numerous than fungal species. Here again, we see a distinct shift in life strategy from that of clonal filamentous fungi and algae ancestors that were mainly networks of multinucleate clonal organisms, with common cytoplasm linking the network of individual cells. In sponge, the network becomes predominantly that of separated individual cells which are not necessarily clonal and also have flagellated interior cells. Gametes are motile, but the adult organism is non-motile animal, resembling sessile fungi in this lifestyle. In most sponges, gametes fertilize eggs and resulting embryos develop within the adult. In a sense, we see the reappearance of flagella as motor organelles and sensory receptors. However, these organelles are no longer used to propel gametes, which now move via pseudopod process. It seems likely, however, that not all of the features of extant sponges were present for the initial evolution of sponges. Although Porifera is the oldest metazoan phyla, it is composed of three classes: Hexactinellida (glass sponge), Desmospongiae and Calcarea. Hexactinellida are considered the oldest class, based on both phylogenomics and fossil evidence left by deposits of its silicate fossils (siliceous spicules). These fossils were found in shallow sea beds from the early Cambrian (in what is now in South China). Extant species, however, are now either restricted to deep or cold waters. Calcarea are calcareous sponges that are more related to higher metazoa phyla. Hexactinellida has the unique and unusual distinction of being the only metazoa that possesses mainly syncytial tissue rather than mainly unicellular tissue. Thus, each body is a giant macrophagic synctium surrounding discrete groups of undifferentiated cells, clearly resembling one giant multinucleated cell. Thus the early sponge resembled the fungi in having communicating multinucleate organization that consequently must also have identity systems compatible with multinucleate states. As discussed with fungi, a multinucleate organism presents a problem concerning how apoptosis can operate to control somatic cell fate in such a situation. Like fungi, a sessile filter-feeding sponge would also seem to present an ideal environment for colonization by genetic parasites. It is thus most surprising that there are no reports of any viruses known for any sponge species. This is particularly intriguing when it is considered that sponges are highly prone to symbiotic relationships with various other organisms. Often, symbionts of sponges are colorful and able to produce light. Since sponges are commercially used and grown, it seems likely that significant occurrences of viral-induced disease should have been seen as it has been in most other marine plants and animals. Viral-induced mass die-offs and virus-induced sexual dysfunction of many other marine lives have not been difficult to observe. However, it remains possible that viral parasites have simply escaped observation, especially if they are persistent infections. Sponges therefore seem to represent another major shift in general viral–host relationship. In addition, sponge cells display a clearly distinct social behavior, in that individual cells of the same type will converge to form an organism. In fact, an adult organism can be dissociated into individual cells that can spontaneously reassociate. Sponges reproduce sexually and will release motile sperms that are able to find and fertilize an egg, resulting in a motile larva. It is therefore curious that individual cells and gametes retain the capacity to move but the adult is immotile. In this reassociation ability, sponge cells also display a capacity to recognize cells of the same type.

Sponge Biological Characteristics

Sponges do make sex cells and other cell types (flagellated collar cells, epithelial skin). They release sperm to internally fertilize an egg and produce a motile larva. However, sponges can be considered not to possess real tissue in the sense that all cells are totipotent. Unlike the algae with motile haploid gametes, there is little evidence that sponge gametes depend on pheromones for mate identification, although they are clearly capable of producing a very large number of pheromone-like and toxic metabolites. Asexual (diploid) reproduction of the totipotent cells can be used to regenerate the adult organism. Such totipotency, as well as the ability of cells to aggregate to form organisms, suggests that different species of sponges must have some type of identification system to recognize themselves. Dissociated adults cells will move like amoebas toward each other to form aggregates, suggesting that some type of quorum-sensing system is operating. A mix of dissociated cells from two species with distinct colors, however, only results in like-species cell aggregations, so motility and quorum-sensing must also be species specific. Sponges have no neurons. Although adult sponges are immotile, they appear to pump water and move in rhythm as if coordinated. This motion, however, is based on a condensation and swelling process of mesenchymal cells to generate a pumping action and is not due to actual cell contractions involving myosin. Motility in adult animals is a characteristic of *C. elegans* that will be presented below.

Genome

Genomes of sponges are not very well studied due to the fact that sponges have an unusual and highly infested life strategy with other organisms. Such rampant symbiosis suggests the likely presence of many T/A or addiction modules that would stabilize the cell association (discussed further below). However, this infestation creates a major problem in the study of sponge DNA due to the presence of contaminating DNA from so many epibiontic, parasitic and symbiotic bacteria, algal and fungal species. The dry mass of some sponges can be composed of up to 40% of these fungi and algae. Sponge DNA is therefore hard to purify, as few sterile regions may exist in adult tissue. Dissociated tissue or the use of gametes can partially address this problem and some progress has been made in the study of sponge genomes. Yet, vertical transmission of symbionts appears to occur adding further complication to sequencing projects. Recently, however, some researchers are proposing a rather remarkable (but controversial and yet to be confirmed) characteristic for sponge genomes. The sponge genome appears to be surprisingly large and may encode as many as 100,000 ORFs, making it the most gene-rich genome so far characterized. Recent CsCl density analysis of sponge DNA indicates it is of uniform density, so these genomes do not appear highly colonized by simple repeat sequences (satellite bands) and thus sponge DNA appears to be gene-dense. This very high number of ORFs compares to the 26,000 ORFs of the human genome. Why would such a simple organism need so many genes relative to much more complex organisms? As there is some uncertainty regarding the sponge genome, it may not be productive to speculate much about this issue. However, similar to what was proposed for sponge genome, the genome of the fungal *Aspergillus* also seems to be unusually large, but without the controversy of possible DNA contamination. Although most fungi have small genomes between 10 and 30 Mbp (encoding about 6,000 ORFs), the A*spergillus* genome was significantly larger (95 Mg) and encoded significantly more genes (33,500 ORFs). Why does a simple fungus, *Aspergillus*, have so many genes and how did this large-scale genetic novelty originate? Genomic analysis indicates the *Aspergillus* has many complex pathways for the synthesis of secondary metabolites. These biosynthetic enzymes account for much of these expanded ORFs but do not appear to have evolved from simple adaptations of previously existing single-function enzymes. Often, evolutionists have suggested that large-scale chromosome duplication has promoted the evolution of new protein function by freeing duplicated sequences to adapt to new functions. In *Aspergillus*, however, this does not seem to be the case. As presented above, the fungal RIP system present early in fungal evolution should generally prevent gene duplication. Consistent with this, analysis of *Aspergillus* fails to see evidence of massive genome duplication. Thus, it seems more likely that these metabolic pathways originated from an exogenous genetic source. However, unlike bacteria, direct colonization by large DNA viruses encoding complex gene sets would not appear to be an option for sponge or fungus since these organisms are not known to support any large DNA viruses. However, the filter-feeding lifestyle of a sessile marine sponge will expose it to many viral parasites present in the oceans. And although sponges themselves do not appear to harbor complex viruses, it is clear that many of their symbionts do (e.g., cyanobacteria which harbor lysogenic DNA viruses). We can expect that a symbiotic bacteria would be protected from lytic viral agents by living within a sponge cell (similar to PBCV-1, which cannot infect its chlorella algal host when symbiotic in its *Paramecium* host). A suggestion can now be inferred from these assembled observations. Sponges, with all their numerous toxic secondary metabolites, provide their epibionts with a cellular T/A habitat that protects them from non-persisting, acute genetic parasites. However, the origin of these protection systems (T/A modules) themselves are likely to derive from the persisting genetic parasites that successfully colonized their epibiont (i.e., bacterial) host and provided them with addiction modules as systems for the stable colonization of their sponge host. Is there evidence that the epibionts are the source of some of these predicted T/A or toxin gene sets?

Symbionts and Toxins

Sponges are notorious for their ability to symbiotically associate with cells of several other species, indicating a general tendency to tolerate or persistently enslave foreign cell types. They are also well known for diverse toxin production. Clearly, they must have distinct systems of group identity that allow such a mixed lifestyle. They also have surprisingly big genomes, able to make many of the secondary toxic metabolites. These two facts appear to be related. The sponge production of secondary metabolites is very chemically diverse and includes derivatives of amino acid, nucleosides, macrolides, terpenoids, prophyrins, aliphatic cyclic peroxides and sterols. Sponges have thus been an especially rich source of bioactive compounds, such as the original source of ara-C, a broadly active antiviral drug. They are also the source of many other antiviral drugs (active against herpes virus DNA pol and retroviral RT, including HIV). Furthermore, these secondary metabolites are often known to be important or crucial for interactions between sponges and their specific epibionts. If such interactions constitute sets of T/A modules, these could define, stabilize and enforce these species-specific co-habitations. One possible example is found in the Diclemnid sponge family, which often has obligate cyanobacterial symbionts, such as prochloron species. These sponge-associated cyanobacteria make complex (non-ribosomal) cyclic peptide neurotoxins originally and mistakenly thought to be encoded by the sponge genome. It is now clear that the cyanobacteria produce the toxic cyclic peptides. However, the sponge produces proteins (channels) that are tolerant of, or resistant to, this toxin. Thus we can consider this as a situation in which a bacteria T and sponge

A combine to provide a T/A module that supports symbiosis. Conversely, desmosponge, when aerated, expresses large quantities of tyrosinase, which synthesizes diphenols from monophenolic compounds. These compounds, which can be toxic to most organisms, are able to be used as a carbon source by the symbiotic SB2 bacteria of *S. domuncula*, which encode a detoxifying pathway that produces acetyl co-A. In this case, the bacteria provides A, and the sponge, T, of a T/A set. These sponge symbionts make lots of clinically active molecules that affect several eukaryotic life processes (such as actin formation, DNA polymerization, ion channel conduction). In fact, sponges produce the most diverse natural products of any organisms known and this feature represents a major novelty for this order. Some well-studied sponge toxins include myclamide A and petaemine. In addition, many brominated organic compounds are produced by sponges and often used as commercial flame retardants and fungicides. In fact, in one field survey in Antarctica, each sponge sample had > 35 brominated compounds (of which 14 were common to all isolates). These brominated compounds also include pyroles, which affect (reduce) voltage-dependent Ca^{++} levels, as measured by sensory neurons in rhinophore of sea slug. Also brominated indols and phenols can be highly produced, constituting up to 12% of the dry weight of some sponges. Consistent with a possible T/A relationship, the symbiotic bacteria are often involved in dehalogenating these same compounds. Since each species of sponge has its own distinct species of cyanobacteria, this further suggests that there exists a match between toxins/antitoxins and symbiont. Also, these cytotoxic metabolites often show intraspecific variation, such as domyclamide A and petaemine. Halogenated fatty acids (C 14–22) are also found in sponges. Thus we might understand the highly diverse capacity of sponges to produce toxins in the context of T/A modules that support highly diverse symbiosis.

Sponge Group Identity and Apoptosis

As the only non-motile animal, the ability of sponges to form an organism from the migration and aggregation of individual cells leads to the question of how sponges control this social interaction and group identification. It was long known that some type of auto- and allografting process is operating, since cells from different species do not mix. Sponges have molecules found in the extracellular matrix/basal lamina that are typical for multicellular organisms, such as adhesion molecules (galectin) and adhesion receptors (receptor tyrosine kinase, integrin receptor). Such molecules are known to be characteristic of metazoan basal groups and associated with tissue identity. It now appears clear that a system able to provide surface identification allows aggregated sponge cells to form an individual. Tysosine kinase receptor (TKR) has been identified as a major sponge recognition system. It has an extracellular Ig-like domain that is highly polymorphic and shows individual variation. Fourteen sponge-specific sequences are known that show genetic variation at intron-splitting Ig domains. These TKR genes seem to be parallel to the MHC class I-specific receptors of the human KIR multigene family of NK cells. The first sponge-grafting experiments were conducted in 1826 and showed that sponge will reject non-self. Attempts to graft non-same sponges resulted in rejection involving the formation of collageneous barrier. However, if individual cells were inserted into a foreign sponge, the cells were destroyed. Modern attempts that try to merge Geodia cydonium and Microciona prolifera at a two cell aggregate (primmorphs) stage result in clear induction in one of the primmorphs leading to apoptotic death. During this apoptosis, one can measure all the usual proteins (caspase, Bcl-2) associated with apoptosis. Sponges encode two Bcl-2 genes that are clearly related to Ced-9 of *C. elegans* (see next chapter). In addition, all Bcl-2 homologs, including those of vertebrate animals, have highly conserved structural features (see Fig. 5.9). This structure is that of a membrane (mitochondrial) pore protein with clear structural similarity to pore proteins of phage-encoded toxins. Sponges also have a (2′–5′) oligoadenylate synthase system, a core feature of interferon in higher animals. Thus in the sponge, programmed cell death via apoptosis involving a Bcl-2 homolog is used for group identification against non-self cells. In terms of sensory receptors, sponges do have a family of three GPCRs, which includes the main receptors for neurotransmitters, pheromones and for odorants used by fish and found in Dyctyostelium, but not in plants. One toll-like receptor, which uses an MyD88-type adapter, has also been reported, but the nfKappaB component of toll transduction has not yet been established. Curiously, sponges do encode receptors that are considered as nerve cell specific (channel proteins), but have no synapse or other neuron markers. This metabotropic glutamate receptor is of special interest in its connection with the origin of the nervous system and will

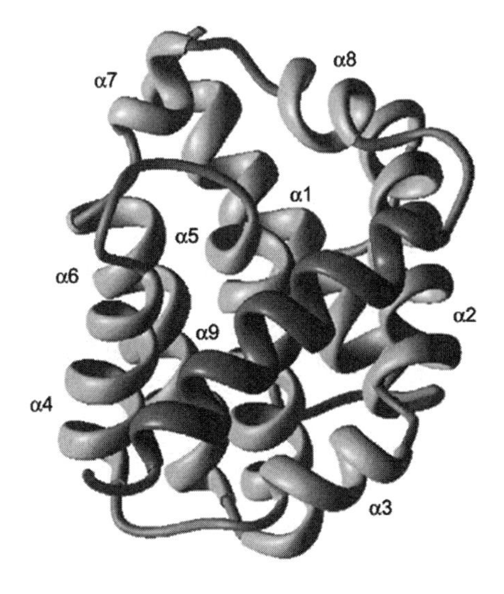

Fig. 5.9 Proposed 3D structure of Bax (an apoptosis protein) (reprinted with permission from: Petras, Oleinic zak, Fesik (2004), Biochimica et Biophysica Acta (BBA) Molecular Cell Research, Vol. 1644, No. 2–3)

be discussed below. It is not clear how sponges utilize this receptor, but it seems likely that it would be involved in some type of basal (non-neuronal) sensory system.

Summary of Sponge Systems

Sponges pose some interesting issues regarding the evolution of group identification. As non-clonal sessile organisms that can be assembled from migrating gametes and individual cells, they are atypical and use molecularly diverse surface recognition systems to specify identity. These surface proteins resemble the surface molecules used by mammalian NK cells. Programmed cell death via an essentially typical process of apoptosis in the sponge is used to recognize and kill cells that are non-self and not to specify the fate of tissue within the organism. Although sponges lack neurons, they encode one surface receptor that is clearly of the neuronal specific class. Sponges appear to be relatively free of lytic viral agents, but have a highly infested (symbiotic) lifestyle with other organisms, mainly cyanobacteria, algae and fungi. Along with this symbiosis, sponges make a very large and diverse array of secondary metabolites which requires a large number of novel biosynthetic genes. Sponges may have more genes than do highly complex animals. These secondary metabolites appear to represent a large number of T/A modules, some of which are clearly involved in maintaining the symbiotic relationship with their partners. In fact, many of these secondary metabolites are encoded by the symbionts.

Motile Eukaryotic Organisms

Before considering the motile multicellular eukaryotes, such as *C. elegans* below, it is worth briefly mentioning some things about motile unicellular eukaryotes, the unicellular protozoa. Protozoa have no cell wall, are motile via amoeba-like pseudopodia motion and show an ability to move towards and phagocytically engulf bacteria. These features seem like typical animal cells. Phylogenetically, however, protozoa are paraphyletic to metazoa and are not ancestral to multicellular animals. It is, however, curious that these motile unicellular organisms did not evolve to become social or multicellular. Protozoa resemble algae and fungi in the types of viruses they will support. Both dsRNA viruses and large DNA viruses are known to infect protozoa. For example, Leishmania is generally persistently infected with LRV1, a dsRNA (5.2 kbp) virus which has an overlapping pol/capsid gene. These viruses are persistent, conserved and congruent with the evolution of their host, like those of the filamentous fungi. Old and New World Leishmania species (seven strains) harbor their own versions of viruses, indicating the conservation and congruence of virus/host. Four of the dsRNA viruses that are known to persistently infect fungi and protozoa appear to have common lineage with UmVH of *Ustilago*, TvV of *Trichomonas*, LRV of Leishmania. Thus clear relationships exist between these viruses and their hosts. DsRNA viruses of *Cryptosporidium parvum*, *Entomoeba histolytica* and Giardia are clearly related to dsRNA viruses of yeast and other fungi. VLPs have been found especially in sporozoite cytoplasm, but not other cells, and have been called hereditary viruses since the host cannot be cured. This relationship, however, is not seen in all the species of this genus. Thus, protozoa resemble both fungi and algae in their virus–host relationship and are in striking contrast to sponges. Like sponges, however, some *Paramecium* are symbiotic and harbor algae (chlorella) which can be virus infected (i.e., PBCV-1). Mimivirus, the largest known DNA virus, was isolated from protozoa (amoebae *Acanthamoeba polyphaga*). Although the biology of mimivirus/protozoa remains unclear, the Sargasso Sea environmental sequencing project indicates mimi-related viruses are highly represented in the open ocean. It seems ironic that this most complex of all DNA viruses is found in the simplest unicellular eukaryote host. However, as we will see below, a sharp transition between viruses and host occurs with the origin of motile multicellular eukaryotes.

Dyctyostelium discoideum

This species represents many of the major functional transitions that have been much examined as a model organism for early animal evolution. Unlike *C. elegans* described below, however, *Dyctyostelium* is not necessarily a clonal multicellular adult, so like the sponge it is clearly aberrant relative to higher metazoans in this feature. *Dyctyostelium* species spend most of their lifetime as free-living individual haploid amoeboid cells that feed on soil bacteria and are thus highly exposed to soil pore toxins and virus. Motility is achieved by pseudopodia formation in a process involving dynamic F-actin polymerization and disassembly at new pseudopods. This process also appears to require ROS. This motility has clear similarity to that of human leukocytes. These free-living haploids can also divide by asexual fission. At this stage, the cells are asocial individuals. When food bacteria become scarce, a starvation response is initiated. Although the mechanism is not known, cells are also phototactic during this single-cell period. One pre-starvation phase change that has been observed is that Dd-TRAP-1 (protein related to tumor necrosis factor receptor) moves to the mitochondria, suggesting some mitochondrial involvement. The possibility of bacterial-like quorum-sensing system or T/A module disruption occurring in *Dyctyostelium* starvation has not been experimentally established. However, as a consequence of this starvation response, individual *Dyctyostelium* cells become social by producing cAMP, which functions as a chemical attractant. Cells move together to become a multicellular (non-clonal) slug, composed of about 10^4–10^5 cells. The slug is able to undergo coordinated movement, which also involves cAMP production at the leading edge. Movement involves the production of an extracellular slime sheath composed of proteins and cellulose, reminiscent of filamentous green algae. This slug is now able to differentiate into two cell types, spores and soma, following exposure to dark periods and calcium. A photoreceptor is clearly involved in this response. Differentiation results in the production of a stalk from about 20% anterior cells which die to produce a hardened cellulose sheath that projects upward. The remaining cells travel up the interior stark to form spores (which can also be of mixed ancestry). Thus here we see the application of programmed cell death as a normal component of multicellular development. This cell death has many similarities to apoptosis, such as chromatin condensation and DNA fragmentation. It differs from apoptosis by not involving the release of cytochrome *c* or the usual inducers (i.e., not via toll receptors or caspase) and is also associated with massive vacuolization (a process that resembles autophagy). However, mitochondria do appear to be involved and undergo ion leakage and a decreased electron potential prior to cell death. It therefore appears that multicellular *Dyctyostelium* uses a process of programmed cell death that is functionally similar to apoptosis (involving mitochondria membrane integrity), but is paraphyletic to it.

Dyctyostelium *Genome*

The full *Dyctyostelium* genome has now been sequenced. Like fungi, it is a relatively gene-dense genome, with 34 Mbp and 12,500 predicted ORFs. Similar to the 'sake' fungal, and also the sponge, genome, it appears to encode many genes involved in secondary metabolism of small molecules, but less so. Therefore, it is a gene-'efficient' genome and does not harbor massive quantities of repeat or parasitic DNA (similar to yeast, 13 Mbp and 5,500 ORFs). No bacteria-like toxin genes are known. Interestingly, membrane depolarization thus does not affect the motility of *Dyctyostelium*, unlike bacteria (their prey). Most *Dyctyostelium* genes resemble those of eukaryotes. *Dyctyostelium* have muscle-like actin. However, in spite of high gene density, it is also clear that *Dyctyostelium* is much more colonized by parasitic genetic elements than were some of the fungal genomes (*Neurospora, cerevesiae*). The nature of such parasitic elements is distinct, and in some cases (i.e., DIR element) unique, to this lineage. This distinctness has led many researchers to label the DIR elements as non-LTR repeats, but this definition can be questioned, see below. *Dyctyostelium* differs significantly from fungi in that it apparently has three systems that target retrotransposons into tRNA adjacent sites. An amazing 75% of all the tRNA sites are occupied by such elements. Interestingly, this tRNA polymorphism is effectively used to identify genetic races of *Dyctyostelium* field isolates and like essentially all other organisms, this parasitic DNA is species specific. Numerically, DIRs are the prevalent retrotransposon and the NC4 lab strain has about 150 copies per haploid genome. As we will see below, however, much larger numbers of other extrachromosomal nuclear genetic parasites are also present in this lab strain. DIRs are clustered and found in two sequence subtypes (5.7 and 2.4 kbp) and can be observed to express seven classes of transcripts, but at very low levels, which can be inducible with mitochondrial poisons. DIRs appear to have undergone multiple insertions into DIR-related sequences consistent with waves of colonization and displacement. Another related retroelement, Skipper (a chromovirus), also shows a similar tendency to insert at random nucleotide junctions into self-related sequences. Yet these complex element patterns are stably maintained. This observation strongly suggests that *Dyctyostelium* has evolved these elements by repeated colonization events that tend to interrupt preexisting but similar elements. As we will see in other chapters, this pattern of genetic parasite insertion and inactivation is seen in many other metazoan genomes.

The Possible Origin of DIRs

This is interesting to consider, since DIRs are much less widely distributed than (and distantly related to) the more basal chromoviruses discussed above. Several researchers have proposed that they must predate endogenous retroviruses, like the chromovirus found in fungi. DIRS1 encode a gag, RT/Rnase ORF as do many other retrotransposons. These RT ORFs are the most conserved sequence and show clear similarity to those of Ty3/Gypsy, which are also related to the chromoviruses. However, DIRS1 also encode a lambda-like recombinase (similar to the *At* of *A. tumifaciens* plasmid), not seen in these other retrotransposons. Also unlike other retrotransposons, the end sequences are inverted split repeats and the integration sites are clearly distinct, not like any retroviruses, leading to many questions concerning DIR origin. However, phylogenetic analysis suggests that a clear DIR-related element is found in Panagrellus redivivus (PAT) that is basal to DIRS1. In PAT, the lambda recombinase is a separate ORF, separated by 'split' direct (not inverted) repeat. Thus, in PAT, the genetic elements resemble a fusion of a chromovirus RT/Rnase, flanked by LTR, adjacent to the lambda recombinase also flanked by a second LTR (recall the chromoviral LTRs are highly variable). One clear implication is that PAT represents the ancestral DIRS1 and was derived from the fusion of a chromoviral RT/LTR to a phage-derived recombinase that acquired the lambda-like integration system and later fused internally to form DIRS1. This event was unique to *Dyctyostelium discoideum.* Other sets of non-LTR transposable elements (some with DIR-like RT/Rnase) are also found (Tdd, TRE, H3R) and together with DRI elements constitute about 10% of the *Dyctyostelium* genome. All of these elements tend to insert into self-similar elements and some are induced with heat shock.

However, as these are poorly studied, we are hard-pressed to say much about their biology. We are left to contemplate some interesting questions. What has been the consequence of this genetic colonization to *Dyctyostelium*? Has it affected host identity? Why was it specific to *Dyctyostelium*?

Epigenetic Parasites

Dyctyostelium discoideum has various other unique genetic parasites. Most notable are the circular nuclear plasmids that are not seen in any of the other eukaryotes we have considered to this point (and are generally uncommon). They have imaginative names like Ddp1, Ddp2, Ddp3, Ddp4, Ddp5, Ddp6. They range in size from 1.3 kbp (Ddp1) to 15 kbp (Ddp5), range in sequence similarity and can encode from as few as none to six ORFs (the latter including rep genes and transcription factors). They can be usually maintained in mixed states, but some appear incompatible with one another. They can readily be found in field isolates but are not uniformly present. Unlike the DNA plasmids of fungi, they are nuclear, not mitochondrial, and appear to replicate as RCR replicons. Their copy levels are stable (even when mixed) with respect to host ploidy and often very high (300 copies per haploid cell). This clearly suggests that these plasmids have some type of persistence module to insure stability. In fact, two of them, Dpd-2 and Dpd-6, have origins and a cognate-binding protein that resembles that found in EBNA-1, a persisting large DNA virus of humans. However, as these systems are poorly studied, we do not know what selective or competitive consequences they might have to their host or other genetic parasite, or if they affect sexual reproduction, as is typical for persisting genetic parasites. They are not associated with killer functions or other known phenotypes. But issues related to genetic competition in natural settings have not been evaluated. One interesting point is that such plasmids have been used to genetically engineer *Dyctyostelium* and have resulted in tandem duplications of inserted sequences. As mentioned above, such duplicated DNA was not tolerated by the *Neurospora* RIP system, so clearly *Dyctyostelium* no longer maintains such a preclusive genetic identifying system. Why these plasmids are peculiar to *Dyctyostelium* is unknown, but I would propose they must define some basic molecular characteristics (immunity?) of their host that allows such stable colonization.

Dyctyostelium *Group ID, Immunity and Sex*

The aggregation-based multicellular lifestyle of *Dyctyostelium* poses some interesting issue with respect to mechanisms of group identification, especially since the multicellular slug can be composed of a genetic mixture of coalesced individual cells. *Dyctyostelium* does not have the sponge IgG-like surface transducers that restricted (by apoptosis) the aggregation of the multicellular form to genetically related individuals. Also, unlike its likely ancestral relatives, red algae, green algae, sponges or even more distant dinoflagellates, *Dyctyostelium* does not appear to produce toxins to any great degree, if at all, so these could not provide diffusible group identification. The oxylipins and the ubiquitous channel toxins of many marine eukaryotes also appear to be absent, as do the T/A modules as seen in prokaryotes or killer viruses of yeast. Yet *Dyctyostelium* must feed on soil bacteria, a well-established source of diverse toxins. It seems likely that *Dyctyostelium* has acquired new communication systems that make it less susceptible to such prevalent toxins. For example, in contrast to *C. elegans*, *Dyctyostelium* does not appear to have any known voltage-gated ion channels, the main target of such toxins. In fact, *Dyctyostelium* motility does not depend on a maintained membrane electronic potential since depolarized cells are still motile, in sharp contrast to the motility of flagellated bacteria and most metazoans. It is thus very interesting to note that *Dyctyostelium* motility is mediated predominately by F-actin assembly and disassembly at the pseudopod, but that such assembly was also one of the major targets of the toxins made by sponges. Instead of voltage-gated ion channels for transduction, *Dyctyostelium* uses G-protein ion channel receptors gated by cAMP that control Ca^{++} influx. Interestingly when first sequenced, *Dyctyostelium* seemed particularly poor in GPCRs (only seven CAR/CRL GPCRs reported). However, it now appears that up to 43 unusual GPCRs (family 2) can be identified, including three glutamate receptors. This is reminiscent of the rice blast genome in this regard. Included in these novel receptors are some that have protein kinase linked to an ion channel. Thus, non-specific chemo-attractant cAMP and cognate receptors appear to have displaced both more specific pheromones and other specific surface receptors as a system of social identification. This system is also involved in the tip (slug) recognition system which coordinates movement by the slug, as the tip is the location of cAMP receptor expression, transduction and assembly of actin. It is not clear if this involves sensory cilia. It is interesting that fusion of the macrocyst (following dark period) results in a cell that will engulf surrounding cells. Other slime molds, such as *D. caveatum*, are able to feed on fellow slime molds, including bigger ones, and some isolates will feed on themselves. Clearly these self-feeding variants have lost the capacity for self-identification. Interestingly, such variants have also lost the ability to complete the multicellular development (macrocyst formation) and remain haploid. Clearly, such a process of group recognition is not representative of most metazoans. For that, we will examine *C. elegans* below.

RNAi and Dyctyostelium

Dyctyostelium does have a functional RNAi system, in spite of the fact that it has no dicer-like gene affecting methylation. The DIRS1 retrotransposon are instead methylated by Dnmt2 gene which results in repressed transcription. It is interesting that *Dyctyostelium* expresses many of siRNA that corresponds to DIRS1 sequence, suggesting that the main role of RNAi is to silence DIRS-like elements (recall that silenced expression is also common in fungi and as a strategy for persistence). Genetic evidence supports this relationship, such as only 0.2% of *Dyctyostelium* C residues are methylated, but these are mostly limited to DIRS1 and Skipper sequences. Mutants in rrpc (RdRp) will increase DIRS1 expression, but not Skipper. In contrast, Skipper does not have siRNA associated with it. Skipper is present at about 30 gag fragments (with 12 predicted full-length gags). Note that Skipper is an LTR retrotransposon complete with gag, separated pro, but no env, so it is clearly a chromovirus similar to those of fungi. There is a link of the RNAi system to sex since RNAi-mediated chromatin silencing is also linked to spermatogenesis. Like the filamentous fungi, genetic parasites appear to be silenced following meiosis. It therefore appears that *Dyctyostelium* has two distinct systems to differentially control DIRS-like and Skipper elements and that these two systems constitute the main systems of genetic identity. The relationship of RNAi to the prevalent epigenetic plasmids is unknown. However, it is interesting that two genes with strong similarity to OpMNPV polyhedrosis capsid size (29.3 kDa domain) are found in *Dyctyostelium* genome. It therefore seems likely that genetic parasites have a basic, but unknown, role in *Dyctyostelium* life strategy and evolution.

Thus *Dyctyostelium* has evolved a form of cell fate group control that operates on somatic cells and involves PCD (terminal differentiation). Some cells (20%) are compelled to die in order to make a stalk for the whole organism to sexually reproduce. This process clearly resembles an addiction module that has been disrupted. It is phototactic and light controlled, but the coordination of systems (slug motility) uses the cAMP receptor as well as an unusual, and slime mold-specific, set of GCRP transducers (NCHM class). In addition, transduction and motility control is not dependent on membrane potential (i.e., not via Ca++ levels). There is little evidence that *Dyctyostelium* slugs are capable of learned group control of behavior. Group behavior appears to be programmed and not developed by associative exposure to environmental stimuli. The *Dyctyostelium* genome is also relatively poor in transcription factors, nor does it have a diverse set of surface receptor proteins that can affect cellular identity.

At Last, Worms and Associative Learning

C. elegans' *Nervous System and Associative Group Identification*

On the surface, the transition from the coordinated crawling 1,000 cell slugs of slime mold (*Dyctyostelium*) to the crawling small transparent nematode worms of similar cell number would not appear to present a difficult evolutionary transition that demands the acquisition of much complexity (relative to the acquisition of the nucleus, for example). Such apparent similarities are, however, misleading. A major transition was indeed required for the evolution of *C. elegans.* The well-studied *C. elegans* provides many examples of systems that represent major acquisitions and remained in all subsequent animals. I would argue that the most prominent of these acquisitions is the development of terminally differentiated nerve cells that provide an entire cell-based sensory and control network. Along with this, numerous associated functions were needed. The acquisition of the sensory GCRPs, the transduction and amplification of signals via alterations in ion-gated membrane potential, the ability of nerves to reach out by invasive processes (dendrites) and control other cells and the conditional delivery of membrane-enclosed chemical packets that penetrate adjacent nerve (or muscle) cells and direct the behavior of other cells and the coordinated and associative behavior of the whole organism were all complex and major acquisitions that are first seen in worms. These were major innovations, establishing an evolutionary path that leads directly to the evolution of human group behavior and intelligence. What do we know generally about the nervous system and behavior of *C. elegans*? Light has been retained as a significant behavioral cue. Sex remains, as always, a major issue regarding group behavior and identification. But both the receptor and transduction repertoire of these systems have been considerably expanded, allowing the development of complex behaviors. With the invention of *selectable* neurons (via apoptosis and experience) and the guidance of their invasive axons and dendrites, nerve cells now provide a platform onto which diverse and highly specialized sensory receptors and transducing systems can be locally and conditionally expressed in response to experience. This invasive and selectable nature of nerve cells now provides the mechanism by which environmental cues can lead to stable alterations in the cellular communication and hence behavior of the whole organism. Along with this, we see the need to prevent or induce programmed neuron death by sensory signaling activity. Thus, sensory input (pheromones, odor, light, heat, mechanical stress) leads to learned behavior, including sex and group behavior. The acquisition of the nerve cell networks also allows the invention of the process of associative learning in which sets of temporally linked sensory cues become associated and encoded in a cell communication network. If such learning can be imprinted (made stable during allowed developmental windows), it also provides an entirely new system for group identification based on such sensory cues. By now the reader should recognize that this system has been described with numerous 'virus-like' qualities: invasive, persistent, capable of killing, even killing itself. Even the very process of axon and dendrite projections to transmit information membrane packets resemble well-known viral (retroviral, poxviral) cell-to-cell transmission processes, as well as transmission processes described in hyphal organisms. In a way, the network formed by the nervous system resembles the multicellular networks of the fungi and algae described above, in that information can be transmitted throughout the entire system, across individual cells. But here, virus or other quantal information packets are not free to diffuse across the entire neuronal network. A synapse provides a gated portal in which a system of cell–cell or group identification must operate. As will be argued below, T/A-like addiction modules will be presented as central to the associative and stable nature of nerve system function. Thus, this addiction character of the CNS is not coincidental, but instead is a fundamental strategy related to acquisition of group identification systems that will allow us to trace a direct path to human behavior.

Our initial subject for this investigation is *C. elegans*, specifically the much studied Bristol N2 strain. This is a free-living, transparent nematode with an unsegmented body plan. It processes a full set of differentiated tissue, neuronal, endoderm, ectoderm and muscle found in 959 somatic cells. Twenty-two races of *C. elegans* are known that can be differentiated (as is common) by parasitic DNA (transposons) content. Some of these transposons seem to have race-specific effects, such as mab-23, which shows severe distortion in male genitalia during development (suggestive of possible involvement in sexual isolation). In terms of behavior, sex and learning behavior are most central and interesting. *C. elegans* uses an XX/XO chromosome system of sex determination. XO males will forage for hermaphrodites via diffusible cues that are not race specific. Mounting, sex and egg-laying behavior result after a partner is located. Mutants are known for many of these activities, such as male osm-5 and osm-6 which do not respond to diffusible cues. The use of odors to train this and other behaviors (i.e., feeding) is discussed below. Other nematodes, however, have distinct sexual genetics. Brugia malayi, a filarial nematode, has XX/XY sex chromosomes. Thus nematodes bifurcate with respect to the evolution of sex chromosomes. It is interesting that Cer retrovirus with env ORF is found only on the X, whereas the Y chromosome has TOY (tag on Y), which is an inactive Y-specific Cer-like element. A Pao-like element, resembling Cer-7 clade, is also present.

Genome; G-Protein/Transcription Factor Expansion

C. elegans has maintained a relatively compact genome of 97 Mbp. It encodes about 20,000 ORFs. Although this number seems excessive for simplicity of a 959-cell round worm (compare to the less than 30,000 human genes), it is well below that estimated for some fungal or the sponge genome mentioned above. A most notable difference from *Dyctyostelium*, sponge and fungal genomes, however, is that 5% of the *C. elegans* genes code for sensory G-proteins, not the unusual G-protein version seen in fungi or *Dyctyostelium*. These sensory G-proteins are mostly gated ion channels and their characteristics are addressed below in considerable detail as they are central to all animal evolution. This represents a large increase in sensory receptors and, as argued before, raises the suspicion of their involvement in group identity and immunity. Also, *C. elegans* code for a lot more transcription factors than seen in *Dyctyostelium*. The evolutionary expansions in two gene types (G-proteins, transcription factors) may be linked.

Genetic Parasites: Cer and Helitron Colonization

The *C. elegans* genome shows the acquisition of two major families of genetic parasites which are clearly distinct to this species; these are the Cer virus/retroelements and Helitron DNA elements. The Cer elements represent a major expansion of chromovirus-like elements. We can recall that in ascomycetes fungi, the Ty3/Gypsy-related clade of chromovirus elements was present, but generally restricted in copy number. Conservation of the Cer element and the Ty3 elements can be seen with RT sequence and within one clade of RT similarity; the LTR sequences are also conserved. Skipper is one member of this clade, but had not undergone expansion in fungi. In *C. elegans*, this family has been considerably expanded. In *Dyctyostelium*, the DIR element had partial similarity to the chromovirus RT, but the presence of a lambda-like DIR integrase makes this element quite distinct from that present in *C. elegans*. *C. elegans* is now known to have 19 families of Cer elements, clustered predominantly on the ends of chromosomes (via RT/IN); 1% of the *C. elegans* genome are LTR elements. However, these numbers are small compared to vertebrate genomes. Most *C. elegans* Cer copies are defective and most are repressed transcriptionally. However, most of these Cer clades also have a full-length version that is highly conserved and this full version is usually phylogenetically basal to the other clade members. For example, Cer-1 is a Gypsy/Ty3 clade, but a full-length version also encodes an *env* gene in two loci. This element appears to be transcribed and, thus similar to Gypsy of drosophila, may be capable of expressing a Cer retrovirus. It is worth noting that an abundant non-LTR version of Cer-1 (at about 1,000 copies) has also been described as the Sam/Frodo elements. Such non-LTR RT-encoding elements are often called LINE elements, and they have received considerable attention in the evolutionary biology literature. However, in the context of a 'virus-first' perspective presented here, the defective elements clearly appear as epiparasites to the full Cer elements. Similarly, Cer-7 also constitutes a separate clade inclusive of other Cers (8, 9, 10, 11, 12, 15, 16, 17, 19). Cer-7 is a full-length endogenous retrovirus and like Cer-1also encodes an *env* gene. It is worth emphasizing unlike the dsRNA-like viruses, the Ad-like viruses and the endogenous chromoviruses seen in the filamentous organisms (both algae and fungi), the endogenous full Cer viruses of *C. elegans* encode an *env* gene which is a conserved virion structural protein. This protein should allow extracellular virion transmission. This is also interesting when we also note that *C. elegans* has lost the multinucleate character of those hyphal organisms.

General Conclusions

C. elegans represents the first animal we have examined that uses neurons and a central nervous system. It has acquired a lot of GPCRs and transcription factors in its genome. Additionally, its genome was colonized with what, for the first time, seems to be a functional endogenous retrovirus (*env* containing) along with a significant increase in endogenous retrovirus-related elements (chromoviruses, first seen in algal and fungal genomes). In addition, a distinct family of DNA elements was also acquired. This worm has a non-network cellar makeup of differentiated tissues. Although these ERVs are highly conserved, their significance is not known. In contrast to the ERVs found in mammalian genomes, the *C. elegans* ERVs are silenced in the germ line. There is evidence of hyperparasitic interactions in that ERVs appear to have integrated into each other and DIR elements, suggesting evolutionary waves of superimposition of multiple genetic parasites.

Helitron Element

Two percent of the *C. elegans* genome is made up of Helitron DNA elements. These are also found in plant genomes, but not in animals. These are quite distinct in that they are DNA elements that do not use the typical 'cut and paste' transposase-mediated mobility mechanism, but instead replicate via rolling circular mechanisms (RCR). Thus they clearly resemble the RCR viruses (Gemini) described earlier, since no host genome replicates by such an RCR mechanism. Accordingly, they encode viral-like Rep gene that functions very much like viral Rep genes (via nicking-dependent initiation of DNA synthesis). Some copies of this element occur in 'epiparasite' clusters of simplified repeats of 200–400 bp, clearly resembling the repeat elements called satellite sequence. The occurrence of so many RCR elements in *C. elegans* is highly intriguing and has not been previously observed in our evolutionary examinations. When we juxtapose the genomic Helitron RCR replicon with the numerous nuclear DNA plasmids described previously for *Dyctyostelium*, an even more intriguing implication presents itself. Ubiquitous, nuclear, episomal and circular plasmids were seen specifically in *Dyctyostelium* host, but are absent from nematodes. That the *C. elegans* genome is now highly colonized by RCR replicons would thus seem more than coincidental. It seems clear that RCR replicon colonization was an event somehow crucial to the evolution of *C. elegans*. I would expect that ancestral or extant genetic identity systems must have been transformed (incapacitated) for this to happen. The superimposing genetic colonizer (Helitron and Cer) would need to override any preexisting immunity/identity systems able to recognize genetic invaders (such as RIP in fungi). But how and why such a colonizing and 'inactivating' process could relate to the emergence of a new identity system (such as a nervous system) would seem to be most mysterious. Neurons would not seem relevant to such genetic colonization. However, as outlined below, this appearance is misleading.

In addition to the Helitron elements, some other more typical DNA (i.e., transposase-mediated) elements are also found in *C. elegans*. Six Tc families are present. Curiously all 32 of the Tcs in the N2 lab strain are invariant in germ line, yet become active with age in soma tissue, such as Tc1, Tc6, Tc7. Tc3 is the most active in the Bristol N2 strain. Tc1 (1.6 kbp, with inverted repeats plus a transposase ORF) is fully silenced in germ line. Mostly all Tcs appear non-functional, but unlike Helitrons, these are also found in animal genomes. Tc7 (921 bp; an epiparasite of Tc1) has a striking pattern of clustered occurrence on the X chromosome. Tc6 is an interesting variant in that it uses covalently closed hairpin DNAs for its transposition. These hairpin DNA structures are of special interest regarding the RNAi system described below but also bear a striking similarity to the V(D)JRS5 junctions used by the RAG gene to generate surface molecular diversity of the vertebrate adaptive immune system.

Genetic Parasites?

A most surprising feature of *C. elegans* is that after many decades of intense research, there have been no reports of viruses (non-genomic) that infect this (or any) nematode. Given that nematodes feed on soil bacteria and slime molds, habitats rich in viruses, and that nematodes are very important vector for numerous RNA viruses of plants, it is clear that they are highly exposed to a broad array of viral agents. It therefore appears that nematodes are indeed rather resistant to most if not all viral agents. As will be presented below, recent experimental results make it clear that the RNAi system is responsible for much of this resistance. Thus in *C. elegans*, we see a major transition in the relationship of host to its genetic parasites. Absent are the nuclear DNA parasites as seen in *Dyctyostelium*. Also absent are prevalent cytoplasmic dsRNA parasites (including killers) and the mitochondrial DNA and RNA genetic parasites of fungi. Instead we see significant chromovirus colonization (19 families) as well as colonization by their defectives and RCR-based DNA elements. Such elements were RIP prohibited in most fungi. Furthermore, both these elements are mostly repressed in germ lines, but can become active in soma. As discussed below, the RNAi system is responsible for much of this Cer repression. Since a major innovation of *C. elegans* is the nervous system, it is highly interesting that the RNAi system is involved in nerve cell differentiation and that the RNAi system does not seem to be functional in nerve cells. This will be considered further below.

Organismal Immunity/Cell Identity: The Impressive Immune System of C. elegans

As mentioned, much of the ability of *C. elegans* to prevent or preclude infection by various viruses has been shown to be due to the RNAi system. This RNAi

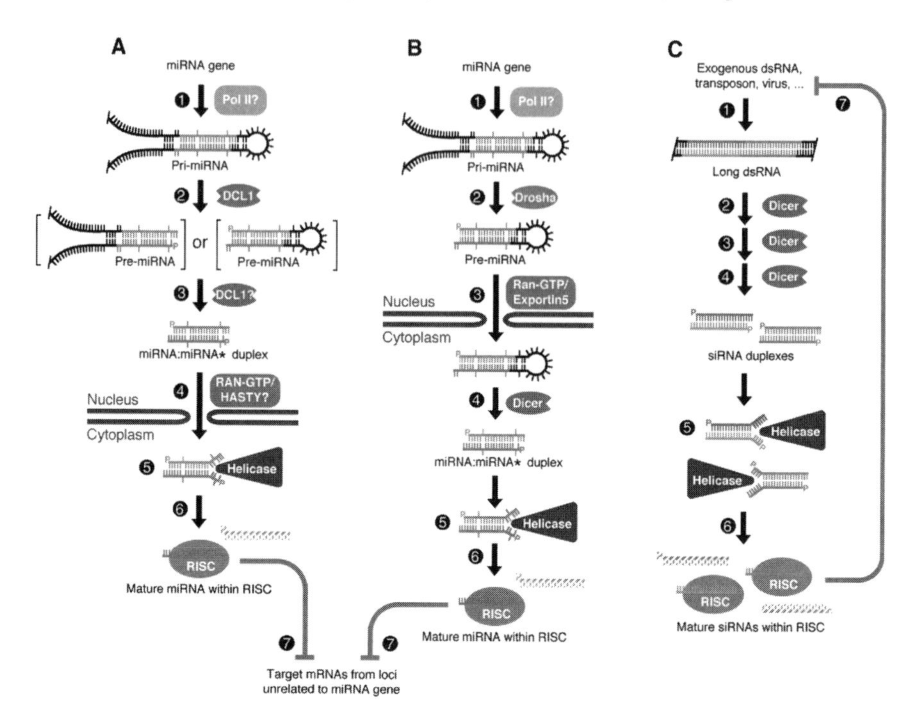

Fig. 5.10 Schematic of miRNA generation, transport and relationship to endogenous virus (*See* Color Insert, reprinted with permission from: David Bartel (2004) Cell, Vol. 116)

response can be considered an 'adaptive' antiviral system in the sense that it will generate a specific protective response to specific and new viral dsRNA sequences. This system, like that in its *Dyctyostelium* relative, is also principally directed at controlling (silencing) the expression of endogenous genetic parasites in the *C. elegans* genome (see Fig. 5.10 for miRNA example). In *Dyctyostelium*, RNAi was also principally used to silence DIR and Skipper expression. It is interesting that Gypsy virus, an env-containing genomic chromo-like retrovirus of drosophila, is also targeted for suppression in ovaries by the RNAi system. However, in *C. elegans*, we now see the involvement of one dicer gene for both siRNA and miRNA responses. This contrasts to drosophila in which two dicers perform these functions. Furthermore, and in striking contrast to the RNAi system of most other organisms, the *C. elegans* RNAi response is systemic and transmissive. That is, by introducing dsRNA against a specific sequence, that RNAi state will be propagated from the point of introduction to the other tissue of *C. elegans* as well as its offspring. This process requires an amplification system that uses RdRp to make more signal (dsRNA) plus some RNA movement proteins (sid-1) to transmit to adjacent cells. It is interesting that unlike dicer, RdRp is found only in fungi, worms and plants, but full RNAi transmission is worm-restricted. It is this combined process that appears to provide the systemic response that provides *C. elegans* with such an effective immunity. The word 'transmissive' should by now conjure up images of a viral-like process. Given the basic role for RdRp (a core viral replicase) in the RNAi response, a strong argument can be made that indeed the origin of the RNAi system itself appears to be viral in character (possibly hyperviral to silence endogenous chromoviruses). The *C. elegans* RNAi response follows the following scheme. dsRNA will elicit RNA-based gene silencing (RBGS) by inducing dicer to cleave dsRNA into siRNAs. siRNA then guides the RNA into a RISC complex that will cleave the specific target RNA. RdRp makes dsRNA that will suppress original DNA template via methylation. The Sid-1 gene is involved in the transmissive state (possibly by providing a ds/siRNA channel between cells), but is not present in neuronal tissue. Genetic studies have allowed the evaluation of the main targets of the RNAi system. Mutants that are resistant to dsRNA induction of RNAi have been isolated, corresponding to the rde-1 and rde-4 genes. These mutants are not defective in growth or development. The main consequence of these mutants is that they mobilize the production of most (but not all) endogenous retroviral sequences. Thus, like *Dyctyostelium*, *C. elegans* RNAi system is also mainly directed at silencing endogenous viral elements. If we consider this as a persistent state of viral silencing, then we would also predict that such silence states could also silence competing genetic elements (other viruses). Indeed, the rde-1 and rde-4 mutants do support the replication of VSV, a negative strand cytoplasmic RNA virus as well as other +RNA segmented viruses (flop house virus, FHV). Given the soil habitat of *C. elegans*, it is likely that these mutants would quickly encounter some lytic virus that would eliminate them. In fact, allowing *C. elegans* to feed on bacteria expressing dsRNA will also result in the induction of the worm's RNAi response, silencing the target gene, and this has provided a most effective method to map and evaluate the somatic (non-neurological) function of *C. elegans* genes.

Thus in *C. elegans* we see the development of a dsRNA-induced systemic RNAi response, able to inhibit expression of endogenous retrotransposons as well as prevent replication of RNA viruses. However, this response does not involve the NFKB transcriptional regulators which are so central to vertebrate immunity. Later, we will see in jawed vertebrates that dsRNA will also acquire the ability to induce an interferon response.

Micro RNA, ERVs and Neurons

In addition to the siRNA response described above, in *C. elegans* we see the evolution of a variation of dsRNA-induced identity system, known as a micro RNA system (miRNA), a system first discovered in *C. elegans*. The retrotransposons of *C. elegans*, when transcribed, also tend to produce hairpin-containing transcripts that can be processed into 20–21 nucleotide micro RNAs. *C. elegans* produces about 100 known micro RNAs, which appear capable of inhibiting translation of matched mRNA via interactions at the 3′ UTR. Overall miRNA, like siRNA, is part of an RNA-silencing response, resulting in methylated (DNA or chromatin) suppression of expression (see Fig. 5.10). As this response is also associated with suppression of retroposons, it could contribute to the silencing and persistent maintenance of these elements. miRNA is also a product of dicer, but unlike the situation in plants, *C. elegans* has only one dicer that is used for both siRNA and miRNA responses. miRNA is the product of non-perfect base pairing of hairpin dsRNA and results in the suppression of mRNA translation also via imperfect paring to 3′ UTR. In contrast, a perfect dsRNA results in RNAi that suppresses transcription. In *C. elegans*, it is estimated that 10% of all genes might be targets of such suppression. The levels of miRNA expression can be very high, between 1,000 and 50,000 copies per cell. Of the known miRNAs, let-7 and lys-6 are most studied. Of special interest is lys-6, which is directly involved in left/right asymmetry of chemosensory (ASE) neurons, and so it is also directly relevant to our focus on the *C. elegans* nervous system. Given this focus, *C. elegans* miRNA involvement in neuronal terminal fate decision via 3′ URL and apoptosis will be examined in detail. In addition, this fate decision also involves the die-1 zinc-finger transcription factors (detailed below), which modify the transcription pattern of the cell, providing a mechanism for stable changes in expression. Other pathways of neuronal differentiation, such as retinoic acid-induced neuronal differentiation, are also regulated by miRNA (-23) operating on transcription factors (Hes1). Thus the ASE neurons appear to represent a general model for controlling neuron fate. Although it seems likely that this miRNA system originally evolved to silence retroviruses (e.g., -poson) that persist in the *C. elegans* genome, this system now also provides the basic mechanism by which neuron fate specificity is affected, promoting the evolution of a central nervous system. The target of the miRNA control is an unstable and undifferentiated precursor nerve cell, a cell that will die unless it receives the appropriate sensory stimulus at the appropriate time. It should be noted that this time-dependent stimulation resembles the process of identity imprinting as described above for addiction modules.

Genomic Clusters of miRNA

The miRNA-producing regions of chromosomes are often clustered, such as the miR-17 cluster in the human genome. Since these miRNAs are often produced from templates of retrotransposons, this clustering also reflects a tendency for these elements to cluster. In *C. elegans*, Tc6 is the major miRNA element. Tc6 expression uses a covalently closed hairpin as part of its mobilization and RNA-processing strategy. Such clusters tend to be somewhat species specific. For example, *Meloidogyne arenaria*, a relative of *C. elegans*, does not have the Tc6 cluster but has the equivalent pMaE element which is present at 16,000 copies per cell and highly conserved in tandem clusters within this species. These are expected (but not yet established) to be miRNA-producing elements. Such species-specific clustering suggests that the retroposon-miRNA loci was the product of genetic colonization that endowed the host with a specific genetic identity. miRNAs are not only clustered but also often expressed from polycistronic loci, as is the human miR-17 cluster. It is interesting to note that a subset of human miRNAs are derived from conventional hairpin-containing RNA via LINE-2 element transcripts. Since these LINE-2 elements can be found in the 3′ UTR of these miRNA-regulated transcripts, this suggest a process by which defective LINE elements can be used to invade and colonize new 3′ UTR of targeted genes. As argued above, LINE elements can be considered as hyperparasites of related retroviruses, such as the chromoviruses or Cer elements of *C. elegans*. Thus *C. elegans* has undergone not only a lineage-specific colonization by Cer elements, but this has also involved the broader activation of LINEs as hyperparasites now able to more broadly affect numerous genes and elements associated with genetic identity and control. Such a colonization process provides a potentially genome-wide mechanism by which a genetic invasion (with something like a retrovirus) can mobilize related or linked defective retroposon elements, colonize sets of transcription units and impart new molecular identity (control) to a broader set of distributed genes. This could provide the initial driving force for the creation of new gene networks that have new but coordinated cues that regulate their expression. It is therefore worth noting that the genomes of all vertebrate lineages appear to be associated with the acquisition of this LINE-2 cluster. If such elements can be used to impart molecular identity, such a use might help explain why non-coding elements, LINE-1, LINE-2 and LTRs, are strongly conserved between divergent genomes, such as the human and mouse genomes. Recently, strong support for this idea has been observed. For example, the ZENK immediate early transcription factor, essential for audio learning (and group identity) in songbirds, has 3′ UTR elements that are the most conserved region of this crucial factor. Also, as will be presented in Chapter 8, human neocortex neurons are now known to differentially mobilize and express LINE-1 elements. This process is also associated with learning but these elements correspond to a recent human-specific evolutionary colonization. It thus seems possible that a colonization by this and/or other genetic parasites has imparted an miRNA-mediated control to a broad set of genes.

A Viral Source of Suppressing Small RNA?

The original pioneering work on RNAi was done by a plant virologist who sought to understand the host suppression of viral gene expression. In 1998, RNAi was also found in *C. elegans*, leading to subsequent discovery of miRNA. It is now clear that *drosha* is a nuclear-expressed gene that works to process pre-miRNA (either viral or host). *Dicer*, on the other hand, is a cytoplasmic protein. Although many host miRNAs are produced by retroposons, extragenomic viruses themselves encode many miRNAs. Such viruses have now been reported to have 40 different miRNAs and 10 RNAi suppressors from diverse viral lineages (including the *TAT* gene of HIV). All of these genes are required for virus replication and are thus not accessory genes. The structures of these viral genes are highly diverse, but mostly all of them bind dsRNA. Generally, viral miRNA and RNAi genes are considered as pathogenicity factors. Although they mostly induce gene suppression, they can also often trigger severe developmental abnormalities in their host cells. Such a large diversity of RNAi-associated gene structure would be consistent with the idea that viruses themselves may have provided the origin of such genes. In herpes virus, the miRNA genes are clustered in region that controls latency and it has been proposed that they silence viral genome to establish and maintain latency. Thus, these appear to be viral identity systems that allow persistence. If this idea is correct, such miRNA genes should be well conserved in virus evolution, but distinct to the different viral lineages. Some retroviruses also appear to maintain miRNA for persistence. For example, primate-specific foamy viruses express miRNAs associated with limiting expression (silencing) of viral genes. RNA viruses, such as RCNMV, also link replication to depletion of RNAi systems such as via production of a viral dicer-like enzyme that prevents siRNA and miRNA biogenesis. Different viral lineages stemming from different host cell sites of latency appear to have evolved distinct patterns of miRNA systems. Thus, we see that EBV expresses several 'EBV-specific' miRNAs during latency. Small DNA viruses also tend to use miRNA for persistence. For example, SV40 expresses viral-specific miRNA that suppresses viral T-Ag production to a level where host CTLs are no longer able to kill them. Thus viruses not only appear to have employed miRNA systems to persist in host, but also have evolved various novel versions of these systems during host evolution.

The Self-Destructive Origin of Neurons: Apoptosis and T/A Systems

The fundamental role of apoptosis in the development and fate of the nervous system was first worked out in *C. elegans*. This accounts for the decrease in neuron numbers with worm development. Apoptosis is a major molecular characteristic of central nervous systems and remains crucial for CNS function in mammals. Recall that the sponge lacks neurons but did use some apoptotic systems to control gametes and cellular identity. The fate and identity of cells thus seem a basal function of apoptosis. In *C. elegans*, a well-studied system of neuronal fate is that of hermaphrodite versus male developmental fate and resulting in group behavior. Here, a resemblance to sponges is more apparent as apoptosis provides a mechanism to make sex-specific fate decisions in response to sensory cues. In both sponge and *C. elegans*, the apoptotic system employs pore proteins and antipore proteins as a T/A system to control the disruption of the mitochondrial membrane. What differs in *C. elegans* PCD compared to ancestral systems (i.e., *Dyctyostelium*) is that in *C. elegans*, PCD involves CED-9, a mitochondrial membrane and pore protein that is the homologue to BCL-2/BCL-Xl. As mentioned, the BCL-2 family of proteins structurally closely resembles colicins and diphtheria bacterial pore toxins and has a prominent hydrophobic groove on the surface that binds antiapoptotic proteins (like BH3). These pore proteins must form heterodimers to make pores (which is a regulated event). CED-4 (BAX/BAK homologue) is an antitoxin protein that acts via surface binding. CED-3 is caspase homologue. Death occurs when pore is formed and open, allowing loss of mitochondrial membrane potential. However, distinct from the apoptotic system of vertebrates, the *C. elegans* system seems simpler and does not need to release cytochrome *c* for caspase activation or other regulators. Also, NFKB is not involved in *C. elegans* neuronal survival as it is in vertebrates. In *C. elegans*, it seems an overlay of the RNAi provides control for control of apoptosis and cell fate, which also controls a fate-determining transcription factor expression. As described below, a main sensory cue for this fate decision is pheromone olfaction as this will affect worm motility via sensory transduction. In this overall logic (small-molecule pheromone, rhodopsin-like receptor, control of transcription factor and ciliated cell motility structure) the system is rather similar to the transduction logic of QS system described for cyanobacteria in previous chapters. This too involved a pheromone signal that eventually affects cilia (flagella) via a G-protein receptor, channel function, Ca++ levels and zinc-finger transcription factor expression (TRA-1A). The big distinction is that in the nerve cells of *C. elegans*, signal transduction to the motility system is via an intervening neuron (not bacterial cytosol) and affects motility via nerve-coordinated muscle function.

Rube Goldberg-Like Sexual Regulation

In *C. elegans*, the sexual chromosomes are of the XX and XO type, in which the XX are hermaphrodites. Within the hermaphrodite, 1,090 cells are initially generated during development but only 959 remain in the adult. Most of those that are lost are neurons, and the male has more surviving neurons than the hermaphrodite. Since there is no male Y chromosome, the sexual fate is essentially an X gene dosage-dependent phenomena that reminds us of the meiotic gene system that was described in *Neurospora*. In *Neurospora*, paired sequences were recognized by RIP (via dsRNA) and elicited a sequence-specific gene methylation reaction (homology silencing). In contrast, with *C. elegans*, some genes are silenced via the actions of an RNAi silencing system. Somehow the X gene dose initiates a gene cascade that ultimately impinges on TRA-1A transcription proteins and induces the activation of apoptosis in those cells destined to die. However, this cascade appears to be overly complex (involving interactions of 16 proteins). A main regulator of male apoptosis is egl-1 (small 91 aa protein which interferes with ced-9, -4) that triggers death in a process (resembling a T/A disruption). Consistent with this idea, living neurons express the killer activities of ced-3 and ced-4, but are protected from death (via ced-9). It has been noted that the system that controls the fate of these sex-specific neurons appears to resemble a Rube Goldberg-like situation in complexity. Most of the interacting 16 proteins are transduction systems that lead to a single control point (ced-9). The origin of this complexity is hard to explain by usual selection and evolution criteria, such as point mutation to increase efficiency or specificity. Since genetic screens frequently find egl-1 mutants, it seems clear that this simple end-point gene is really responsible for neuronal fate regulation. At this point it is worth recalling that persisting viruses frequently invent small efficient regulatory molecules that are used to overtake existing host identification and transduction systems. The small ultimate target (ced-9) gene clearly resembles such a virus-like gene. Furthermore, the ced-9 gene was initially hidden amongst an intergenic parasitic 'junk' DNA, and not found in the initial EST database searches. Also of relevance, other viruses clearly make ced-9-like genes. For example, baculovirus p35 can functionally replace the ced-9 protective activity in *C. elegans* neurons. In fact, many of the large DNA viruses encode some of the simplest cytokine receptors known (discussed below), consistent with a possible viral origin of such genes. If persisting viruses were indeed the source of the above cell fate regulatory genes, it would explain how a Rube Goldberg-like situation could ultimately evolve to control a relatively simple end point due to the overlay of sequential, parasite-superimposed ID-modifying modules.

A Summary of C. elegans PCD

With *C. elegans*, we see the evolution of programmed neuronal cell death as a main system to control the development and fate of nerve cells. Cell death becomes an important system that uses sensory input in order to control the development of the main sensory tissue. There is evidence to suggest a possible viral role in the origin and adaptation of this PCD system. A main trigger is the RNAi-mediated process that initially recognizes retroposons and viruses. The distribution of this control system to other genes appears to have been mediated by further genetic colonization at 3′ UTR regions with related (LINE-like) hyperparasite elements. *C. elegans* also has a highly effective antiviral siRNA system (described below), which is not fully expressed in neurons. The entire process of PCD has the clear characteristics of a multilayered T/A addiction system. In order for neurons to stay alive, they need an 'A' signal that prevents the 'T'-mediated death (via mitochondrial membrane/pore disruption). This system closely resembles that seen in mammal neurons, although one difference is the absence of cytochrome *c* as released in vertebrates.

There is also some evidence that elements of earlier PCD systems still remain in *C. elegans* neurons, similar to the oxylipin-mediated oxidative burst of the innate immune system from red algae. For example, sphingolipids ceramide (*N*-acetylsphengosine) is used as a second messenger lipid, activates ROS and evokes mitochondrial oxidation damage and apoptosis in neuronal systems. In *C. elegans*, n-3 PUFA (made from n-6 by fat-1 genes) can exert antiapoptotic effect on neurons. Further, expression of fat-1 makes neurons less dependent on growth factor (prostaglandin E2) and inhibits apoptosis. Thus these lipid-based signals still seem to exert some control over T/A systems of *C. elegans* neurons.

GPCR Sensory Expansion

The expansion of GPCR sensory receptors in the *C. elegans* genome seems especially important for the evolution of the sensory function of the nervous system. This expansion is not only characteristic of *C. elegans* but has continued in all animals so we will also examine this issue in detail. As noted, *C. elegans* has an estimated 1,006 GPCR-related ORFs, of which 800 appear to code for full GPCRs. Of these, 550 appear to be functional genes of the chemosensory type, mostly of the rhodopsin class. There are some interesting differences comparing these genes to organisms we have already considered. For example, absent in *C. elegans* (and mammals) are the NCHM receptors seen in *Dyctyostelium* genome. Recall that NCHM-like receptors are also found in insects and are the target of various bacterial toxins. Interestingly these rhodopsin-like receptors are themselves the primary target of many pore receptor toxins found in pathogenic bacteria. This allows the use of *C. elegans* to study bacterial pathogens of humans due to this similar toxin sensitivity, as will be presented below. This large number of rhodopsin-like receptors in *C. elegans* is also interesting to compare to much more complex organisms. Mouse has only a little more GPCR genes than does *C. elegans* – 1,106 mouse receptors, of which 44 are of the VR (vomeronasal receptor) type. Humans have only 752 such receptors, with only 18 VR-like ORFs, of which essentially all seem to be pseudogenes. Curiously, these VR-related pseudogene sequences are highly represented (1,800 hits) in the human EST database. This fact will be much discussed in later chapters. Of special interest with respect to the nervous system is that the *C. elegans* rhodopsin receptors are in four groups. The largest group is the olfactory group, but this also includes the monoamine group. Within this group are the receptors for serotonin (5-HT), dopamine, histamine, adrenalin and acetylcholine. *C. elegans* uses serotonin along with olfaction to control many behaviors as presented below.

Transient Receptor Proteins (TRP) and Sexual Group Behavior

TRPs are a superfamily of Ca++ permeable cation channels first observed in Drosophila. TRPs exist in six families of which TRPC was first described. TRPs are found in most eukaryotes, from yeast to mammals. In *C. elegans*, they control olfaction, mechanosensation and osmosensation, and thus are primary sensory receptors associated with sexual and other behaviors. *C. elegans* has five TRP families, all of which are terminal neuron expressed. In *C. elegans*, the TRPV ion channel protein regulates 5-HT biosynthesis and is therefore of special interest for the function of the nervous system. TRP activity itself is regulated by G-protein signaling and polyunsaturated fatty acids. Some of the *C. elegans* TRPs are store-operated channels (stored CA++), thus representing a new nerve cell function in that the use of store-operated calcium release is an important innovation that provides a potential for a more rapid nerve cell response to stimuli (see also myelin sheaths next section). Specifically, *C. elegans* pkd-2 is a TRP family of calcium channel that is necessary for male mating behavior (male aggregation). This gene is also an intracellular Ca++ release channel and furthermore is homologous to the human PDK-2 polycystin gene (a cilia-associated protein). In many animals, TRPs appear to be basic to sex behavior. In *C. elegans*, TRPPs (related to human polycystic kidney disease) are expressed in male-specific neurons. *C. elegans* also has nph-1 and -2, both with similarity to two human polycystic kidney disease genes (part of a highly conserved set of cilia genes, as noted earlier). One of these, TRPM, is ced-11 involved in programmed cell death of these neurons as noted above. Another TRP, GON2, is associated with mitotic division in gonads. OSM-9 is another TRP, but this example is associated with mechanical sensation detection (nose-touch sensor, see below). Thus, if we can consider *C. elegans* as an example of an early nervous system, we see TRPCs are involved in sex cell development, sex cell identification and group sex behavior. Furthermore, the *C. elegans* TRP-3 is sperm specific and required for fertilization. Thus the role of TRPs in group identification is basic to sex cells as well as to the behavior of groups of organisms (this dual characteristic appears conserved in evolution). For example, later we will see that mouse TRPC2 is involved (as a cilia-specific pheromone receptor) in the mouse VNO (vomeronasal organ, also via stored Ca++) and is also crucial for sexual group behavior but remains associated with sperm/egg cellular fusion. These TRP-related VNO receptors are distinct from vertebrate olfaction and visual receptors in that they are cyclic nucleotide-gated channels, but used by all terrestrial vertebrates to elicit stereotypical sexual group behaviors. However, the origins for the TRP role in sexual behavior appear to trace to *C. elegans*, as the worm gene represents the phylogenetically basal version of the whole gene family. Yet, still later in the chapter on human evolution, we will see that TRPC2 has become inactivated, although still a transcribed pseudogene, and that it was made so due to invasion by transposable elements (an HERV K env fusion). In fact, it seems all human VNO TRP receptors were incapacitated. Thus at the very origins of the nervous system and its control over group behavior as well as at the most recent events in human evolution, we see evidence that retroviruses and their epiparasites have left large footprints.

Origins of 7 TM Genes?

With the evolution of the *C. elegans* nervous system, we have emphasized the role of complex regulatory channel proteins as sensory receptors, ligand and transducing proteins. Rhodopsin sensory receptors are considered as the phylogenetic predecessor to all the 7 TM G-coupled proteins. In *C. elegans*, a major expansion and adaptation to neurotropic function was seen with these genes. Can we say anything concerning the likely source of such gene novelty and diversity? I have already noted above that point selection and recombination do not seem to adequately explain the complexity of sensory and sex regulation we have seen in *C. elegans*. Phylogenetic analysis also fails to offer an explanation as to the likely source of these genes, as the various gene families and clades do not slowly evolve from one another, but rather enter the genome as a family at the base of the dendrogram. They thus seem to have originated from exogenous genetic sources that stably colonized *C. elegans*. Where then might we find an exogenous source of such complex genes? In the previous chapter, I have already noted the surprising ubiquity and diversity of rhodopsin sensory receptor from environmental clones of the Sargasso Sea. This same source also has a very large number of mimiviral-related sequences. Can we then propose that some large, complex DNA virus might be the source of 7 TM genes? Mimivirus infects amoebae. Other large DNA viruses infect algae (PBCV-1 and EsV-1). The coccolithovirus was especially noteworthy as a source of novel signal transduction proteins. Could such viruses present in the oceanic habitat have then provided the early source of these 7 TM genes? We do know that some large DNA viruses, such as the poxviruses, have an inherent capacity to induce infected cells to produce cytoplasmic processes that move the virus from one cell to its neighbor. Such a complex dendrite-like function would indeed seem similar to that needed for a nervous system. Yet as emphasized, large DNA viruses are not known in *C. elegans* or any other nematode. Other large DNA viruses, such as herpes virus and poxvirus, do encode many 7 TM receptors with unknown ligands. Some of these viral receptors are simple, having no ligand or show a broad ligand capacity. We know that *C. elegans* was colonized by Cers and Helitrons, not found in algae, and that these retroviral-derived systems are directly associated with the effective antiviral systems (siRNA) of *C. elegans*. The relationship, however, between these elements or siRNA and the absence of DNA viruses is unknown. Yet shrimp and insects species clearly do support related large DNA viruses. I have already mentioned the baculovirus ced-9 sex-determining homologue, a virus family infecting both crustaceans and insects. In this regard, it is interesting that insect entomopoxvirus also has within it a LINE-type RT, suggesting the possibility of more complex virus host colonization patterns. It is interesting that entomopoxvirus also encodes apoptosis inhibitors as well as a phosphatase 2C (PP2C) protein, unusual for other DNA viruses, but similar proteins (FEM-2) are known to be important for *C. elegans*' male sex development and determination. Although these observations are tantalizing, they cannot provide a clear conclusion regarding a possible viral role in the origin of these host 7 TM genes. Their source thus remains a mystery.

Eating Toxins: A Lineage of 7 TM Receptors

One interesting difference between *Dyctyostelium* slugs and *C. elegans* worms is that although both eat bacteria, only *C. elegans* is sensitive to the toxic and pathogenic characteristics of the bacteria it eats. Pathogenic *E. coli* including the enteropathogenic, hemorrhagic, invasive and enterotoxic types will often affect *C. elegans*. Other toxin-producing pathogenic bacteria are similar, including *Staph aureus*, *Streptococcus pyrogenes* and *Vibrio cholerae* toxins as well as other highly prevalent cytolytic pore toxins (i.e., Cry5B) in affecting *C. elegans*. These bacterial pore toxins can often cause paralysis in *C. elegans*, thus making them neurotoxins. The *C. elegans* G-proteins can also be the targets for such toxins (such as cholera toxin). Thus it appears that neurons of mammals and of *C. elegans* have common receptors and signal and transduction systems that make them both susceptible to these bacterial toxins. Besides toxins, *C. elegans* is also affected by pathogenic *E. coli* that colonize localized epithelial cells. Such localized adherence is mediated by QS-controlled bundle-forming pills and flagella. However, as an experimental model for pathogenesis, *C. elegans* is limited by not being able to grow at body temperatures, which frequently affects bacterial pathogenicity. Nematode panagrellus redivivus, however, will grow at 37°F. It is interesting, therefore, that this nematode seems less susceptible to these bacterial toxins (this is possible relevant to morphine receptors described below). *C. elegans*, however, is susceptible to insect-specific pore toxins made in large quantities by plasmid-bearing *Bacillus thuringiensis* during sporulation (Cry5B, a cation-selective channel protein). Although this is also mediated by *C. elegans* receptors and transduction pathways, such receptors are found only in insects and not vertebrate animals, making insects (but not vertebrates) also susceptible to these ubiquitous toxins. The paralysis noted above is observed following the eating of bacteria by *C. elegans*, which normally induces pharyngeal pumping triggered by the amount of bacteria via a serotonin-dependent nerve response in its gut. Thus *C. elegans* appears to represent a bifurcation point in the types of sensory receptor and nerve cell-mediated transduction systems it uses. Both vertebrates and insect receptors are present.

Olfaction

Olfaction in *C. elegans* is also mediated by GPCRs (7 TM channel proteins) and can be found in five neurons, as well as specific synaptic sites. *C. elegans* can sense at least five attractive (plus other repulsive) odors via a single pair of asymmetric neurons (AWC). The olfactory receptors (Str-2) are only expressed in one AWC in each animal (either left or right, but not both) and are localized to sensory cilia on olfactory dendrites. These cilia must undergo complex assembly, then be transported to the dendrites. ODR-3 is needed for cilia morphogenesis but is also an odor receptor subunit. It is interesting that such chemosensory neurons are also the targets of oxidative stress. In *C. elegans*, olfaction can be used for group identity (quorum sensing via Dauer pheromone) to initiate mate identity and sexual behavior. As will be presented below, this behavior can be learned and provides much insight into the learning mechanisms. Thus olfaction receptors, transduction and expression of transcription factors are all involved. The major behaviors affected include finding mates, egg laying and feeding on bacteria. Recall that *Dyctyostelium* had distinct set of olfaction receptors (NCHM), not present in *C. elegans*. *C. elegans* instead has many (1,106) rhodopsin-like receptors.

Olfaction Transduction: Invented and Conserved

C. elegans will respond to dozens of odors via mostly ciliated head neurons. The AWCs alter expression via transcription factor Tbx2 and Tbx3. Olfaction affects phosphorylation of these factors allowing cytoplasmic to nuclear transport (reminiscent of the NFKB response). Odor transduction is via a voltage-gated Ca++ channel, which is involved in switch decision to express Str-2. This transduction involves a cGMP signal pathway, needed to maintain Str-2 expression. If the sensor is inactivated, the coupled receptor kinase (Ce-grrk-2) and sensory perception is lost along with alterations in Ca+ + levels. Note that mammalian olfaction receptors (rat17) have been shown to function in *C. elegans*, and thus these transduction pathways appear to be conserved in vertebrates. The ACWs also respond to a 'QS-like' fatty acid-based Dauer pheromone detected when juvenile populations are dense as described below. *C. elegans* also has one TOL-1-like receptor which is strongly expressed in neuronal tissue. This is the tir-1 (Toll-interleukin receptor) gene involved on on/off decision at the time of synapse formation, so it is also part of the AWC (on/off) decision. This response is though to be involved in pathogen avoidance behavior. However, this toll does not signal via NFKB/My88 as in mammalian organisms. The presence of this receptor in *C. elegans*, however, suggests that siRNA first evolved in fungi and algae, adapted to the Toll-like receptor in *elegans*, but further adapted the NFKB transcription factor control in vertebrates.

PCD Cell Biology of C. elegans Neurons

Neuronal cell biology of *C. elegans* involves selection for neuronal growth versus death, as well as adaptations of neuronal plasticity. As mentioned above, many neural-specific genes are resistant to RNAi, so an RNAi role in PCD is distinct and neurons seem to represent a segregated system of identity.

PCD is mainly seen in the context of male/hermaphrodite development (hermaphrodites 959 cells, males 1,031 cells). Of these, the hermaphrodites have 302 neurons and 56 glial cells, whereas males have 381 neurons and 92 glial cells. One half of the neurons are found in the head; 39 classes of sensory neurons have been predicted; 87 neurons are sex specific, so numerically sex-associated neurons predominate. This implies that a major selective force is sex behavior regarding the origin of the nervous system. Each neuron has between 1 and 30 synaptic partners and the total synaptic connections of all neurons are about equal to one hippocampal pyramidal neuron of a mammal. The ciliated head neurons are associated with pores to the outside of the animal, so this represents a primitive nose-like sensory organ. As mentioned above, AWA and AWC neurons, which control sexual attraction, operate via TRPV receptors. The ASH and ADL neurons are associated with repulsion behavior. These neurons use immediate early genes to control expression (i.e., *Homer 1a* in *elegans*). Thus this control represents a highly conserved strategy of CNS systems of higher organisms. A 7 TM G-protein sensory channel receptor, altered Ca++ levels, rapid activation of transcription factor, alterations in stable expression and protection from PCD of properly stimulated neuron are all features found in *C. elegans* nerve cells. Genes eor-1 and eor-2 also specify neuronal fate, leading to egl-1-mediated death in male lines. Behavior is thus set by neuronal death. Alterations in worm motility behavior are the most common consequence and involve both attraction and repulsion. In the case of hermaphrodites, males' foraging activity is affected. In hermaphrodites, egg-laying behavior is controlled (HSN neuron fate). A second level of neuronal identity also occurs via connections between neurons. Axonal path finding uses distinct channel subtypes. *C. elegans* makes various neuropeptide-like proteins (NLP-29, NLP-31), which are curiously induced by fungal infection and also appear to be antifungal toxins. This might suggest an original role as toxins for these neuropeptides (and a role in T/A states). *C. elegans* can make up to 60 neuropeptides. *C. elegans* netrin, UNC-6, is a secreted protein that guides growth, cone, dorsal and ventral migrations, and mutant in this prevents branching of axons. In vertebrates, netrins are in the DCC superfamily of receptors that are immunoglobulin-like. There is a clear link between neuronal connections and PCD. Some PCD mutants that fail to die can make cells that also fail to make normal connections with targets. With *C. elegans*, zag-1 zinc-finger transcription repressor is required for some aspects of axonal path finding and can misexpress chemosensory receptors. In vertebrates, the process of programmed cell death also applies to glial cells, which is not a significant process in *C. elegans*.

Group QS Behavior, the Dauer Hormone

Besides olfactory cues associated with finding hermaphrodites and mating, *C. elegans* also produces a Dauer pheromone This is a short-chain fatty acid (heptanoic acid) with a rhamnose ring derivative whose density-dependent production by larvae induces starvation-stressed young larvae to arrest development as Dauer larvae. Arrested larvae do not progress to sexual maturation. Such a pheromone, and its density function, is clearly reminiscent of the fatty acid-modified lactone pheromones of bacterial QS systems. Interestingly, *C. elegans* retains a capacity to respond to bacterial QS pheromones and will modify gene expression of many gut genes following lactone exposure. Some have argued that the Dauer pheromone (for density detection) would have no fitness advantage for an individual *C. elegans* and suggested that this should not therefore be called a pheromone. Similar logic, however, applied to lactones and bacteria, where genetic studies clearly indicate pheromone involvement in group (colony) formation, would not support this suggestion. The possible role of the Dauer pheromone as a T/A module involved in *C. elegans* group ID seems strong, but lacks experimental evaluation. Unlike the bacterial lactones in which the sensory receptor directly affects the cell motility (flagella) system, the target of the *C. elegans* pheromone is a sensory neuron. Dauer pheromone binds to a subset of chemosensory ciliated neurons. Signal transduction is via two G-proteins alpha subunits (GPA-2, GPA-3) which affect promoter activity (via FOXO transcription factor). Mutants affecting Dauer formation can also affect male mate searching (apparently via another uncharacterized pheromone). Dauer acts during critical period (L1 and early L2 larval stages) and thus has the characteristics (developmental time window) of an identity imprinting process. Besides inducing a behavioral change from solitary feeding to social feeding of the juvenile male, it also induces a recycling of cellular components through the process of autophagy. This can be considered a process of 'self-eating'. This situation reminds us of the *Dyctyostelium* sex mutants, discussed above, which lost the ability to identify slug group membership and became cannibals. Why the Dauer pheromone would affect social feeding (clustering together in liquid drops) is odd, since the Dauer larvae are themselves non-feeding. Their aggregation may instead be a form of social identification. The Dauer larvae become long lived, resistant to oxidative stress and no longer respond to nose touch (prodding) with reversals of motility. Also, Dauer induction can also be done by high-temperature stress, independent of pheromones. All this clearly suggests that the Dauer larvae are arrested in the development and expression of some type of T/A or identity modules. With their resistance to oxidative stress, they seem stuck in the early A half of a T/A module. As discussed below, Dauer pheromone also has an intriguing ability to antagonize the activity of volatile CNS anesthetics associated with the interruption of consciousness in mammals.

Serotonin Receptors, Opiates and Sex Behavior

C. elegans uses serotonin (5-HT) as a nerve signal to control many behaviors, such as motility towards food. Since GABA, acetylcholine, dopamine, glutamate, noradrenalin are all biochemically derived from this one amino acid, the early use of serotonin in *C. elegans* nerves is of significant interest from an evolutionary perspective. *C. elegans* has four distinct 5-HT receptors. One of these receptors (SER-7) has clear similarity to mammalian receptors, but its agonist affinity is different. A main behavioral consequence of these receptors affects two feeding behaviors, one of which is social feeding and the other of which is solo feeding as mentioned above. 5-HT receptor mutants can also affect both pharyngeal pumping and egg laying, although some mutations are specific. Genetic analysis has identified a single aa in NPR-1 involved in this behavior, which is a GPCR of the neuropeptide Y-type receptor. In this case, a cell-specific regulation of 5-HT is regulated via TRPV. TRPV is expressed in serotonergic ADF neurons but not other serotonergic neurons, and thus these receptors can specify neuron-specific expression. Other mutants, such as osm-9 and ocr-2, can dramatically downregulate 5-HT biosynthesis. Mutants of serotonin receptors generally affect nerve cell development, not nerve cell plasticity. Although 5-HT is used to control egg-laying behavior in the hermaphrodite, other hermaphrodite-specific neurons are responsive to acetylcholine and will inhibit egg laying. Nicotine stimulates egg laying which suggests that nicotine-sensitive receptors are basal elements in the evolution of nervous systems and control of behavior. Along these lines, it is interesting to note that biologic amines, which include LSD and mescaline, operate via 5-HT receptors. It seems that 5-HT receptors and transduction systems of some early lineages have maintained certain chemical susceptibilities. For example, the monoamine CNS toxins of some fungi, which are associated with affects of human higher CNS function (audio and visual hallucinations), appear to operate via such receptors. Also interesting is that volatile anesthetics (VAs) disrupt higher nervous system functions (pain detection, consciousness) by ill-defined mechanisms. In fact, these VAs can disrupt behaviors of all metazoans, such as volitional movement and consciousness as well as block memory formation. These affects are not antagonized by any known drug, suggesting non-specific mechanisms of action (such as membrane fluidity). As mentioned above, it is therefore very curious that the Dauer pheromone will antagonize the effects of VA (halothane) on *C. elegans*, possibly by affecting G-protein-mediated neurotransmitter release. The implication is that many molecular processes associated with higher brain function were already in place in this simplest of nervous systems.

As a species, *C. elegans*, and nematodes in general, does not have much of a social character to its populations and behaviors. Absent are pair or group bonding/imprinting, the death of individual for the benefit of the population or the presence of non-reproductive caste as common in social insects and even some vertebrates. However, it is interesting that Dauer larvae do resemble these non-reproductive castes in several ways (arrested juvenile development, non-mating individuals). *C. elegans* male mating (which is sexually dimorphic) is the most complex behavior in nematodes and is pheromone induced via sensory neurons. With the evolution of a nervous system able to respond to sensory cues and control complex mating behavior, *C. elegans* provides a possible foundation for the evolution of even more complex (overlaid) social-based group identification systems and behavior. However, *C. elegans* does not have all the regulatory systems we see in vertebrate nervous systems. For example, it lacks the large numbers of olfaction receptors seen in mammals. It also lacks the nerve cell-based NO (oxidation) system, as nitric oxide synthase is absent from the *C. elegans* genome. This NO system is also crucial for the innate and adaptive immune system of vertebrates. Of special interest from our perspective of social group behavior is the absence of opiate and oxytocin receptors, since, as will be developed later, these play significant roles in vertebrate CNS-mediated group behavior (social bonding). It is worth noting that oxytocin is a neuropeptide that can keep neurons alive, thus preventing death (resembling the A of a T/A state). Later we will consider how such neuropeptides are used in many mammals during sex and birth to provide neurological systems for social bonding. In vertebrates, there frequently appear to be links between opiate receptors and the NO system, such as magnocellular oxytocin neurons of rats that can rapidly develop profound morphine dependence (within 5 days of exposure) and display major excitation on withdrawal. These neurons undergoing withdrawal also strongly express NOS. Opiates generally stimulate dopamine release. The absence of opiate receptors and NOS would seem to limit the application of these systems and mechanisms for the control of behavior in *C. elegans*. But *C. elegans* does have other receptors that are highly susceptible to addicted states. Nicotinic acetylcholine receptors are amongst the most addictive receptors known. In *C. elegans* nicotine can be used to affect (stimulate) egg-laying behavior. And chronic exposure to nicotine results in dramatic and long-lasting effects on behavior associated with long-lasting decrease in abundance of nicotinic receptor that controls egg laying. Although *C. elegans* lacks opiate receptors, it is most interesting that parasitic nematodes, such as Ascaris suum, produces opiate alkaloids morphine and morphine-6-glucuronide, especially in nerve chords and in the female reproductive organs (uterus – males produce much less). Here, morphine stimulates the release of NO, and this release is blocked by naloxone, similar to vertebrates. So both morphine and NO production are present in these parasitic worms. The receptors involved, however, are distinct from vertebrate ones. Interesting, however, is that other worms, such as Mytilus edulis, do have mammalian-like opiod receptors, which are also involved in NO release. Note that parasitic helminthes (Ascaris suum) are biologically intriguing in that they do not induce strong immunity in their host. Morphine and NO appear to be involved in this immune evasion as NO is a regulator and the relative absence of worm parasitization in developed populations appears related to the increase in asthma worldwide. Morphine can suppress respiratory burst induced by NO in neutrophils as well as a complement receptor expression. This suggests that some orders of nematodes have opiods and receptors similar to those of mammals that strongly affect both nerve and immune functions.

Cell Basis of C. elegans *Associative Learning*

Learning is a change in behavior that follows experience (a process that includes habituation). Associative learning is the changed behavior in response to conditioned stimulus following its pairing to a temporally coincidental condition. Associative learning can also be thought of as the temporal-based memory of linked events. For many years, it was assumed that simple worms were not capable of associative learning and this required the evolution of higher brain function. However, associative learning has now been experimentally observed in *C. elegans* using olfaction, thermotaxis, chemotaxis and mechanical sensation as the sensory cues. Early on, benzaldehyde was used to positively train worms concerning the presence and content of food. With such training, prolonged olfaction exposure was seen to lead to habituation, and the odor becomes repellent. Similarly, mechanosensation (taps) is also susceptible to habituation. Males will normally move towards hermaphrodites, but a plate tap will cause inhibition of this movement. Although males habituate to this plate tap, they can also recover from this habituation. This demonstrates a simple form of associative learning. This learned state, however, is not persistent, but in other cases, experience can modify chemotaxis in more persistent manner. Specific sensory experience can cause changes in chemosensory receptor gene expression which will be stable. For example, in ASI neurons, srd-1, srd-2, srd-3 are repressed by exposure to Dauer pheromone (as a quorum sensor). The expression of two of these receptors is regulated by pheromone levels well below those needed to induce Dauer formation, so this identifies an early molecular response to chemosensation. It is thus likely that the modified behavior is a main target of this pheromone. *C. elegans* can also associate unconditioned stimulus (starvation) with a conditioned stimulus (i.e., NaCl concentration), a more complex form of associative learning. Such learning indicates the existence of a plasticity of the learned state, since NaCl is normally suppressive of chemotaxis and this can be overcome by learning. Olfaction-directed behavior can also be used to isolate mutants in the process, such as the *lrn-1* and *lrn-2* mutants. Therefore, it seems clear that *C. elegans* provides the basis of experience-based behavioral adaptation and that a main objective of such behavior is sexual. *C. elegans* nerve cells, however, are not myelinated and thus fire at relatively low speeds. This problem is considered further in the next chapter

C. elegans has an XX/XO system of sex chromosomes but we are most interested in understanding XX/XY systems as found in vertebrates and their role in group identity. In this, Brugia malayi is most interesting as a filarial nematode since it has an XX/XY chromosome system for sex determination. Brugia malayi has a 100 Mbp genome, much larger than *C. elegans* so it has acquired a lot of non-coding DNA. Thus it is very interesting that it has a prominent TOY (tag on Y) genetic marker which is a Cer-like repeat element. Y chromosomes, in general, are especially colonized with lineage-specific endogenous retroviruses and related epiparasites (including LINEs). Although most copies of this genetic parasite are not active, the X chromosome retains a full Cer copy with an intact *env* gene which is expressed. As we will see later, these retained ERVs are typically expressed in reproductive tissue of most animals. It is further interesting that a few parasitic nematode species may use these repeat elements to identify and developmentally eliminate regions of DNA from soma tissue, a process called gene diminution. Yet, all the parasitic elements are conserved in germ lines. A similar phenomena is seen in hagfish which suggests that the evolutionary path to human evolution resembles these parasitic nematodes (next chapter).

Summary

In this chapter, we have traced the evolution of metazoan group identity from algae to nematodes. The main objective is to understand the origin of the nematode nervous system and how it controls group identity. The basic mechanics of sensation and group behavior, however, were mostly adapted from systems that had been used by ancestral microorganisms. These common molecular systems include the rhodopsin-like sensory receptors, detection of small-molecule pheromones, gated ion channel proteins, signal transduction systems that target flagella to control motility behavior. These systems all evolved into the main elements of sensory detection and group behavior in metazoa. The strategy of using toxin/antitoxin gene sets to define group identity has also been mostly maintained in early metazoa. With the red, green and brown algae, we presented the known mechanisms that affect group identity and immunity and noted the innovation of gamete-associated oxidative burst that uses reactive oxygen species for innate immunity. In red algae, the role of PUFA (oxylipins) as toxins and in group identity was presented. The multinucleate (hyphal) eukaryotes (algae and fungi) also employed ROS-based innate immune responses. Here we see the emergence of a programmed process of cellular self-destruction (PCD), mostly associated with sexual cell identity and vegetative compatibility, but still also associated with resistance to genetic parasites. The system clearly resembles a T/A module that must prevent mitochondrial disruption via toxic pore proteins (Bcl-2-like). The main constituents of this PDC system were the basal elements of apoptosis and were maintained and further evolved in early animals. The nematodes represent a clonal, non-hyphal, non-sessile animal that moves under the control of a central nervous system responding to sensory cues. Evidence for the involvement of genetic parasites during the evolution of *C. elegans* is extensive. Congruent with the development of nerve cells, we see the large-scale colonization of worm genomes by ERV (chromovirus) and other genetic parasites. These colonizers contributed to the development of new systems of genetic identity and control (via RNAi) that were also used in the development of sensory-dependent PCD of neurons. In *C. elegans*, an expansion of sensory receptors also occurred. This combination of sensory mechanisms controlling neuronal PCD and interconnections and provided control neurons over movement laid the foundation for CNS-determined group behavior in all animals. It also provided the foundations for systems that learn identity (including simple associative learning) from their sensory experience. I have asserted that a main driving force for the evolution of the CNS has been for the purpose of defining sensory-based group identification. Since group identification normally requires some form of T/A or addiction module to create new group identities, the CNS itself must be prone to addiction states that set behavior. Nerve receptors associated with addictive behavior are thus expected and observed in this simple worm CNS. As we will see, these basic systems are conserved and further adapted to provide group identification and behavioral systems for higher eukaryotes and the foundations that lead to human evolution.

Recommended Reading

Red Algae: Its Biology, Infectious Nuclei, Virus and Immunity

1. Archibald, J. M., Rogers, M. B., Toop, M., Ishida, K., and Keeling, P. J. (2003). Lateral gene transfer and the evolution of plastid-targeted proteins in the secondary plastid-containing alga Bigelowiella natans. *Proc Natl Acad Sci U S A* **100**(13), 7678–7683.
2. Cole, K. M., and Sheath, R. G. (1990). "Biology of the red algae." Cambridge University Press, Cambridge [England], New York.
3. Douglas, S., Zauner, S., Fraunholz, M., Beaton, M., Penny, S., Deng, L.-T., Wu, X., Reith, M., Cavalier-Smith, T., and Maier Uwe, G. (2001). The highly reduced genome of an enslaved algal nucleus. *Nature (London)* **410**(6832), 1091–1096.
4. Goff, L. J., and Coleman, A. W. (1990). Red Algal Plasmids. *Curr Genet* **18**(6), 557–565.
5. Goff, L. J., and Coleman, A. W. (1995). Fate of Parasite and Host Organelle DNA during Cellular-Transformation of Red Algae by Their Parasites. *Plant Cell* **7**(11), 1899–1911.
6. Goff, L. J., Ashen, J., and Moon, D. (1997). The evolution of parasites from their hosts: a case study in the parasitic red algae. *Evolution* **51**(4), 1068–1078.
7. Moon, D. A., and Goff, L. J. (1997). Molecular characterization of two large DNA plasmids in the red alga *Porphyra pulchra*. *Curr Genet* **32**(2), 132–138.
8. Nagasaki, K., Ando, M., Imai, I., Itakura, S., and Ishida, Y. (1994). Virus-Like Particles in Heterosigma-Akashiwo (Raphidophyceae) – a Possible Red Tide Disintegration Mechanism. *Mar Biol* **119**(2), 307–312.
9. Pueschel, C. M. (1995). Rod-shaped virus-like particles in the endoplasmic reticulum of Audouinella saviana (Acrochaetiales, Rhodophyta). *Canadian Journal of Botany-Revue Canadienne De Botanique* **73**(12), 1974–1980.
10. Witvrouw, M., and De Clercq, E. (1997). Sulfated polysaccharides extracted from sea algae as potential antiviral drugs. *Gen Pharmacol* **29**(4), 497–511.

Chromoviruses in Eukaryotes

1. Gorinsek, B., Gubensek, F., and Kordis, D. (2004). Evolutionary genomics of chromoviruses in eukaryotes. *Mol Biol Evol* **21**(5), 781–798.
2. Gorinsek, B., Gubensek, F., and Kordis, D. (2005). Phylogenomic analysis of chromoviruses. *Cytogenet Genome Res* **110**(1–4), 543–552.
3. Kahn-Kirby, A. H., and Bargmann, C. I. (2005). TRP Channels in *C. elegans*. *Annu RevPhysiol* **68**, 719–736.

4. Kordis, D. (2005). A genomic perspective on the chromodomain-containing retrotransposons: Chromoviruses. *Gene* **347**(2), 161–173.
5. Miller, K., Lynch, C., Martin, J., Herniou, E., and Tristem, M. (1999). Identification of multiple Gypsy LTR-retrotransposon lineages in vertebrate genomes. *J Mol Evol* **49**(3), 358–366.

Green Algae and Their Genetic Parasites

1. Lawrence, J. E., Brussaard, C. P. D., and Suttle, C. A. (2006). Virus-specific responses of Heterosigma akashiwo to infection. *Appl Environ Microbiol* **72**(12), 7829–7834.
2. Nagasaki, K., Shirai, Y., Tomaru, Y., Nishida, K., and Pietrokovski, S. (2005). Algal viruses with distinct intraspecies host specificities include identical intein elements. *Appl Environ Microbiol* **71**(7), 3599–3607.
3. Rico, J. M., and Guiry, M. D. (1996). Phototropism in seaweeds: a review. *Sci Mar* **60**, 273–281.
4. Short, S. M., and Suttle, C. A. (2002). Sequence analysis of marine virus communities reveals that groups of related algal viruses are widely distributed in nature. *Appl Environ Microbiol* **68**(3), 1290–1296.
5. Van Etten, J. L. (2003). Unusual life style of giant chlorella viruses. *Annu Rev Genet* **37**, 153–195.
6. Yamada, T., Onimatsu, H., and Van Etten, J. L. (2006). Chlorella viruses. *Adv Virus Res* **66**, 293–336.

Brown Algae; Pheromones and Virus

1. Allen, M. J., Schroeder, D. C., Holden, M. T., and Wilson, W. H. (2006). Evolutionary history of the Coccolithoviridae. *Mol Biol Evol* **23**(1), 86–92.
2. de Carvalho, L. R., and Roque, N. F. (2000). Halogenated and/or sulfated phenols from marine macroalgae. *Quimica Nova* **23**(6), 757–764.
3. Delaroque, N., Maier, I., Knippers, R., and Muller, D. G. (1999). Persistent virus integration into the genome of its algal host, Ectocarpus siliculosus (Phaeophyceae). *J Gen Virol* **80**(6), 1367–1370.
4. Delaroque, N., Muller, D. G., Bothe, G., Pohl, T., Knippers, R., and Boland, W. (2001). The complete DNA sequence of the Ectocarpus siliculosus Virus EsV-1 genome. *Virology* **287**(1), 112–132.
5. Pearson, G. A., and Serrao, E. A. (2006). Revisiting synchronous gamete release by fucoid algae in the intertidal zone: fertilization success and beyond? *Integr Comp Biol* **46**(5), 587–597.
6. Pohnert, G., and Boland, W. (2002). The oxylipin chemistry of attraction and defense in brown algae and diatoms. *Nat Prot Rep* **19**(1), 108–122.
7. Reisser, W. (1993). Viruses and virus-like particles of freshwater and marine eukaryotic algae – a review. *Archiv fur protisten kunde* **143**, 257–265.

Fungi: Genetic Parasites, Mating-Type Sex and RNA Silencing

1. Bagasra, O., and Prilliman, K. R. (2004). RNA interference: the molecular immune system. *J Mol Histol* **35**(6), 545–553.
2. Cogoni, C. (2001). Homology-dependent gene silencing mechanisms in fungi. *Annu Rev Microbiol* **55**, 381–406.

3. Fritz, J. H., Girardin, S. E., and Philpott, D. J. (2006). Innate immune defense through RNA interference. *Sci STKE* **2006**(339), 27.
4. Ghabrial, S. A. (1998). Origin, adaptation and evolutionary pathways of fungal viruses. *Virus Genes* **16**(1), 119–131.
5. Glass, N. L., Jacobson, D. J., and Shiu, P. K. (2000). The genetics of hyphal fusion and vegetative incompatibility in filamentous ascomycete fungi. *Annu Rev Genet* **34**, 165–186.
6. Hammond, T. M., and Keller, N. P. (2005). RNA silencing in Aspergillus nidulans is independent of RNA-dependent RNA polymerases. *Genetics* **169**(2), 607–617.
7. Karaoglu, H., Lee, C. M., and Meyer, W. (2005). Survey of simple sequence repeats in completed fungal genomes. *Mol Biol Evol* **22**(3), 639–649.
8. Pickford, A. S., and Cogoni, C. (2003). RNA-mediated gene silencing. *Cell Mol Life Sci* **60**(5), 871–882.
9. Rohe, M., Schrage, K., and Meinhardt, F. (1991). The linear plasmid pMC3-2 from Morchella conica is structurally related to adenoviruses. *Curr Genet* **20**, 527–533.
10. Smith, M. L., Bruhn, J. N., and Anderson, J. B. (1992). The fungus armillaria-bulbosa is among the largest and oldest living organisms. *Nature (London)* **356**(6368), 428–431.
11. Sogin, M. L. (1991). Early evolution and the origin of eukaryotes. *Curr Opin Genet Dev* **1**(4), 457–463.
12. van der Gaag, M., Debets, A. J., Osiewacz, H. D., and Hoekstra, R. F. (1998). The dynamics of pAL2-1 homologous linear plasmids in Podospora anserina. *Mol Gen Genet* **258**(5), 521–529.

Dictyostelium: Genome and Genetic Parasites of a Social Microbe

1. Eichinger, L., Pachebat, J. A., et al. (2005). The genome of the social amoeba Dictyostelium discoideum. *Nature* **435**(7038), 43–57.
2. Golstein, P., and Kroemer, G. (2005). Redundant cell death mechanisms as relics and backups. *Cell Death Differ* **12**, 1490–1496.
3. Gonzales, C. M., Spencer, T. D., Pendley, S. S., and Welker, D. L. (1999). Dgp1 and Dfp1 are closely related plasmids in the Dictyostelium Ddp2 plasmid family. *Plasmid* **41**(2), 89–96.
4. Goodwin, T. J., and Poulter, R. T. (2001). The DIRS1 group of retrotransposons. *Mol Biol Evol* **18**(11), 2067–2082.
5. Metz, B. A., Ward, T. E., Welker, D. L., and Williams, K. L. (1983). Identification of an Endogenous Plasmid in Dictyostelium-Discoideum. *Embo J* **2**(4), 515–519.
6. Noegel, A., Metz, B. A., and Williams, K. L. (1985). Developmentally Regulated Transcription of Dictyostelium-Discoideum Plasmid Ddp1. *Embo J* **4**(13B), 3797–3803.
7. Samuilov, V. D., Oleskin, A. V., and Lagunova, E. M. (2000). Programmed cell death. *Biochemistry (Moscow)* **65**(8), 873–887.
8. Schioth, H. B., Nordstrom, K. J. V., and Fredriksson, R. (2007). Mining the gene repertoire and ESTs for G protein-coupled receptors with evolutionary perspective. *Acta Physiol* **190**(1), 21–31.
9. Urushihara, H. (2002). Functional genomics of the social amoebae, Dictyostelium discoideum. *Mol Cells* **13**(1), 1–4.
10. Winckler, T. (1998). Retrotransposable elements in the Dictyostelium discoideum genome. *Cell Mol Life Sci* **54**(5), 383–393.

C. elegans: *Genetic Parasites, Apoptosis, Neuron Development and Olfactory Behavior*

1. Abrusan, G., and Krambeck, H. J. (2006). Competition may determine the diversity of transposable elements. *Theor Popul Biol* **70**(3), 364–375.
2. Antebi, A. (2005). The prepared mind of the worm. *Cell Metab* **1**(3), 157–158.
3. Bartel, D. P. (2004). MicroRNAs: genomics, biogenesis, mechanism, and function. *Cell* **116**(2), 281–297.
4. Bagasra, O., and Prilliman, K. R. (2004). RNA interference: the molecular immune system. *J Mol Histol* **35**(6), 545–553.
5. Bowen, N. J., and McDonald, J. F. (1999). Genomic analysis of *Caenorhabditis elegans* reveals ancient families of retroviral-like elements. *Genome Res* **9**(10), 924–935.
6. Buss, R. R., and Oppenheim, R. W. (2004). Role of programmed cell death in normal neuronal development and function. *Anat Sci Int* **79**(4), 191–197.
7. Chisholm, A. D., and Jin, Y. (2005). Neuronal differentiation in *C. elegans*. *Curr Opin Cell Biol* **17**(6), 682–689.
8. Ganko, E. W., Fielman, K. T., and McDonald, J. F. (2001). Evolutionary history of Cer elements and their impact on the *C. elegans* genome. *Genome Res* **11**(12), 2066–2074.
9. Jeong, P. Y., Jung, M., Yim, Y. H., Kim, H., Park, M., Hong, E., Lee, W., Kim, Y. H., Kim, K., and Paik, Y. K. (2005). Chemical structure and biological activity of the *Caenorhabditis elegans* dauer-inducing pheromone. *Nature* **433**(7025), 541–545.
10. Kahn-Kirby, A. H., and Bargmann, C. I. (2005). TRP Channels in *C. elegans*. *Annu Rev Physiol* **68**, 719–736.
11. Kapitonov, V. V., and Jurka, J. (2001). Rolling-circle transposons in eukaryotes. *Proc Natl Acad Sci U S A* **98**(15), 8714–8719.
12. May, R. C., and Plasterk, R. H. (2005). RNA interference spreading in *C. elegans Meth Enzymol* **392**, 308–315.
13. Melkman, T., and Sengupta, P. (2004). The worm's sense of smell. Development of functional diversity in the chemosensory system of *Caenorhabditis elegans*. *Dev Biol* **265**(2), 302–319.
14. Robert, V. J., Vastenhouw, N. L., and Plasterk, R. H. (2004). RNA interference, transposon silencing, and cosuppression in the *Caenorhabditis elegans* germ line: similarities and differences. *Cold Spring Harb Symp Quant Biol* **69**, 397–402.
15. Roth, K. A., and D'Sa, C. (2001). Apoptosis and brain development. *Ment Retard Dev Disabil Res Rev* **7**(4), 261–266.
16. Schott, D. H., Cureton, D. K., Whelan, S. P., and Hunter, C. P. (2005). An antiviral role for the RNA interference machinery in *Caenorhabditis elegans*. *Proc Natl Acad Sci U S A* **102**(51), 18420–18424.
17. Uchida, O., Nakano, H., Koga, M., and Ohshima, Y. (2003). The *C. elegans* che-1 gene encodes a zinc finger transcription factor required for specification of the ASE chemosensory neurons. *Development* **130**(7), 1215–1224.
18. van Roessel, P., and Brand, A. H. (2004). Spreading silence with Sid. *Genome Biol* **5**(2), 208.
19. Vastenhouw, N. L., and Plasterk, R. H. (2004). RNAi protects the *Caenorhabditis elegans* germline against transposition. *Trends Genet* **20**(7), 314–319.
20. Viney, M. E., and Franks, N. R. (2004). Is dauer pheromone of *Caenorhabditis elegans* really a pheromone? *Naturwissenschaften* **91**(3), 123–124.
21. Wiens, M., Krasko, A., Muller, C. I., and Muller, W. E. G. (2000). Molecular evolution of apoptotic pathways: cloning of key domains from sponges (Bcl-2 homology domains and death domains) and their phylogenetic relationships. *J Mol Evol* **50**(6), 520–531.

22. Wilkins, C., Dishongh, R., Moore, S. C., Whitt, M. A., Chow, M., and Machaca, K. (2005). RNA interference is an antiviral defence mechanism in *Caenorhabditis elegans*. *Nature* **436**(7053), 1044–1047.
23. Yeo, W., and Gautier, J. (2004). Early neural cell death: dying to become neurons. *Dev Biol* **274**(2), 233–244.
24. Zwaal, R. R., Mendel, J. E., Sternberg, P. W., and Plasterk, R. H. (1997). Two neuronal G proteins are involved in chemosensation of the *Caenorhabditis elegans* Dauer-inducing pheromone. *Genetics* **145**(3), 715–727.

Chapter 6
Group Identity in Aquatic Animals: Learning to Belong

Goals of this Chapter

Immunity and Group Identity in Marine Animals

The goal of this chapter is to trace the continuing development of group identity and immunity from worms to vertebrate fish, a path that leads to the terrestrial tetrapods and eventually to humans. The origins of animal group identity systems in the oceans will be our focus. The oceans present a specific habitat and ecology that were initially crucial in molding the systems used for animal group identity and immunity. A water media promotes various specific types of mechanisms used for group identity, such as water soluble molecules used for olfaction, vision and sound; all transmit information in the oceans. Our objective is to trace the origins and evolution of social identity that led to the evolution of bony fish. Teleost fish show some significant changes in group behavior compared to their jawless ancestors. They often live in large shoals, thus they display expanded tendencies toward social or group behaviors. Accordingly, they appear to have evolved various mechanisms of group identity and behavior. Bony fish have also developed the adaptive immune system which also plays a role in group identity and was also maintained in all higher vertebrates. This adaptive immunity was also absent from their jawless ancestors. As presented in Chapter 5, the control of *Caenorhabditis elegans*' group behavior (as in most animal species) is generally mediated by the central nervous system which responds to sensory cues (e.g., olfaction) that modify neuronal cell function and communication (via cell viability and interneuronal connections). *C. elegans* had also evolved an effective RNAi system that provides immunity to genetic parasites. This RNAi system is also involved in CNS modifications that result in stable alterations (the product of apoptosis), associated with learning and memory and have lasting affects on group membership (i.e., sexual behavior). As the basal behavior is generally sexual reproduction, affecting the identification of mates is a main outcome. In this chapter, I will examine the simplest such animal sexual identity and behavior, starting with cnidarians, to annelids, mollusk and echinoderms. Much of this behavior will be shown to be mediated by biochemical (or olfactory) mechanisms, mediated by small diffusible molecules (i.e., pheromones and toxins). Overlaps

L.P. Villarreal, *Origin of Group Identity*, DOI: 10.1007/978-0-387-77998-0_6,

between social systems of group membership and biochemical systems (immunity systems) will be frequently noted. However, with the evolution of shrimp and crabs, we also see the introduction of highly elaborate visual systems (involving UV and light polarity-based vision) which is used for social communication. The octopus and cuttlefish also have highly developed visual systems that are briefly mentioned. However, for the most part, these extended visual capacities were not maintained or developed in the bony fish, although fish clearly use visual patterns as part of a shoal membership process. With bony fish, we see the emergence of olfaction via a predecessor of the vomeronasal organ (VNO) as a crucial sensor of small molecule pheromones. We also see the emergence of the use of MHC locus (peptides) of the adaptive immune system for olfactory group identification purposes. These systems take on a central role in sexual group identification that were preserved in all terrestrial tetrapods, although lost in the great apes. The basic mechanisms of how sensory detection (e.g., olfaction) can set behavioral tendencies and group identities were thus initially evolved in CNS of oceanic animals.

Genetic Parasites as Driving Force of Identity

As in the prior chapters, the premise of the role of genetic parasites in the origin and evolution of immunity and group identity will continue to be examined. Thus, in this chapter, the emergence of major new host identity systems, such as adaptive immunity of bony fish, will be introduced from the perspective of the possible involvement of such genetic parasites. Both genomic and extragenomic genetic parasites and their persistence will be considered. The origin of the adaptive immune system will especially be of interest in this context and a likely role for ERVs can be evaluated from this perspective. Indeed, considerable evidence will be presented that supports the concept that the core components of the adaptive immune system originated from the stable colonization and cooperation of various genetic parasites which were not present in the genomes of the immediate ancestors. Ironically, following the evolution of the adaptive immune system, we also see an expansion of types of viruses that infect the bony fish (especially retroviruses and rhabdoviruses). Since vertebrate fish can live in large shoals that are strongly affected by both disease causing and persisting viruses, the effects of such virus relationships on group identity will be explored. In terms of genomic parasites, significant transitions have occurred in the genomes of these marine animals. Although the ancestral fungal genomes were essentially devoid of retrotransposons and satellite sequences, as presented in Chapter 5, major lineage-specific colonization and expansion of these elements is seen in marine animal genomes. These events will be presented and examined to evaluate their possible role in host identity. An overlap (interaction) between the acquisition of genetic parasites and the development of host sensory odor (pheromone) detection will also be presented.

Advent of Addictive Behavior from Addiction Modules

A central theme that persisting genetic parasites often succeed in colonizing their host due to the use of addiction modules will be also be considered in the context of marine animals. Such modules are deemed essential for providing the stability and precluding competition by other genetic parasites. However, as presented with the P1 exemplar, addiction modules also promote the creation of new systems of group identity that can both harm non-group members and protect members. The role of toxins and antitoxins (T/As) in group identity will be evaluated from the context. The general concept of an addiction module, however, will also be evaluated for their role in apoptosis and other forms of cellular group identity and differentiation. Of special significance is the introduction of T/A addictive circuits in the central nervous system that directly affect behavior. In vertebrate fish, group identity (shoal, sex behavior) often involves behavioral modifications mediated by the central nervous system. With the evolution of vertebrate fish (and extending to all terrestrial tetrapods), we now see a basic role for the central nervous system for controlling such group behavior. This required the advent of general emotional states involving aggression, obsession, fear and protection to promote both positive (beneficial) and negative (harmful) behaviors which became crucial for controlling behavior. Since this behavioral-based identity is mediated by the CNS via sensory input, vision and olfaction (pheromones) assume crucial roles defining group membership, and such sensory input must generally occur during crucial windows of susceptibility (often in juveniles). The role of endorphins in the central nervous system and establishment of stable addictive behavioral circuits will thus be of special interest to evaluate since, as will be argued later, these same circuits are often used to allow CNS-based group identity and membership. Addiction-based behavior not only results in stable group identification and social bonding, but can also promote individual self-destruction or altruism (as often associated with sexual reproduction). The role of endorphins, pheromones and olfaction in this process is thus presented in some detail, especially relating to oxytocin and morphine receptors. In addition, links between mechanisms of group identity and the immune system (such as MHC makeup) will be considered. Also, the VNO (vomeronasal organ) receptors for pheromone make their first evolutionary appearance in vertebrate fish, and the links between this olfaction and addictive sexual behaviors will be examined. This system exerts much control over sexual and social identification of most vertebrates. This VNO circuit presents a basal system for group identity and will be henceforth traced to the evolution of human social behavior. Endogenous retroviruses appear to be of relevance to the evolution of this organ in fish. This chapter will develop and present various specific exemplars that illustrate key innovations in the evolution of group identity. For example, in shrimp species, highly elaborate visual systems have developed, which are used for social purposes. Some shrimp species have also developed myelinated nerve cells which allow much faster and more compact brains. Worms (annelids) appear to be the first organism to use endorphin (morphine receptors) circuits for both the CNS and immune system control. Toxic sea snails are considered and appear to be the first example of an animal that has developed oxytocin/vasopressin-like molecules. In these snails, these appear to have been used as toxins. These molecules are much used by vertebrates for social regulation. Vertebrate fish will also be presented as examples of the first organism we consider that has used steroid-like molecules as pheromones. In addition, some fish species are also able to establish pair bonding associated with the care and upbringing of offspring. The mechanisms by which such social bonding occur will receive much attention in subsequent chapters.

Zooplankton: the Massive Foundation of Marine Animals

In the prior chapter, considerable attention was given to marine phytoplankton, their mechanisms of innate immunity, sexual identification and their relationship with viruses. These organisms often undergo light-associated spring blooms that include the cyanobacteria, diatoms, dinoflagellates, red, green and filamentous brown algae. The virus association was seen to be a major issue regarding host evolution, blooming and survival. However, besides these spring, light-associated blooms, this phytoplankton also feeds zooplankton, the massive quantity of small marine animals which also undergo especially pronounced spring blooms in the polar oceans. These zooplankton blooms are also associated with reproduction of major shoaling fish species (such as herring and pollack in Alaskan waters). This massive zooplankton community thus presents us with many of the same issues regarding group identity, immunity and viral interactions. Coordinated behavior is well established in many instances (such as mass diel vertical migrations or nocturnal migrations). Like phytoplankton, zooplankton behavior is also much affected by light. And also like phytoplankton, zooplankton species must exist in a marine viral soup which would seem to expose all marine animals to potentially high viral loads. However, unlike phytoplankton, the viral biology of zooplankton is poorly studied. The animals that make up zooplankton represent diverse life histories that includes many larval and adult forms. Planktonic larval forms are essentially microbiological animals that swim, but within this community, we can also find the larval forms of most larger marine animals, such as bivalves, shrimp or spawn from fish. Most larvae show small incremental structural changes from their corresponding hatchling morphology. These larger animal forms, however, are often associated with sometime complex patterns of metamorphosis (such as shrimp). As metamorphosis involves the turnover of most cells in the organism, apoptosis is frequently an essential process for such dramatic change. Thus the altruistic-like death on many cells must be involved to accomplish the wholesale change of cellular identity associated with metamorphosis, and this suggests the involvement of T/A modules (as presented with *C. elegans* neuronal apoptosis).

The specific makeup of all the animal species within zooplankton is not well characterized. It is clear, however, that various copepod species are a numerically dominant component of this population. As copepods are clearly related to shrimp, they represent early and highly successful marine animal lineage. Copepods are also known to exhibit group behaviors. Much of this behavior is light associated, but sex-based behavior is also established. However, in some respects, copepods represent some rather advanced neurological features. For example, some species have myelinated neurons capable of high-speed conduction of impulses through nerves. Because of these more advanced features, copepods will be considered in greater detail later in this chapter. However, it would probably be best to first consider organisms that represent the early development of a nervous system, since nerve-based behavioral control will be a central issue to be examined for all marine animals.

As presented in Chapter 5, basic mechanics of sensory systems and of motile group behavior were invented by microbes. Here, rhodopsin-mediated light detection and signal transduction to the flagella were common pathways linking senses to motility. Also, bacteria invented quorum-sensing systems that use small diffusible molecules, and these were sometimes linked to these same light sensory/production-based circuits, but also linked to motility. As we will see, most of these basic bacterial elements (rhodopsin G-coupled receptor, cilia proteins) are conserved and provide the same sensory input function needed for animal group identification. However, in marine animals, the sensory input is intercepted by the nervous system which provides a behavioral system for the control of all higher eukaryotes. As presented in Chapter 5, with the *C. elegans* exemplar of neuronal control, sensory detection for quorum sensing was via a G-coupled receptor present in olfactory neurons that controlled sexual behavior. Sensory neurons, interneurons and motor neurons were all involved. Thus, in spite of the fact that *C. elegans* represents a rather simple version of a eukaryotic nervous system, it still has sufficient complexity to indicate that it is not representative of the first nervous systems.

Hydra

Hydra represent sessile eukaryotic animals that are considerably simpler than *C. elegans*. Thus, hydra provides a better example from which to trace the first nervous system. As hydra are Coelenterata, they have only two cell layers: ectoderm and endoderm. This provides the most primitive cell structure of any animal with neurons. Hydra do not have full nervous systems, but instead have simple nerve nets. Only 3% of all hydra cells are neurons, so it has many fewer neurons than does *C. elegans*. In addition, Hydra neurons are not differentiated into various cell types. Only one nerve cell type provides all the needed functions: sensory neurons, motor neurons, interneurons and secretory neuropeptide production. In hydra, bioamines are not used; all nerve signaling is via neuropeptides. Thus it is very interesting that such a simple nerve system has both oxytocin- and vasopressin-like neuropeptides. Given how relatively few signal peptides known for hydra (about six in total), the presence of oxytocin/vasopressin-like peptide is most intriguing. One of the most abundant hydra neuropeptides is substance P. In contrast, mollusks show both presence and absence of oxytocin/vasopressin-like peptides, suggesting a bifrucation in the conserved evolution of this system. As discussed below, apylsia lacks these peptides, but all coelenterata and annelida have preserved these genes. Curiously, Conus snail express oxytocin/vasopressin (conopressin) as venom toxins (discussed below). Platyhelmenthes and echinoderms also do not have these peptides, and most arthropoda (except Braratha brassica) similarly lack them. Thus, the presence of oxytocin/vasopressin-like peptides early in animal evolution seems to provide an interesting pathway that leads toward vertebrate evolution. In those mollusks species that have conserved this receptor (see below), expression is restricted to reproductive organs as well as the brain. This corresponding cognate receptor is also related to the rhodopsin 7 TM G-protein. Furthermore, this mollusk receptor gene is phylogenetically basal to those found in vertebrates. What then is the function of oxytocin/vasopressin in hydra? This is not clear. The expression pattern in mollusk suggests a role in sexual identification, but this has not been experimentally evaluated in hydra. Cellular expression patterns are distinct in various animal neuronal lineages. In hydra, cyclostomes and insects, only one population of neurons make both peptides, whereas in birds, reptiles, amphibian and fish, these peptides are made in separate hypothalamic neurons. In mammals, which also use bioamines as nerve signal molecules, a conserved region of the brain stem makes both peptides. Like hydra, jellies (cnidarians) and cyclostomes also have simple nerve nets of one cell type, and all use neuropeptides. Thus, these simple nerve systems must represent the earliest example of nerve systems. The role of neuropeptides regarding group identity is unknown. If neuropeptides play a role in early nerve systems and sexual identification (as the Dauer hormone does for *C. elegans*), it would be expected that they serve a role in sexual identification. The cognate receptors for these neuropeptides would thus be considered as the earliest example of nerve-specific 7 TM receptors able to detect small peptides. In red algae, we saw that reactive oxygen and peroxidase were involved in innate immunity. Like the fungi and algae presented in Chapter 5, hydra also appear to make peroxidases that can be used to generate ROS and potentially provide immune functions (Fig. 6.1).

Hydra Genomes and Genetic Parasites

In terms of potential virus, very little is known regarding hydra (or any cnidarian species). Viral-mediated mass die-offs have not been reported, so it seems that large-scale acute infections are not common in these species. In this paucity of

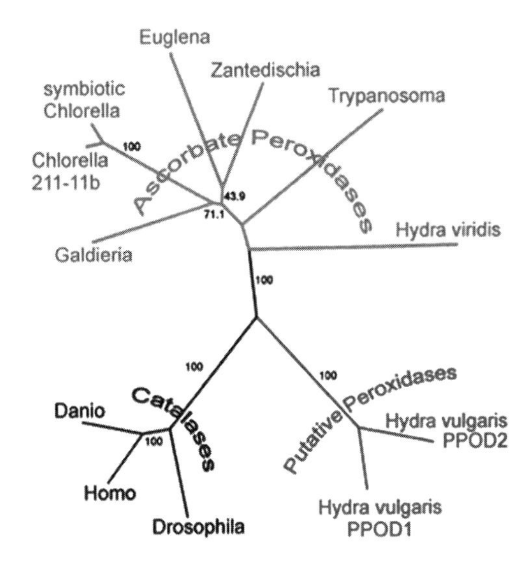

Fig. 6.1 Dendrogram of putative peroxidase of hydra (reprinted with permission from: Habetha, Bosch (2005) The Journal of Experimental Biology, The Company of Biologists, Vol. 208, No. 2157–2164)

virus, they appear to resemble nematodes. Since hydra is a well-studied laboratory model organism, however, it seems likely that prevalent acute viral agents would have been identified by now. There is, however, one well established and very interesting viral relationship that involves hydra. Many hydra are symbiotic host for unicellular algae that provide photosynthesis for the pair. These algae, however, can frequently be infected and lysed by phycodnaviruses, very similar to PBCV-1 described previously. How this viral relationship may relate to symbiotic relationships is presented below. Essentially, nothing is known in hydra about antiviral immune systems. It seems likely that some persisting viruses may be present, especially since VLPs have been observed in hydra by EM studies. However, the nature of putative virus or its host biology has not been explored. In terms of genomes, more information is available. Hydra genomes range from 380 to 1,400 Mbp. Thus, unlike higher eukaryotes, hydra have rather compact DNA and are expected to have a more limited load of genetic parasites (such as ERVs, LINES, satellite elements). However, hydra genetic parasites have not been well characterized. It, however, has been established that hydra have significant numbers of the polinton DNA transposon in their genomes. This transposon is rather unusual in that it closely resembles the RCR of several ssDNA viruses. They are not conserved in the genomes of many other eukaryotes. Such RCR DNA replication is not a characteristic of host DNA synthesis; this clearly suggests that hydra underwent a successful colonization by some virus like ssDNA parasite that now persists as dsDNA in the genome. Thus, it is of great significance to note that this transposon is transcribed and codes for a transposase which is clearly similar to the RAG1-like proteins. These transposase proteins play a crucial role for the origin of the adaptive immune system of vertebrates (discussed below). However, the role of such transposase genes in hydra biology is not clear. According to principles presented earlier, successful colonization by new genetic parasites requires superimposition of new stable identity system generally mediated by a T/A addiction module. It is, however, not known if these genes have any role in hydra group identification or if they affect recognition of other genetic parasites.

ID Systems of Hydra

It is clear that hydra (and cnidarian species) have the capacity to recognize themselves and other organisms. Hydra are frequently in specific symbiotic relationships with other organisms, especially algae as mentioned above. The well-studied example of this relationship is the freshwater *Hydra viridis*, which is symbiotically colonized by one algae cell (a chlorella-like species). As mentioned, this Chlorella algae will be lysed by a hydra-specific chlorella virus (HVCV-1) if the algae are grown outside of the hydra host. All symbiotic hydra so far examined (eight species) also have a virus that will lyse their corresponding photosynthetic symbiotic chlorella algae. Furthermore, the algae-specific phycodna virus that lyse these algae are prevalent and abundant in their natural water habitat. One implication of this situation is that the hydra are providing a protective habitat for the algae that prevents viral lysis and maintains the symbiotic relationship. As also mentioned, cnidarians can exist in both colonial or non-colonial species. It is clear that the colonial species have mechanisms that allow the recognition of self and non-self, even differentiating clones from each other. Colonial species (including corals) are able to kill neighbor organisms via phagocytic epithelial cells and an induced apoptosis response. Related colonial choral organism (i.e., Bitryllus) also distinguishes individuals and will fight via cellular killing when clonal colonies make contact. The phagocytic cells of one (or the same) species will reach out and kill neighboring non-identical organisms. This is clearly colony (group) recognition, and death to non-members is therefore an established feature of these organisms, and this involves apoptosis. The ability of these epithelial cells to phagocytize non-self cells, however, suggests the origin of cell-based immunity and the early evolution of the first blood cells. As we will see, blood cells of essentially all marine animals have phagocytic cells. It is very interesting to note that non-colonial, solitary species that have lost this cell feature are unable to kill neighbors and do not reject foreign tissue that is introduced into polips. Thus, phagocytic cells are not needed to protect these solitary species against parasites, but seem instead to provide colonial organisms with a basal system of group recognition active against similar cells. Here, apoptosis is also being used for group recognition.

Apoptosis in hydra has several elements that are characteristic of most higher eukaryotes, compared to *C. elegans* which is a curiously simpler situation as presented in Chapter 5. Specifically, caspases and bcl2 proteins are present and expressed in hydra, although interferons are not seen. Besides killing non-self cells, apoptosis in hydra is also used for sex cell development. During hydra oogenesis, the oocytes are surrounded by 1,000s of nurse cells. These nurse cells are in suspended state of apoptosis in which caspases are expressed and DNA has already been cleaved, but cells stay alive, feeding the egg and awaiting an oxidative burst. With hatching of the egg, high levels of peroxidase activity are induced, and these cells die. This sex cell associated killing by ROS is reminiscent of that described earlier for red algae sex cells. Hydra sperm precursor cells also undergo apoptotic cell death. Thus, in hydra, sexual reproduction is associated with apoptotic cell death and with ROS. It is worthwhile to retrace the likely evolution of the components of apoptosis. As mentioned in Chapter 5, the basal pore protein that disrupts the mitochondrial membrane potential has clear structural similarity to pore toxins found in viruses and plasmids of bacteria. The antipore proteins that prevent function have the clear hallmarks of a T/A addiction module which will destroy the cell with disruption of molecular identity. In red algae, ROS bursts were used to kill sex cells that had lost their group identity via parasites. In hydra, we see the employment of apoptosis to kill neighboring colonies (groups) that have non-self identity. In addition, in hydra, we see a suspended state of apoptosis used for the generation of sex cells, but here new components (caspase) are also used. ROS, however, is still used to ultimately kill suspended cells. In contrast to *C. elegans*, hydra neurons are not undergoing sensory (olfactory)-associated high-level apoptosis nor are there any blood cells that might similarly undergo apoptosis or selection (as will be presented later in the adaptive immune system, which also links to olfaction via MHC peptide detection). Hydra does not appear to have been colonized by the same retroviral elements (cer) associated with the RNAi response and apoptosis control seen in *C. elegans*. Also absent in hydra neurons are either endorphins or morphine-like receptors (discussed below).

Curiously, other cnidarians (i.e., box jellies) have highly elaborate eye structures, but still retain the simple (brainless) nerve nets, suggesting that any eye-derived pattern recognition (commonly light/nocturnal/lunar-based group behavior) is not CNS determined and must operate by more biochemical mechanisms. This detection must reside in the cell biology of eye itself, not the nervous system. Light does seem to play a special role in the life and reproductive strategies of many jelly fish. Siphonophore, a deep sea relative of jellyfish, also uses light for social communication, and it is one of the very few species to produce red light to communicate movement. Some organisms also have blue patches suggesting color-based patterns for social recognition. There are clearly also other systems in cnidarians that control group behavior as well. For example, young coral reef larvae return to their natal reefs. Here, it is thought that sounds produced by nearby fish and shrimp are involved. Although sensory information is clearly used, it is not significantly processed by a central nervous system.

Urchins (Echinoderms)

Echinoderms are lower deuterostomes of which the sea urchin is the best-studied organism. Like nematodes, with the sea urchins (and echinoderms in general), there seems to be a paucity of acute virus associated with these species. Because sea urchins are filter feeders and have been commercially farmed from larvae in Japan for decades, there has been ample exposure and opportunity to have observed acute viral disease. No such viruses have been observed. Although the mechanism by which sea urchins might resist virus infection have not been evaluated, it is likely they are very similar to those described above for *C. elegans*. Thus the suspicion is that RNAi system derived from retrotranscripts is likely to provide a response that can recognize and degrade foreign RNA. However, this prediction lacks experimental evaluation. It is clear, however, that the sea urchin genome is colonized by Ty3/Gypsy class element (Chromovirus). This virus is also similar to viruses produced in drosophila eggs. In sea urchins, these agents are called SURLs, and the large majority of these elements are defective. With the completion of the sea urchin genome, it has been established that there are 37,000 total copies of such repeat elements, so colonization has been massive. Furthermore, these elements are mostly species limited and do not move between species. However, some SURL copies are intact and even conserve the *env* gene. These SURL-like elements are now found in all Echinoidea yet examined, which includes 33 urchin species as well as some sand dollars and heart urchins. None of these elements appear to move between species; thus it seems they were acquired at the time of speciation, not later. Using the RT sequence as a probe for evolution, phylogenetic tree based on the SURLs have been generated. From this it appears that rate of synonymous mutation in these intact SURLs is equal to that of single copy genomic DNA, indicating that the intact SURLs are under purifying selection that maintains their sequence stability. This analysis also suggest that four cases of horizontal transfer occurred in the common ancestor to all extant echinoid species. Thus colonization by these genetic parasites is highly associated with their origin.

Sea Urchin Identity Systems and Origin of RAG

Due to its reproductive strategy of producing large numbers of sperm and eggs in the oceans, sea urchins are faced with many issues regarding sex cell identification. Thus the study of sperm and egg identity has received much attention. For the most part, specific surface-recognition molecules are involved. Sea urchins do have an innate system of immunity. One organism will generally reject the cells of another organism when grafted tissue is transferred between different individuals. A main component of this innate immunity is via phagocytosis and cytotoxic cells found in blood. Thus sea urchins represent an early example of blood cell evolution which is also involved in blood-based immunity. However, this cell-based immunity is neither adaptive nor specific. In this feature, most marine animals have blood cells that include amoeboid phagocytic cells. Sea urchin also has a homologue of the C3 compliment component. This is a central component of the complement system that creates the attack pore. In sea urchin, it is called spC3 and is expressed in amoeboid phagocytes.

This C3 is retained in coral and is also in shrimp species, described below. It should be noted that complement acts by producing attack hole structures in membranes. This is normally held in check by association with another protein that blocks the attack complex (clearly resembling a T/A addiction module). The ERV sequences present in the sea urchin genome are highly expressed early in development of sea urchin embryos. The function of such high-level ERV expression, however, remains mysterious, but I would suspect some role in genetic identification. Recently, it has been observed that sea urchin genome has two genes that show clear similarity to the core genes of the adaptive immune system, the RAG1 and RAG2 genes needed to generate the Ig sequence variation. RAG1/2 are the core molecular elements of the vertebrate-adaptive immune system (see below) that are needed to generate the diversity of T-cell receptor and immunoglobulin-hypervariable region essential for adaptive immunity. In sea urchin, these homologues are transcribed from a DNA transposon superfamily (transib DNA transposon superfamily, of which the retroviral integrase gene and Mu phage transposase are the basal members). Their function in the sea urchin genome, however, is completely unknown. Given the lack of any known viruses, it has simply been assumed that these transposases represent selfish colonizing elements, but the patterns of conservation below clearly suggest that is likely to be an oversimplification. Clearly, sea urchin (but not other lower animal genomes) was colonized by a DNA or retroviral parasite that provided an early version of RAG1/2 transposase. Normally, I would expect such parasite genes to be involved in host identity systems. These RAG1 and RAG2 are complexed to each other, resembling a TA gene pair. But they are not found in Drosophila or *C. elegans*, so they were not under selective pressure to be maintained in those animal lineages. Instead, they are found in deuterostomes invertebrates and urochordate, but not in the direct ancestors to bony fish (i.e., lampreys, hagfish). The implication of this colonization pattern is that related genetic parasites that encodes RAG1/2 colonized both sea urchin and, later, bony fish genomes, but not the intervening animal species. In the bony fish, these parasite genes, along with additional functions, provided the foundation for a blood-based adaptive immune system (discussed below).

By now it should be apparent that a basal role of CNS is to allow for the evolution of more complex social cohesion, identity, via the behavior of higher animals. Light, sound and pheromone sensory detection must be processed by a central nervous system and result in stable alteration to nerve cell function and identity which lead to stable alterations of behavior. In this regard, *C. elegans* (presented in the previous chapter) provides an exemplar of the role that neurons can play in group behavior and how sensory (olfactory) input affects neuron biology. However, many basal adaptations needed for higher CNS function were also first developed in arthropods. Since arthropods are numerically dominant constituents of zooplankton described above, we will now consider CNS developments in copepods and shrimp as representing a pathway to higher animal CNS function and social recognition.

Social Identity Mechanisms and CNS

Many marine animals live in large schools or shoals. We want to know if this social capacity is specific and learned from sensory input. Complex behavior of copepods is also influenced by pheromones, light and sound. How is such group membership and behavior established? Copepods are the most abundant arthropods on planet and grow to huge numbers during spring blooms in the polar oceans. Marine arthropods represent a basal animal lineage that evolved both insects and vertebrates. Some of these species have myelinated nerve cells, which significantly improved CNS function in animals, although brains remained small in arthropods. Many viruses are known to interact with both neurons and myelin-forming cells of these species. Many species show diel vertical migration, surviving in deep, dark oceans. Some surface-dwelling species, however, are restricted as surface dwellers and do not show these migratory patterns. The mobbing of lights at night by krill indicates a strong influence by light on group behavior of some species. Sex behavior is also pheromone regulated. For example, *Temora longicornis* (copepod) males show chemically mediated female tail-following behavior. Sea louse (parasitic *Lepeophtheirus salmonis*) similarly uses olfaction of pheromones for mate location as adult males move toward females. Shrimp are particularly interesting in their social behavior as they have also developed pair bonding for care of their young in some species.

In contrast to nematodes or echinoderms, shrimp biology is clearly affected by various DNA and RNA viruses, such WSSV and numerous RNA viruses (bunya-like). Although copepod virology is much less studied relative to shrimp, WSSV can also infect some copepods. The WSSV–shrimp relationship will be examined as an exemplar below. Other virus–host relationships will also be presented, such as herpes-related viruses and the nervous systems of bivalve species. Recent assessments have reported that the oceans harbor a large variety of +ssRNA (picorna-like) viruses thought to infect marine protozoan species, but some of which clearly infect shrimp species. Later we will also examine the emergence of RNA viruses (rhabdoviruses) in bony fish, a virus family which was not prominent in ancestral animal species. We will also see the first occurrence of significant autonomous retroviruses in fish, later in this chapter.

Copepods CNS and Immunity

The development of myelinated nerve fibers in CNS was a major innovation, but only in some arthropod species. Calanoid and cyclopoid copepods are the dominant species in zooplankton. Some (but not all) members of these species have lamellated glial sheaths surrounding axons of up to 60 layers. Similar myelination is found in palaemonid and penaeid shrimp, discussed below. These sheaths allow high-speed conduction of trains of impulses (10× velocity) resulting in a remarkable saving of space and energy. This innovation allows the development of fast and compact brains characteristic of all higher animals. They are found in both sensory and motor neurons, as well as in the first antenna interneurons, the primary source of sensory input. This nerve feature occurs in widely differing phyla, including most vertebrates and annelids, but it is rare in other invertebrates as well as absent in Agnatha (lamprey). Thus it seems to represent a major bifurcation point in CNS evolution. Shrimp gial sheaths differ from that of vertebrates in that sheaths are continuous and circular, not spiral. As a result of this innovation, copepods can show extremely high-speed escape jumps (800 mm/s, 200–500 body lengths) in response to rapid decreases in light intensity or brief hydrodynamic disturbances from fish predators. This feature seems to be a more recent evolutionary acquisition in shrimp species in that Calanoid evolution is monophyletic and the split with the myelinated members of this species appearing more recent. These species are not light migratory or as widespread as the non-myelinated species which have more light-dependent behavior (diel behavior, deep water during the day and shallow at night).

Immunity in copepods is not well studied. Although we can assume it is related to the innate immunity found in shrimp species, specific details are not available (see below). However, with shrimp, virus biology is highly linked to host metamorphosis, and since copepods have much less complex metamorphosis, it is likely they also have more restricted virus relationships. However, since neither their genomes nor their genetic parasites have been characterized, little can be said on this issue. If they are indeed shrimp-like in their innate immunity, we would expect to see the presence of an RNAi. The presence of transposons and satellite sequences has not been characterized. It is known, however, that copepods have an efficient reaction to previously encountered parasites (such as tape worm larvae) so it seems they have some system of immunity that is sensitive to prior exposure.

It is worth mentioning something about Decapods. As their larvae resemble their adults, they are distinct from shrimp in developmental complexity. Like most marine animals, they use olfaction for sex identification, for ovary maturation, as a juvenile hormone and for molting and reproduction. However, instead of peptides or lactones, they use MF (methylfarnesoate), which is also used in insects. Thus, the pheromone chemistry has undergone a significant shift in these species.

Bivalves: Oysters, Mussels and Virus

Mussels are the biomass-dominant species of the deep sea. They are also highly successful in terms of species, and 110,000 species of Molluska are currently estimated to exist. There is overwhelming evidence that mollusks have many intimate but poorly understood relationships with viruses of various types. In deep sea species, for example, gut tissues can often be seen (by EM) to harbor virus-like inclusions. Some of these agents appear to further be associated with pathogenesis of big shelled individuals. It is clear that some viruses have asymptomatic relationships with their host, but these same viruses can also be pathogenic. This is especially true for viral members of the herpes virus family. In particular, OsHV-1 (Oyster herpes virus in adult Pacific Crassostrea) is a prevalent infection found in natural populations. Outbreaks by this virus were observed to account for a high mortality in larvae and juvenile oysters, especially in farmed shellfish. Yet, curiously, surveys of natural adult populations frequently show that an asymptomatic OsHV can be found in gonad tissue of 90% of the individuals. It is worth noting that gonad tissue is the main tissue that shows ongoing differentiation in bivalves. It seems that in natural populations, most adult oysters are survivors of virus colonization. This view is supported by the observations by investigators of commercial oyster farms, such as Tomales Bay, CA. Most adult oysters here have OsHV, and the version of the virus appears to be specific to the host population. It is thought that persistence in gonads may allow vertical transmission. Other herpes viruses of bivalve species are also known to show high mortality in farmed populations. For example, a herpes-like virus causing ganglioneuritis destroyed farmed abalone species in four farms of the Victorian coast of Australia. Here, the virus had an affinity for nervous tissue. The OsHV virion is similar to mammalian HSV-1 and CMV (i.e., capsid T = 16 characteristic of herpes), thus it seems to be clearly related to animal herpes viruses. It has many of the usual HSV-like proteins in addition to some unusual ones, such as ion channel membrane proteins (used by CSV to preclude other viruses from infected host). These host-like viral proteins often appear to be phylogenetically basal to their related host proteins. Besides herpes viruses, a lot of aquatic birnaviruses are known to infect shellfish (and fish). Clearly, marine viruses have major consequences to bivalves, but the viral ecology of this situation remains poorly understood.

Mircosatellites: Mussles and Gastropods Suggest a Retroposon Origin

Bivalve species are of special interest with regard to the presence of microsatellites DNA sequences. As mentioned previously, microsatellites are simple repeat sequences, sometimes found in large numbers in the genomes of some eukaryotes. In contrast, some lower eukaryotes, such as fungi, have low satellite DNA content (40× reduced relative to many higher eukaryotes). Recall that many fungi genomes also have almost no LTR transposons. In *C. elegans* (nematode), ERVs and LTR made a significant appearance in their genomes. In the bivalves and gastropods, satellites show both a clear presence and absence. Satellites are of unusually low abundance in many bivalve species, although bivalves are over-represented with introns. Yet closely related bivalve species can differ significantly in their specific satellite makeup and have large quantities. In spite of this variability, some satellite sequences are also highly conserved in all bivalves. A specific example of this situation is found in commercial oyster, *Ostrea edulis*. It has a conserved AT-rich satellite that is related to satellites of other bivalves. This element is also clearly related to part of a mobile (retroposon) element (the Hind III family conserved in all oysters). This observation, of both high variability and high conservation, does not fit the currently accepted model in which satellites and their variation are a product template slippage during DNA replication. Instead, an alternative model has been proposed which suggest that satellites are the result of retroposon gain and loss (via unequal cross-over). Accordingly, Lopez-Flores et al. (2004) concluded that the ancient transposable element generates the satellite units via unequal cross-over (not replication slippage). This would be in keeping with earlier proposals (Batistoni et al., , 1995; Kapitonov et al., 1998) that satellites can originate from retroelements via interspersed retrotransposition and unequal crossover. Although this retroposon-based idea was originally proposed by Smith, in his original 1976 proposal on satellite origins, evidence to support this view had been lacking until these recent oyster results provided direct evidence to support a retroposon origin. Why are bivalves so uneven in satellite DNA content? In some species, satellites are whopping 5–30% of the genome. If indeed these elements originate from retroposon colonization and activity, it would be suspected that they must have altered host molecular genetic identity systems at the time of colonization. This is a large difference with respect to bivalve genomes and seems to indicate a bifurcation in the nature of parasitic DNA tolerated by these two animal lineages. Compared to other related animals, freshwater gastropods (snails), for example, also show lots of variation in microsatellite composition, and here 28 satellite families are known. In these species, the pattern of variation is characterized by repeat length expansion, and multistep mutations predominate. These changes are also species specific. Some worm species, such as meal worm, can also have up to 36% of DNA as satellite sequence. Although it is suspected these satellites are involved somehow in species identity, there is little evidence that relates to this suspicion.

Shrimp: Innovations

There are over 30 extant invertebrate phyla, and they have an interesting tendency to form symbiotic relationships with other organisms. This suggests that many symbiotic species have T/A-based strategies linking them to their host. Copepods, like shrimp, are arthropods and can live in very large social populations, and we have assumed they have shrimp-like immune systems and viral parasites. However, shrimp have coordinated heart beat regulated by cardiac ganglion, indicating a more advanced nervous system. Also, shrimp are significantly more evolved than copepods and can also have highly sophisticated visual systems as well as complex social interactions. Within shrimp, we find our first example of pair bonding associated with rearing offspring. This will be a crucial social bond as presented in the vertebrates. I suggest that the more efficient and compact CNS of the myelinated shrimp brain was an important innovation for the emergence of such complex social behavior and group identity.

Viruses, Immunity and Blood

Shrimp viruses are prevalent and there is much evidence indicating that an array of RNA and DNA viruses can infect shrimp. Much of this is an acute infection. However, some of these infections are clearly persistent and asymptomatic. For example, healthy wild caught *Penaeus monodon* brood stock were over 98% positive for lymphoid organ RNA virus (LDV). Other studies of healthy natural populations have reported that three distinct virus families were generally present: WSSV, monodone baculovirus and hepatopancreatic parvovirus (HPV). Yet other studies report frequent co-infection with WSSV and TSV. Thus shrimp clearly harbor many inapparent viral agents. However, in terms of viral agents that infect shrimp populations, although DNA viruses and some types of picorna-like RNA viruses abound (and are linked to metamorphosis), no reports of negative strand RNA viruses are known, which are so important for vertebrate fish (see below).

Shrimp Immunity

Shrimp are colonial multicellular organisms. Like most colonial animals, they have well-developed defense systems. However, their corresponding defense systems appear to vary widely even within one phyla. They do have allograft rejection via complement and compatibility factors, but these factors are not related to parasite defense. Mostly, it appears that graft rejection occurs via phagocytic blood cells. However, shrimp do not have CTL or NK cells, nor are the hemocytes undergoing ongoing proliferation or differentiation as the vertebrate blood. Shrimp do not have the elements of the interferon system as interferon genes are mostly absent from invertebrate genomes, although they do have a toll receptor. Shrimp do have various RNA-responsive immune systems, and dsRNA will induce both sequence specific and non-specific antiviral response. Here, it appears that long dsRNA (not short siRNA) are most involved. Some shrimp have alpha 2 macroglobulin of unknown function. Phagocytic blood cells are known to be able to express C3-SpC3 and RCA (regulator of complement activation) which seems to be involved in graft rejection. RCA provides a self-protection system that prevents complement damage by inhibition of complement conversion. These proteins have the SCR (consensus) domain. These membrane proteins are highly expressed in gonads, suggesting a role of sexual function as well. ROS is part of the shrimp innate immunity. Some invertebrates use phenoloxidase as a defense system. Shrimp have additional, poorly characterized antiviral activities. Penaeus shrimp have activities that inhibit a variety of DNA and RNA viruses in culture (sinbis, vaccinia, VSV, mengovirus poliovirus). Some of these inhibitory activities appear to function at virus attachment. In addition, maternal factors can also provide antiviral function, such as with WSSV in penaeus shrimp which can be inhibited by maternal factor. Specific hemocytes seem involved in this protection. Recently, several reports indicate that shrimp can be vaccinated against virus. Mostly, DNA-based vaccines have been used, but it also seems env protein of WSSV and other viruses can be used for vaccination. Oral exposure to such env proteins results in humoral antiviral activity that lasts about 25 days, so memory is not lifelong. Shrimp do not have the classical adaptive immune system, so how this vaccination is functioning remains unknown.

Apoptosis: Caspases, Metamorphosis and Virus

The apoptotic process in normal adult shrimp tissue is mostly associated with sex cell development (oocyte nurse cells, sperm) as mentioned above. Apoptosis is also prominently used during metamorphosis as most of the preexisting cells of a shrimp larval state will die following apoptosis. As also mentioned, shrimp apoptosis has additional genes, such as caspase-3, cytochrome *c* release and bcl2 proteins that were not seen in *C. elegans*. In *C. elegans*, apoptosis was mainly used for the development of neurons and not sex cells, metamorphosis or antiviral states. However, virus infection of shrimp can also induce high-level apoptosis that involves these additional elements and can be visualized by staining for the cleaved DNA that results in infected hemocytes (see below for WSSV). The apoptosis in shrimp involves DNA fragmentation as seen in higher animals. Thus it is curious that WSSV replication both induces apoptosis and is also dependent on the specific larval stage that normally undergoes apoptosis. With some shrimp, it can have especially complex patterns of morphogenesis. Curiously, in insects (see next chapter), CNS neurons are not apoptotic during metamorphosis which can allow memory learned in a larvae stage to persist into next morphological stage. Thus shrimp have more complex apoptosis that is used for additional developmental events, but the apoptotic process is also highly manipulated by prevalent acute viruses of shrimp.

Genome and Genetic Parasites

With the major growth of worldwide shrimp aquaculture, interest in shrimp genomics and viral disease has increased significantly. The population crashes in commercial shrimp farming that occurred in the 1990s due to virus induced mortality, however, compelled an intense study of shrimp–virus relationships. Thus, we now know much about the genomes of various shrimp viruses (see below). However, shrimp genome itself and its sequencing projects are not very developed so we have limited understanding of the overall shrimp genome. Yet several distinct characteristics are still clear.

Shrimp DNA is much less colonized by endogenous retrovirus and their related sequences than are the genomes of bony fish. Nor has the shrimp genome undergone the large-scale genome duplication proposed for bony fish. Yet some elements appear relatively distinct regarding shrimp DNA. One distinct shrimp genetic element partially resembles a non-LTR retrovirus and is known as the Penelope element. Although this element has an RT gene, it also has a 'Uri-' like endonuclease domain that is distinct from all known retroviruses or retroelements. As will be presented later, however, the Penelope RT element has interesting biology in other animal species, such as being highly iterated on the fish X chromosome, or being involved in hybrid disgenesis in crosses between *Drosophila* species. Other genetic parasites of shrimp DNA appear to resemble the Dir elements, described in Chapter 5 for Dictyostilium (an non-LTR RT element). Overall, the shrimp genome seems to more resemble the genome of *C. elegans* (such as having low ERV levels, high tc-1 elements) than the genomes of bony fish (with many ERVs and LINES). However, we must await the completion of the genome project to be confident about such generalizations. There is however one clear general distinction between shrimp and *C. elegans* DNA. Shrimp genomes seem highly prone to the presence of large quantities of satellite DNA sequences (similar to what was described above with annelids). These shrimp satellites can be highly abundant (up to 100,000 copies/genome) but are often species specific. Generally, they appear to reside within the heterochromatic component of the DNA so their expression should be silenced. Thus it is most curious that satellite sequences can be highly transcribed during shrimp development. Satellites are expressed in the mRNA fraction, initially at low levels in the larvae stage, but at high levels in post-larvae stages for unknown reasons. This situation is reminiscent of (but inverse to) high-level transposon expression early in sea urchin development and suggests some role for this non-coding RNA in programming during shrimp development. Penapus shrimp have been observed to show high polymorphism in their microsatellite makeup, and this can be used to evaluate the genetic polymorphism of natural populations. Clearly, shrimp genomes are being much altered by their own peculiar genetic parasites, but the major elements here seem not to be the ERV (or LINE) elements as we see in vertebrate fish genomes. Instead, non LTR, DNA types of transposons seem to have been more active (similar to Dictyostelium). It is thus very curious, given this paucity of genomic LTR transposons, that the mitochondrial genomes of shrimp species have unusually large number of pseudogenes. This clearly suggests a peculiar and high activity of retrotransposons in these organelles. Other 'viral' curiosities are also seen in shrimp genomes, such as the presence of sequences clearly related to HHNV, a common DNA parvovirus of shrimp. Also, these HHNV sequences are flanked by satellite sequences that resemble LTRs of retroviruses. Thus shrimp genomes have been much affected by genetic parasites of unknown biological consequence, and the makeup of these genetic parasites are distinct from those seen in vertebrate fish.

WSSV: the Shrimp/Virus Exemplar

WSSV (white spot syndrome virus) was discovered due to large population crashes that occurred in commercial shrimp farms mentioned above. It is a well-characterized 300 kb dsDNA virus that has many distinctive genes as well as a distinct ovoid virion resembling baculovirus. It has a wide host range and can be isolated throughout the world. However, although it was discovered due to its highly pathogenic nature, it also resembles oyster herpes virus (OsHV) in that it can often be isolated from adult host (shrimp) as asymptomatic persistent infections. The natural biology of WSSV is not understood. Besides shrimp, WSSV can infect 20 species of marine crabs, but some species are resistant to disease. The genomes of natural isolates range from 293 to 312 kbp, mainly due to variation in one general polymorphic loci. The natural isolates of WSSV thus have very similar DNA sequence except for 54 bp repeat element and two polymorphic ORFs (ORFs 14/15, 23/24) that are highly specific to the isolate. In fact, the satellite-like 54 bp repeat pattern can even be 'pond' specific relative to freshwater isolates. Curiously, the most pathogenic version of WSSV is also the one with the smallest genome. Furthermore, phylogenetic analysis supports the idea that the larger genome is ancestral to the smaller, pathogenic genome. As I will argue later (with poxvirus and CMV), persistence of a DNA virus in its natural host generally requires additional genes that are needed to control potential host responses to persistence (such as innate immune reactions, apoptosis, ROS, etc.). These additional genes are associated with host-specific molecular identity systems and are not needed for acute replication life strategies. Thus I suggest that these larger WSSV isolates represent virus being made by a persistently infected but specific and unknown host. In terms of WSSV-induced shrimp disease, replication is tightly linked to development and larval stage of its host. Thus, it is interesting that these larval stages also specifically control satellite expression. Since WSSV is unusual in having a satellite-like 54 bp repeat, this presence might relate to host control systems during metamorphosis. Larval development of *P. monodon* is amongst the most complicated in crustaceans. In the early stages, nauplius, protozoae and mysis show no viral disease even when WSSV is present. In contrast, other viruses, such as TSV, show preference for early post-larvae (PL1-10) and late post-larvae (PL1-20) (sex maturation) stages. With WSSV, pathology is mostly seen in late post-larvae and in juveniles. Here, the pathology is fast (3–7 days) and complete (100% mortality). Disease is mainly visible as high-level apoptosis in lymphatic cells (as seen via caspase-3 increase or tunnel assay). This is curious since the lymphatic cells are not virus-producing cells; thus pathology is not resulting from virus replication, but from the induction of apoptosis in the 'immune' phagocytic cells that do not normally undergo apoptosis. Thus, in acute WSSV infection, the levels of apoptosis are high and results in large decreases in hemocyte numbers (10-fold). ROS (oxidative stress) is also strongly induced in infected hemolymph. Since these affected cells are neither making virus nor providing a clear antiviral defensive function, neither the ROS nor the apoptosis appears to be protective in this case. The responses are instead pathogenic. We can say that WSSV has triggered a self-destructive response present in the host, apart from its own replication. Similar pathology is seen in other marine crustacean species, marine and freshwater crabs, copepods and prawns. As noted, shrimp adults show no WSSV pathology. Some specific shrimp species (4/20 field isolates) are refractive to WSSV disease, but WSSV may persist. In marine crabs, WSSV can be found in 60% of benthic larvae in natural populations with no disease. Polychaete worms have also been suggested to be asymptomatic carriers of WSSV. It is likely that these persistent, asymptomatic relationships are the species specific virus–host interactions that will strongly affect related (competing) species that do not harbor persistent virus. WSSV codes for three LATs (latency-associated transcripts, first described in herpes virus-ganglion persistence). Thus persistence appears crucial for WSSV natural biology, but this is poorly studied. Thus it seems plausible that the WSSV shrimp virus is providing group identity in persistent host and inducing the destruction of non-persistent host. Interestingly, the shrimp eye stalk (or its removal) is also associated with disease (discussed below). Thus this sensory organ has direct consequences to viral disease.

Social Consequences of Persistence and Virus

The above exemplar of WSSV shrimp biology suggests some potentially major consequences to the group biology of a shoal species. A shoal represents a population structure (density) that would seem highly susceptible to acute viral agents, as experienced in commercial shrimp farms. However, a shoal persistently infected with virus (potentially destructive to non-members) would seem to also be at a considerable advantage for survival of the shoal. In addition and equally important, if the resident persistent virus expresses addiction modules that preclude other related viruses, the biology of the shoal is transformed also to resist related acute agents. Could such persistent and acute viruses affect social organization and survival of marine animals? The shrimp industry is currently threatened by about 20 distinct shrimp viruses of which WSSV poses the greatest threat. WSSV will show 100% mortality in shrimp infected at vulnerable periods of development. But what is the natural situation of wild populations which do not undergo phased development? Do other viruses alter the WSSV outcome? It has been reported that pre-exposure to hypodermal and hematopoietic necrosis virus will protect shrimp against WSSV. Coinfections with WSSV and TSV are also frequently observed. Some healthy wild caught brood stocks of *P. monodon* were found to be 98% positive for lymphoid virus (LOV). In other studies, healthy wild post-larval *P. monodon* shrimp were seen to be infected with up to three viruses: WSSV, monodone baculovirus and hepatopancreatic parvovirus (HPV). Thus, multiple viral persistence in natural settings is well established. The relationship of these persisting viruses to infection by other viruses or larvae development, however, is unknown. Nor is it known if social density affects virus outcome in natural settings. The mass-reared shrimp that are highly vulnerable to virus and WSSV do show a strong density-dependent mortality, ranging from 76% to 1% as a function of host density. Such density dependence also reminds us of quorum-sensing systems described previously. Might we think that persisting virus is providing the equivalence of a quorum-sensing system that is able to induce self-destruction if thresholds are surpassed? There is little direct evidence regarding this idea. Clearly, resident and acute genetic parasites have considerable potential to provide determinants of group survival.

Shrimp and Fancy Sight

Shrimp appear to use an elaborate visual detection system for the purpose of group identity and membership. The relationship between shrimp eye stalk and virus reproduction noted above is thus interesting to consider. Is this relationship a trivial coincidence or does it have some relevance to virus–host biology? On the surface, the question seems preposterous. What could virus replication possibly have to do with host visual sensory detection? The obvious answer would seem to be that this virus linkage must be coincidental and that visual perception is unrelated to virus biology. Even if visual perception first evolved solely for the purpose of group detection and identification, it would still seem unrelated to the presence or activity of any genetic parasites. But these same visual systems are also directly linked to hormonal, developmental and sexual biology of shrimp. And these developmental systems are indeed highly relevant to viral biology. As we have seen, virus biology is also highly related to self-destructive (apoptotic) systems such as seen in infected hemolymph and that used in the CNS. If visual detection is used for group identification, the resulting sensory input could induce lasting changes in the neurons of CNS, and apoptosis would be a likely mechanism by which to achieve these lasting changes. A link between neuronal apoptosis and visual sensory input would seem much more likely to have had a viral consequence. But such speculation lacks any experimental analysis. Shrimp, however, show many social behaviors, such as snapping shrimp, or the sponge-dwelling *Synalpheus*, which have complex social interactions and complex socially based group membership involving vision. In addition, some shrimp are monogamous and are pair bounded, such as seen with the nesting behavior of *Hymenocera picta* Dana. This is an issue of great interest with respect to the evolution of human social behavior. How did some shrimp come to evolve such social bonding? This question is mostly unanswered. Shrimp eye stalks are the anatomical structures that are involved in shrimp visual, olfactory and hormonal responses. We will now examine the links between these structures, host development, group membership and virus biology. As in other animals we have presented (i.e., *C. elegans*), the most basic group behavior is usually associated with olfaction and sexual identification for mating. In shrimp, mating is a complex social activity and is not the product of random encounters. Many shrimp females, for example, are choosy and will pick specific males by various criteria including visual criteria. Thus we see clear evidence of more advanced sociality in some shrimp.

Eye stalk of shrimp contains several tissues associated with development. In snapping shrimp, the X organ (a sinus gland) is found, which releases peptide hormones (neuroendocrine factors) into hemolymph. These hormones induce changes in extralarval stages associated with metamorphosis. Thus, removal of the eye stalks will ablate the molting of the shrimp. The eye stalk also has the optical ganglion, which controls very fast neurons which control surface color changes associated with social communication. These changes are mediated via nerve conductance and erythropores. Thus, in these species, there exists a fast nervous control of color change associated with social communication. Albation of the eye stalk results in the loss of this color change capacity as well. Stalk ablation also stimulates gonadal development, which is used in shrimp aquaculture, and the ablation of molting can have big consequences regarding susceptibility to viruses.

Complex Social Behavior

Snapping shrimp are also studied for their use of light and sound as sensory cues for social systems. Snapping shrimp are of special interest due to their extremely fast strike, which produces the snapping report. This strike is so fast that it produces vapor bubble collapse that results in cavitations along with a light flash which is not the product of enzymes (such as luciferase). This strike is probably the fastest mechanical motion in any biological system. However, like the fast escape response of copepods described above, this requires rapid nerve conduction via myelinated neurons in both sensory and motor neurons pathways. Snapping shrimp are also studied for the highly complex social behaviors, some of which are clearly learned. Social relationships can extend to other species. Over 100 snapping shrimp species are found to live in coral reefs and have various social relationships with other species. One clade of snapping shrimp, for example, is an obligate sponge-associated lineage. In some cases, a mutualistic partnership is formed, such as with Gobiid fish which construct burrows in which fish and the shrimp both reside. The fish will not burrow without shrimp. The mechanisms for such social symbiosis and communication are not well understood. However, since about 120 fish species and 30 shrimp species are involved, this is not an uncommon relationship. The social interactions of snapping shrimp can also be eusocial, showing both altruistic and multigenerational behaviors. In the case of *Synalpheus* species, shrimp form monogamous pair bonding of mates that can also share in the rearing of offspring. Thus, some shrimp can show surprisingly advanced social interactions, but the neurological mechanisms involved are poorly understood. Another complex social trait of snapping shrimp is their ability to see and produce complex patterns of light for social purposes. These shrimp can see the vector of light (polarization) via highly aligned visual pigments in microvilli. This feature is absent from mammals, but present is some bird, amphibian and reptile species. Most invertebrates, insects, crustaceans, cephalopods and mollusks can also detect polarized light. Shrimp antennae of some species can also produce striking patterns of polarized light which are used for social communication. Such visual complexity is reminiscent of cuttlefish, which also make polarized light that is produced into dynamic body patterns for social communication purposes. Highly complex visual sensory capacity is also seen with mantis shrimp (stromatopods). These are aggressive diurnal predators that show complex social behavior involving displays of colored and polarized body markings. Their use of color in communication is especially observed by the displaying of bright colors in forward appendages, and they also produce striking patterns of color and polarized light in their antennas. Recall that antenneas are also the main target of pheromone binding which is associated with sexual and group identity (olfactory communication). Thus visual and olfactory sensory detection are both communicated via antennas. These mantis shrimp are especially well studied with respect to visual system and are the record holders of all organisms in having 16 types of visual pigments. They have remarkable color vision with up to 12 channels of color, 8 photoreceptors in visible range and 4 receptors in UV range. They also see vector of light but use distinct neurological data stream to CNS for communicating vector versus color information. Their compound eyes have an elaborate trinocular design, with three sections in which specific regions are adapted for specific color and polarized light. In contrast to most other species, in mantis shrimp, the long-wavelength color receptors are individually tuned with filters that vary in individual animals. Thus they have a phenotypically plastic individually based visual system which seems able to learn to see specific wavelengths. The social reason for such a remarkable visual capacity and resolution is not clear. These shrimp would seem to be able to visually differentiate individuals. This would contrast with the biochemical-based (e.g., MHC peptide) olfactory detection of individuals, as present in many vertebrates (see Chapter 7). It is generally thought that such visual sensory capacity must be involved in sexual and other social group identification. However, the evolutionary pressure that would have created such visual complexity are not clear, nor is it clear why other, 'more advanced' species (e.g., mammals) do not also retain the highly developed visual characteristics seen in these shrimp.

Crabs

Like shrimp, with the increased aquaculture of crabs, it has become apparent that many viruses can also infect them. These include large dsDNA viruses (WSSV,

WSBV), parvoviruses (NPHV) and reoviruses. However, rhabdoviruses and retroviruses have not yet been observed to infect crabs. With respect to the above discussion on vision, it is interesting that WSSV infects many crab species and is found to replicate in crab eye stalks. Of special note, however, is the appearance of crab Bunya-like viruses, which in higher organisms are often associated with CNS disease. Although they can also infect shrimp, Bunyaviruses represent a new viral lineage compared to those that have been presented so far. They are segmented ambience RNA viruses which package tri-segmented ssRNA (+/−) into a membrane-bound capsid. The natural biology of crab Bunyaviruses is not well known. They were initially isolated in mixed infections with WSSV. The details of the pathology of viral-induced disease or persistence in crabs have been well worked out. Little is known regarding Bunyavirus persistence in crabs. One observation is noteworthy, though. Unlike various DNA and retroviruses of vertebrate fish, few crab viral infections are associated with hyperplasia. In Bunyavirus infected diseased crabs (*Cancer pagurus* with systemic bunya-like virus, CpSBV), virus can be found in very large numbers in hemolymph and is observed to bud from hemocytes. This contrasts with WSSV in shrimp hemolymph described above and clearly indicates acute pathogenic replication in these cells. In vertebrates, Bunyaviruses, such as LaCrosse virus, show a general and clear tendency for neuronal damage, via demyelination and neuronal apoptosis. This vertebrate disease is the result of mosquito borne transmission. Thus it is very interesting that in mosquito insect host, no insect disease is seen and transovarial virus transmission occurs, suggesting a sex-linked persistent infection. Consistent with this, no genetic changes are observed in LaCrosse RNA when passed in mosquitoes, suggesting this is natural persistence. It is not known if some crab species might similarly harbor persisting Bunyaviruses. Crabs can also be infected with highly pathogenic WSBV (white spot bacilliform virus). This is a large dsDNA virus with clear resemblance to insect baculovirus, although with little sequence similarity between them. WSBV has a 300,000 bp DNA with 181 ORFs. These ORFs are mostly unique and have few matching sequences in the genetic databases, although weak similarity to herpes virus can be seen. It is interesting, however, that WSBV has a full collagen-like gene, a gene that is characteristic of animal genomes. WSBV is also unusual in having 47 repeat elements of various types, some of which clearly resemble satellite sequences, but of unknown function.

It is clear crabs also have various social and sexual identification systems. Female hermit crabs, for example, are known to produces two premolt pheromones (probably peptides) in their urine. Peptide-based sperm-attracting pheromones are also known for crabs. One pheromone is detected by male outer flagella antennules and induce male to grasp and copulate with female. In addition, female crabs can clearly distinguish male crabs via olfaction, tactile and visual cues, resembling shrimp in these characteristics. It appears, however, that males depend more on olfaction and less on visual cues to distinguish females, suggesting some sexual dimorphism with respect to partner identification. Male behavior is also distinct from female behavior. Male horseshoe crabs for example will aggressively guard females against other males. As it is not know if any sex-mediated virus transmission is occurring in crabs, so little can be said of potential viral roles in this process.

Mollusks, Snails and Slugs and Tunicates

Mollusks represent highly successful marine species that is also commercially farmed. It has become apparent that viruses can also have large impacts on cultured mollusk populations. Some of these viruses, however, seem to represent basal members of important animal viral lineages. Possibly the most interesting of these is OsHV (oyster herpes virus), as mentioned above, a novel class of herpes virus harboring 124 unique ORFs. This virus represents the first example in the evolution of the animal herpes virus lineage. They came to the attention of aquaculture due to devastating losses in farming of oysters and abalone. Although the biology of this virus is not well investigated, it seems to have several interesting characteristics. In Pacific oyster isolated from California waters, it appears that essentially all adult oysters are OsHV infected. Thus all surviving adults carry the virus suggesting a naturally high prevalence of virus persistence. In these crabs, virus seems to reside in nervous tissue and retrograde neuronal transmission has been proposed (a feature conserved in animal herpes viruses). Some evidence suggests that infected female oysters are able to protect their young from OsHV by unknown mechanisms. In acutely infected young (larvae), disease is seen via hemocyte apoptosis and connective tissue disease. Iridoviruses (common insect viruses) are also known to infect oysters.

Clearly, social and sexual behavior of mollusks is simpler than that of shrimp. However, they still represent some important developments regarding CNS function in social associations, learning and behavior. Of special interest is the ability of morphine to stimulate nervous and immune functions in mussels. Both Ca^{++} and NO are stimulated by morphine in mussel immunocyte. Mussels also use amine neurotransmitters, unlike hydra neuropeptides. However, with respect to sexual identification, as mentioned above, mollusks still retain the use peptide hormone associated with egg laying. One of these is a sperm-attracting peptide, SepSAP, which is released by oocyte to promote external fertilization. Crabs and cuttlefish are similar in this and both produce sperm-attracting peptide pheromone.

Mollusk: Aplysia

Aplysia has been well studied as a model for learning, memory and social behavior. It has two chemosensory organs: the rhinophore that is used for egg response and osphradium not involved in egg response. *Aplysia* uses pheromones to attract conspecifics for mating. These are cyclic, waterborne peptide pheromones. However, these pheromones are not species specific and similar molecules are found in hydra. Normally, *Aplysia* are solitary animals most of the year. However, they gather together during breeding season, transiently expressing a social system. Egg masses are known to provide social cues by releasing hormone which affect feeding behavior. Hormone release is regulated by numbers (quorum), reminiscent of the *C. elegans* dauer hormone. Memory is also involved in social behavior as *Aplysia* can learn (associatively with other stimuli) to identify food. But memory is also affected by quorum levels as *Aplysia* must be together in a group with others in order for long-term memory to be established (also via pheromone and affected by NO). Thus, memory itself can be considered as a component of a quorum system. *Aplysia* then form large mate aggregates. A sperm-attracting peptide hormone (temptin) is then released by oocytes. Temptin is made of four genes, one of which is Alb, which is highly expressed. Two of these proteins are membrane associated. These proteins are 90% identical between species. The cues for mating do not appear to be species specific as multiple species will aggregate.

Aplysia californica is a well-studied system for neuroplasticity and learning. Short, intermediate and long-term changes in sensory neurons are all seen in this organism. Its nervous system resembles that of vertebrates in that it uses serotonin as a transmitter, via two G-protein receptors, five HT receptors, as found in sensory neurons. These receptors are also linked to egg laying behavior. *Aplysia* thus resembles higher animals in that the NGF family of neurotrophins are used, which were absent in *C. elegans* and *D. melanogaster*, but present in mollusks and coelenterates. It is interesting to note some other aspects of neurotransmitters found in Annelids, slugs and worms. In contrast to *C. elegans*, annelids have CNS-like neuropeptides free in hemolymph, which function as hormones. This seems to represent a bifrucation in the evolution of mollusks. One such hormone, crucial to many social interactions, will receive considerable attention in the remaining chapters below, that of oxytocin. Interestingly, octopuses have conserved the use of oxytocin, although its biology is mostly unknown. The virology of *Aplysia*, however, is not well studied.

Elysia chlorotica: a Photosynthetic Slug with ERVs

This is a sea slug that is interesting due to its transient symbiosis with algal chloroplasts. Thus it represents a highly unusual 'photosynthetic animal' in which chloroplasts have been acquired from an alga. Hatched slug larvae are free of chloroplast but are able to home and attach to specific algae and eat one species of plankton, the filamentous Chromophytic algae, *Vaucheria litorea*. These are giant cellular algae that display light-associated mitosis, with growing tips capable of undergoing fusion. Tip fusion suggests the likely presence of some form of a MAT system (as previously described for filamentous fungi). Although filamentous brown algae and fungi frequently harbor persisting viruses, this is not established for *V. litorea*. The algal chloroplast genomes are large circular DNA and have recombinogenic inverted repeats. Although such repeats are characteristics of non-cellular genetic parasites, the evolutionary origin of this plastid sequence has not been evaluated. After 9 years of observations, it has become clear that the slug life cycle is odd and is of a 10-month duration. After larvae eat algae, they take up the chloroplast which continues to function for photosynthesis for a period of 8–9 months. How the slug prevents O_2 cytoplasmic damage from these active chloroplast, however, is not clear. However, when the juvenile slug undergoes metamorphosis, it must be attached to the algae so their development is linked to chloroplast. Following the cool winter temperatures, a slow warming of water occurs in the spring, which is associated with the laying of egg mass. The role of pheromones in this process has not been studied, although it seems likely. However, following this phased egg laying, all the adult slugs die. During this period, slugs produce an unusual virus in large quantities. The virus appears to be a retrovirus, but rather odd in showing nuclear icosahedral particle assembly. Also highly unusual, viral particles are also seen in the chloroplasts, occurring as icosahedrel paracrystaline array. This within plastid assembly of such VLPs would be highly unusual for any virus, and is unknown for a retrovirus. In the cytoplasm, VLPs are also seen to be budding as enveloped particles into vacuoles (reminiscent of IAP of mouse embryos described in Chapter 7). Thus all individual slugs have this endogenous virus. Since the VLPs can be physically purified and shown to have RT activity, they seem to be authentic retroviruses. Consistent with this is an increase of RT activity by 100× in cellular extracts made during the spring. It thus seems likely that the life strategy and mass die-offs of *E. chlorotica* are associated with the unusual production of endogenous retrovirus. There is also some evidence that DNA (photosynthetic genes) are moving between algal and slug genomes which could be mediated by these endogenous retroviruses. The generality of this relationship is unknown, but the slug–algal relationship appears to be young on an evolutionary timescale. Whether other, non-photosynthetic slugs (such as *Aplysia*) harbor any endogenous retroviruses has not been reported. It is interesting to note, however, that baculoviruses of insects also harbor retroviral agents (TED), suggesting the possibility that the consortia of various virus types could work together to generate and move genes. Baculoviruses are known to infect oysters, but slug infection is not known.

Cone Snail, Conotoxins and Species Identity

Another apparently young evolutionary relationship found in Mollusk is that of the cone snail which produces conotoxin. This toxin production is diverse and has the hallmarks of a group identity system (e.g., diverse species-specific production, along with antitoxins in apparent T/A gene pairs). These numerous snail species are relatively young in mollusk evolution and about 500 species are known, all of which are venomous predators. Curiously, and unlike other venom-producing predator, cone snail prey are highly generalized. Each snail species expresses about 100 distinct peptides that are highly active neurotoxins in which there is little overlap in the peptides. The toxins are specific neuropeptides of about 10–30 aa. They can also be highly selective, mostly against voltage-gated ion channel protein targets but also including some 5-HT receptor targets. Thus, they are very discriminatory. This specificity includes all the major classes of neuroreceptors: omega, Ca channels, alpha, nicotinic receptors, etc. A schematic of the proposed mechanism of neurotoxicity for conotoxins is shown in Fig. 6.2. Curiously, conotoxins always occur in paired genes: the Alpha conotoxins (which excite channels, channel activation) and the Sigma conotoxins (which block channels). Thus both receptor activators and blockers are made and appear to be in matched sets that would seem to counteract each other. One example of this, Na + channel inhibiting cone venom peptide also has a second peptide that blocks this effect. This second

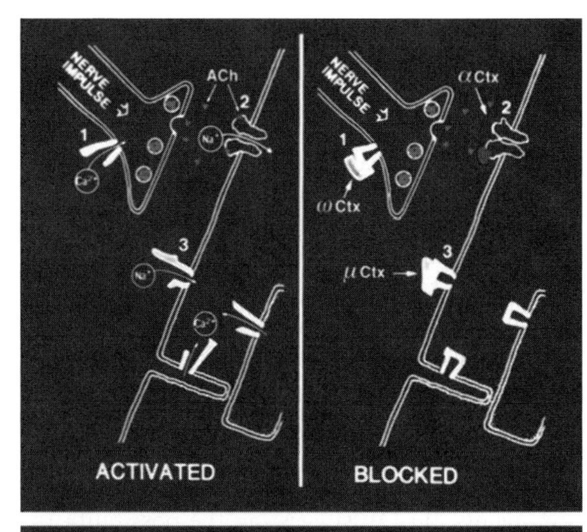

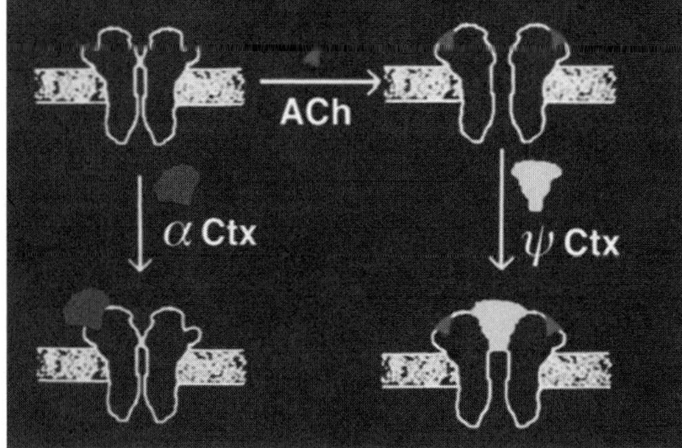

Fig. 6.2 Mechanism of conotoxin neurotoxicity (reprinted with permission from: Baldomero M. Olivera, Molecular Biology of the Cell, Vol. 8)

peptide inhibits channel inactivation. Although both peptides are individually neurotoxic, they antagonize each other. This is a general situation that clearly has the characteristics of a T/A pair. Accordingly, I suggest that the original purpose must have been to provide a snail species-specific toxin-based ID system that became adapted for use as venom for prey. However, the current literature does not experimentally address this proposal very well. Yet some evidence suggests that these toxins can indeed be used between competing species. The origin and evolution of such diverse toxins is a mystery. Toxin evolution seems to have occurred mainly by the action and variation of introns, so genetic invasion seems to be relevant. Also, there appears to be some focal hypermutation in toxin genes, which might suggest a role for satellite DNA or other error-prone elements. In addition, toxin evolution is staltatory, involving acquisition of gene sets, not point changes. There is also an interesting link between venom production and snail shell color pattern. In addition, these shells show specific patterns under polarized light. The variable shell patterns are thought to relate to sexual and egg laying behavior of the female (see Fig. 6.3). The inference is that toxin production and visual identification might be linked in cone snails, but no current evidence directly links snail visual perception or neurological sensory activity to these toxins. It is interesting, however, that in annelida (earthworm), annetocin (related to conotoxin) is known to be involved in reproductive behavior.

Fig. 6.3 Cone snail patterned shells (*See* Color Insert, reprinted with permission from: Baldomero M. Olivera, American Society for Cell Biology, Vol. 8)

Emergence of Major Social Receptors: Oxytocin, Cannabinoid and Morphine

The possibility that conotoxins were initially used for group identification purposes leads to some interesting topics. One of considerable interest to higher animals is the use of oxytocin and vasopressin and their receptors in the context of social or group identification. It should be recalled that oxytocin-like peptides and receptors were first seen in hydra (above) as a peptide neurotransmitter. The receptor is not found in platyhelminthes or *Aplysia*, or in other mollusks that have more complex nervous systems than does hydra. Thus, its occurrence in some mollusks is of special evolutionary note, especially since those mollusks that do have this receptor have a version that is phylogenetically basal to that found in vertebrates. Furthermore, in these mollusks, the receptors are expressed in brain and reproductive organs, clearly suggesting a role in sexual (social) identification. Interestingly, the receptor is also found in octopus, but here it is without introns (suggesting a retroviral-mediated hop into this lineage). In the case of the venom of *Conus geographus*, both vasopressin- and oxytocin-like peptides are made by the conotoxin-producing systems in significant quantities. Thus here, oxytocin appears to be used as a toxin. Furthermore, this toxin retains its neurological activity in a male mouse brain. Oxytocin (or apomorphine) will induce penile erection and yawning in mouse males, as does the related conotoxin. This induction is associated with increased NO production in the hypothalamus. It is interesting that penile erection can also be induced by agonist to the cannabinoid CB1 receptor. In *Aplysia*, two cannabinoid receptors are found, CB_1 and CB_2, both lack introns and are G-proteins. These receptors were not found in echinoderms nor are they present in *C. elegans* or Drosophila. Mollusks thus seem to represent an entry and bifrucation point in the evolution of these important receptors in that $CB_{1.}$ CB_1 was retained in most fish, amphibians and birds, whereas CB_2 is found in mammals, but also curiously in puffer fish. The morphine and cannabinoid receptors are also of major significance regarding social evolution and will be presented later in the context of addictive CNS circuits used for the purposes of group identification. Mollusks also appear to represent a significant evolutionary juncture in the introduction of the morphine receptor. Although no morphine receptor is found in hydra, jellies or free-living nematodes (*C. elegans*), like the cannbinoid receptor, it is found in mollusks as well as parasitic nematodes and parasitic helminthes. Overall, mollusks also appear to represent a shift in the pattern of neurotransmitter, from that of the predominantly peptide-based transmitters in hydra to amine-based (cholinergic) transmitters. For example, snails have 10% of their neurons as nicotinic acetylcholine receptor (which contrast with *C. elegans*). Thus cholinergic neurons are 10% of snail CNS and ganglia.

I suggest that the evolution of CNS circuits that use neurotransmitters and receptors associated with addictive behaviors provides a foundation for a stable

CNS system which is able to create complex social group membership based on sensory experience. Although not established in mollusks, later we will see that these opioid circuits are present in hagfish and central in the evolution of vertebrate social interactions and membership.

Tunicates; Crawling Toward Vertebrates

Tunicates represent an important and basal ancestor to the vertebrates. Thus their genomes and immune systems would seem to provide important clues that allowed the evolution of more complex organisms and group identities.

Tunicate Genome and ERV Colonization Patterns

Tunicate genome is the smallest of all animal genomes with only about 15,000 genes with 65–75 Mbp. It has been proposed that tunicates have undergone a seemingly massive elimination of retroelements in comparison to jawed vertebrate genomes. However, as presented above, more ancestral genomes, such as shrimp and annelids do not maintain a high load of LTR- or RT-based elements, so it seems more likely that the vertebrates underwent a significant colonization of such elements relative to tunicates. Tunicate *Oikopleura dioica* has the most compact genome of any animal. Its DNA has about 20× the gene density of human DNA. Within this, there is little heterochomatin. This genome has six clades of non-LTR retrotransposons, yet most other known families of retrotransposons are absent. It is therefore very curious to note an astonishing level of diversity for the Gypsy/Ty3 elements (a Chromovirus related new family called Odin), most of which are not corrupted. Examples would be Tor3 and Tor 4b, both of which appear to have maintained an intact env ORF, but oddly, this *env* resembles the *env* of paramyxovirus, not other retroviruses. Both the non-LTR and ERV-like elements of tunicates are thus distinct from their mammalian counterparts. But in contrast to lower eukaryotes, no Ty1/copia or Bel-like elements are present, but curiously DIRS1 elements are present. This can be compared to the puffer fish genome which is also a compact genome with a low percentage of repetitive DNA, but has a much higher diversity of autonomous RT elements.

Other Non-vertebrate Animals and Chromovirus

ERV colonization is also seen in flatworms (trematode) such as *Clonorchris sinensis*. Similar to most non-vertebrate animals, flatworms also show clear, but limited patterns of colonization by specific ERVs and their derivatives. Again the Chromoviruses (Gypsy LTR containing elements) are present as uncorrupted copies, but their function is also unknown. One such element (CsRn1) is present at about 100 copies per haploid genome, clearly representing an expansion relative to ancestral genomes. Other LTR retroposons, such as Pao/Bel, were maintained in the nematode, echinodermata, chordata and insecta animals, but not in most vertebrates. However, since these same elements were absent from fungi, lower eukaryotes and plants, their acquisition appears to be animal associated. In sea urchin, starfish and tunicates, we can also see that each of these lineages have distinct ERV groups that are also well conserved. So far, eight classes of Gypsy/Ty3 are recognized in these animals. However, in these genomes, there are also non-LTR repeats that are of low abundance (less then 1%) relative to vertebrates. In *Ciona intestinalis*, for example, only one class of novel non-LTR element is present; the Tor element is a Ty3/Gypsy-related element (Chromovirus) that also encodes an env, but is present at an astounding diversity. These ERV copies are generally intact, and many are not corrupted. Tor3 and Tor4 are both phylogenetically distinct from vertebrate retroviruses, and they resemble sequences found in insect and flowering plants. As presented below, a significant expansion of some LTR elements is also seen in fish species, such as Gypsy in Salmonidae species, but these Gypsy-like viruses were not maintained at significant numbers in genomes of mammals and birds (see next chapter). Curiously, at least one such element was conserved (zebrafish ERV) in human DNA (discussed below). Thus, these non-vertebrate animals have a limited expansion of and conservation of specific versions of Chromovirus-related elements that were present at low levels in genomes algae and fungi as described in prior chapters.

Non-LTR Colonization and Methylation

Tc1 and related elements are particularly prevalent in invertebrate animals. Tc1/mariner elements are related to MITES and are found in high copy number (about 15,000) in *Ciona intestinalis* (sea squirt). Tc1 elements were also found in *C. elegans*, mosquito, fish, amphibians (xenopus) and human DNA, so this lineage of repeat elements has been maintained. Some loci of these elements encode *C. elegans*-like transposase. In addition, Ciona has five clades of non-LTR retroposons at low copy (<100 c/c). As in all species, these elements (and their specific introns) differentiate highly related Ciona species from each other (i.e., European and Pacific species). In amphioxus, BfCR1 (non-LTR, CR1-like element) is monophyletic and present in only about 15 c/c. This same element (like LTR elements) became highly expanded in vertebrate genomes (to greater then 10^6 copies, constituting > 20% of the genome). In Atlantic salmon, 1,200 Tc1/cell were maintained, and 68% of these elements were transcribed. This is 20–30× greater level of Tc1 from that seen in other vertebrates. These expressed copies are stable, showing little sequence divergence. In zebrafish, Tdr1 is a Tc1-related element (at 1,200 c/c) that appears to be specific to the zebrafish genome. The corresponding Tc1 transposases are mostly non-functional in vertebrates. So whatever their role was in lower animals, they became inactivated in most vertebrate genomes. Still, some selection for gain and loss of specific DNA elements seems to have occurred. In contrast to the non-colonial multicellular Dictystioium genome presented in Chapter 5, no DIRS1-like elements are present in Ciona. However, Ciona repeat DNA does show some clear overall distinctions from that of vertebrates. Most significantly, Cephalochordate repeat DNA is unmethylated. Cephalochordates such as Amphioxus is considered sister group to vertebrates. These genomes have the invertebrate pattern of DNA methylation (mainly unmethylated). This is in contrast to that seen hagfish and lamprey, which have vertebrate pattern of methylation (post-embryo heterochromatin formation). Ciona has a mosaic (mixed) pattern of DNA methylation. Such variations in patterns of heterochromatin cannot be due to the simple concept that DNA methylation is needed for genome defense against genetic parasites since similar patterns of ERV colonization (chromovirus) are seen in these organisms. In addition, as presented below, vertebrates which do methylate and suppress these repeat elements have a large expansion of ERV and related retroposon colonization. Curiously, some jawless vertebrates (hagfish) have large quantities of simple repeat satellite DNA (~50% of their total DNA), which is maintained in the germ line but eliminated from somatic chromosomes. For example, Japanese hagfish have species-specific C-bands (heterochomatin domains) composed of simple repeats (85 and 172 bps) that are eliminated from somatic chromosomes (via chromosome diminution) but maintained in germ line chromosomes. Chromosome diminution is known for some lower animals (ciliated protozoa, parasitic nematodes, crustaceans, insects), but essentially unknown in higher animals. Clearly, this situation poses a problem for the concept that satellite DNA is simply 'junk' DNA. That it is both species specific and maintained in the germ line specific remains unexplained. In Chapter 8, we will see that similar C-bands also distinguish human from chimpanzee DNA suggesting an ongoing role to speciation.

In general, it seems clear that the Gypsy-like LTR elements (but not LINES or SINES) and Tc1-related DNA elements show a distinct pattern of colonization and expansion in lower animals (non-vertebrate). Overall, a clear, but low-level ERV and RT element colonization occurred as well as significant expansion of Tc1 elements. However, these specific elements are lineage specific and broader DNA rearrangements were not uniformly seen. In amphioxus, for example, we do not see a massive duplication of genome nor has genome undergone major rearrangement like Ciona.

Immunity/Identity and Group ID in Pre-vertebrates

Ciona intestinalis (sea squirt) as a non-vertebrate chordate (Urochordate) has received much attention recently as a representative predecessor of the chordates. Since it has none of the pivotal genes of the adaptive immune system (i.e., RAG1/2, T-cell receptor, IgGs, MHC I/II/III), it is considered to represent an animal that was present prior to the evolution of the adaptive immune system. Colonial tunicates, however, such as Botryllus, do show a capacity to differentiate self from non-self colony members as mixed tissues are rejected. This capacity is associated with an MHC-like polymorphic gene locus, Fu/HC, but these genes have no sequence similarity to the vertebrate MHC locus. When cells with different Fu/HC loci are mixed, they will reject all tissues not sharing at least one allele at this highly polymorphic fusibility locus. Rejection is via the induction of massive phagocytosis. Thus the cellular response is not a T-cell selected process or due to an apoptosis process as presented with other invertebrates. Nor does this Fu/HC system appear to represent a pathogen-response system. It does, however, provide a group identity system. In addition, a C3-like protein (as well as other complement proteins) are present, suggesting the presence of some complement function, but it is missing almost all the complement regulators associated with the MCH III region. Although not well evaluated, it is assumed to have an innate RNAi-based immune system but lacks components of the interferon system. Some minimal domains related to adaptive immune function can be found. These include the presence of two Ig-like domain proteins and also an MHC-like peptide-binding domain. But these domains have no known role in immune recognition or tissue rejection, and the adjoining genes are very distinct from those of vertebrates. Ciona does have a relative to JAM/CTX (junctional adhesion molecule). These are 'virus' membrane receptor which resembles an Ig antigen receptor. The absence of RAG in the sea squirt is especially interesting, since it was found as a DNA transposase (parasites) in the echinoderm genome. Yet no invertebrate uses RAG genes as a component of an immunity system. Thus this RAG-related element seems to have independently colonized the echinoderm and vertebrate genome, but not their respective ancestors. This clearly supports the concept that the basic elements of the adaptive immune system were acquired via complex genetic colonization by parasites (see below). Thus the Ciona genome and its immune system seem to represent the genetic lull before the storm of colonization associated with the evolution of vertebrates and their adaptive immune system. However, we know very little concerning autonomous or endogenous viruses of sea squirt, so we are hard-pressed to identify likely candidates for this colonization.

Ciona Social Identity

In terms of CNS and sensory-mediated group identity, Ciona, unlike Aplysia presented above, has minimal sensory, memory and social systems. Ascidian (Ciona) adults are mostly sessile, solitary species that do not react much to their environment, so they do not seem to exert much complex social control of individual behavior (in contrast to shrimp). Sexual ID behavior, however, should retain some capacity to differentiate mates, but this may be mostly at a cellular (sperm/oocyte) level. What memory does operate in Ciona seems to be via pigmented brain cells. In terms of CNS receptors, like the mollusks, Ciona retains both the CB_1 (CNS associated) cannabinoid receptor and the CB_2 (immune system endogenous ligand) cannabinoid receptor, which is expressed on sperm. Morphine receptors (methionine–enkephalin) have also been reported via immunocytochemistry studies. Neurons in the ganglion and visceral nerves are positive during spawning season and induced by peroxides to high level expression in oocytes. Thus, morphine-related receptors are involved in sexual cell behavior. Ciona lacks hemoglobin, and instead, it uses hemocyanin for O_2 transport. This protein is also used as a phenoloxidase, which could be involved in an ROS innate response. Ciona show intense interspecies competition by blocking polyspermy, thus displaying a biochemical, not 'sensory' group ID system. Thus it is very interesting that sperms are chemoattracted to egg via sulfated steroid pheromone. This represents the first use of steroids as pheromones we have yet seen, which become prominent in fish. These steroids are not very species specific, however, but may be family specific. They are released via store-activated Ca^{++} channels in sperm. Thus ciona group identification (memory, social, sexual) is not very developed, but does introduce the use of steroids for sexual cell identification. Ciona also introduces a polymorphic genetic loci system to control surface recognition and cellular group (colony) identification.

From Tubes to Bones. Sex Behavior and Vertebrae

Although many aquatic animals are social, as noted, Ciona was relatively devoid of social capacity. Lampreys and hagfish have more advanced social/sexual interactions and cellular ID systems, although still somewhat simplified relative to other animals. Lampreys are unusual animals in that they undergo true metamorphosis from a larval filter feeder to a more mobile adult that is predatory or parasitic. Adults show both sexual and migratory behavior. Unlike Ciona, here, behavioral control is primarily on individual animals, not sex cells. Spawning behavior is mediated by sulfated steroids (petromyzonol sulfate, PS), given off by larvae in spawning streams, similar to Ciona. For sexual behavior, female lampreys, when ovulating, will excrete a sex pheromone from their gill epithelia that will induce males to initiate a searching behavior. This pheromone is not a peptide, but is 3 keto-petromyzonol sulfate. Reception of this cue is via olfaction, along with bile acids, and thus resembling *C. elegans* in sexual strategy. In the silver lamprey, the nasal cavity olfactory region is innervated by a cranial ganglion from the forebrain. How CNS circuits respond to sensory input and compel lamprey behavior has not been studied. Lamprey adult tissues lack gonadotropin-releasing peptide hormone, but a GRH peptide is released by ripe gametes which induces spawning behavior. Also, lampreys make only conopressin – which is a vasopressin-like peptide that controls male behavior. Clearly, lampreys depend on olfactory cues to communicate via the nervous system and control individual sexual behavior, and this is significantly more advanced than Ciona. However Lampreys, lack oxytocin-like peptides, unlike jawed vertebrates as presented below.

Lamprey Cellular ID and the VLR System

Lampreys have also further developed the ability of their lymphocytes to recognize foreign cells. For this, they use a system that can rearrange DNA to express variable lymphocyte receptors (VLRs), which are leucine-rich surface-binding proteins. A germ line copy of the VLR DNA undergoes somatic DNA rearrangement in the lymphocyte allowing the insertion of a leucine-repeat region into an incomplete VLR copy. This is followed by DNA amplification with recombination during cellular differentiation to form an expressing lymphocyte cell population. The process has been called a suspended apoptosis, reminiscent of the nurse cells described above. The system can potentially express up to 10^{14} distinct VLRs. The origin of this system is mysterious. Components required were not present in representative ancestral genomes, such as Ciona. It does, however, have clear attributes of a DNA parasite, such as amplification, site-specific invasive recombination and apoptosis. Yet none of these components are similar to those immune genes in the jawed vertebrates (e.g., RAG1/2, MHC I, II, III). Thus it appears lampreys underwent a distinct type of genetic colonization that resulted in their VLR system. They did not, however, undergo the major DNA rearrangements or ERV/retroposon colonization of jawed vertebrates. Their DNAs do have a high abundance of AT-rich satellite repeats, of unknown function, but their natural populations show low genetic variability. They also have high levels of potentially recombinogenic V-SINEs (vertebrate 5' tRNA-related elements). It is interesting that lampreys do support some herpes viruses, but if this virus has a nerve or lymphoid cell affinity is not know. No retroviruses (ERV) or other viruses have yet been reported for lampreys.

A similar situation exists for hagfish. In addition, Pacific hagfish (Eptatretus) does make C3, C4 and C5 complement components, but they do not have multiple copies of these genes (C3) as seen in bony fish, which have about 10 c/haploid genome that is species specific. In contrast, tetrapods have only one c/haploid genome. Hagfish, like vertebrates, also use hemoglobin. Interesting that in contrast to lampreys, hagfish have both the vasopressin and oxytocin hormones. However, these genes have unusual intron/exon junction flanked by an ITR element (Tos1) that resembles a Tc1 element. It seems that a genetic colonization was involved in the acquisition of these hormones. In general, Tc1 elements of hagfish are highly species specific, but also species conserved.

Thus an enhanced social capacity is seen along with expanded olfaction, CNS sensory processing (including social electroreception) and sexual behavior all occurred in lampreys relative to sea squirts. In addition, lampreys developed a sophisticated cellular immunity system that uses highly variable surface protein expression and somatic DNA variation in a suspended apoptosis-like state, for the purposes of leukocyte killing. But this system is completely distinct from that of jawed vertebrates. Although this system also has many of the hallmarks of a genetic colonization event, such an event is unrelated to that occurred in bony fish. There is no progressive and diverse leukocyte differentiation in lampreys, as seen in vertebrates, nor did imprinted social identity develop (such as shoals or pair bonding).

Jawed Vertebrates: Complex Adaptive Immunity and Social Identity

The transition to jawed vertebrates from lampreys represents some major adaptations in both vertebrate behavior and immunity systems. Bony fish show major changes in the immune system, major alterations of the genomes and major shifts in relationships between hosts and viruses. Overall, fish have developed much more sophisticated social interactions, group identities and cognitive functions. The Teleosteans represent about 38 orders, which are composed of about 23,600 species. Although this is a large number, this is not enormous compared, for example, to the estimated 200,000 species of only parasitoid wasp (see Chapter 7). These fish include many shoaling species, such as herring and pollard, which are the most numerous vertebrates on Earth. The first vertebrates to evolve adaptive immunity were probably early sharks and rays, which have much smaller extant species counts. Most fish that live in large shoals undergo mass behaviors (migrations) for spawning and other reproductive purposes. Accordingly, they have clearly developed sensory and nervous systems that support such large group interactions, and both olfaction and visual systems are known to be involved. In addition, some species show pair bonding and shared rearing of offspring, social characteristics that are of great interest for human evolution. The acquisition of a complete adaptive immune system, however, is particularly enigmatic. This is made curious by the fact that no extant species has a clear set of subcomponents of this complex immune system. The adaptive immune system seems to have been acquired essentially *en toto*. These components include the T-cell receptors, RAG 1/2-mediated DNA variation in Ig molecules, the MHC I, II, III loci and related T-cell selection via apoptosis. Related to this acquisition, we also see the development of jaws and calcified bones. Bone marrow is the site of proliferation and differentiation of the hematopoetic system, which allows the differentiation of diverse leukocytes via apoptosis-dependent leukocyte selection and differentiation. An enigma of this acquisition is to understand the selective pressures that led to it. Consider the following points. Pathogens were clearly present and must have been an issue for the ancestors of bony fish (e.g., lampreys). It appears that these organisms deal well with their corresponding pathogens as do their extant decedents. It thus seems clear that the origin of the adaptive immune system was not simply directed at pathogens. Instead, it is more reasonable to propose that this adaptive immune system represents a highly sophisticated self-identity system, whose initial selective pressure now seems obscure. In conjunction with this immunity acquisition, major genome changes also occurred. These include the large-scale colonization by ERVs, expansion of RT-dependent elements, such as LINES, SINES, large-scale duplication of the genome and a large-scale elimination and/or inactivation of Tc1elements. There was also a major shift in virus–host relationships. Bony fish support a noticeably expanded set of viruses, relative to ancestral species. Most of these fish virus families are observed to also infect mammals, including autonomous retroviruses, herpes viruses, baculoviruses, parvoviruses, many + RNA viruses, Bunyaviruses and especially the negative strand rhabdoviruses (which have not been previously discussed). Thus it is most ironic that following the origin of the highly sophisticated adaptive immune system, we see such a corresponding increase in viral families that infect the bony fish.

Chromoviruses, LTR Elements and Fish Genomes

In bony fish genomes, a large increase in chromoviral diversity as well as in related LTR elements is observed. However, for the most part, these chromoviral elements were not conserved in the genomes of mammals. All fish species examined to date have 30 novel ERVs families (via pol/RT sequence similarity). Mart and Jule elements are better known examples of the chromovirus found in fish. However, they are absent from mammals or birds, except for one conserved uncorrupted copy (see below, Dario). The Gypsy elements are LTR- and RT-containing relatives of chromoviruses that are lacking gag similarity (hence defective, see Fig. 6.4). These are especially high in fish and plant genomes (absent in other vertebrates). Jule is a prominent Gypsy ERV of fish and has a gag, pol, RnaseII, integrase, but no *env* gene. Puffer fish (tetradon nigrovividis) is of considerable interest from an ERV perspective since it has a compact genome. Yet it too has conserved most, but not all of the fish ERV families. Specifically, puffer fish have 25 LTR-containing retrotransposon elements that include 5 major groups. This level of ERV diversity is not seen in mammals, so it very surprising to find that in such a compact genome, that presumably deleted unnecessary DNA, they have been retained. Some, but not all of these elements are transcribed. What might be selecting the maintenance of such elements?

In terms of transition from ancestral genomes, we can compare to the fish RT elements to those that were clearly present in tunicates, sea urchins and *C. elegans*. As noted above, chromoviruses, SURL, CER (of the Mag RT

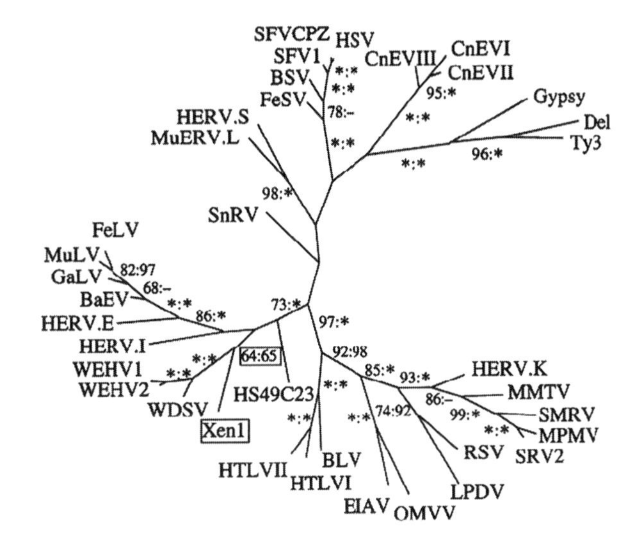

Fig. 6.4 Dendogram of Gypsy-like (chromoviral) retroviruses showing placement of amphibian Xen1 (reprinted with permission from: Kambol, Kabat, Tristem (2003), Virology, Vol. 311)

family) were all seen in these invertebrate animal genomes, but most such elements were at relatively low copies, even though full-length elements were often conserved. Gypsy-related elements were found in sea urchins, starfish and tunicates, including Ciona which had 100 c/c of CsRn1 (a well-conserved Ciona-specific version). The related chromoviral *gag* encodes a capsid gene and its conserved structural domain is clear. This *gag* show a clear relationship to MLV gag, hence it can be considered as a gammaretrovirus and related to elements in mammals. But most chromovirus sequences are defective for gag (e.g., Gypsy). The original Gypsy retroposon, first observed in drosophila; had earlier been defined as a non-viral transposon (via RT and LTR analysis, a defective retroposon). However, it is now clear that even in Drosophila, Gypsy is actually a complete retrovirus virus (including *gag*) that can be expressed in reproductive tissue. In terms of more distal or ancestral genomes, we can recall that in pathogenic fungi, Pyret was a gag containing chromovirus which unlike most viruses of filamentous fungi, encoded an *env*-like gene. Clearly, versions of complete chromoviruses were present and affected invertebrate animal evolution. However, the phylogenetic analysis of RT and LTRs of such elements is not always congruent with that of *gag*. Thus, invertebrate animal genomes show many specific but non-congruent associations with their ERVs. As described below, there are some overall clear distinctions in this with the genomes of vertebrates.

ERV Differences with Mammals

It is important to compare general genomic and ERV differences between mammals and lower vertebrates, especially fish. Mammals have thousands of copies of endogenous retroviruses. Although they are mostly inactive, their diversity is relatively limited compared to that lower absolute number found in fish. Mammals also have almost no functioning LTR retrotransposons, in spite of the large numbers in their genome. In general, mammals lack the various chromoviral elements noted above, but expanded in fish. Mammals have no Penelopy retroposons (i.e., Poseidon and Neptune) as found in shrimp and sea urchins, therefore these specific and abundant ERV lineages were lost. There are also significant differences with DNA transposons between fish and mammals (discussed below). Here too it appears that the higher diversity of such elements in lower vertebrates was lost, supplanted by expansion of a limited, very abundant but non-functional subset. What selective forces might then have favored such general shifts in genomic makeup or parasitic DNA?

ERV Role in Bony Fish Evolution

The significant increase of ERVs and LTR elements in fish DNA is well established, but is seldom thought to have been involved in the development of immunity or group identity of fish. The previously accepted explanation has been that such elements were simply the result of the accumulation of selfish DNA, with no phenotypic or real evolutionary consequences. According to the central theme of this book, however, stable colonization by exogeneous genetic parasites is not simply the result of a selfish replicator. Such successful colonization will generally involve addiction modules that have strong affects on immunity (especially against related genetic parasites) and to group identity. Furthermore, they can provide the mechanisms of new host identity (immunity) systems. Is there then any evidence that fish ERVs might have such affects in the origin of fish immunity or group identity? As will be presented below, there is much evidence that indicates a basic role of ERVs in the origin and evolution of the fish adaptive immune system. Evidence will also be presented that suggests a role in sexual reproduction. It is known that zebrafish do encode an 11.2 kb ERV with intact env and LTRs. Curiously, this element is one of the very few fish ERVs conserved in the human genome. This is an MLV-like virus whose lineage is distantly related to the HERV Ks found in the human genome. In fish, a major site of the ERV expression is in thymus, which is developmentally restricted after 2 days of larval development. EST screens also find expression of this ERV in brain and retina, but especially olfactory rosettes. Unlike the human HERV K, an autonomous retrovirus relative of this fish ERV does exist. This zebrafish ERV has *gag*, *pol* and *env* sequences that are related to SSSV, an autonomous retrovirus that causes bladder sarcomas in juvenile Atlantic salmon. Additionally, a second related autonomous retrovirus infects Chinook salmon. This is the plasmacytoid leukemia retrovirus (discussed below). This second virus is especially a problem for pen reared fish, but interesting in that differentiating leukocytes are the targets of virus infection. Thus both lytic and endogenous versions of related retroviruses infect fish.

Zebrafish germ line, in contrast to that of mammals, is susceptible to infection and colonization by modified retroviruses. Such virus capacity provides the currently used technology to make transgenic fish. To make MLV mediated transgenic fish, the VSV envelope protein has been used to coat and MLV recombinant retrovirus (MLV gag, pol, integrase) as a pseudotype virus. This mixed virus can then be used to efficiently infect differentiating spermatogonia resulting in sperm with multicopy integration of MLV. However, such germ line infections does not seem to frequently happen in nature as most fish LTR transposons are functionally silent. Yet the MLV capsid (gag) is similar to chromoviral capsid found in fish, suggesting that there are no fundamental barriers to retrovirus integration into fish germ lines. Yet there is strong evidence of recent germ line ERV activity associated with sex evolution in fish (discussed below).

A Curious and Ancient Link of ERVs to X and Y Chromosome Evolution

Teleost fish are of special interest with respect to genetically determined sex. Given the central importance of sexual behavior and identity to group behavior and identity, this issue also is of major significance for this chapter. Some species of platyfish (*Xiphophorus maculatus*) have been the focus of studies that seek to understand the origin of sex chromosomes. Teleost fish have a remarkable variety of ways by which they determine sex (SD), not all of which are genetically determined. In this particular genus, SD systems vary both depending on species but also within one species depending on the specific population (I. Nanda2000). *Xiphophorus maculates* thus appears to represent an organism that has evolved several sex-determining systems that may represent the origin of genetic SD systems. The genetic variations include XX females with XY males, WY females and YY males. It has been hypothesized that a male-determining locus is present on the X, Y and W chromosomes, but only the Y chromosomal locus is active. It is therefore very interesting to note that in this model system for the evolution of sex chromosomes, we also see close link to color patterns. *Xiphophorus maculates* strains show variation in melanophore patterns (spotted, stripped, extended, etc.) as well as color patters (dorsal red, anal red, brown). These strain-specific color patterns are tightly linked to the sex determining (SD) locus. Such a close linkage would be consistent with my thesis that visual group identity (via pattern and color) is most fundamentally associated with sexual identity. Unexpectedly, this SD locus is also related to ERVs.

Viral Connection for the Origin of Y

Mart (aka Mar) is a genus-specific genetic locus of fish that contains a chromovirus with gag gene-related to Sushi (a LTR-containing defective retrovirus).

This locus is also tightly linked to sex determination. Unlike most chromoviruses, this is an ERV genome that has been maintained by purifying selection (not as pseudogenes) in the evolution of mammals. However, in mammals, it is no longer able to function in transposition but still maintains the gag ORF (including unique introns). This chromovirus is found in both mouse and human genomes (both with 11 copies, 8 on the X chromosome). This ERV is expressed in mouse embryo. Curiously, the majority of Mar genes are expressed in brain, and some are subjected to molecular imprinting (paternal-specific embryonic expression). For example, the two autosomal copies show paternal embryo and placental expression in mammals. As noted, in fish some *X. maculates* strains do have stable X/Y sex determination. It is thus very interesting that *X. maculates* has retained four nearly intact copies of Jule (a MAG family chromovirus) on its X chromosome. Even more intriguing, however, when interspecies hybrids are formed with other *Xiphophorus* species, spontaneous melanomas (pigmented tumors) will develop. It is now clear that the Jule chromovirus is involved in duplicating and activating the Xmrk EGF receptor (a tyrosine kinase receptor), which promotes growth of melanocytes onto tumors. Intact Jule is found in the first intron of the Xmrk gene and a second defective copy is only 56 nt downstream of the end of the gene. Hybrids are somehow allowing these otherwise repressed Jule copies to rearrange and activate the Xmrk gene, which controls pigment production and proliferation. Thus, these silent fish chromoviruses are responding to disturbed sexual identity in the F1 by loss of suppression and inducing lethal tumors in hybrid progeny. This situation has the hallmarks of a T/A addiction module that responds by activating a suppressed ERV. As will be presented in the next chapter, mice (from Lake Casitas) are also able to induce silent endogeneous retroviruses to produce lethal cancers (lymphomas) in sexual hybrids. However, the Xmrk gene has led to some very useful markers for the X and Y chromosome. Although Xmrk is found on both X and Y, the 5' region of the Y copy contains a repeat LTR retroelement (XIR) adjacent to a Rex3 element. This repeat is clustered, highly reiterated and specific to the Y chromosome. Here, simple repeats (satellites) are apparently absent. Thus, this LTR appears to have amplified along with the origin of the Y chromosome. Estimates, based on mtDNA, suggest that the *X. maculates* Y chromosome is much younger than the other chromosomes and originated only 10 million years ago, making it extremely young and still evolving (compare to an estimated 200 million years for the mammalian Y). Since the XIR expansion is not seen in other maculates populations from other rivers, this strongly suggests that this retroposon colonization and expansion is an early step in Y chromosome evolution. Given the diversity of Y chromosome composition, that X and Y can have very different composition, and that Y is generally composed of highly repeated, heterochromatic non-coding DNA, the process seen above may represent that usual process involved in Y chromosome evolution. In the following chapters, special attention will thus be paid to understand the new retrovirus and retroposon acquisitions associated with Y chromosomes.

Another way of thinking about the above results is to propose that fish sexual identity is being mediated by the presence of persisting genomic retroviruses. These viruses show activity in immune cells, olfactory cells, pigmented cells and sex cells, all tissues associated with group identification. These chromoviruses are sex associated, affect pigment expression and are also brain expressed and this is noteworthy from the context of their proposed role in group identity.

Non-LTR Elements, or ERV Hyperparasites?

Puffer fish have 14 families of non-LTR retroposons (encoding or using RT). The level of these specific elements differ significantly, from a few to thousands, but this is generally well beyond what was present in ancestral genomes. One such element is Rex1, a CR1–LINE-like element that is also found in birds, but not mammals. The levels of this element also vary considerably, from 5 to 500, but were present in Acanthopteergii, the common ancestor to main teleost lineage. Curiously Rex1 is not present in trout, pike or zebrafish DNA, so it must have been lost from some fish lineages. Unlike ERVs, however, LINES do not move between species. Nor is it clear how they amplify given that mostly all are inactive. Fish also have some SINE elements. Puffer for example has two SINES. It is interesting that different clades of fish have distinct copy numbers of these elements. In total, puffer fish have 3,000 such elements that constitute about 2.6% of its genome. Although this represents a significant expansion, it is still a small fraction compared to that of mammalian genomes. The Rex3 elements are also conserved in the otherwise compact puffer fish genome (up to 1,000 related copies). Some fish (including Japanese puffer) also have Rex6 – a non-LTR retroposon but with a type II restriction enzyme-like endonuclease, thus more resembling hyperparasite elements found in prokaryotes. This specific element has undergone several burst of expansion in teleost species (including *X. maculates*, see above for sex evolution), which is likely associated with ERV activity, consistent with hyperparasite activity (invasive of specific parasites). Rex-6 is also absent from human DNA. Given such levels and changes during evolution, it is curious that in zebrafish, no active natural transposon is found. Yet resident elements can be genetically re-engineered back into transposon activity, suggesting they are selectively suppressed. Thus I suggest that most of these non-LTR elements are subjected to influences of ERV activity. Since some such elements do not move between species, but can also be found in the genomes of DNA viruses, such viruses would seem to provide the more likely source of their genomic colonization. That both ERVs and non-LTR elements are tolerated at high levels, however, clearly indicates that unlike some filamentous fungi, vertebrate genomes do not preclude repeated DNA. Genomic identity is relaxed, yet curiously few DNA viruses colonize the genomes of any vertebrate. However, unlike invertebrate genomes, ERVs and LINES are mostly repressed by DNA methylation. Since such repression is a silencing mechanism, it is consistent with a persistent mechanism associated with colonization by genetic parasites.

Lower vertebrates had many more DNA type transposons that are seen in mammals. In mammals, such DNA elements constitute about 4% of the genome, but this corresponds to a few, very abundant but inactive transposons. Both the diversity and percentage of such elements is higher in invertebrates. Ciona, for example, has 15,000 copies Tc1 alone. Compared to Drosophila, fish species also have diverse transposons but show evidence of relatively recent evolutionary activity. Insects, fish, birds and mammals differ dramatically from each other in these patterns of DNA and LTR elements. It has been suggested that the RNAi system may contribute to this difference but how this happened or any potential role for ERVs or other exogeneous virus in this DNA transposon colonization is seldom evaluated.

RCR–DNA Transposons

Fish genomes have maintained or been further colonized by additional specific DNA elements. Politoron is a DNA transposon found in protist, hydra, nematodes, sea urchin, fish and lizard genomes. Some of these elements were conserved in fish. Yet these same elements were also not retained in mammalian DNA. The Politron elements appear to have evolved from protein-primed linear plasmid and uses DNA pol B. This very much resembles the adenovirus family replication and also clearly resembles the mitochondrial parasites of filamentous fungi presented earlier. However, Politorns are distinct from these viral-like lineages by having acquired a retroviral integrase (involving a 6 bp target site duplication) as well as some other genes. Thus, like DIRS-1 elements, Politorns have hybrid characteristics (RT/DNA-transposase). Why some elements but not all were lost is unknown and difficult to explain by the selfish DNA hypothesis. I would suggest that these elements initially provided some form of colonized genomic identity that were subsequently displaced by newer genetic parasites (via horizontal displacement). In terms of RCR replicons, we can recall that the RCR-based helitrons were seen in ancestral genomes, but were also retained in compact fish genome of *Danio rerio*. Note that even earlier major colonization by non-LTR DIRS-1 was seen in some animal lineages (dyctyostellium, fungi, nematodes, sea urchins, fish and amphibia). This non-LTR element shows some relationship to RCR elements, but it is clearly also a hybrid (chromovirus RT and lambdoid YR-recombinase). It too is largely absent in mammals. The Tc1 DNA transposon element is also found in fish (between 300 and 1,200 c/c), including Fugu (puffer fish) which have 1/8th the genome but similar gene content to mammals. In zebrafish, Tdr1 is a species-specific version of Tc1 family present at about 1,200 c/c. Tdr1 has an *elegans*-like N-terminal transposase. Similar elements are found in flatfish

(~300 c/c) and salmon (~1,200 c/c). In salmon about 68% of these copies are transcribed, and these transcribed copies are genetically stable, consistent with positive selection. Tc1-related elements are retained in amphibians and in numerous insects. This element can be modified and shown to be an active transposase in cell culture transfection studies. Its distribution pattern in various lineages strongly suggest that it has been acquired via horizontal colonization, an event also most likely mediated by a virus since these non-LTE elements cannot otherwise move between species. This implies a role in virus colonization. Although Tc1-related sequences can be found in vertebrate genomes, all such elements are non-functional and have been inactivated. Satellite DNAs (simple repeats like the Pst1 element) are also highly preserved in all fish species, often at centromeric positions. The link between these repeats and other genetic parasites is not understood, although as noted earlier, they may result from entry and excision of LTR elements.

As we will see below, the issue of transposase colonization and activity in vertebrates is of central importance for the origin of the adaptive immune system. The IS4 family of transposase (a Mu-like family) is clearly related to the RAG 1/2 genes, which provide a core function for adaptive immunity, but was absent from the ancestors to the jawed vertebrates. Large DNA viruses infecting lower eukaryotes are known to harbor large numbers of transposase and other genes (which I have asserted provide hyperparasite function). Mammals have generally lost the DNA elements, but acquired specific ERVs and LTR-based elements. All these observations are consistent with a role of viruses and other genetic parasites in the development of vertebrate-identity systems.

The Viruses of Fish and Fish Group Identity

The possible relationship between bony fish, their systems of immunity/identification and the viruses they support has not been previously explored. Since bony fish represent various significant developments in immunity, group membership and group behavior, they provide many of the foundation mechanisms that were used for the evolution of mammalian identity systems. Beyond the adaptive immune system, this includes chromosome determined sex, pheromone/olfaction pattern recognition for reproduction, visual pattern recognition of sex partners and shoal membership and associative learning for social bonding. At the onset, it would not seem that viruses could be related to such issues. However, as already presented above, a clear association has been noted between the presence of ERVs and the origin and evolution of fish sex chromosomes. Since the biology of a persisting virus is frequently linked to (and can limit) sexual reproduction, this ERV linkage to sex chromosome is not theoretically unexpected, and it functionally resembles lambda dysgenic effects during a sexual exchange of its *E. coli* host. Given the likely dynamic (exclusion) between ERVs and exogeneous viruses, group membership can also be expected to be influenced by such virus interactions. Bony fish are noteworthy due to the large shoals they often establish. The sensory cues associated with this group behavior are not fully understood, but clearly involve both olfaction and visual pattern recognition. We now examine the potential links between viruses of fish and these systems.

Transitions

In Lampreys, representative of fish ancestors, very few viruses have been reported. Essentially, only one herpes virus has been characterized. Since herpes viruses are known for other invertebrate marine animals, its occurrence in lampreys is not unexpected. However, since lampreys are not commercially important, it is possible that this viral paucity results from a sampling bias, not a general capacity for virus resistance. Yet, there is reason to suspect that early vertebrates were indeed less prone to virus infections. A basal version of jawed fish with a fully adaptive immune system is the sharks and rays, which have mainly cartilage-based bones. Curiously, here too there are almost no reports of viruses of these species outside of one Dogfish herpes virus. As sharks are commercially harvested, viral pathology should be more apparent in these species (especially since many viruses of fish induce visible surface lesions). The suggestion would seem to be that sharks are less prone to visible virus infections and pathology. Additionally, some sharks, such Dogfish (*Mustelus canis*), are noteworthy in that they are placental species, since live birth presents a problem for the adaptive immune system (see Chapter 8). Until we are more certain that this apparent paucity of virus reflects actual biological phenomena, however, we will be unable to explore its potential significance.

Teleostean species (bony fish) are very numerous ($\sim$ 23,600) and are the most species diverse of all vertebrates. This diversity well surpasses that of the phylogenetically older sharks and rays. In addition, bony fish often exist in very large populations, as found in the North Atlantic and Alaska-Pacific oceans. Most bony fish are shoaling species, and herring appear to represent the largest populations of vertebrates on Earth (into the trillions). It is clear that some of these large populations are strongly affected by viruses. In terms of large-scale effects on natural populations, fish rhabdoviruses and nodaviruses seem especially prominent (see below). For example, with commercially harvested Atlantic halibut, infections caused by Nodavirus that result in viral encephalopathy and retinopathy have caused major losses (especially with fry populations). However, the existence of an adaptive immune system in fish means they can be vaccinated against many viruses. Fish reproduction and development is often of relevance to fish–virus relationships. Bony fish are mostly produced from eggs which hatch into fry, although a few species are viviparous, placental species that give live birth (including sharks).

Fish Retrovirus Are Clearly Autonomous

As mentioned above, with the evolution of bony fish, we see the first clear example of an autonomous retrovirus (env containing) that is disease causing in a natural host population. Given this together with the major expansion of fish-specific ERVs and related LTR elements in all fish genomes, it appears that a major shift occurred in retrovirus–host interactions along with the evolution of bony fish and adaptive immunity. The best-studied autonomous fish retrovirus is WDSV – Walleye dermal sarcoma virus. Although as mentioned, this virus resembles both chromoviruses and MLV in some respects, it is a more complex family of retrovirus. WDSV is a 13 kd retrovirus that encodes several additional genes, including a viral cyclin D, which is involved in inducing proliferation of infected cells, resulting in benign skin tumors. These viral induced skin tumors appear to provide a site for virus persistence. This virus is prevalent in natural settings. In terms of the evolution of WDSV, the viral cyclin gene is clearly unlike host cyclin genes. The presence of this unique gene, however, makes several important points. This virus did not 'steal' a host protooncogene (cyclin) in order to induce cellular proliferation. This cyclin is a unique viral version that is highly conserved in this viral lineage. Furthermore, it is clear that WDSV did not emerge from LINES (or other non-LTR RT-encoding elements) present in the ancestors to fish genome (a common misconception). In fact there is no established example of any autonomous retrovirus with such an evolutionary history, despite several such early assertions in the literature based on incomplete phylogenetic analysis. Instead, such autonomous viruses are related to the very ancient Chromoviruses lineage (via similarity to *gag* as described above). However, there is evidence that a descendent of WDSV did undergo lineage-specific endogenization in another species. Xen-1 is a complete endogenous retrovirus specific to African clawed toad (*Xenopus laevis*) (see Fig. 6.4). This very large ERV is highly unusual and has retained the additional complex genes of WDSV. Another related fish retrovirus is WEHV-1, Walleye epidermal hyperplasia (skin growth associated) which also retains a more complex genome. This virus is seasonal, observed during the spring spawning run (Lake Oneida). Almost all fish eventually get infected, so it is highly prevalent. Sea Bass retrovirus is another fish retrovirus that affects blood cells. However, although all these viruses may persist in individual host, they do not normally establish germ line infections. Finally, SSSV (swim bladder sarcomas virus) is a retrovirus that infects in juvenile Atlantic salmon poses a major commercial problem. Chinook salmon are also subjected to infection with a similar plasmacytoid leukemia retrovirus (especially pen-reared salmon). Thus autonomous retroviruses are of considerable natural and commercial importance and have big impacts on their host. However, as noted above, WDSV is clearly related to Chromovirus or ERVs of zebrafish via gag/pol/env similarity. This particular ERV is noteworthy as it is the only fish ERV that is known that has been conserved in the human genome and is also expressed in immune and olfactory tissue and linked with sex determination and Y chromosome evolution in fish.

Viral Proliferation

The autonomous retroviruses of fish tend to cause cellular proliferation, often on skin. Leukemia is an immune cell proliferation that is a retroviral induced characteristic in birds, placental and marsupial vertebrates. Leukemia is further associated with ongoing 'endogenization' of autonomous virus (i.e., koala virus). Thus, the adaptive immune system with its proliferating and differentiating lymphocytes seems especially suitable for autonomous retrovirus infection, but this outcome is also frequently affected by endogenous viruses. This immune association may date back to the very origins of adaptive immunity and T-cell proliferation. It is also interesting that SSSV induces cell proliferation in swim bladders. Bladder is of interest from an evolutionary perspective since this organ is thought to be ancestral to the origin of lung. Transition from bladder to lung required enlargement of the organ and transformation of an epithelia to a mucosa (which occurred in two fish orders). Mucosal reactions to viral infections were observed in tunicates; thus, retroviruses could well have been involved in these transitions.

Behavior, Olfaction and Virus

Fish Nodavirus represent another prevalent RNA virus that can also affect behavior and olfactory neurons. One of these, viral nervous necrosis (VNN of halibut), induces erratic behavior that results from spongioform damage to the central nervous system. This CNS infection and damage is related to incomplete differentiation of nerve cells. Disease is especially a problem in hatchery-reared larvae of juveniles. However, it is curious that many wild caught halibut are sub-clinically infected. Yet young fish grown in virus-free settings are highly susceptible. In these juveniles, the perinasal route of infection involves rather specific nerve cells in the olfactory lobe, which becomes extensively necrotized. Also thalamus involvement (at the innervation point) is seen. Infected nerve cells show viroplasmic inclusions, loss of dendrites and myelin-like structures (not apoptosis-like WSSV-infected shrimp). This curious link to fish olfaction is also seen with SGNNV which shows large numbers of infected cells and damage in the olfactory lobe. This results in altered behavior, including loss of balance, enlarged swim bladder and altered skin coloration. Thus, major systems of fish social communication are being targeted by these viruses.

VEE (Venezuelan equine encephalitis)-related viruses also infect fish. This line of virus has interesting tendency toward demyelination in vertebrates. Two fish versions are SPDV (salmon pancreatic disease virus) and SDV (sleeping disease virus). The natural biology and pathological mechanisms of these viruses are not understood. However, the mammalian versions of this virus family also conserve a tendency to CNS damage, and demyelization is most curious. Other viruses also show a tendency to maintain peculiar patterns of host cell type infection through long periods of evolution. Retrovirus infection of immune cells was noted above. Also alpha herpesviruses have shown a tendency to persist in ganglia. These biological patterns have been maintained even into human evolution.

Rhabdovirus

The negative strand unsegmented RNA viruses have not been previously discussed in any ancestral organisms. Versions of such virus that infect any prokaryotes or lower eukaryotes are unknown. As they are non-recombining RNA, their evolution is limited to point mutation and deletions. They are also thought to be ancestral to the other negative stranded animal viruses (paramyxovirus, orthomyxovirus). As mentioned, rhabdoviruses have major effects on large fish populations. One of the best-studied viruses is VHSV (viral haemorrhagic septicemia virus). This virus is clearly related to mammalian rhabdoviruses (VSV, rabies). VHSV is mostly seen in freshwater stages of rearing trout and salmon. It is also observed to infect fish in marine net pens. Of particular interest is that it can also occur in natural population. It has reportedly been responsible for severe epizootics in wild populations, especially shoaling species such as herring, sprat and mackerel. Thus VHSV appears to be responsible for large fluctuations in natural populations (up to 90% crashes in Alaskan waters). This virus is in three distinct genetic populations. Curiously, phylogenetic analysis suggests that the north American and European viral populations diverged only 500 ybp. Such recent divergence suggests some human involvement, such as commercial fishing, is likely involved. The natural biology of VHSV is not fully understood since its origin and or any naturally persistent host is not known. It seems ironic that bony fish with their highly developed adaptive immune system seem so prone to such a large-scale acute viral disease. It also seems clear that such agents can and will have profound effects on host population structures.

Can such major viral effects on fish populations have any relevance to host group or sexual genetics? The ERV/sex chromosome story mentioned above clearly supports this idea. It is certainly expected that fish not harboring a protective ERV or persistent RNA virus will have a distinct group identity that can promote survival relative to groups of fish that lack these extragenomic elements. Thus we can fully expect that persisting and acute viral agents will have major impact on such group interactions.

Fish DNA Viruses

The oceans are known to harbor immense quantities of various types of large DNA viruses. The majority of such viruses appear to be phage of various prokaryotes (especially cyanobacteria). However, large quantities of mimi-like virus of eukaryotic amoeba and phycodnaviruses of green algae can also be found in specific aqueous habitats. The phycodnaviruses show clear relationship to the herpes viruses of vertebrates and were one of the few viruses observed to infect lampreys. Herpes viruses are also known to infect other marine invertebrates. For example, Abalone farmed off the south west Victorian coast of Australia recently underwent a die-off due to herpes viral ganglioneuritis. The origin of the virus was unknown, but it appears to likely have been introduced from a wild host. In the Pacific, most adult Pacific oysters harbor similar virus. It is interesting that this virus shows an affinity to persist in nerve tissue (ganglions), a characteristic of mammalian herpes viruses as well, suggesting an unbroken lineage of large DNA viruses from marine animals to mammals. Various DNA viruses of fish are prevalent and include herpes viruses, iridoviruses and adenoviruses. All these viruses tend to display a biology that is peculiar to fish, such as a tendency to target integument tissue, which has no traditional basal lawyer, and can induce proliferation.

Iridoviruses

Fish iridoviruses are noteworthy in that they show a wide host range (unlike herpes viruses). This virus family also infects shrimp and insects. LMBV (large mouth bass virus) is a well-studied member that shows lots of strain variation in natural isolates. The reason for this variation is not understood, but is suspected to be associated with its wide range of host species. LV (lymphocystitis virus) is another fish iridovirus that can induce skin hypertrophy, a papilloma-like lesions of skin, fins and tail. This virus has a worldwide distribution, but it is especially found in tropical coral fish, affecting 140 different species. Thus in sharp contrast to most of the large DNA viruses of mammals, these fish DNA viruses tend to infect many species. This associated viral mortality is also mainly seen in juvenile fish. Another fish iridiovirus is ISAV (infectious salmon anemia virus) that was first seen in Altantic salmon in the 1980s and 1990s after the fish transfer to sea from freshwater stage. Although this virus causes acute disease in the predatory Atlantic salmon, its prey fish (wild herring, *Clupea herengus*) are asymptomatic carriers of the same virus. This carrier state is species specific as the related Pollock are not carriers. Iridoviruses can also infect the olfactory epithelia of some fish, preventing amino acid detection and feeding. Clearly, these viruses have inherent capacities to alter host behavior and survival. Overall, it appears that iridioviruses tend to cause much disease in many species, but may also persist in limited numbers of species.

Fish Herpes viruses

Although there are various types of fish herpes viruses, these represent rather diverse virus families, some of which show little direct relationship to extant mammalian herpes viruses. One such virus is Channel catfish virus (CCV). CCV clearly has a herpes-like capsid structure (T = 16), but it shows little or no sequence similarity with HSV genes. Other fish herpes viruses, however, show greater relationship to HSV, via DNA pol and capsid structure. For example, CHV (herpes virus cyprini, an acute disease in carp) more closely resembles HSV. CHV DNA persists in basal cranial nerve ganglion, similar to HSV and produces recurrent virus (via papillomas) with reactivation of the same persisting virus. Another fish herpes virus, Koi herpes virus is a lethal carp infection that was responsible for massive viral epidemic in 1995, 1998 in Australian Pilchards. Herpes virus 2 was also responsible for heavy mortality that occurred in Coho salmon, as well as an epizootics and mass mortality of Plichards in Australia. It too produces hypertrophic lesions similar to those of LV described above. Thus there seems to be several distinguishing characteristics between the herpes viruses of fish and mammals. The fish viruses show a clear tendency for much greater epizootic acute disease of large populations, and the disease tends to have proliferative lesions. Clearly, we expect these viruses to have major impacts on host population structure and identity.

Viral/Host Identity: Disease and Persistence

The above virus–host relationships appear to identify a selective circumstance that could be expected to affect host group identity. Are there field studies that support this implication? Although we do not know of many specific fish species that might harbor persistent infections associated with acute outbreaks, this issue has been somewhat evaluated in farmed eels. HVA is a herpes virus of *Anguilla anguilla*, a farmed species of Japanese eel and has been responsible for massive disease in farms. However, it now appears that HVA originated from European eels. HVA can be isolated from healthy stocks of European eel, indicating that these eels are latently infected. HVA infection and virus production does not appear to be harmful to the European eel. Since this HAV production can be induced by treatments with dexamethasone, such induction has been used to evaluate virus pathology but did not produce any clinical disease. Thus HAV appears to be in an essentially asymptomatic state in its persistent host. Although HVA is prevalent in European eels, it is lethal to Japanese eels. Clearly, HVA is behaving as a group (species)-specific harmful virus against the Japanese eel. But might it also provide protective function in the European eel? However, we lack sufficient knowledge regarding viral ecology of the European eel, to be able to answer this question. Oceanic herpes viruses can also affect group identities across species. For example, Green turtle herpes virus induces a fibro papilloma disease with significant mortality in turtles. This acute infection, however, is sporadic. But the cleaner fish that are associated with these turtles show latent infections with this same virus but with no disease. Although the viral role in cleaner fish–turtle relationship is not clear, it does seem apparent that persistence in the cleaner fish will have a significant effect on such relationship. Thus the tendency of herpes viruses to be latent in a species-specific host, but potentially harmful to related and unrelated species, has been observed in several marine herpes viruses. Related relationships are also known for mammalian herpes viruses. For example, human HSV-1, primate HBV and elephant herpes viruses all retain such a biological characteristic (see below). However, in mammals, herpes family of virus does not tend to infect broad species as we see in the oceanic herpes virus.

Fish also support adenovirus, such as the White sturgeon (SnAdv). The biology of these viruses is not well studied, and they have not yet demonstrated the large impact on commercial fishing as the viruses noted above. However, it is clear the fish adenoviruses are phylogenetically basal to the adenoviruses of both birds and mammals.

Overall, we see that natural populations of bony fish can often support persistent infections with various RNA and DNA viruses. However, these same or very related viruses are also often able to cause widespread disease with high mortality in similar natural populations. Although iridoviruses and +RNA viruses were previously presented as prevalent infectious agents of marine invertebrates, with bony fish we now see the emergence of several new viral agents, such as autonomous retroviruses, rhabdoviruses and herpes viruses, all associated with large-scale fish disease. It is thus most ironic that in bony fish, with the evolution of their highly sophisticated adaptive immunity, we see a congruent expansion of these prevalent viral parasites.

Blood Cell Evolution and Virus

As previously presented, phagocytic cells were seen in early animals such as sponges and hydra, but were maintained in all invertebrate marine animals. In lower animals, innate immune systems (such as RNAi), not blood cell phagocytosis appears to be most important for an antiviral response. Tunicates have evolved several complement components (C3, C4, C5), but this does not appear to be linked to adaptive immunity. When we consider how the blood cells have evolved in mollusks, arthropods, echinoderms and tunicates, we see that they are essentially similar, and how these cells respond to pathogens is also similar. The main action is that of phagocytic blood cells (non-proliferative amoebocytes) that will wall off invaders and are cytotoxic with lytic granules. These cells resemble vertebrate NK cells in activity. However, complex surface recognition of foreign cells (such as seen in tunicates) is not a uniform characteristic of these invertebrate blood cells, and it does not seem to have been initially associated with pathogens as these organisms deal effectively with most pathogens. Some viruses (e.g., WSSV) do induce considerable damage via apoptosis in invertebrate blood cells. In contrast, vertebrate blood cells are significantly more complex. They undergo antiviral-induced proliferation and progressive complex differentiation in bone morrow and lymphoid organs. They employ ongoing apoptosis and selection to generate an array of cytotoxic and other cells, and they now recognize virus infected 'self' cells and produce antiviral antibodies. However, given clearly the effective nature of invertebrate immunity and their blood cells, it is not obvious that selective pressure against pathogens was crucial in the evolution of such complex blood cell biology. Clearly an effective immunity system was already present and remains present in the vertebrate ancestors. Adaptive immunity, therefore, does not seem to result from more efficient pathogen exclusion process as its major selective pressure. Why then did such complex adaptive immunity evolve? It is noteworthy that it is now active against virus-infected host (self) cells. Like the P1 colonized *E. coli*, when these vertebrate cells are perturbed by infection of a foreign genetic parasite, they are identified as foreign and killed by blood cells. As we will present below, numerous components of the adaptive immune system also are constituents of group and sexual identification systems (such as MHC olfactory peptides). In addition, as presented below, most of the novel components of the adaptive immune system have been externally acquired (were absent from ancestral genomes). Thus, they must have been the products of genetic colonization process (horizontal transfer). The entire system thus represents an acquisition of new molecular genetic identity system. The thesis that this likely originated from stable colonization by genetic parasites will now be presented.

Adaptive Immunity, Group Identity and Addiction

A central assertion has been that new systems of group identity can be superimposed permanently onto host by the action of persisting genetic parasites. Such persisting parasites generally require addiction modules or strategies to attain both stability and group identity, and both harmful and protective components are essential for such addiction. The propagation (transfer) of the new molecular identity to offspring (or new cells) will then typically require a temporally limited window during which protection against the long-lasting harmful component is transferred (identity transfer), prior to activating the destructive systems. The adaptive immune system essentially has all these elements and can thus be thought of as one highly elaborate system of cellular addiction and group identity. In an outline, destruction, in the form of apoptosis and cytotoxicity, is a prevalent characteristic of the immune system that must be controlled or selected against. During development, the immune system goes through a limited period of education, in which proper MHC expression and self-identification of peptides is allowed by thymic killing (apoptosis) of active CTLs. Once this education (tolerance) is complete, any cell not expressing appropriate MHC markers or presenting foreign antigens (peptides) will be destroyed. Addiction and self-identity have thus been both inseparably attained. Below, observation in support of this thesis is presented in greater detail.

Origins of Adaptive Immunity and Role of Genetic Parasites

The adaptive immune system can be evaluated from the perspective of a new and highly complex cellular identity system. As in all the identity systems so far presented, it too has both inherent destructive and protective components that must work together to provide identity and prevent self-harm. In addition, the sources of these components bears clear hallmarks of originating from a genetic colonization. Also, since all the basal components of adaptive immunity appeared to have been acquired together, *en toto* as no extant representative of an ancestor exists with the possible intermediates of this system, this too suggests a colonizing event was involved. In terms of blood cell biology, these characteristics include the ongoing, renewing and effector-induced proliferation of blood cells, ongoing apoptosis, surface receptor selection and ongoing peptide signals that are needed for cell survival. In addition, adaptive immunity produces long-lived cells that serve as memory cells. None of this cell biology was seen in any invertebrate or the lampreys or hagfish, yet hagfish had a highly efficient (but completely different) VLR-based immune system that also used somatic DNA variation for surface expression. A most basic component of the adaptive system is the T-cell receptor (TCR), which controls both cellular (T cell) and humoral (B cell) immunity. TCR is an IgG-like surface molecule that spans the membrane. Related viral receptors (JAM/CTV) also have IgG-like surface domains. One of these proteins (CTX) is phylogenetically basal to other related T-cell receptors. CTV is also a basal viral receptor (in having a V domain). This CTV receptor was present in the genomes of protochordates, Branchiostoma and Ciona. In these organisms, CTV shows variable expression, but its binding targets and potential role in immunity are not apparent. It seems likely that peptides played some role in ancestral immunity, since peptide toxins and peptide sex-hormone binding to receptors are used in identity systems of all early animals. Thus it seems possible that the TCR could have evolved from such a receptor ID system that was initially based on binding processed peptides. The peptide ligand for the TCR is processed for presentation by MHC. Thus the ligand (MHC-peptide) and receptor (TCR) are matched systems that must both co vary to evolve resembling T/A functional set. Another basic and novel component of the adaptive immune system is the gene needed for recombination of Ig genes and its role in generating Ig surface diversity (via RAG1/2). This gene set is also absent from the ancestors of the jawed vertebrates, but as presented below, it has clear characteristics of originating from a genetic parasite. Finally, and also fundamental to the adaptive immune system is the acquisition of the gene dense MHC I/II/III loci found in all vertebrates. Most of the genes within this loci are also absent from the ancestral genomes. The acquisition of these genes as a tight, dense loci of complex, interacting but novel genes is similar to the genetic colonization mediated by phage in prokaryotes. In addition, clearly much gene duplication is also involved in the later evolution of these loci. All of these characteristics regarding the origin of the adaptive immune system are consistent with the involvement of genetic parasites.

Origin of RAG

The adaptive immune system generates diversity during the V(D)J DNA (variable, diversity, joining regions) recombination in the Ig gene which is catalyzed by the vertebrate RAG1/2 proteins. As mentioned, RAG1/2 appear to be of external origin (i.e., in lacking introns but with inverted repeats in RSS region), and these genes are absent from ancestral-jawed vertebrates. The recombination reaction involved is similar to both a retroviral integration and the transposition by Tc1 family transposases, all of which involve DDE metal binding domains at the active site. RAG is thought to act like a transib family of transposases (cut and paste). RAG expressed in vitro indeed shows transposase activity. Of all the IS4 family of transposases, the phage Mu transposase is the most active and the best characterized and also appears to be phylogenetically basal to other family members. Thus these transposases appear to have a viral heritage. Recently, RAG homologues, SpRag1L and SpRag2L, have been found in an echinoderm (sea urchin) genome, but not ciona, lamprey or hagfish. This puzzling absence has led some to argue for loss of RAG via genome reduction and that RAG containing ancestors are now extinct (or not yet sampled). But ciona has retained 15,000 copies of Tc1 transposon elements (salmon have 1,200, of which 1/3 are expressed, but none are functional). This transposon was also seen as intact ORF in *C. elegans* and was retained even in the compact puffer fish genome. Ciona has also retained various LTR 'transposons'such as chromoviruses (Gypsy), which are also retained in puffer fish. Why should so many copies of this these two types of transposons be retained during a proposed genome reduction, but the RAG ancestor lost in tunicates only to be regained in jawed fish? The genome reduction hypothesis is thus problematic. Furthermore, it is clear that SpRag1L and SpRag2L are not functioning as components of an immune system in sea urchin. They do physically bind to each other, resembling a T/A system of a stable genetic parasite. It therefore seems much more likely that RAG was never present in common ancestors to vertebrates. Its presence in echinoderm and other invertebrates (but not *C. elegans*) indicates a related but separate colonization from a similar infectious source, which serves a distinct purpose (not for adaptive immunity, but likely some other ID function). But what might that be? The widely expressed view that a DNA 'transposon' was the source of RAG is not at all clear, since DNA transposons cannot move between species. The conserved diagnostic domain of RAG (the DDE domain) is also found within the RNAse H-fold that is part of this same catalytic domain of a retroviral gene. Thus, it remains possible that an RNAse H-like ancestor (from and colonizing retrovirus) could have equally been involved. That RAG is IS4-like, a 'cut-and-paste' transposon, has been used to argue against retroviral involvement. Transposons act via two major mechanisms. Simple proteins (i.e., Tn5, hermes) and complex proteins that are also site specific (Tn7, transib, IS4). Others, like IS917 are distinct RCR transposons. For use in Ig gene variation, RAG must be highly specific to the Ig gene. Most transposons lack such gene specificity. However, homing introns (transposons) are abundant and conserved in crucial viral genes (such as DNA polymerase) and can clearly have gene-specific interruption capacity. Also recall that the Bordetella T7-like phage system uses a phage-encoded RT to generate gene-specific surface receptor variability (in variable region 1), clearly reminiscent of Ig variation. Such receptor expression is an example of lysogenic conversion by a prophage. Other prophages (such as SF370.1 of streptococcus) encode superantigens (exotoxin C) as part of their lysogenic conversion and host addiction. Thus viruses have established the capacity to interrupt and generate somatic variation in specific proteins (including surface receptor proteins) and alter immunological regulation. Such capacity involves a pre-selected set of highly specific, complex interacting genes, found within a genetic parasite. Thus, I argue that the current concept that the RAG 'transposons' fortuitously colonized the germ line of jawed vertebrates, which was then co-opted for the complex gene-specific immunity function as outlined above, is inadequate. First, how this initially happened presents a problem since transposons are almost always silent or defective in host genome and do not move between species: let alone understanding what selective pressures promoted the two RAGs to become involved in host immunity. How then might a specific transposon set enter the vertebrate genome? The only apparent solution to this problem is that the original source of the RAG (and other immune) gene sets must be external, non-ancestral; that is viral. But this was a virus (or virus mixture) that persisted permanently and permanently altered host antiviral immunity. We should thus look to the various complex viruses that were and remain prevalent in the oceans as the most likely source of the entire Ig variation system. As we have presented, the oceans are an especially good source of highly diverse large (integrating T-7 like) viral genomes that infect invertebrates and lower eukaryotes. A recent screen of oceanic water for viral DNA indicates that many novel transposases, integrases, homing introns (of all classes) are amongst the most numerous and diverse of such viral genes (along with receptor proteins). Some large DNA viruses also have complete retroviral genomes within them (viruses within viruses), so a mixed virus colonization is also possible. Such a colonization would be able to supply not only the complex, sequence-specific RAG1/2 function, but other complex, pre-selected, interacting genes not found in ancestral genomes which would provide new molecular identity systems with inherent antiviral function. However, for such a complex colonization to succeed, these new viral genes must provide a stable colonization phenotype, as an addiction or identity module that precludes competition. It is also likely that such a new set of genetic parasites would have needed to displace prior competing or excluding host identity modules (such as the VLR system). Thus the extensive colonization by ERVs and their LTR derivatives in the genomes of all jawed vertebrate could have been the colonization event that led to wholesale reorganization of the host genome along with emergence of adaptive immunity. This would explain why such a widespread genetic invasion was associated with the origin of the adaptive immune system.

Origin of the MHC Locus as Group Identity

The major histocompatibility region (MHC) is a gene-rich loci (>200 human genes) with a high density (40%) of immunity associated genes found in all jawed vertebrates. It has within it the TRC receptor and rearrangeable Ig genes, especially genes involved in antigen processing and presentation. Class I and II MHC genes are found in all jawed vertebrates, but are mostly Ag presentation genes. MHC I is highly gene dense, and lacks many introns, suggesting recent horizontal origins. Class I genes are responsible for proper folding, translocation and peptide processing, ER transport and MHC assembly (e.g., TAP1/2, beta2 microglobulin). Class II binds and presents Ag, but it is not ER processed and therefore employs a distinct pathway. Some of these genes can be identified in the tunicate genome. These I/II loci are separated by about 700 kb which has class III genes (an intragenic, dense region with ~60 genes); these genes are less broadly conserved, but are very similar between mouse and human. This complex organization in mammals is simplified in fish. In fish, the conserved core function still encodes glycoproteins that deliver peptides to the cell surface and initiate signal transduction to promote cellular proliferation and differentiation. These cells create a system of self-identification or histocompatibility (tissue rejection), for the immune system. Cells and pathogens are recognized as foreign cells that have either lost or gained molecular markers and are destroyed. The MHC loci are the most polymorphic loci in both fish and humans. It also shows a complicated evolutionary history of rearrangements, duplication, insertions and deletions which are often genus specific. In tetrapods, MCH locus is highly clustered. However, in fish, it is much less clustered and many genes are unlinked. Yet even here Ag presentation genes (IA) remain linked.

If the protection against pathogens cannot account for the origin of the MCH locus, then how are we to explain it? If adaptive immunity represents an elaborate and new cellular identity system, can we suggest an origin for the MHC locus? Antigen presentation is a complex process of cell biology and membrane-associated assembly. Some (but not most) of the genes involved can be found in vertebrate ancestors. However, several are novel and vertebrate specific. We have already presented the case for the origin of RAG1/2 from genetic parasites above. In addition, a viral receptor, the RAG target, could also represent the origin of TCR. What about other novel elements of MHC? Could the novel MHC genes also be of viral origin? Chemokines are essential small signaling proteins (8–12 kd) that bind specific receptors of various immune cells. Viruses encode many chemokines, including some that are not host derived. In some cases, such as with IL-10, it appears that viral versions have independently originated on at least three occasions. Although the neurotropic alpha herpes viruses do not generally encode any such chemokines or their receptors, all immune tropic beta herpes viruses (such as cytomegalovirus) do. But because some of these viral chemokine genes can show high similarity to their human gene counterparts, it is often thought that viruses acquire host chemokine and receptor genes. Yet in other cases, such arguments cannot be supported. For example, herpes viruses can encode several broad spectrum CC chemokine, such US28 from HCMV and the CXC ORF74 chemokine from HHSV8 (active via MAP kinase pathway). These viral genes have no similarity to host genes. US28 is interesting for its novel ability to undergo constitutive endocytosis and recycling to the plasma membrane and also for activating several different signal transduction pathways. Some viral genes, like EBV BILF1, are constitutively active, requiring no ligand and have simpler basal protein structures. Genes related to BILF1 are conserved in all gamma herpes viruses, indicating an ancient origin. Again, BILF1 shows no similarity to any known chemokine receptor, suggesting it is a unique viral creation. In fact, EBV seems to control GPCR settings at many levels for the purpose of immune evasion during persistence, and also for virus dissipation. Some viral genes, like latently expressed vGPCR of KSHV, appear to be crucial for inhibition of cellular apoptosis. In terms of TCR-like molecules, European fowlpox encodes both a V-type Ig domain protein as well a G-protein coupled receptor. Lumpy skin disease virus, LSDV, also encodes a G-coupled CC chemokine receptor, not found in other poxviruses. Since phylogenetic analysis does not place any of these viral genes at tips of host gene clades (such as CCR1), this suggest that they are not derived from host genes, but are all old viral creations. DNA viruses, like vaccinia, also control immune response by synthesizing steroid hormones (3beta-HSD). Finally, with regard to vertebrate lymphocyte apoptosis and possible viral origins, we can recall the ability of WSSV to induce high-level apoptosis in haemocytes of shrimp, a cell that doesn't normally proliferate nor does it seem to replicate WSSV. Other DNA viruses of insects (such as TnBV1 of parasitoid wasps) express a viral protein that induces host immune suppression by inducing a programmed apoptosis. Thus persisting DNA viruses have many genes, active during persistence, that could provide numerous basal functions needed for an early simplified MCH locus and adaptive immune response. Since persistence normally also requires some form of host addiction, the inherently addicted character of the adaptive immune system (presented below) would also be consistent with an early viral origin.

MHC Locus Architecture. Fish, lower vertebrates and some birds show a simpler genetic organization with respect to MHC loci and its expression. Also, some birds only have 20 total linked MHC genes, whereas zebrafish have 41 linked MHC genes (of possibly 149 scattered *en toto*). Often, in lower

Fig. 2.1 A planktonic bloom (cyanobacteria) in the ocean as seen from space

LEFT: Living stromatolites at Shark Bay, Western Australia, showing the laminated structure of a single column and the microscopic cyanobacteria that construct them. Each cyanobacterial filament is about 100th of a millimetre in width. These organisms are rarely preserved in fossil stromatolites, although some have been discovered in chert (silica-rich rock) at other localities in the Pilbara that are about the same age as the Trendall locality.

The bacteria precipitate or trap and bind layers of sediment to make accretionary structures, which can be stratiform, domical, conical or complexly branching. They can range in size from smaller than a little finger to larger than a house. Some branching stromatolites resemble modern corals.

Fig. 2.2 A current stromatolite, similar to those 2,500–2,590 million ybp

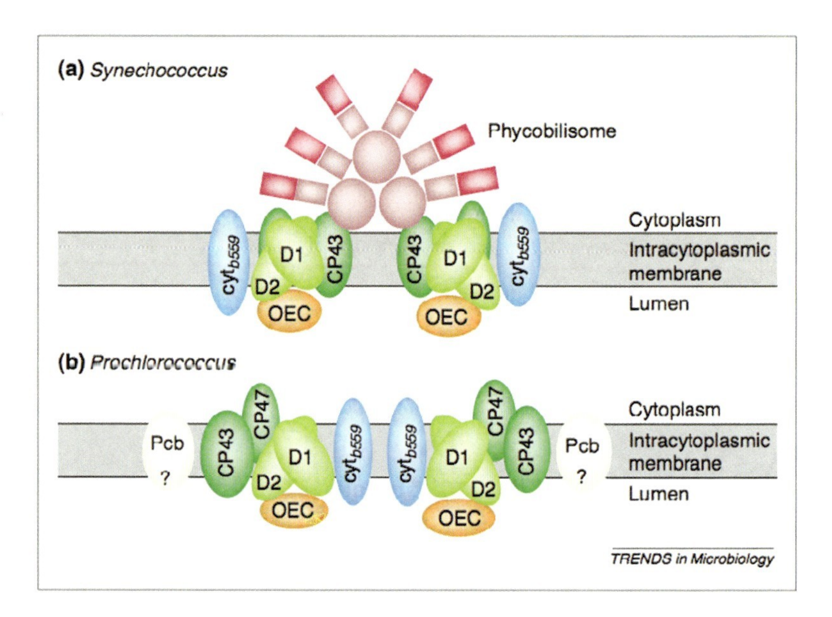

Fig. 3.4 Cyanobacterial photosynthesis

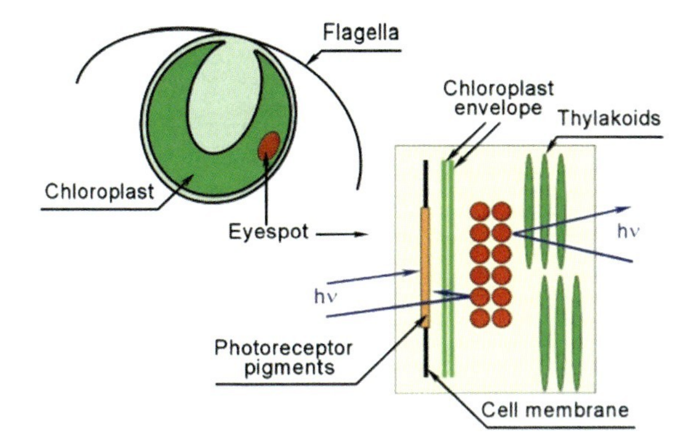

Fig. 3.6 The light antennae and flagella of green algae

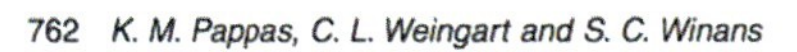

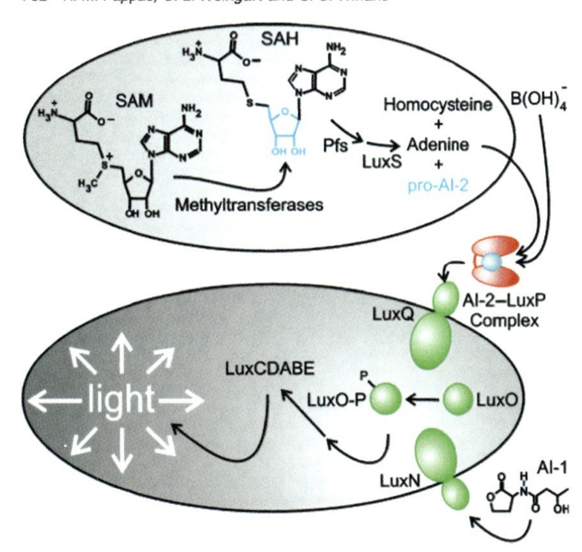

Fig. 3.8 Multiple control of ALH

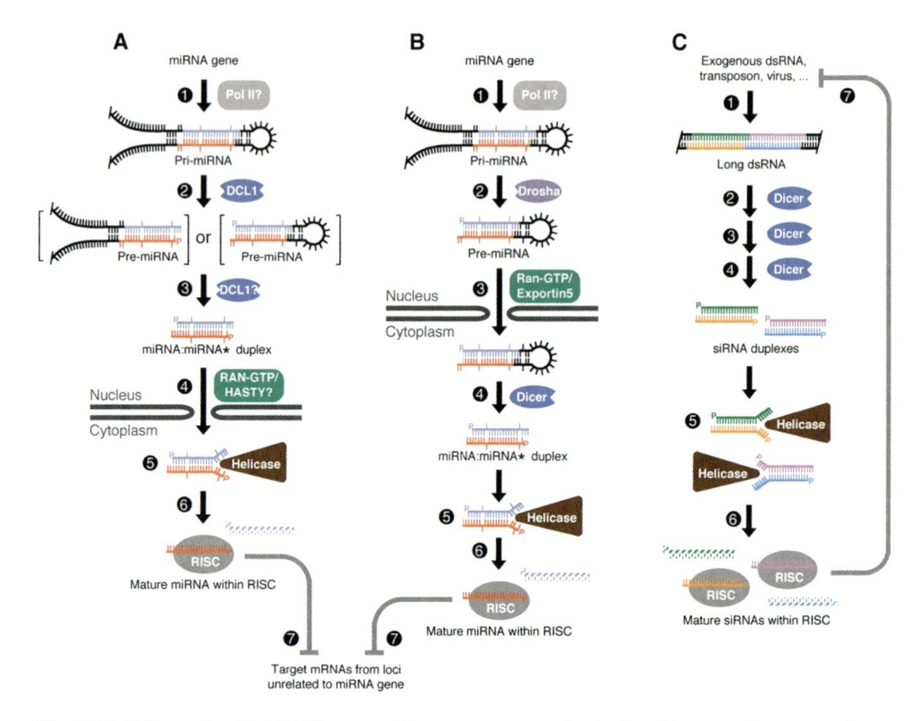

Fig. 5.10 Schematic of miRNA generation, transport and relationship to endogenous virus

Fig. 6.3 Cone snail patterned shells

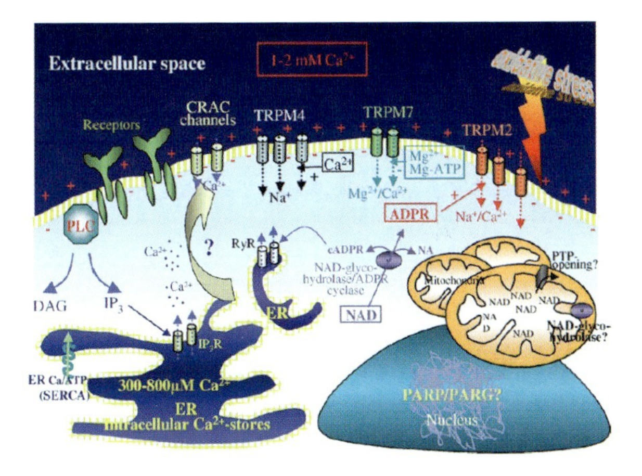

Fig. 6.6 Role of TRMP ion channels in immune cell function

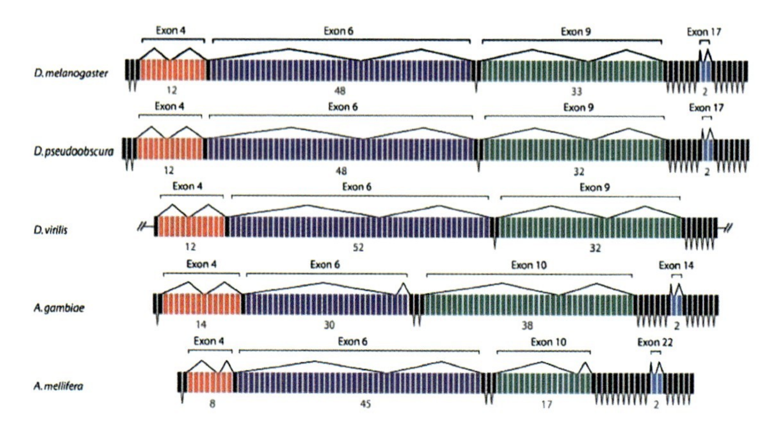

Fig. 7.1 Insect DSCAM gene map

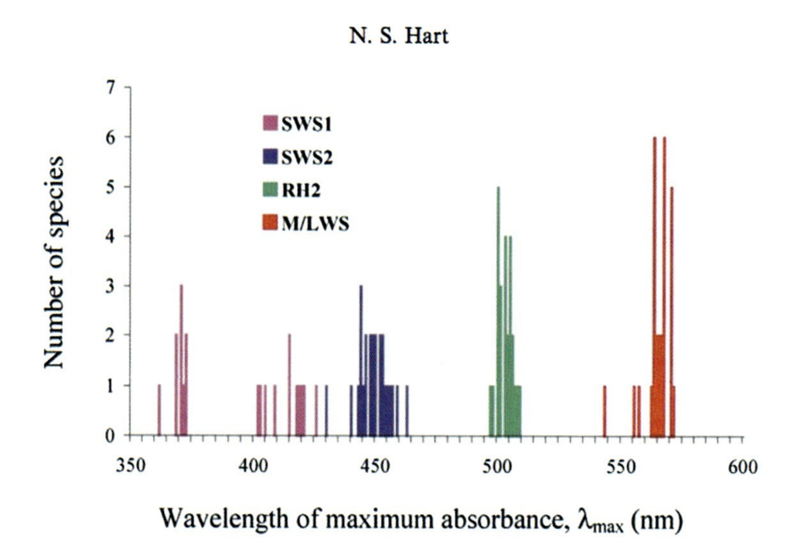

Fig. 7.4 Spectral response of avian visual receptors

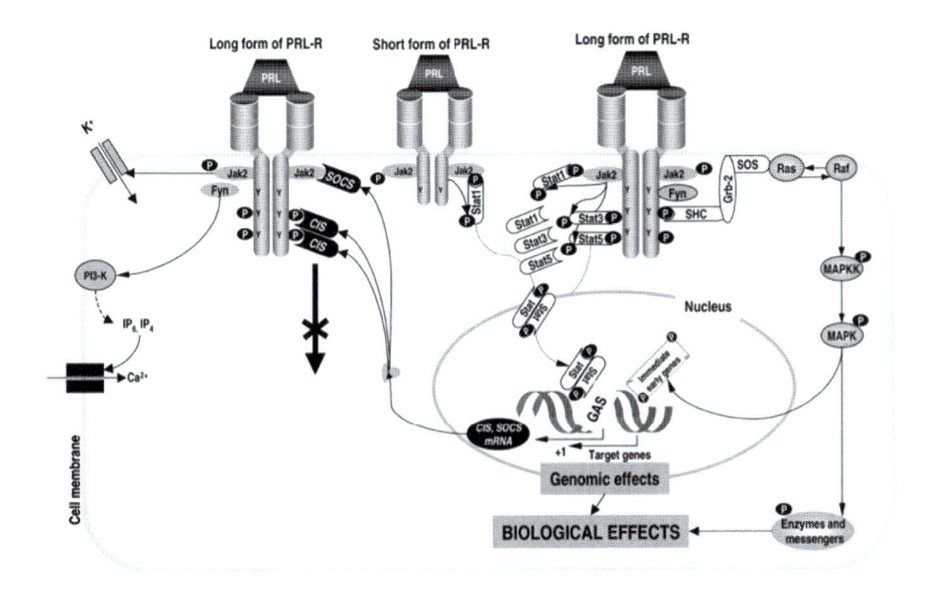

Fig. 7.8 Prolactin receptor transduction

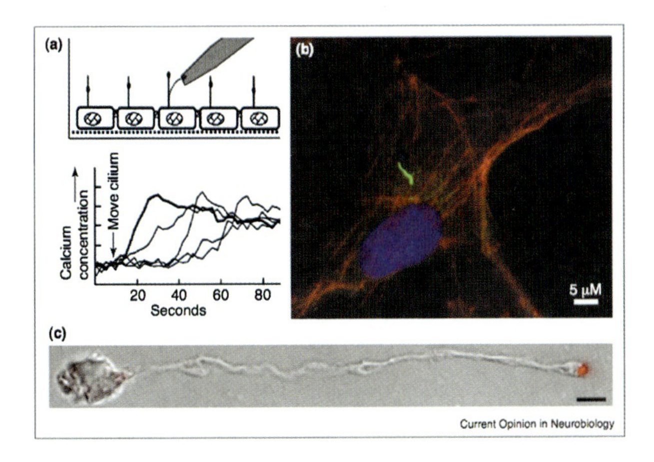

Fig. 7.9 Mouse VNO neuron TRPC expression

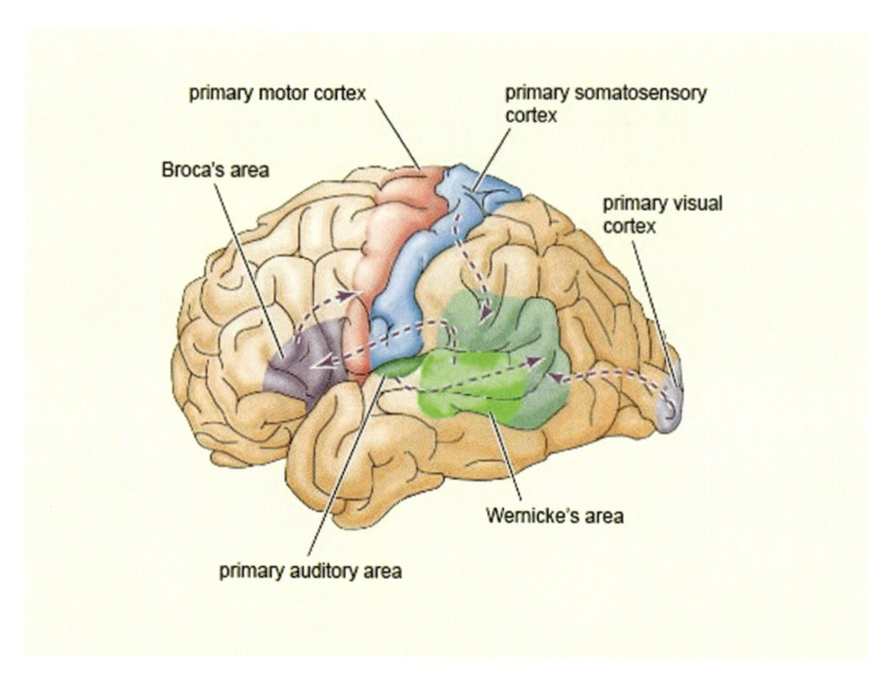

Fig. 9.3 Broca's area identified onto human brain

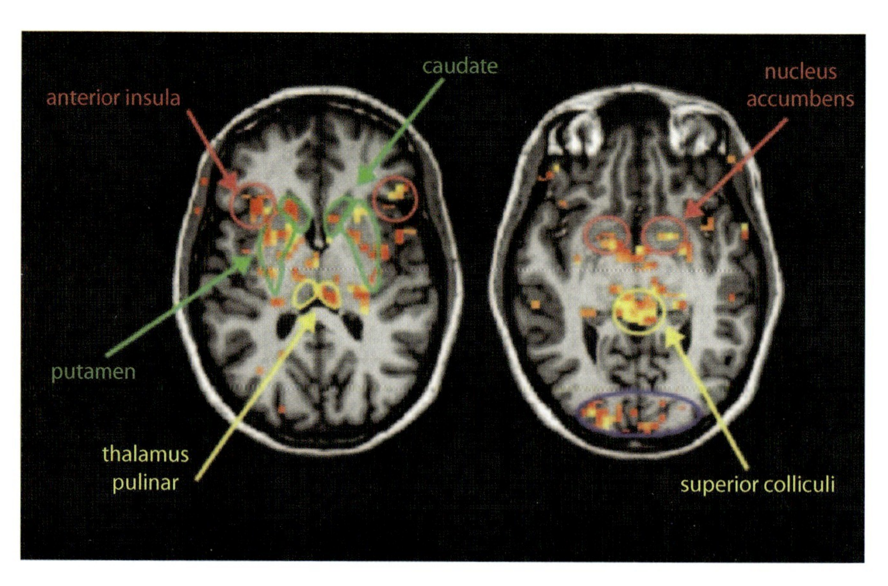

Fig. 9.4 fMRI study of brain response to fearful body positions

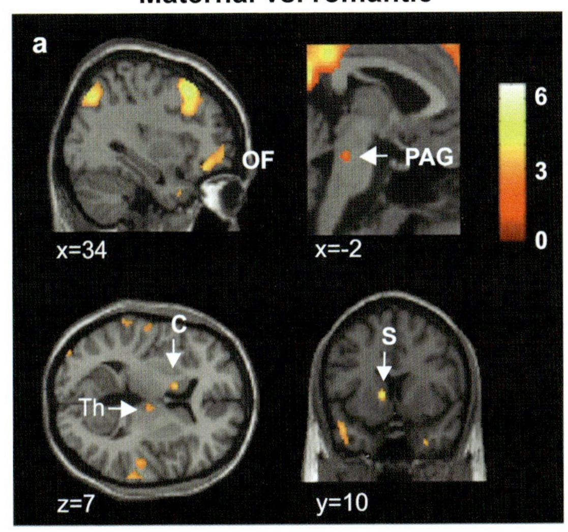

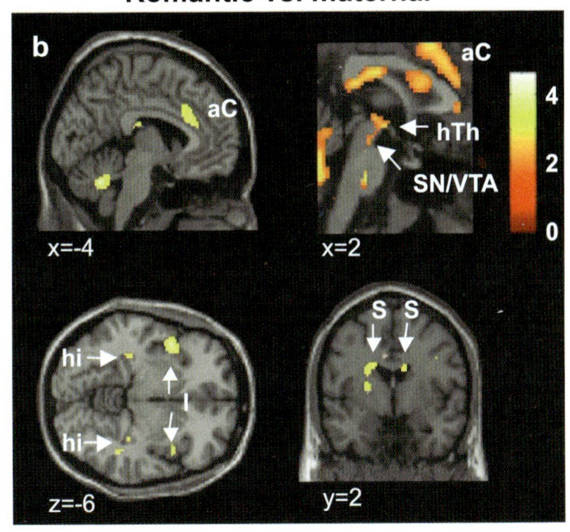

Fig. 9.6 fMRI brain study of maternal and romantic love

vertebrates, only one class I MHC gene (allele) is expressed. In fish, 2–12 classes of MHC genes can be expressed but individual fish (Atlantic salmon) tend to express 2 or less alleles. The class IA genes remain linked in fish. Why would such variation exist in the number of MHC genes? Given the virology of the oceans and fish presented above, it is not apparent that reduced pathogen diversity or loads can account for such limited variation. The increasing diversity of MHC alleles in mammals relative to lower vertebrates is thus hard to explain. Since the lower vertebrates have efficient immune systems, it is not clear how mammals benefit from a higher diversity of MCH alleles or why this was selected. In addition, given this considerable genetic variability in MHC, it is most curious to note that this locus also encodes a significant number of odor receptor (OR) genes, and that these OR genes are conserved, often 5' to the TCR in all jawed vertebrates. In fact, the total extended human MHC locus has more OR genes (34, all in MHC I) than HLA I (26) or HLA II (24) genes. Why does the human MHC still retain so many OR genes?

Mouse MHC I Has a Cluster of OR Genes at the Distal End

One cluster of mouse OR genes is just 5' to the TCR genes. Fifty-nine OR genes are found here as well as another 20% pseudogenes. In the same loci of human DNA, 25 OR genes with 50% pseudogenes are seen. Most of these OR genes have independently evolved with some mouse subfamilies version lacking human homologues. Thus mice have greater quantity and diversity of OR genes relative to humans and less gene inactivation. The OR genes do not seem to undergo somatic recombination as do the TCR genes. ERV (LTR elements), SINES and alu elements are abundant in MHC loci and their presence is associated with its dynamic evolutionary history. As presented below, fish also show a link between MHC and OR genes. Furthermore, another distinct type but much smaller class of odor receptor, the VNO (vomeronasal receptor), is also linked to the MHC loci (described below). Since the VNO receptor binds to pheromones and affects sexual behavior in all tetrapods, it is especially curious to note that all five VNO ORFs within the human MHC I locus (and throughout the genome) have been converted to pseudogenes. These are the only ORFs in the human MHC with this feature. What then is the link between MHC, OR and VNO receptors? Why is this link so basal yet so conserved? Why did humans lose MHC-linked VNO receptors but not mice? On top of that, how can we understand the diversity of OR genes or their inactivation as pseudogenes? If these receptors simply detect environmental odorants, why have so many been lost in humans? It has been proposed that olfactory-driven mate selection might explain the presence of the OR genes in the MHC locus. As argued earlier, mate identification or selection is really a form of sexual group identity. The link between the MHC locus, OR and VNO receptors might therefore be understood if they are all components of a linked group identification system invented by the jawed vertebrates, but partially modified during primate evolution. That is, as primates no longer depend on pheromones or many odorants for group or sex identification, VNO has been lost. This thesis is presented below.

The Origin of the Odor/MHC Linkage

OR genes are the largest known mammalian gene family and show strong evidence of recent evolution in most mammalian (and vertebrate) lineages. Mice, for example, are estimated to have about 1,500 OR genes. Compared to prokaryotes, this would place OR genes along with the corresponding restriction/modification enzymes and holins as the most diverse genes characterized. The latter genes are clearly playing a role in prokaryotic cellular identity systems. The rapidly evolving OR genes also include many examples of recent gene loss (via pseudogenes) in most vertebrate lineages. If the main purpose of OR genes is simply for odor detection, why might so many rapidly evolving OR receptors be needed? Thus the gain and loss of so many odor-detecting genes seems illogical. Recall, however, our suspicion that gene diversity itself is suggestive of involvement in an identity system. If we apply this reasoning to OR genes, we can consider how such genes might be used to make and ID system. What is the ligand for these receptors and how might this relate to the MHC locus? Why would the VNO receptor also be involved in this locus? Furthermore, since odor detection is a sensory process that necessarily involves peripheral neurons and CNS, how might this result in a group identification process? Does stable behavior modification create group identification? If so, dose this also involve addiction modules as I have previously asserted? Much of the rest of this book will attempt to address these very questions and trace how they relate to the evolution of human group identification. For now, however, let us examine these issues in the context early evolution of group identity in jawed vertebrates.

Lampreys lack the MHC locus, although they do have some of the genes related to MHC function. In terms of OR genes, however, lamprey has six OR genes. The number and types are similar to those found in *C. elegans*. These OR genes are 7 TM GCRPs whose main purpose appears to be for the sensory detection of smell. Recall that in *C. elegans*, a similar receptor was also used to detect the Dauer sex pheromone and modify *C. elegans* sexual and feeding behavior. The role of these six OR receptors in lamprey sexual biology has not been explored, but would be expected, since lampreys use olfactory-based bile acid detection such as 3-keto petromyzonol sulfate for sexual identification. However, these ORs are distinct from those of the jawed vertebrates in that they do not belong to either class I or II OR genes (described below). Thus in the lamprey, the type and number of OR genes is distinct from the jawed vertebrates, and no linkage to MCH is present.

A significant change in OR gene type and number occurred with the evolution of teleost fish but has been expanded in all jawed vertebrates. Bony fish have 50–100 OR genes. All fish and jawed vertebrate OR genes are clearly related to each other with 20–95% amino acid similarity. These ORs are classified as type I (aqueous ligands, found in fish) or class II (volatile ligands, found in amphibians and tetrapods). Most fish have only type 1 ORs. Humans have mostly type II ORs, but dolphins have lost their type II ORFs as pseudogenes, reverting to type I ORs.

The fish OR genes have been evaluated to have evolved from about seven prototype genes. Extant OR genes are found in about four subfamilies, such as with Japanese loach (24 ORs and two pseudogenes in four subfamilies). Curiously, these subfamilies show strong evidence of recent and species-specific variation. In addition, sequences that appear to correspond to the original prototype OR genes can be found, but have been converted to pseudogenes. Overall, the pattern of OR evolution in fish seem overly complex for the sole purpose of odor detection, given the long evolutionary age of these receptors. However, if ORs are also used for group identification, such a dynamic evolution involving gene displacement might be understood. Although most teleost fish have class I ORs, there are some exceptions. Coelacanth (considered a living fossil) represents a minor branch of teleost, but one that was ancestral to the terrestrial vertebrates. Thus it is very interesting that Coelacanth have both class I and II OR genes.

What then is the relationship between fish olfaction and immunity as seen via the MCH locus in group identification? Clearly fish use olfaction for group and sexual behavior and link this to MHC. All teleost fish have a single olfactory epithelium that, unlike tetrapods, contains both ciliated and microvilli cells. Bipolar neurons in this epithelium express individual OR genes. One of these OR genes (5.24 of goldfish) is a G-protein with clear similarity to pheromone receptors in mammals (see Fig. 6.5). In tetrapods, the olfactory tissues and types

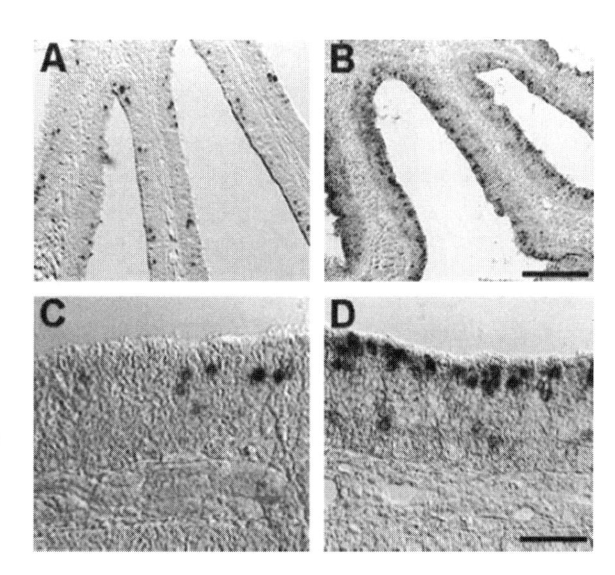

Fig. 6.5 Expression of odor receptor (OR) in goldfish olfactory epithelium (reprinted with permission from: CaO, Oh, Stryer (1998), National Academy of Sciences, USA Neurobiology, Vol. 95)

of receptors expressed are more complex. Here, microvilli epithelium in a separate tissue (the VNO) which expresses the VNO receptors for pheromone detection. The ciliated cells are the main olfactory epithelia (MOE) which express the more numerous OR genes used more for odor but also some pheromone detection. Tetrapods have two multigene receptor families, GFA, which is MOE-like, and GFB, which is VNO-like, expressed in these distinct epithelia. Goldfish also have both GFA and GFB (V2R class) GPCR olfactory receptors, but their expression is in vertically distinct regions of the same olfactory epithelium. These receptors may be detecting steroid and peptide pheromones respectively, as well as other odorants and affecting ion homeostasis. Thus it seems that teleost fish have a simpler epithelial tissue that combined OR and VNO functions but expressed in distinct sensory neurons. The relationship between VNO receptors and sexual/social identity and immunity is presented below. Of special interest will be the role of ion channel proteins in olfaction, nerve cell and immune system function. Indeed as shown in Fig. 6.6, Ca++ ion homeostasis in immune cells, like neurons presented earlier, is of central importance for their evolution and function. This will be a main focus of the next chapter. Given the very early history of these olfactory receptors and their retention during the evolution of all tetrapods, why they were lost from the human (great ape) genomes becomes an exceedingly interesting topic that will be developed in detail in the subsequent chapter.

It has recently become clear that fish (such as female sticklebacks) display an odor preference for mates with dissimilar MHC locus, possibly via the peptide binding domain. Female Atlantic salmon also choose male mates which express increased MHC heterozygoticy. Although Altantic salmon have 12 total MHC alleles, individuals express 2 or less, suggesting a selective pressure to expand MHC alleles. A similar link in mice between sexual-based olfaction, MHC and MCH peptide presentation is presented in much more detail in the following

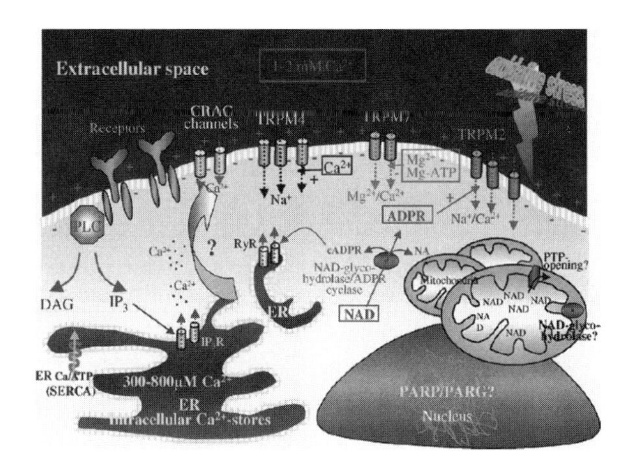

Fig. 6.6 Role of TRMP ion channels in immune cell function (*See* Color Insert, reprinted with permission from: Knowles, Schmitz (2004), Molecular Immunology, Vol. 41, No.6–7)

chapters. However, fish olfaction is complex, involving interactions beyond MHC. Four classes of olfactory molecules are known, including amino acids, sex steroids, bile acids (also seen in lamprey) and prostaglandin peptides. Several of these molecules can have dramatic effects reproductive and sexual behavior. In goldfish, pre-ovulatory and peri-ovulatory females secrete distinct pheromones (steroids and prostaglandins respectively) that strongly affect male courtship and sperm production. In fact, goldfish behavior in general can be best understood via the action of a blend of pheromones that illicit different behaviors including courtship and aggression. Much of this behavior can also be considered as group (shoal/sex/mate) identification. Other fish additionally utilize visual cues and other learned sensory patterns to further control group identity. Olfactory-controlled reproductive behaviors of fish can be exceedingly powerful and clearly harmful to the individual fish. Schooling and spawning behaviors are often associated with seasonal patterns of OR expression and odorant detection. But during spawning, fish often use olfaction to find spawning grounds and induce reproduction that can result in the death of the individual fish. Such self-destructive behavior has the clear hallmarks of a behavioral addiction module (presented below). However, the involvement or overlay of MHC in such olfaction allows the targeting of conspecifics with respect to such group and sexual behavior. And if the MHC alleles are numerous, even individual animals can be identified by their MHC composition (as seen with mice/voles). This link between MHC and OR was retained in all jawed vertebrates, but it underwent an expansion in the mammals.

Summary

In this chapter, we have traced the evolution of marine animals from cocopods to vertebrate fish. We have considered their immune and group identity systems and possible relationships with exogenous and endogenous viruses. An especially well-conserved endogeneous retrovirus of zebrafish, for example, shows high thymic and olfactory bulb expression. These issues appear always to be intertwined and appear to link immune cells with receptor-expressing sensory cells. We have examined the role of colonizing genetic parasites and seen the major changes associated with the origin of the adaptive immune system. I have outlined various lines of evidence that support such a role. With the emergence of the MHC locus of the adaptive immune system, we also see evidence that olfactory-based detection and cellular-based immunity have been combined into one genetic locus and are under some common selection. It has been asserted that the adaptive immune system is really a highly complex system of cellular group identity that has the addictive properties inherent to most identity systems. The inclusion of odor receptors in this identity system extends the cellular system to apply to reproductive or sex-associated group identity. This linkage also makes clear the role the central nervous system will have on group identity in the jawed vertebrates. The chapter that follows will therefore continue to trace the OR/immune linkage from fish to primates. An emphasis will be to understand how CNS system has developed an ability to provide group identity and promote reproductive behavior. As always, the role of genetic parasites in this genesis will also be considered.

Recommended Readings

Hydra Genome and Immunity

1. Bonnefoy, A. M., Kolenkine, X., and Vago, C. (1972). [Virus like particles in hydrae]. *C R Acad Sci Hebd Seances Acad Sci D* **275**(19), 2163–2165.
2. Grimmelikhuijzen, C. J. (1983). Coexistence of neuropeptides in hydra. *Neuroscience* **9**(4), 837–845.
3. Steele, R. (2006). Hydra – Small animal, big genome, lots of surprises. *Dev Biol* **295**(1), 343–344.
4. Steele, R. E. (2005). Genomics of basal metazoans. *Integr Comp Biol* **45**(4), 639–648.

Urchins, ERVs and RAG Genes

1. Batista, F. M., Arzul, I., Pepin, J. F., Ruano, F., Friedman, C. S., Boudry, P., and Renault, T. (2007). Detection of ostreid herpesvirus 1 DNA by PCR in bivalve molluscs: a critical review. *J Virol Methods* **139**(1), 1–11.
2. Batistoni, R., Pesole, G., Marracci, S., and Nardi, I. (1995). A tandemly repeated DNA family originated from SINE-related elements in the European plethodontid salamanders (Amphibia, Urodela). *J Mol Evol* **40**(6), 608–615.
3. Arzul, I., Renault, T., and Lipart, C. (2001). Experimental herpes-like viral infections in marine bivalves: demonstration of interspecies transmission. *Dis Aquat Org* **46**(1), 1–6.
4. Britten, R. J., Mccormack, T. J., Mears, T. L., and Davidson, E. H. (1995). Gypsy/Ty3-Class Retrotransposons Integrated in the DNA of Herring, Tunicate, and Echinoderms. *J Mol Evol* **40**(1), 13–24.
5. Cruz, F., Perez, M., and Presa, P. (2005). Distribution and abundance of microsatellites in the genome of bivalves. *Gene* **346**, 241–247.
6. Elston, R. (1997). Bivalve mollusc viruses. World J Microbiol Biotechnol **13**(4), 393–403.
7. Goodwin, T. J. D., and Poulter, R. T. M. (2001). The DIRS1 group of retrotransposons. *Mol Biol Evol* **18**(11), 2067–2082.
8. Hibino, T., Loza-Coll, M., Messier, C., Majeske, A. J., Cohen, A. H., Terwilliger, D. P., Buckley, K. M., Brockton, V., Nair, S. V., Berney, K., Fugmann, S. D., and erson, M. K., Pancer, Z., Cameron, R. A., Smith, L. C., and Rast, J. P. (2006). The immune gene repertoire encoded in the purple sea urchin genome. *Dev Biol* **300**(1), 349–365.
9. Kapitonov, V. V., and Jurka, J. (2005). RAG1 core and V(D)J recombination signal sequences were derived from Transib transposons. *PLoS Biol* **3**(6), e181.
10. Kapitonov, V. V., Holmquist, G. P., and Jurka, J. (1998). L1 repeat is a basic unit of heterochromatin satellites in cetaceans. *Mol Biol Evol* **15**(5), 611–612.

11. Kapitonov, V. V., and Jurka, J. (2006). Self-synthesizing DNA transposons in eukaryotes. *Proc Natl Acad Sci U S A* **103**(12), 4540–4545.
12. Kuznetsov, S. G., and Bosch, T. C. (2003). Self/nonself recognition in Cnidaria: contact to allogeneic tissue does not result in elimination of nonself cells in Hydra vulgaris. *Zoology (Jena)* **106**(2), 109–116.
13. Lopez-Flores, I., de la Herran, R., Garrido-Ramos, M. A., Boudry, P., Ruiz-Rejon, C., and Ruiz-Rejon, M. (2004). The molecular phylogeny of oysters based on a satellite DNA related to transposons. *Gene* **339**, 181–188.
14. Renault, T., and Novoa, B. (2004). Viruses infecting bivalve molluscs. *Aquat Living Resour* **17**(4), 397–409.
15. Saavedra, C., and Bachere, E. (2006). Bivalve genomics. *Aquaculture* **256**(1–4), 1–14.
16. Smith, G. P. (1976). Evolution of repeated DNA sequences by unequal crossover. *Science* **191**(4227), 528–535.

Shrimp, Viruses (WSSV, TSV, LDV) and Immunity

1. Bonnichon, V., Lightner, D. V., and Bonami, J. R. (2006). Viral interference between infectious hypodermal and hematopoietic necrosis virus and white spot syndrome virus in Litopenaeus vannamei. *Dis Aquat Organ* **72**(2), 179–184.
2. Chen, L. L., Lo, C. F., Chiu, Y. L., Chang, C. F., and Kou, G. H. (2000). Natural and experimental infection of white spot syndrome virus (WSSV) in benthic larvae of mud crab Scylla serrata. *Dis Aquat Organ* **40**(2), 157–161.
3. Hameed, A. S. S., Sarathi, M., Sudhakaran, R., Balasubramanian, G., and Musthaq, S. S. (2006). Quantitative assessment of apoptotic hemocytes in white spot syndrome virus (WSSV)-infected penaeid shrimp, Penaeus monodon and Penaeus indicus, by flow cytometnic analysis. *Aquaculture* **256**(1–4), 111–120.
4. Lo, C. F., and Kou, G. H. (1998). Virus-associated white spot syndrome of shrimp in Taiwan: a review. *Fish Pathol* **33**(4), 365–371.
5. Loker, E. S., Adema, C. M., Zhang, S. M., and Kepler, T. B. (2004). Invertebrate immune systems – not homogeneous, not simple, not well understood. *Immunol Rev* **198**(1), 10–24.
6. Mohankumar, K., and Ramasamy, P. (2006). White spot syndrome virus infection decreases the activity of antioxidant enzymes in Fenneropenaeus indicus. *Virus Res* **115**(1), 69–75.
7. Rameshthangam, P., and Ramasamy, P. (2006). Antioxidant and membrane bound enzymes activity in WSSV-infected Penaeus monodon fabricius. *Aquaculture* **254**(1–4), 32–39.
8. Sanchez-Martinez, J. G., Aguirre-Guzman, G., and Mejia-Ruiz, H. (2007). White spot syndrome virus in cultured shrimp: a review. *Aquac Res* **38**(13), 1339–1354.
9. Suttle, C. A. (2007). Marine viruses – major players in the global ecosystem. *Nat Rev Microbiol* **5**(10), 801–812.
10. Suckale, J., Sim, R. B., and Dodds, A. W. (2005). Evolution of innate immune systems. *Biochem Mol Biol Educ* **33**(3), 177–183.
11. Tirasophon, W., Roshorm, Y., and Panyim, S. (2005). Silencing of yellow head virus replication in penaeid shrimp cells by dsRNA. *Biochem Biophys Res Commun* **334**(1), 102–107.
12. van Hulten, M. C., Witteveldt, J., Peters, S., Kloosterboer, N., Tarchini, R., Fiers, M., Sandbrink, H., Lankhorst, R. K., and Vlak, J. M. (2001). The white spot syndrome virus DNA genome sequence. *Virology* **286**(1), 7–22.

13. Withyachumnarnkul, B., Chayaburakul, K., Lao-Aroon, S., Plodpai, P., Sritunyaluck-sana, K., and Nash, G. (2006). Low impact of infectious hypodermal and hematopoietic necrosis virus (IHHNV) on growth and reproductive performance of Penaeus monodon. *Dis Aquat Organ* **69**(2–3), 129–136.
14. Wu, J. L., and Muroga, K. (2004). Apoptosis does not play an important role in the resistance of 'immune' Penaeus japonicus against white spot syndrome virus. *J Fish Dis* **27**(1), 15–21.

Shrimp Vision, Eye Stalk and Social Behavior

1. Cheroske, A. G., Barber, P. H., and Cronin, T. W. (2004). Ecological requirements determine the expression of a phenotypically plastic color vision trait in three species of Caribbean mantis shrimp within the genus Neogonodactylus (Stomatopoda, Gonodactyloidea). *Integr Comp Biol* **44**(6), 535–535.
2. Cheroske, A. G., and Cronin, T. W. (2005). Variation in stomatopod (Gonodactylus smithii) color signal design associated with organismal condition and depth. *Brain Behav Evol* **66**(2), 99–113.
3. Cowles, D. L., Van Dolson, J. R., Hainey, L. R., and Dick, D. M. (2006). The use of different eye regions in the mantis shrimp Hemisquilla californiensis Stephenson, 1967 (Crustacea: Stomatopoda) for detecting objects. *J Exp Mar Biol Ecol* **330**(2), 528–534.
4. Schiff, H., Dore, B., and Donna, D. (2002). A mantis shrimp wearing sun-glasses. *Ital J Zool* **69**(3), 205–214.
5. Wagner, H. J., and Kroger, R. H. H. (2005). Adaptive plasticity during the development of colour vision. *Prog Retin Eye Res* **24**(4), 521–536.

Crabs Virus

1. Hameed, A. S., Balasubramanian, G., Musthaq, S. S., and Yoganandhan, K. (2003). Experimental infection of twenty species of Indian marine crabs with white spot syndrome virus (WSSV). *Dis Aquat Organ* **57**(1–2), 157–161.

Aplysia and ERVs

1. Cummins, S. F., Nichols, A. E., Schein, C. H., and Nagle, G. T. (2006). Newly identified water-borne protein pheromones interact with attractin to stimulate mate attraction in Aplysia. *Peptides* **27**(3), 597–606.
2. Cummins, S. F., Xie, F., de Vries, M. R., Annangudi, S. P., Misra, M., Degnan, B. M., Sweedler, J. V., Nagle, G. T., and Schein, C. H. (2007). Aplysia temptin – the ' glue ' in the water-borne attractin pheromone complex. *Febs J* **274**(20), 5425–5437.

Elysia Chlorotica

1. Mondy, W. L., and Pierce, S. K. (2003). Apoptotic-like morphology is associated with annual synchronized death in kleptoplastic sea slugs (Elysia chlorotica). *Invertebr Biol* **122**(2), 126–137.

2. Pierce, S. K., Maugel, T. K., Rumpho, M. E., Hanten, J. J., and Mondy, W. L. (1999). Annual viral expression in a sea slug population: life cycle control and symbiotic chloroplast maintenance. *Biol Bull* **197**(1), 1–6.

Conotoxins

1. Corpuz, G. P., Jacobsen, R. B., Jimenez, E. C., Watkins, M., Walker, C., Colledge, C., Garrett, J. E., McDougal, O., Li, W. Q., Gray, W. R., Hillyard, D. R., Rivier, J., McIntosh, J. M., Cruz, L. J., and Olivera, B. M. (2005). Definition of the M-conotoxin superfamily: characterization of novel peptides from molluscivorous Conus venoms. *Biochemistry* **44**(22), 8176–8186.
2. Norton, R. S., and Olivera, B. M. (2006). Conotoxins down under. *Toxicon* **48**(7), 780–798.

Tunicate/Hagfish ERVS, Tc1 and VLR

1. Alder, M. N., Rogozin, I. B., Iyer, L. M., Glazko, G. V., Cooper, M. D., and Pancer, Z. (2005). Diversity and function of adaptive immune receptors in a jawless vertebrate. *Science* **310**(5756), 1970–1973.
2. Kasamatsu, J., Suzuki, T., Ishijima, J., Matsuda, Y., and Kasahara, M. (2007). Two variable lymphocyte receptor genes of the inshore hagfish are located far apart on the same chromosome. *Immunogenetics* **59**(4), 329–331.
3. Pancer, Z., Saha, N. R., Kasamatsu, J., Suzuki, T., Amemiya, C. T., Kasahara, M., and Cooper, M. D. (2005). Variable lymphocyte receptors in hagfish. *Proc Natl Acad Sci U S A* **102**(26), 9224–9229.
4. Rogozin, I. B., Iyer, L. M., Liang, L., Glazko, G. V., Liston, V. G., Pavlov, Y. I., Aravind, L., and Pancer, Z. (2007). Evolution and diversification of lamprey antigen receptors: evidence for involvement of an AID-APOBEC family cytosine deaminase. *Nat Immunol.*
5. Shintani, S., Terzic, J., Sato, A., Saraga-Babic, M., O'HUigin, C., Tichy, H., and Klein, J. (2000). Do lampreys have lymphocytes? The Spi evidence. *Proc Natl Acad Sci U S A* **97**(13), 7417–7422.
6. Sorensen, P. W., Fine, J. M., Dvornikovs, V., Jeffrey, C. S., Shao, F., Wang, J., Vrieze, L. A., and erson, K. R., and Hoye, T. R. (2005). Mixture of new sulfated steroids functions as a migratory pheromone in the sea lamprey. *Nat Chem Biol* **1**(6), 324–328.
7. Tweedie, S., Charlton, J., Clark, V., and Bird, A. (1997). Methylation of genomes and genes at the invertebrate-vertebrate boundary. *Mol Cell Biol* **17**(3), 1469–1475.

Fish Sex (X/Y) Chromosome and Viral Tumors

1. Froschauer, A., Korting, C., Bernhardt, W., Nanda, I., Schmid, M., Schartl, M., and Volff, J. N. (2001). Genomic plasticity and melanoma formation in the fish Xiphophorus. *Mar Biotechnol* **3**, S72–S80.
2. Schartl, M., Hornung, U., Gutbrod, H., Volff, J. N., and Wittbrodt, J. (1999). Melanoma loss-of-function mutants in xiphophorus caused by Xmrk-oncogene deletion and gene disruption by a transposable element. *Genetics* **153**(3), 1385–1394.
3. Volff, J. N., Korting, C., Altschmied, J., Duschl, J., Sweeney, K., Wichert, K., Froschauer, A., and Schartl, M. (2001). Jule from the fish Xiphophorus is the first complete vertebrate Ty3/Gypsy retrotransposon from the Mag family. *Mol Biol Evol* **18**(2), 101–111.

4. Volff, J. N., and Scharlt, M. (2002). Sex determination and sex chromosome evolution in the medaka, Oryzias latipes, and the platyfish, Xiphophorus maculatus. *Cytogenet Genome Res* **99**(1–4), 170–177.
5. Volff, J. N., and Schartl, M. (2001). Variability of genetic sex determination in poeciliid fishes. *Genetica* **111**(1–3), 101–110.

Fish ERVs, Retroviruses and Other Viruses (WDSV, VHSV, LMBV, CCV, CHV)

1. Britten, R. J., Mccormack, T. J., Mears, T. L., and Davidson, E. H. (1995). Gypsy/Ty3-Class Retrotransposons Integrated in the DNA of Herring, Tunicate, and Echinoderms. *J Mol Evol* **40**(1), 13–24.
2. Casey, R. N., Quackenbush, S. L., Work, T. M., Balazs, G. H., Bowser, P. R., and Casey, J. W. (1997). Evidence for retrovirus infections in green turtles Chelonia mydas from the Hawaiian islands. *Dis Aquat Organ* **31**(1), 1–7.
3. Eaton, W. D., and Kent, M. L. (1992). A Retrovirus in Chinook Salmon (Oncorhynchus-Tshawytscha) with Plasmocytoid Leukemia and Evidence for the Etiology of the Disease. *Cancer Res* **52**(23), 6496–6500.
4. Einer-jensen, K., Ahrens, P., Forsberg, R., and Lorenzen, N. (2004). Evolution of the fish rhabdovirus viral haemorrhagic septicaemia virus. *J Gen Virol* **85**, 1167–1179.
5. Kawakami, K. (2005). Transposon tools and methods in zebrafish. *Dev Dyn* **234**(2), 244–254.
6. Kurita, K., Burgess, S. M., and Sakai, N. (2004). Transgenic zebrafish produced by retroviral infection of in vitro-cultured sperm. *Proc Natl Acad Sci U S A* **101**(5), 1263–1267.
7. Leaver, M. J. (2001). A family of Tc1-like transposons from the genomes of fishes and frogs: evidence for horizontal transmission. *Gene* **271**(2), 203–214.
8. Lilley, J. H., and Frerichs, G. N. (1994). Comparison of Rhabdoviruses Associated with Epizootic Ulcerative Syndrome (Eus) with Respect to Their Structural Proteins, Cytopathology and Serology. *J Fish Dis* **17**(5), 513–522.
9. Martineau, D., Bowser, P. R., Renshaw, R. R., and Casey, J. W. (1992). Molecular Characterization of a Unique Retrovirus Associated with a Fish Tumor. *J Virol* **66**(1), 596–599.
10. Paul, T. A., Quackenbush, S. L., Sutton, C., Casey, R. N., Bowser, P. R., and Casey, J. W. (2006). Identification and characterization of an exogenous retrovirus from atlantic salmon swim bladder sarcomas. *J Virol* **80**(6), 2941–2948.
11. Poulter, R., and Butler, M. (1998). A retrotransposon family from the pufferfish (fugu) Fugu rubripes. *Gene* **215**(2), 241–249.
12. Shen, C. H., and Steiner, L. A. (2004). Genome structure and thymic expression of an endogenous retrovirus in zebrafish. *J Virol* **78**(2), 899–911.
13. Volff, J. N., and Schartl, M. (2001). Variability of genetic sex determination in poeciliid fishes. *Genetica* **111**(1–3), 101–110.
14. Zhang, Z., Du Tremblay, D., Lang, B. F., and Martineau, D. (1996). Phylogenetic and epidemiologic analysis of the walleye dermal sarcoma virus. *Virology* **225**(2), 406–412.

Eel Herpes Virus, HVA

1. Haenen, O. L. M., Dijkstra, S. G., van Tulden, P. W., Davidse, A., van Nieuwstadt, A. P., Wagenaar, F., and Wellenberg, G. J. (2002). Herpesvirus anguillae (HVA) isolations from disease outbreaks in cultured European eel, *Anguilla anguilla* in The Netherlands since 1996. *Bull Eur Assoc Fish Pathol* **22**(4), 247–257.

2. Shih, H. H., Hu, C. W., and Wang, C. S. (2003). Detection of Herpesvirus anguillae infection in eel using in situ hybridization. *J Appl Ichthyol* **19**(2), 99–103.
3. van Ginneken, V., Haenen, O., Coldenhoff, K., Willemze, R., Antonissen, E., van Tulden, P., Dijkstra, S., Wagenaar, F., and van den Thillart, G. (2004). Presence of eel viruses in eel species from various geographic regions. *Bull Eur Assoc Fish Pathol* **24**(5), 268–272.
4. van Nieuwstadt, A. P., Dijkstra, S. G., and Haenen, O. L. M. (2001). Persistence of herpesvirus of eel Herpesvirus anguillae in farmed European eel *Anguilla anguilla*. *Dis Aquat Organ* **45**(2), 103–107.

Fish MHC, OR Genes and Pheromones

1. Boehm, T., and Zufall, F. (2006). MHC peptides and the sensory evaluation of genotype. *Trends Neurosci* **29**(2), 100–107.
2. Dulka, J. G. (1993). Sex pheromone systems in goldfish: comparisons to vomeronasal systems in tetrapods. *Brain Behav Evol* **42**(4–5), 265–280.
3. Freitag, J., Ludwig, G., and reini, I., Rossler, P., and Breer, H. (1998). Olfactory receptors in aquatic and terrestrial vertebrates. *J Comp Physiol A* **183**(5), 635–650.
4. Hamdani, E. H., and Doving, K. B. (2007). The functional organization of the fish olfactory system. *Prog Neurobiol* **82**(2), 80–86.
5. Johnson, P. T. (1976). Herpes-Like Virus from Blue-Crab, Callinectes-Sapidus. *J Invertebr Pathol* **27**(3), 419–420.
6. Kawakami, K. (2005). Transposon tools and methods in zebrafish. *Dev Dyn* **234**(2), 244–254.
7. Kobayashi, M., Sorensen, P. W., and Stacey, N. E. (2002). Hormonal and pheromonal control of spawning behavior in the goldfish. *Fish Physiol Biochem* **26**(1), 71–84.
8. Kulski, J. K., Shiina, T., Anzai, T., Kohara, S., and Inoko, H. (2002). Comparative genomic analysis of the MHC: the evolution of class I duplication blocks, diversity and complexity from shark to man. *Immunol Rev* **190**, 95–122.
9. Matsuo, M. Y., and Nonaka, M. (2004). Repetitive elements in the major histocompatibility complex (MHC) class I region of a teleost, medaka: identification of novel transposable elements. *Mech Dev* **121**(7–8), 771–777.
10. Ottova, E., Simkova, A., Martin, J. F., de Bellocq, J. G., Gelnar, M., Allienne, J. F., and Morand, S. (2005). Evolution and trans-species polymorphism of MHC class II beta genes in cyprinid fish. *Fish Shellfish Immunol* **18**(3), 199–222.
11. Shi, P., and Zhang, J. Z. (2007). Comparative genomic analysis identifies an evolutionary shift of vomeronasal receptor gene repertoires in the vertebrate transition from water to land. *Genome Res* **17**(2), 166–174.
12. Sorensen, P. W., Pinillos, M., and Scott, A. P. (2005). Sexually mature male goldfish release large quantities of androstenedione into the water where it functions as a pheromone. *Gen Comp Endocrinol* **140**(3), 164–175.
13. Spehr, M., Kelliher, K. R., Li, X. H., Boehm, T., Leinders-Zufall, T., and Zufall, F. (2006). Essential role of the main olfactory system in social recognition of major histocompatibility complex peptide ligands. *J Neurosci* **26**(7), 1961–1970.
14. Stacey, N. (2002). Hormonally derived pheromones in fish: exogenous communication from gonad to brain. *J Physiol (London)* **543**, 6S–6S.
15. Ward, A. J. W., and Hart, P. J. B. (2003). The effects of kin and familiarity on interactions between fish. *Fish Fish* **4**(4), 348–358.
16. Ward, A. J. W., Webster, M. M., and Hart, P. J. B. (2007). Social recognition in wild fish populations. *Proc R Soc B-Biol Sci* **274**(1613), 1071–1077.

Chapter 7
Development of Tetrapod Group Identity, the Smell of Self

Overall Chapter Objectives

Starting with the teleost fish and tracing a path through amphibians, to reptiles and to placental mammals, the various mechanisms of sexual and social group identity and social bonding will be explored. The capacity of sensory (social) inputs to set group identity, including sexual identity, will provide a central theme for this chapter. As in other topics, the possible role of genetic parasites will continue to be evaluated. We will also consider how species attain individual as well as group identification in conjunction with MHC-based biochemical detection in vertebrates. Fish in particular exist in large shoals and have expanded the use of CNS and olfaction to control group identity. Thus, the basal importance of odor-based systems and their various receptors as used for individual, sexual and group identification will be examined. The issue of olfaction is also of particular relevance to insect social systems, so a short segue into insect-based pheromones will be presented. In addition, visual- and audio-based group identification as used by fish and terrestrial animals will also be presented. The role of the CNS-mediated learning and group imprinting in these identity systems will be discussed. As previously argued, stable group identification generally requires addiction module-based strategies. Genetic parasites will be considered as possible sources for the origins of such addiction modules. In addition, the social systems of avians which involve prevalent pair bonding will provide another segue into group identity systems. Here, the role of neurotransmitters, dopamine, opioid receptors and especially oxytocin as mediators of bonding is considered. These social relationships will be traced to the related systems as used by placental mammals. The maternal–offspring bond of mammals, in particular, will provide a major and basal example of social bonding systems that will relate to human evolution. This chapter ends byconsidering how stable pair bonds in vole mating pairs are formed and how this relates to maternal bonding mechanisms. Since our main objective is to understand the evolution of primate and human group identity systems, the next chapter will trace such non-maternal mammalian systems of social bonding in primates.

L.P. Villarreal, *Origin of Group Identity*, DOI: 10.1007/978-0-387-77998-0_7,

The Continuity of Group Identity Systems

Fish, relative to lampreys or mice, show major differences in olfactory tissues and their corresponding central neurons (olfactory bulb). Fish show a considerable expansion of this capacity relative to lampreys, and rodents underwent another large expansion. Along with this, we can also see the expansion of specific sets of odor receptors as well as the emergence of a stable link between olfaction and the MHC locus. In the mouse, major brain adaptations in sensory systems (VNO tissue and receptors) that sense and process this sensory group (pheromone) information can also be seen. Curiously, main elements of this olfactory-based pheromone identity (VNO tissue and receptors) were lost in the great apes (see next chapter). In terms of vision, major transitions are also seen. The prior chapter presented the elaborate visual systems of shrimp, which were mostly not maintained in fish or other vertebrates. However, the visual acuity and spectral resolution of the mouse has reduced relative to that of fish as most mammals have dichromic vision. Audio group detection, however, has undergone significant adaptations in mammals. Yet as presented in the next chapter, African primates acquired both color vision and major adaptations in CNS processing of vision.

A Quick Retracing

At this point, it would be worthwhile to briefly summarize the mechanisms of group identity that have been presented in the prior chapters in order to better understand the context of adaptations that occurred in mammals. The aim is to place the development of group identity systems used by tetrapods within the context and continuity of how lower organisms have addressed this issue. Specifically, we will be interested to examine how the central nervous system as evolved in teleost fish from such beginnings and has set the stage for the evolution of an extended identity that includes and links a group of individuals. A main focus of this chapter will therefore be the evolution of CNS-based mechanisms (such as behavior) of sexual and social identity.

Unicellular Bacteria and Eukaryotes (Chapters 1 and 2)

We can recall that bacteria employed sensory receptors, involving signal transduction (via intracellular Ca^{++} levels), as components of the cilia complex that could also be involved in light detection and cellular motility. This ciliate-based sensory apparatus generally uses Rhodopsin-like receptors, which are 7TM GCRPs that induce signal cascades, and will often affect cell motility. Strategically, the overall molecular logic of this sensory control has been conserved in higher eukaryotes. It generally involves small molecules (or light) and surface recognition that affect individual cell movement toward or away from other group members. In addition, group members often produce toxic products active against non-members or members with altered identity. These can be parts of T/A modules that can induce self-destruction following colonization by foreign genetic entities. This is the basic unicellular sensory apparatus, controlling both movement and sex, and is also able to establish group identity and survival in these simple organisms. Bacteria also exist in large communities and the toxins/peptides they produce can also frequently be involved in group identity. Thus, these small molecules in bacteria appear to provide the origins of small molecule group identifiers, such as pheromones, retained in most organisms. An example was the QS molecules of bacteria which small molecule pheromones provide group detection. Phage and other genetic parasites such as plasmids are also frequently involved in bacterial group identity. Here, bacterial surface identity is frequently affected by phage (lysogen) conversion or bacteriocins.

Caenorhabditis elegans (Chapter 4)

In the model nematode, *C. elegans*, similarities to early pheromone-based bacterial QS systems appear to have been maintained, such as the Dauer pheromone. *C. elegans* also retains some capacity to recognize and directly respond to bacterial pheromone remains (i.e., AHL QS molecules which will alter 5% of *C. elegans* proteins expressed). *C. elegans* has also acquired a high diversity of channel proteins. However, the development of multicellularity in *C. elegans* represents a major transition in cell identity systems that fundamentally required the development of social control over individual cells and tissue growth. Such social control of individual cell fates frequently involves a self-destruction-like or apoptosis process. In *C. elegans*, this apoptotic system is especially observed in the development of the central nervous system. And the Dauer pheromone (a social control pheromone) also operates on individual behavior by affecting sensory neuron survival (apoptosis) to set pheromone detection and sexual behavior. Asymmetric neurons are differentiated in response to the pheromone and receptor transduction by a process that involves both RNAi and apoptosis. The resulting sensory neurons can now control sexual identity by controlling behavior. This examplar thus provides us with a foundation for how sensory learning and CNS function can control animal group behavior. In terms of immunity or biochemical self-identity, *C. elegans* uses a highly efficient and transmissive RNAi-based immunity system to control viruses. Some, but not all, of the components of this system have been conserved in higher animals. *C. elegans* lacks the cellular-based immunity (phagocytes), characteristic of most higher animals. However, although all invertebrates have phagocytic blood cells, only in the jawed vertebrates did these cells acquire the capacity to differentiate altered genetic identity of its own cells and became able to recognize virus and kill infected cells.

Lampreys and Hagfish (Chapter 5)

The last chapter presented the pheromone/olfaction situation as seen in Lampreys with regard to group and sexual identity. Lampreys do have the class A (rhodopsin-like) OR receptors which show similarity to serotonin, dopamine, protaglandin and opioid receptors in mammals. These mostly novel lamprey OR genes consist of a small family (9) of genes that differ significantly from the fish OR genes. Some of these receptors are clearly used for sex-cell chemotaxis, a significant biological strategy in lampreys, indicating their involvement in cellular identity. In lampreys, ORs can bind bile acids and link this detection to sexual behavior. For example, migratory adult lampreys use OR to detect bile acid-derived pheromone released by larvae in order to locate spawning rivers for conspecifics. Some solitary lamprey species are hermaphrodites that can release both sperm and egg simultaneously, but can be self-sterile. This interaction is regulated by surface interactions. Ciona intestinalis (a self-fertilizing species) has sperm that is chemotactic via sulfated steroid which will move toward eggs using flagella and store activated Ca^{++} channels. This is similar to sea urchin, which is also known for chemotactic sex cells and produces sperm that have polycystin receptor and a canabinoid receptor, apparently involved in chemotactic activity. It was interesting to note that lamprey also has specialized organelles associated with oxidative burst, which would appear to also contribute to cellular identity. Thus, lamprey sexual reproduction has a much more single 'sex-cell' motility character to it than does reproduction in the jawed vertebrates. With jawed fish, we see the emergence of a greater reliance on whole organism motility (not simply egg or sperm cell movement) which involves reproductive group behavior. In terms of immunity, like other invertebrates, lampreys also have motile phagocytic blood cells. But there is no evidence that such cells can recognize virus infected self. In contrast, in jawed fish, such blood cells evolved to become both self-renewing and clonally selected to recognize virus infection of self. Curiously, however, lampreys do not seem as prone to general virus infection relative to jawed fish. These cells of the adaptive immune system became subjected to 'social control' by other (presenting) cells and undergo apoptosis during the establishment of cellular identity (immunity). In addition, jawed fish are noteworthy relative to jawless fish for their propensity to live in much larger communities as large social (shoal) populations. How this expanded group identity evolved will now be considered below.

The Transition to Fish and Tetrapods

Many, but not all, biological features seen in lampreys were retained in more complex organisms, such as jawed fish and tetrapods. Chief among them is the retention of a compact and rapidly functioning CNS able to control behavior and motility, although lampreys are mostly not very social species. Jawed vertebrates also show many other evolutionary acquisitions. For example, the use of pheromones has been expanded in jawed vertebrates, and steroids are now commonly employed in addition to the earlier peptides. This expanded set of pheromones is used for a more complex and differentiated social identity that entails small molecule olfaction as well as internal use of peptides. Neuropeptides, in particular, now become more prominent components for internal CNS control. This CNS development is also accompanied by the emergence of cholinergic neurons and nicotine, acetylcholine receptors. Morphine receptors (not present in jellies or *C. elegans*, which use both peptide and some steroid pheromones) were also developed. Curiously, much of the lamprey immune system (VSR) is fully distinct from that of jawed vertebrates. In jawed vertebrates, additional immune-based peptides (MHC) are used as a biochemical basis of group and individual identification. The CNS of the jawed vertebrates also now employs the use of internal vasopressin and oxytocin peptides in social control. In addition, the use of external olfaction in jawed vertebrates was significantly expanded, allowing the CNS to recognize sensory patterns that can become imprinted at crucial times (such as spawning grounds, a learned state) allowing the recognition of reproductive sites or membership in shoals. In contrast to prokaryotes (and cone snails), toxins are no longer common systems of external group identification in vertebrates, but toxins are internally used for cell identity and social control, such as in ROS and apoptosis. Visual recognition in particular is enhanced in jawed vertebrates. This necessarily requires a more developed CNS capacity for acquired memory of sensory pattern recognition. Although steroids become important agents for sex and group identity, many tetrapods (amphibians, reptiles) still employ environmentally determined sex strategies, in contrast to the genetically determined X/Y sex chromosomes of mammals. The highly effective innate immunity system as seen in *C. elegans* (RNAi and NO) becomes less efficient in vertebrates (losing its intracellular transmissive capacity) but is now involved in CNS memory and behavioral control. A major and novel acquisition in jawed vertebrates is the origin of adaptive immunity. This system also involves the application of apoptosis to differentiate leukocyte in blood. Links between this adaptive immunity system and group ID become set and maintained via olfaction of MCH peptides, but, as mentioned, this also requires stable learning (imprinting) of olfaction as a regulator of group identity (as will be seen in maternal and pair bonding of mammals). A genetic link between MHC and olfaction (via OR/VNO receptor genes) becomes established in vertebrates. However, in contrast to vertebrate animals, insects (especially social insects) adopt distinct olfactory systems to from group identity and rely heavily on the uses of hydrocarbon as an olfactory basis for such recognition. In vertebrates (and insects), a link of immunity regulation to morphine receptors is also developed. In some fish species, self-destructive (altruistic) behavior becomes linked to reproduction en mass. These are all features associated with expanded scale of group membership, mostly mediated by CNS and behavior. Thus, the essential features of higher animal group recognition function involve the learning sensory cues (olfaction and vision) as used for group identity. This feature is mostly present in bony fish but has been maintained in all tetrapods.

Ironically, the development of this highly complex adaptive immunity of vertebrates is accompanied by a much greater diversity of virus in bony fish (e.g., the first autonomous retrovirus able to infect the immune cells, the first rhabdovirus). Huge populations and crashes in shoal species mediated by virus are also seen in large natural populations. In addition, virus persistence becomes prominent in some circumstances, and essentially all adult species as survivors of persistent viral infections.

Overall Patterns of Virus and Genetic Colonization: From Fish to Tetrapods

An implication is that ERV colonization and retrovirus infection seem associated with the evolution of the adaptive immune system. Previously, I presented the overall patterns of genetic colonization that occurred with the evolution of teleost fish. Significant changes relative to jawless vertebrates were seen. Although fish genomes show considerable variation in size, these patterns of genetic colonizers are generally well conserved. Most notable was the colonization by the chromovirus-related retroviral elements (Gypsy, penelopy, neptune) and the absence of the MLV-related ERVs as found in tetrapods and especially seen in mammals. As presented below, chromovirus-related sex chromosome colonization is also associated with the evolution of sex chromosomes in some fish species (platyfish). Also notable was the relative decrease in the presence and activity of DNA-based elements, i.e., DNA transposons, compared to invertebrate animals such as mussels.

Both the autonomous retroviruses of fish and the large DNA viruses of fish are associated with the capacity to induce cellular proliferation, but especially interesting is the ability of some of these viruses to induce such proliferation in the immune cells of fish. Some of these autonomous retroviruses of fish control cell proliferation using viral (non-cellular) regulatory genes (such as Walleye retrovirus). Since bony fish have also evolved a renewing (stem cell based) adaptive immune system, that such stem cells are also the targets of these transforming viruses seems especially significant. This transition in virus–host biology seems relevant to the bony fish-specific transition regarding their genomic makeup and Gypsy viruses noted above. In bony fish, we see major transitions in relationships to autonomous retroviruses and genomic ERVs. Could such genomic ERVs have also contributed to the evolution of systems that preclude other viruses (antiviral immune cells)? In bony fish, large-scale apoptosis now underlies the differentiation of the adaptive immune cells (from stem cells) leading to the attainment of an antiviral state (able to kill self). However, the adaptive immune systems seem to be complex (i.e., involves too many novel genes) to have evolved only from the possible action of ERVs. Yet both lampreys and sharks (as well as bivalves) still appear to support infection by herpes viruses. Therefore, it is likely that related large DNA viruses were present and prevalent prior to the evolution of the adaptive immune system and remained adapted to those vertebrates that evolved adaptive immunity. It is also known that various large DNA viruses of invertebrates (and vertebrates) can themselves be colonized by intact retroviruses. Thus, clear links between DNA viruses and retroviruses are known. Such links also appear much later in vertebrate evolution, such as avian and mammalian evolution (see primate evolution, next chapter). Overall, the viruses that infect mammals and the genetic parasites found in their genomes both very much resemble the viruses and genetic parasites found in teleost fish, but do not resemble those of other invertebrates. The specific prevalence of especially retrovirus elements on the sex chromosomes (especially Y), and their tight association to speciation will be of special interest and is presented below from the perspective of evolution of group identification.

Adaptive Immunity as a Viral Colonization Product

The emergence of adaptive immunity in jawed vertebrates such as fish appears to represent a very successful development since such fish make up 50% of all vertebrate species on Earth and greatly outnumber ancestral tunicate (lamprey) related species. However, the adaptive immune system does not appear to be a better defense against viruses. I suggest that it might be better to think of it as a more complex and elaborate system of group identity, able to recognize and preclude viruses, but a system that likely originated from the stable colonization by a mixture of endogenous RNA and DNA viruses and other genetic parasites. The resulting adaptive blood-based cellular system, unlike prior blood-based phagocytic defense systems, is now able to recognize and exclude specific viruses or virus-infected tissues. Let us consider the evidence and arguments in support of this idea. The nurse shark has been considered to represent the earliest vertebrate that has all the hallmark genes of adaptive immunity, RAG1/2, T-cell receptor, antigen processing and presentation genes (TAP, LMP; combined MHC I/II). Comparative genomics indicates that the MHC region is the most rapidly evolving and diverse genetic loci in the vertebrate genome which has become increasingly complex. It is clear that this evolution occurs by a process of duplication of 'frozen' gene blocks (alpha, beta), followed by diversification. The MHC region is also unusual in being densely colonized by ERVs and retroelements. It is now clear that the basic unit of MHC gene block duplication (leading to MCH II/III evolution) includes an ERV (HERV-16 in the case of human genomes). Thus, ERV colonization and mobilization seem most associated with all MHC evolution (see next chapter for primate MHC). I have already discussed the RAG1/2 genes as likely evolving from colonization by a Mu-like viral transposase (IS4 family) and its unusual gene-specific (IgG) transposition, which also resembles many viral transposes. RAG1/2 also conserves the DDE domain as conserved in the RNAse H fold of retroviral integrases. I have mentioned the JAM/CTV (found in prechordates) is a viral receptor that also appears to be phylogenetically basal to the TCR gene. And in terms of the MHC I genes (ancestral to II/III), this gene-dense region mostly lacks introns and is absent from tunicates (hence appears horizontally acquired). Large DNA viruses (known to infect tunicates) are well known for the many MHC I-like genes they encode (chemokines, receptors, processing) that regulate MHC in non-host-like ways. One invertebrate baculovirus (granulovirus cydia pomonella) also encodes a Tc1-like transposase element (TCP3.2) that closely resembles the V(D)J recombination system. This element, when present, is used to preclude and outcompete similar viruses that lack the element. Related granulovirus can be vertically transmitted in some species. Tc1-like elements are prevalent in the genomes of *C. elegans* and jawed fish. Baculoviruses are prevalent in oceanic invertebrates. Taken together, it is clear that we can account for all the elements of the adaptive immune system as possibly originating from viruses. What then would be required is that these parasites (possibly acting cooperatively) should be able to attain a stable colonization that transformed the host blood cells and endowed them with new preclusive and antiviral activities.

If the adaptive immune system is indeed a component of a more general group identity system, perhaps we can explain other curious aspects of adaptive immunity. For example, this would also suggest why immunity should operate via addiction modules. The large-scale cellular apoptosis that is inherent to development of adaptive immunity can be considered from the perspective of an addiction module. In the next chapter, the link between addiction modules and adaptive immunity will further be traced to the origin of CNS (behavior) based addictive systems (i.e., opioids) and their involvement in social identity. Addiction modules have also been proposed to generally promote and mediate symbiotic relationships and could relate to why vertebrate animals are host to more complex gut flora relative to invertebrates. Such symbiosis would seem to require a nuanced and complex system of group identity. However, the most basic state of group identity is self, kin and sexual identity. Thus, we might expect that at its origin, the adaptive immune system should link cellular (antigenic) identity to sexual or kin identity. Indeed, an association between immunity and olfaction involving OR genes, VNO genes and MHC genes is seen in teleost fish and all tetrapods. Tetrapods, including most mammals, depend on olfaction via OR/VNO detection of MHC peptides for social and sexual purposes. These olfactory systems are used for mate and kin identification. However, in jawed vertebrates, the CNS is a main target of such olfactory identification which mediates the resulting and sometimes stable changes in behavior. This chapter will thus focus on the evolution of such CNS adaptations in vertebrates with regard to group identity. The next chapter will continue that focus with the primates and hominids. Curiously, the biochemical (antigenic) and olfactory-based group identification system of tetrapods became incapacitated, by genetic colonization, in the great apes. This allowed the evolution of their social identity systems to becoming a much more brain-based process (cognitive or sensory-based learning). In the next chapter, we thus consider the role of genetic parasites in the loss of this immunity-olfaction system of group identity in the African primates.

Vertebrate Brain Evolved to Mediate Group Identity

A major objective is to understand how sensory information can program group identity via behavior. Therefore, the evolution of the CNS and its sensory apparatus will be prominent in our evaluation. In general, vertebrate animal brains have evolved by a lamination process, in which more complex species acquire new brain layers that control more complex sensory systems or behavior. In teleost fish, the development of brain structures associated with odor sensation seems to have undergone a major evolutionary transition. For example, the vagal lobe is 20% of the brain in cyprinid species. In jawed vertebrates (gnathostomes), like bony fish, the vagal lobe has 15 laminae, thus identifying a significant structural adaptation during their evolution. In representatives of ancestral species (jawless agnathans like lamprey), this lobe is unlaminated. Also, in agnathan species, such as hagfish, the pineal gland (the source of oxytocin) and thalamus are absent. Agnathostomes lead relatively simple and passive social lives, spending much time in mud as larvae before transforming to swimming forms (lampreys) or scavenging as bottom feeders (such as hagfish). Jawed vertebrates became much more social and developed complex group membership. The lamination that is seen in the jawed vertebrate olfactory lobe can be considered to result from a new, hypertrophic layer that is controlling the brain via processes it sends to older layers. Evolution by lamination thus has both a hypertrophic and an invasive character to it; features common to virus affects onhost tissue. For example, poxvirus-infected cells make and extend cell processes to transmit virus to adjacent cells via transfer of small membrane-bound viral packets. Why did olfaction expanded expand to such a degree in bony fish and why did it continue its evolutionary expansion in tetrapods? The olfactory lobe has continued to evolve in complexity in animals, and increasing layers are seen in reptiles, birds, mammals and primates. This olfactory bulb expansion was especially noteworthy in the transition of mammals relative to fish. In terms of fish olfaction, fish have a palatal organ that lines the roof of oral cavity and gill rakes and have huge vagal lobes used to process odor/taste within the mouth. The fish forebrain is noticeably less complex than that of reptiles and birds. In most vertebrates, the olfactory sensory neurons bypass the cortex and connect directly to the limbic system, an area associated with unconscious innate behaviors. Such a connection could allow olfaction to control basic emotional behaviors without a conscious sensory or intervening CNS process. In teleost fish, this innervates the hypothalamus which is relatively large and differentiated. The hypothalamus is involved in homeostatic behaviors (feeding). Other brain structures also differentiate fish from their ancestors. For example, lampreys have few astrocytes. In addition, the cerebellum is also absent in hagfish as are myelin-producing oligodendroglia and microglia. Microglia are marcophage-like cells that remove damaged cells (such as those killed by apoptosis). Both these structures are present in ray-finned fish and all jawed vertebrates. However, hagfish and lampreys have more extensive projections on olfactory bulb mitral cells, which appear relatively 'pruned' in most vertebrates.

In fish, olfactory neurons are responsible for the only recorded nerve tumors ever observed. This may be related to the fact that the olfactory sensory neurons are one of few neurons that are produced throughout life in all adult vertebrates (including mammals). This ongoing differentiation may also explain the curious feature in that the olfactory neurons from fish to humans retain a susceptibility to virus infection and are often used by viruses to gain entry into the central nervous system. In adult mice and humans, neurons of the olfactory bulb are born as distant neuroblast in the lateral ventricles and must then migrate large distances to reach the olfactory bulb. Such migration clearly requires that some type of cellular ID system must be at work. Recent results suggest immune stem cells retain a similar 'neuron-like' migration capacity. Brain cell identity is now known to involve Dscam surface expression as discussed below.

As will be presented below, fish depend heavily on olfaction for social communication, and it will be asserted that such communication provides a strong selection for brain function. Interestingly, electric fish have developed a highly laminated gymnotoid organ associated with electric perception, but their vagal lobes are reduced. It is thought electric fish communicate via this electric organ. A similar situation appears to apply to platypus which has a tiny olfactory bulb, but a large electrosensory organ. Thus, it seems the evolutionary demands that increase fish olfactory brain capacity may be associated with more communication than gustatory detection.

Neuropeptide (Oxytocin) Evolution

Hagfish and lampreys seem not to have the neuropeptides oxytocin or vasopressin. However, they do have vasotocin, which is a vasopressin homologue. Vasotocin is also clearly related to the conopressin, discussed in the prior chapter as a highly toxic peptide produced by toxic cone snails and is related to similar neuropeptides that are found in all mollusca and annelids. These specific neuropeptides provide a most interesting trace of all animal evolution. For example, basal fish (elasmobranchii; such as ratfish, skate, rays, sharks) underwent a transition in which vasotocin was lost, but isotocin-related peptides (cyclic nonapeptide oxytocin homologues) were acquired in place. These fish orders are also associated with the early evolution of adaptive immunity.

Thus, it is particularly noteworthy that the elasmobranches have a most unusual degree of diversity among these peptides, not seen in any other animal. Can this molecular diversity suggest involvement in group identity? One of these diverse neuropeptides is oxytocin itself. The ratfish (*Hydrolagus colliei*) produces oxytocin, but such capacity does not make a reappearance until the evolution of mammals. It is very interesting to note that the vasopressin homologues differ noticeably in action from the oxytocin homologues. Vasopressin homologues are associated with regulation of ion flow and water balance (hence conotoxins effects), whereas oxytocin is associated with reproductive physiology and behavior (smooth muscle, milk letdown, prostate production). As will be presented at the end of this chapter, receptors for these two neuropeptides have much to do with vertebrate social interactions (see rodent pair bonding). Clearly evolution favored the use of such peptides for central aspects of mammalian group behavior. However, bony fish exclusively make isotocin, which was displaced by mesotocin in lungfish. Amphibians retained the mesotocin (MT) but many also acquired hydrin (a vasotocin-like peptide). Reptiles also retained the mesotocin, but did not acquire hydrin. Mammals (and platypus) have all lost mesotocin and all express oxytocin (OT) and vasopressin but suiforms (pigs) also express lys-vasopressin (LVP). Marsupials also have distinct neuropeptides (all express OT, some retain MT, all also express either LVP or PheP). These sharp transitions in oxytocin-related peptides appear to correlate exactly with sharp transitions in vertebrate lineage identity. Such striking phylogenetic congruence between neuropeptides and major animal lineages is not seen with other neuropeptides, such as substance P or neuropeptide Y, even though the latter peptide is involved in sex behavior. It was largely conserved from lampreys to mammals. These specific neruopeptides (VT, OT) thus have all the hallmarks of providing important markers of group identity in the vertebrates.

Opioid Neuropeptides

Another group of neuropeptides that are also involved in sex and group behavior are the opioids and their receptors. Although these receptors were discovered due to their interaction with analgesic alkaloids such as morphine, the endogenous neuropeptides (opioids such as enkaphalins, endorphins) that bind to them are found in nervous tissue of early animals as well as all vertebrates. Most prevalent are the Leu-Enk-like and the Met-Enk-like peptides found in crustaceans, mollusks, annelids and insects. Although their functions are not well established, expression is almost always limited to central nervous tissue and reproductive tissue (ovary and testis) in these species. Endorphins can directly affect reproductive tissue (ovary) function. In some instances (crustaceans), they are also expressed in eye stalks and control chromatophores by pigment dispersal, thus affecting visual pattern expression, presumable in response to visual stimuli. Recall in the prior chapter I presented the elaborate use of pigments and color in crustaceans for visual pattern recognition. Endorphins also affect crustacean defensive behavior, such as responses to passing shadows. Most interesting crustacean response to passing shadows can show habituation with repeated stimulation and that this appears mediated by endorphins since nalaxone increases this habituated defense response. In this case, learning and endorphins appear to be linked. As presented below, opioid peptides are also linked to mammalian immune cell function, a feature also seen in insects.

Learning to Belong, Shoal Membership and CNS Imprinting from Sensory Inputs

Social fish, like herring and minnows, are more numerically successful than non-shoaling species and can exist in huge populations. In terms of the characteristics that determine shoal membership, odor and visual cues appear to predominate; thus the CNS and learning is involved in both. In jawed fish, the MHC locus also becomes linked to odor detection and sexual behavior. Thus, CNS circuits accepting sensory input from OR (and VNO) and retinal receptors are impacting the memory systems that modulate group membership. Steroids and neuropeptides both play a role in this. In the case of the VNO receptors, oxytocin is also involved (discussed below).

Fish Emotions

The memory systems involved in fish group identity are likely to be emotion based (not cognitive) and elicit general (emotional) behaviors (discussed below). Thus, behavioral group responses such as fear, anxiety, aggression, defense and pleasure are the likely mechanisms by which emotional memory compels group behavior that defines group membership. In some cases, these emotional (and reproductive) responses must be able to override even the most basic instinct of self-preservation, such as with salmon's lethal return trip to spawning grounds. Thus, these extant CNS mechanisms that control such emotions and behavior must be capable of exerting complete control over the fate of individual fish (resembling behavioral apoptosis). The vertebrate CNS has thus acquired a capacity to learn sensory-based identity that which can determine the fate of individuals. The brain, and what it learns, has become a central system for group membership.

Mechanisms of group membership based on visual and olfactory cues probably have upper limits concerning the numbers of familiars that are possible to recognize. Yet Pacific herrings, which can form very large shoals, retain individual associations for several years over considerable distance, so some individual recognition capacity clearly exists. Some shoals, such as rainbowfish, are sex specific (females with females), thus showing broader or general patterns of group membership. Although group membership is common in fish species, pair bonding or monogamy is rare in fish, but does occur. Species that have external egg fertilization usually have promiscuous behavior, and are not pair bonded. One example that is pair bonded is the midas cichlids. In this species, not only is pair bonding attained but feeding and care of young also occurs. Female parents will feed young with mucus from skin (a likely predecessor of a milk gland and interesting opportunity for virus transmission). Thus, mechanisms of tight social bonding clearly developed in some fish, but the relevant biochemical processes that control these behaviors are not known. The mechanisms by which social environments affect sexual development are not well understood in this example. Recently, however, it has become clear that some fish can visually determine the social status of other fish. Thus, a considerable capacity for visual learning in the context of fish-specific identity must exist. Social-based sensory input is clearly used and in some cases can determine sexual differentiation and identity. For example, the goby and bluehead wrasse can change sex (female to male) in response to social circumstances. Other cichlids (African) also show social control over sexual maturation. Thus, even this most basic sexual identity can be under social control in some fish. Since sexual identity is a most basic form of group identity, this issue is of great general significance. Most fish use odor and visual-based cues for group membership. However, some South American freshwater fish species also use audio-based mechanisms, such as the Dorado which will lay eggs in response to male audio signals (clicks). Little is known concerning the mechanisms by which this has evolved.

The common existence of socially determined sex in fish thus makes a major point; sensory and CNS-processed information can determine this most basic form of group identity. In such species, the brain can override and control genetic and biochemical systems that determine sex.

Specifics of Social (Sensory) Determined Sex

It is worth recalling the situation in *C. elegans* regarding hermaphrodites and the Dauer hormone response. The sensory trigger for this hormone is an odor receptor. Here too, the sex of hermaphrodites was not determined by the presence of a sex chromosome. Rather a small sensory odor molecule initiates sex development, thus defining sensory-based sex determination as a basal and common process of many animals. It should come as no surprise then that many fish species also use sensory-determined sex and that in teleost species, hermaphrodite species abound. But the existence of some basically female species that can differentiate into male (or hermaphrodites) raises some significant evolutionary problems. In most animals, males compete for females, and this situation has been explained by the capitalist-inspired concept of a cost-investment model. Here, females have clearly more investment in the next generation than do males; hence males compete for females. Thus, it is reasoned that males will often also display sexually selected visual patterns to attract females. The often fierce male to male competition and aggressive behavior are rationalized by such models. But if all males are females that will become males only in the appropriate social circumstances, such as in many fish species, this simple cost-based model cannot be applicable. In addition, the existence of female competition for males poses a problem. For example, in two-spotted gobys, the case for female competition as well as female-specific coloration is clear as females have bright yellow-orange bellies during breeding season that attracts males. This is clearly not consistent with established sex role theory regarding competition or the function of female ornaments in attracting males. It has been proposed that melanin-based ornaments function in male–male competition, whereas more energetically costly carrotenoid-derived ornaments are involved in female mate choice. These explanations seem contrived to me, especially if gender can be reversed. For example, in brightly colored tropical marine fish of shallow water, sex change is common. All fish start off as females, only some of which become males, but the converse sex change can also happen. These harems are composed of one large male (displaying sexual body coloration on dorsal fin) to about eight females. The top female changes into a male and shows alterations in external coloration, behavior, gonad structure and steroid hormones when resident male is lost. The top female will also induce expression of H-Y antigen (sex determining, discussed below). In this situation, female and male sexual selection are conditional on social standing. The female must also become male. How then does sex selection or cost–benefit operate here? This seems the wrong question to ask. What I suggest is a better question: How has sexual and social identity been attained? And why did it change from sensory determined to become genetically determined in some fish species? These two states differ by their stability. Genetically determined sex is much less conditional and more stable. In the case of socially determined sex, how is the stability of gender attained if it is also plastic and behavior controlled?

In many cases, socially determined fish sex can operate through visual cues, but is controlling gonadal development. Turbidity, for example, is known to be able to block sex change in gobys, as they fail to perceive male coloration patterns. Therefore, visual input in this specific species is needed to set (or reverse) sexual identity. Here, sexual identity is not dependent on presence of gonads, but rather social perception induces gonad growth. Behavior and physiology are both stably altered by sensory input. Courtship, aggression and offspring care can all result. For example, black male gobys build and defend their nest to then attract females. They also provide parental care of eggs (although younger males may cheat at this). Endocrin profiles have been examined in bluebanded goby, and androgin (11-ketotestosterone) is high in experienced male parents. Both grass goby and black goby can also have older males that defend the nest and provide parental care of young but vary from each other concerning pre-optic forebrain expression of gonadotropin-releasing hormone (GnRH) in that the grass goby is sexually dimorphic. These forebrain structures appear to control behavior and may also relate to differences in migratory and nest type behavior in these two species. The forebrain shows other changes during sex change. As noted above, all teleost fish have conserved isotocin, and this too seems involved in social modulation. For example, *Lythrypnus dalli* showed elevated isotocin immunoreactive cells in females as well as early during sex change that were lost after the dominant female became male. The implication is that isotocin is directly involved in early regulation of sex change. In mammals, the arginine vasopressin/vasotocin (AVP/AVT) system is strongly androgen dependent and mediates sociosexual responses. Also, with bluehead wrasse, neither development nor sex change depends on presence of gonads. Rather optic sensory input to the preoptic area of hypothalamus shows high level of AVT mRNA in sex changing females, even in ovariectomized females. Furthermore, castration of resulting dominant males had no effect on AVT mRNA levels or any sexual behavior, nor did implants of 11-ketotesterone into socially subordinate ovariectomized females (although it could induce male coloration in females). In addition, AVP receptor antagonist prevent females from gaining dominance. In saddleback wrasse, sex change can be reversed by blocking dopamine or serotonin (monoamine neurotransmitters). Increasing norepinephrine will also block sex change. Together, these observations demonstrate the basal importance of sensory social cues and CNS neurochemistry (via neuropeptides) in sexual phenotype of these fish. It also suggests that the resulting social behaviors (defense, aggression) are also emotionally learned responses to set group identity. As a side note, it is interesting to recall that Aristotle first developed many principles of scientific observation based on his studies of goby, including sexual behavior, in the Mediterranean Sea around the island of Lesbos in that some members of the goby family show socially determined sex.

H-Y antigen of fish is now known as Sxs (sex-specific antigen) and it is thought that H-Y also controls sex determination in fish. H-Y expression is induced by behavior in these socially controlled fish. This H-Y (Sxs) antigen is the synthetic product of the sex-determining gene. In most non-mammalian vertebrates (including some fish), sex determination can also be chromosomal, either xx/xy or zw/zz. In ZW/ZZ species, Sxs is expressed in the female, not males. However, in socially determined sex (homogametic species), females (e.g., protogynous wrasse) are Sxs negative but positive (in gonads) when they become secondary males. When sex is reversed (to female) by androgen administration, for example, all tissue becomes Sxs positive. For this to work, it is likely that visual receptors are able to transduce appropriate signals to the CNS that then controls Sxs gonadal gene expression. This response must also be subjected to memory control in order to become set (learned) by social situations. It is worth noting H-Y antigens were originally observed in mice via MHC-restricted lymphocyte rejection of male tissue in allogeneic female mice (discussed below). This rodent Yc-ag cross-reacts with (but is distinct from) fish Sxs antigen; therefore, it appears these sex regulators are highly conserved. In mammals, both Yc and Sxs antigens are encoded on the Y chromosome (therefore it is male specific).

Genetically Determined Sex, a Viral Role

The social (non-sex-chromosome) system described above is more prevalent in teleost fish than is genetically controlled sex. The implication is that this state (like that seen in *C. elegans*) is ancestral to that of the fish sex chromosome determining system. Thus, XY system in some fish as presented in the last chapter would appear to represent a newer evolutionary development (consistent with phylogenetic data), leading to the later evolution of the mammal XY sex system. How might this change evolve? Since Sxs expression is conditional in some fish, colonization with an ERV that encoded a viral version of Sxs (e.g., H-Y) could allow genetically determined sex to be superimposed onto species that had socially controlled sex. Since the SD locus is precisely within and adjacent to a Y-specific ERV, this proposal is consistent with genetic data. Furthermore, these ERV-related sequences are known to be rapidly expanding within newly evolved Y fish chromosomes. This also appears consistent with the involvement of endogenous retrovirus with Y chromosome evolution. If a main purpose of such ERV colonization was to create new group identity (especially sexual identity), this could also account for why melanin production and body coloration patterns were also linked to Y and to ERV evolution on these fish species. Furthermore, since viral-derived identity systems are often addictive and preclusive of related viruses, the melenoma tumors of F1 crosses between species with different ERVs would be expected to result from derepression of silent persisting ERV. Thus, the melenoma tumor induction might also be understood as the result of an interrupted viral addiction module.

If ERV colonization was involved in the establishment of the sex chromosome and sex determination in some fish species, then along with sex switching, male behavior must have also been stably changed by ERV action. This also suggests ERV-mediated stable alterations to CNS function. This would be especially expected for the brain regions involved with the relevant sex neuropeptides/neurohormones. For example, with the loss of visually determined sex, peroptic areas and its isotocin responses would likely have been permanently altered as they no longer exert control over gonad development. In XY-determined sex, such as in mammals, hormones and peptides produced by the gonads now take on the primary function of setting sexual identity as determined by sex chromosomes, not optic input. Thus, many sex-specific emotional and behavioral states would now be the product of the gonads (i.e., hormones) to some degree. Thus, the gonads will now be expected to affect brain development as it relates to sexual identity. Also, ERVs in the sex chromosome can take on an added role in speciation as they are expected to become likely barriers (by preclusion or by tumor induction) to sexual exchange with species hosting different ERVs.

Female sex partners would need to harbor ERVs that were compatible with those of the males to prevent disgenesis. Such a state would also promote a tendency for ERV (and defective) to expand in host genome due to selection for defective-mediated persistence (and dispersed LTR-mediated regulation) allowing stable colonization. Thus, with the evolution of X/Y chromosomes, new species will tend to result from the successful colonization of the sex chromosome by new exogenous genetic colonizers that will themselves tend to inactivate resident endogenous elements that resist colonization. If so, all mammals will show their own species distinct ERV colonization on the Y. This is indeed the case as described below.

Olfaction as the Basal Model in Fish

Humans tend to overlook the olfactory senses in favor of visual and audio senses. Human olfactory lobe is reduced from that of chimpanzee and even more reduced relative to other mammals. But in most other vertebrates, olfaction is crucial for group and sexual behavior, and the evolution of this characteristic can be directly traced to aquatic vertebrates, which have water as a soluble media for olfactory detection. Fish have a single olfactory epithelium with both ciliated and microvilli cells that appears basal to the differentiated VNO and MOE of tetrapods. Tetrapod VNO is typically composed of ciliated cells, whereas main olfactory epithelia (MOE) are composed of microvilli cells. The olfactory epithelium has the most polymorphic cells which express a high diversity of odor receptors. Goldfish, for example, have two multigene receptor families that are all G-coupled proteins. The two families are GFA and GFB. GFA has MOE-like receptors and GFB has VNO-like receptors. Goby, our social model fish, for example, have typical teleost style of a compact olfactory epithelium (single lamella with both microvillar and ciliated neurons) that expresses both receptors. How does odor detection link to sex partner identification or social bonding in fish? This is an active area of research, but one which has not yet been fully resolved. Odor detection and motility remain important characteristics even for larval fish. Young larval fish can learn to return to their habitat based on odor memories and display directional swimming relative to these odors. This represents an ancient process that resembles that odor-based motility seen from cyanobacteria to hagfish sperm cells previously described. In all these cases, pheromones, G-couples receptors and directional motility are involved. Thus, odor detection remains crucial for fish social behavior. For example, salmon display species-specific homing (spawning) behavior via olfaction. Salmon return to same spawning river after 2–5 years in open ocean. This memory is acquired as juveniles. Vision does not seem important for the ocean phase of migration, but the olfactory system is necessary and seems to use layer-dependent detection and swimming when in open waters. Below, we will examine how olfaction affects migrations, sexual identity, group imprinting and learning in fish.

Ligands and Receptors of Fish Olfaction

As previously mentioned, a significant adaptation in fish relative to lampreys is the use of peptides in addition to steroids (or bile acids) for sex pheromones and hormones. For example, goldfish release 3 steroidal and 2 prostaglandin pheromones for conspecific control of reproductive behavior via olfaction. However, some internal neuropeptides have received special attention above, such as prolactin, vasopressin and oxytocin homologues. These peptides also use similar receptors.

7TM Receptors, Prolactin and Opioids

In fish, oxytocin-related (isotocin) receptor is 7 TM protein that is similar to that found in mammals. Fish oocytes have been observed to express and release this peptide, suggesting a role is sexual cell development. [Arg8] vasotocin AVT is also produced by some fish species and is the most primitive vertebrate neurohypophyseal peptide in this family and generally occurs in pairs (i.e., vasotocin/isotocin like). Fish can show an overlap in the expression of beta endorphin and olfactory receptors. In catfish, receptors for endorphin are seasonal, and beta endorphins in female olfactory neurons have been observed using antibody reactivity. This reactivity changes with annual female reproductive cycle, increasing during pre-spawn, decreasing with post-spawning. This receptor expression varies with four distinct reproductive phases: a preparatory phase in which little expression is seen, a prespawning phase in which a robust increase in expression is seen, a spawning phase in which receptor expression continues than a post-spawning phase in which receptor expression is significantly decreased. Such changes may also be involved in fish feeding behavior. In zebrafish, ZFOR1 is a putative opioid receptor, shown to bind opioids in laboratory studies. Zebrafish have opiod receptor delta, which is phylogenetically basal to the vertebrate clade (showing 68% similarity). This gene is ZFOR1 (a G-coupled receptor), which has distinct ligand-binding patterns (beta endorphins show good affinity). However, ligand binding is distinct from that of most endorphin receptors, so its significance is not clear. In gold fish, both endophins and enkaphalins have been reported to be involved in fear habituation. Interesting that these peptides will also modulate goldfish macrophage activity (which is also inhibited by naloxone). In social confrontations between aggressive fish (i.e., Arctic charr), the immune system of subordinate fish will be suppressed as measured by non-specific CTL response (here skin darkening also occurs). Naloxone will also alleviate this response. The endogenous opioids also appear to modulate fish immune systems. This is seen via social contact (shoals) and also affects sexual and reproductive behavior. Goldfish, unlike zebrafish, are a more social species. In goldfish, vasotocin (VT) inhibits approach to visual stimuli without aggression, consistent with a role in shoaling behavior. This is the opposite of the affects of isotocin. In puffer fish, which have highly compact genomes, vasotocin and isotocin are closely linked genetically (within 3–12 kb with tail/tail orientation). Vertebrates maintain a similar genetic linkage, suggesting a strong selection has maintained a close genetic coordination between these two genes (a possible T/A module). In relationship to receptors and possible behavioral addiction modules, it should be noted that CB1 is the mammalian canabinoid receptor which is seen only in vertebrates, including fish, amphibians and birds. *C. elegans* does not have CB1. However, Ciona (a urochordate) has a CB1 that is related to vertebrate clade. Thus, various receptors that we relate to behavioral addiction in mammals occur early in the evolution animals. However, the role in these early animal species has not been established. Although these above observations are tantalizing, their behavioral significance remains to be fully resolved. They do, however, suggest that neuropeptides and opioids in fish may be involved in sexual and shoaling behavior.

Zebrafish are the main laboratory model for teleost fish, but are less social than many other fish species. They also lack sex chromosomes. Zebrafish are a fresh water species (from the Ganges River) with simple social structures and do not display parental egg care. Although some shoaling of adults is seen, females generally appear to prefer living with one or two males in a non-paired state, but do not live with other females. Males also do not pair bond. Adult zebrafish seem especially visual (relative to goldfish) in response to their environment and to social (sexual) cues as indicated by maze studies. Recognition by females of sexually mature zebrafish appears to be due to their color striped patterns as females prefer male-like horizontal striped patterns over vertical patterns, which is not seen with males. Thus, it is clear that significant visual pattern recognition occurs. Also, females do not develop the sex-specific body colorations as do other species. Although adult zebrafish seem to be a highly 'visual' species with respect to sexual and shoaling behavior, young fish appear to depend heavily on olfaction. Zebrafish have a complex olfactory epithelia that have several lamella forming oval-shaped rosette. This type of epithelia is seen in many other and more social teleost (salmon, goldfish, stickleback). Bile acids appear to be main stimulants of this ciliated olfactory sensory neurons, whereas nucleotides activate microvillous OSNs (recall bile acids were sex pheromones in lamprey). Amino acids appear to also stimulate both types of sensory neurons. It is most interesting that the olfactory rosette is also the main site of zebrafish ERV expression. Although visual stimulus is clearly important in adults, mate selection also seems to have a strong olfactory component since males clearly respond to female pheromones. In terms of young fish, however, it is interesting that the zebrafish olfactory neurons differentiate very early in development, and odor receptors are expressed on the first day post-fertilization (dpf) and respond via nerve connections by 3 dpf in development. This is prior to taste bud formation and swimming development. Thus, amino acids and bile acid responsiveness is established within 24 h post-hatching. This early activation suggest a role for olfaction in early development, most likely odor-based imprinting on spawning sites.

Olfaction can clearly serve to identify spawning sites in many fish species, but has also been used for mate selection in some fish species. For example, with rose bitterling, amino acids are also functioning like sex pheromones. In addition, female sticklebacks use an MHC-based odor detection to choose mate, and its mate choice appears based on diversity of MHC allele. In char, juvenile fish are also attracted to conspecific via MHC-based odor types, whereas sexually mature adults are not. An MHC-based odor detection system presents an interesting identity system that could potentially provide individual identity if sufficiently diverse MHC alleles are also present. It seems clear that some type of high-resolution sensory mechanism (such as neuron specific or diverse receptor expression) would be a necessary precondition in order to allow olfaction to attain individual specific social identification to occur.

As mentioned above, some fish species do show considerable social bonding and even parental care of young. In some well-studied cases, it seems clear that prolactin (PRL) is causally involved in such social bonding and caregiving. The North American bluegill (*Lepomis macrochirus*) males will construct nests in colonies. Although females are attracted to these nest to spawn, they leave the males to care for the offspring (fanning eggs, defending larvae and fry against predators). Bromocriptine is a known antagonist to dopamine receptors that inhibits PRL secretion in fish. Males treated with bromocriptine were less aggressive toward predators and fanned their eggs less, which clearly resulted in a lower reproductive success, although they showed other strong nesting behavior. Similar results were seen with the three-spined stickleback (*Gasterosteus aculeatus*), which also construct nest by males, but these studies used pituitary implants. However, some fish, such as the monogamous blue discus (*Symphysodon aequifasciata*), also show PRL-induced mucocyte skin production in which these epidermal cells produce mucus as food for the young (aka 'dicsus milk'). Such a social bond in which parents produce food for their young is clearly similar to milk production in mammals, which is also the product of epidermal secretion and is PRL regulated (discussed below).

Neuron Identity, Connectivity and Memory

With *C. elegans*, the central role of apoptosis in CNS development was presented. This process allows sensory information to affect neuronal survival and thus stably affect CNS cell structure and function, as seen in the olfactory neuron development associated with *C. elegans* sexual behavior. With the increasing CNS complexity of the brains of teleost fish, we now see that the connectivity between neurons and how they identify cell–cell connections becomes complex and prominent in addition to apoptosis. It now appears that there exist molecular systems that provide the basis for some of this connectivity and identity. One such system appears to be due to Dscam surface proteins (Down sydrome cell adhesion molecule) that are involved in axon guidance and branching. Although these proteins were initially described in relationship to human CNS genetic disease, their biological roles have been best worked out in Drosophila. Dscams are IgG-like cell surface protein which have 38,016 potential alternative spliced isoforms in the fly. This situation is reminiscent of the variable surface expression seen in the lamprey. Thus, Dscams posses a molecular diversity that could provide considerable identity information. It now appears that Dscams are involved in isoform-specific homeopathic binding in neural cell adhesion (via extracellular Ig domains). Dscam appears to provide Drosophila (and the zebrafish, NCAM – neural cell adhesion which has five extracellular Ig domains) with an important system for neuronal cell–cell recognition used for pathfinding. Dscam is absent from *C. elegans*, which appears to use the much less variable Syc-1 for neuronal pathfinding. In *C. elegans* let-2 is SynCam, an alternatively spliced SyG-1 (IgG-like surface protein). Dscam is also absent from slime mold or algae. The existence of a complex pathfinding and surface identity system in the brains of insects and fish provides the foundations of complex sensory-determined brain structures. It thus seems likely such systems will also have a role in group behavior. In addition, other molecular systems, such as micro RNA, also appear to regulate brain morphogenesis (in zebrafish), since mutants that cannot silence micro RNA expression fail to differentiate some brain tissue. However, it also appears that micro RNA are not involved in early zebrafish brain development, consistent with a more recently evolved role in the fish brain.

Learning, Alarm and Innate Behavior

With fish, we can see that group membership can be learned from olfactory and visual input. The capacity to learn group membership is one that has been retained in all mammals and provides the main basis of their social cohesion. Such a CNS-mediated group membership requires stable memory to form from appropriate sensory cues at the time of group membership imprinting (typically as juveniles). Sensory information must be transduced to new members, affecting gene expression and neuronal cell biology that ultimately affects cognition (recognition) most likely via emotional (non-conscious) based memory mechanisms. Although some cues may be learned, other responses are innate, not requiring learning. For example, the olfactory signals that can induce spawning behaviors in goldfish and zebrafish do not require learning to spawn. However, memory is important for many other social activities. Goldfish appear to have two distinct memory systems. One is an emotional memory that appears to operate via medial telencephalic pallia (MP) and the second is a type of rational memory associated with spatial recognition involving lateral telencephalic pallia (LP). Memory consolidation would thus be a significant issue which appears to be associated with the emergence of primitive REM sleep in fish. With the evolution of mammals, controlled body temperature and advanced REM sleep are further associated with consolidation of memory. Olfactory imprinting appears to apply to many animal species. In fish, salmonids of the Pacific Coast were well studied for their ability to imprint on odors, i.e., being able to make stable memory of odors near their natal streams. As mentioned, fish larvae show innate directional swimming with respect to amino acids. Many adult fish are also known to have an innate (positive) response to amino acids (L-alanine). However, such fish can also be trained to discriminate binary mixtures of amino acids, establishing a learned component to even such innate recognition. Such learning fits an associative conditioning paradigm and is especially evident with regard to alarm states. Fish can give off alarm pheromones via club cells that release pheromone when the skin is damaged (such as isoxanthopterin in Danio or hypoxanthine 3-N-oxide in black tetra). Although a fear response to alarm pheromones is innate, fish can learn to pair this pheromone odor with predator cues. In this case, it seems an innate emotional response becomes associated with learned visual and odor cues. However, such learning can also be modified to respond to either additional visual or odor cues, such as unthreatening species that will be learnt as an alarm. Rainbow trout also show fear response to novel objects and odors. Interesting that morphine prevents this response in groups of fish. In terms of shoal behavior, a fear response is particularly interesting as it has an infectious or transmissive character to it that can affect an entire shoal. Clearly, emotional states can be learned to not only affect innate behavior but can also affect the integrity of extended groups.

Visual-Based Fish Identity

In prior chapters it was noted that marine invertebrates have frequently used visual patterns, including light production, as a method for sex and group identity. Squid, for example, showed much light-based physiology and sexual and group behavior. Some squid have three visual pigments and can show an innate light reflex (side incident light causes head and trunk roll toward light). Some squid will also use light flashes for communication, such as Humboldt squid (producing red and white light). In contrast, teleost fish are not known for their light production associated with group behavior, but have clearly preserved color vision and many teleost species have colorful body and sexual markings. Thus, color vision was clearly present in the earliest vertebrates. Visual perception is also clearly a part of fish social behavior, such as mentioned with zebrafish above. Visual-induced aggression is also known for other fish species. One well-studied example of this is *Betta splendens* male Siamese fighting fish. These males show stereotype aggression when confronted with the visual profile of another male (via fin markings) and are even aggressive to their own mirror images, indicating a lack of visual self-identity. Males will show intense deepening of fin color as they are approached by other males. Curiously, morphine potentiates this aggression whereas other drugs (antihistamines) suppress it. The role of visual perception in shoal membership appears to be more complex. For example, Danio fish can apparently discriminate individuals based on their pigment patterns and also between shoals. Such discrimination requires an early (imprinting) experience between the specific fish. Some fish habitats seem more prone to visual-based interactions. In the great barrier reef, for example, most fish have elaborately patterned bodies that reflect a range of colors from UV to far red (these are highly diurnal teleost fish species). Some of these patterns are clearly species specific. In one example, 36 Wasser species were examined for corneal color phenotype, but only one had yellow corneas, suggesting the existence of a group-specific color recognition. Clearly fish can use color and visual patterns as sensory source of group identification. Given this use early in vertebrate evolution, it is thus curious that vertebrate tetrapods are all dichromic for vision essentially lacking color vision.

Audio and Electric-based Fish Identity

Other sensory systems are also used by some fish for the purpose of group identification. Although rare, both sound production and biologically generated electric fields have been thus employed. For example, Gulf toadfish will produce a boatwhistle call for male–male competition and to attract females. These are sets of short duration grunts generated by sexually mature males. Here it also appears that fish learn to recognize the snap-reports of nearby pistol shrimp, indicating the development of complex audio pattern recognition. African electric fish use electric organ discharge for differentiation in mate selection and reproductive behavior. Thus, in teleost fish, olfactory, vision, audio and even electric field production and sensation are all used for group membership purposes, although clearly olfactory process predominate. In addition, it is clear that complex CNS-mediated learning is important with all these sensory systems and the acquired learned state can exert profound effects on individual fish physiology, group and sexual identity and behavior. Essentially, teleost fish appear to have developed all the main sensory systems used by all vertebrates for group and sexual identity. However, as presented below, we will see that tetrapods, mammals in particular, expanded their olfactory system and linked it to their immune system for the purposes of group and sexual identification. Why and how did this happen?

Social T/A Modules and Emotional Memory

How specifically can sensory-induced emotional memory provide the basis for group identity and membership? An acquired identity must compel retention (stability) of the new ID if it is to survive selection and preclude identity competition. This will generally be accomplished by an addiction module or addiction strategy that involves the linkage between the production of stable but potentially harmful element to the production of a protective or beneficial but less stable element. In bacteria and simple eukaryotes, toxins and lytic viruses can provide the harmful element, whereas antitoxins or viral immunity elements (including viral defectives) can provide the protective elements (together constituting a T/A module). The apoptosis involved as a cellular ID system has these same essential T/A characteristics. However, what might this mean exactly in terms of behavior and emotional memory? For behavior to provide the elements needed for group identity, it too must have inherently harmful (toxic) and beneficial (antitoxic) elements that are linked to the new identity state. For this to be satisfied, the resulting behavior states must have *general* characteristic of being potentially harmful (T) and potentially beneficial (A) in linked sets in which the beneficial component is less stable. Such general behavioral characteristics can be satisfied by states of emotional memory. Fear and anger, and aggression, would be examples of general but harmful emotional states that affect behavior and group identity and direct harm to non-members. Of these, possibly the two most basic emotions involved are aggression and protection (defense) as a behavioral T/A set. Non-members are met with aggression and members are provided protection. It thus seems clear that emotional states can indeed provide the behavioral T/A elements needed for group identity. Although fish social and sexual groups clearly employ aggression and protection, it is less clear if they also use additional emotional phenotypes to define group identity. For example, mammals (especially humans) have developed emotions that include the negative emotional states of anxiety, frustration, compulsion, obsession, insecurity and depression. Humans also have positive (pleasurable) emotional states that would include arousal, joy, excitement, contentment and satisfaction. It seems most unlikely that fish can experience the full range of emotional states that mammals and humans experience (see Chapter 9). However, some fish clearly experience at least some of these states. We will now seek to understand the relevant mechanisms and how such states evolved with an initial emphasis on olfaction memory and cognitive group imprinting.

Aggression

The overall objective is to trace sensory input (such as olfaction) to the establishment of behaviorally mediated (emotional) group membership in fish and other model organisms. Since our ultimate objective is to understand these systems in humans, we will focus on pathways that are known to be retained (or curiously lost) in human physiology. We start with a potentially harmful emotion – aggression, since this appears to be a basal emotional state. Aggression is a common fish defense mechanism that can apply to offspring, spawning sites, mates, shelters and food. Social aggression is thus well established in fish and has clear physiological consequences. The majority of such aggression is from males, but some female aggression is also known. Besides the direct physical consequences of aggression on the health of fish, other physiological effects are also seen. As noted above, states of social aggression modulate sexual state (identity) in some species. States of social aggression can also modulate immune state in some fish populations which can involve opioids, since naloxone. For example, plasma cortical levels, a central immune modulator, also appear to be subjected to social control as seen in highly crowded fish (i.e., Arctic charr). Individual fish that are the subjects of social aggression (subordinate fish) can show suppression of PHA stimulated lymphocyte proliferation. This PHA reaction probably also involves cortisol levels as a general immune suppressor, but here too nalaxone will suppress this PHA response. In rainbow trout, plasma cortical levels are known to be elevated in stressed fish, especially immature fish. In Arctic charr, such stress and social subordination result in skin darkening (affecting visual social patterns) which is also mediated by elevated plasma cortisol. In terms of sexual identity, it is worth noting that for successful sexual encounter to occur, there must be an absence of aggressive behavior during approach. Clearly sensory recognition would allow a successful sexual approach and must modulate an aggressive emotional state. Neuropeptides, such as vasotocin, are known to inhibit visually stimulated aggressive response in some fish. In terms of aggression in larger groups (shoals), salmonids fish are probably the most studied and show that detection of familial fish reduces aggression. Overall, we see that fish aggression is a common and often learned social state associated with identity and group membership.

Protection/Prolactin/Aggression

Fish can also display positive responses, such as care, toward young fish. One example is with the cichlid family, which will often provide parental care of young and can discriminate between related and unrelated offspring. This care is also seen to be provided by non-reproductive adult males that have no genetic link to the young fish being protected. To establish such a relationship, it appears that daily exposure to young fish will sensitize adult to recognize and defend the young. Thus, this is a learned state, but little is known concerning the neural mechanisms involved. However, the adults can also establish a bond to their mates and will also show aggression to a non-paired mate. This non-mate aggression can also be prevented by morphine. Thus, opioid neuropeptides appear to be involved in this emotional state. Prolactin (neuropeptide that uses a 7 TM receptor) is also highly associated with social behavior, especially caregiving to young. Relevant to this, two fish species are known to show exclusive paternal care of young. The North American bluegill male constructs colonial nest in which females will visit to spawn, then disappear, leaving males to provide all care of young. Prolactin expression level in these males correlates with caregiving and a correlation of male caregiving to prolactin levels is also seen in many other vertebrate species (including avians, presented below). In terms of fish and evolution, it is most interesting that the lungfish prolactin gene is distinct from that of other teleost fish and is phylogenetically basal to that found in all tetrapods. This prolactin differs markedly from that of other teleost fish. Overall, prolactin shows a very uneven pattern of evolution. It appears to have been stable and slowly evolving in teleost fish, reptiles and avians, but underwent four bursts of evolution in mammals. Clearly, the name and association with milk production is a misnomer, given its early presence in vertebrates. Prolactin expression in fish clearly suggests that it predates evolution of hypothalamus–pituitary where it is regulated by a releasing peptide (PrRP). Affecting prolactin production also affects aggression in most species.

Care of young by fish can also be linked to sexual maturation, such as with bluegull sunfish, in which some males delay maturation, yet build nests and provide parental care for their own fry. In contrast, neighboring fry will instead be aggressively pecked. Some fish will cannibalize fry of their own species that are not kin, such as live-bearing Poecilid. In stiklebacks, the ability to discriminate siblings form non-siblings is stably maintained by both visual and olfactory cues. The discrimination ability occurs early in development, at egg hatching, and is seemingly an innate capacity at this time. It has been proposed that such kin recognition is actually learned (as a template) during sensitive early developmental periods and once learned, the memory persists up to 5 months without reinforcement. Coho salmon also depend on early association to learn to recognize kin. If salmon are raised in a mix setting (with non-kin salmon), they do not discriminate between kin and non-kin. Similarly, Arctic charr raised in isolation are subsequently unable to discriminate kin. Here, kin recognition clearly must have considerable cognitive (memory) character. In some cases, the learned template crosses species barriers. For example, guppies raised with non-kin species can learn to associate with these species; thus they have acquired a new species 'template'. Such templates, when they exist, often seemed to be learned from prevailing sensory cues during crucial windows of development that result in stable group behavior. Thus, developmentally restricted but learned patterns of sensory (olfactory, visual) input determine subsequent group and sex identity. Behavior and group identity is thus set by (learned) emotional, but non-conscious states in which aggression and protection are the most basic emotions, but both states may involve neuropeptides and the opioid systems.

In fish, we thus can accept that sensory input, such as olfaction, operating via VNO-like receptors is communicating with the hypothalamus to become learned and control group identity and group behavior. The details of the neurobiology and neurochemistry of such fish group behavior, however, are not well studied, although clearly neuropeptides such as oxytocin and prolactin can be involved. Much more information on this topic will be presented below in the context of rodent and primate social behavior.

Drosophila and Insects

Insects offer many important insights into social biology, although not on the same lineage as mammals. Since *Drosophila melanogaster* is such an important biological model and also since there are numerous insect species that are highly social, the mechanism of their group identity will be briefly considered. Insects are the most numerous terrestrial animal species. It currently seems likely insects evolved from an aqueous arthropod that resembles the extant fairy shrimp. Most social insects (wasps, bees, ants) have evolved from solitary parasitoid wasp species (hymenoptera), so this lineage will be presented in greater detail. As noted previously, some shrimp species have elaborate visual systems. Many insect species have also maintained relatively elaborate visual systems that can sense UV, color and far infrared light. Insects are not found in the oceans for reasons that are not clear.

What's the Stink About Bugs?

Insects represent highly successful terrestrial animals that also developed elaborate social systems that include sex and group identification. They lack an adaptive immunity and their central nervous system is significantly less complex than that of the vertebrates. Although they represent an evolutionary pathway that is distinct from that leading to humans, they have clearly retained and developed sophisticated pattern recognition, especially olfactory and visual recognition, using similar molecular mechanisms. Overall, insect olfaction seems especially important for group recognition, and the genes involved are clearly similar to those of vertebrates (e.g., receptors and transduction pathways). Learning in insects is also associated with sensory input. Like the oceanic animals, pheromone and small molecule odor detection remains a dominant system used for sexual and group identification. In most insects, males process female sex pheromones in antennae sensory neurons to the olfactory lobe of their brains. Also, like some vertebrate fish presented above, social cues will determine the reproductive biology of many social insects. Some insects also employ audio sensory detection in association with group and sex identity. *D. melanogaster*, our main model, is known to use odor detection in linkage to sexual behavior in which cuticle hydrocarbons are the most common odorants. Memory in drosophila is important for this, although insect memory is much less capable than that of vertebrates. And as presented in the prior chapter on vertebrate fish, the insect endogenous/exogenous virus relationship is also relevant to sexual and species membership.

Genetic Transitions/Colonizations

Insects have their own peculiar relationship with genomic genetic parasites, which is quite distinct from that of vertebrates. This relationship has been best evaluated in those insect genomes that have been sequenced, the first of which was *D. malanogaster*. Most numerous are the elements that resemble components of endogenous retroviruses (especially LTRs). *D. melanogaster* has 178 full-length LTR retroposons and have been classified into 12 families. Overall, these families are found in three larger groups (Gypsy/chromovirus, TY1/copia and Pao-like). The copia group is related to pseudoviridae that expanded significantly in the genomes of higher plants, but is absent from the genomes of tetrapods. Curiously, this copia family is also relatively absent in other fly genomes. LINEs are rare in drosophila, in contrast to mammals as only two copies of a BS LINE-like element, for example, are present. Thus, unlike vertebrates (especially mammals) LINE elements have not undergone expansion in insect genomes. In contrast, acquisition of insect retroviral elements is an early event in drosophila evolution, but curiously the specific LTR composition also differentiates recent hymenoptera evolutionary events. For example, a majority of LTR elements have been transposed after divergence from *D. melanogaster* from *D rosophila simulans* (about 2.3 million years ago). These different drosophila lineages thus have their own distinct LTR makeup, but the LTR retroposons that are found in the euchromatic portion of genomic DNA are much younger than the host species, whereas some of the heterochromatic elements are much older. This link to speciation is also true for acquisition of the DNA-based P element. Thus, the maintenance of genetic parasites in drosophila is a rather specific relationship. Some of these genetic parasites are involved in sexual isolation. For example, hybrid dysgenesis can also be mediated by both P-factor (via differential splicing) and alpha LINE-1 retroposons. But as indicated above, LINE-mediated alterations are uncommon in insects.

Intact Insect ERVs

Besides numerous LTR-related elements, the genome of *D. melanogaster* also harbors and conserves several intact ERVs. DmeGypV is an element that is now recognized as an env-containing endogenous retrovirus related to the Gypsy element or chromoviruses, as presented previously. Expression of DmeGypV is controlled by an X-linked flamenco allele, which normally maintains repression. Mutants of this allele will induce virus production in ovaries. Drosophila Gypsy virus is expressed in follicular epithelia of the ovaries and can also mediate maternal effects. ZAM is specific allele of an env-containing Gypsy virus and can be overexpressed in a small set of follicular cells surrounding oocyte. All the retroviral-encoded genes are expressed in these cells and assembled into virus-like particles. RevI of *D. melanogaster* is a mobilized ZAM that was caused by a spontaneous retroposon insertion. These ERVs can also be associated with hybrid dysgenesis. Similar elements can be found in *Anopheles gambiae*, and these elements can also be lineage specific with clear evidence of recent genome colonization. The relationship of a silent endogenous virus to hybrid dysgenesis following mating with an uncolonized mate seems reminiscent of the situation described in the last chapter for melenoma induction XY-sex-determined fish. Thus, overall, in insects we see a link between speciation and genome colonization with specific ERVs. Recent comparative genomics of various fly species clearly confirms this point. A linkage of endogenous virus production to reproductive tissue is also noted below with some parasitoid wasp (hymenoptera) and polydnavirus. The general significance, however, of such EV associations to sex and other social identity has not been evaluated.

Bombyx mori (silk moth) has also been sequenced so we can compare its composition of genetic parasites to those of Drosophila. *B. mori* has a large compliment of LTR elements with at least 29 separate LTR families which constitute 11.8% of the total genome. Within these LTR families, chromovirus (Gypsy-like elements) is the most numerous. It is interesting that the W sex chromosome of *B. mori* in particular has a peculiar pattern of LTR retroposon colonization. The copia-like Yokozuna elements have colonized the W chromosome in an interrupted manner, unlike the LTR transposons seen in drosophila. Little is known concerning the significance of this sex-associated colonization to genetic isolation or speciation.

Viral footprints are thus prominent in insect genomes and linked to speciation. But insects also have peculiar relationships to various autonomous viruses as well. For example, baculoviruses and ascoviruses are prevalent in many insect species, and in some cases can be sexually transmitted. Curiously, some moth baculovirus are known to contains a TED element, a complete chromovirus (Gypsy retrovirus) including an intact and functional *env* gene. Some DNA viruses, such as iridoviruses, can infect crane flies and mosquitoes, but can sexually (transovarially) be transmitted and many types of related viruses are known. However, the biological consequences of such viral host relationships or these mixed viral consortia are not clear.

Drosophila OR Genes and Neuron Development

As olfaction provides our central focus from which to evaluate sexual and group behavior systems in insects, the molecular details interest us. Olfaction affects many insect behaviors, such as oviposition and locating mates. In social species, early olfactory cues can determine the caste development of the individual larvae. Most insects determine male sex behavior by the action of pheromones. In some cases (parasitoid wasp), this characteristic is manipulated by plants which can produce the same pheromones and induce wasp to copulate with plants flowers. Social insects also have alarm pheromones and many insect species show pheromone-mediated aggregation. For example, cydia cocoons are aggregated in response to quorum odors. Thus, the role of odor in insect behavior goes well beyond sexual identity. Genomic analysis indicates that the *D. melanogaster* genome has 60 identified OR genes. It appears that different classes of OR genes are expressed on specific neurons and these neurons innervate distinct targets in antennal lobe of the brain. It now appears that specific classes of neuron OR require Dscam expression to properly from symapse between communicating nerves. Thus, OR diversity and Dscam diversity in neurons are linked. Mutants in Dscam will form ectopic connections in and out of antennae, affecting their ability to differentiate odors. As noted below, there are multiple spliced forms of Dscam and these are expressed in developing OR axons of developing antennae. Dscam is clearly an Ig superfamily with Ig-like external domain receptor protein which appears to be involved in neuron self-recognition and axon guidance, probably via isoform-specific binding. In Drosophila it is expressed in hemolymph and brain. Unlike mammalian nervous system, in drosophila brain glia cells are few in number. Thus, glias will not be able to contribute much complexity to insect neuron identity or plasticity.

Dscam hemolymph expression has been measured at 18,000 isoforms and has been used to argue that Dscam is involved in drosophila immunity. Figure 7.1 shows a schematic of DSCAM from various insect species that generate gene diversity by exon shuffling. Dscam mutants can indeed affect bacterial phagocytosis. However, such mutants have no known antiviral affect. Curiously, social insects, like honey bees, have significantly reduced the number of genes involved in innate immunity (such as the toll receptors), yet, as described below, they can have very long-lived queens (ant queens up to 28 years, bee queens 12 years) and live in colonies that clearly support various persistent and pathogenic viruses. It is thus clear that insect hemolymph is distinguished from that of vertebrate lymphocytes with respect to antiviral activity and denotes a major difference in animal immunity. Although Dscam undergoes alternative splicing, it bears no relationship of lymphocyte TCR rearrangement via splicing. It has frequently been argued that insects and other invertebrates generally have short life spans, hence do not need adaptive immunity but there are clear exceptions to this statement, such as colony queens, cicadas, 250-year-old giant

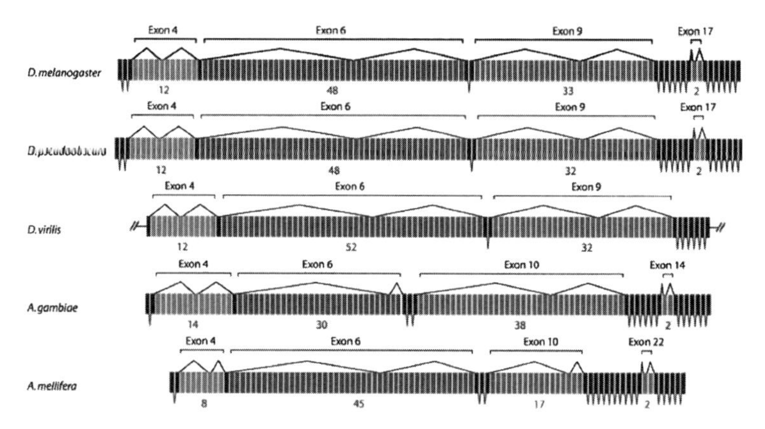

Fig. 7.1 Insect DSCAM gene map (*See* Color Insert, reprinted with permission from: Graveley, Kaur, Gunning, Zipursky et al. (2004), RNA, Vol.10)

hydrothermal clams and others. Recall that sea urchins are also long-lived invertebrate species and lack adaptive immunity. Sea urchins, however, express an array of 222 toll-like receptors (not Dscam) in their hemolymph, also with no known antiviral consequence. Sea urchins, however, have conserved RNAi systems (like *C. elegans*) which are clearly an antiviral system, and sea urchins, like *C. elegans*, do not support any known autonomous viruses. Given the established ability of viruses to affect insect biology, Dscam role in drosophila hemolymph immunity is therefore unclear. It appears more likely to me that its role is in the recognition of cell surfaces or cellular identity, especially neuron identity. That different species of drosophila show surprisingly similar alternative Dscam exons, suggests this system of neuron variation is highly selected. How then does neuron identity (plasticity) relate to odor-mediated behavioral identity?

Links of Sensory OR Detection to Neurons and Learning and Behavior

In Drosophila, courtship conditioning is an associative learning process. Male behavior is modified by experience with unreceptive previously mated females. Female cuticle hydrocarbons (9-pentacosene) are clearly a conditioning stimulus for this. That this is learned state is shown by newly emerged males that will court other males, but modify their subsequent courtship to only females. Neuron plasticity and communication with olfactory lobes are likely involved. Mutations in Drosophila learning have been studied. Dunce is one well-studied learning mutant that competes poorly with other males for female mating. Dunce males do not appear to learn or establish persistent memories. This is a recessive mutant in which males are sterile. Dunce is mutation in cyclic A 3′5′ phosphodiesterase (increase in cAMP) which resulted from insertional inactivation by retrotransposon. An inserted sequence of 8.2 kb can be recovered at 5′ end of the gene, corresponding to a roo-copia-like element. It is interesting that elevated cAMP levels in dunce mutants induce copia-like RNA transcription as shown by differential display. Clearly, this learning mutant has also disturbed ERV expression and likely resulted from ERV action.

Drosophila females can also learn to associate odors with respect to the substrate choice site for oviposition. Quinine has been used as a chemical cue for this training. This association requires both learning and memory. It is also possible to habituate drosophila with ethanol not to startle. Drosophila larvae have simpler brains than adults, yet they too can also learn to associate odors with avoidance. Thus, Drosophila larvae can be trained with electroshock to avoid specific odors. Surprisingly, such memory is retained after metamorphosis. Larvae of other species, such as manduca sexta larvae, can also be trained to associate odor and such training also persists into the adult phase. Given the massive tissue remodeling during metamorphisis, memory persistence is not expected. The Drosophila nervous system also undergoes profound changes during metamorphosis, so the retention of odor-based memory here is also striking. Almost all adult motor neurons derive from larval motor neurons but are respecified for adult function. Many adult interneurons also derive from larval interneurons. However, those neurons related to complex adult sense organs derive from larval cells persisting as neuroblasts. Larval sensory neurons mostly die during metamorphosis and adult sensory neurons arise from the imaginal discs. How then was odor memory retained? It is known for Drosophila that behavioral feedback can delay the programmed death of some neurons during metamorphosis, so perhaps these specific trained sensory neurons survive. It is curious then that few larval cells survive metamorphosis (most undergo apoptosis), yet memory harboring neurons seem to survive. If such memory is essential for group identity, we might understand why its retention is so exceptional.

Drosophila, Odor Detection and Sex Isolation

Odor attraction and repulsion both seem to be used for mating in drosophila and can lead to sexual isolation. *Drosophila elegans* does not use body coloration as a sexual cue, yet black and brown variants do occur and appear to be undergoing sexual segregation. This segregation is due to large difference in the cutile hydrocarbons produced. Why this production is linked to color production is not clear, although oxidation of colored products may be involved. It is very curious to note that there is a nearly universal production of black-brown pigment in arthropods that can accompany cellular innate immunity. Such pigmentation is not seen in vertebrates. In the mealworm beetle, cuticular color (tan to black) has been linked to pathogen resistance and is also associated with higher phenoloxidase activity in black beetles.

Virgin drosophila females are known to express a sex-peptide gene at high levels during ovulation, but are unresponsive to males. Male mutants defective in learning (dunce) or olfaction indicate that virgin females can repel males via volatile pheromones, indicating both an attractive and a repulsive capacity. In drosophila, a single population of sensory neurons appears to mediate avoidance behavior. Male olfactory-dependent courtship behavior is known to be dependent on Fru(M) expression, a male-specific fruitless product (transcription factor) that controls neural substrates. It appears that expression of Fru(M) is necessary and sufficient to hardwire male courtship behavior (as suggested by genetic studies). Fru(M) expression is seen in only 2% of male neurons. It is not involved in female courtship so it is clearly sexually dimorphic. Drosophila also uses sound during courtship. A mutant, cacophony, encodes a calcium channel subunit and causes courtship song defects or abnormal response to visual stimuli. Such males generate longer cycles of song pulses. Thus, although many details of odor-dependent sexual behavior in drosophila are known other aspects are less clear.

Moths (Other Bugs)

Moths in particular seem to use a large diversity of pheromone structures for long distance signaling. These are mostly composed of C10-C18 unsaturated acyclic aliphatic compounds with an oxygenated functional group (reminiscent of early oxidation products) and are made by females to attract males. Another major class of sex pheromones are polyene hydrocarbons and epoxides (characterized by C17-C23 straight chains, double bonds and epoxides). This attraction is very sensitive and can occur over large distances. The receptor is male-specific G-protein-coupled olfactory receptor. In *B. mori*, this receptor gene is on Z sex chromosome. The synthetic genes involved are associated with much genetic change in moth species in which new genes evolve and old genes disappear, by unclear mechanisms. It seems clear that these genes provide the basis of much molecular identity in moths. Learning is clearly involved as moths can learn and imprint on specific host and habitat via molecular odor cues.

Wasps Social Mechanisms as Basal to Social Insects

Parasitoid wasps are the most numerous and successful of all insect and terrestrial animal species. Most such wasps are not social but many wasp species are and they are also the ancestors to many other social insect species (ants, bees). Wasps like most insects depend heavily on olfaction for sex and group identity. Thus, the Vespidae show a wide range of social complexity, from solitary to eusocial colonies. In eusocial colonies, reproductive casts are determined early in larval development at a pre-imaginal period. Nutritional variables, juvenile hormone and pheromones appear to be involved in programming development. It is unknown if memory or group imprinting is also involved in this early period. In adult wasps, odor detection is clearly important as can be seen by stereotypical antennae cleaning behavior which employs specially adapted forelimbs. Parasitoid wasps oviposit their fertilized eggs into other insect host species, where the larval wasp grows and develops. The parasitoid host are generally instar larvae of other insect species, just prior to metamorphosis. When attacked by wasp, these host larvae will often attempt to defend themselves against the wasp, and clearly seem angry to a causal observer. What appears to be anger can also be seen in many other insect species and can be induced by alarm pheromones in social species eliciting a defensive colony aggression. This behavior clearly suggests the existence of some type of emotional memory that is also used for group aggression. Parasitoid wasps typically use odor cues to find their host. This odor-based host relationship can also be modified by other species, such as plants, the prey of the larvae or aphids, all of which can give off the same olfactory cues that allow the wasp to find its host larvae. Some parasitoids also use the sex pheromones of their prey in order to find them. Some parasitoid wasps attempt to fertilize specific orchards, which produce a single wasp sex pheromone. Such orchards often rely exclusively on this specific wasp for pollination. Clearly the world of the wasp is highly olfactory, but since olfactory mechanism are best understood in drosophila, the relevant neurological mechanisms will not be further developed. Some examples of visual-based group identity are also known in wasps. For example, *Polistes fuscatus* wasps have highly variable facial and abdominal markings that are used for individual recognition within their society. This visual recognition is associated with a flexible nesting strategy in this particular species. The parasitoid wasp also provides much information regarding how virus–host interactions may affect group recognition, as presented below.

Polydnaviruses and Hymenopteran Parasitoid Wasp

Ichneumonidae and Braconidae are both orders of parasitoid wasp that are also known to harbor both endogenous and persistent viruses. These parasitoids are endoparasites that use mostly larval flies and moths as their host. These wasp orders also show low incidence of acute viral disease, yet viral persistence (especially in reproductive tissue) by various virus types is common. These viruses include rhabdovirus, poxvirus, baculovirus, ascovirus, polydnavirus (PDV) and some unidentified viral forms. One example is a baculo-like virus (HZ1V) which is a persistent ovarian virus. HZ1V encodes a JH inactivating esterase expressed especially in ovary and venom gland. Persistent viruses like HZ1V can sometimes be found in all individuals in a population, and viruses are frequently injected during oviposition and along with endogenous (PDV) virus particles. Polydnaviruses (PDV) are endogenous small circular DNA viruses found in the genomes of their wasp host, produced only in female reproductive tissue. PDV replication is hormonally triggered by ecdysone during egg production. These PDV endogenous viruses can clearly immunosuppress the host cellular immune response, modify host neurochemistry and alter developmental programs. PDVs were first observed to alter host immunity when viral material was removed from the eggs, which resulted in host hemocyte encapsulation and rejection of the wasp egg. The presence of PDV appears to be genus specific and is absent in some specific genus, such as *Alysla*, *Bracon* and *Aleiodes*. PDVs are found in two distinct classes that entered their respective host genomes early in their evolution; at the divergence of Bracoviruses and Ichnoviruses orders. They are thus species specific. Parasitoid wasps are also known to harbor other persisting viruses. For example, many can also harbor a reovirus (dsRNA virus) which also appears to suppress host larval defense and also contribute to parsitoid egg development. Five such reoviruses have been reported and all are non-pathogenic for wasp. One such reovirus is DpRV-1 (*Diadromus pulchellus*), which is common in field isolates and always found associated with wasp ascovirus (DpAV4a, a large DNA virus of insects) which is also essential for host immune suppression. This DNA virus is maintained as a nuclear DNA episome in all cells of the female wasp, with virions only produced in the oviducts (similar to the endogenous PDV). It appears that the reovirus retards ascovirus replication in the host wasp, clearly suggesting virus–virus interactions, but of unknown biological consequences. Thus, we see here a three-virus consortia (PDV, DpRV-1, DpAV4a) associated with host reproduction and larval development and immuno-suppression; hence all are highly linked to host sexual and group identity. Other specific populations of wasp harbor distinct reovirus (DpRV-2 in which DpRV-1 and DpAV-4 are absent), suggesting a viral role in population identity. Besides the viral relationships noted above, parasitoid wasps also function as primary vectors for granulosis virus (GV), an insect baculovirus. GVs are numerous, mostly restricted to lepidopterous insects and are phylogenetically congruent with their host. This virus host can be highly efficient in natural settings. For example, in some California field populations, greater than 90% of all examined larval host species had persistent GV infection due to parasitoid action. Ascoviruses can also be vectored by hymenoptera parasitoid. Phylogenetically, ascoviruses cluster with iridoviruses and phycodnaviruses, but show no homology to PDVs. Yet ascovirus biology is rather similar to PDV (DNA, persistence, egg transmitted). Ascovirus is sex transmitted, such as SfAV1 and HvAV3 which use lepidopteran host parasitoid wasp and transmit at oviposition. Some ascoviruses, such as DpAV4 noted above, are very poorly infectious in vitro, but highly efficient in natural populations. This natural efficiency appears to be due to vertical transmission in *D. pulchellus*. Only females are infected, and here DpAV4 is not integrated but it is an immunosuppressant that enhances wasp development and also prevents larval host development, very much like PDV. Thus, the silent ovary associated biology of DpAV4 clearly resembles PCV. Recent sequencing of ascoviral genome indicates that they have a most interesting set of apoptosis modifying enzymes (caspase, cathepsin B, several kinases) as well as some lipid metabolizing genes associated with viral vesicle formation. Why is it that parasitoid ovaries seem so prone to a wide variety of silent viruses and what consequence has it had for wasp evolution and group identity? We start to see common links between endogenous viruses and ovary biology in numerous species.

Genomic Adaptations for Virus?

The presence of PDV in the genomes of numerous parasitoid wasp species and the even more frequent colonization by other persisting viruses suggest that parasitoid wasps have special relationships with these viruses and must also have genomic adaptations that are consistent with these prevalent genetic parasites. Wasp genomes do indeed show interesting patterns of genetic parasites. Genomes of Diadromus and Eupelmus species are unusual for having a substantial amount of satellite DNAs that consist mostly of single family repeats and make up 5–7% of the genome. In other parasitoid species, satellites can be up to 20% of genome, well in excess of other animal genomes (except bivalves as presented in the last chapter). The presence of this repeated DNA seems to be under positive selection, since evolutionary conservation of non-random patterns of variability is observed as well as the absence of sequence divergence within a species. This observation is clearly not consistent with the prevailing view that satellites are hypervariable sequences due to DNA polymerase strand slippage with no phenotypical or selective consequence. However, these conservation patterns do suggest that the formation of hairpin structures via dyad symmetry of the inverted repeats may be needed for satellite sequence maintenance. Given the importance of hairpin RNA in the function of the RNAi antiviral system of *C. elegans* presented earlier, the large presence of potential hairpin RNA in the genomes of parasitoids is very interesting with regard to persistent virus maintenance. Recall, no viruses were known for *C. elegans* (or any nematode), but viral persistence in parasitoid wasps is common. It is also interesting to note interspaced short palindromic repeats (CRISPR) have recently been shown to be major mediators of virus immunity in bacteria and archaea, but these elements themselves appear to have been virus derived. One implication is that parasitoid wasp genomes may be uniquely adapted to support species-specific sex-associated persistent viruses.

Given the importance of parasitoid wasp in the evolution of social insects, it is worth noting a few other relevant issues. One is that the oviposition system of parasitoids evolved to become the sting apparatus of most social insects. The poison gland of a stinger has also evolved from its association with ovipositor-related glands. In the context of the evolution of placentas in mammals with corresponding live birth and immune modulation that will be described below, it is worth mentioning that a parasitoid wasp faces similar biological conundrum when it injects its egg into a host larvae. The egg must establish a food supply, alter host developmental programming and evade immune rejection to overcome the host identity systems. The concentrated polydnavirus injected along with the egg directly contributes to all these functions. However, wasp teratocytes (large polyploidy hemolymph wasp cells) are also injected along with the egg. These cells also protect the egg from host immune response and appear to aid in feeding the egg. The similarity of this biological function to that of a mammalian placenta is clear.

Social Insects, Odor Identity and Virus

Some parasitoid wasps are social. Mostly, social insects use cuticular hydrocarbons to provide social identity and differentiate reproductive status within a colony. These odorants are produced from ovaries of reproductive females. A simple social structure is that of the polistine wasp, which can be in either associative (polygynous) or solitary (monogynous) colonies. The establishment of the colonies appears to be linked to level of cuticle hydrocarbon production. Social environment of insects strongly affects the reproductive physiology and odorant production. In some wasps (*Polybia sericea*), colony migration is also controlled by odorant production and swarming to new nest is mediated by trail pheromones (volatile aldehyde compounds), produced by Dufoure gland. Clearly, social control is olfactory here. Some parasitoid wasp, virgin females deposit sex pheromones that provide trail scents to males. The polyembryonic parasitoid wasp (*Copidosoma floridanum*), which is parasitic of moth larvae, will clonally produce offspring of two distinct castes: a reproductive wasp and non-reproductive soldier wasp for defense. The soldier wasp can differentiate kin from non-kin wasp, by unknown mechanisms. However, it appears that the very immunosuppressive system that protects the wasp larvae from host immune response is also involved in and provides cues for soldiers to identify kin. Although odor or pheromone mechanisms seem likely to apply, they have not been characterized in this case. Given, however, the prevalent viral role is such immune suppression, a viral role in kin recognition also seems plausible. In fact, some parasitoid viruses are known to be able to modify behavioral and olfactory cues in parasitized larvae. The parasitoid *Leptopilina boulardi* uses fruitfly young as its host. This drosophila parasitoid normally avoids superparasitism by an odor-based avoidance of host larvae that have already been injected with a wasp egg. However, this avoidance trait is highly variable in field populations and is not determined by nuclear genes. The superparasitism trait was shown to be due to a horizontally transmitted infectious agent and viral particles were seen in affected female (not male) reproductive tissue. It seems likely this virus is modifying wasp behavior, allowing superparasitism that favors virus transmission to other wasp eggs. Thus, parasitoid group behavior and kin identity can clearly be modified by virus. Endogenous parasitoid viruses can also clearly modify host larvae behavior. There are two major groups of hormones regulating insect growth, development and reproduction: ecdysteroids and juvenile hormone (JH). PDV can induce very high levels of ecdysiotropic activity in brains of parasitized larvae which has large effects on larval neurochemistry, cellular immunity and behavior. Also, pea aphids parsitized by braconavirus PDVs show developmental arrest with total ovary disruption. In addition, transient paralysis of the aphid can occur. Parasitoid will induce permanent termination of host feeding just prior to wasp emergence. It is clear that parasitoids very much manipulate these two hormone systems. Clearly, these infectious agents have the capacity to completely alter host identity hormones. As dopamine is one of the primary neurotransmitters associated with behavioral control, it would be interesting to know if this transmitter is also altered, but this issue is poorly studied. It is known, however, that parasitoids of cockroaches contain dopamine that affects grooming behavior and immune systems.

Flower/Parasitoid Olfactory Overlap

In terms of hymenoptera, a most basic use of the odor cues relates to sexual control and identification and mating. It is accepted that insects and flowering plants evolved in conjunction with one another. Orchards are representative of early flowering plants and water lilies are perhaps even earlier to evolve. Some orchards are known to emit hymenoptera sex pheromones that induce copulation of male wasp with the flower. These highly specific hymenopteran sex pheromones which induce male copulation with flower structure are extremely species specific for plant and wasp, raising questions regarding its evolutionary origins. Female oviposition can also be controlled by odorants from plants. The fig wasp in particular is interesting in this regard. Wasp use the fig fruit as an oviposition site of wasp egg development. Fig gives off a sex pheromone that attracts specific wasp to pollinate fruit. After pollination, the pheromone is no longer released by the fig. Female wasps will lay eggs in fig flowers and embryos develop within the seeds as galls in the plant. The wasp young are later liberated by males that find them via olfaction, excavate a hole, inseminate then release the nubile female inside. Males can also recognize and attack other males around the galls. Clearly, in this example, the odor-based sexual and social identity of male and female wasp and fig fruit and fig flower are all interlinked. How did such complex interlinking of two species with vastly different ancestral systems of identity evolve such reproductive synchrony? This requires coordination of the synthesis of various odorants, receptor expression, behavior modification and developmental control. Such a coordinated situation seems overly complex. It is usually explained by suggesting that the plants evolved insect pheromone expression by Darwinian processes to control their fertilization by wasps. However, in a prior book I suggested such insect–plant relationships more likely originated by egg-laying insects that used viruses to aid directly in the colonization and evolution of flowers in plants. The simplest flower resembles a modified development of a plant gall. Galls can also be the product of the endoparasitic wasp. As noted above, wasps can use endogenous and exogenous viruses made in female ovaries to control host development and immunity (identity). Drosophila, for example, can produce large quantities of Gypsy virus (a chromovirus) in ovarian tissues. The evolution of flowering plants is also associated with massive colonization of higher plant genome with Gypsy-related genetic parasites. This dramatic retroposon expansion is specific to flowering plants. Also, the endogenous Gypsy retroviruses, as seen in Drosophila genome, are phylogenetically basal to the retroposons of flowering plants, but are absent from the genomes of fern plants. In this scenario, the genetic parasite (ovarian wasp viruses) would have also colonized plant genomes and provided insect pheromone genes (synthetic and receptors), leading to the evolution of flowers. Regardless of how plant came to evolve flowers and express insect pheromones, clearly insect sexual behavior can be manipulated externally by other species.

Overall, the hymenoptera (ants, bees, wasps) have diverse sex pheromones. However, many species appear able to discriminate among mates and are even able to identify individual kin or colony mates. For this, insects do not depend on innate recognition. They can clearly learn to identify odors. Honey bees can also be trained to associate an odor with food. For example, orchard bees are not attracted to nectar (food) of the flower, but are instead attracted to orchard-specific odor molecules that they have learned. Interesting that bees can also have a mutation in the dunce locus, but here dunce phenotype (learning deficit) expression depends on age and social status of the bee. Much of social insect group identification is also learned via odors. Sterile worker bees can be considered as part of a super-organism that will protect the group against outsiders. Workers learned to recognize kin via cuticular hydrocarbons. However, the social recognition of a colony can also be disrupted by parasitoid wasps that are social parasites and can mimic or adapt to colony-specific olfactory identity. Such social parasites can also disrupt established identity and induce colony infighting.

Hymenoptera Lack Sex Chromosomes

Sex determination in hymenoptera is via haploid/diploid system, in which males develop from unfertilized haploids and females from fertilized diploid. Since reproductive caste are restricted in social wasps and other insects, clearly there is also an environmental overlay that provides sensory cues that will generate non-reproductive caste. Reproductive status is not simply determined by ploidy. How this basic group ID is attained is not well understood. Clearly, sensory input seems essential. However, the sex-determining genes that affect bee brain function and behavior are distinctly different genes for sex determination in Drosophila. Thus, it seems clear that these two social and non-social insects have distinct gene set for sex determination. The evolution of this difference is not yet understood. Nor is it known whether virus persistence or sex-based transmission has had any role in insect sex evolution or determination, although these orders clearly differ in regard to the types of viruses they support or endogenous viruses made by reproductive tissue.

The DNA sequence of the honey bee genome is now known. As mentioned, one surprising difference with drosophila is the reduced number of genes associated with innate immunity. In terms of pheromones and nerve receptors, bees have 36 brain peptide genes that encode pheromones for a total of about 200 neuropeptides. This is similar to the numbers found in drosophila, so it does not seem that the bee social life style has significantly affected these signaling genes. But nine of these genes are unique bees, so some adaptation must have occurred. Apis, like drosophila, also lacks BDNF and its signaling machinery, so this nerve factor is also the same. This factor is of significant interest with regard to mammalian brain function since it seems to be involved in mammalian memory. Although some of the neuronal mechanisms of bee memory seem to be distinct from those of mammals, they do not appear to differ much from that of drosophila. However, a significant difference between the bee and fly genome is seen with regard to OR genes. It seems likely that OR sensory neuronal link to brain function is crucial for bee social function.

Accordingly, bees show a remarkable expansion of the OR receptors in their genomes relative to drosophila (170 versus 62). This OR expansion is also often clustered; 142 of these ORs are found in 14 tandem arrays. These OR proteins are reflected by neuronal OR expression in the bee antennae. Thus, a major genetic adaptation from drosophila to social bees is in olfaction, but how olfactory sensory input is linked to the social life style of the honey bee is not well understood.

Genetic Parasites of Social Species

The social living style of bees is highly crowded, genetically homogeneous and would seem to provide a most favorable environment for virus and parasite transmission. Viruses are indeed well known and prevalent in bee colonies (such as *Apis mellifera*), but they are mostly silent, persisting infections. Due to major commercial interest of using bees as crop pollinators and for honey production, viral-mediated crashes in bee populations have occurred and have received considerable attention. Thus, the virology of bees provides the best virus–host system of any social insect. Curiously, the viruses so far found in bee populations are distinct from those discussed above for parasitoid wasps. By far, the most prevalent bee viruses resemble picornaviruses. No ascoviruses or polydnaviruses (prevalent in parasitoid wasps) have yet been reported for bees. Also, these picornaviruses are mostly present as persistent, non-pathogenic infections, often in reproductive and nervous tissue. Curiously, the bee genome has significantly fewer proteins implicated in insect innate immune function relative to fruitfly (i.e., JAK/STAT-like genes are reduced by 2/3). Why is there such a sharp contrast between parasitoid wasps and honey bees in virus–host interactions? Bee genomes are also noteworthy in their relative lack of most major transposon families. Yet some retroposons are present but these are mainly from the mariner family (not Gypsy) of retroviral-like elements. These are seen in six families (AmMar), with copy numbers ranging from 70 to 360 mostly of about 900–1300 nucleotides long. Of these, 4 AmMar1 appears phylogenetically the youngest and have many nearly intact copies. In contrast, AmMar 6 appears much older but most copies are degraded. Chromovirus (Gypsy/ty3) are absent and copia-like, and BEL12 elements also much reduced (with some degenerate remnants). DIRS elements are also highly degraded. Overall, the bee genome appears to represent colonization by waves of AmMar family retroposons that displaced other (Gypsy, BEL12, DIRS) elements present in ancestral species. The AmMar family itself also seems to have undergone several waves of self-similar colonization and displacement. Bee genome also resembles more vertebrate genomes relative to drosophila genomes in their RNAi and DNA methylation genes. The bee genome is evolving more slowly relative to the fly genome. As mentioned, less gustatory receptors but more odor receptors are present relative to the fruitfly. How can such changes account for the development of a social species with non-reproductive altruistic worker caste? How was this alternate bee identity specified? One possible clue to this issue is that bees express a caste-specific and unique set of miRNAs. The implication is that miRNA expression is somehow involved in regulating the caste and social identity of bees. A possible relationship between miRNA expression and hairpin transcripts from retroposon elements, as seen in *C. elegans*, has not been investigated. One other point is worth noting. Queen bees typically have 10 times the life span of other bees and in some cases can have 100 times the life span (living up to 13 years). This is longer than the life span of most mammals in natural populations, yet bees lack the adaptive immunity that has often been invoked to explain why vertebrates evolved adaptive immunity.

Many RNA viruses of bees are known and six of these have been most commonly studied. Bee viruses were first seen in the 1960 s with an outbreak of sacbrood paralysis in Britain. This virus is now known to be prevalent in European honey bee larvae. These viruses are mostly picorna-like viruses that persist as inapparent infections and most colonies support a mixture of viruses with no disease. However, these otherwise persistent viruses can sometimes multiply with high mortality, destroying a colony. The reasons for the emergence of pathogenic virus replication in some colonies are not well understood. In some cases, it seems that other parasites, such as mites (vorra destructor), are perturbing persistent states and mediating the induction of virus pathology. These viruses may also depend on parasites of bee for their transmission as well as sometimes for pathology. For example, APV commonly persists as an inapparent infection of adult bees, but following mite infestation, mites induce APV multiplication and bee disease. Mites themselves also seem to support a mix of viruses but mite virology is not understood. These viruses are common throughout the world and often endemic in healthy colonies. However, both protective and destructive relationships are known. For example, queens infected with CPVA appear to protect the hive against other viruses and also have protective reproductive effects. CPVA (chronic paralysis associated virus) is a satellite virus that persists in female reproductive caste (queens) and depends on CPV for its replication, modifying CPV pathology. These viruses often infect all caste members of a bee colony, but persist in the colony via infection of the queen and are also frequently egg transmitted. This type of protective and destructive virus–host relationship has the hallmarks of a virus-mediated addiction or identity system. The genetic and population structure of these viruses, however, has not been evaluated, but it appears that the viral sequence is stable, as expected for a persistence life strategy. Some viruses, such as BQCV (black queen cell virus), which although common in field colonies, can specifically kill queen larvae under some conditions. The natural biology of such mixed virus infections in bee colonies is not understood. Any potential affects on bee fitness are also unclear, although likely. However, these viruses can show rather specific effects on host. For example, the CPV virion is highly produced in the head of the infected bee but it is infection of the abdominal ganglion that leads to paralysis. Such relationships would also seem able to differentiate different hives that harbored distinct viral infections and could provide a form of colony selection. One such possible example was seen in Costa Rica in which a colony of Africanized bees suffered significant virus pathology in the field, presumably via a virus from a native colony. These viruses also seem capable of modifying group behavior. For example, in one case KV was found in the brains of aggressive worker honey bees and appears to have induced high aggression to wasps. How or why a virus would modify aggression was not clear, since the neurological basis of aggression in general is poorly understood. Overall, we see a striking shift in the exogenous and genomic virus biology of this social insect the honey bee.

Other social insects are much less studied with regard to virus–host interactions, since commercial interest in such species is low. Thus, we have only a few examples from which to attempt any generalizations. Viruses of termites are known, such as termite paralysis virus, which appears to be a picornavirus, but the study of their natural biology is lacking. Ants are also poorly studied with regard to their virus biology. It is known that fire ants can be persistently infected with SINV-1, a picorna-like virus that not only infects all caste members but also transmits and persists in a colony via queen infection and egg transmission. This relationship very much resembles that described above for honey bees. Curiously, however, although no viral symptoms are seen in field isolates, broods from infected queens will all die within 3 months in laboratory growth conditions. This implies that some natural but unknown conditions are preventing virus pathology (reminiscent of the CPV–CPVA relationship above). Normally, for persistent viral infections, a violated addiction strategy would be expected to produce the type of uniform pathology seen in these lab broods. This seems similar to mouse lab colonies dying when grown in the presence of mouse hepatitis virus, a very common inapparent field virus (see below). Recall that viral addiction strategies, when they exist, can also differentiate population identities. Some ant queens are known to have very long life spans, up to 28 years. It would be interesting to know if virus persistence is prevalent in such long-lived ant colonies.

Learning of Group Behavior in Social Insects

Insects, like the vertebrates, also learn from sensory input as a principle mechanism to establish social group and sexual identity. Insect learning, however, is most remarkable in its sophistication, resolution and pattern identification considering that insect brains generally have only about one million neurons and that the neuronal interconnectedness of these cells is much less compared to vertebrate brains. Genetically, bees have reduced genomes relative to non-social insects and seem only to have overrepresented OR genes from which to create their social life style. Yet social insects, like bees, learn to associate flower color, shape, scent and location with a food reward and communicate source location with dance language to other bees. I would suggest that the most basal of these social learning functions relates to olfaction in group and sex identity. In social insects, early sensory and gustatory input can determine caste which results in a complete subjugation of the individual producing individual insects that have lost all functional reproductive potential, except for queens and a few males. In cellular (tissue) based group identification, we can recall that most tissues subjugate the fate of individual cells using terminal differentiation and cellular self-destruction (apoptosis) which were often a part of an addiction strategy that controlled cell fate. In a social context, a similar concept of individual subjugation continues to be applicable but must be mediated by learned CNS functions. The result is an insect that has not only undergone a caste-specific cellular differentiation program but also acquired certain behavioral states. For example, soldier ants will behave aggressively toward ants that are not colony mates. This aggressive and harmful behavior is complemented by protective behaviors that promote feeding and care by worker bees of larvae young. Clearly, an insect colony has both protective and harmful general behaviors that apply to colony social identity. These characteristics therefore adhere to the concept of a behavioral-based addiction module as the basis of social identity. Since the resulting behavioral states will have inherently but general harmful or protective characteristics, we can also consider if these acquired behaviors are equivalent to 'emotional' states or memory. When social insects learn such 'general states' of behavior that respond to specific sensory cues, does this correspond to an emotional psychology? It seems clear that terms like alarm, aggression, anger and defeat appear to characterize defensive colony behaviors. How then does sensory input to the brain and group imprinting or learning set such generalized behaviors?

Aggression is a key feature of social insect behavior, colony identity and defense. Sterile workers of a colony will protect the nest and attack invaders via the recognition of cuticle hydrocarbons. Ants are also known to use cuticular hydrocarbon blends to differentiate nest-mate recognition and have a sensory sensillium in their antennae specifically used for this purpose. Such detection is colony specific, and in some cases, such as with argentine ants, the colony-specific hydrocarbons are affected by the diet. Thus, a communal colony diet (via insect prey) can define current colony membership and can also distinguish prior nest mates that have been fed distinct diets as these nest mates will be met with aggression when reintroduced into the colony. The relative rank of queens in multiqueen colony can also result from distinct hydrocarbon patterns, made by the queen's Dufour gland. Queen-derived hydrocarbons also mark queens egg and protect them against worker destruction. When a queen dies, unrelated workers can move into the nest and lay their own eggs. Normally worker ovaries are reproductively inactive (a type of reverse Bruce effect) due to queen-produced odorants. If workers do lay eggs, they lack odor protection and are eaten by fellow workers. Different caste have distinct odor patterns. Thus, aggression, odor and social status are tightly linked in most insect societies.

Bug Opioids

At the beginning of this chapter, the role of neuropeptides such as opioids in group behavior was emphasized. For example, aggression in fish could be modified by opioids. The opioid receptors are clearly also present in insects and opioid peptides, closely resembling those of vertebrates, have also been identified. But such studies of insect opioid are scattered and do not provide a coherent picture. Yet there is some reason to suspect they have a role in female reproductive systems. Young ovarian follicles of locusta show immunoreactivity to methionine(met)-enkephalin, consistent with a role in reproductive physiology. Also, melanotrophins are observed in locusta CNS tissue. Endorphins are also present in sperm and young ovarian cells of locusta. Here, Met-enkphalin is seen in oocytes and trophocytes, which may be involved in reproductive physiology. In American cockroach, opiates induce increased feeding which naloxone decreases. Here, Kappa opiate agonist increases the ingestive feeding response in cockroaches, whereas naloxone blocked this augmentation. In addition, it seems clear that opioids affect various other types of insect behavior. Morphine increased cockroach locomotor activity. Thus, it seems possible that the opioid system may control insect feeding behavior analogous to that proposed for vertebrates. Beetles' nervous tissue also shows reactivity to beta endorphins and ovine prolactin. In terms of social insects, honey bees show a morphine-dependent inhibition of stinging response that was very sensitive to blockage by naloxone. Although these limited observations do not allow many firm conclusions, clearly opioid circuits are present in insects and could contribute to addiction-based behavior resulting from sensory input. Insects also seem to have some links between opioids and immune systems.

Audio-Based Insect Groups

The above section has emphasized mainly the olfactory systems for insect social and sexual group identity. Although not nearly as widespread as olfaction, songs produced by various insect species, such as cicadas, crickets and stridulation in grasshoppers, are also well established as a mode of group identification. Such songs, when used, are species specific and JH is often involved as a modulator of nerve cells associated with song-based behavior. Cicada singing is stereotyped in response to playback of conspecific songs and populations can show chorus (group) singing. However, female oviposition has remained primarily an olfactory-based behavior as it is responsive to chemical cues. Clearly, these insects have a set pattern recognition, but it is not generally adaptive or individual specific. Interesting that some parasitoids also use these audio cues to find their cricket host. Furthermore, some crickets will alter their singing patterns when parasitoid wasps are in the vicinity. Thus, there seems to be some degree of audio learning going on in both parasite and host. Clearly this parasite–host situation could provide a strong selection on nature of audio-based identity. How do wasp and singing host acquire such matched audio pattern recognition? Clearly, audio pattern recognition must also link to behavior. Generally, male crickets sing to attract mates and the singing often rhythmic. Audio pattern requires brain-specific capacity for such rhythmic patterns. Sexual identification and audio production can also occur in flies. With *Drosophila virilis*, the males produce courtship song of unpaused pulse trains. Other species (Montana phylad) produced paused songs. Montana females crossed with virilis males show that pause length was largely X chromosome determined, although crosses between other species did not observe an X chromosome effect. Audio and olfactory recognition can also be linked in some insects. In some true bug species, males will emit sex pheromones after detecting female vibrations on the same plant that promotes completion of mating behavior. In terms of aggressive behavior, olfactory recognition may still be required in audio insect species. For example, male crickets, which clearly use audio signaling for mate identification, will fail to display male–male aggression of their antenna lobes that have been removed. Thus, olfactory cues remain needed for full mating behavior. Interesting that this loss of antennae also results in lowered serotonin levels (5-HT), likely associated with decreased aggression. Little is known about any possible virus role in these audio insect species. Crickets clearly are prone to virus infection by viruses similar to those of bees (cricket paralysis, producing CNS disease), but the natural biology of virus–host relationship is not known.

Visual Systems, Sexual Isolation and Body Colors

Insects, like shrimp, have remarkable visual capacities regarding the spectra of light they can detect. Unlike terrestrial vertebrates, insects mostly have trichromic vision that can include UV, blue and green light detection and resolution. Most hymenopteran species also have trichromic color vision. It is interesting that many insects have darkly colored eyes with distinct patterns and colors, especially metallic colors, as seen in flies for example, suggesting some role in group recognition. Butterflies are known to use ultraviolet reflections for mate recognition, as do dragon flies and some other insects. This tri-color capacity in pterygote insects is due to three sets of 7 TM receptors. Curiously, these three receptor genes do not have the same phylogenetic pattern of evolution, thus do not seem to have been the products of adaptation from a common ancestor gene. It thus seems color vision was present in common ancestor to all insects. Besides color, Drosophila can also visually discriminate and remember landmarks. Such discrimination is due to five types of pattern discrimination that have been characterized: size, color, elevation, vertical and contour layouts. In terms of visual receptors, TRPC in drosophila is expressed in photoreceptor cells and form heteromultimers. The evolutionary relationship of these insect TRPC genes relative to those of fish and human and mouse vertebrates is shown in Fig. 7.2. Insect genes are clearly basal to those of higher vertebrates. These channel proteins affect Ca^{++} influx and output (from store-operated Ca^{++}, via diacylglycerol). The original TRP mutant (TrpP365) was involved in massive degeneration of neuronal photoreceptors (via Ca^{++} influx), a phenotype that could be rescued by RNAi. Visual and olfactory memory in the fly seem to have distinct memory traces as well as distinct mechanisms. Although absent in drosophila, BDNF in mammals is associated with TRP channels, dendrites and memory formation, especially with regard to VNO sensing and sexual memory in the mouse (described below). Drosophila dunce mutants (two X-linked alleles) do not perform well with regard to visual memory. In some cases, visual group recognition appears to have displaced much of the olfactory mate recognition, such as with the tsetse fly (*Glossina morsitans morsitans*) which shows no olfactory component by male antennae for mate recognition. Here, attraction of male to female is visual (but not color). With this fly, female sex pheromone are still expressed but appear to promote male mating strikes, not attraction. Clearly, visual pattern recognition can serve a social function in insects. Fireflies, like many oceanic invertebrates, not only are attracted to light for mating purposes, but will also produce and emit light in patterns to attract mates. Fireflies can also display social responses to light in the formation of swarms. Such swarming is also sex associated. Both female and male fireflies use bioluminescent signals for sexual communication and can discriminate conspecific light signals with the aid of eye light filters. Some firefly larvae also emit light, presumably for social reasons. Although not insects, glowworms also use chemiluminescence for mate phototaxis at night, a process that involves color light production and detection.

Aphids and Mites/Sex and Virus

Most aphid (ascari) species bare live young and most are females, although able to sexually reproduce. Thus, they have a high tendency to use parthenogenic and clonal reproduction. Aphids are often primary prey of parasitoid wasp that use olfaction to hunt this prey. Overall, however, the ascari have much less diversity in their odor-based behavior relative to insects. Different ascari species tend to use the same or similar molecules for olfactory communication. However, olfaction is still used for clustering and mate finding. Aphids use volatile phenols to regulate this courtship behavior. However, neither aphid oviposition nor kin recognition shows much olfactory specificity. It seems any species-specific detection that may remain may reside mix of volatiles that are detected. Most aphids have large cells that harbor a bacterial symbiont (such

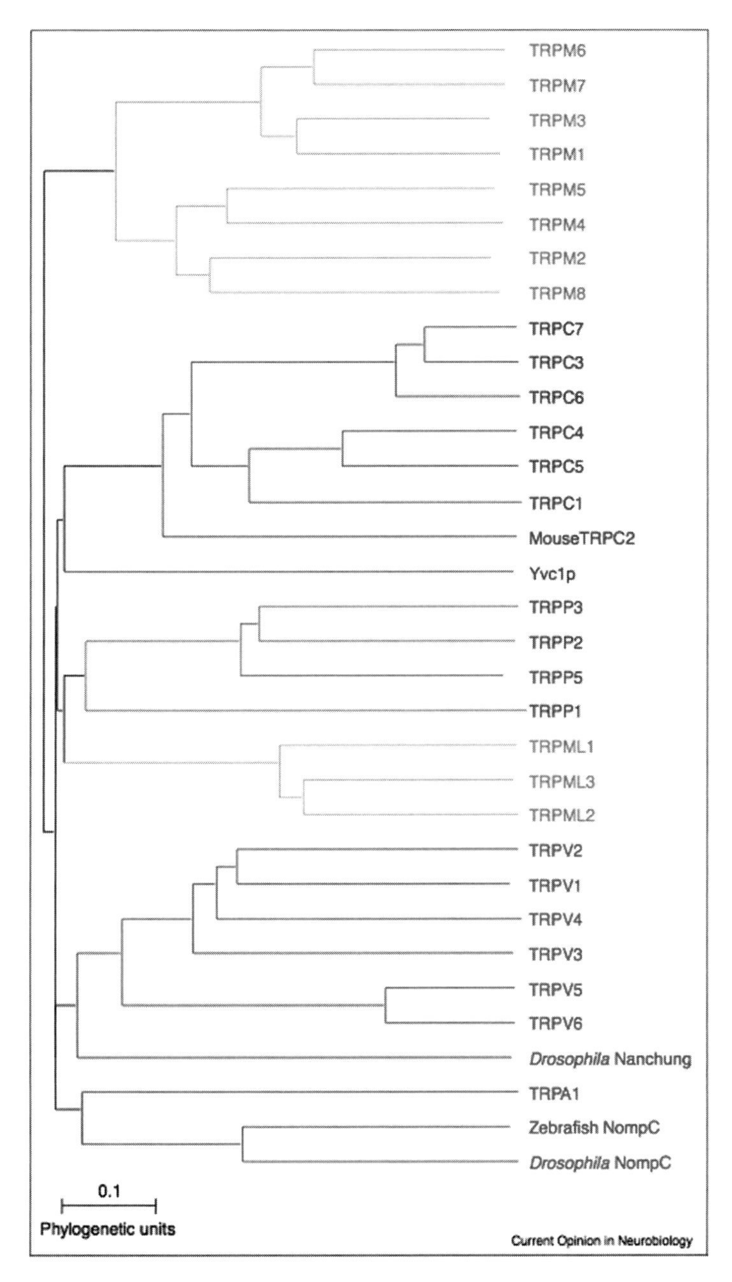

Fig. 7.2 Evolutionary relationships of insect and all TRPC genes (reprinted with permission from: Moran, Xu, Clapham (2004), current opinion in Neurobiology, Vol. 14)

as Buchnera) that can provide the aphid with metabolic functions. These symbiont bacteria have received much attention as they also appear able to sexually isolate aphid populations. Pea aphid also harbors a secondary bacterial endosymbiont. This second bacteria itself is colonized by prophage that is too associated with symbiont function. Given that a primary predator of pea aphid is parasitoid wasp, it is striking to learn that this prophage produces an endotoxin, but this endotoxin is lethal to the larvae of parasitoid wasp and not harmful to the aphid. This relationship (endosymbiont and prophage) is similar to the wolbachia-like rickettsia found in ovaries of hymenoptera, such as *G. morsitans* species. In this parasitoid hymenoptera, parthenogenesis and cytoplasmic incompatibility is associated with symbiotic bacteria. This symbiont is also a maternally transmitted cytoplasmic factor associated with sexual incompatibility. Wolbachia is thus an obligate symbiotic rickettsial bacteria, but this bacteria also harbors and produces bacteriophage particles and also expresses phage genes for lysogenic conversion, modifying bacterial surface expression. In *Culex pipiens* species, cytoplasmic incompatibility also mediated by specific symbiotic Wolbachia, but the Wolbachia is distinguished by a bacterial prophage (orf7) locus. Variation in this phage can correlate with variation in cytoplasmic compatibility. Given the complex virus–host and reproductive biology of parasitoid wasp species noted above, the addition of this symbiont–virus story is truly dizzying! Furthermore, the story does not end here. In the adzuki bean beetle, some of the Wolbachia surface proteins (lysogenic conversion?) appear to have been transferred to the beetle X chromosome via non-LTR retrotransposons, suggesting a role for a retrohyperparasite as well. Clearly, in aphids and hymenoptera, many viral footprints are found in their reproductive biology. Interesting that mites also have a lot to do with the biology of honey bees. As mentioned, in California, for example, a major crash in agricultural honey bee population is due to mite (varroa distructor) infestation.

What did We Learn from Bug and Their Societies?

It is clear that the insect lineage adapted a separate but related evolutionary pathway toward the establishment of systems of group and sex identification. The most dramatic departure from that of vertebrates is seen in the highly successful social insects which evolved essentially clonal societies that include non-reproductive caste. Such societies generally lack individual or family pair bonding seen in vertebrates, but have biochemically coherent colony identity and eusociality, not seen in vertebrates. These social species mostly evolved from parasitoid wasps ancestral species. These insects have maintained olfaction as a principle mechanism for sex and group identification, but mainly use hydrocarbons, not peptide or steroids pheromones for this purpose. There is much evidence of virus involvement with these species (especially persistent infections of reproductive tissues) and the viral relationship is typically peculiar to the specific host. Some insects also use visual and audio signaling for group membership purposes. Insect vision resembles that of their arthropod ancestors (color vision, compound eyes). The CNS mechanisms that control insect group behavior, however, are not well understood. For example, the existence of opioid addiction circuits and their relationship to group behavior, such as aggression, seems likely, but is poorly explored. Clearly, insects learn at least odors but also song and visually in relationship to group membership so in this they resemble the vertebrates. However, it needs to be acknowledged that insect learning is rather inefficient in comparison to the vertebrates. For example, training fruitflies in the laboratory is a difficult undertaking relative to that of vertebrates.

Amphibians and Reptiles: Back to the Human Track

The amphibian and reptile vertebrates are members of the vertebrate lineage that leads to human evolution. They represent the transition from a water-based habitat of bony fish to land-based tetrapod species. Overall, however, amphibians and reptiles are not known for their elaborate or higher order social interactions or structures. Although sex-based aggregations will often occur within species, stable learned group associations such as shoaling and pair bonding seen in some fish are not characteristic of these species. Thus, they are not highly informative concerning the development of higher odor group identity in animals. These tetrapods have undergone some general changes in their sensory organs. In terms of olfaction, they have developed a distinct sensory epithelia in that the VNO that is now a separated tissue from MOE, expresses distinct OR proteins and has distinct innervation into the brain. The VNO organ is important for sexual identification, detecting mainly peptide or steroid molecules. Their visual systems, however, have lost spectral capacity as all terrestrial tetrapods are dichromic; hence color vision is generally absent. Tetrapods have, however, developed more elaborate hearing which is often used for sexual identification. Curiously, sex chromosomes are generally absent in amphibians and reptiles. Sex determination is often environmental, resembling that of many teleost fish. The majority of species are egg laying, but importantly some members have also developed live (viviparous) birth in multiple lineages. Most species do not care for their young, although some female reptiles do.

Genomic Transitions/Parasites

The evolution of land-based animals seems to have involved large-scale genetic alterations with respect to genetic parasites. Thus, tetrapod evolution development correlates with a major event in genetic colonization beyond the earlier retroviral invasion characteristic of fish genomes discussed above. It seems clear that lungfish represent a transition species between bony fish and tetrapods. In terms of genes associated with group identity, we have previously noted that in lungfish, isotocin as found in vertebrate fish was displaced by mesotocin, a molecule involved in sex and group identification. Mesotocin is maintained in amphibians and reptiles, but was again displaced in mammals. Lungfish also conserve the duplicated proenekephalin genes similar to those found in all gnathostomes but are in contrast to the diverse and evolving genes as found in teleost fish. Lungfish also have simplified MHC I gene set, expressing only one locus in blood cells. As noted below, frogs have VNO receptors that are intermediate between fish and mammals. Along with this shift, lungfish underwent a dramatic expansion of their genomes. Lungfish have 80X the DNA content of zebrafish (40X human) genome, consisting mostly of parasitic DNA elements. Genome duplication also is thought to account for some of this expansion, but duplication cannot explain the highly expanded mix of genetic parasites and occurrence of many inactivated pseudogenes. Since lungfish have similar tissue makeup relative to teleost fish, it is clear that a massive colonization and expansion by DNA has occurred in their genomes but was not for the purpose of creating gene or tissue complexity. The specific makeup of this DNA is not well characterized. A similar genome expansion was seen in urodele amphibians (basal amniotes such as salamanders), so amphibians and lungfish likely harbor similar elements. Ty1/copia elements are found in fish, amphibians and reptiles, but not mammals. However, their numbers are low, mostly range between 10 and 100 c/c, but numbers in lungfish are unknown. Related elements did undergo a massive expansion in some higher plant genomes (i.e., representing half of maze genome) for unknown reasons. Amphibians are known to harbor large numbers of spumavirus/MLV-related elements, not found in later vertebrates. In addition, large numbers of chromovirus (Gypsy) related elements are also known for amphibians (e.g., salamander Hsr1 elements), which are a major part of their genome (also absent from mammals). In addition, non-LTR elements related to chicken repeat-1 (CR1) were also introduced into amphibians (urodele) genomes and this amphibian version (nf-CR1) is found in significant numbers (0.05% of DNA), but also absent from most vertebrates. This element is highly represented only in avian genomes. Reptiles, birds, fish and mammals all have endogenous viruses with clear similarity to HERV-I, which are a monophyletic group. This set of ERVs is also more related to MLV class (41% identity in RT) than to other retroviral genera or the elements discussed above. As previously noted, ERVs of *Xenopus laevis* are closely related to walleye fish retrovirus. Poison dart frog also has ERVs (DevI, DevII, DevIII) that are clearly distinct from those of mammals and birds. These are distinctly related to avian viruses and seven recognized retroviral genera and are more like MLV. These appear to be present at high copy (>250) and amphibian wide with some sequence similarity to spumavirus and walleye dermal virus. African clawed toad also has Xen1, mentioned before as intact ERV clearly related to WDSV and WEHV epsilon retroviruses of Walleye fish. Reptiles have their own peculiar ERV patterns, often distinct from amphibians. Of special note are the endogenous retroviruses of pythons (PyERV), which do not classify with other retroviruses (B,C, D types), but show relationship to placental associated ERVs of mammals.

In summary, a very large scale colonization by ERVs and other genetic parasites occurred in the genomes of lungfish and basal amphibians. This genome colonization underwent a great reduction later during the evolution of the tetrapod genomes. The specific patterns of ERV and other elements that colonized amphibians (i.e., chromovirus, Xen1, satellites) were mostly not retained in subsequent mammals but were also distinct from that found in birds. The reasons for this massive colonization and expansion in lungfish and salamanders and subsequent loss (and displacement) in tetrapods remain obscure. Such a genetic colonization would seem to have been a highly disruptive event. Curiously, the social insects show a reduction in the level of genomic parasites and genes associated with innate immunity, although they increased the number of OR genes.

DNA Viruses of Amphibians

Fish and amphibians show considerable overlap in viruses they support, especially iridoviruses (which share antigenicity). Iridioviruses are linear dsDNA membrane-bound capsid viruses that code for about 100 proteins and show some relationship to the large viruses of algae. They are grouped according to geographic and taxonomic origins and are specific to cold-blooded species. No iridioviruses of birds and mammals are known, which instead support the large DNA herpes viruses and poxviruses. When first discovered, it was thought that amphibian iridioviruses were not having adverse effects on their host in natural populations, as many infections were subclinical or asymptomatic. However, in the last 10–15 years this view has changed following catastrophic mortality and population crashes of commercial frog farms in China, followed by various field reports of iridovirus-mediated die-offs of pond breeding in natural frog and salamander populations. Yet the early asymptomatic view seems also correct in that most natural populations are frequently infected with little disease. The mortality seen in frogs is associated with the metamorphosis of tadpoles as adults show little disease and may provide passive immunity to larvae. Clearly, some environmental variable, such as close aggregation of frog populations or altered parental immunity, is allowing mass mortality to occur. By far the best studied of these frog viruses is FV3, related to LCDV that causes cutical warts in fish. It has also been proposed that FV3-related viruses may be contributing to the worldwide die-off of frog populations that is currently happening. In salamanders of the Sonoran desert, for example, it is clear that RRV (a close relative of FV3) has caused a mass die-off of the Tiger salamander. Thus, there is reason to think both persistence and population die-offs are associated with these DNA viruses. The possible effects of such population-based disease host population structure or group identity have not been studied.

RNA Viruses

Other viruses of amphibians and reptiles are also known, such as RNA viruses. For example, 16 reptilian paramyxoviruses have been reported (a rare virus in fish). These viruses have a high tendency to infect lungs and immune tissue. A well-studied paramyxovirus member is Fer-de-Lance Virus (FDLV), which is pathogenic in some species. These reptilian paramyxoviruses are clearly similar to mammalian viruses and appear to be phylogenetically basal to those of mammals. They are found in two clades but with relatively low divergence. These viruses are also prone to cause infection across species. Persistence in specific reptiles has been reported, but this issue is not well studied in reptiles.

Curiously, no reports of autonomous retrovirus for any amphibian are known. However, various amphibian ERVs have been described, but interestingly, none of these retroviruses group to the five members of established non-avian, non-mammals retroviral families that are currently recognized. As mentioned above, amphibians (Rana) are known to harbor ERVs that are clearly related to autonomous proliferative disease-causing retrovirus of fish (Walleye DRV). These are unusually complex retroviruses not found in any other vertebrate. However, it does appear ERVs of frogs are functional to some degree and associated with sexual compatibility. Hybrid frogs resulting from crosses between several species of Asian pond frogs (*Rana nigromaculata, Rana plancyi, Rana brevipoda*) will usually develop tumors, especially pancreatic-derived peritoneal tumors. These tumors had high levels of C-type particles in them. Such tumors are never seen in wild populations, thus appear to be the direct product of hybrid formation between species. Thus, these amphibians and their resident ERVs appear to share common species identity.

Olfaction and Vocal Group ID in Amphibians

Amphibians and reptiles are not known for the social complexity or ability to learn stable social identity and few species are gregarious. Thus, compared to teleost fish they have much less diversity in their social structures. Pair bonding, parental care of young, family associations and large stable gatherings are all rare (although not fully absent). Accordingly, courtship, copulatory and territorial behaviors are also much reduced. Eggs are generally laid and abandoned by amphibians and reptiles so that parents tend never to meet their offspring, thus minimizing parental care. Frogs make up 90% of all amphibian species (about 4,200 species). Most frogs go through tadpole stage of metamorphosis (usually aqueous), although some frog species vary with respect to metamorphosis. Some anuran frogs, for example, have lost the aqueous tadpole stage and develop directly out of water. As mentioned, frog virus has a tendency to cause lung and kidney tumors, apparently taking advantage of these emerged terrestrial organs. In terms of group and sexual identity, frogs mainly use auditory senses for bioacoustic communication; thus, this is by far the most examined topic for frog social studies. It thus seems frogs underwent a major shift away from olfaction to audio-based social identity relative to fish. It is not clear if this was accompanied by specific changes in OR or VNO receptors or genes. In frog brains, acoustic stimuli are able to alter hypothalamus connectivity similar to that as done by olfaction. This auditory process also involves regulation by vasotocin (vasopressin-like) and mesotocin (oxytocin-like) neuropeptides, which both affect sexual behavior. Thus, this basic neurocircuit is very much like that of olfaction but has apparently been adapted for audio-based group identity. Ancestor species, such urodele amphibians, however, do maintain the use of olfaction for most of their social communication and even appear to have adapted lung morphology and physiology (buccal oscillations) for the purpose of olfaction, not respiration. Salamanders and basal frog species still use scents to locate males and females and will both odor mark and prefer sites of conspecifics. They thus maintain this significant level of olfaction for group identification. In salamander, a female sex hormone, which is a 10 aa peptide, allows conspecific recognition. This peptide is under prolactin regulation and acts via the VNO. Male salamander skin glands also secrete a female-attracting peptide. Xenopus has both OE and VNO tissue. Clearly attractive sexual behavior is under olfactory control in these amphibians. However, unlike the snake skin lipids used for sexual attraction (described below), peptides provide most of the olfactory sexual identifiers in amphibians. In terms of learning olfaction, pond frogs do appear to learn their native spawning pond based on olfaction, so at least this fish-like system has been conserved. But why was there an overall and major shift in identity mechanisms from olfactory to audio cues in so many frog species? Possibly, an acoustic-based identity mechanism was made possible by evolution of lungs in amphibians along with their ability to pressurize air and to control sound patterns as well as the evolution of ears. Insects are also adapted to terrestrial habitats, but with only a few exceptions, they mostly retained their olfactory-based social identity as described above. What was the advantage of audio-based social identity over that of olfaction for frogs? One possibility is that audio identity could provide a highly species-specific identity system that competitors or predatory would not recognize if they remained olfactory based. Some visual identity in frogs did develop, such as used by dart frogs, but such species are not common. Also, some social bonding can occur in specific amphibian species, but these are also uncommon. For example, Porto Rican coqui frogs will provide parental care of young and show territorial behavior. Amphibian social monogamy is also known, although rare, such as with the red-backed salamanders that will form male–female pair bonds. The mechanisms underlying these social associations, however, are not known. It does seem clear, however, that social imprinting (addiction) is not a prominent feature of the biology of amphibians and reptiles and individual recognition is uncommon.

In terms of reptiles, lizards are the most species diverse, although snakes are close in numbers (about 2,900 species). These numerous lizard species, however, mostly tend not to show complex social structures. In contrast to amphibians, bioacoustic communication is not characteristic in reptiles. Some reptile-based social gatherings, however, are known. Snakes, for example, will aggregate at denning sites with conspecifics for sexual exchanges. In these gatherings, VNO and olfaction are known to be involved via an odor-based recognition. This is a robust phenomena mediated by pheromones which correspond to skin lipids. This pheromone/olfaction chemistry is mainly based on cuticular organic compounds, which are similar to those of insects as described above. The snake gathering behavior, however, seems innate and does not appear to be a learned state. Overall, and unlike birds and some frogs, reptiles are not very colorful species, so visual recognition is not highly prevalent. Some lizards do show visual sex and territorial signs, such as head bobbing and push ups. Some lizards are also sexually dichromatic and can have sex-based color markings. But clearly, visual identification is not highly used for sexual or group purposes, although it has occasionally evolved. Although most reptiles tend not to use acoustic social communication, there is one prominent exception to this in that crocodilians clearly employ acoustic communication for group identity purposes. All crocodilians can produce sex-associated low-frequency sonic oscillations. Interesting that relative to other reptiles, crocodilians also show considerable parental care. A few lizard species also show parental care, such as some Australian lizards that learn stable family social structures including monogamy in mating. Some Iguana females will migrate to sites to lay eggs together, showing phased reproduction. How such relationships evolved, however, is a mystery.

Reptiles, Sex and Placentas

Reptiles and amphibians use both genetic mechanisms and environmental conditions for the determination of sex. As cold-blooded species, they resemble teleost fish in this situation. Most commonly, ambient temperatures are used for sex determination, which is similar to some fish species. However, the consequences of ambient temperature vary with the individual species. For example, alligators will produce males from eggs incubated at warm temperatures, whereas most turtles will produce females from eggs incubated at warm temperature. Curiously, snapping turtles are the converse of this, producing males from eggs incubated at warm temperatures. It is puzzling that such variation has evolved in sex determination, yet other developmental plans have been highly conserved. In terms of egg development, snakes are of particular interest in their ability to be either egg laying (oviparous) or giving live birth (viviparous). Live birth requires the ability to grow genetically distinct offspring within the tissues of the mother. Since these vertebrate mothers have an adaptive immune system, this poses numerous problems. The mother's immune system must not recognize and reject her allogeneic embryo. It seems clear that the placenta has solved such problems. Thus, the origin of the placenta and the solution to this immunity/identity problem are of fundamental interest from the perspective of organismal identity and the evolution of placental species. Reptiles form the basal amniotes and are monophyletic. As they adapted to a full terrestrial life style, with no aquatic stage, they utilize internal fertilization. Thus, the males must be motile to find females. Snakes are rapidly diversifying reptiles, for unknown reasons. All are strict predators, mostly of rodents. Their genomes, however, are poorly studied. Most snakes make shelled eggs with large yolk sac. However, viviparous snakes (such as *Virginia striatula*) show complex embryonic nutrition with chorioallantoic placenta that is distinct from other squamates. In fact, and unlike the placental mammals, squamate reptiles show multiple independent origin of placentation and reproductive diversity. Thus, reptiles show both oviparity and viviparity, but have evolved viviparity several times. They also show both genotype-determined sex and environmental sex determination. Normally, viviparity occurs in genetically determined sex reptile species. Thus, the co-occurrence of viviparity and environmental sex determination is very rare. One exception to this seems to be a species of basking viviparous lizard, whose body temperature determines sex of offspring (males being cooler). The reasons for this dislinkage of vivipary to temperature-determined sex are not clear. Pythons and boas are related snake lineages. However, all pythons are oviparous whereas the related boas are mostly viviparous. Thus, the acquisition of live birth appears to distinguish these two snake lineages. Although some lizard species are parthenogenic, this is a rare biology in snakes. It is thus most interesting that one exception to this is the Burmese python which can reproduce parthenogenically (as observed in zoos). Burmese pythons also expresses python endogenous retrovirus (PyERV) at high levels in uterine tissue. This virus is apathogenic to Burmese pythons but can be pathogenic to boid species. Its role in the biology of Burmese pythons, however, is unknown. Most interestingly, PyERV is phylogenetically related (basal) to the endogenous retroviruses associated with placental functions of all placental mammals (discussed below).

Venoms and Sex

Snakes are known for their ability to make venoms used to incapacitate prey. The evolutionary origin of this capacity in vertebrates is not clear, although some venoms clearly resemble the channel toxins described previously for oceanic organisms. Other venoms are proteases or resemble elements of the adaptive immune system, such as complement. Above, we noted that the venom apparatus of the social insect mostly evolved from the ovipositor of parasitoid wasp, as a component of the wasp sexual biology. Mammals (and fish) generally lack venom production. However, there are some exceptions from early representatives of mammals, such as the duck billed platypus and a few shrew species (Soricidae). There is also recent evidence that early extinct mammals had similar capacity for venom production from salivary glands. In platypus, this venom is made at hind spur of males, used mainly for sexual competition, not to incapacitate prey. Thus, there is reason to suspect that toxin production was involved in the early evolution of sexual identity in mammals, but was lost during their evolution. The reasons for this loss are not clear. However, it is clear that group and sexual identity mechanisms underwent a significant shift in the early evolution of mammals. It is also interesting that the platypus also makes a single oxytocin-like peptide in pituitary gland, which distinguishes it from reptiles and amphibians.

Avians

Avian species are numerous (9,000) and although they diverged from amphibians and reptiles (as a sister clade) they are not on the lineage that led to humans. In this regard, the mechanisms they have adapted for group membership may not be directly relevant to human evolution. Therefore, I will focus on general issues in avian social evolution that may suggest how complex social structures evolved for the purposes of comparing this to primate evolution in the next chapter. As will be presented in the next chapter, one of the main distinctions that allow more complex social identities to develop is the evolution of general intelligence, especially social learning. It is clear that avian species did evolve social intelligence and have many complex social behaviors. Mammals and birds diverged about 310 million ybp. One general but distinct characteristic of many bird species is their tendency to live in large densely packed flocks. Also, the majority of bird species (93%) tend to form pair bonds, in contrast to mammals in which only 3–5 % of species form pair bonds. Thus, birds are highly prone to form monogamous lifelong mating pairs and will often share in offspring upbringing. These flock-based social associations are in sharp contrast to amphibians and reptiles, but similar in scale to herring shoals in the oceans. Both herrings and bird, such as finches, can live in enormous populations (in the tens of millions). Clearly, birds have a capacity to recognize groups and individuals and also have mechanisms that stably link their common behavior. Although the molecular basis of this bird group identity is not well studied, with the sequencing of the chicken genome can summarize and consider overall changes to bird genomes.

Birds also provide examples of group identity systems that are multiply adapted. These adaptations include the development of colorful plumage and visual recognition, the complex audio learning associated with songs, pair bonding between mates, parental care of eggs and young and imprinting of young into social groups. It is interesting that both mammals and birds have also evolved three primary, discrete and related emotional systems in their respective brains used for mating reproduction and parenting (group ID). These are lust (sexual urge), attraction (sexual interest) and pair bonding which also provide individual level resolution. How or if such avian neural processes might be in common with that used in mammals is poorly understood, although some examples (e.g., prolactin and caregiving) appear to exist. Significant differences between avian and mammalian brain structures, however, suggest parallel rather than common solutions were used. Relative to the amphibians and the reptiles, avians show much more social complexity and diversity but also curiously have genomes significantly reduced in genetic parasites. This contrasts to the human (primate) genomes which underwent an expansion of genetic parasites (see next chapter).

Although brain structures differ, overall in both birds and mammals, general learning ability correlates positively with brain size. Many birds are excellent general learners and have various additional brain adaptations. Unlike primates which have expanded isocortex, birds have expanded neostriatum (nidopallium) and hyperstriatum ventrale (NeoHV, hyperpallium) complex, along with the Wulst (a bird-specific sensory projection area for visual and other sensory input). For brain size, we can compare the larger and more capable brains of hawk or crow brains to relatively smaller and less able brains of doves. Vertebrate brains are much larger than those of insects. Although we noted that drosophila could learn simple associations, training flies is difficult and their learning is rather restricted and lacks any general intelligence. The much enhanced learning capacity of birds and mammals allows for much more flexibility in learning group and sexual behavior. Learning that is stable (imprinted) further allows the establishment of a social identity system based on sensory patterns as processed during critical developmental periods by the CNS. Such enhanced learning ability would also provide an inherently more capable cognitive capacity for learning other forms of sensory pattern recognition, including more complex general learning (general intelligence). Thus, learning-based group identity and general intelligence should be linked. We can see specific examples of highly expanded learning capacity of some songbird species, such as song thrush which have been observed to learn 171 distinct songs. I have previously asserted that phenotypic diversity that is peculiar to any species will often identify the existence of an identity system that is also peculiar to that species. Accordingly, in these songbirds, we would thus expect song learning to be a prominent system for group identity.

Loss of Avian VNO

Fish, amphibians and reptiles all use VNO and olfaction as part of sexual and group identity system and for controlling sex behavior. However, like the great apes (next chapter), avian species have also lost their VNO (both the tissues and receptors) during their evolution. This would suggest avian species are no longer dependent on these associated olfactory pheromones for sexual and group identification. Yet, curiously, avian genomes appear to have expanded the number of genes for odor receptors. Also, it has recently become clear that some avian species can use olfaction to recognize individual nesting sites and offspring in large communal nesting sites. The Antarctic prions, seabirds that from monogamous pair bonds, for example, can differentiate their own burrows in large breeding flocks and have been reported to show partner-specific odor recognition. Possibly, olfactory recognition has shifted from VNO to MOE OR receptors in some avians.

Genomes and Genetic Parasites

Relative to mammals and other vertebrates, avian genomes are small and less variable. In addition, their rate of genome evolution (via indels) seems slower, although single nucleotide polymorphism rate is high (3–6X relative to mammals). Like mammals, avians represent a transition to genetic control of sex determination. There are no avian species that retained environmentally controlled sex such as crocodilians or turtles. It is interesting that about 50% of bird species are hard to sex; thus many species are not very sexually dimorphic. In avians, genetically determined sex is by female being heterozygous for sex chromosomes (ZW). Males are homozygous, ZZ. In contrast to mammalians, the male determining Z chromosome is a large gene-rich chromosome, whereas the heterozygous female (ZW) has a W chromosome which is small and heterochromatic, resembling the heterochromatic Y-like chromosome of mammals. Ancient or early birds, such as emus, have large W chromosomes. Overall, bird genomes did not obviously evolve due to genetic expansion or large-scale duplication, like proposed for fish. Since they are significantly smaller (40%) than other vertebrates, they harbor many less repeat elements than ancestral species. Most repeated element is CR1 (chicken repeat 1, non-LTR) family which is present at about 200,000 copies, but most of these are degenerate. A dendrogram of CR-1 element as found in various species is shown in Fig. 7.3. This element is absent in most vertebrate genomes, but interestingly present in lungfish DNA. It is remarkable, however, that the chicken genome has only one full-length CR1 locus. Retroviral-like elements are also clearly present, but the types and patterns of occurrence are distinct from that in mammals or fish. Most surprisingly is that all copies of chromoviruses (Gypsy) and copia elements are lost. These elements were abundant in lower animals, including fish. Most abundant chicken ERV is GGERVL, which is most closely related to ERV-L found in mammals (discussed below). ERV-L and GGERVL appear to have diverged from each other early in the evolution of avians and mammals. However, GGERVL shows limited genetic drift in avian genomes for unknown

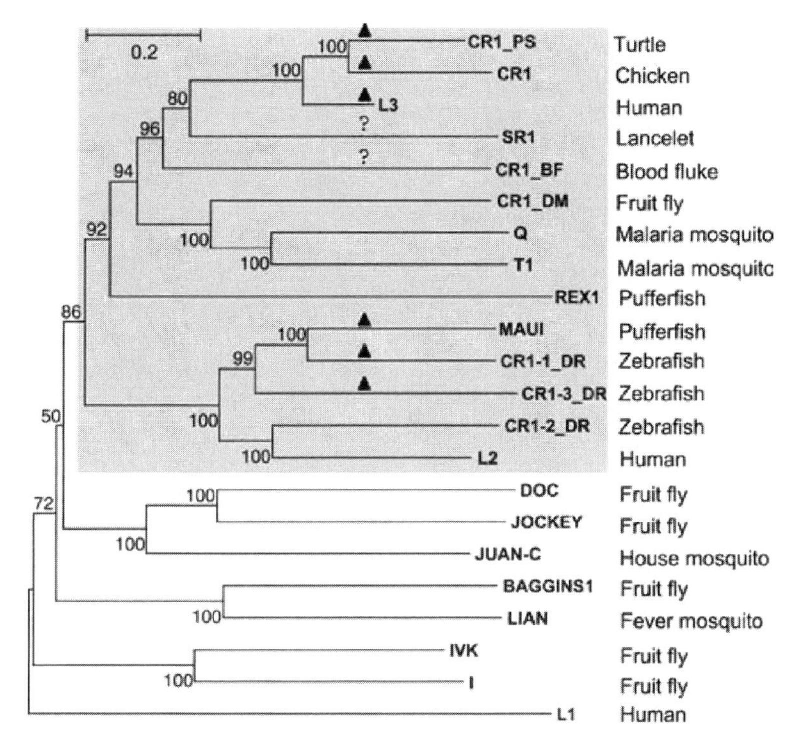

Fig. 7.3 Evolutionary dendrogram of chicken CR-1 repeat element (reprinted with permission from: Kapitonov, Jurka (2003), MolBio Evol, Vol. 20, No. 1)

reasons, suggesting positive selection and one class II GGERVK10 may still be active. SINEs (LTR derivatives described below) are absent from the chicken genome. The overall colonization and loss patterns of genetic parasites suggest that a major genetic disturbance occurred in avian genomes that appears to have allowed CR-1 colonization, but also displacement of prior chromovirus and copia-like elements.

In terms of gene content, chicken genes associated with immune function are least conserved relative to mammals. Interesting that chickens, like mammals, also show the largest expansion of zinc-finger transcription factors (known for brain activity). The most overrepresented chicken genes (by a factor of 200×) are various receptors. Chicken DNA is especially represented by genes 5U1 olfactory receptor genes and GPR43, Ig-like receptor. Overall, olfactory receptors are at about 200 genes, Ig-like receptors 25 genes and other related types about 40 genes. However, as mentioned, VNO receptors (V1R) are absent from avians. Curiously, the chicken genome is underrepresented by taste and other olfactory receptor (such as 53EX). This curious pattern of receptor gain and loss is hard to explain. It has been accepted that birds have poor sense of smell, so the loss of VNO seemed to fit this perception, but such a large gain in OR

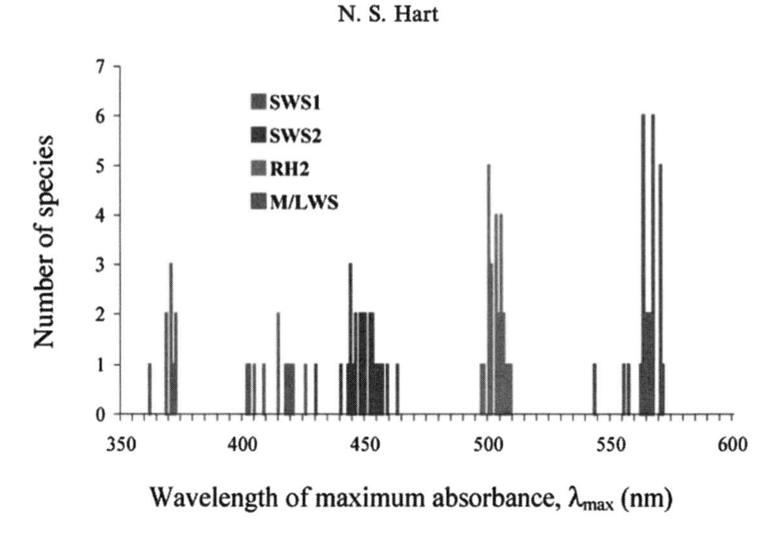

Fig. 7.4 Spectral response of avian visual receptors (*See* Color Insert, reprinted with permission from: Nathan Hart (2001) Progress in Retinal and Eye Research, Vol. 20, No.5)

genes would not make sense. VNO receptors are associated with sexual and group identity, leading some to think that avians did not retain olfaction as a system of social identity but as argued above, OR expansion may be related to this social identification. There is one set of sensory genes in which birds excel relative to mammals. These are in the visual receptors and their corresponding pigments in which birds can see in extended wavelengths, including ultraviolet (see Fig. 7.4).

Bird, Retroviruses and Tumors

Avian species support many avian-specific DNA and retroviruses (MDV REV, LLV). These are noteworthy for their general tendency to cause proliferative lymphoid and hematopoetic disease in domestic flocks. The very first tumor inducing virus was observed in the early 1,900's in Peyton Rouse associated with domestic chickens. In terms of avian retroviruses, three main types are recognized: avian leucosis/sarcoma (ASLV, mostly of chickens), riticuloendothelial virus (REV, mostly of turkeys and chickens) and lymphoproliferative (LPDV turkey specific). Although commercially important, the natural relationship of these viruses to birds is not well studied, since it is clear that these domesticated birds may represent aberrant virus–host relationships. In all, 10 avian retroviral species are known for Galliform species. REV (reticuloendothelial virus) is a well-studied infection of domestic fowl which is virulent (tumor inducing) but can be found in some wild bird populations also as a tumor. However, most evaluated wild bird populations are asymptomatic but viremic for REV, yet curiously do not seem to develop antibody. An example of this are the Attwaters prairie chickens, in which 50% of captive wild flocks are REV positive but healthy. REV distribution is peculiar in terms of species and geography. No ALV-related viruses, for example, are found in geese nor are RAV-related viruses in this species. But REV-related viruses are found in other gallinaceous birds. It is interesting that most of these REV related viruses are more similar to mammalian C-type viruses than the other major avian retrovirus, ALV. For example, REV env gene is 50% similar to env of simian D retrovirus, suggesting these avian viruses may originate from mammalian sources (a view also supported by various phylogenetic studies). REV infection of embryonic or neonatal ducks can result in persistent infection that induces no antibody response, but is eventually lethal. This retroviral–host embryo situation is distinct from that seen in mice where the virus becomes silenced by methylation and fails to express in adults (see below). The reasons for such difference are mysterious, but probably have to do with overall differences between avian and mammalian systems for controlling ERVs. REV uses the same receptor type as related simian retrovirus (SRV), a host receptor used to control cell-mediated immunity. REV also encodes a v-rel, a highly efficient NFKB B family transcriptional regulator and regulates a variety of immunoregulatory molecules. REV is clearly attuned to the host immune systems. REV is rather common in wild turkeys, prevalent in third-world chicken flocks, which has been increasing in recent years. Some of this may be due to the occurrence of REV as locus within the DNA of fowlpox virus (FPV), a large DNA virus that is a common pathogen of domestic chickens. FPV may be helping to transmit REV, but in FPV, the REV region occurs as a pathogenic locus for this DNA virus. It is also most curious that a REV sequence (also as a full REV element) similarly occurs in Turkey herpes virus, a distinct lineage of avian DNA virus, but not in other avian herpes viruses. Clearly, some complex host–retrovirus–DNA virus biology is happening in avians that has not been seen in mammals. The biological consequences of an endogenized REV are not fully clear. It is known that chicken REV-A env sequences will protect birds against REV disease. It is also known that defective REV vectors will transmit DNA into chicken germ lines in vivo, suggesting selective pressure for this defective ERV. Vertical REV transmission in breeder turkeys has been observed, via hens but not toms, so some link to sexual biology is known. In chickens, congenital ALV transmission is through egg production (ev21 locus assisted). Both REV- and ASLV-related sequences (via gag sequence) can be found in mammalian viruses, which suggests that these ASLV-related viruses have been recently distributed into avians mainly by horizontal transmission and most likely from mammalian sources. Thus, the capacity for REV to have major effects on bird immune identity seems clear.

There are clearly ERVs in the genomes of avian species, but the ERV–host relationship is distinct from that seen in mammals. Overall, the number and diversity of ERVs are much smaller relative to placental mammals. EV21 is an ERV locus of domestic chickens, associated with early feathering. It is of some interest as this is an endogenous virus that affects the ability of exogenous ALV infection to induce tumors. An F1 cross of $EV21^+$ male to $EV21^-$ female results in a most enhanced susceptibility to tumor induction by ALV. Such loss of tumor control in the offspring of parents that had distinct ERV makeup is very similar to the melenoma induction that was described earlier in the previous chapter for platyfish. In the fish situation, the responsible ERV was also associated with the sex chromosome and possibly contributing to the evolution of this sex chromosome. The chicken ERV21 phenotype (delayed feathering) is also sex linked and EV21 can be found in the Z sex chromosome of some breeds (e.g., broilers). The net effect is that EV21 is male transmitted and it is in such males that it makes eradication of ALV difficult, since they allow silent ALV infections that can then readily be transmitted via eggs. This relationship clearly has the characteristics of a viral addiction state in which protection is provided by persisting EV21 and destruction is provided by acute ALV. ALV selects against birds lacking ER21 by efficiently inducing tumors in them, whereas EV21 selects for birds that can be persistently infected with ALV, preventing ALV extermination. The sex chromosome ERV link to tumor formation is most interesting, resembling both the platyfish situation noted above and a mouse ERV F1 tumor situation that will be described below (i.e., Lake Casitas). Domestic chicken populations appear prone to epidemic retroviral-related acute disease, much more so than most domesticated mammals (except possibly sheep). Some of this disease susceptibility is known to have an MHC component (described below). An interesting side note is that chicken Z chromosome also has a high level of CR1 element (constituting 10% of this DNA). CR1 is thought of as an avain version of a LINE-like element, although this is clearly not strictly correct since it has several clear distinctions. CR1 has a distinct 3′-end UTR which terminates with a microsatellite sequence (found in all CR1 families). It is not a LINE element. As we will see below, ERVs, LINEs and sex chromosomes are also frequently linked in mammals and associated with their speciation.

Other Avian Viruses

Besides retroviruses and ERVs, birds have their own peculiar relationships with other viruses. Birds appear to support various specific types of RNA viruses, such as influenza and sinbis virus reovirus, as well as specific DNA viruses such as NDV, avian poxvirus and avian polyomavirus. Avian species are the main host for these viruses. Often, however, natural relationships between virus and host are poorly understood. Most observations come from affects on domestic or commercial bird populations. For example, we think of influenza as a human disease, but it is now clear all human versions originally evolved from avian versions that adapted to humans. Every year, the numbers of birds dying from influenza (especially domestic birds) vastly exceed the deaths of any other species. The original source of most of these influenza viruses appears to be various specific species of water fowl. However, even within bird species, virus–host relationships are generally lineage specific. In contrast to ancestral vertebrates, some clear differences are seen. Some of this difference seems to be associated with birds being warm blooded. For example, the iridoviruses, common to cold-blooded insects and vertebrate and invertebrate oceanic animals, are not found in birds. More curious, however, is that few, if any, rhabdoviruses are known of birds. Recall that rhabdoviruses were responsible for massive die-offs of herring fish. Rhabdoviruses are also common to various placental mammals, especially bats. Rhabdoviruses can infect insects and plants and even infect RNAi-incapacitated *C. elegans*. As they are distantly related to influenza virus, their absence in birds is a mystery. Avians do support numerous types of herpes viruses, which were well represented in fish, oysters, and amphibians. The herpes viruses of avians and mammals show common descent, with the turtle herpes viruses possibly representing the common ancestor. These avian herpes viruses, such as Marek's disease virus (MDV), have big and virulent impact on domestic flocks (as do influenza and avian pox). These herpes viruses cause malignant lymphomas and also arthrosclerosis by altering cellular lipid metabolism. Interesting that genetic resistance to Marek's disease virus is strongly associated with B-F region of MHC. This MHC determinant has a big effect, resulting in mortality between 0 and 100%. Such a dramatic dependence of viral pathology on MHC is not seen in mammals. Avipoxviruses are complex large DNA viruses that are both more complex and more diverse than their mammalian counterparts and are phylogenetically basal to the mammalian poxvirus. These avipox are notable for the acquisition of large and interesting gene families. Many of these genes are involved in immune evasion, including eight natural killer cell receptors, four CC chemokines, three G-protein-coupled receptors, two beta nerve growth factors and a transforming factor beta. Most interesting is the presence of viral Bcl-2 homologue, basic to apoptosis, as well as viral genes for steroid biogenesis and antioxidant functions. Therefore, the potential of these genes to regulate most aspects of host immune and nervous system is considerable. Many avian viruses also show a capacity to move (retrograde) through neurons and the CNS by poorly understood mechanisms and can affect behavior. For example, avian influenza often causes CNS disease in non-domestic bird species. In mice, LCMV RNA virus establishes lifelong CNS persistence infections which impair spatial–temporal learning.

Avian Immunity

Birds have some interesting differences with mammals regarding their immune system, especially how they create their B-cell repertoire. In contrast to teleost fish and some well-studied mammals, chickens have a single and unusual V_H1 gene. In addition, and unlike the mouse and human genome which use V(D)J site-specific recombination to form a single gene function, chicken B cells diversify by a pseudogene templated gene conversion of the heavy chain which occurs in the bursa and then uses somatic hypermutation to attain IgG diversity. Pseudogenes are donors for the needed recombination as full IgG genes are not usually encoded in the germ line (see Fig. 7.5). Curiously, a

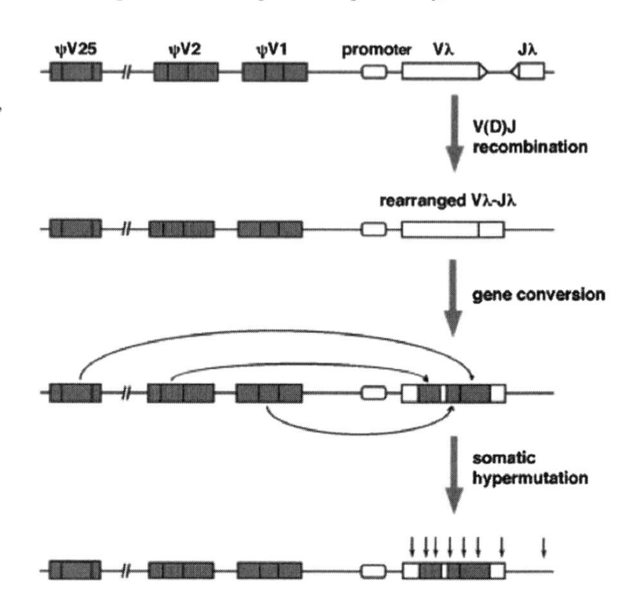

Fig. 7.5 Avian V(D)J gene map (reprinted with permission from: Arakawa, Buerstedde (2004), Developmental Dynamics, Vol. 229, Pg. 458–464)

very similar process is also used by domesticated farm animals. Phylogenetic analysis supports the view that this system evolved after the original V(D)J site-specific recombination system present in fish. The chicken system seems to have been derived from just one of the three Vh genes present in other vertebrates rather recently. Since this one chicken Vh1 gene also has a D region fused onto it, it seems clear that retroviral-like reverse transcription was involved in the origin of this chicken immune gene. Avians diverged rather early from the mammalian lineage. Yet some later vertebrates (mouse, primates) conserved the original V(D)J recombination process. It is difficult to understand the selective pressures that might have allowed such a shift in immune recognition systems, given how crucial immunity seems to be for survival. What then might have led to this shift in immune systems for birds and some mammals? I suggest that is most likely that the pre-existing V(D)J system was invaded and displaced via the action of some ERV. According to assertions in this book, I would suggest that this transition was the result of a colonization by a new genetic identity system, which incapacitated the prior V(D)J system, eliminated most resident ERVs (i.e., chromoviruses) and was likely related to avian-specific CR-1 acquisition. At this time, the avian VNO OR receptors (and MHC-linked receptors) were also eliminated by pseudogene formation. Thus, there was a major shift in genetic identity systems of avians. Bird genomes are noteworthy for the relative paucity of ERVs (and LINEs and SINEs) compared to mammals. Within mammals, the murid rodents and the primates are significantly more ERV (and LINE) colonized than other mammals (e.g., farm animals).

In mammals, mutated or dysfunctional MHC loci are most associated with autoimmunity diseases. Some associations with viral disease are known, but these are less prevalent or pronounced. With birds, however, there is much stronger link of MCH to viral disease. For example, the chicken MHC B locus is tightly associated with survival to MDV. However, such viral mortality is mostly the consequence of virus-induced tumor outgrowth, not direct tissue destruction as seen with most mammalian viruses. Thus, it seems that the chicken MHC locus more closely affects virus survival. Yet, curiously, chicken MHC locus seems inordinately simplified. Many common chicken lineages express a single dominant class I (and II) molecule relative to multigene expression seen in mammals. The chicken expression is similar to that described earlier with lungfish. Mammals also use their MHC locus for odor-based identification of sex partners and close relatives. This function is absent in birds.

The Prolactin Story: The Neurochemistry of Caregiving

Bird relationships to neuropeptides are of special interest due to their capacity to form tight social and family associations. The role of prolactin in forming family and caregiving associations was initially presented above in teleost fish in which prolactin was associated with sexual and nesting behavior, such as with the three-spined stickleback. A similar prolactin relationship also applies to caregiving associations with birds. Prolactin is a small circular peptide with a 7 TM receptor protein. In birds, prolactin is made in both the CNS and the immune system (although its role in immunity is unknown). Production is subjected to control by light, audition and olfaction. The bird prolactin receptor is mainly expressed in the preoptic region of the CNS, which also specifically binds prolactin peptide. This link to optical processing nerve cells is interesting, given that birds have particularly light-dependent breeding, known as photorefractoriness. Bird breeding and parental care is mostly seasonal, and their reproduction is tightly linked to light duration. Most free-living birds in temperate zones have such reproductive photorefractoriness. In contrast to mammals, this light-dependent reproduction is easily manipulated to keep reproductive system inactive. Clearly, in birds, sensory input controls gonad development and reproduction. For this, birds do not use retinal photoreceptors (like with all mammals), but use extra-retinal photoreceptor cells. Also unlike mammals, the bird pineal gland does not modulate reproduction. Thus, photosensory input can generally repress gonad development, via hypothalamus and hormone action in birds. Some of this effect appears to be mediated by prolactin since day length affects prolactin production and an increased prolactin level appears essential for refractory reproductive state. Prolactin binding to preoptic areas also seems directly related to caregiving behavior in both males and females. Elevated prolactin binding in this optic region is especially true for males of those species that are known to provide paternal care, and lesions here will prevent caregiving. In species with maternally behaving female birds, prolactin levels are also elevated. In Wilson's red-necked phalaropes, males provide parental care and also show elevated prolactin levels, females do not provide care and similarly do not have high prolactin levels. A striking example of male caregiving is seen with Harris' hawks in which male nest helpers also show elevated prolactin levels. These birds are interesting in being social hunters. Thus, prolactin levels generally correlate with caregiving in both male and female of numerous bird species. In some nesting birds, other inputs, such as physical contact or sight of nestlings, are also needed to induce caregiving. In these species, the parent must retain some contact with chicks to retain parental care. In terms of its evolution, prolactin appears to have diverged from other growth hormones. Bird prolactin appears to have been stably maintained during evolution, unlike rodents and primates which show evolutionary burst and variation prolactin. Although prolactin was originally discovered in connection to mammalian milk gland, clearly its name must be a misnomer from an evolutionary context. Its role in group formation and caregiving in fish and avians (non-mammals) is clear.

Sensory Learning Modes

A main assertion of this chapter is that the large complex brain of mammals and birds allows learning to be used as a major system for group/sex identification and membership. In birds, visual and audio sensory input has been especially adapted for this purpose. Birds are highly visual animals and depend most on their visual input for numerous group assessments and reproductive capacity. Thus, many birds have extravagantly colored and even fluorescent plumage, which was the main basis for the original Darwinian concept of sexual selection. Most birds appear sensitive to ultraviolet wavelengths, which is also used for mate choice, thus they surpass mammals in visual capacity. They have very complex retinas and four classes of retinal cone cell with visual pigments that resolve UV, S, M, L wavelengths (see Fig. 7.4). In contrast, humans have three and drosophila has six visual pigments. Birds also use retinal oil droplets to filter and resolve light frequencies. But numerous coloration patterns as found in birds, such as proximity-associated colors seen in zebra finches, cannot be readily explained by sexual selection. Thus, it seems that color patterns are used for group identity and membership as well. This demands that bird brains provide excellent visual pattern recognition.

Audio learning (vocal learning in bird literature) is another complex form of learning that promotes group membership. Learning to reproduce conspecific audio patterns has evolved only in a limited number of species: three birds, three mammals. It is used in birds mainly for sexual group identity in which males mostly learn and sing to attract females. Canaries, parrots and hummingbirds are all vocal learners. It follows the pattern of an identity system in that it also has a limited period of identity transfer in which sexually immature males are able to learn songs. Some birds clearly perform a version of vocal mimicry as a basis of their song learning. Interestingly, songbirds are also one of the few species that is known to have mirror neurons. Mirror neurons were initially observed in monkey brains in Broca's area that were active both by excitation and observation. Songbirds are monophyletic. These species also show movement imitation. Structures within avian forebrain control audio learning. Interestingly, early RNA hybridization reassociation (c_0t) based estimates established that canary forebrains express about 100,000 unique transcripts, much greater number than total genes (about 20,000) or transcripts expressed in other tissues. The bulk of these forebrain-expressed RNAs are expressed at low level, although its functional significance remains unknown. It seems clear that most of this RNA must represent transcribed parasitic regions of DNA. In both human and songbirds, CNS generally expresses the greatest number of RNA transcripts for unknown reasons. Of those transcripts identified as genes in the finch, receptors and transcription factors are the most commonly expressed in the forebrain.

The songs learned by songbirds are species and even region specific. Some flocks can have distinct and geographical specific song dialects. Such regionality makes it clear that songs can provide high-resolution group identity information as applied to mates, individuals and larger social groups (flocks). Cowbirds (*Molothrus ater*) can also show regional cultural transmission of such vocal traditions. Such song differences can be culturally learned in that transplanted birds will acquire the new song tradition and transmit them to offspring. These acquired traditions affect courtship patterns in that transplanted males preferentially courted females with same vocal tradition. Social (sexual) ID is thus clearly learned via song. Zebra finches are well-studied songbirds which also show group living, are socially monogamous and form lifelong pair bonding. Although the physiological link between song and pair bonding is not yet clear, songs can clearly attract mates and repel rivals. It is also known that female hearing mates' song will modify their nesting behavior. The male zebra finches learn song before 90 days post-hatching (as juveniles), after which songs are stable and they no longer learn or alter their song. During this learning, songbirds must first hear a new song, then after 3 days they can reproduce it. This learning is unlike that of primates (hominids) which can immediately reproduce a novel song. Bird song learning is thus more biologically specialized than that of human but by unknown mechanisms. Transcriptional control is clearly involved in song learning as novel songs can rapidly induce altered forebrain transcription. Zenk-mediated fast transcription in the finch auditory forebrain is involved as are cfos and cjun expression (all immediate early genes). Zenk is a single copy gene that also provides deep molecular genetic marker for songbird phylogenetics. Thus, zenk seems central to the evolution of song learning, although zenk also is conserved in non-song species. Interestingly, this zenk gene also has 3′ UTR that is even more conserved than is the coding region. Although the reasons for this conservation have not been established, it is likely to involve audio sensory regulation of expression, suggesting that regulatory elements (transcripts) might be central to the evolution of such sensory control. Differential display of transcribed RNA specific to song-affected forebrain regions suggests similar transcriptional mechanisms are likely involved in controlling numerous other transcripts. Thus, the fast transcriptional control of expression seems central to learning and brain function.

Evolution of Vocal Learning as a Paradigm

Understanding how vocal learning evolved to provide birds with group identity is thus of great general interest. Clearly regulatory DNA evolution appears central to this process. But how can a complex coordination of CNS gene function be attained by evolution? Regulatory adaptations that have allowed audio sensory input to control complex neuron expression and behavior must impose a form of coordination (i.e., enslavement) onto a set of distributed and previously uncoordinated genes. Some of these target genes are known. The role of immediate early genes, and transcription factors (such zenk) in vocal learning, for example, appears to identify gene sets that link sensory input to neuronal control and behavior that could be used to evolve such CNS coordination. Interestingly, immediate early genes were first discovered in and are used by many retroviruses. Songbirds are monophyletic and the typical passeriform songbird has a 1–2 billion bp genome with about 40 haploid chromosomes. Their coding genes do not differ much from non-song species. As it appears that brains and the vocal learning in male and female birds are distinct, the needed evolutionary coordination to acquire song learning is sexually dimorphic. Since this song learning is mainly done by males during sexual maturation, it would be logical for the sex chromosomes to be involved in creating the needed regulatory coordination. All vocal learning animal species have sex chromosomes. I have already presented the arguments that in both fish and avian species, sex chromosomes (which are often gene poor) are mostly evolving due to the action of (colonization by) ERVs and related genetic parasites (CR-1). As persisting genetic parasites must attain stability, displace resisting resident molecular identity (via gene interruption by defectives) and generally employ addiction strategies (also via defectives), we can now present an evolutionary scenario that can explain the origin of such coordination. ERVs and defectives (CR-1 and its satellite elements) colonized sex chromosomes of ancestral song learners (via the usual selection for persistence to resist acute disease from similar agents). In doing so, they superimposed a form of group identification that selected for sexual and group behaviors. This involved the distribution of defective LTR, CR-a and satellite elements that share common regulatory origins. Some of these elements colonized the regulatory regions of immediate early genes, bringing them under the coordinated control of the new genetic parasites. Selection for group and sexual identity then promoted behavioral adaptations (social addiction) that preserved the new parasite-derived group identity and led to sexual selection for vocal learning.

Social addiction has been asserted by me to provide a general strategy that is used for social bonding. How then might song learning contribute to such putative addiction? First, song learning would need to establish a stable pattern recognition which links a specific sensory audio pattern to addictive (and emotional) behavior. How can such stability attained in the CNS of songbird compel sexual or nesting behavior? In other vertebrates, such as fish and other bird species, PRL can clearly be involved in such social states. In terms of song learning, neuronal survival clearly appears to be involved as neurotropin expression is induced in juveniles and expression persists into adult. Singing is known to enhance BDNF transcription in high vocal center (HVC) of brain in relation to the number of times the song is heard. It seems likely this allows new neuron survival. Thus, a role for brain-derived neurotrophic factor seems apparent. Neurons expressing this factor are also the CNS sites for expression of non-gonadal steroids. In songbirds, the ability to audio learn correlates with testis mass and testosterone production which also relates to vocal plasticity. Dopamine (and PRL?) also appears to be involved as receptor antagonist will reduce female mating calls (courtship singing) and mating behavior. This activity involves a new region of the songbird brain, the song control system, that differs markedly between male and female birds. Thus, it is a product of sex chromosome (SDR). In classical evolutionary biology, such large sex-dependent differences have been explained by the action of sexual selection. But here, it is asserted song learning is part of a complex addiction module that provides group identity. Similar addiction strategies would also apply to other bird social bonding. And like the song response, female doves also induce cFos and ZENK when brooding their chicks consistent with common pathways for circuits of nesting behavior proposed above. Together, these observations suggest common mechanisms operate to establish social addiction.

A social species can also be socially parasitized. As presented above, social insect can have insect parasites. Some bird species are social parasites of other nesting birds. Cowbirds are one social (nest) parasites that can also discriminate between parasitized potential host birds by unknown neurochemical mechanisms.

Visual Intelligence

Cowbirds are additionally interesting in that they integrate their song to a visual display. Male cowbirds will synchronize elaborate wing movements during atypical long silent periods of their songs. This suggests a strong visual component to group identity in cowbirds. Other bird species appear to more fully depend on visual cues for group membership. Domestic birds, like chicks, do not show any auditory imprinting but visual imprinting is most important for them. The ability to recognize complex vocal and visual patterns implies a possible linkage in birds to general intelligence. Although songbirds can be very intelligent, song learning is not generally required for higher bird intelligence. Crows, for example, are highly intelligent, show tight family groups, but are not vocal learners. Black birds are probably the fastest learners, but are also not vocal learners. Although bird intelligence appears to be generally correlated with brain size, most birds still show tight social bonding, regardless of brain size or general learning ability. For example, doves, considered less intelligent and with small brains, nevertheless form that tightly bonded pair (imprinted) in which both male and female have a role in offspring caregiving. Although pair bonding can be strong in birds, it is not necessarily permanent as suggested by divorce among gulls in a 1991 book by William Jordan. The physiological basis of such dissociated pair bonding is unknown. Thus, we can see that bird social group identity often uses elaborate complex visual and audio pattern recognition (generally correlated with intelligence) but is not dependent on general intelligence.

Social and Cooperative Breeding/Hunting

Our interest in intelligence and social bonding is ultimately aimed at understanding human evolution. Thus, the topic of social cooperation in birds, especially in relationship to breeding and hunting, is of special interest in this regard. Although uncommon, some bird species do show social and cooperative hunting and breeding. The occurrence of cooperative avian breeding shows a strong phylogenetic bias and strong family associations, suggesting that it is genetic or hardwired characteristics. Such a genetic basis would be inconsistent with cost–benefit theory that is generally used to explain cooperation, see below. For example, cooperative breeders tend to lay smaller clutches, which has been used to argue they have a large parental cost investment in a few offspring. However, non-parental investment in young is well established and is problematic for cost–benefit theory. However, such cooperative social situations are seen in only 3% of avian species. Cooperative hunting is even more rare. A most interesting example of this, however, is seen with the Harris' hawk (*Parabuteo unicinctus*), which are social predatory birds found in the American south west. Harris' hawks (and other raptors) show both cooperative breeding along with their cooperative hunting in that more than two individuals (both male and female) can be observed living in groups and raising young. However, these non-parental helpers need not be genetically related individuals. It seems clear that kinship effects have been overemphasized in the scientific literature for males in such situation. The absence of a required genetic linkage to the helper male birds thus presents a significant dichotomy for kin selection theory. According to Hamilton's 1964 concept of inclusive fitness theory (i.e., cost–benefit analysis), the helpers must gain some indirect benefit from their altruistic action. Their helping must improve the likelihood that genes related to theirs will survive. This cannot apply to Harris' hawks. An alternative idea is that helping by non-breeding birds via exposure to critical stimulus at early period (begging young, before nest departure) establishes imprinted cooperative behavior (more like social addiction). Although this idea is more consistent with observations, it fails to explain the traditional evolutionary forces that create and maintain it. I suggest that the group identity, via behavioral addiction modules, offers a better explanation not only for the occurrence of cooperation, but also for its origin and phylogenetic pattern of conservation. Harris' hawks do show clear signs of other common forms of group identity and display both territoriality and aggression to conspecifics of other genetically related social groups. Thus, the unit of identity is the cooperative breeding group which although genetically programmed is determined by critical sensory input, and is not strictly genetically determined. I have emphasized the role of pheromones and hormones in establishing behavioral social bonding through social addiction. Prolactin was causal in several instances of fish and bird nesting behavior. Prolactin also appears to play a role in Harris' hawks social interactions. Its levels are conspicuously high in helper males (a situation also seen in Mexican jay, relative to non-helping Western scrub jay). Also, prolactin levels in the Harris' hawks breeders declined immediately after the eggs hatched, in contrast to the pattern seen in many other altricial species. However, following hatching, prolactin levels in the adult-plumaged male helpers rose significantly (to 9.1 ng/ml blood). At this time, these non-parental helper birds bring more food items from group kills to the nestling than any other group members. The elevated prolactin levels in the adult-plumaged helpers thus appear to facilitate the helping behavior exhibited toward the nestlings and fledglings. Other cooperative breeders (e.g., scrub jay) also show elevated prolactin levels in parental mates, as do some songbird males that also provide biparental care (i.e., red-eyed vireo). These observations clearly support a role for prolactin in bird helping behavior. The mechanism by which prolactin might promote such stable behavior is not yet clear, but it is interesting to note that in mammalian cell cultures, the prolactin receptor can also induce apoptosis. We know little concerning the virology (or ERV biology relevant to sex) of Harris' hawks or if this is of any relevance to group identity or sex chromosomes. They are known to harbor lethal, species-restricted but unique adenoviral infections. But the biological significance of this is unknown. One thing seems clear; the origin and evolution of cooperative hunting and breeding as observed in these few avian species is consistent with the role of group identity in social evolution.

Rodents as a Mammalian Model

The Transition to Mouse

The mouse is our most important exemplar for understanding the group identity of mammals. In the mouse, we see numerous adaptations, some associated with identity and immunity, that are characteristic of mammals in general. These characteristics include endothermy, hair production, early nourishment of young via yolk sac and placenta, nourishment of newborns via milk and larger brains. In terms of some of the more basic developments of identity systems, mammals all have adapted genetically determined sex based on X/Y sex chromosomes (not the avian ZW). Unlike avians, large conspecific congregations (shoal or flock-like), although well known in some herding mammals, are not the norm. Also in contrast to avians, social bonding between mating pairs is not highly prevalent in mammals, but social bonding is always present between the mother and her offspring. However, similar to that described above for avians, mother–offspring bonding involves related mechanisms (i.e., prolactin, oxytocin affecting brain areas). The mammary gland was an early development in the evolution of mammalian lineage and most likely predated the evolution of the placenta. Early mammals were egg laying, but still provided nourishment for their newborns. With the evolution of mammary glands we see the origin of the basic mammalian social unit in which much more parental bonding and care of young have emerged relative to that which was present in ancestral reptiles and amphibians. In addition, placental mammals (which displaced oviparous mammals) have even more investment in offspring in that the placenta provides nourishment for the early embryo. In terms of immunity and identity systems, vivipary presents a dilemma (noted above in reptiles) in that the mother's adaptive immune system must not reject the embryo, which is antigenically distinct. Along with these adaptations, major changes also occurred in genome colonization by genetic parasites, especially with regard to ERVs. What is especially evident (as compared to avian genomes) is the relative loss of chromovirus elements (Gypsy related) and the gain of various other ERV (MLV-related, HERV K) elements. Along with this colonization, a major expansion of additional elements (LINEs, SINEs, alu elements) also colonized the mammalian (but not avian) genome. This provides direct evidence of significant changes in genome identity which are common to all mammals and will be evaluated below.

In terms of social interactions, mice also have much to tell us. Like most mammals, mice do not normally live in large, shoal-like groups, although occasional population bursts and aggregations have been seen. Mice, like all mammals, do show strong maternal–offspring bonding and care of young. Also like most mammals, most species of mice do not show pair bonding or social monogamy between mates (present in only 3% of mammals), although some voles and hamsters can clearly pair bond, as presented below. Mice are predominantly olfactory creatures, and often use olfaction for discriminating individual and social identity. Olfaction (along with elements of the immune system) is employed to control social behavior, such as kin and mate recognition, and even embryo development can be affected by olfaction which is set into maternal memory via an imprinting process. Mice have also retained some audio and some visual recognition capacity. But like most other mammals, mice have a reduced visual spectrum, maintaining only two photoreceptors. Some visual imprinting may occur, but this is not the most crucial sensory input. Clearly, mice do not show the strong capacity for visual-based individual discrimination as do primates (or some fish and birds). Some audio recognition occurs in mice, especially during sex, but this is limited and not individual specific.

Olfaction and Maternal Bonding

Olfaction has a robust role in mouse behavior. It provides direct links between mate recognition, immunity and reproduction and is involved in imprinting mother to offspring and offspring to mother. In addition, it is involved in specific mate recognition. The mouse has undergone a general expansion of its CNS capacity, especially with respect to its olfactory system, as seen by an enlarged olfactory bulb. The main mouse olfactory epithelium (OE) expresses many OR genes on microvilli tissue that provide a pathway to the olfactory bulb. Olfactory receptors (ORs) are expressed on olfactory neurons (all are GPCRs with rhodopsin-like structures). As previously mentioned, the evolutionary dendrogram of the VNO V2R and V1R receptors is shown in Fig. 7.6. Mice also show a clear VNO tissue, not just receptors, that is distinct from MOE and use VNO for the control of sexual behavior (see Fig. 7.7 All tetrapod (non-hominid) mammals have also acquired the V1R receptors (all 7 TM GPCRs) which detect pheromones and are specifically expressed in the VNO. These receptors are composed of roughly 150 family members that show no sequence homology to other receptors. Thus, mice, like amphibians and reptiles, have maintained a clear differentiation of MOE and VNO tissues, although there remains considerable functional overlap of these receptor types (as seen for fish). In terms of OR ligands, volatile molecules and urinary proteins are clearly involved in rodent sexual and territorial behavior. In mice, many VNO receptors are also genetically linked to MHC locus and provide the capacity to differentiate individuals based on these peptides. Non-volatile peptides thus appear to convey sex-specific and individual-specific chemical identity. This contrasts to the bile acid and steroid-based olfaction system present in lamprey.

In order for olfaction-mediated maternal–offspring bonding to attain the needed stability and specificity, the mouse memory systems must be capable of detecting transient, post-partum but developmentally restricted odor cues that become set into memory and tightly control maternal behavior. All mammals must learn to bond to their offspring. Such social bonding must be maintained by enhanced long-term (and emotional) memory systems. Mammals indeed have enhanced learning and memory systems. There also appears to be a clear link between evolution of endothermy, REM sleep and long-term memory maintenance, all found in mammals. Teleost fish have a primitive form of sleep, although many teleost fish can clearly imprint on odors. But REM sleep (involving muscle atonia) is tightly associated with endothermy. In mice, olfaction systems clearly promote learning and bonding to young, but learning can conversely control the mother fetus (offspring) immunity and terminate a pregnancy by preventing embryo implantation. For example, fetal rejection by the mother (of specific mouse breeds) can be induced by non-congenic male odor molecules in a process known as the Bruce effect. Thus, olfactory sensory cues not only affect CNS-mediated learning but also induce innate (macrophage-mediated) prevention of embryo implantation.

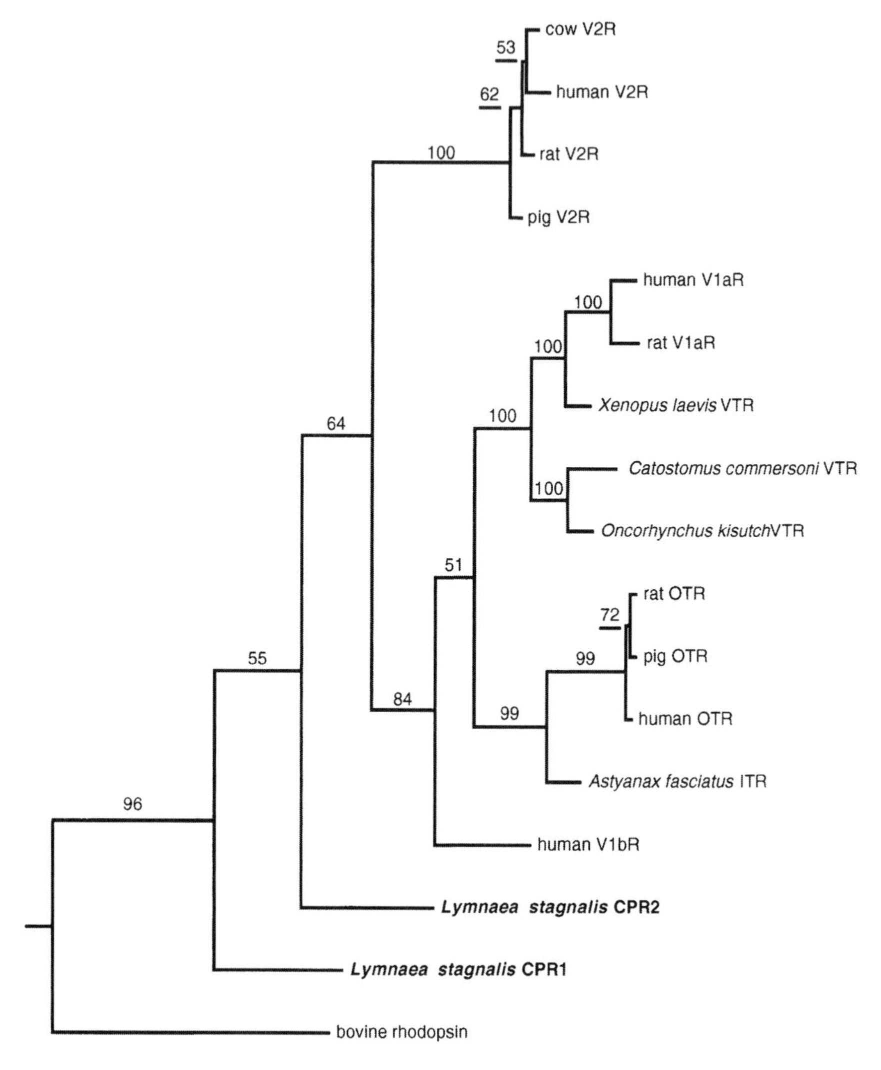

Fig. 7.6 Evolutionary relationship of V1R and V2R receptors relative to oxytocin receptor (reprinted with permission from: Kesteren et al. (1996) Journal of Biological Chemistry, Vol. 271)

Emotions and Social Bonding in Mammals

It is usually thought that emotional states are used by mammals to control social interactions and social bonding. Thus, the maternal–offspring bonding inherent to all mammals engages various emotional states which must be acquired (learned), following birth. Emotions are general states of behavior that are clearly the products of a more complex CNS, although under considerable hormonal and sensory control. It is therefore important to consider what evolutionary developments in emotional capacity occurred in mice (mammals)

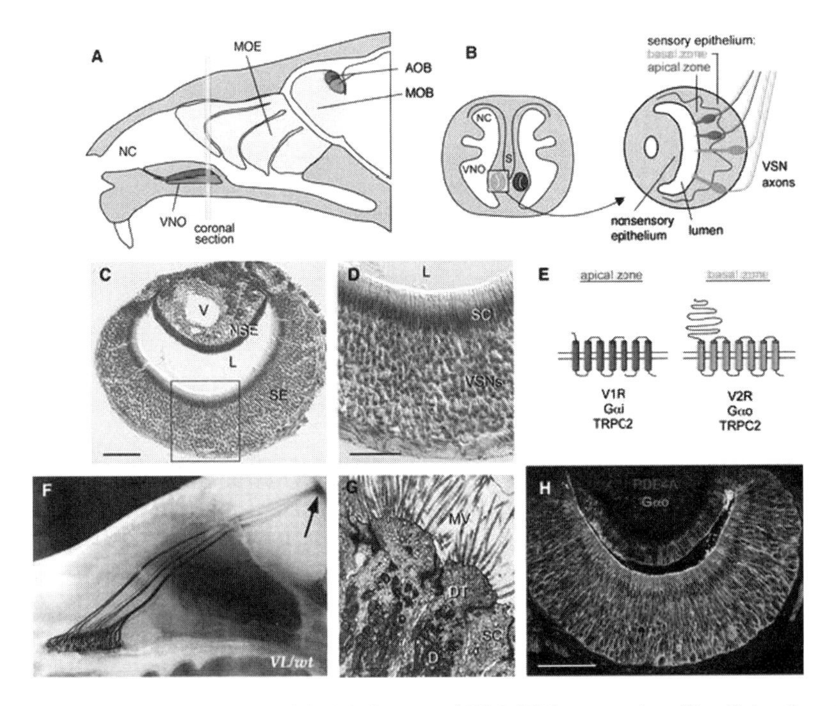

Fig. 7.7 Rodent VNO/MOE epithelial tissue and V1R/V2R expression (*See* Color Insert, reprinted with permission from: Freeman, Kanyicska et al. (2000) Physiological Reviews, Vol. 80, No.4)

and how they might be regulated. As mentioned previously, reptiles and amphibians (and possibly insects) show fear, flight, anger and aggression as general emotional states. These are the basal emotional states found in most animal life forms. However, in mammals, we see acquisition of additional and more complex or nuanced emotional states. These additions appear to include the states we call lust, obsession, anxiety, joy and contentment (an antianxious state). Most of these emotional states can be observed as applying to a maternal mouse bonded to her offspring. In addition, all mammals display social pleasure (playing), an acquired characteristic that was absent in all other tetrapods. Social displeasure (rejection) also seems to have developed as an emotional state in mammals and can induce some clear physiological effects. There are also clearly major hormonal links that can control these states but rodents are by far the best understood concerning how hormone/pheromone affects general behavior. Thus, we will consider in some detail the role that prolactin and oxytocin have had in such social bonding. Of special interest will be the prolactin receptor and its transduction system (see Fig. 7.8). Recall in birds and fish, these two peptides were highly relevant to and in some cases causal to social bonding. With the evolution of mammary glands, we continue to see a central role for prolactin in milk production and oxytocin in

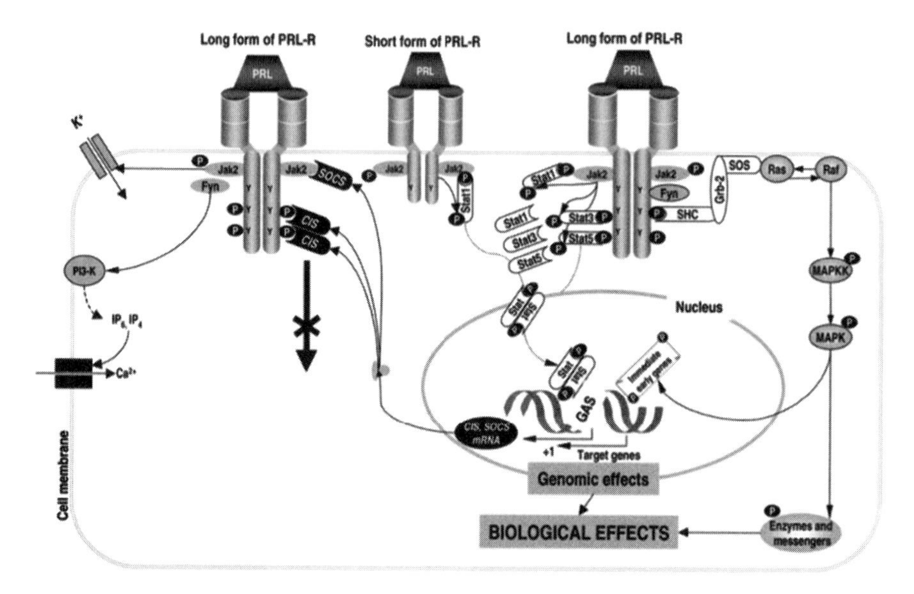

Fig. 7.8 Prolactin receptor transduction (reprinted with permission from: Frank Zufall (2005) Pflugers Archiv, Vol. 452, No.1)

bonding. We can also see a bewildering array of distinct types of prolactins specific to the placenta, suggesting a diverse role in this tissue. Also, as presented below, we will see a robust role of opioids in social bonding and pair formation.

Shrews as the Basal Mammals

Early mammals were likely to have been egg laying, which inherently requires less maternal investment prepartum. As egg shells are generally impermeable to molecular cues, social bonding between oviparous mammals and offspring must have occurred with hatching and mammary glands would likely have been involved as a biological cue. Fossil evidence indicates that these early mammal species were small shrew-like predators (such as the extinct multituberculates). These basal egg-laying mammals were displaced by placental mammals along with emergence of viviparous birth. Vivipary clearly required major compensatory changes to the immune systems to prevent live embryo rejection. Marsupials appear to be sister group to mammals, and as they mostly lack placental tissues they can help us understand the nature of the placental-specific genetic changes that occurred.

The first placental species are estimated to have evolved between 35 and 65 million ybp and probably most resemble extant shrews in morphology and their predatory life style. Thus, shrews best represent the likely ancestor to all placentals, including primate and rodents. Shrews have a functional VNO tissue, such as seen in the house musk shrew. It thus seems likely that VNO tissues were also present in the basal placentals, making it likely that olfaction was involved in offspring and sexual recognition as had been conserved in all terrestrial tetrapod animals. Afrotheria (elephants, aardvarks, manatees, anteaters) appears to represent a lineage that diverged early from the other placental species. The human and rodent lineage appears to also have diverged early from the lineage that also led to bat/carnivore. Shrews are placental with the smallest brain, suggesting a limited capacity for general intelligence. It is interesting, however, that shrews, squirrels and primates show similarities in forebrain organization that are distinct from rodents. Shrews also differ from rodents in food finding patterns, and will search for food more like primates than do mice and appear to identify a conserved learning style. Shrews block polyspermy; thus, like lampreys, some sexual identity is maintained at a cellular level in these species. Also, penile penetration by shrews induces ovulation, so it is clear that sexual reproductive control differs considerably from that of rodents.

Tree shrews support many types of virus, including many latent viruses, such as orthopox virus, parvoviruses, paramyxoviruses, rhabdoviruses, adenovirus, herpes virus 2. Most of these viruses can lytically infect cells of other species or young of related shrew species. Such virus–host relationships can be highly species specific. For example, tree shrews can have THV and THV2 adenovirus infections which are epizootic and lethal in colonies of some shrew species but are latent in healthy colonies of other shrew species. Shrew social identity systems are poorly studied. But clearly the potential for viral-mediated group membership exists in shrews. Mostly, these shrew viruses are phylognetically basal to those of other mammals. For example, rhabdoviruses of shrews are phylogenetically basal to rhabdoviruses (including rabies virus) found in all mammals. Many (but not all) of the tree shrew viruses are also similar to those in murid rodents. Oddly, there is no rodent alpha herpes virus, which is so well conserved in many other mammals. Interesting that the much feared ebola virus has also been found persisting in two species of shrew, although bat host may also be involved in persistence. In terms of retrovirus, only a few species have been studied. The house musk shrew (*Suncus murinus*) has an endogenous retrovirus, Sm-MTV (an MMTV-like virus), which has been isolated from and released by spontaneous mammary tumors, common in this species. So this basal representative, with the evolution of mammary glands we also see a productive retroviral association. This ERV (Sm-MTV) of musk shrews identifies a family of retrovirus that is distinct from established B, D and C-retroviral types characteristic of other mammals. It is most interesting that Sm-MTV produces a spikeless extracellular particle (IAP) that resembles a cytoplasmic form of MMTV. In rodents, IAP production (of unrelated sequence to shrew) is also seen and is highly associated with placental development. The function and evolutionary significance of the shrew IAPs are not known. IAPs are unknown for marsupials. However, as presented below, ERVs and IAPs are of special evolutionary interest regarding placental reproductive biology and speciation of all mammals.

In mice, production of MMTV-related virus is associated with spontaneous mammary tumors in mice, similar to mammary viruses of shrews. Mouse MMTV is a well-studied retrovirus (B type) that is infectious and milk borne. The virus can induce mammary tumors via insertional inactivation of protooncogenes. Mammary glands start dividing cells during lactation (via prolactin induction), and it is the resulting tissue differentiation that induces the gland to become receptive to viral infection, allowing selection for tumor production. MMTV also shows interesting links to the immune system. The MMTV LTR encodes an ORF which is also super antigen (Sag). Super antigens are interesting in that they are able to induce B-cell differentiation in an antigen-independent way. Such an induction process seems essential for the normal life cycle of some B cells. Some wild mouse species (from Taiwan and Malaysia) were completely free of enMMTV. Thus, MMTV viral colonization need not be maintained in all species, for unknown reasons.

ERV General Expansion and Rodent Evolution

Rodents have diverged drastically from shrews, although phenotypically, rodents and shrews resemble each other. Rodents are the most numerous species for all mammals, and it is estimated that there are about 2,300 rodent species (half of all current mammalian species). Most of these rodent species have diverged and evolved rather recently. The murid family makes up half of all the rodents. Within murids, a New–Old World split in lineages is estimated to have occurred about 9.8 million ybp. This event was early in the divergence of murids. Genus Mus has about 30–40 murid species native to Europe, Asia and Africa, but probably originating in northern India. Evolution of murids seems to be by a sporadic process with nested pulses or radiation of new species (suggesting involvement of colonization events). Rats have diverged from Mus rather recently. Since murids are the best studied mammals, they have much to tell us about the processes involved in speciation. To evaluate this speciation story, I will concentrate much attention on the role of endogenous retroviruses in rodent evolution, especially in evolution of vivipary.

In the prior chapter, the colonization of the vertebrate fish genome with various types of chromoviruses (Ty3/Gypsy-related elements) was presented. Such chromoviruses were also abundant in the genomes of invertebrates and lower vertebrates. However, mammals have mostly lost these elements, although a residual but inactive set of nine chromoviral elements have been conserved. One of these elements is a Fugu element (one, sushi-ichi is on the X chromosome and undergoes X inactivation). Other versions are not on X and are not inactivated. Thus, it appears that ancestral vertebrate chromoviruses were mostly displaced or made inactive in the placental genomes. But along with this loss, there was an acquisition of various additional ERVs.

All placentals genomes have been colonized with ERV-L-related DNA. In most cases, these ERV-L elements are present at about 10–30 copies per genome. This ERV-L pol sequence is clearly related to the pol found in extant foamy virus. Foamy viruses are considered to be the oldest lineage of all the genus retroviridae, but are curious in that unlike all other retroviruses, they are all silent infections with unusual RNA replication strategy. Although lytic and fusogenic in culture, no foamy virus is known to cause disease in vivo and they evolve extremely slowly in their primate host. Since foamy viruses are especially able to infect non-human primates, this is discussed in detail in the next chapter. However, both mice and simians underwent expansion of ERV-L to about 200 copies per genome. This simian ERV-L bust was an early event in the simian lineage and appears to have occurred shortly after they diverged from prosimians (about 60 million ybp). The burst in the mouse ERV-L, in contrast, is much more recent, and has likely occurred after rat split from the mus lineage since rat DNA lacks ERV-L expansion. Some other expansions of ERV-L may also have occurred, such as in the wooly mammoth. Clearly, some colonizing version of ERV-L was present at the origin of placental lineages and seems to have been maintained well into the placental radiation. The human genome has about 200 copies of HERV-L. However, in contrast to the murine ERV-L, human HERVL element is less defective and displays full open reading frames in gag and pol genes. This HERV-L gag is 43% identical to foamy virus 1 (Fv1) gag, indicating similarity in both pol and gag. These patterns of infection and colonization are consistent with the idea that simian species were colonized early by ERV-Ls, but may also provide a source of exogenous ERV-L via foamy virus that were able to colonize genomes of other species, such as human and mus. Thus, it is interesting that mus species also have a fully coding MuERV-L element, absent in rat.

All placentals have additional relationships with their ERVs. Most species are colonized with complete ERV copies that are rather unique to each lineage and tend to be highly expressed in placental and other reproductive tissue (ovaries, uterus). Mostly, these ERVs are present at low copies (sometimes unique copies), but have ORFs that have been conserved. As discussed below, in mice, human and sheep genomes, some of these ERVs are encoding functional env genes that provide essential functions (syncytin; cell fusion) for the placental tissues (syncytiotrophoblasts). The ERVs found in pig genomes (PERVs), like most ERVs, are congruent with their host and show phylogenetic correlation with time of speciation between closest American native pig species, but have unknown function. Feline ERV (RD114, and env-containing FERV) is similarly present in all feline species (and is placentally expressed), whereas the autonomous version (FeLV) is not common in wild feline populations. Recently, JSRV and enJSRV of sheep have also been reported to be congruent with their host (including the most recent domesticated breeds). Baboons also have specific ERV (baboon endogenous retrovirus), but with no homology to the human HERV Ks. Some defective ERV elements (such as IAPs, RD114, VL30) are highly expressed in placental tissues such as trophoblast. In some cases, these ERVs or their autonomous relatives can affect reproductive success. For example, feline leukemia virus (FeLV) is an exogenous relative of the endogenous RD114 virus which is conserved in all felines. FeLV can disrupt embryo implantation in cats, but it rarely occurs as an autonomous infection in feral cats (1% prevalence) and is not seen in wild felines. Rodents seem to be particularly associated with ERVs. All mus species have IAP sequences that have been shown to be highly expressed in trophoblasts, when examined. Yet rats lack any IAP-related sequences, but conserve ERV-encoded syncytin genes. Other rodent species, such as Syrian hamsters clearly produce IAPs in reproductive tissue, but these IAPs differ significantly from those of mus (i.e., hamster and mus LTRs have clear similarity, but not their RTs, discussed below) and the hamster sequence shows little polymorphism. Interesting that these same ERVs are frequently present in abundant numbers on the Y chromosome. For example, the 3′ ends of mouse and Syrian hamster IAPs are present at 950 copies in Syrian hamster genome, but mostly on the Y. Yet in the close relative, the Chinese hamster, these IAPs are completely absent. The Syrian hamster IAP is of special significance relative to human evolution since its pol gene is MMTV-like and was used as a probe to discover and clone HERV K (K for lysine tRNA RT primer).

What is the significance of all these diverse, confusing, peculiar and species-specific placental ERVs? More specifically, why do all placental species have their own peculiar ERV composition, including humans and why the frequent association with reproductive tissue? As I have been asserting throughout this book, I suggest they represent successful genomic colonizers that helped create new group identities within their host. I will now consider this idea in the context of rodent evolution below and primate evolution in the next chapter.

Rodents especially seem to have a dynamic relationship between their ERVs and related exogenous retroviruses as compared to most other mammals. Mus species in particular are prone to produce what are known as xenotropic retroviruses, viruses that can infect other rodent species, but not their own species. Such xenotropic viruses can be derived from endogenous mouse sources. These viruses are especially associated with embryos, immune systems and placental tissues. Rodents thus seem to represent a particularly large potential source for the production of exogenous retroviruses that are able to infect other species. In contrast, non-human African primates are known to harbor numerous species-specific exogenous retroviruses (such as foamy virus and SIV), but are not known to produce any retrovirus from a related endogenous copy. In humans, for example, there are no known autonomous version of endogenous viruses infecting other species (such as HERV K). All known ERVs of primates are inactive. Thus, it is very interesting that many of the ERVs as found in most mammalian genomes most resemble rodent viruses, especially the autonomous MuLV of mice.

Mouse Speciation, Autonomous Virus and a Very Complex ERV Story

Murid species evolved rather recently, compared to other mammalian lineages. It is thought that one reason is that the main food source for murids is seeds, so their development depended on the evolution of seed-producing higher plants, such as grasses, to provide a sufficiently energy-rich food that allows maintenance of body temperature in such a small mammal. Murid rodents are also the best studied mammals and their speciation shows many links to retroviruses and ERVs. There is an interesting historical link between rodent and human distribution. The four species of *Mus musculus* are commensal populations, living in proximity to humans and have invaded the Americas and Australia along with human colonization. Most inbred lab mouse strains derive mainly from *Mus domesticus* (shown via mtDNA) with some genetic contribution (especially Y chromosome) from *M. musculus* and *Mus castaneus* (via outbreeding often with males). It is interesting to note that fossil remnants of mice species and other rodents can be used to date hominid bones and estimate prevalent weather during human evolution. The rodent–hominid association goes back at least to the time of homo erectus, establishing a relationship (about 1 million ybp) that predates modern human evolution. This will be of further interest given the curious link between genomic ERVs found in humans and rodents discussed below. Voles are the most recently diverging of rodents and receive special attention (regarding pair bonding) below. As a side note, it is also interesting that rodents also differ from other mammals in conserving BC1 RNA, a retroposon-derived neural-specific small cytoplasmic RNA whose expression is confined to rodentia nervous systems. Clearly, the highly studied inbred lab mice are best understood regarding their retroviruses and such strains differ significantly in susceptibility to retroviruses. BALB/c, for example, are not highly susceptible to MLV unless they are fostered by C3H (MLV producer) mouse mother. Such fostered BALB/c mice that become infected, however, also become highly susceptible to MLV in establishing a multigenerational transmission chain. Thus, the virus status of the mother (not necessarily its genetic makeup) determines offspring susceptibility. BALB/c males are also interesting in that they are socially aggressive compared to other breeds and often lack corpus callosum. Thus, MLV virus–host outcome has a clear link to maternal (mammary) relationships and social settings (such as shared maternal care). Similar MLV virus–host links are not confined to inbred strains. The European wild mice, *Mus specilegus*, for example, has a species-restricted susceptibility to MLV-like ectropic virus. This European mouse also shows a distinct biology to these MLV infections (i.e., syncytial formation) relative to other species (such as M dunni). Sensitivity to mammary tumors is also seen in some wild mus hybrids between different mus species, suggesting a link to sexual isolation. Asian wild mouse (*M. castaneus*), however, normally resist polytrophic mouse gammaretrovirus. This resistance is via the XPR1 locus (a surface receptor for P-MLV) and is main factor in hybrid viability between *M. castaneus* and XPR1 variant mice that allow MLV infection (such as with *M. specilegus*). This resistance locus itself is due to X-MLV env gene that is interfering with P-MLV, which is part of a full-length X-MLV-like provirus. Thus, exogenous MLV resistance is mediated by a strain-specific MLV-like ERVs. It is thus curious that most lab strains of mice also encode an MLV-like env sequence. In some wild strains (i.e., *Mus spretus*) the MLV ERV elements appear to be recently acquired, but in some instances this can be detrimental. The presence of endogenous MLV-like viruses can have clear significance in field situations. For example, on a field study at Lake Casitas (Ventura County, CA), a colony of east Asian castaneus mouse was resistant to virus via the FV-4R locus (a defective receptor blocking MuLV provirus with gp70). Field crosses between this species and a neighboring *Mus musculus domesticus* results in some offspring that do not retain FV-4R locus, which leads to reactivation of endogenous virus and kill the F1 hybrid mouse due to resulting lymphomas. I suggest that one way to think of this result is that the F1 hybrid violated a state of endogenous MLV addiction that was present in castaneus parent, losing the protecting version of defective ERV, resulting in reactivation of MLV and disease. This proposal clearly resembles the original concept of virus addiction as proposed for P1 phage persisting in *Escherichia coli* that will kill mating pairs that are 'cured' of persisting P1. In the F1 hybrid mice, they were 'cured' of the persisting defective ERV.

Clearly, MLV-like ERV elements can confer resistance to exogenous MLV. The resident provirus prevents autonomous virus development and infection within the same or similar species. Can we infer from this observation a general explanation for the peculiar presence of ERVs in the genomes of all placental species? Probably not. For example, as mentioned, primates are one of several lineages of placental species in which there are no exogenous versions of endogenous viruses (i.e., HERV Ks). Yet primate genomes recently underwent HERV K colonization in their genomes. How then might we explain these prevailing ERV colonization patterns in species that lack autonomous versions? I suggest this requires cross-species-mediated colonization events. A potentially strong example of this proposal may be found in leukemia viruses of koala bears, which clearly resemble a rodent virus. Some wild mice (mus dunni terricolor, not native to Australia) have conserved a full endogenous virus, MEDV, that is present at only 1–2 copies per genome. This MEDV is hydrocotisone activated, implying a possible link to behavior and social stress. This mouse and MEDV are interesting for other reasons. MEDV harbors an MLV-like LTR that is similar to the highly expressed VL30 of *M. musculus*, suggesting long evolutionary link to mouse biology. However, this MLV-like virus also has a coding region (non-env) that is related to Gibbon ape leukemia virus. This MEDV is of special interest since it seems to best represent the source of an autonomous retrovirus that is responsible for the ongoing endogenization of koala bears (described in detail below). One implication is that rodents may have become a major source of exogenous viruses that colonize the genomes of other placental species.

Like the platyfish, there is also an apparently strong association between mouse species, EVSs and their sex chromosomes (Y). A significant portion of mus ERVs is associated in a species-specific way with the Y chromosome (see below). For example, *M. musculus* IAP sequences on Y (MuRVY) are present at about 500 copies per genome, most of which is on the Y chromosome of all mus species. One such ERV, IAPE-Y, has an intact env gene that flanks the more numerous defective copies. MuRVY is a normally in a silenced state, but can be induced by 5-aza-cytidine. Curiously, IAPE env lacks immunosuppressive CKS 17, conserved in the env genes of essentially all retroviruses. But MuRVY is specific only to mus species. Within mus, *M. spretus* has MuRVY-related sequences but lacks IAPE-Y sequences which all other mus species conserve. *M. musculus* has both autosomal and Y versions of this ERV. In fact comparisons of *M. musculus, M. domesticus, M. spretus, Mus hortulanus and Mus abbotti* indicate that all have their own peculiar pattern of MuRVY on their respective Y chromosomes. As mentioned above, crosses in the field between distinct mus subspecies (i.e., European *M. musculus* X Asian *M. castaneus*) results in F1 offspring that can have a high incidence of lymphomas, due to recombinant reactivation of an endogenous AKV-1 that is no longer repressed. Thus, the abundance of ERVs on the rodent Y becomes most interesting to understand with respect to sexual isolation. Hamsters have their own IAP elements distinct from those of mus species. In Syrian hamsters, the entire Y chromosome exists as heterochromatin and over half of genome is IAP sequence. However, unlike some of the rodent MLV-related elements, rodent IAPs are more like primate HERV Ks in that they do not seem to be able to produce exogenous virus.

The IAP-HERV K Link

All mus species, like most rodents, have IAPs. However, IAPs have not traditionally been considered as ERVs since they have distinct LTR classes and although they make particles, they do not make transmissible virus. Although their purpose is obscure, by any general definition of ERV, IAPs would certainly be included. VL30 s are defective versions IAPs that are highly expressed by many mouse cells, but lack any functional coding. Both IAPs and VL30s are expressed in mouse embryo and germ lines at high levels. However, there is a curious historical and sequence relationship between IAPs and the HERV K ERVs, so characteristic of great ape but especially human genomes. In 1982, Masao Ono used a plaque hybridization technique to first clone IAP (using a phenyl-ala primer sequence) from *M. musculus* and showed that it was present at about 900 copies per cell. The same probe was also able to clone IAP sequences from Syrian hamsters (indicating related LTRs). No homology to other retroviral LTRs or human DNA was seen with this primer. But although LTRs were similar, the mouse and hamster IAP genes were not conserved outside of LTR. Also, this hamster IAP was not found in closely related Chinese hamsters. Most IAP genes have lots of stop codons. But some IAP gag and env copies are intact. The IAP gag is MMTV-like and the hamster IAP pol is RSV-like. This hamster pol probe was then used as a probe to initially find human HERV K (HERV K and MMTV pol are 70% identical). These HERVs are not present in any fish or ancestral genome and are characteristic mostly of primates. In characterizing this HERV K, the right LTR was seen to be unusually long (>200 bp relative to typical 100 bp). It now appears that the human HERV K LTRs are A-rich sequences that have been the source of 3′ end for many SINEs, linking HERVs and SINEs. Like IAPs, HERV K proteins (gag/env) are expressed in early embryos and are also diagnostic for human germ cell tumors. Curiously, HERV K env expressed in mouse embryos disrupts mouse development. HERV K transcripts are also induced by female steroids. As will be presented in the next chapter, HERV Ks are of special relevance to human evolution and the distinctions between human and chimpanzee genomes.

It is likely that any reader, including most dedicated aficionados of retroviruses and ERVs, will struggle to understand all the complex biology and interrelationships between these rodent retroviruses, the ERVs and host described above. Below, that complexity is even further exacerbated by considering relationships between ERVs, LTR, LINEs, SINEs and other retroelements. Then there are links to sex chromosomes, reproductive isolation and speciation. However, it is important to remember the big picture and not drown in this daunting complexity. These agents have clear and major consequences to host identity and the survival of their host as well as how the host competes with related species. These elements are most certainly not simply selfish DNA. Nor are they acting in isolation. Mammalian genomes resemble a quasi-species swarm of interacting, often defective elements that are both within and external to the genome and it is the swarm behavior that matters, not one specific ERV element. We are not disposed to think of the forces of evolution from such a perspective and our brain struggles to follow a linear logic of single elements.

ERVs and LINEs and Mammal Speciation

At one time, the view was widely held that retroviruses likely evolved from their LINE-like relatives present in early eukaryotic genomes. As developed earlier in this book with the discovery of chromoviruses *gag* and phylogenetic evaluation of DRIS elements, this argument, although still popular among many evolutionist, is no longer tenable. Also, in the context of the well-studied mouse models which are so prone to producing xenotropic viruses, no example of a retrovirus evolving from a LINE-like element has ever been observed. All known exogenous retroviruses have evolved from other retroviruses or ERVs. Comparisons of rodent genomes, however, allow us to further consider this LINE–ERV issue and its relevance to speciation. One telling example is found with the South American rodent, *Oryzomys palustris*, which has undergone a major loss of LINE-1 elements along with speciation. In contrast, the closely related *Sigmodon hispidus* South American rodent retains LINE-1 elements and activity. However, like all other rodents, these two related species have distinct ERVs (e.g., MysTR) that have been independently and highly amplified in their respective genomes. *O. palustris* has 10,000 MysTR copies whereas *S. hispidus* has 4,500 copies. Thus, ERV colonization, not LINE amplification, associates with this recent speciation. The sequenced mouse genome has about 80,000 LINE-1 (L1Md) elements. This contrasts with the much larger number in humans (L1h @ 500,000, which is 26% of the Hu-X chromosome). The great majority of these elements are defective for retrotranscription and transposition. Humans also have highly expanded alu (7S RNA derived) and SINE (tRNA derived) retroelements, previously thought by many to be the products of LINE activity. All these elements can potentially be mutagenic via transposition resulting in gene interruption and seem to have undergone waves of amplification during primate evolution, possibly in linkage to ERV colonization. In mouse, the L1 RNAs have distinct UTR and ORF within them that are seen to be expressed male and female germ cells. Although this expression may provide some selection, its function is not understood. More recently, a role for LINE RNA expression in primate brain function has also been proposed (see next chapter). The LINE expression pattern in reproductive tissue resembles that of ERVs. For example, human L1 is expressed in undifferentiated EC, but not differentiated EC cells (similar to that of IAPs and HERVs). Curiously, although much more abundant in human DNA, mouse L1-mediated retroposition is 20 times more likely to be mutagenic to mouse genes relative to human genes. My suspicion is that this might be related to the much higher activity of mouse versus human ERVs which I suggest triggers this mutagenesis. Why do mouse and human differ this way with regard to L1? Unlike the primate genome, in the rodent genome, L1 does not appear to be highly dynamic. When L1 sequences were examined in 30 rodent species from nine genera they showed low levels of variation. Thus, there seem to be clear constraints on the sporadic L1 amplification that sometimes occurs in evolution. In rapidly evolving microtus species, L1 variation is not seen at all. This is especially telling with regard to recent speciation of microtus as it has undergone a very recent explosive radiation (in the last 1.2–1.6 million ybp) generating 60 extant species. Thus, although microtus evolution is recent and ongoing, L1 variation is not involved and unlike Y-associated ERVs, L1 phylogenetics does not resolve these species. In fact LINE-1 elements seem to have undergone reduction in some of these species. *Microtus montanus* compared to *Microtus ochrogaster* (prairie vole) shows that they have same LINE-1 makeup relative to each other, but are missing many other LINE-1 elements found in other rodents and primates. Although the link between ERVs and LINEs is not clear, it is clear that pattern LINE amplification cannot explain ERV colonization or host speciation nor can they explain retroviral evolution.

Y and ERV and Speciation

The uniform use of the X and Y chromosomes for sex determination distinguishes mammals from reptiles and avians. In the mouse, I have asserted that the brain is a terminal target of sexual identity in order to control behavior, which would seem necessarily linked to sex chromosomes. A mouse brain (as in all mammals) will develop into a set female behavior if not in the presence of testosterone producing fetal testes. Estrogen receptors are essential for this as receptor mutants clearly result in mascularizing (de-feminizing) the developing brain. Thus, a genetic control of gonadal tissue and sex hormone production results in stable brain development and behavioral alterations. In contrast to some fish and amphibians, sensory exposure (temperature, social circumstance, pheromonal) no longer directly determines mammalian sexual development and identity. In most mammals, the X and Y chromosomes are very different in size and content from each other. The Y is much smaller, mostly in a constitutive heterochromatin (condensed, late S-phase replicating) state with few genes and lots of highly repeated ERV DNA. The human Y has only 78 genes, and the SRY locus is at the tip of Y, ~1,000 bp coding for 204 aa domain that binds DNA. The mammalian Y evolved at time when ferns and reptiles prevailed. This seems rather late in evolution of animals. The Y chromosome does have some euchromatic portions, but even this is mainly made up of repeat DNA. The mouse Y chromosome seems to have only about 30 genes within this euchromatin. In terms of evolution, the Y chromosome shows an interesting dichotomy. Within one species, the Y is more stable than autosomes (presumably due to less recombination) and can be used to trace paternal lineages for many generations relative to autosomes. However, between species, the Y chromosome is the most rapidly evolving relative to autosomes and is diagnostic for any placental species. Thus, the mouse Y chromosome is evolving faster than autosomes mainly due to acquisition and amplification of repeat elements, such as ERV-derived elements as discussed above with MuRVY in various mus species. For example, a significant portion of IAPEs is found in tail to tail arrangement on the *M. musculus* Y chromosome, which is not seen in the *M. spretus* Y chromosome. Although the mouse X and Y are very different in size and content, a 65 Mb region of the Y is homologous to the X chromosome over two small regions, which contains about 13 genes (including the SRY genes involved in sex/testis determination). Mammalian SRY genes are evolving 10X greater than other gene families. However, this potential pairing region is not uniformly conserved in all rodents. In some microtus species, for example, no pairing between X and Y is seen. Also some microtus species have an X chromosome with extended Y-like heterochromatin regions. The most extreme example of Y variation is found in the mole vole, which has no Y chromosome or SRY region. In some microtus and other rodent species, it seems sex determination has been significantly altered. In some South American mice, half of the females have XY chromosomes but are functional females that bare young and are biologically fine. Some species, such as lemmings, can have large numbers of XY females as well as XX females. Sex reversal seems to have occurred by unknown mechanisms in these species. However, in eight known mouse species (*Microtus cabrerae*), the sex reversal in XY females seems to be Y linked. These eight species can have extremely large sex chromosomes, with large blocks of heterochromatin consisting of highly repeated DNA. In field voles (*Microtus arestis*), some individuals have an alternative version of the Y chromosome, the Lund Y, which has a much longer short arm and can be distinguished by cytogenetic staining patterns (G-bands) visible on the Y. It is interesting to note that human Y chromosome also shows only 20% of the nucleotide diversity of autosomes but that G-bands similarly distinguish human from chimpanzee Y chromosome, our closest relative. We might now ask why the Y chromosome is so stable relative to the autosomes, given its non-coding selfish 'junk-DNA' makeup. We can also ask why the Y chromosome varies so substantially and reliably between species given its stability and its selfish or presumably dysfunctional genetic makeup. Without exception, in all mammal species, the Y has the most species-specific makeup of all chromosomes. A direct proposal from these facts is that the Y chromosome is providing new sources of sexual and species identity. It would appear that placental mammals are using ERV-like and other repeated DNA on the Y chromosome to somehow create new species identity. Recall our prior discussion on the evolution of vocal learning in songbirds along these lines. If so, this could explain why the Y chromosome has been observed to resemble an ERV 'graveyard'. However, I would expect that not all Y-associated ERVs are dead. They must maintain genetic identity, probably by addiction strategies. Thus, the highly defective ERV-like genomic makeup of the Y can also be thought of as a viral quasi-species that provides the host with a new swarm or group selection and competition against other related elements. The placental Y contrasts with the marsupial Y chromosome which is tiny (only about 10 MB) and lacking any cytogenetic relationship to mouse Y. For example, the dunnart has a tiny Y with no region of X homology. The marsupial Y does seem to preserve SRY function. This must represent a distinct lineage of marsupial genetic identity.

In terms of rodent-specific ERV makeup of the Y chromosome, we have noted that MuRVY (murine repeat on Y) is present at about 500 copies per cell, mostly on the Y chromosome of all *mus* species, but absent from non-*mus* species rodents and that within *mus* species the specific MuRVY makeup varies. However, this MuRVY colonization also seems to be associated with displacement of older ERVs. Recall that the more ancient Gypsy-like chromovirus (LTR-ERV) with *gag, PR, RT, RNaseH* and *IN* ORFs has been conserved by purifying selection in humans, sheep, mice and rats. But the copy numbers are greatly reduced and uniformly low in mammals and other vertebrates. These elements, however, are found in great numbers in invertebrate genomes. Thus, it is very interesting that the majority of remaining Gypsy-like elements in mice are found on the X chromosome, and some of these X elements undergo X chromosome inactivation. These elements, similar to IAP-E, are also placentally expressed as are ERV3 (R) and RD114. ERV3 is closely related to the SY1,2,3,4 elements. All these elements encode an *env* ORF and are found on the Y. Of these, the SY2 homologues are also found in all other mammals so it appears this specific ERV lineage was acquired early in mammal evolution, but later supplemented by SY1, 3 and 4 acquisition in *mus* species. Such patterns of ERV acquisition and displacement support the concept that ERV-mediated Y colonizations in placentals are providing new species-specific and sexual identity.

Y chromosome, Sexual Identity and MHC Odor

What is the role for the placental Y chromosome in group or sexual identity? Clearly, male sex and testis determination function is usually a Y-encoded function via SRY as previously mentioned. SRY regulation also appears to control sex reversal in some species and SRY is often duplicated or ERV associated in specific rodent lineages. Thus, sex, the most fundamental of group identifiers, is clearly subjected to Y chromosome control. SRY appears to be under positive selection during evolution and is rapidly evolving. The phylogenetic patterns of the Y chromosome are highly concordant with speciation. In addition, the Y chromosome can also contribute to antigenic identity. In placental mammals, embryos are surrounded by an embryo-derived placenta, which must necessarily express paternal proteins not present in the mother. If the embryo is male, the genes encoded by the Y (such as SRY) will necessarily be foreign to the mother. It was such Y antigens that led to the initial discovery of the major histocompatibility locus (MHC). The H-Y (human Y histocompatibility) was originally discovered as transplantation antigen when male skin was grafted onto allogeneic female. This resulted in an MHC (in mouse called HLA) controlled T-cell reaction that rejected the male skin. It was also observed that females that had been previously pregnant with the male embryos would make long-lasting T cells against male antigens. This also led to the discovery of minor H antigens, such as the TCR – Db-specific antigens. The locus for this is the H-Y which is also on short arm of Y chromosome in mice. This locus can be deleted in the Sxrb mutation and is separate from the SRY testes-determining genes. Thus, the discovery of the MHC locus is associated with sexual differences. However, we have since come to learn that the mammalian MHC locus is of central importance for the whole adaptive immune system (described previously). In addition, MHC is also linked to odor detection and antigenic identity. Mice clearly show odor-based male–male recognition and aggression stemming from interstrain cues transmitted by VNO neurons. Mother mice also clearly recognize their own and specific offspring via VNO-based odor cues. Mice show a clear link between Y chromosome and agonistic behavior, a behavior that can be of two types: offensive and defensive. Offense can be seen as bite and kick attacks to opponent rumps. Defense behavior consists of lung and bite attacks. Offense is heritable and one or more of these genes reside on the Y chromosome. Genetic crosses indicate male-specific non-recombining parts of the Y are involved in such behavior. Offense behavior can respond to urinary odor types from the Y chromosome.

Production of these odors seems to be due to testosterone-dependent pheromones in male urine. It is also possible that minor histocompatibility genes on the Y are involved (via the SRY locus). Overt aggression can also be seen between ICR males to BALB/c males, mediated by VNO-dependent increase in c-fos expression. This process appears to activate a separate population of neurons compared to other sex-related chemosensory cues, suggesting multiple modes of Y-mediated sex behavior. Other VNO circuits are associated with attraction of male to estrus females. In addition, as noted previously, it has long been known that pregnant mice display odor-dependent Bruce effect to foreign males. This is a selective pregnancy block mediated by peptide ligands of MHC. Clearly, Y-associated odors (antigens) have much influence on mouse social recognition and behavior. Thus, the Y chromosome is not only central to sex determination, but also crucial to communicating sexual identity via odors and initiating behavior. As the Y chromosome allows specific odor-type perception within the same species of mice, it clearly provides individual-specific resolution of identity.

Mouse MHC/OR/VNO Genes

Mice, like most mammals, have both VNO and OR receptor genes within their MHC I locus. With respect to the VNO receptors, we can recall that mouse V2R VNO receptors specifically interact with M1 and M10 gene families of MCH 1 (which have no human orthologs). M1 and M10 are non-classical MHC proteins expressed only in VNO that facilitate V2R expression. Ligands that bind these are not typical MHC peptides and are currently unknown, but are likely to be involved in social recognition and aggression. This VNO system must also contribute to formation of persistent memory as mice can acquire persistent memory in one trial of learning (i. e., the Bruce effect). This learning also likely involves MHC-like peptides. OR genes are also found within the mouse MHC 1 locus. In mouse DNA, the cluster of OR genes is at the distal MHC I locus and appear to have been co-duplicated along with the MHC genes. This constitutes the largest OR cluster in the entire mouse genome, although the reasons for this duplication are not clear. Other mammalian species have retained the link between MHC 1 and OR genes. In humans, the OR cluster in the MHC region is the largest OR cluster of any organism. Both mouse and human have a cluster of OR genes 5′ to the TCR (alpha/delta locus). Mice have six OR genes in this cluster (of 46 total genomic clusters encompassing 1,500 total genes, 3X that of human). The coding regions of OR genes can show especially strong sequence identity between species, in contrast to their regulatory regions which vary. Yet OR genes are among the most rapidly evolving of all genes (new and lost versions are common, see avian section above). In mice, the link between odor receptors and MHC1 is thought to be due to olfactory-driven mate selection. However, experimental evaluation does not always support the view that MHC odor detection is primarily for mate selection. I have argued the mother–offspring recognition is a more fundamental and conserved social bond for all mammals. However, as discussed in the next chapter, humans do not depend very much on odor detection for mate of offspring identification, so clearly an MHC role in a mate selection cannot apply to all mammals. Thus, it is most interesting that all the human VNO genes and many human OR genes found within the MHC cluster have become pseudogenes. Mice and other non-hominid mammals retain functional VNO and OR gene clusters and very much use this system for social recognition (see below).

Vertebrate OR/VNO Receptor Evolution

OR receptors appear to play a central role in establishing rodent and tetrapod group and sexual identity. In tracing their evolution, we can see that the last common ancestor between fish and tetrapods had nine groups of ancestral OR genes (each of which is monophyletic). Currently, OR genes are classified into two classes (I and II) and all are GPCRs of about 310 aa. All fish ORs are class I, whereas mammals are mostly class II (with a few class I). Class II is composed of nine families. Eight of these families are still found in teleost fish. Frogs have almost the same receptor families (with the addition of a gamma family). However, only two of these early OR families are found in mammalian genomes. Fish (and amphibian) OR families are therefore much more diverse than that of mammals. The one OR family that expanded enormously in mammals was the gamma family, which contains 90% of this entire gene family. This family was nearly absent from fish. Why did such major shifts in OR receptors occur if as assumed they were mostly adapted for small molecule food-odor detection? Such molecules would not seem to be changing so quickly during evolution. In addition, mammals (tetrapods) also acquired the V1r receptors; these are 7 TM GPCRs which detect pheromones and are expressed in the VNO. There are roughly 150 members of the VNO family that show no sequence homology to other receptors. Recently, a fish version of a single V1r sequence was reported and is exclusively expressed in punctate chemosensory neurons of olfactory rosette. Unlike other fish OR genes, this unique V1r shows a remarkable sequence variability between fish species, suggesting role in species group or sexual identity, which I suspect is most likely associated with the ability of fish to identify MHC-associated sex partners. This likely is the original V1r that radiated into VNO 14 families composed of about 150 receptors. A second family of receptors, V2r, are also expressed on olfactory sensory neurons of fish and many vertebrates. This family is closely related to calcium sensing and glutamate receptors and detects basic amino acids.

Odor receptors are the largest mouse gene family and make up 2% of all mouse genes and mice have a total of about 1,400 OR genes of which 25% are pseudogenes. Many of these OR genes lack introns, suggesting an RT-mediated acquisition during evolution. The expanded repertoire of mammalian OR genes retains certain biological or expression characteristic. Typically, only one of the OR genes is transcribed in an individual neuron. Thus, OR expression resembles the clonal expression and exclusion similar to that of TCR genes as seen in immune cells. In addition, OR expression pattern is a mosaic of 50% paternal, 50% maternal, with no biallel expression. How this cell specificity is set is not understood but it would seem to provide a system capable of high-level resolution and recognition. As mentioned, the1,400 OR genes in haploid mouse DNA are found to reside in 49 genome clusters of up to 100 genes. This contrast to the 60 total OR genes in found in drosophila. As previously asserted, gene diversity on this scale is suggestive of an identity system, peculiar to placental mammals. Most of these clusters predate the primate–rodent divergence so OR diversity appears to be a mammalian-wide characteristic. Some clusters are species unique. These clustered genes mostly lack homology in their regulatory regions in spite of similar regulatory control. Oddly, the human MHC-linked OR cluster is larger than that found in mouse (36 versus 14) and seems to have complicated transcription and splicing.

In contrast to mouse, humans only have about 800 ORs, of which 50% are pseudogenes. VNO tissue and its receptors as well as many MOE pheromone-related receptors have been lost in the hominid, especially human, lineage. The reason the human OR MHC linkage was so diminished (via pseudogenes) has not previously been understood. Even harder to explain is the complete loss of VNO ORFs in the human MHC locus. This extensive OR pseudogene occurrence, however, does not apply to those OR genes that are expressed in the testis. Testis express greater than 50 OR genes, but these ORFs lack pseudogenes. Given the antiquity of this VNO system, its loss presents a major dilemma regarding human-specific systems of group identity (discussed in the next chapter).

Mouse VNO Receptor Evolution

The existence of VNO sensory neurons, as a small epithelial tissue patch in the nasal/oral mucosa, was initially described in 1813 by Danish veterinarian Ludwig Jacobson (see excised mouse neuron Fig. 7.9). With time, its role in sexual behavior became clear. As noted above, mouse VNO expresses mainly two classes of receptors (V1R and V2R) in segregated epithelial zones (one basal). The mouse has about 100–200 V1R genes and about 100 V2R genes (expressed in the basal VNO structure). These genes appear to have originated from about seven ancestral VNO gene families that have expanded to about 73 members. A third VNO family is also known, V3Rs. VNO neuronal tissue also expresses a third class of receptors known as the H2-MV-9 receptors, non-classical MHC class I genes. It appears that about 10 MHC Ib genes expressed in basal VNO sensory neurons. The mouse V1R superfamily has 187 and rat has

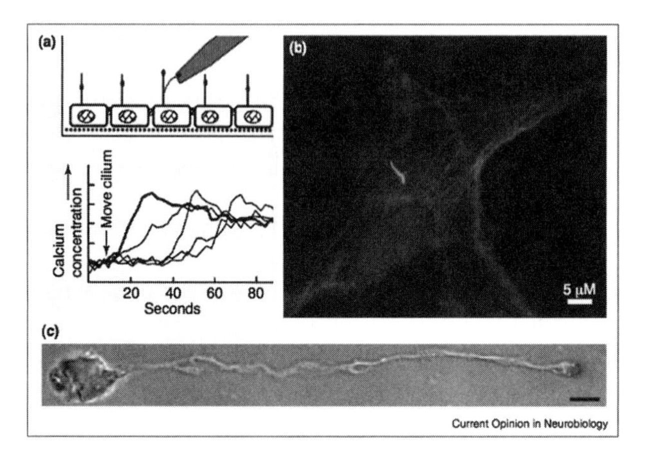

Fig. 7.9 Mouse VNO neuron TRPC expression (*See* Color Insert, reprinted with permission from: Moran, Xu, Clapham (2004), Current Opinions in Neurobiology, Vol. 14)

102 members. In contrast, dogs have only have 8 and cows 32 V1r genes. Thus, it seems that there was a loss of VNO receptors in some carnivores, but a gain in rodents. The large V1r gain in rodents seems to have been mainly by gene duplication. Interesting that V1R also shows dramatic variation between placental and marsupial mammals. Opossum, for example, has 49 V1rs, but this gene repertoire is independent of mammals and suggests a sophisticated pheromone communication in marsupials. As mentioned, humans underwent a massive V1r pseudogene formation that started just prior to separation of hominids from Old World monkeys. In contrast to hominids, New World monkeys show lots of OR and VNO diversity, although they also have many pseudogenes in these ORFs. Bats also display a unique and highly diverse set of VNO receptors. Bats are thought to have radiated rather recently in mammalian evolution. Although the humans have about 200 VNO receptor-like sequences (seven are V1r-like), they are not intact ORFs and this is predominantly due to alu element insertions. The V2R receptors can be seen to be part of the dendrogram that includes the more derived V1r cluster (Fig. 7.6). Most interestingly, the OTR gene (oxytocin receptor) is a sister clade to the V1r cluster. Also, basal to this OTR clade is found the isotocin receptor from ray finned fish (*Astyanax fasciatus*). However, basal to the entire dendogram (including V2R, V1R, ITR, OTR) were CPR2 and CPR1 as found in mollusk pond snail (*Lymnaea stagnalis*). It is worth recalling my prior discussion on origin of oxytocin as likely related to conotoxin in conesnails which would be consistent with this dendrogram.

VNO Receptor Variation Between Mouse and Rat

Given the differences between mouse and rat, the large V1r gene expansion in murid rodents appears to have been relatively recent in evolution. This variation in VNO receptors is puzzling, given the importance we assign to MHC linked to odor detection for group behavior and its conservation in evolution. Within the V1r family rats have about 40 genes, with no pseudogenes. Only in rodents do we see so many V1r genes with so few pseudogenes. Curiously, rodent version of VNO pseudogenes appears to be distinct from human VNO pseudogenes in that the rodents are prodominantly the result of LINE-1 element insertions, not alu insertion, suggesting pseudogene formation can be due to a different kind of lineage specific genetic colonization. For example, the mice L1 insertions (L1_MM) are mainly found in 6D V1R (occurring after mouse/rat split), whereas an Lx insertion is found in V1r loci prior to mouse/rat split. Thus, mouse and rat have diverged with respect to the specific VNO receptors. Rats have about 100 V2R that are pseudogenes.

Other VNO and MOE Genes

VNO neurons also specifically express other associated genes. One of these is the Trp2 gene, which is thought to specifically transduce VNO sensory signals in neurons. The Trp2 gene also appears to have been lost from the hominid genome along with VNO receptor pseudogene formation. Some of the mouse VNO neurons also express H2-Mv genes. These are nine non-classical MHC I receptors. MCH 1b proteins (M1 and M10 s) appear to interact with V2Rs which are exclusively expressed in VNO. Similar to V1r and Trp2 genes, there are also no human analogue to the M10 s. These genes are clearly related to MHC molecules as M10 s show 50% sequence identity to classical MHC receptors. In addition, M10 is structurally similar to classical MHC but differs in that their peptide binding groove is open and unoccupied. This is the first protein structure with an empty class I molecular groove, suggesting that it is normally occupied. However, the ligand for this receptor in not clear and does not seem to be a class I peptide. These M10s are thought to play a direct role in initiating reproductive and territorial behaviors, and likely providing peptide-based identity information. It also seems likely that in addition to the VNO, the MOE can also be stimulated by MCH I-derived peptides. In fact, this MOE response appears to be more involved in Bruce effect than are the VNO receptors. It undergoes induction at early puberty, but also controls social aggression. However, not all mouse VNO-based odor identity is via the urine. Peptide-based receptor recognition also appears to operate from other sources. For example, some peptides originate from the male lachrymal gland and require physical facial contact in order to transfer to the female. Here, the male extraorbital lachrymal gland will make and transfer a male-specific 7-kD peptide to stimulate the female V2R expressing VNOs. Clearly, mice depend on various receptors and ligands to transmit group identity.

TRPs and VNO Transduction

As mentioned above, Trp2 is specifically expressed in the mouse VNO and is involved in V2R signal transduction. Trp2 is also involved in sex behavior, including sex discrimination, male–male aggression, courtship, as well as in maternal aggression toward males in lactating females. These phenotypes can be seen using Trp2 (–/–) mutant females which show much less aggression toward intruding males. Trp2 thus has a central role in behavior-based mouse social group identity. With respect to gender discrimination, male TrpC2 –/– mice are more sexually interest in other males. TPRC2 is also essential for pheromone-evoked male–male aggression. Trpc2 mutant males also fail to establish dominance hierarchies with other males but do display urine marking behavior (presumably MHC associated) like other subordinate males, but mutant males do not show lowered spiking response to urine pheromones. The Trpc2–/– mutation increased male–male sexual behavior (mounting). However, this affect is curiously not seen in other mice in which their VNO was ablated, suggesting TRPC2 can also operate outside of the direct VNO nerve connections. Although TRPC2 is well conserved in most mammals, TRPC2 was much altered in primate evolution, especially the great apes. It thus seems that TRPC2 transduction was a central component of a behavioral identity involved in recognition and aggression in rodents but the great apes have lost this system.

TrpC Characteristics

Trp2 is one member of a family of a diverse group of cation-selective channel proteins with six transmembrane domains that can form diverse heteromultimeric channels (see Fig. 7.2). These channel proteins are considered members of sensory 'signalplex' protein complex and are mostly expressed on neurons. Some of these proteins can be activated by receptor-coupled PLCs. DAG (polyunsaturated fatty acid) activates many TRPCs. The TRPC family are also a Ca^{++} and Na^{++} channel protein activated by GCP transduction via DAG. Trp-related proteins were initially identified by similarity to mutants found in drosophila that affected various sensory modes; vision, odor, taste, vision, hearing, mechanosensation and thermosensation. These proteins are now known to be highly diverse and able to form a large variety of heteromeric channels. The formation of some heteromers is only seen in embryonic brain. Although most TRPC expression is in the CNS, some variable and specific expression in peripheral tissue in known. For example, Trps are also important for T-cell-dependent activation and apoptosis via altered Ca^{++} levels. This expression is associated with store-operated Ca^{++} influx induction, which can kill cells if it is excessive. Some Trps are also expressed in reproductive cells to high levels, such as seen in human sperm. With the completion of numerous

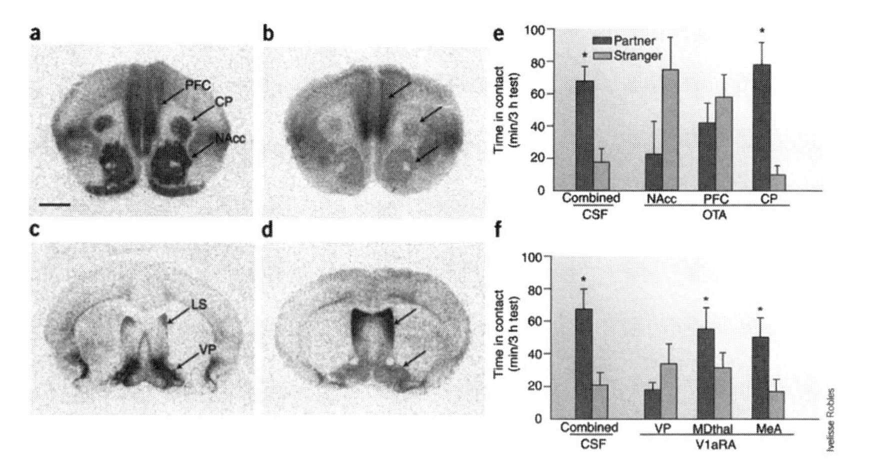

Fig. 7.10 Vole brain, pair bonding and V1aR regulation (reprinted with permission from: Elizabeth A.D. Hammock and Lorry J. Young (2006), Phil. Trans. R. Soc. B, Vol. 361, pg 2187–2198)

genomes, it has been realized that the overall size of the Trp family is much larger than had been anticipated. Trp genes are highly spliced and spread out on the chromosome and they can encompass from 20 to 306 kb of DNA, yet code for only about 1000 aa. Of all the TRPs, TRPC is the simplest member (see Fig. 7.10). *C. elegans* has three of these members but mammals (including mice) have seven members. Humans have conserved six TRPCs. In a total of all the TRP subfamilies, *C. elegans* has 17, mice 28 and humans 27. The *C. elegans* TRP-3 is the phylogenetically basal member of C subfamily and is involved in sperm fertility in hermaphrodites and in males. Thus, early TRPCs were clearly providing a sexual identity function at a cellular level. In *C. elegans*, TRPV (osm-9) is involved in the control neurons involved in 'social feeding'. The mouse TRPC2 is phylogenetically basal to the other TRPC1-7 mammalian members. TRPC2 is found at the dendritic tip of VNO sensory neurons. Since TRPC2 was retained in prosimians and New World monkeys, its loss in humans, by pseudogene formation (ERV fusion), was puzzling.

Although TRPC seems to be a crucial component of VNO-mediated olfaction, TRPC2 function is not synonymous with VNO function as surgical removal of the VNO causes effects beyond TRPC2 deletion. For example, VNO transection will strongly prevent the persistent memory needed for Bruce effect (selective pregnancy failure), but TRPC2 deletions did not. Also, TPCR2 deletions still reject pregnancy when new males are introduced. Yet TRPC2 is clearly involved in olfactory imprinting in which pregnant females must retain individual-specific pheromonal memory of male mates. Another difference is that VNO removal prevents male sex-associated ultrasound vocalizations in response to female chemosignals, whereas TRPC2 deletions do not. VNO ablated mice lose their strong preference to investigate scents from estrus females and will now investigate males equally, although female mounting preference persists. In male guinea pigs, VNO ablated males also fail to investigate female odors, although they remain able to discriminate male from female odors and retain normal sexual behavior. Thus, VNO removal affects preference for urinary estrus female odorants, but not sex discrimination.

MOE and Social Behavior

In addition to the VNO, MOE receptors also contribute to mouse sexual behavior, not just for the detection of volatile odors as is often thought. In contrast to the VNO (and TRPC2), the MOE uses adenylyl cyclase for signal transduction. Mutants in this transduction gene (AC3–/–) mice can still respond to VNO pheromone detection. However, mutation is another MOE-specific gene mutant; the cyclic nucleotide-gated channel alpha2 (CNGA2) results in mice that fail to mate or fight. This one gene has affected both aggression and mating. Clearly, this transduction system has a major affect on sexual behavior and identity. The mouse MOE also expresses trace amine-associated receptors (TAARs), a second family of odor receptors, related to serotonin receptor. These recognize volatile amines as found in mouse urine, including a stress-related compound and pheromones that distinguish male from female mice. Besides volatile urine-based odorants, non-volatile MHC I peptide molecules also function as olfactory cues for MOE (via CNGA2, CNGA4) and also affect social preference. Thus, the MOE has several features that are similar to the VNO/MHC response with respect to odor-based sexual recognition and aggression. This again attests to the redundant and robust olfactory systems used for controlling sexual behavior in rodents.

MHC and the Fundamental Maternal Social Bond

It is often proposed that a main selective reason for the occurrence of MHC-restricted odor detection in mammals (i.e., ORs in MHC locus) is for the purpose of mate selection which allows Darwinian sexual selection to operate. However, some animals, even some with an adaptive immune system (fish, amphibians, reptiles), do not have VNO and OR receptors within the MHC locus, yet they still select mates by odor. In addition, as noted above, mice mutated in various relevant VNO and MOE genes involved in pheromone detection and transduction do not always affect mate recognition. As mentioned, most mammal species do not form stable pair bonds between mates. Thus, in most mammals, mating pairs interact only intermittently so mate recognition is at best a transient affair that requires little long-term memory. However, all mammals bind mother to offspring for extended durations and in this feature differ fundamentally from most reptiles. In addition, even those few species that do pair bond between mates must still establish a mother–offspring bond. As all mammals provide extended care and nursing for their young, I propose this type of social bond that is invariant and fundamental. Furthermore, I suggest that the mechanisms that were adapted for such maternal–$infant bonding would most likely provide the sensory, transduction, neuronal and behavioral systems that could also be adopted by mammals for other, more extended social bonds. Some of these mechanisms are known. In mice, it has long been recognized that mothers can preferentially recognize their syngeneic pups from other pups that differ with respect to MHC. The resolution of this recognition is very high and it has been established that single amino acid changes in an MHC genes can be detected by mothers that will reject the foreign pups. This recognition and the necessary imprinting of memory that results is also the product of post-partum lactation, involving both prolactin and oxytocin. Oxytocin is released in the maternal brain during post-partum suckling and stimulates brain olfactory systems and reward processing areas. Pup suckling is more rewarding to post-partum mothers than is cocaine and dams will chose suckling over cocaine when presented a choice (relative to virgin females). Recall that various novel forms of prolactin were made by the placenta. Suckling also induces prolactin in the brains of dams. Mothers that are mutant in this response (both OT and PRL receptors) are also profoundly deficient in providing maternal care. Also important, nalaxone, a receptor inhibitor of endo-opiates, completely suppresses the normal pre-estrus prolactin surge. Clearly, endo-opiates are involved in the maternal prolactin response. In rats, ultrasounds from hungry pups will also stimulate prolactin secretion in virgin female, so it is also clear that various alternative sensory systems (non-olfactory) can affect maternal PRL states. The relationship between maternal bonding and brain addiction may define a fundamental system for social bonding and is developed further below and in the next chapter.

Maternal bonding also involves the formation of a strong bond of newborns to their mother. Here, like their dams, pups also recognize their familial environment via specific MHC makeup and learn these odors. Even when foster mothers are used, pups appear able to learn this early experience and identify their new family. It is worth noting that these pups have neither placenta nor lactation to provide PRL or OT-based bonding. Male pups bind effectively to their mothers. But this must still be learned. Such a pup state would seem less biologically restricted compared to a pregnant or lactating female. The resulting maternal–pup family (or group identity) is not only enforced by aggression (emotion) but also maintained by emotion (nurturing). And the VNO is important for, but not the sole determinant of such behavior as VNO transections inhibit, but does not prevent some maternal behaviors, such as latencies to maternal behavior.

I have asserted that all group identity systems must be established during limited, crucial periods of development and transfer. In a mouse impregnated by an MHC-specific male, if she encounters sensory input that is incoherent with this MHC identity, or is presented by an alternate peptide identity (via urinary MHC peptide from a mismatched male), the mother–embryo recognition can be thwarted leading to induction of embryo rejection (the Bruce effect). Post-partum, however, the rejection is no longer simply cellular, but becomes behavioral or emotional. Mothers that encounter MHC mismatched pups now react aggressively (emotionally) to reject the foreign pup. For this, behavior and emotional learning become crucial. But in rodents, this learning involves MHC. Neither the VNO, its transducers, nor the link between MHC and pheromone olfaction was maintained in humans. Nor do humans have anything like the Bruce effect. Yet humans do form strong and lasting social bonds without involving MHC.

Clearly, the role of olfaction has changed substantially during vertebrate evolution. Mice also express TAAR1 receptor which is involved in social detection used (urine odor detection). This is 33% identical to serotonin receptor 4 and there are 15 TAAR genes in the mouse genome. However, there are many more (57) such receptors expressed in the zebrafish olfactory rosette. This suggests that TAAR genes had a much more important role early in vertebrate evolution that has been significantly diminished in mammals. Humans retain and express only five trace-amine-associated receptors and their role in human behavior and social identity is not clear. Thus, even with this system, humans no longer depend to any major degree on olfaction for social identity.

CNS, Micro RNA and Sexual Behavior

In rodents, a main purpose of olfactory identity detection is to establish social group membership and to accordingly affect behavior. How then does olfactory sensory input set a CNS-behavior-based social identity? Clearly, this requires lasting changes to neurons, memory that control behavioral (i.e., emotional) states. However, neuroscience is not yet sufficiently developed so as to be able to inform us of the specific mechanisms involved as we do not understand the relevant mechanisms that link to sensory perception and identity transfer. This is especially true with regard to the behavior (emotional) mechanisms and the possible role of behavioral addiction modules might play in group identity. Still, it is worth considering some plausible scenarios until such information is known, especially if we can extrapolate from the better studied simple animal models how things might work.

We can recall in *C. elegans* that chemosensory neurons were involved in establishing sexual behavior. Something is known regarding the mechanisms that are involved here and a role for micro RNAs has been established. For example, it is known that lsy-6 miRNA expression is required following sensory Dauer pheromone exposure for the left–right asymmetric expression of chemosensory neurons (ASE L/R) essential for behavioral differentiation. Here, miRNAs prevent the expression of specific mRNA targets and function much like the *C. elegans* innate immune system to silence mRNA. In *C. elegans*, the chemosensory neurons' asymmetric (left/right) expression of chemoreceptors is controlled by a zinc-finger transcription factor. This expression is controlled by mir-273 (miRNA). miRNA involvement in neuronal development is also known for various invertebrates, and programmed cell death can also be miRNA controlled. Clearly, such a process can set lasting changes into nerur-onal function and alter behavior following crucial sensory input. However, the general applicability of such a process to mammalian social behavior is not established. In the mouse, the neurons of main olfactory epithelium eventually project into and through the accessory olfactory bulb to hypothalamus and amygdala. These are brain regions where basal behaviors, such as aggression and sex discrimination, appear to originate. So according to this basic architecture, the proper linkages seem to be present. However, how neuronal connections are set, the link to memory mechanisms and/or specific immune-based olfaction, is not known. Let alone how any of these processes alter stable behavioral (emotional) changes. There are, however, reasons to suspect similarities to what was seen in *C. elegans*. For example, in developing rat and monkey brains, micro RNAs are expressed in temporal waves through the CNS and waves of micro RNA expression have been observed during rat development. In P19 EC cells, miRNA (lin-28) appears involved in differentiation into neurons. It seems clear that at least a subset of miRNA are expressed during mouse neuronal differentiation. There is also reason to suspect miRNA involvement in memory. In rat hippocampal neurons, miR-134 is localized to the synaptodendritic compartment. These RNAs provide negative regulation via inhibition of protein kinase, thus miR-34 negatively regulates the size of dendritic spins, where excitatory synaptic transmission occurs. Extracellular stimulation (e.g., neurotransmitters and BDNF) relieves this suppression. Since it appears that local protein translation is involved in stable communication (memory), miRNA could be providing a mechanism for establishing this.

The possible origin of such putative RNA-based CNS control is also interesting to consider. In *C. elegans*, CER elements (*C. elegans* ERVs) provide the bulk of snap-back RNAs used to form stem-loop precursors of the RNAi reaction. As presented earlier, the entire *C. elegans* RNAi system has clear virus-like characteristics in its transmission and suppression activity (curiously absent in *C. elegans* neurons). It has been proposed that specific neuronal mRNAs are transported to specific sites of synaptic contact within dendrites, where translation awaits specific appropriate extracellular stimuli (such as via BDNF). In zebrafish, miRNA is crucial for early brain development. As we will see in the next chapter, there is much HERV expression associated with human neocortex function, not seen in other primates. Along these lines, it is interesting that rodents also differ from other mammals in conserving BC1 RNA, a retroposon-derived neural-specific small cytoplasmic RNA confined to rodentia nervous systems. Although we cannot now interpret the significance of all these observations, they clearly suggest some role in neuronal function.

Endo-opioids and Extended Social Learning

Above, I introduced some observations linking oxytocin (OT), prolactin (PRL) and maternal behavior to endo-opioids. I have asserted that the main and invariant social units in mammals involve the bond between a mother and her offspring and its converse with the pups. Such bonds must set appropriate, specific and lasting emotional states (via stable learning). In some species of mammals, adult male mates may also be included in social bonds (as in pair-bonded voles, see below). Some mammal species also have larger social organizations involving hunting packs and herds, although unlike avians, these larger social identities are less common and the underlying neuroscience is seldom understood. My premise is that the mother–offspring bond will likely identify most of the mechanisms that have been adapted to these other more extended social bonds. Mice will thus provide our examplar for understanding the basal maternal bonding. In order to result in a stable social bond, an emotional bond must persist after it is acquired from the appropriate sensory cues (mostly olfactory) following birth. Although the onset of maternal behavior is hormone mediated, it still requires appropriate sensory experience to complete the bonding. The first few weeks post-partum is crucial for dam interaction with pups. Pup identities must be learned at this time or dams will fail to establish a bond. Pups separated from dams for more than 10 weeks during this initial period will be rejected. This defines the window of group membership transfer or imprinting and as asserted above, and clearly MHC, oxytocin and prolactin are all involved. I suggest that the acquired emotional stability will be the result of a social addiction module. The dam and pups have become emotionally addicted to one another. The acquisition of this state is also the acquisition of a new behavior (emotion) based group identity. We can also consider that this acquisition was learned; thus such emotional addiction also represents a form of social learning. The result is a committed and stable state of codependency or cooperation. This underlying mechanisms of this acquired behavioral stability most likely resembles that described above for the *C. elegans* Dauer pheromone, but here would result from stable changes to central neurons (i.e., neuron connections, differentiation, apoptosis) that communicate with or control emotional states and memory (i.e., amygdala). The amygdala in particular seems to affect social behaviors such as sexual, pair bonding, mate guarding and parental care behavior in rodents (and emotions such as love and fear in humans). How then do we create such emotional addiction?

We noted above OT and PRL links to endo-opioids in maternal behavior. The transduction of a prolactin signal is complex (see Fig. 7.8) and has the potential to interact with many cell communication systems. Recall that mice mutant in OT receptors fails to develop maternal care or social memory. Also, suckling mice induced PRL in dams via dopaminergic neurons in hypothalamus. Dopamine is the main prolactin inhibition factor and withdrawal of dopamine induces prolactin release. PRL also stimulates beta endorphin release and this release is blocked by nalaxone. It has indeed been proposed that the effects of oxytocin resemble those of addiction. The prolactin receptor also is known to be associated with human female sexual behavior. In rodents, PRL is known to promote parental behaviors, such as nest building, gathering, cleaning, crouching and nursing, but such behaviors are not initiated by PRL. Clearly, this topic needs more study as such observations remain inconclusive.

An assertion I made above is that the maternal–pup rodent bond represents a basal social bond that will be similar to other social bonds in mammals. There is some evidence that supports this idea. In the California mouse (*Peromyscus californicus*), males show the same amount of paternal care as females, with the obvious exception of lactation. These rodent fathers partake in the building of nests, carrying young, grooming young and spend large amounts of time in physical contact with pups. These males have significantly elevated levels of plasma PRL, relative to non-fathers and equal to those of mothers. Interestingly, non-fathers will often commit infanticide when they encounter strange pups. In terms of extended social membership, some carnivores, such as wolves, show communal breeding and pup caregiving systems that also relate to PRL levels. The wolf's mother (an alpha female) nurses her young but also other females that have undergone a pseudopregnancy with elevated PRL levels also nurse the pup of the alpha female. Even males (e.g., alpha male father) provide caregiving by regurgitation of food, licking, playing with and defending their pups. Such parental care occurs in spring which is also coincident with a seasonal PRL peak in both sexes.

Emotional Sets

It seems apparent that emotional states are often linked into sets which I suggest can constitute emotional addiction modules. In mice, the sets of emotions linked to maternal bonding include not only compulsive care of the young, but also fear of threats to them and aggression to intruders. It is known that olfaction, prolactin and oxytocin are also associated with such aggression in addition to being involved in parental care. Fear and aggression are probably the most commonly linked emotional states that can be found present in most all animals, including fish, insects, amphibians and reptiles. Compulsion (a version of obsession) regarding care of young (avians and mammals), however, is not found in many species (i.e., most reptiles, amphibians). In mice, anxiety would seem to be a central component of pup care as separation anxiety is a common emotional response and is seen in most mammalian mothers when separated from their newborns. In mice, some of the genes involved in this anxiety reaction have been identified. They have also been shown to causally contribute to the emotional state via local CNS expression during an anxiety reaction. These same genes are also involved in oxidative stress, reminiscent of early immunity genes in animals. The mechanism by which such gene expression sets an emotional state, however, remains obscure.

Positive and Negative Emotional States as Addiction

In the context of a mammalian social bond, group identity established by emotional mechanisms should require both a long-lasting negative component and a less stable positive component, the basic elements of an addiction module. As just outlined, it seems clear that maternal bonding entails both essentially positive and negative emotional states. Positive emotions would include those of pleasure, joy, love, protectiveness, happiness, contentment. Clearly, we cannot be certain that all mammals experience all these states as known to us. In most addiction modules so far examined, the positive element tends to be acquired early in the establishment of new group identity. The stable negative component may not be apparent until after this initial acquisition. It has long been a curiosity that young mammals (but not other tetrapods, amphibians or reptiles) participate in play, a state clearly associated with social joy and pleasure. It seems likely to me that this must be part of the establishment of social group identity via acquisition of a positive emotional state. Negative emotional states include the emotions of fear, anxiety, depression, jealousy, anger and aggression. Again, we are not certain if all mammals experience all of these states, but it seems clear that most of them can be observed. However, it also seems clear that together these identified positive and negative emotional sets can be learned and applied to both sexual and social group identity. For example, as we have noted, sexual approach and male aggression in mice are learned states subjected to manipulation by olfaction and neurochemistry. In addition, it is well established that the dopamine pathway, in particular, is involved in emotional response to the environment. This pathway is also involved in learned social affiliations and social outcomes associated with stimuli leading to the expression of either appropriate approach or avoidance behavior. It is thus of considerable interest to realize that dopaminergic neurotransmissions are constituents of reward circuits (via mesotelecephalic dopamine system) which are activated by opiates and other drugs and are mainly seen in the basal forebrain, which also mediates CNS effects of oxytocin. Along these lines, recall that the opioid system also specifically regulates prolactin secretion. Clearly, the prolactin/oxytocin system and opioid system are linked and posses the mechanistic capacity to provide an emotion (behavior) based addiction strategy applicable to social membership.

Brain Structures Involved in Oxytocin and Opioids

It is thought that forebrain dopamine plays a casual role in the opioid drug addiction process. These forebrain and limbic structures include the olfactory tubercle nucleus accumbens and the hippocampus. Oxytocin is a neurohypophyseal peptide (as is vasopressin) that resembles some conotoxins. OT is normally synthesized in the brain and released at the pituitary gland. However, the microinjection of oxytocin directly into these forebrain structures induces various stereotyped behaviors, including a sniffing behavior. This reaction would be consistent with a proposed but ancient olfactory role for oxytocin action, such as would be needed for dam–pup bonding. There are also other links between oxytocin and beta endorphins. For one, beta endorphins response is attenuated in a dose-dependent way by oxytocin, and direct brain microinjection studies show the hippocampus and basal forebrain to be most sensitive to this attenuation. In rats, oxytocin neurons (magnocellular) develop strong morphine dependence after just 5 days of exposure. Oxytocin acts to directly inhibit the development of morphine tolerance (addiction) and attenuate withdrawal symptoms in mice. Oxytocin also decreases cocaine-induced hyperlocomotion and serotype compulsive grooming behavior. There is also a link between responses to conditioned stimuli, prolactin and oxytocin. For example, in rats, conditioned fear to a novel environmental stimuli shows a suppression of vasopressin and an augmentation of oxytocin and prolactin release. As mentioned, oxytocin appears similar to some conotoxins (such as omega-conotoxin) that are known to selectively bind specific neuronal receptors (such as acetylcholine). These toxins retain sexual activity in rat CNS as microinjection induces penile erection and increases NO production in hypothalamus. Such links between these peptide hormones and CNS addiction are far from definitive. However, they provide plausible pathways by which oxytocin, olfaction and opioids could work in concert to establish addictive behaviors.

Addiction as Social Learning

I have asserted that the maternal bond is a form of social learning that uses addiction systems. It has previously been suggested by several researchers that the functional tolerance to opiates can also be considered as a form of learning, as a repeated sensory stimuli leads to long-term changes to nervous system function. In an odd sense, drug addiction is a highly stable form of behavioral learning since repeated sensory exposure has led to stable behavioral adaptations. That oxytocin can also affect addiction and withdrawal reactions noted above would appear to provide it with a means of regulating learning that might occur via such a process. Consistent with this idea, naloxone not only induces withdrawal from opioids but also induces profound excitation in oxytocin neurons along with this withdrawal. Other drugs that can be precipitated by withdrawal can also enhance oxytocin secretion. Oxytocin neurons also strongly express nitric acid synthase and administration of nitric oxide inhibits oxytocin neuron activity. This NOS response could suggest a linkage between OCT to immune cell modulators. Interestingly, other links between opioid and immune responses have been reported. For example, morphine causes immunosuppression and inhibits NF-kappaB nuclear binding in neutrophils and monocytes. NF-kappaB is the main transcriptional regulator of the adaptive immune response. Morphine also seems to act like NO in inhibiting phagocytosis and oxidative burst in neutrophils via mu(3) receptor. Morphine will also affect NO production to prevent oxytocin-induced penile erection noted above. Interesting to note that antagonist to the cannabinoid CB1 receptors also induce penile erection, distinct from the oxytocin receptor response. Although opioid and morphine reactions have been mostly studied in relationship to drug abuse, this system may well be essential for psychosocial experiences and membership via associative learning, such as pair bonding, maternal attachment and imprinting.

Adverse social experiences (aggression, subordination) can also lead to learned states that also act via the dopamine and opioid systems. In mice, for example, it is known that social conflict can be associated with two forms of analgesia: opioid and non-opoid. The non-opioid form is also scent associated as well as associated with aggressive conspecific defeat experience (both mate and VNO associated). The opioid form is not scent associated and occurs in response to extended aggressive conspecific attack. This produces a social defeat stress, resulting in profound aversion to social contact. This aversion is a stable conditioned state. BNDF expression is required for the development of this long-lasting behavior, suggesting induced changes in neurons. This was shown by using a recombinant virus expressing inhibitor of brain-derived neurotrophic factor (BDNF); thus BDNF expression is required for development of social aversion. Mice experiencing repeated aggression thus have long-term behavioral and neuronal consequences. We might call this the toxic outcome of an addiction module. In this case, social contact has lost its pleasure but retains toxicity, so mice become aversive to social contacts. This aversion state can be alleviated by long-term administration of antidepressants that affect the dopamine pathway.

REM Sleep, Learning and Imprinting

Sleep seems to be important for memory consolidation. Thus, we would expect it to be involved in the establishment of social group membership, such as maternal bonding in mammals. A nearly universal property of new memory formation is that it requires protein synthesis. Memory consolidation is a thus a biochemical process, but it is unknown how sleep links to the other processes for social bonding and learning outlined above. It is thought that memory consolidation requires activity-dependent translational control resulting in long-term alterations to dendrites. Thus, local (dendrite) protein synthesis is suspected to be involved. Eutherian mammals developed NREM sleep from primitive REM sleep. NREM sleep thought to be especially important for consolidation and maintenance of long-term memory, which is also associated with tonic immobility. Tonic immobility as seen in REM sleep can also be elicited by hypnosis. Hypnosis is not often considered in the context of memory or learning. It is important to introduce hypnosis now as we consider sleep and memory since in the last chapter we will again consider that hypnosis can induce false memories.

Also interesting in relation to addiction is that irritative pain will prolong periods of hypnosis, but can be blocked by naloxone. Morphine also potentiates the duration of hypnosis, suggesting an opioid mechanism may be active during hypnosis. We do not understand this situation well enough to allow a clear evaluation of its significance. However, since I will again consider sleep, memory and hypnosis and their relationship to dreams in the last chapter regarding their involvement in human group (social) identity, hypnosis needs to be remembered in the context of sleep and memory. The REM link to memory consolidation that emerged in eutherian mammals, however, is clearly both consistent with and congruent with the evolution of the maternal social bonds as a major system of group identity.

Vision-Based Group Identity in Mammals and Primates

Most mammals (including rodents) are basically dichromatic with respect to their color vision. This contrasts with insects, fish and birds which all have an expanded capacity regarding the spectra of their vision, and this also applies to group identification. The reduced spectral sensitivity in mammals has been rationalized as being due to the biological strategy needed during the early evolution of mammals. As mammals are endothermic, unlike reptiles and dinosaurs, they would have the capacity to be active at nocturnal periods. Given the dim light of the night, it has been reasoned that these early mammals did not need full spectral color detection, but rather needed to see low light with greater sensitivity and resolution. Thus, most mammals have only two, M/LWS and SWS (medium/long and short wave), visual opsins. Prosimians (which are nocturnal) lack a functional SWS opsin gene but lemurs (which are closer to human branch) have this receptor. However, primates have trichromic vision due to the acquisition of a red receptor (from gene duplication and divergence on the X chromosome). The selective pressure that led to primate color vision, however, has been obscure. It has often been asserted that being able to see red fruit, for example, was important for arboreal primates, but clearly other arboreal fruit eaters did not similarly acquire red vision. Another idea is that the adaptation of primate color vision was for sex and group membership purposes. For example, many primates (but not rodents) frequently do have colorful female sexual displays during estrus, consistent with an X-linked color vision. Most mammals are not colorful, unlike birds, and do not seem to use color for sex or group markings. Although mice do have some degree of visual recognition ability, this is clearly less developed than that seen in primates. Rodents are like most mammals and still depend much on olfaction for group and sex identification. Thus, unlike humans, mouse group membership is color blind. Indeed, in all the Eutherian mammals, only primates possess trichromic color vision and here it is linked to the sex chromosome. However, in order for color vision to be used for sexual identity, I would expect that it would also need to coevolve links to learning and social bonding, or some form of vision-induced behavioral addiction. This idea is developed in the next chapter. As will be presented, vision indeed seems to be especially important for social identity in the African primates. New World primates are intermediate between other mammals and African primates and show much species-specific variation in the occurrence of color vision. Some species have three alleles of distinct spectral sensitivity in single X-linked locus (gene) that can undergo random X inactivation. Thus, heterozygous females can have trichromic color vision (opposes to males with dichromic vision via random X inactivation and mosaic expression in eye). Therefore, even in New World primate species, color vision is sex linked. Since African primates have also loss of much of their sex-linked olfaction, together, this strongly suggests primates underwent an evolutionary shift from olfactory to vision-based group identification that was uncommon in other mammals.

Family Identity, Pair Bonding in Rodents

Prairie Voles, Pair Bonding and Emotional Addiction

The evolution of the placental life strategy also required close coordination between the mother's adaptive immune system and that of the embryo-derived placenta. The allogeneic placenta must not be recognized as foreign and rejected by the mother's adaptive immune system. Mothers use both olfaction and MHC recognition to immunologically and individually identify their offspring. These two identity systems (OR/VNO olfaction and MHC immunity) are also genetically linked in placental mammals. In rodents, maternal behavioral bonding is mediated MHC olfaction and by the action of prolactin and oxytocin on the CNS, resulting in learned stable emotional states. As the placenta is also known to produce many placental versions of prolactin, it is not hard to envision a placental role in maternal behavior. However, male bonding to female mates or offspring could not similarly have a placental role. Such bonding is not common in mammals, and the great majority of placental mammals (97%) are promiscuous and do not form pair bonds between mating pairs. Those few species that do undergo pair bonding, however, have been highly informative. There exist closely related vole species that differ with regard to pair bonding phenotype and both pair-bonded and promiscuous species are known. The prairie voles will form pair bonds between mating pairs during their first mating. They engage in extended matings that allow bonding to develop during a 24-h period. The resulting pair bond will last through lactation of young and also establishes biparental care for the young. In the case of prairie voles, it is clear that olfaction exerts crucial control over reproduction and mating. Male vole urine odor will not only initiate sexual behavior in virgin female (to seek sex partners) but also induce virgins to grow ovaries and develop mature eggs. Glucocorticoids and fos expression are involved in this response. In addition, VNO removal in females prevents estrus induction by male odors. These transected VNO females also failed to pair bond. Thus, olfaction is a crucial sensory signal for this bond. In contrast, in polygynous solitary meadow vole (such as microtus), removal of VNO in estrus female did not affect mating for animals housed in short photoperiods, rather for voles housed in long photoperiods, a significant increase in mating was observed. These promiscuous voles have spontaneous estrus and lack the induced estrus of the monogamous, group-living prairie voles. It has become clear that both oxytocin and vasopressin are centrally involved in this pair bonding. CNS administration into females of oxytocin and antagonist, respectively, facilitate and inhibit partner preference formation. In males, vasopressin facilitates bonding whereas V1a receptor antagonist (for vasopressin) inhibits pair bonding. It is most interesting that the prairie vole V1a receptor is located and expressed within the reward pathway of vole brain (see Fig. 7.10). These receptors are G-couples 7 TM receptor proteins.

Given that these social differences occur in two highly related vole species (99% similarity in coding regions), we can explore how such differences might have evolved. Genetic analysis has established that differences between bonding and from non-bonding vole species can be seen within promoter regions of receptor genes. There has been a large expansion in V1a receptor 5′ region in prairie vole, which is seen by the presence of satellite elements one of which also has a LINE-1 element next to it. Somehow this change has made the CNS Vr1 receptor expression subjected to sensory (olfactory) regulation in the prairie vole (by unknown mechanisms). Although many think such satellite expansions are due to DNA polymerase slippage, as I argued in prior chapters, a more likely explanation is due to retroposon activity (genetic invasion and unequal excision). Thus, I would argue that the altered Vr1 regulation is a likely example of regulatory evolution by genetic invasion. Along these lines we can note that a major genetic distinction between these vole species can be found in their sex chromosomes in that they differ significantly by LINE-1 and ERV element makeup, especially in their Y chromosome. This is very much in keeping with the distinctions in Y makeup between murid species described above. It is also interesting to note that by using a recombinant virus that overexpresses the Vr1 receptor, this gene can be delivered into the brains of voles and other non-pair bonding rodents and induce increased states of pair bonding without mating. Pair-bonded voles clearly have several characteristics of an addiction state, similar to that of opioids, and dopamine plays critical role in male vole partner preference formation. The brain regions involved in increased receptor expression include the nucleus accumbens, which is associated with reward learning and addiction. Furthermore, apomorphine can also help induce partner preference formation. With this, the rD2, D2 receptors are involved. These receptors are also involved in opposing control of cocaine-seeking behavior as well as certain behavioral aspects of copulation. It also seems clear that both positive and negative emotional states have also been engaged for the pair bonding, such as sexual pleasure, caregiving, anxiety and aggression to conspecifics all develop. For example, voles given vasopressin receptor in the ventral pallidum were more anxious. Bonded males become aggressive after bonding, whereas they were gregarious prior to bonding. In terms of female aggression to non-bonded males, olfaction seems important and visual recognition would not seem to likely be involved as wrong male must get close to initiate aggression. It also seems that in the females, oxytocin is more involved in pair binding whereas in males vasopressin appears more involved. There is also some evidence that the presence of the bonded mate provides an opioid-like state. For example, in prairie voles, exposure to stressor (via swimming, injection of corticisterone, or corticotropin-releasing factor, CRF) facilitates social preference of male for its female partner (a preference blocked by nalaxone). The converse reaction is not seen in females consistent with a sexually dimorphic neurological circuit in these voles. Other pair-bonded species seem similar. For example, in male guinea pigs, a stress hormonal response (cortisol) is ameliorated by the presence of bonded females, but not by other females. However, unlike voles, female guinea pigs are similar to males in this response.

Voice and Vision in Rodent Social Bonding

Although olfaction seems to be the most basal sensory input for most mammals to establish social identity, it is clear that non-olfactory sensory cues have also been evolved for the purposes of identity and pair bonding. One such example is with the Pygmy marmosets which also undergo pair bonding. Here it is known that a vocal call is specific to the pair bond. Mate pairs have a stable audio call that is retained after 3 years separation from their mates. The male trill also appears able to preclude other males from mating with its paired mate. The calls are also linked to social status and, like odor cues in the prairie voles, may directly affect female reproductive receptivity and ovulation cycles. Although this has not been experimentally evaluated, I can propose how this might develop. It is likely that very similar neurological systems will be used to establish this pair bonding (i.e., oxytocin/vasopressin/endo-opioids). A main difference will likely be instead of the sensory and central olfactory neurons; sensory and central audio neurons will provide or augment the signal that initiates emotional addiction. This will further require enhanced CNS processing of audio patterns in order to attain sufficient resolution and recognition as to provide individual-specific identity. In male and female prairie voles, individual variation in nesting behavior appears linked to variation in microsatellite (GT repeat) composition in vasopressin receptor at the 5′ regulatory region, which was highly expanded in pair-bonded voles. Socially indifferent vole species also lack this satellite sequence in this promoter. Social indifference suggests a lack of feeling (emotion or compassion) for other individuals. So such non-bonded males fail to establish emotional addiction. In a Pygmy marmoset, we might expect that similar circuits have become sensitive to audio neuron regulation, probably also by colonization or alteration of regulatory regions via genetic parasites (likely Y chromosome associated). It is interesting that humans and bonobos also differ from chimpanzees in such repeats (GTs) on their Vr1 receptor promoter. Humans have the bonobo pattern of genetic parasite. However, here, most of the promoter sequence variation between humans and chimps does not seem to have been derived from a possible strand slippage event as the conserved repeat element has also conserved both clean break points and large stretches of non-simple repeat. I suggest this difference is easier to explain as having been due to the product of a distinct genetic colonization and unequal excision. In terms of social bonding, chimpanzees show higher basal aggression to each other than do bonobos or even humans. This issue will be considered in the next chapter.

It should by now be clear that some significant distinctions exist between humans and other mammals regarding mechanisms of group membership. For one, humans do not depend much on olfaction as a main sensory cue for sex, family identity or territory marking. Nor does our MCH locus or immune system seem much involved in this. Humans do show considerable similarity to the great apes in how social identity and membership is attained and this will be the focus of the next chapter. Visual and audio learning, especially language, are much more important for cohesive human social structures. This will be the focus of the last chapter. It also seems clear that humans are very much using cognitive systems to establish group membership, a characteristic that distinguishes us from all other species. Yet still, there is much evidence that the underlying neurological systems, such as the basal mammalian-wide materna-l–infant bond, have been retained in humans. For example, oxytocin and prolactin remain clearly involved in the human mother–child bonding experience. Human males have elevated vasopressin levels during sexual arousal. Also, the basic use of addiction modules as a strategy to create social bonds still seems to operate in humans. For example, the ventral pallidum has been directly implicated to be involved in romantic human love and humans shown pictures of loved ones will activate this same brain region. This region is also involved in drug addiction. Drugs targeting this system are also used to treat various behaviors and disorders associated with social attachment as well as addictive disorders. The chain of reasoning and evidence that leads us from the olfactory-based group identity in mammals to the cognitive-based group identity in humans will now be presented.

Recommended Reading

Jawed Fish: ERVs, Genomes, MHC and Olfaction

1. Boehm, T., and Zufall, F. (2006). MHC peptides and the sensory evaluation of genotype. *Trends Neurosci* **29**(2), 100–107.
2. Britten, R. J., Mccormack, T. J., Mears, T. L., and Davidson, E. H. (1995). Gypsy/Ty3-Class Retrotransposons Integrated in the DNA of Herring, Tunicate, and Echinoderms. *J Mol Evol* **40**(1), 13–24.

3. Cao, Y., Oh, B. C., and Stryer, L. (1998). Cloning and localization of two multigene receptor families in goldfish olfactory epithelium. *Proc Natl Acad Sci USA* **95**(20), 11987–11992.
4. Collin, S. P., and Trezise, A. E. (2004). The origins of colour vision in vertebrates. *Clin Exp Optom* **87**(4–5), 217–223.
5. Dulka, J. G. (1993). Sex pheromone systems in goldfish: comparisons to vomeronasal systems in tetrapods. *Brain Behav Evol* **42**(4–5), 265–280.
6. Freitag, J., Ludwig, G., Andreini, I., Rossler, P., and Breer, H. (1998). Olfactory receptors in aquatic and terrestrial vertebrates. *J Comp Physiol [A]* **183**(5), 635–650.
7. Gottgens, B., Barton, L. M., Grafham, D., Vaudin, M., and Green, A. R. (1999). Tdr2, a new zebrafish transposon of the Tc1 family. *Gene* **239**(2), 373–379.
8. Hamdani, E. H., and Doving, K. B. (2007). The functional organization of the fish olfactory system. *Prog Neurobiol* **82**(2), 80–86.
9. Kambol, R., Kabat, P., and Tristem, M. (2003). Complete nucleotide sequence of an endogenous retrovirus from the amphibian, Xenopus laevis. *Virology* **311**(1), 1–6.
10. Kobayashi, M., Sorensen, P. W., and Stacey, N. E. (2002). Hormonal and pheromonal control of spawning behavior in the goldfish. *Fish Physiol Biochem* **26**(1), 71–84.
11. Kulski, J. K., Shiina, T., Anzai, T., Kohara, S., and Inoko, H. (2002). Comparative genomic analysis of the MHC: the evolution of class I duplication blocks, diversity and complexity from shark to man. *Immunol Rev* **190**, 95–122.
12. Leaver, M. J. (2001). A family of Tc1-like transposons from the genomes of fishes and frogs: evidence for horizontal transmission. *Gene* **271**(2), 203–214.
13. Nanda, I., Volff, J. N., Weis, S., Korting, C., Froschauer, A., Schmid, M., and Schartl, M. (2000). Amplification of a long terminal repeat-like element on the Y chromosome of the platyfish, Xiphophorus maculatus. *Chromosoma* **109**(3), 173–180.
14. Niimura, Y., and Nei, M. (2005). Evolutionary dynamics of olfactory receptor genes in fishes and tetrapods. *Proc Natl Acad Sci USA* **102**(17), 6039–6044.
15. Spehr, M., Kelliher, K. R., Li, X. H., Boehm, T., Leinders-Zufall, T., and Zufall, F. (2006b). Essential role of the main olfactory system in social recognition of major histocompatibility complex peptide ligands. *J Neurosci* **26**(7), 1961–1970.
16. Stacey, N. (2002). Hormonally derived pheromones in fish: exogenous communication from gonad to brain. *J Physiol (London)* **543**, 6S–6S.
17. Thompson, R. R., George, K., Dempsey, J., and Walton, J. C. (2004). Visual sex discrimination in goldfish: seasonal, sexual, and androgenic influences. *Horm Behav* **46**(5), 646–654.
18. van Ginneken, V., Haenen, O., Coldenhoff, K., Willemze, R., Antonissen, E., van Tulden, P., Dijkstra, S., Wagenaar, F., and van den Thillart, G. (2004). Presence of eel viruses in eel species from various geographic regions. *Bull Eur Assoc Fish Pathol* **24**(5), 268–272.
19. Volff, J. N. (2005). Genome evolution and biodiversity in teleost fish. *Heredity* **94**(3), 280–294.
20. Volff, J. N., Korting, C., Altschmied, J., Duschl, J., Sweeney, K., Wichert, K., Froschauer, A., and Schartl, M. (2001). Jule from the fish Xiphophorus is the first complete vertebrate Ty3/Gypsy retrotransposon from the Mag family. *Mol Biol Evol* **18**(2), 101–111.
21. Volff, J. N., and Scharlt, M. (2002). Sex determination and sex chromosome evolution in the medaka, Oryzias latipes, and the platyfish, Xiphophorus maculatus. *Cytogenet Genome Res* **99**(1–4), 170–177.
22. Volff, J. N., and Schartl, M. (2001). Variability of genetic sex determination in poeciliid fishes. *Genetica* **111**(1–3), 101–110.
23. Ward, A. J. W., and Hart, P. J. B. (2003). The effects of kin and familiarity on interactions between fish. *Fish Fish* **4**(4), 348–358.
24. Ward, A. J. W., Webster, M. M., and Hart, P. J. B. (2007). Social recognition in wild fish populations. *Proc R Soc B-Biol Sci* **274**(1613), 1071–1077.

Viruses of Fish

1. Eaton, W. D., and Kent, M. L. (1992). A Retrovirus in Chinook Salmon (Oncorhynchus-Tshawytscha) with Plasmocytoid Leukemia and Evidence for the Etiology of the Disease. *Cancer Res* **52**(23), 6496–6500.
2. Essbauer, S., and Ahne, W. (2001). Viruses of lower vertebrates. *J Vet Med Series B* **48**(6), 403–475.
3. Hoffmann, B., Beer, M., Schutze, H., and Mettenleiter, T. C. (2005). Fish rhabdoviruses: Molecular epidemiology and evolution. *World of Rhabdoviruses* **292**, 81–117.
4. Martineau, D., Bowser, P. R., Renshaw, R. R., and Casey, J. W. (1992). Molecular characterization of a unique retrovirus associated with a fish tumor. *J Virol* **66**(1), 596–599.
5. Zhang, Z., Du Tremblay, D., Lang, B. F., and Martineau, D. (1996). Phylogenetic and epidemiologic analysis of the walleye dermal sarcoma virus. *Virology* **225**(2), 406–412.

Evolution of Adaptive Immunity, MHC and ERVs

1. Alder, M. N., Rogozin, I. B., Iyer, L. M., Glazko, G. V., Cooper, M. D., and Pancer, Z. (2005). Diversity and function of adaptive immune receptors in a jawless vertebrate. *Science* **310**(5756), 1970–1973.
2. Amadou, C., Younger, R. M., Sims, S., Matthews, L. H., Rogers, J., Kumanovics, A., Ziegler, A., Beck, S., and Lindahl, K. F. (2003). Co-duplication of olfactory receptor and MHC class I genes in the mouse major histocompatibility complex. *Hum Mol Genet* **12**(22), 3025–3040.
3. Flajnik, M. F., and Kasahara, M. (2001). Comparative genomics of the MHC: glimpses into the evolution of the adaptive immune system. *Immunity* **15**(3), 351–362.
4. Kulski, J. K., Anzai, T., and Inoko, H. (2005). ERVK9, transposons and the evolution of MHC class I duplicons within the alpha-block of the human and chimpanzee. *Cytogenet Genome Res* **110**(1–4), 181–192.
5. Kulski, J. K., and Dawkins, R. L. (1999). The P5 multicopy gene family in the MHC is related in sequence to human endogenous retroviruses HERV-L and HERV-16. *Immunogenetics* **49**(5), 404–412.
6. Kulski, J. K., Gaudieri, S., Bellgard, M., Balmer, L., Giles, K., Inoko, H., and Dawkins, R. L. (1998). The evolution of MHC diversity by segmental duplication and transposition of retroelements. *J Mol Evol* **46**(6), 734.
7. Kulski, J. K., Gaudieri, S., and Dawkins, R. L. (2000). Using alu J elements as molecular clocks to trace the evolutionary relationships between duplicated HLA class I genomic segments. *J Mol Evol* **50**(6), 510–509.
8. Kulski, J. K., Shiina, T., Anzai, T., Kohara, S., and Inoko, H. (2002). Comparative genomic analysis of the MHC: the evolution of class I duplication blocks, diversity and complexity from shark to man. *Immunol Rev* **190**, 95–122.
9. Ohta, Y., Goetz, W., Hossain, M. Z., Nonaka, M., and Flajnik, M. F. (2006). Ancestral organization of the MHC revealed in the amphibian Xenopus. *J Immunol* **176**(6), 3674–3685.
10. Olson, R., Huey-Tubman, K. E., Dulac, C., and Bjorkman, P. J. (2005). Structure of a pheromone receptor-associated MHC molecule with an open and empty groove. *PLoS Biol* **3**(8), e257.
11. Spehr, M., Kelliher, K. R., Li, X. H., Boehm, T., Leinders-Zufall, T., and Zufall, F. (2006a). Essential role of the main olfactory system in social recognition of major histocompatibility complex peptide ligands. *J Neurosci* **26**(7), 1961–1970.

Evolution of the Vertebrate Brain

1. Jaaro, H., Beck, G., Conticello, S. G., and Fainzilber, M. (2001). Evolving better brains: a need for neurotrophins? *Trends Neurosci* **24**(2), 79–85.
2. Pfaff, D. W. (2002). "Hormones, brain, and behavior." 5 vols. Academic Press, Amsterdam; Boston.
3. Roth, K. A., and D'Sa, C. (2001). Apoptosis and brain development. *Ment Retard Dev Disabil Res Rev* **7**(4), 261–266.
4. Striedter, G. F. (2005). "Principles of brain evolution." Sinauer Associates, Sunderland, Mass.

Oxytocin and Prolactin

1. Argiolas, A. (1999). Neuropeptides and sexual behaviour. *Neurosci Biobehav Rev* **23**(8), 1127–1142.
2. Grimmelikhuijzen, C. J. (1983). Coexistence of neuropeptides in hydra. *Neuroscience* **9**(4), 837–845.
3. Kanda, A., Satake, H., Kawada, T., and Minakata, H. (2005). Novel evolutionary lineages of the invertebrate oxytocin/vasopressin superfamily peptides and their receptors in the common octopus (Octopus vulgaris). *Biochem J* **387**(Pt 1), 85–91.
4. Smeltzer, M. D., Curtis, J. T., Aragona, B. J., and Wang, Z. (2006). Dopamine, oxytocin, and vasopressin receptor binding in the medial prefrontal cortex of monogamous and promiscuous voles. *Neurosci Lett* **394**(2), 146–151.

Insects, Virus, Bees

1. Fujiyuki, T., Ohka, S., Takeuchi, H., Ono, M., Nomoto, A., and Kubo, T. (2006). Prevalence and phylogeny of Kakugo virus, a novel insect picorna-like virus that infects the honeybee (Apis mellifera L.), under various colony conditions. *J Virol* **80**(23), 11528–11538.
2. McPartland, J., Di Marzo, V., De Petrocellis, L., Mercer, A., and Glass, M. (2001). Cannabinoid receptors are absent in insects. *J Comp Neurol* **436**(4), 423–429.

Avians; Genomes, Viruses, Immunity

1. Fadly, A. M., and Smith, E. J. (1997). Role of contact and genetic transmission of endogenous virus-21 in the susceptibility of chickens to avian leukosis virus infection and tumors. *Poult Sci* **76**(7), 968–973.
2. Herniou, E., Martin, J., Miller, K., Cook, J., Wilkinson, M., and Tristem, M. (1998). Retroviral diversity and distribution in vertebrates. *J Virol* **72**(7), 5955–5966.
3. Tristem, M., Herniou, E., Summers, K., and Cook, J. (1996). Three retroviral sequences in amphibians are distinct from those in mammals and birds. *J Virol* **70**(7), 4864–4870.
4. Vleck, C. M., Mays, N. A., Dawson, J. W., and Goldsmith, A. R. (1991). Hormonal correlates of parental and helping-behavior in cooperatively breeding harris hawks (Parabuteo-Unicinctus). *AUK* **108**(3), 638–648.
5. Witter, R. L. (1997). Avian tumor viruses: persistent and evolving pathogens. *Acta Vet Hung* **45**(3), 251–266.

Mouse: Genomes, ERVs Olfaction, and MHC

1. Beraldi, R., Pittoggi, C., Sciamanna, I., Mattei, E., and Spadafora, C. (2006). Expression of LINE-1 retroposons is essential for murine preimplantation development. *Mol Reprod Dev* **73**(3), 279–287.
2. Jones, S. (2003). "Y: the descent of men." Houghton Mifflin, Boston.
3. Levy, J. A., Oleszko, O., Dimpfl, J., Lau, D., Rigdon, R. H., Jones, J., and Avery, R. (1982). Murine xenotropic type C viruses. IV. Replication and pathogenesis of ducks. *J Gen Virol* **61 (Pt l)**, 65–74.
4. Ono, R., Nakamura, K., Inoue, K., Naruse, M., Usami, T., Wakisaka-Saito, N., Hino, T., Suzuki-Migishima, R., Ogonuki, N., Miki, H., Kohda, T., Ogura, A., Yokoyama, M., Kaneko-Ishino, T., and Ishino, F. (2006). Deletion of Peg10, an imprinted gene acquired from a retrotransposon, causes early embryonic lethality. *Nat Genet* **38**(1), 101–106.
5. Yoon, H., Enquist, L. W., and Dulac, C. (2005). Olfactory inputs to hypothalamic neurons controlling reproduction and fertility. *Cell* **123**(4), 669–682.

Voles and Pair Bonding

1. Aragona, B. J., Liu, Y., Curtis, J. T., Stephan, F. K., and Wang, Z. (2003). A critical role for nucleus accumbens dopamine in partner-preference formation in male prairie voles. *J Neurosci* **23**(8), 3483–3490.
2. Bales, K. L., Kim, A. J., Lewis-Reese, A. D., and Sue Carter, C. (2004). Both oxytocin and vasopressin may influence alloparental behavior in male prairie voles. *Horm Behav* **45**(5), 354–361.
3. Curtis, J. T., Liu, Y., and Wang, Z. (2001). Lesions of the vomeronasal organ disrupt mating-induced pair bonding in female prairie voles (Microtus ochrogaster). *Brain Res* **901**(1–2), 167–174.
4. Hammock, E. A., Lim, M. M., Nair, H. P., and Young, L. J. (2005). Association of vasopressin 1a receptor levels with a regulatory microsatellite and behavior. *Genes Brain Behav* **4**(5), 289–301.
5. Hammock, E. A., and Young, L. J. (2005). Microsatellite instability generates diversity in brain and sociobehavioral traits. *Science* **308**(5728), 1630–1634.
6. Landgraf, R., Frank, E., Aldag, J. M., Neumann, I. D., Sharer, C. A., Ren, X., Terwilliger, E. F., Niwa, M., Wigger, A., and Young, L. J. (2003). Viral vector-mediated gene transfer of the vole V1a vasopressin receptor in the rat septum: improved social discrimination and active social behaviour. *Eur J Neurosci* **18**(2), 403–411.
7. Meek, L. R., Lee, T. M., Rogers, E. A., and Hernandez, R. G. (1994). Effect of vomeronasal organ removal on behavioral estrus and mating latency in female meadow voles (Microtus pennsylvanicus). *Biol Reprod* **51**(3), 400–404.
8. Pennisi, E. (2005). Genetics. In voles, a little extra DNA makes for faithful mates. *Science* **308**(5728), 1533.
9. Pitkow, L. J., Sharer, C. A., Ren, X., Insel, T. R., Terwilliger, E. F., and Young, L. J. (2001). Facilitation of affiliation and pair-bond formation by vasopressin receptor gene transfer into the ventral forebrain of a monogamous vole. *J Neurosci* **21**(18), 7392–7396.
10. Wang, Z. X., Liu, Y., Young, L. J., and Insel, T. R. (2000). Hypothalamic vasopressin gene expression increases in both males and females postpartum in a biparental rodent. *J Neuroendocrinol* **12**(2), 111–120.

11. Wysocki, C. J., Kruczek, M., Wysocki, L. M., and Lepri, J. J. (1991). Activation of reproduction in nulliparous and primiparous voles is blocked by vomeronasal organ removal. *Biol Reprod* **45**(4), 611–616.
12. Young, L. J. (2002). The neurobiology of social recognition, approach, and avoidance. *Biol Psychiatry* **51**(1), 18–26.
13. Young, L. J., Murphy Young, A. Z., and Hammock, E. A. (2005). Anatomy and neurochemistry of the pair bond. *J Comp Neurol* **493**(1), 51–57.

Chapter 8
Origin of Primate Group Identity: Vision and the Great ERV Invasion

Overview

This chapter focuses on the evolution of group or social identity of primates, the great apes and humans. One aim is to outline the genetic changes that underlie the development of human group identity and to trace the origin of cognitive (social)-based group identity in humans. Our starting point will be the biological basis of pair-bonded rodents, as presented in the last chapter with voles. I will consider what similarities and dissimilarities can be seen to primates and hominids. In the vole model, the formation of a stable pair bonding also involved biparental care. The systems involved included the VNO-based olfaction and a CNS role for imprinting. Also, the neuropeptides oxytocin and prolactin were much involved operating through their receptor expression in brain structures associated with addiction. The resulting bond is mediated by social behavior. Many of these same systems also appear to be involved in avian social bonding (especially prolactin). Thus, in the vole model, olfaction has remained basal and it seems likely that a form of CNS-based addiction is operating. The vole provides the basal exemplar of CNS-based learning in which imprinting group (mate) identity. Although creates the role of vision in group imprinting in the vole is limited, in avian's we see a much expanded role of vision in group and sexual identities (via colorful markings and plumage). Yet, there are some rodents that have developed senses other than olfaction as the sensory input associated with pair bonding. For example, vole species that use specific vocal calls for this purpose. In the vole, we can also note that variation in satellite DNA in the regulatory region (promoter) of the prolactin receptor accounts for the genetic changes that affect pair bonding. In primates, complex, strong and long-lasting social bonding is prevalent, although mate pair bonding is not common, with the exception of gibbons presented below. However, in primates, the role of olfaction in social recognition is much reduced, and it appears that much of this core social sensory system has been lost by genetic interruption. African primates show that most of their social olfactory genes (VNO) and systems (transducers and tissue) were lost. However, African primates also show a gain in visual (color) and audio (vocalization) capacity used for social purposes. In humans, vocal (language) learning

L.P. Villarreal, *Origin of Group Identity*, DOI: 10.1007/978-0-387-77998-0_8,

systems in particular were gained and with their emergence, such systems are much applied to human social recognition. These adaptations were also reflected in corresponding primate-wide gain of the brain capacity, especially regions dedicated to visual, and in humans, audio processing. The next chapter considers how such biological adaptations provided learned or cognitive states of group membership.

About 35 million years ago, primates underwent some clear morphological adaptation. The ancestral primate snout became smaller (likely associated with reduced sense of smell), plus the eye orbits became forward facing allowing enhanced stereoscopic vision, indicating an increased dependence on vision. Later, color vision was acquired in the Old World primates. The visual cortex similarly expanded in these primate brains. As mentioned, catarrhine species uniformly have trichromic color vision, whereas most platyrrhine primates (snout containing) have polymorphic (mostly absent) color vision. Thus, we will consider how these primate and hominid adaptations, that enhanced their visual capacity, might have affected social mechanisms. The coincidental loss of both VNO-based and MHC-linked social identification, especially given its role in the establishment of mother–offspring social bonding in all other mammals, will also receive much attention. Clearly, prolactin remains involved in human maternal imprinting as is oxytocin. This loss in the VNO receptors is coincident with a gain in color vision and the acquisition of complex, brain-based color visual processing. This transition also correlates with the emergence of more complex social bonds and social groups in primates. Thus, the emergence of enhanced visual recognition has been accompanied by an expanded role for CNS-based visual learning used for social membership. This expansion of the large social/visual brain appears central to primate, especially hominid, evolution. The chimpanzee brain is noticeably larger than that of Howler monkey, but the hominid brain is noticeably larger than that of chimpanzee. In comparison to other brains and visual centers, we can note that most small mammals have less than 20 distinct neocortical areas (lateral geniculate nucleus). But macaques have over 30 in their visual system alone (of a total of about 50 overall) and their color-sensitive retinal neurons project to distinct, deeper layers. Thus, primates clearly have higher order visual cortical systems as a phylogenetic addition. Primate brain connectivity, not just size, also evolved. A brain from a lower invertebrate or vertebrate animal has an average of about 500 connections per neuron compared to 50 trillion in the human brain. This was also a big change. We can recall that amphibians generally show less social intelligence and correspondingly had essentially no neocortex. Reptiles do have a more developed forebrain but can also show more complex social behavior. Most of the larger brain adaptations are found in both the great apes and humans, but are absent in monkeys. In the great apes, we now know that facial expression and body gestures are primary systems for social communication. The recognition capacity needed for face-based social (emotional) communication is highly developed and hardwired. This face recognition system also has both a symbolic and a high resolution character to it. Furthermore, it is individual specific. Thus, the facial–social recognition capacity appears ingrained into the function of primate brains and operates through specific neurons. The great apes also have enhanced some vocal and audio recognition capacity, but this is not dramatically different from that of other mammals. For example, the howls, grunts, alarm calls, aggression barks of apes are all used socially but are not so different from those used by other mammalian predators. Humans, however, have acquired the capacity to learn and communicate language, and how language served a group identity function is presented in the last chapter.

What Only Hominids Have and Lack

The biggest biological differences between human and great ape species are found in reproductive systems, and their brains or their minds. Human organs are very much similar to those of the other great apes, so why do our brains differ so much? The doubled size of the human brain is mostly due to the expanded neocortex, relative to that of the chimpanzee. However, the overall patterns of transcription in the various organs are very similar between apes and humans, including the brain. Yet genes active in the human brain appear to have accumulated more changes relative to chimpanzee. The human neocortex layer is relatively expanded, providing a 'learning layer' that is more invasive, by innervation of other, deeper brain structures. In humans, learning after birth is more important than for many other mammals and is needed for the acquisition of many basic functions. In fact, humans are the most helpless of all newborn mammals and seem only able to eat and defecate at birth without additional learning. The human newborn thus seems functionally incapacitated relative to other mammals. Learning has taken on a much more fundamental role in our survival. Also, although other mammals have impressive physical abilities and coordination of movement, such capacity is mostly ingrained or innate in them. Humans, in contrast, can learn many highly complex physical functions, from basic crawling, walking, talking to riding a bike, swimming, gymnastics or flying a high performance aircraft. Clearly, humans start from a position of diminished capacity but can learn highly complex tasks. The most fundamental form of human associative learning is that of social attachment between mother and offspring and this too is learned. In addition, the human brain has other specific functional and physical adaptations regarding learning. Our brain has the capacity for learning complex recursive language that has correspondingly required asymmetric adaptations to our brains. Language acquisition depends on these split brain features. Language acquisition can also be considered an imprinting process that promotes group-specific communication and membership as will be presented in the next chapter as such. Furthermore, since language imprinting is a culturally transmitted system of identity, it allows culture to take on a much more important role in social identity. Human culture owes much to early vocal traditions for its maintenance and transmission. However, human culture has also developed to the point where it can also assume a role in social attachment. Primates, with their enlarged social brains, all show considerable and expanded capacity for extended social attachment, from the fundamental mother–offspring attachments to broader (troupe) attachments. In addition, variability in such social structures is seen in our close relatives as both male-dominated social structures (i.e., chimpanzee) and female-dominated social structures (i.e., bonobo) exist. Humans, however, also form family-based as well as pair-bonded social attachments, in which the father is significantly involved, which goes beyond attachments to mothers or tribes (troupes). Thus, the participation of males (i.e., father figures, tribal chiefs) in social attachment to their offspring distinguishes human from most primate social structures.

Big Social Brains: Their Cost and Social Stability

The earliest ancestors to humans had chimpanzee-sized brains, such as *Homo erectus* and *Homo ergaster*, which appeared about 1.9 million ybp. About 1.9 million years ago, the brain in *H. erectus* underwent a more than doubling in size. Then about 500,000–200,000 ybp, another brain expansion occurred in the human lineage. *Homo sapiens* and Neanderthals both had these larger brains and appeared about 200,000 ybp. However, such a large increase in brain size represents a very expensive resource. A human brain consumes about 25% of total calories compared to 8% on average for apes. In resting newborn, this is even more energy demanding in that 60% of total calories is needed for the infant brain. This seems most wasteful given how the newborn brain is hapless and helpless. What selective pressures might favor such a seemingly unfavorable energetic state? The evolutionary success of humans appears directly related to their more advanced social structures as their physical abilities are modest and/or less capable than that of their great ape ancestors. *H. sapiens* originated in Africa less than 1 million ybp and migrated out of Africa to eventually form thc caucasoids and mongoloids races. This chapter focuses on biological basis and origins of the features of human group identity. The next chapter will focus on the culturally learned social identity systems that can result from the emergence of the human mind, including a role for reading the origin of the modern mind and culture. However, it is argued that the evolution of language as a group identity system appears to have been a basal development in the Homo lineage that led to culture. The biological basis of this language origin will be the focus of this chapter. Along with language, our brains have developed the capacity to create and store abstractions. The stability of language thus relates to the stability of social identity. In keeping with the thesis of this book, the stability of identity will best be maintained via the action of addiction modules. In the context of language, the meaning of language must attain stability if it is to be used either for group identity or to convey thought. The requirement for stable meaning can be equated to the requirement of a belief state, which is a retention system for meaning that has a substrate in the human brain. From this premise, the next chapter will also trace the evolutionary origins and role of belief systems as an extension of language-based group identification system.

Genetic Outline

As the main focus of this chapter is to understand the evolution of human language, we are most interested to consider the genetic events and selective pressures that lead to the enlarged human brain and its capacity for vocal learning and speech. Comparative genomic analysis now allows us to present this issue from the perspective of global changes in chimpanzee and human DNA (presented below). Historically, evolutionary biologists generally adopt the perspective that selection for more efficient brain function results from better sensory and behavioral function. For example, color vision in the great apes, it has been argued, can allow for more efficient foraging for colored fruit, hence greater feeding success and selection. Similar proposals for speech adaptation have also been made (an assertion that is addressed below). Another prevalent idea is that sexual selection (mate selection) may also contribute. In contrast, my main assertion is that selection for group identity is a basic force in evolution that underlies the above situations. The enlarged social human brain should be considered to have resulted from the selective pressure of group identity. According to this thinking, genetic parasites are likely to be much more involved as elements that both disrupt and create genetic identities and promote the emergence of new group identities. It is clear that the primates' lineage was indeed subjected to major genetic disturbance that has continued into the human genome. The most apparent of these disturbances was the widescale colonization by an array of specific endogenous retroviruses (ERVs) and their more numerous defectives (LTRs) as well as expansions of other 'hyperparasitic' elements (LINES, SINES, alu elements). These were all primate (and human)-specific colonizations which are also associated with selective pressures created by exogenous genetic parasites (viruses). This genetic disturbance also incapacitated the extant system of group identity (odor, VNO and MHC based) that had originated in teleost fish and had been maintained in all tetrapods. Thus, these African primates were then released from the selective pressures of odor-based group identity which allowed the expansion of vision-based group identity. In the lesser apes (gibbons), in addition to vision-based identity, we see a form of genetically determined vocal identity (species-specific songs) that is also associated with pair bonding. In the great apes, vision-based group identity required major brain adaptations for sophisticated visual-based recognition. Later, humans also adapted audio-based group identity that, unlike the gibbons, was learned, not genetically determined (human vocal learning and language). This further required major brain adaptations needed for complex audio pattern recognition and group membership. Thus, starting from a seemingly destructive genetic colonization, genetic parasites created the foundation which eventually led to the cognitive (visual, audio)-based group identity in primates and humans. This enhanced cognitive capacity then allowed the evolution of complex human social structures. Let us now start presenting the evidence in support of this assertion by first focusing on the patterns of genetic colonizations associated with primate and human evolution.

A Genetic Colonization and Disruption Leads to Hominids

The biggest difference between humans and chimpanzees is in their behavior and social structures, which suggests that this is most likely due to brain differences. However, in terms of ORFs or coding genes, including those that are expressed in the brain, there are relatively few genetic distinctions between human and chimpanzee as the nucleotide differences in these coding regions correspond to only 1.23%. These modest alterations in coding regions would seem unable to explain the changes involved. Non-coding DNA, however, shows a much greater difference between human and chimpanzee genome. It is also possible that changes in non-coding regulatory DNA, especially of brain genes, might also be involved. Overall, the major human–chimp DNA variation corresponds to about 68,000 INDELS (insertions and deletions). Most of these alterations average about 300 bp and often involve alu and LINE elements. However, there are also large differences in the makeup of ERVs and LTR elements, including many that are 2–3 kb long. This difference in ERVs, LTRs and alu's is especially seen in the Y chromosome, and the patterns of repeat DNA on the Y are human or chimp specific. It is also interesting that these numerous alu elements are derived from nuclear receptor superfamily (estrogen) and that these are only found in higher primates. Forty-five percent of the human genome is composed of repetitive arrays derived from admixtures of these transposable elements. These repeats often correspond to gene-poor regions (non-coding) yet display remarkable conservation, consistent with selection. However, this material has mainly been considered as a genetic wasteland of little phenotypic significance to human survival and evolution. As previously mentioned, early in the evolution of all placentals, it was clear that some ERVs were of direct functional (placental) significance. Human genomes have a curiously more uneven distribution of genes relative to other animals in that its genes tend to occur in more lumped regions. Contributing to this lumpiness, humans have the most LINES of all primates, about twice that found in chimpanzees, which are themselves greater than that present in gorillas and orangutans. Human LINES show evidence of recent activity, but these active elements are human specific and all the older (primate-like) elements are inactive. In terms of ERVs, the human genome has 25,000 HERV-K-related LTR elements. A genetic map of the full HERV-K genome is shown in Fig. 8.1. It differs from many other ERVs in also containing dUTPase. Only 20–50 full proviral copies of HERV-K are present, although these sequences seem to be conserved. Unlike rodents (and sheep), there are no replication-competent

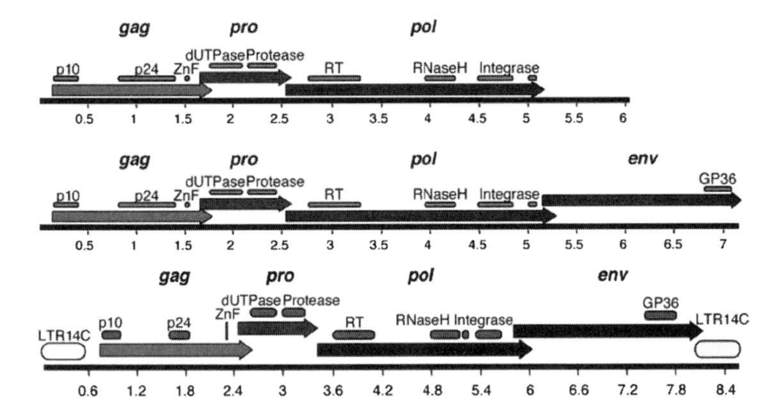

Fig. 8.1 HERV-K gene map (reprinted with permission from: Flockerzi, Burkhardt et al. (2005), Journal of Virology, Vol. 79, No. 5)

retroviruses derived from HERV-K nor are any autonomous versions of this virus seen today. Thus, it seems the original source of this vast number of HERV-K must have been exogenous to the primate genomes. As mentioned in the last chapter, the closest relative to HERV-K is MMTV of mice and also the Jaagsiekte sheep retrovirus. Sheep genome has about 20 copies of enJSRVs that like primate HERV-W and HERV-FRD are needed for placental function.

On a cytogenetic level, more than half of the 48 chimpanzee chromosomes have a distinct terminal C-band, absent from human chromosomes. C-bands will also be of considerable interest regarding gibbon evolution, the lesser apes discussed below. These stainable regions contain heterochromatin repeat sequences that has many satellite, HERV-K and HERV-W sequences in a large block acquired by retrotransposition. Why this has been generated and retained in the chimpanzee genome is not clear. It seems humans and chimpanzees may have been under distinct viral selection schemes. Since humans differ from all the African primates in their relationship to foamy and SIV viruses (see below), it seems clear that humans underwent a significant adaptation regarding these exogenous agents.

It seems clear that there was an early colonization of the ancestral placental genomes by various ERVs, which is also coincident with the displacement and loss of chromovirus ERVs (Ty-1-related) elements. ERV-L elements are found in the genomes of all placental species but their relationship to this radiation is not clear (see Fig. 8.2). A genetic map of the full ERV-L virus is shown in Fig. 8.3. These are present, however, in low copy number, but did undergo expansion in primate and some rodent (mus) lineages. With primate placentals, both HERV-FRD and HERV-W colonized their genomes. These latter ERVs have a clear involvement in placental biology. HERV-FRD is present in all simians. HERV-W is found in all catarrhines (Old World apes). HERV-W (syncytin 1) is a single complete open reading frame including *env* on chromosome 7 and is mainly trophoblast expressed (paternal gene). HERV-W predated New–Old World split but was not conserved (became inactive) in New World

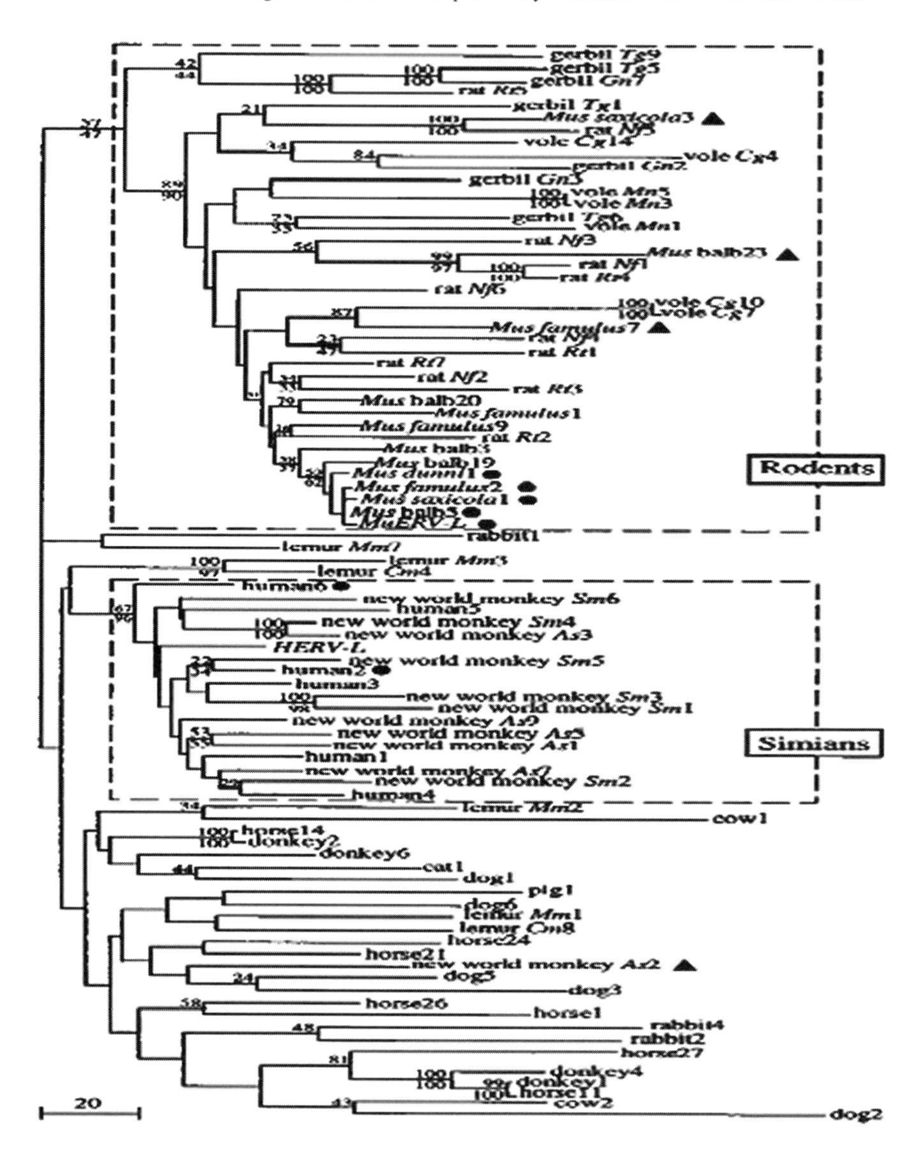

Fig. 8.2 Dendrogram of ERV-L evolution in placental mammals (reprinted with permission from: Benit, Lallemand, Casella, et al. (1999), Journal of Verology, Vol. 73, No. 4)

primates, presumably due to relaxed selective pressure. Thus, many HERVs were fixed into the genomes of Old World monkeys immediately after separation from New World primates (about 35 million ybp). A burst of HERV colonization that is also related to alu retrotransposition and amplification thus characterizes the major genomic changes in evolution of Old World primates. Curiously, much of this genetic material appears to have been maintained for unknown reasons in hominids as testis-specific RNA that is transcribed.

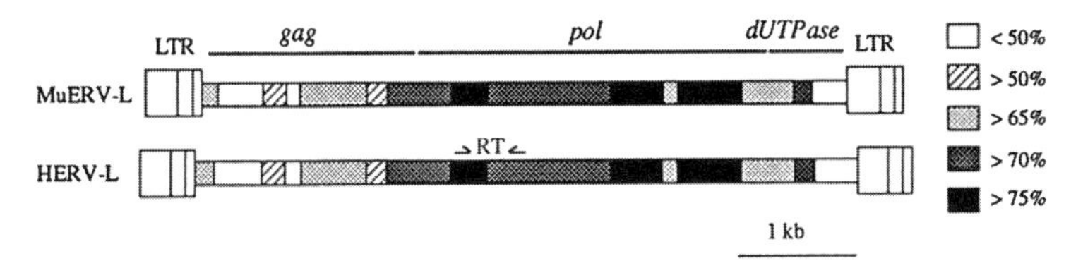

Fig. 8.3 Genetic map of HERV-L (reprinted with permission from: Benit, Lallemand, Casella, et al. (1999), Journal of Verology, Vol. 73, No. 4)

The HERV-H elements are numerous in the great apes and one locus is adjacent to the HERV-W locus (syncytin 1). HERV-H also entered the primate genome prior to the New/Old World separation but is numerous only in Old World primates, remaining low copy New World primates. It is interesting that chimpanzees have retained this specific HERV-H locus, but not baboon (which conserves HERV-W syncytin) or other lower primates. Thus, HERV-H is highly abundant (almost 1000 c/c, mostly defective) in and shows a high rate of evolution in hominids for unknown reasons. HERV-H has also affected various cellular genes (such as zinc finger transcription factors) and like HERV-W and HERV-FRD also expresses *env* sequences in placenta and testis. It is also very interesting that HERV-H is highly expressed in brain regions involved in various disorders, such as multiple sclerosis. We do not currently understand the selective pressures that promoted these various patterns of primate ERV colonization and expansion. The most likely reason would seem to be that exogenous versions of these endogenous viruses were prevalent during hominid evolution and selected for persistence, which is also likely to have affected group identity. It is known that the HERV-W *env* gene will form pseudotype virions with SIV, suggesting that some interactions with exogenous viruses are still observed in African monkeys. However, the most dramatic ERV colonization and expansion in the Old World great ape genomes were due to HERV-K elements. Intact copies of HERV-K (including *env*) are found only in human, chimpanzee and gorilla genomes. In addition, these HERV-K elements conserve a functional integrase. Human evolution, in particular, shows specific colonization by HERV-K family members and an associated expansion of LINE and alu elements (discussed below).

How might such major and lineage-specific genetic changes have affected hominid group identity? At first glance, proposing ERV involvement in human-specific adaptations would seem unlikely (if not preposterous). Given our firm consensus regarding the central importance of gene function in the emergence of a complex phenotype such as cognitive-based behavior, ERVs and their LINE/alu epiparasites would seem to offer no solutions. Yet the gene changes between chimpanzee and humans appear unable to explain such complex functions. Clearly, these ERV colonizations were big genetic events. But they seem much more likely to present opportunities for a destructive outcome rather than events that might promote higher complexity. But, as I have argued, stable colonization by new genetic parasites must confront existing host molecular genetic identity (immunity) systems, incapacitate them and/or displace them in order to persist. In so doing, prepresent the opportunity for coordinated regulatory changes. And if they persist successfully, they create new addiction strategies that promote new group identities.

Sources of ERVs (Foamy and MLV)

Comparative genomics now establishes that most ERVs as found in placental genomes are closely associated with their host species. There is little evidence that ERVs undergo species jumps. This applies to ERV-L, HERV-FRD, HERV-W, enJSRV, HERV-H and the numerous HERV-K family members. These viruses must have originated from mixed sources and therefore are from polythetic lineages of exogenous viruses. Yet in the case of hominid ERVs, no known examples of any such exogenous viruses exist. As mentioned previously, there are some exceptions. Exogenous viruses able to infect other species are especially well known in rodents (i.e., MLV). In addition, species-specific persistent infections with exogenous retroviruses are highly prevalent in some cases (foamy virus and SIV in African monkeys). A genetic map of foamy virus is shown in Fig. 8.4. Its genome shows greater complexity relative to simple retroviruses and encodes several small regulatory genes. In terms of old lineages of retroviruses, the foamy viruses represent the oldest known genus of Retroviridae. Foamy viruses also have several distinct characteristics that resemble other ancient lineages of retroviruses (such as Ty-1 like chromoviruses and pararetroviruses via DNA metabolism). However, their most unusual biological feature is that they are exclusively persisting asymptomatic viruses in their natural host. They were initially discovered from primate primary kidney cultures in the early 1950 s during efforts to use primate cells for poliovirus vaccine studies. In cell culture, foamy virus will induce clear syncytia and drastic lytic pathology; yet, in vivo, no pathology is seen in any primate. In addition, unlike MLV-related viruses, no tumor induction is seen with foamy viruses. Foamy viruses are highly prevalent in non-human primates, but do not naturally transmit or persist in human. Thus, there was a clear evolutionary shift in human–foamy virus relationship relative to the non-human primates. In addition to non-human primates, foamy viruses are found in feline and cattle.

Foamy viruses (FV) display an essentially pure persistent life strategy and induce lifelong persistent infections with no pathology. Transmission is likely via saliva through the oral mucosa. Latent infection is seen in CD4+ leukocytes. Persistence is established via defective interfering particles along with a unique internal promoter that inhibits FV replication. This internal promoter controls expression of the *Bet* protein which is highly expressed and mediates the establishment and maintenance of persistence. Thus, FV have an additional early gene that interferes with early entry and integration of FV and MLV. *Bet* product is most unusual in that it is secreted to and internalized by adjacent uninfected cells

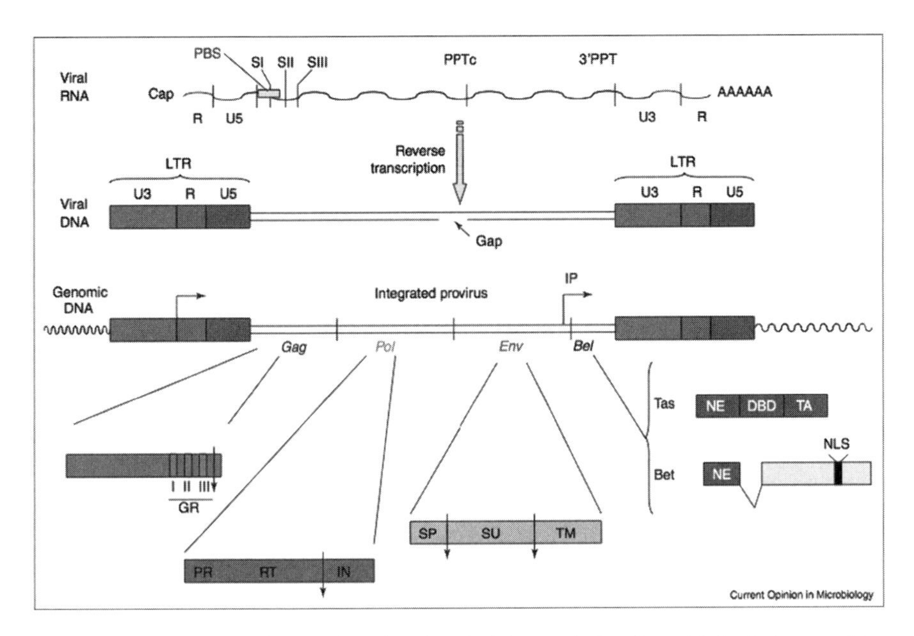

Fig. 8.4 Genetic map of foamy virus (reprinted with permission from: Delis, Lehmann-Che, Saib (2004), Current Opinions in Neurobiology, Vol. 7)

making them also resistant to FV (and MLV). Thus, persistent infections with FV would seem likely to affect the outcome of infection with MLV-like viruses. FV also show alternate DNA metabolism and use asymmetric Ty-like integration at non-coding regions, which presumably explains their low tumor incidence. The *bet* gene product clearly has protective properties consistent with a possible role as part of an addiction module in that it can allow FV to occupy its host and preclude competing retroviruses. Thus, in contrast to mouse in which resistance to exogenous retrovirus infection is encoded by genomic retroviral-derived elements (like mouse Fv1–Fv6 *gag* elements), all FV encoded their own functions for persistence. This seems to define a basic difference between primate and mouse retrovirology. The xenotropic mouse viruses that can infect other mouse species are all MLV-related exogenous viruses in which resistant strains harbor ERVs that block the same virus. Primates lack such a xenotropic–ERV relationship, but instead harbor FV and SIV.

Like most persistent virus infections, FV show an extremely low rate of evolution and also show host congruent evolution and does not appear to be under systemic immune control or selection. FV distribution is so tightly linked to its host that it has been used to reconstruct host (i.e., orangutan) habitat colonization on specific islands. Also, in primate host, there is a direct overlap between FV and SIV distribution for unknown reasons. Many questions are raised but not answered by these observations. Why is this virus–host relationship so prevalent in all primates, but absent in humans? How has it affected primate group identity? What consequences did it have for human evolution?

ERV-L and Primates

Can foamy virus in primates allow us to better understand the likely events leading to the colonization by ERV-L elements? For one, the RT gene of ERV-L is distinct from other retroviruses and shows clear similarity to that of foamy virus. Foamy viruses have several features similar to both chromoviruses and pararetroviruses, thus it seems clear that foamy viruses are ancient and its relatives were present prior to the evolution of placentals. However, Old World simians underwent an expansion from 10–30 c/c of HERV-L to about 200 c/c (also HERV-K, see below). A schematic dendrogram outlining this ERV-L radiation is shown in Fig. 8.5. This expansion was also coincident with brain enlargement in prosimians. It thus seems likely that foamy viruses,

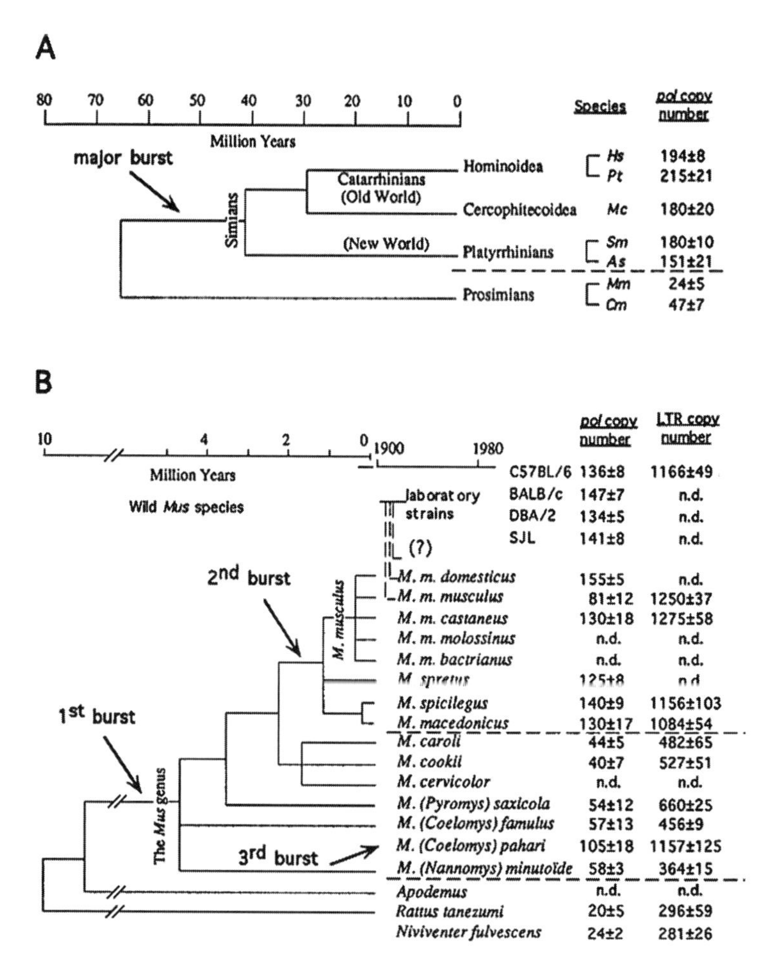

Fig. 8.5 ERV-L copy burst in primates and rodents (reprinted with permission from: Benit, Lallemand, Casella et al. (1999), Journal of Virology, Vol. 73, No. 4)

present when primates evolved, were involved in ERV-L endogenization. HERV-L generally has low expression activity in most human tissues, although curiously it is expressed in brain. It is also interesting that ERV-L-related pol has also undergone amplification in the mouse genome. In fact the species-specific and variable MuERV-l as found on the Y is ERV-L related. However, the source of this ERV is unknown. As this element is absent from the rat, it seems it too must have been derived from some non-rodent exogenous virus, most likely a prosimian source of FV. Thus, it is most interesting to note that the Fv1 gene locus of mouse is a resistance gene for MLV-like xenotropic viruses and this Fv-1 gene is clearly similar to the gag protein of HERV-L and foamy virus. The implication is that early FV (most likely primate) may have provided resistance to MLV-related xenotropic viruses of mice.

MLV-Like Virus and HERV-K

An almost converse relationship to what was described above for FV may exist between primate ERVs and exogenous retroviruses of rodents. Old World primate genomes (but not rodent) are highly colonized by HERV-K-related sequences. However, there are no primate infecting exogenous viruses that resemble HERV-K viruses. Instead, HERV-K is related to MMTV-like viruses well known to infect rodents. Xenotropic rodent viruses seem to matter to natural mouse populations and appear to provide some selection. For example, the Asian wild mouse, *Mus castaneus*, resists polytrophic mouse gamma retroviruses (P-MLVs), which involves a defective XPR1 receptor in resistance. Other factors, such as the XPR1 variant that permits MLV infection, are also involved in virus susceptibility of interspecies hybrids. This resistance is also due to a residual proretroviral gene, the env glycoprotein, in this case a spliced X-MLV env that is found in resistant mice. Although we cannot now identify the specific ancestor to the various and diverse primate HERV-Ks, the rodents clearly host MLV-related ERVs that can form exogenous viruses and infect other species. Thus, we would expect rodent viruses to be likely candidates to have provided the infection that underwent endogenization to become the HERV-K ancestors. And as discussed below, the primate-specific acquisition of MLV-resistant genes (such as APOBEC3) is also coincident with the HERV-K colonization. Indeed, it is precisely such a rodent-derived virus that currently appears to be undergoing endogenization in Australian koala bears (also related to GaLV discussed below). The idea that rodent viruses were ancestral to HERV-Ks is also consistent with phylogenetic analysis in that the HERV-K clade are within the broader MMTV clade. In this cluster, however, MMTV is the only exogenous virus in this basal clade, consistent with MMTV-like viruses providing exogenous virus to form new endogeneous states. Also consistent with this idea, enJSRV appears to be a descendent virus of the basal MMTV clade, as is marsupial enTDV. The non-primate lentiviruses are a sister clade to the other two (HERV-K and MMTV). The most parsimonious explanation of these combined observations is that an MMTV-like rodent virus was ancestral to primate HERV-K.

HERV-K, Hominids and Brains

Besides the link between the expansion of HERV-L and HERV-K in African primates, there was also an expansion of HERV-I that occurred at the time of New/Old World primate split. Why then was there such intense and coincident ERV, alu and LINE activity in Old World primates? Do mixtures of viruses and hyperparasites provide some distinct evolutionary consequence? What did primates get from this invasion? Humans have 30–50 HERV families (all of which appear to have resulted from independent HERV colonizations), 10 of which are HERV-K families. Yet, in spite of these colonizations being roughly coincident, HERV-I, ERV-L and HERV-K are all distinct retroviral lineages and only HERV-K has continued to be active in human evolution. The HERV-K lineage is also distinct in that primate version also contains and conserves a dUTPase gene. A dendrogram of dUTPase evolution is shown in Fig. 8.6. In fact sequence analysis with combined dUTPase and RT supports the idea that HERV-K- and MMTV-related viruses are sister clades which are both descendents from earlier HERV-L-related virus population (which also encodes a dUTPase). In spite of all the evolutionary activity of ERVs during primate evolution, human ERVs do not show polymorphism in integration sites in any extant human populations. Thus, these various integration events appear basal, punctuated and not ongoing but seem associated with the divergences of hominids. Most HERV-Ks are not found in New World monkeys, with the exception of low copies of HERV 5 that are found in both New and Old Worlds. That hominid HERV-K colonization is ongoing, even after the human–chimpanzee split, is supported by 8/10 full-length HERV-Ks being human specific. HERV-K expression was originally observed in human teratocarcinomas, which are able to make HDTV (HERV-K viral-like particles), especially in placental trophectoderm tissue. These VLPs have core structures that contain RT activity and HERV-K RNA but no *env* protein. The reason for their expression and/or function in these tissues remains a mystery as unlike enJSRV, HERV-W and HERV-FRD, they are yet to be experimentally evaluated.

As presented in the last chapter, the original cloning of HERV-K stemmed from the discovery of RT genes of hamster IAPs, which had similar in LTRs (but not RT) to mouse IAP sequences. Thus, ERV evolution is inherently complex and of a mixed viral lineage and such observations are at the very foundations of HERV-K studies. The HERV-K designation is due to the lysine tRNA primer used to initiate replication. However, HERV-K nomenclature, like ERV nomenclature in general, has been most confusing due to the mixed nature of their evolution. Currently, various superfamily or groups of HERV-K are recognized (typically defined by RT similarity). Within these families, HML

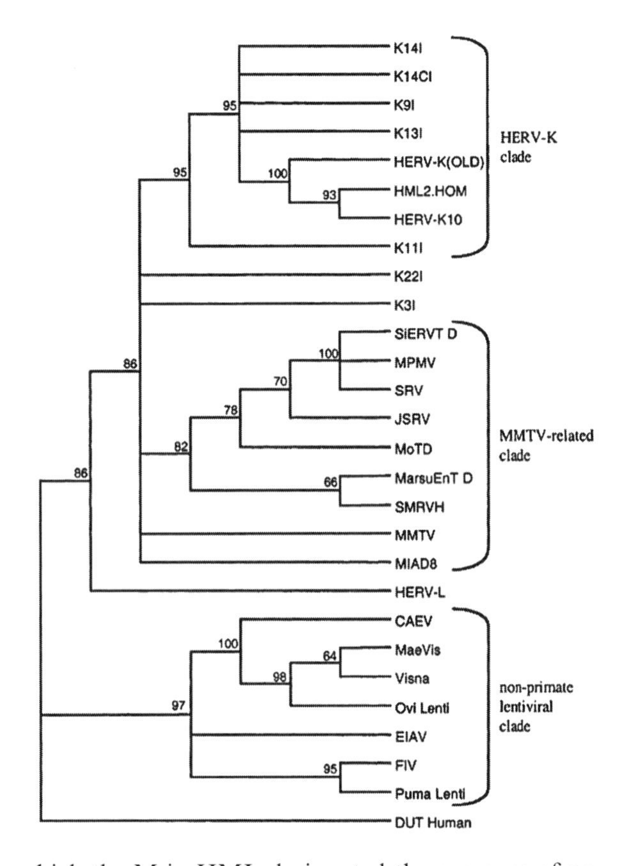

Fig. 8.6 Dendrogram of dUTPase evolution (reprinted with permission from: Jens Mayer (2004), Journal of Molecular Evolution, Vol. 57, No. 6)

designations are used, in which the M in HML designated the presence of an MMTV-like RT (for example HML-1 to HML-6) and 10 HML families have been proposed. The HML-2 and HML-3 families also encode a full dUTPase gene (noted above for foamy virus and ERV-L). dUTPase fragments are further found in HML-6, -5. With one possible exception, phylogenetic analysis suggests that all HERV-K families may have initially encoded a dUTPase gene. One member within the human HERV-K clade, K11I (HML8), appears to be the phylogenetically basal member. One well-studied HERV-K group is the HERV-K(10), which is considered as the parent of several other HERV-Ks. HERV-K(10) has six HML groups (all have dUTPase). All HERV-K groups seem to be monophyletic, including the older HML5 (via dUTPase analysis). However, they are not monophyletic via LTR sequence analysis. Although the HML5 group seems older than other clades, it does not appear to be ancestral to the more recent HML2 group. This observation supports the independent (exogenous) colonizations by HML5 and HML2 ERVs which is also a general characteristic of most ERV families. According to arguments made above, I would also expect a likely rodent ancestor for the exogenous predecessor of HML2. In keeping with this, rodent genomes do harbor a group of beta retroviruses with *env* genes that are HML-2 like. Also, IAPE-like viruses also include the HERV-W-like group (MuERV U1), which would also support a rodent origin of HERV-W (crucial for primate placenta function).

HERV env *genes*

Humans have retained 18 intact HERV *env* genes. One *env* containing HERV-K includes an ERV that seems very recent in human evolution, the HML2 element (found on the human-specific cluster 9). The alu insertion patterns of this HERV are also consistent with the out-of-Africa hypothesis of human evolution. Also, these elements are more diverse in African populations. In addition, heterozygotes of these recent HERVs appear to be under negative selection (75% below expected). Human DNA has 74 HML2 sequences, of which 18 are near complete, but some show allelic variation (mostly in LTR). However, older HERVs also can conserve *env* genes, such as HERV 5 which is found in both New and Old World primates but conserved in an intact *env* (on the Y chromosome in multiple copies). What might conserve HERV colonization and *env* selection? It is most interesting to note that the recent HML2-specific *env* will function as a structural gene to form pseudotype virions with SIV. The implication is that endogenous viruses maintain *env* function to interact with exogenous primate retroviruses. If so, these exogenous viruses could be providing the selective pressure for endogenization and *env* maintenance. HERV-K91 is a human-specific HERV, most likely entering our genome after the split with chimpanzee. One of the most recent evolutionary HERVs in the human genome is the full-length HERV-K113 and HERV-K115. HERV-K113 is found in only 20% of the human population (distributed in African, Asian, Polynesian). HERV-K115 is found only in 15% of the human population mostly in sub-Saharan Africa. It has been estimated that HERV-K113 entered the human genome about 100,000–200,000 ybp, near the time modern humans and Neanderthals emerged. Curiously, HERV-K113 has two LTRs that appear to have diverged at a much older time, 1.2 million years ago. Such divergence might reflect evolutionary events in the ancestral exogenous version of these viruses, prior to human colonization. In terms of LTRs, HERV-K LTR cluster 9 is found only in humans whereas LTR cluster 1 is much older and is found in Old World monkeys as New World monkeys either lack or have much reduced copies of most ERVs. However, this makes clear that HERV LTR and HERV ORF ancestry can often be distinct and unlinked. This African distribution of numerous ERVS, such as HERV-K115, suggest some selective relationship to prevalent African primate retroviruses (such as FV or SIV).

Brain Virus?

Although the evidence of intense HERV activity during the evolution of the genomes of African primates is incontrovertible, what this tells us generally about evolution of the hominids, let alone hominid-specific features such as brains, is at best confusing. On the face of it, surely such widespread DNA integration and amplification must have been bad for the integrity of the ancestral genome. In addition, as presented below, these waves of primate HERV colonization and amplification are also correlated with waves of amplification of other (hyperparasitic) retroelements (LINES, alu's, SINES). How can all this genetic disturbance generate any good for primates? More specifically, how could such colonization affect complex brain development and group behavior as seen in primates? As emphasized, a main characteristic of the human brain is that it has developed an enlarged, hypertrophic and invasive neocortex that controls more and other CNS regions, relative to that in chimpanzee. Such tissue enlargement and invasiveness must have required widespread and stable alterations to the cellular programming and to cellular identity systems, most likely involving the ancestral neuronal stem cells. Retroviruses indeed have the capacity to alter precisely such programming. For example, MLV viruses do integrate into and transform both mouse and human hemopoietic stem cells. These stems cells can differentiate not only into immune cells but also into human neurons following mouse brain implantation. Thus, MLV has the inherent capacity to manipulate and transform CNS cell programs at this level. But did they? For that, we currently have little evidence. Yet some observations seem relevant. ERVs were clearly present in non-human prosimians and primates and each species has its own specific makeup. And although placental and reproductive expression is often seen for many of these ancestral ERVs, the most striking difference in their hominid expression is in brains. For example, HERV-E is highly expressed in human brain, but not in brains of OWM. Various HERV-Ks are also highly expressed in human, but not Old World Monkey brains, such as HML-10 and HML2. In contrast, other HERV-Ks, such as HML3 and HML4 are expressed in OWM brains, whereas HML6 is not. Thus, HERV-E, HERV-Ks(HML 2,6), HERV-W, HERV-FRD and HERV-L were all generally more expressed in human brain (especially in the prefrontal cortex) than in OWM. Why? We do not currently understand the significance of this human brain HERV expression. Also, for the most part this expression corresponds to RNA transcripts, not proteins. Although such RNA expression does have a high potential to provide regulatory functions, it appears unable to provide much direct gene function. Along these lines, it is most interesting that the human-specific HERV-K10(HML2) is significantly overexpressed in active regions of schizophrenic brain, regardless of clinical condition. Since schizophrenia is a disease of higher brain function and social capacity, this seems a most intriguing but currently unfathomable link.

The Basal Hominid; Gibbons, Their Songs, Sex and Virus

As a rule, non-human apes are not vocal learners and do not depend heavily on vocalization for social structures. However, the lesser apes, such as gibbon (genus Hylobates) and siamang, are of special interest regarding group identity.

They are the basal hominid primate (Hominoidea) and thus reflect several core evolutionary developments applicable to group identity of all the great apes. These include the loss of VNO- or MHC-mediated group identity and the use of color visual and vocal systems for social communication. They are also the most species diverse (about 12) of all the ape species. As many Hylobatid species are found in island populations around Indonesia (i.e., Sumatra, Java), they appear to represent ecologically isolated groups that have been maintained since the raising of the ocean level in the Pleistocene period (about 1.8 million ybp). Thus, they may present a snapshot of hominid evolution that has remained isolated. In contrast to most mammal and other primates, they also have a high tendency toward monogamous pair bonding and family-sized unit social organization (about 90% of observed species). As they are arboreal, fruit eating primates they would not offer an ecology expected to favor monogamy. In addition, and also in sharp contrast to most apes, they all heavily use species-specific and sex-specific vocalizations for social communication and pair bonding. From the perspective of the evolution of human group identity, they offer many interesting insights regarding the evolution of language and social bonding.

The persisting viruses of gibbons, for the most part, are very similar to those found in humans and the other great apes. For example, there is evidence that gibbons harbor all major classes of herpes viruses (HSV-1, HSV-2, EBV, CMV). Like essentially all primates, it is known that gibbons harbor herpes B-related viruses in wild populations and that these viruses seem to be species specific. However, field and molecular studies of such viruses have not been done, thus the natural biology or species specificity of this relationship is not known. It is, however, clear that the human type I and II herpes virus can cause lethal infections in gibbons and has at times caused zoonosis in especially young animals of captive gibbon colonies, thus species determined viral pathology is clear. Given the strong general tendency of herpes viruses to cause severe CNS disease in related species, it would not be surprising to learn that gibbon herpes viruses (especially type II) might contribute to sexual isolation via cross species infection. Gibbons also appear to be infected by prevalent, inapparent and geographically restricted versions of hepatitis B virus, which also seems to be sexually transmitted. In terms of retroviruses, essentially all non-human primates seem prone to latent infections with various retroviruses and viral-mediated adaptations seem to have occurred. Like most primates, gibbons appear to be frequently infected by foamy viruses. SIV and HIV can also both infect gibbons in laboratory settings and lead to virus production and disease. However, the natural distribution or consequence of SIV in gibbon populations has not yet been evaluated. Yet, it seems clear that SIV-like viruses have had major impact on gibbon evolution. Gibbons (but not other apes) for example have a RANTES gene mutation associated with HIV disease progression in humans. Gibbons do not have TRIM 5 alpha as do the great apes. Also, gibbon serum normally has anti-retroviral activity (antibody) in natural populations, suggesting common exposure to unknown retroviruses. Most compelling, gibbons are susceptible to gibbon ape leukemia virus (GaLV, a gamma retrovirus) in both captive and natural settings. This was first observed as an emerging leukemia from a pet gibbon, due to reactivation of persisting virus. GaLV is found in many natural gibbon population, but the natural biology or species specificity of GaLV has not been studied. The lesser apes clearly have a distinct pattern of ERV colonization compared to both the monkeys and the great apes. For example, as presented above, we can see that chimpanzees underwent major colonization by gamma retrovirus (MLV/HERV-W related). And we can recall that endogenous MLV-related viruses are prevalent in amphibians and reptiles and rodents. The gibbon GaLV virus closely resembles ERV as found in Mus dunni (MDEV, a VL30/GaLV chimeric env), but most mouse lines are resistant to GaLV. GaLV is also related to koala virus discussed above. In contrast to chimpanzees, humans did not undergo this same gamma retrovirus colonization but instead underwent a beta retrovirus colonization via HERV-K HML-2 (an MMTV/JSRV-like virus). Thus gibbons appear to host many species-specific persisting viruses (including primate herpes viruses, retroviruses, hepatitis B viruses and papillomaviruses) which could all contribute to group identity.

Gibbon genomes show strong evidence of major viral-mediated affects. Indeed a distinguishing feature of the Hylobates is their extreme variation in chromosome karyotypes and size. Hylobate chromosomes vary widely in makeup (rearrangement) and size relative to other apes. Most of this variation is due to variation in C-chromatin content (stainable repetitive heterochromatin). The C-band pattern and chromosome composition is highly species linked in gibbons. Thus, it is most interesting that the chromosomal break points associated with these rearrangements are conserved (as hot spots) in all great ape chromosomes, including human. These break points are LTR flanked, and the LTR corresponds to those of HERV-K family. In gibbons, this break point domain appears to be the simplified (non-duplicated) ancestral version or a 'preduplicated' unit. This corresponds to a 12 kb DNA region in gibbons. It consists of repeat elements of which LTR (HERV-K derived) are the most numerous, followed by SINE-R and simple beta satellite elements (SINE-R has an HERV-K LTR). Most likely, rearrangements are mediated by action retrovirus or their defectives. This hot spot was reused during evolution of great apes. Gibbons (like monkeys) lack the expanded HERV-K colonization seen in the great apes, which includes the HERV-K-LTR colonization of the various (chromosome 3,4,7,11) clusters of olfactory receptor genes (seen in all great apes). Many Hylobatides have an extremely rare placement of C-band and nuclear organizer onto the Y chromosome, suggesting an interesting link of LTRs to sexual evolution. C-bands are usually considered to be made up mostly of satellite DNA. The specific repetitive elements of C-bands, however, appear to correspond to ERV colonization events. For example, early primates (monkeys) underwent both an ERV-L and alpha satellite expansion. Apes underwent an additional HERV-K and beta satellite expansion (duplication). This expansion seems to have been RT (HERV-K) mediated since junctions are alu elements (requiring RT action). Thus, that Gibbons have the preduplicated

C-band hot spot and a low copy of beta satellite duplicon is consistent with their ancestral relationship to the great apes and also consistent with a heavy role for ERVs in this evolution. Indeed, the pattern of SINE-R evolution is directly congruent with the evolution of all great apes. For example, humans have a specific SINE-R.C2 found in their C2 complement component of their immune systems. This element is not found in gibbons, but only in the African great apes. This element not only matches that of hominoid evolution, it also traces the recent evolution of MHC genes (and identity) in the hominids (described below).

Social group identity in gibbons. A very apparent feature of Hylobates is the presence of conspicuous and brightly colored sexual swellings shown by all females. This colorful, visual display is clearly linked to the acquisition of color vision by the apes and defines a basal role for vision in the maintenance of group or sexual identity in the lesser ape. Most likely, it is part of a social (pair) bonding system but the neurochemistry of this situation has not been studied (i.e., possible role of addiction, prolactin, oxytocin or vasopressin). Such visual display presents big problems for theories of sexual selection based on mate choice in a monogamous, pair-bonded species. A one male mating system (pair-bonded monogamy) should not result in conspicuous display of estrus. Similarly, as mentioned above, an arboreal habitat would also not seem suited to such visual sexual display. This cannot be due to male competition for female but must instead be part of bonding communication between mates. It is unknown if any odor detection remains involved in the social structures of the lesser apes. It is, however, interesting to note that a reddened swollen sexual tissue resembles that which might develop from a viral infection. Lactation appears to prevent these sexual displays.

Gibbons clearly have various (non-sexual) forms of social communication. These have been observed to include tactile, visual gestures and facial expressions. Gestures predominated numerically, but facial expression was next most common. These visual communication skills were acquired up until age 6 years. Clearly, visual communication is a major process for gibbons. A strong gazing response to mirror image has been observed, although specific social response elicited to mirror images varies greatly with the individual and appears to be learned, not instinctively set. Gibbons will also align themselves with gaze of others, including humans. Thus gibbons do respond socially to their mirror images and both associative and aggressive reactions can be seen. Gibbons, however, do not recognize themselves in the mirror and uniformly fail the modified makeup mark test. Thus, it is clear that gibbons have good visual and social recognition, but not self-aware as are the great apes.

A particularly intriguing and uniform feature of Hylobate social communication is their species-specific and sex-specific vocalizations or songs. All species produce such songs, thus it seems to be a basal feature of these lesser apes. Most of this singing is associated with mating pairs, and males and females will participate in duetting in most (but not all) species. Such paired vocalization rates are linked to copulation rates. These songs are a loud and complex assembly of notes into species-specific patterns. Mating pairs sing in non-identical duets, and the songs appear to have a clear sexual emotional valiancy, increasing in duration and intensity in duetting mated pairs. It thus appears that these songs serve to link the mated pairs. In considering how this might have developed, I propose it is likely that songs participate in pair bonding during initial copulation. Such vocalizations during gibbon copulation would function like the MHC olfactory/VNO peptides during the first vole copulation described in the last chapter. In gibbons, the song would establish an emotion-based addiction module during initial sex and set pair identity via very similar addiction brain systems as used in the vole (i.e., vasopressin receptor expression). The difference is that the olfactory lobe would not be (or much less) involved (i.e., no MHC or VNO requirement). Instead, the gibbon brain needs to recognize song patterns and link this initial recognition to emotional addiction systems. Later, this system evolved to extend vocal addiction that no longer required ongoing sex, leading to the duet singing between mating pairs. Other apes have retained some tendency toward copulatory vocalizations, especially baboons and to some degree humans, but such vocalizations do not appear linked to pair bonding in the great apes. However, in contrast to songbirds and humans, gibbons are not good vocal learners. The great call produced by gibbon females is mostly under genetic control, not learned. It is acquired between 5 and 32 months of age which indicates a brain-related developmental competence. Such development would require the acquisition of audio processing brain regions able to recognize patterns of specific tone and rhythm. The genetic character of song acquisition is made clear by mixing young female gibbons with adults of different species which clearly establishes that the great call was not learned from parents. Thus, in gibbons, vocal-mediated identity is directly linked to genetic (species) identity. Songs are also used for more extended group identity. Adjacent groups, for example, will not respond to each other and remain silent. Yet when intergroup conflicts do occur, there is a strong vocal nature to it. Although we see above much evidence that genetic parasites are directly involved in gibbon speciation, how these two phenomena (genetic/vocal) might be linked is a mystery. However, some limited vocal learning does occur in gibbons. Variation in duration and intensity of song along with some increase in complexity are associated with mating. In addition, there is some evidence for combinatorial (possibly recursive) use of notes in altered songs as observed during the presence of predators. Possibly the dissonance between these altered songs and expected pattern detection in the brain serves as a warning system. Clearly, in gibbons we see the beginnings of vocal learning and a possible link to emotional systems associated with social bonding in apes. Unlike humans, however, vocal learning has remained genetically (not culturally) determined in gibbons (see next chapter). In this link between group identity and genetic or epigenetic composition, gibbons are like most other animals. Gibbons, unlike humans, show little if any link between brain lateralization, such as handedness to vocal learning. As will be presented in the next chapter, humans may be unique among all animals in unlinking much of their group identity from genetic or epigenetic markers.

Overview of Human Immunity/Identity System ERVs

The adaptive immune system depends on the MHC locus which is the most dynamic genetic locus in our genome and is associated with identity, immunity, and possibly mate identity. ERVs show a very strong and ongoing influence on this system and account for most of the dynamic nature. As presented in the last chapter, the origin, evolution and diversification of the MHC I and II genes are all clearly mediated by duplication and divergence involving an ancestral ERV duplicon. Even recent events marking the differences in MHC evolution of chimpanzee and human are also ERV mediated. But in the MHC, HERVs are not acting alone. There are clear interactions between the colonization and expansion of HERVs and hyperparasites; alu's LINES and SINES. The interactions are complex. In some cases hyperparasites can act to incapacitate genetic functions associated with earlier genetic parasites and ID systems (i.e., earlier HERVs and MHC ID proteins). LINES themselves can also become incapacitated by HERV colonization. Thus, HERVs, alu's and LINES constitute consortia of interacting genetic parasites but consortia that must work in the context of prevalent exogenous viruses to establish the selective determinants of group selection. There are striking distinctions in retrovirus biology of Old World non-human primates and humans. All Old World primates support FV and SIV persistent infections. Humans do not. Yet humans show major transitions in anti-retroviral genes, especially various APOBEC genes not seen in chimpanzees. In addition, pseudogenes within the MHC locus are unusually numerous in the primate genomes and some recent formation is apparent.

At this time it is important to re-examine another significant development in Old World primates, all of the VNO and many OR genes (associated with extended MHC) have become pseudogenes via the action of hyperparasites. This book has previously emphasized the central role of VNO, OR and MHC in vertebrate group identification as seen in all terrestrial tetrapods. In mammals, the basal need to socially link the mother to its offspring was also VNO and MHC regulated. Within the mouse MHC cluster, 59 OR loci (20% pseudogenes) are likely used for odor-based identification. The mouse MHC I genes are interspersed in this cluster. But in humans MHC loci have only 25 OR genes (50% pseudogenes) and the human OR clusters are also highly dense with LINEs. In the mouse, the OR/MHC gene cluster underwent co-duplication, suggesting that selection has kept these regions together. The mouse OR genes are more diverse than human and show evidence of recent expansion. Three mouse OR families have no human orthologs. It thus seems clear that the mouse OR families have evolved independently of human OR families. However, the most dramatic difference is in the human xMHC locus in which all five VNO genes have become pseudogenes. Given the importance of VNO receptors to social and sexual behavior in other mammals, how can we understand this change?

The argument I now present is that Hominids, in particular, lost most of their odor-dependent capacity for social and sexual detection (although some residual olfaction may still remain). This was caused by massive HERV colonization that also activated various hyperparasites (alu's/LINES) to incapacitate gene sets needed for both prior ERV identity and MHC-based group identity (VNO, some OR). Such a colonization was mostly due to exogenous (MLV-like) viruses not present in primate ancestors and hence not carried over to New World primate. The resulting Old World primates, however, were altered in their retroviral-host biology becoming permissive persistent host for FV and SIV. Such a host-specific persistence set up a situation for group identity mediated by retroviral (and possibly other) agents. The loss of odor-based group identity, however, presented a major dilemma for Old World primates. Since group identity is an essential survival characteristic, the surviving primates were under strong selection to expand other mechanisms of group and sexual identification. The visual sensory system was adapted and expanded for this role. To accomplish this, the great apes underwent a major expansion of brain regions especially those associated with visual recognition. The X-linked color vision was one of these visual adaptations. This complex vision adaptation was also likely mediated by genomic HERV colonization of their lymphoid (or neurological) stem cells, allowing hyperproliferation, expansion of cellular identity systems and invasive control of adjacent cells along with expanded vision-mediated behavioral control. This now required visual sensory input to control those identity systems that regulated emotional memory (such as prolactin and oxytocin) and linked vision to CNS opioid-based behavioral addiction circuits. These emotional memories had previously been under mostly olfactory control and imprinting. Much of the section below presents the details of observations that support the above scenario. Since these details necessarily involve reference to arcane terminology (i.e., MHC immunology, retroposon and HERV nomenclature), it is likely that non-expert readers will find this evidence burdensome and may wish to skip to the section on brain changes.

The Human Y Chromosome and ERVs

The Y chromosomes are the most species-distinct chromosomes in mammals and as presented in mice usually show the presence of the most recent HERVs (and numerous other retroposons) to have colonized their host genome. The Y chromosomes have been called ERV 'graveyards' due to this feature. However, I suspect the term ERV nurseries might be more applicable since they appear to represent ERV entry points into chromosomes. The great ape and human Y chromosomes retain this recent HERV characteristic. The human Y chromosome has more than 14 HERVs per MB, well in excess of the other chromosomes (but not the MHC locus). One particular 5 MB Y hot spot window had more than 120 HERVs which were often inserted right next to and into existing ERV locus (consistent with ongoing genetic displacement). The human-specific part of Y chromosome is about 3–4 million bp long. This region is rich (60% of sequence) especially with LINES, HERVs and some alu's. It also has an overabundance of pseudogenes. However, one distinct full HERV (HERV ID 40701) seems ancient since it was rather similar to full-length ERV of zebrafish (ZFERV) and is the only element that seems to have been conserved from that ancient chromovirus colonization. The majority of HERV-K14CI proviruses were also on the Y. Autosomes can also show distinct patterns of recent HERV colonization. ERVS can also affect gross chromosome structure, such as cytogenetic stains that can clearly distinguish most chimpanzee from human chromosomes based on staining characteristics of repetitive DNA. As previously mentioned, chimpanzee chromosomes have characteristic C-bands absent from human chromosomes. These regions also have a high density of HERVs and other retroposons (alu's, LINES), so it is clear that significant differences in colonization exist in great ape chromosomes. It is interesting to note that the Y chromosome provides the best genetic marker of male ancestry and has been used to trace ethnic and religious ancestry (such as Jewish rabbi ancestry, see next chapter). The relationship of the Y chromosome to sexual or group identity, however, has not been well studied. Yet it is known that mice do encode sexual odor markers on the Y. However, there is no strong evidence that humans (or chimpanzees) similarly use the Y chromosome for any odor-dependent sexual identification. In terms of odor detection, the HERVs (K14I) are interesting in that they are fused with ion channel TRPC6 (and FAM8A1) and express a hybrid mRNA. In fact, it seems proviruses have moved this TRPC6 hybrid to 14 new positions including various human autosomal chromosomes for unknown reasons. As TRPC genes are directly involved in sensory detection and TPRC2 is the basal family member expressed in rodent VNO olfactory neurons, this fusion with TRPC6 is most suspicious. TRPC2 can physically associate with TRPC6 and is also involved in the guidance of nerve growth cone in cerebellar cells. However, the functions of these hybrid TRPC6 genes remain unknown.

Hyperparasites (LINES and alu's)

There were clear bursts of LINE and alu expansion in primate and human evolution. One LINE burst occurred early in eutherian evolution and the other was specific to the Old World monkeys. Alu elements (described below) also underwent sharp amplification at the primate expansion, prior to OWM emergence, but also recently in human evolution. These have differentially affected human and chimpanzee chromosomes. Human chromosome 21, for example, has around 21,000 young alu elements that account for most insertions that differentiate human from chimpanzee DNA. This compares to only 2,600 alu inserts in the equivalent chimpanzee chromosome. In terms of LINE elements, these are less distinct as humans have 82,000 LINE-1's relative to 78,000 in chimp. The number of human and chimp LTR elements is similar although humans do have more ERVs. Indeed, all primate radiations are associated with peculiar ERV colonizations. HERV-Ws make an interesting case that supports the primacy of ERV colonization leading the other primate alterations. In humans (Hominidae) HERV-W (syncytin-1) encodes an *env* ORF that has been maintained by purifying selection (shown by indel analysis). However, in non-ape primates (Ceropithecidae), they use distinct ERVs for trophoblast fusion as their HERV-W (syncytin-1) gene has progressively been colonized by other HERV and LTR elements.

HERVs, especially those on the Y, often appear to be associated not only with other HERVs but with other retroposons as well (LINE, alu or SINE elements). These hyperparasites appear to have been used by some ERVs as agents active against competitor ERVs. For example, in primate evolution, we see a frequent coincidence of HERV colonization, HERV LTR expansion, alu amplification and specific LINE/SINE expansion or extinction. Since LINES, SINES and alu's do not generally move between species, it seems HERV colonization must provide the initiating event for such genetic changes. Yet LINE displacement seems ongoing in mammalian (including human) evolution. Although placental mammals underwent an early and significant ERV and LINE colonization, LINE and HERV colonizations are not always in synchrony. In human DNA, about 30% of LINES are intact and 26% of the X chromosome is LINE (in contrast to the HERV-rich Y). Similarly, the mouse X has 80,000 LINE elements. Human DNA has been estimated to have about 500,000 L1's (Kpn1 repeats), although Wu-BLAST reports only about 100,000. LINE colonization and displacement seem to have occurred in episodic waves. Human genomes have five major families; L1PA5, L1PA4, L1PA3B, L1PA2 and L1PA1, but only L1PA1 remains active. Yet L1PA1 is human specific, being absent in great apes, whereas L1PA2 is present and remains active in ape genomes. Clearly, a major amplification of L1PA2 occurred only in African apes, followed by LPA1 in humans. From these patterns it appears that LINEs have colonized primate genomes over a period of 25 million years in which old versions of LINES seem to get cleared out of genome. Prior L1 elements have also become silenced by DNA methylation. Clearly, new LINEs do not mobilize older versions. What then mobilizes the new LINEs? Since most full-length LINE elements are to be found on the Y and also X chromosome (i.e., LP1A on the X) and further testis determining factor gene (SRY) that can modulate L1H promoter activity, LINE amplification also seems sex linked. Thus L1 families have succeeded each other as sequential families but the sex chromosome link suggests that specific ERV colonization is involved. L1 is active in germ lines and translation of RT appears to use viral-like IRIS elements. The needed transcription factor for LINEs expression is absent from most human tissue, except for a Y factor which allows testes-specific expression. As mentioned in the last chapter with South American rodents, LINE-1 colonization is clearly not essential for speciation. Yet, in these rodents, an endogenous retrovirus family differentially amplified in the various species (MyTR, present 4,500 copies, 1,000, 10,000 copies in related species). This ERV colonization appears recent but independent of amplifications linked to LINE-1.

Genomic Effects of Genetic Parasites (HERVs and Pseudogenes)

LINE-1 ORF II encodes an endonuclease but also encodes an RT. Endonucleases clearly have the potential to be genotoxic. The LINE RT has 10 conserved regions which clearly resemble those of the gypsy element RT (chromovirus), the phylogenetically basal version of this gene. Chromovirus and LINES use distinct insertion mechanisms that are less destructive than those used by other retroviruses (avoiding coding regions). In the MHC locus, recent HERVs have often interrupted prior LINES and alu's can interrupt recent HERV LTRs. The triggering event for these rearrangements appears to be ERV colonization. In terms of consequences to host genes, LINES (especially L1) and RT ERVS are thought to generate the majority of pseudogenes. Curiously, the occurrence of pseudogenes in different genomes is highly variable. For example, as mentioned previously, chicken genomes are surprisingly pseudogene poor (only 51 total pseudogenes), yet ironically, chicken B cells use pseudogenes to generate Ig L and H diversity of their adaptive immune system. In contrast, mammal genomes have about 15,000 pseudogenes (including all human VNO receptors), yet neither humans nor rodents use pseudogenes to generate diversity of the adaptive immune system. Although LINES often seem involved in making HERV-W pseudogenes, LINES do not seem to provide the same outcome to the more recent HERV-Ks. One example is the HERV-K integrase which has been functionally maintained in primate genomes. As discussed above, these HERV-Ks are clearly not the result of 'remarkable' RV captures, but are instead pervasive and recurrent events in primate evolution that have also affected LINE patterns.

Alu elements can be considered as defective parasite of active RT elements. Alu's are about 300 bp sequences in length present in up to 1.5 million copies and have two components. They have evolved from 7SL RNA (signal recognition particle), which conserved two-dimensional RNA structure, thought to be involved in retrotransposition. They also have an internal RNA pol III initiation site. They depend on exogenous RT (such as LINE expressed) for transposition. Alu's are mostly found in AT-rich regions, compared to GC rich for LINES. Interaction between alu's and ERVs seems clear; as mentioned in numerous cases alu's insertion will interrupt LTR and can lead to antisense LTR ERV expression and immunity to related ERVs. SINES are also dependent on exogenous RT sources and most are derived from tRNA, which sometimes function as HERV LTR primer tRNAs (such as SINE-R). Curiously, given the prevailing view that SINES are strictly selfish genetic agents, they are absent from chicken genomes for unknown reasons. It is possible that this SINE paucity may also relate to the paucity of pseudogenes present in chicken genomes. Primates and avians thus seem to differ fundamentally in the parasitic elements they host. It seems primates have utilized HERVs, LINES and alu's much more for sculpting their recent evolution than have avian genomes.

An idea that might explain all the above HERV–retroposon associations with primate evolution was put forward by Kim, Hong and Rhyu. They argued that there has been periodic explosive expansion of various LINES and alu elements associated with distinct LTR colonizations and such events are coincident with and underlie hominid evolution. Here too, the Y chromosome seems especially prone to alu insertion and is around three times denser on the Y chromosome than the X. The question is raised as to why humans in particular expanded alu's so much. All these elements appear to have a significant capacity to affect gene regulation (via LTR, promoter integration, RNAi, antisense regulation, pseudogene formation, etc.), not by simply adding new genes. Colonization by such consortia could result in highly complex and potentially coordinated regulatory capacity able to control dispersed genes. With the loss of social olfaction and gain of visual group identity in the great apes, a much more complex regulation of vision-based brain function was needed. Thus, we see that the rapidly changing DNA between human and our chimpanzee relatives is to be found in the non-coding repeat sequence and is most associated with genes involved in brain function. Genetic colonizations, pseudogene formation and complex interactions of hyperparasites are thus proposed to have incapacitated olfaction and set the stage for the evolution of cognitive (initially vision)-based human group identification.

Can Complex CNS Regulation Result from ERVs and Repeats?

The above hypothesis will not be easily dissected due to the consortia or population character of the genetic elements involved. If repeat transcripts, such as LINES, HERVs, alu's and LTRs for example, indeed have a clear role in higher human brain function, this will be very difficult to causally establish by typical genetic approaches, since such elements are numerous and dispersed. It is possible, though, that strong associations of repeat element expression to brain activity might be observed. In addition, other genetic alterations in regulatory regions, such as satellite DNA, might also result from the action of genetic parasites. For example, in prior chapters the evolutionary origin of simple satellite sequences in mollusk genomes was presented. In mollusks, it seemed clear that uneven insertion and excision of parasitic genetic elements (retroposons) contributed to the large-scale satellite variation and conservation. In vertebrates, however, a model is generally accepted that most satellite variation occurs by a DNA polymerase strand slippage, not by the action of genetic parasites, which should not conserve satellite elements during evolution. Direct evidence regarding the DNA strand slippage model, however, is generally lacking. As mentioned in the last chapter, satellite DNA in promoters controlling crucial brain expression was of special interest from the perspective of social bonding. From a social perspective, humans resemble bonobos more than chimpanzees in that humans and bonobos both form social structures in which the role of an alpha male and aggressive relationship between males and males and females varies and need not dominate. Chimpanzees are typically more socially aggressive. We can recall from the vole model of pair bonding that the presence of satellite DNA in the VR regulatory region correlates with species susceptibility to social (mate pair) bonding. We noted chimps differ in the satellite makeup of their VR receptor promoter from that seen in humans and bonobos in that the latter conserve three specific satellite elements found in the promoter of the AVPR1A receptor (one of which is a simple GT repeat). The other satellites are more complex, thus do not seem likely to have been the product of strand slippage. I propose that these conserved promoter regions are more likely to have resulted from retroviral and/or retroposon activity via multiple retrotransposition, with uneven excision, not DNA pol strand slippage. Thus, I would further propose that a related genetic invasion occurred in the ancestor of both human and bonobo that generated the altered AVPR promoter which made humans and bonobos more prone to extended social bonding. Along these lines, other regulatory regions of crucial brain genes are also very interesting regarding satellites. The human dopamine transporter gene, for example, has an array of tandem repeats that are polymorphic, possibly associated with learning disorders. More recently and more relevant to the role of endorphins, prodynorphin (PDYN) is a precursor to a number of endorphins. Human PDYN promoter has 68 bp repeat that carry one to four copies. This contrasts sharply to non-human primates that have only one copy and multiple copies expressed at elevated levels. This is consistent with the view that human-specific evolution involved complex regulatory adaptations involving simple repeat elements that control genes important for social interactions. And I further suggest these adaptations resulted from hyperparasitic responses to ERV colonizations. In general, behavior is the most rapidly evolving characteristic of mammals. General changes in behavior seem to result from rapid, parasite-driven regulatory shifts, especially with respect to RNA metabolism, promoter regulation, RNAi, siRNA and micRNA. Indeed, such a relationship between RNA metabolism and behavior may actually be ancient. Recall that miRNA also appears to regulate brain development in zebrafish. Also, consider for example, a central element of interferon, RNAse L, which used ankyrin repeats to bind 2′5′ A as a signal for RNA decay. This gene was also involved in Drosophila as the dunce sexual learning mutant.

Other Primate Viruses: What Made Us Different from Monkeys?

There is much evidence that suggests hominids have been strongly selected by retroviruses and possibly other viruses. Of the 40 different simian species native to Africa, all support SIV. Yet humans, gorillas and bonobos are all normally

SIV free. Even chimpanzees, which can support SIVcmp in some populations are mostly SIV free in the wild. However, we can expect that exogenous viruses will provide strong selection regarding endogenous viruses and that the high prevalence of SIV and foamy viruses in African monkeys would be relevant. However, neither SIV nor FV is closely related to the HERV-Ks (or other HERVs) found in hominid genomes. The recent HIV-1 pandemic could represent a potentially transforming human agent. However, it is clear that HIV is now not able to cause a selective sweep of humans as public health interventions (a product of culture) have limited its spread. Yet, had HIV emerged during an earlier historic period, prior to modern medicine or culture, it might have had a broad evolutionary consequence to the human population. Both HIV-1 and HIV-2 have evolved from primate sources of SIV. The SIV of the red capped mangabey bar monkey (SIVrcm) was the first SIV to be discovered and many versions of SIV are now known. Yet, neither HIV-1 nor HIV-2 originated from a single SIV or simian species. Both required a recombination prior to human adaptation (e.g., recombinant virus from chimpanzee and sooty mangabey SIV). SIV smm appears to be tightly linked to its natural host, the sooty mangabey monkey, and to have provided a genetic component of HIV-2. In chimpanzee, SIVcmp itself appears to be a recombinant with SIV gsn-like virus. However, none of these SIV–primate relationships seem to directly relate to HERV-K as HERV-Ks are not derived from any of the prevalent exogenous primate viruses (SIV of foamy). However, as presented above, HERV-Ks clearly resemble MMTV-like viruses of rodents and it is now known that Old World primates (and some human cells) do express anti-retroviral proteins, such as TRIM5 alpha protein. However, it does not appear that TRIM5 evolved to target HIV but instead it seems most active and potent against MLV-like viruses. New World monkeys also express TRIM5, but its activity differs and most do not block HIV post entry as do Old World TRIM5 proteins. But New World TRIM5 does resist entry by N-tropic MLV and can also be active against SIV. In the New World monkeys, only owl monkey species also show a restriction to HIV post entry via TRIM5 and resembles the Old World function. However, this TRIM5 is distinct form those of other New World monkeys and has been interrupted by a LINE-1 element that also catalyzed retrotransposition-mediated fusion with cyclophilin A gene. This fused gene is active against HIV-1. This LINE-1 insertion occurred at the base of the owl monkey radiation.

How then might SIV or FV contribute to the evolution of Old World primates and why would the hominids differ so much in this relationship? It appears that SIV is species congruent and stable in its natural African monkey host, but can be highly adapting and disease inducing in non-African primates. This situation clearly provides a potentially lethal system of species (group) identity. For example, the four distinct species of African green monkey are all infected with SIVagm in the wild; each has a phylogenetically distinct gag and env sequence. SIVsmm9 in sooty mangabey is also non-pathogenic in natural host, and in this SIV, genetic variation is limited. Old World primate immune cells respond sluggishly to SIV (and HIV-1) infection and are inherently resistant to SIV. However, in contrast, in Asian rhesus macaque monkeys, an AIDS-like disease will result following SIV infection. And the rhesus macaque with SIV-induced AIDS shows lots of sequence variation. This clearly has the features of a species-specific persistent virus that has adapted to acute lethal viruses in related host species (virus addiction).

The absence of both human foamy viruses or a human-adapted SIV indicates that the human (and other hominid) lineage underwent a significant adaptation in it virus–host relationship relative to the other Old World primates. The question we would like to answer, however, is if such 'anti-viral' adaptations involved colonization and exclusion by HERVs. One human-specific anti-viral adaptation appears to be the expansion of APOBEC3C gene (an anti-FV gene mentioned previously). This expansion strongly supports the idea that humans did indeed undergo a sweep and selection by retroviruses. But these anti-viral genes are most active against MLV-like (HERV-K-related) viruses. As I have asserted, HERV-Ks are likely to have evolved from an MLV-like exogenous virus, as frequently found in rodents (xenotropic virus). Fossil evidence (camp sites) indicates that since the time of *H. erectus* the human lineage has indeed had close interactions with rodents. The retroviral-mediated selective sweep and ongoing endogenization by MLV-related retrovirus in koala bears in Australia may be a good model of the likely circumstance that led to human HERV-K endogenization. If so, the likely pathology associated with a related human viral selective sweep would also have been in the immune system disease (via infectious lymphomas), which would impose a direct selection on progenitor human hemopoietic cells, the target of MLV-like transformation. Thus, immune system functions would be under the strongest selection by these viruses (consistent with significant HERV-mediated variation between human and chimpanzee MHC locus). Intriguingly, as mentioned, these very same immune progenitor cells can also differentiate into fully functional neurons of the CNS. Thus the transformation of immune cells by MLV-related viruses also has the capacity to alter brain cell development and might help HERV association with human-specific brain tissue.

The primate HERV-K viruses can also be evaluated by the presence of their dUTPase gene (not common to many retroviruses). This gene provides relatively unique trace of primate ERV evolution and it appears that the ancestral progenitors of HERV-K conserved an expressed copy of dUTPase. Such dUTPase-HERV-K conservation is not seen in the mouse genome. dUTPase appears to provide a function that overrides an anti-retroviral situation in that it allows (detoxifies) cytoplasmic dUTP, which will otherwise poison cytoplasmic cDNA synthesis. Cytoplasmic DNA synthesis would also make dUTPase relevant for and conserved in various DNA viruses, such as the alpha and gamma herpes viruses described below. The lentiviruses of non-primate animals have both conserved and lost HERV dUTPase activity for unknown reasons. dUTPase is conserved in non-primate lentiviruses but is absent in primate lentiviruses, where HERV-K dUTPase is conserved. Thus, it is also most interesting that primates and rodents also differ in this regard.

Besides primate-specific retroviruses in natural primate populations, most primates are frequently infected with other viruses, including many DNA viruses. Humans, for example, are colonized by eight human-specific herpes viruses (i.e., HSV-1, HSV-2, CMV, VZV, EBV, HHV6, HHV7, HHV8) and these herpes viruses are mostly phylogenetically congruent with their primate host. We have already discussed the broad occurrence of herpes viruses, hepatitis B virus and GaLV in gibbons, the basal lesser apes. In natural monkey populations herpes viruses are also prevalent. It is known, for example, that temple monkeys in Nepal show a high prevalence of RnCMV (cytomegalovirus), SV40 (polyomavirus), CHV-1 (herpes virus) and SFV (simian foamy virus) all together. It is likely this mixed situation has prevailed during primate evolution. There is also reason to think such viral mixtures are linked to host identity and can affect host species. For example, in rhesus macaque monkeys, interaction between CMV and SIV has been reported in which CMV limits or prevents SIV antibody formation. The CMV-like viruses are known to have many genes that regulate immune cells, especially T cells. Humans, like rodents, have maintained CMV-like viruses that infect immune cells. Both HCMV and MCMV downregulate expression of conventional MHC I at the surface of infected cells. Both also encode MHC heavy chain homologs, which bind to light chain (beta2 microglobulin). It is also interesting that 11 microRNAs are encoded in latent regions of their genomes indicating substantial regulatory capacity to support latency. VZV is remarkably stable in the human population. These viruscs have reiteration (repeat elements) regions that are associated with origins of DNA replication and differentiate geographically distinct clades. This viral variation can also differentiate human populations. The human alpha and beta herpes viruses persist in nervous tissue. Although humans lack FV- and SIV-like viruses, it is interesting that human herpes viruses appear to have evolved along with the interaction of retroviruses, since these herpes viruses lack introns, and thus they appear to be products of cDNA retrotransposition. This lack of introns is not true of many other human DNA viruses (such as adenovirus, polyomavirus). Also, some of the ancient ancestors of herpes virus, such as phycoDNA viruses, conserve introns, such as in their DNA polymerase gene. We can also recall that avian herpes viruses often contain entire retroviruses in their genomes. However, a specific human evolutionary significance of potential DNA virus–retrovirus association is not understood.

Could retrovirus–DNA virus mixtures possibly contribute to primate group identity? We cannot clearly answer this question. Herpes viruses are all species specific and can clearly damage related species and would thus seem to have good potential for enforcing group identity. For example, bovine herpes virus is lethal in sheep as is wildebeest herpes virus. Herpes viruses of Asian and African elephants are lethal to the respective offspring of each other species. In Asian primates, herpes B virus (Cercopithecine herpes virus) is asymptomatic in nervous tissue of native primates but often causes lethal encephalitis in humans

(70% lethal). Human to human transmission with this virus is rare however, so it is clearly not human adapted. This B-like virus appears to be an ancient alphavirus and likely was also present in prosimians (related to marmoset herpes virus). In terms of recent human evolution, we do not know if there might have been a Neanderthal-specific herpes virus or if they were susceptible to the *H. sapien* herpes viruses. It is curious that humans, unlike all other primates, harbor two close relatives of alpha herpes family (in which B virus is basal): HSV-1 and HSV-2. Given that HSV-1 is an oral virus and HSV-2 is genital, we might wonder what human-specific sexual selective pressures or behavior might have led to the divergence of these two viruses. It has been proposed that face-to-face mating might have allowed selection for the segregation of oral and genital mucosal habitats. Besides face-to-face mating, human mating biology has several distinctions with non-human primates, such as unapparent estrus and persistently large mammary gland, which could also affect sexual behavior and virus transmission dynamics. Thus, there is some evidence that human-specific sexual behavior has affected human-specific virus evolution. Human sexual behavior has clearly also contributed to the adaptation and evolution of HIV-1 in its new human host.

Viral Alterations to Primate Identity and Immunity – MHC

From a biochemical or cellular perspective, the MHC locus of the human immune system represents the most dynamic gene set. As mentioned, vigorous rearrangements in MHC were seen in transitions of fish to tetrapod and later in the emergence of primates. The human MHC locus is the most polymorphic of all human gene regions and differs in several aspects from that of other primates, including chimpanzee. In addition, distinct MHC differences are also apparent between New and Old World primates. Much of the change in MHC has been mediated by the action of ERVs, LINES and alu elements but the overall evolution of the MHC locus is from simple to complex. MHC I appears to be ancestral to MHC II, and MHC III seems to represent a latter addition. MHC III was acquired along with origin of placentals and is the most gene dense intron-lacking region (62) in the human chromosome. Some have proposed that the MHC III locus is an inflammation region. Why was this locus acquired with placentals? I previously argued that the MHC III locus resembles the product of colonization by a DNA virus. MHC III encodes C4, a core compliment gene, which in humans was clearly affected by insertion of HERV-K(C4) into the first intron, making the C4 long. Curiously, this HERV has an antisense alu element inserted into the HERV LTR which implies highly complex but also conserved regulation of C4. In mammals, C4 (core) genes also show unusual size in polymorphism which is also due to HERV-K(C4) LTR colonization. Such colonization events also differentiate human and chimpanzee C4, but even account for a human-specific polymorphism. Chimpanzees and gorillas both have a short C4 locus. Why do humans have HERV-K polymorphism in a core gene of compliment?

Human MHC locus is organized into three regions (class I, II, III) plus some extended (x) region. Humans have 19 HLA class II genes (11 coding and 8 pseudogenes, which are often reverse oriented). Human and chimps differ in class II, human being more complex. MHC I of human has about 18 HLA class I genes (6 coding and 12 pseudogenes). MHC I has within it three recognizable gene blocks: beta, kappa, alpha. Within HLA-1, humans have six coding genes (-A, -B, -C, -E, -F and -G) and 12 pseudogenes (found in the above three distinct blocks). The human alpha block has 10 duplicated MHC I genes and these genes are shared in primates, but those within the beta block are less shared. The C block is found only in gorilla, chimp and human and not in other Old World primates. In contrast, Asian rhesus macaque has a markedly different MHC I with an additional 20 genes in the alpha block and 17 in the beta block. This compares to the wild Indian Ocean macaques which have a surprisingly simple MHC haplotype makeup (but are susceptible to SIV infection). Thus, distinct versions of these MHC I blocks are found in most mammals but their gene content can differ significantly and non-human versions generally have fewer pseudogenes. All the MHC I and II genes are proposed to have originated by a duplication from MHC I, and an ancient 'preduplication' version of HLA has been proposed which was composed of an HLA gene, a PERB11 gene and an HERV-16. Why humans, in particular, have so many pseudogenes in this locus is unknown.

Chimpanzee MHC I also differs from human in that it has large deletion (95 kb within beta block). This deletion has lost HERV-L and HERV-16, elements which are thought to be important for the fact that chimps do not develop AIDS from HIV-1 infection. Of special interest in this is that HERV-16 seems to be in an antisense orientation and should express an anti-RT pol RNA. In general, the MHC locus is surprisingly dense with retrovirus (ERVs). Human MHC I has 16 HERVs (mostly HERV-16, but also a basal HERV-L). Many HERV-16 s are in the alpha and beta block. However, in the HLA-b or HLA-c regions, few if any HERV-16 s are found. Instead HERV-I is present (which colonized Old World primates at the New/Old World split). Thus, HLA-b/c seems to represent waves of distinct ERV colonization associated with evolution of this locus. MHC is also dense with LINES, SINES and alu elements. Yet there are some blocks in which these elements seem frozen in evolution. Within these regions, ERV-16 is most often found (part of the basal MHC duplicon). However, although MHC is dense with ERVs, most are interrupted by alu inserts and only HERV-K91 (in HLA-c) has conserved pol and protease ORFs. This HERV is also the most recent insertion into class I locus. It is interesting that the biggest difference with chimpanzee and human MHC I locus involves both ERVK9 and L1 (human genome has an ERVK9 insertion). In contrast, in the mouse MHC locus, MuERV-L is found at this primate ERVK9 position and has also conserved gag, pol and dUTPase. Besides being interrupted by alu's, many ERVs (such as HERV-L) have themselves interrupted and inserted into LINE-1 elements, such as L1MD1 and the primate-specific L1PA2. Thus, it is most interesting that all LINES in MHC I are fragmented. Human MHC also encodes the multicopy gene family of P5 that is specifically transcribed in immune cells. As P5 is related to HERV-L and HERV-16, and since P5 is transcribed into an mRNA that is RT pol antisense, it seems clear that retroviruses (ERVS) and retroposons (alu's/LINES) have been most responsible for shaping the regulation and evolution of the MHC locus, including the human locus. Below, we consider the purpose of this locus and why it had been linked to VNO receptors which were lost as pseudogenes in hominids.

Purpose of MCH: Self and Group Identity

Although it is well accepted that the MHC locus is a principal component of the adaptive immune system, the majority of mutations within MHC result in unexpected phenotypes. Rather than affecting susceptibility to infectious agents, such as viruses, by far the most common consequence of MHC mutations is autoimmune disease, especially in women. As I have previously asserted, immune systems should properly be thought of as identity systems which provide molecular self-recognition. As identity systems can also provide group identity and infections (such as virus) can alter host group identity this can induce a toxic immune response. In addition, an MHC mutation that misidentifies self would also be expected to induce autoimmune disease as a common phenotype. If indeed this is the underlying role of MHC, we can now offer a more coherent explanation as to why ERVs are so associated with MHC evolution, why MHC would also encode VNO and OR genes and also why pseudogenes are so common in the human MHC. My assertion is that the most common source of new systems of molecular genetic identity is from persistent and genomic viruses and hyperparasites. The waves of ERV colonization in primates must have thus superimposed a new set of genes and/or RNA regulators associated with their persistence. Generally, host resident identity systems will oppose such colonization by a new genetic parasite. Thus for the new colonizing virus to succeed, it will also often need to incapacitate these resident molecular systems (by integration and mutation). As such resident systems are often redundant or numerous, a dispersive mechanism of incapacitation will be needed. Thus the integrating HERVs and the dispersion of the defectives and LTRs pose an ideal system with inherent distributive and coordination capacity that can affect multiple loci. Along with a particular HERV's defective and LTR colonization, alu's, SINES and LINES can all also react as an identity consortia. And since these agents are hyperparasitic to and active against each other, they are ideal manipulators of each other. Thus, HERVs can initiate a more complex but interacting set of elements and rearrangements resulting in new identity strategies as well as incapacitating prior versions of hyperparasites. The net result is a tendency to incapacitate old genetic parasites with the acquisition of new ones that promote a net evolutionary increase in the complexity of host identity. Colonization by such consortia can also affect and coordinate other gene systems. For example, the zinc finger gene ZNF177 is a single copy gene associated with nerve signaling that incorporates alu, L1 and HERV env segments into 5′ UTR. This resulting RNA expression is now differentially regulated via the action (coordination) of these hyperparasites. In this example, the alu is in antisense orientation which may affect alternative splicing. Is this complex regulation typical of the types of changes that result in consortia of parasites? Possibly. In all tetrapods, prior to the great apes, the MHC region also encodes OR and VNO surface receptors that are used for social recognition. However, in contrast to the molecular genetic and cellular memory systems used by MHC I/II, as we presented earlier, these odor receptors establish group identity via CNS-mediated recognition (memory). However, the unique HERV-K colonizations that occurred in hominids were also associated with the pseudogene-mediated loss of all the VNO ORFs within the human MHC locus as well as many other OR pseudogenes. This loss, as presented below, represents a major shift in group identity strategy in which hominids depend little on odor recognition for social or group recognition.

ERV-Driven MHC Genesis and Human Evolution

The above scenario makes clear why we should expect a basic link between the evolution of ERVs, exogenous retroviruses and host identity or immune function. It suggests why the origin and evolution of MHC loci was so ERV (and hyperparasite) dependent. An inherent link between ERVs and host identity is germinal for the immune and possibly other identity systems. Thus when we observe, for example, that HERV-K18 encodes a superantigen with a broad capacity to stimulate the immune system, we might consider this from a germinal perspective, not simply a recent viral adaptation to immunity. Both HERV-K18 and MMTV can stimulate T cells via receptors without antigen. MMTV stimulates Vbeta7CD4 T cells and this viral antigen can stimulate 10^3–10^5 more T cells than conventional antigens. Such stimulation is via selective binding to Vbeta chains, thus they are not dependent on their antigen specificity. How might this relate to the evolution of host identity, such as the evolution of tolerance? The studies with this MMTV superantigen were indeed crucial in mice to work out much of our understanding of how immune tolerance works. HERV-K18 also stimulates Vbeta7 and Vbeta13 T cells which are constitutively expressed in thymus, resulting in central tolerance (these cells are also associated with the establishment of EBV persistence). I would suggest such less-specific T cell stimulation might better represent how T cell evolution was initially promoted. Other basic links between immune system function and retroviruses are also clear. The process of somatic hypermutation, for example, is used by the adaptive immune system to expand diversity and depends on genes that perform site-directed cytidine deamination to generate high-affinity Ig. But cytidine deamination genes can also be thought of as part of an ancient viral lineage (found in T4 and *Bacillus subtilis* phage) that are active against other phage (such as PBS2).

We already mentioned that humans have APOBEC3C which is an anti-retroviral gene and shows distinct anti-foamy virus activity. The expansion of APOBEC3C to five additional genes indicates that these genes were under the strongest positive selection in recent human evolution. However, humans also have many APOBEC pseudogenes in a tandem array. Thus, both APOBEC3C gene acquisition and pseudogene formation have been ongoing and active in human evolution. Curiously, APOBEC3C expression is restricted to neurons. Why might this be? Does APOBEC3C serve some identity role in the human brain that has been the target of retroviral invasion? In mouse cells, APOBEC3G strongly inhibits mouse IAP retrotransposition. But such IAPs are absent from the human genome and APOBEC3C is absent from the mouse genome. Also, MLV (mouse retrovirus) can replicate in APOBEC3C-producing cells, so it would not seem that resistance to such rodent viruses was the main selective pressure for APOBEC3C expansion. Somehow, human brain seems to have become a principle target of such identity adaptations but we lack sufficient understanding of the brain to evaluate this relationship.

Decreased Importance of Smell When Great Ape Color Vision Develops

The VNO Pseudogenes

The role of ERVs in MHC evolution is now clear. We can now focus on the details concerning the loss of the great ape VNO genes, including a set within the MHC locus, and the consequence to human group identity. In addition to the traditional MHC locus, MCH locus also includes an extended region, xMHC, which is adjacent to class II and III genes. The human xMHC I locus has seven VNO ORFs within it in a tight cluster, but all these ORFs (along with those outside of MHC) are pseudogenes. A related, but less dramatic, pattern also applies to human xMHC OR genes in which 34 OR are intact ORFs and 20 OR are pseudogenes. Also in the human xMHC are 36 zinc finger ORFs with 10 related pseudogenes. These VNO and OR genes were mostly maintained as ORFs in the mouse xMHC. In the whole human chromosome, there are about 200 V1Rs, and most (if not all) are also inactivated. It seems there are no remaining functional VNO genes in the human genome since library screening fails to find any. Other cloning screens have found 56 BAC clones with 34 distinctive VNO sequences (V1R and V2R related), but all were pseudogenes mostly by alu inserts and frame shifts and all were present on unexpressed chromatin. This is in striking contrast to the rat MHC which has about 40 VNO

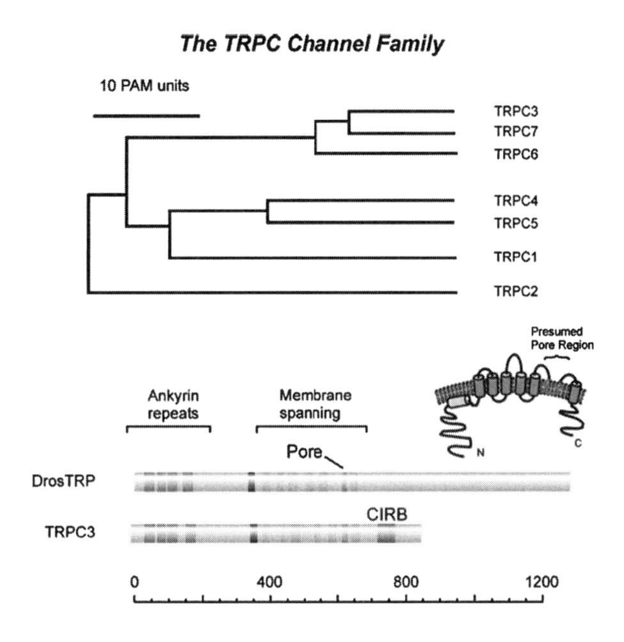

Fig. 8.7 Evolution and gene structure of TPRC genes (reprinted with permission from: Mohamed, Guillermo, Gary, Putney (2003), Cell Calcium, Vol. 33, No. 5–6)

genes with no pseudogenes. We can recall that mouse had 137 intact V1R genes. In contrast to mouse, rat V2R do have many pseudogenes, but these are mostly due to LINE L1 inserts, and distinct from both mouse and also human V2R pseudogenes which are mainly due to alu inserts. The primate VNO pseudogene formation was shortly before separation of hominoids. It thus seems that the great apes specifically experienced tremendous selective pressure to inactivate all VNO genes and many MHC-linked OR genes. In addition, TRPC2 is an essential ion channel transducer for VNO gene function (exclusively expressed on VNO microvilli), also became a pseudogene in humans but was well preserved in prosimians. TPRC represents the basal copy of the TPR gene family as shown in Fig. 8.7. As mentioned in the last chapter, its loss in mice shows altered sexual and aggressive behavior. Hominids thus show a striking loss in VNO genes and function. In contrast New World primates have well-developed VNO. What then could have been the strong selective pressure to degenerate the numerous hominid VNO genes?

The OR Pseudogenes

Humans also have large numbers of OR pseudogenes relative to mice and prosimians. There are about 1000 human OR genes of which half are pseudogenes. This corresponds to an astounding 1% of all human genes. In contrast, Drosophila has 60 OR genes and vertebrate fish has only 10% of mammal OR genes. OR genes tend to be clustered into tandem blocks for unknown reasons. For example, human class I OR are a single extended cluster with 17 OR sequences (10 intact, 6 pseudogenes, with 1 dimorphic pseudogene intact in some individuals). In this same cluster chimps have 15 OR genes (as do other apes), including 4 pseudogenes conserved in apes (2 related to human). The human OR cluster on 17p13.3 has a high proportion of pseudogenes (60%), in which interruption by LINE elements predominate. As mentioned, human MHC locus of 34 OR genes has 14 coding and 20 pseudogene (all LINE interrupted). In contrast the mouse xMHC I region has 59 OR genes of which 20% are pseudogenes. Thus, by all measurements mice have more MHC OR genes relative to humans but humans have more pseudogenes. In addition, for most OR subfamilies, mice have several members whereas humans have only one member and in four out of the nine OR subfamilies, humans have only a single locus that is a pseudogene. Also, mice have three OR subfamilies absent in humans. Yet the OR gene order and transcription are conserved between mice and humans, clearly supporting a common ancestry. Thus it seems mice underwent an OR expansion during their evolution whereas humans lost several basal OR members and accumulated more pseudogenes. In contrast, New World monkeys have an especially high number of OR pseudogenes.

Overall Gene Changes

The above VNO genes can directly trace their ancestry to teleost fish and have been maintained for social and sexual identification by all other terrestrial tetrapods. What possible selective pressure could have caused this to change in only hominids and why was it not strongly counter selected? ERVs and their hyperparasites do have the potential to integrate widely and interrupt many genes. And clearly widescale and multiple HERV colonization and expansion occurred at this phase of primate evolution. Considering that the VNO, TRPC2 and many OR genes all were lost together in the great apes, it appears clear that these OW primates were under intense selective pressure to loose their olfactory-dependent social identity. In contrast, rodents and other mammal species (bats) were selected for expansion of these same genes and functions. And ERVS, LINES and alu's appear to have been involved in both the expansion of rodent odor receptors and the loss of primate VNO receptors. We have noted the basal importance of group and sexual identity to the mammalian lifestyle, especially the mother–offspring bonding. The involvement of the VNO in this in ancestral mammals would seem to pose a huge problem for hominids. How can we rationalize such a evolutionary transition in a social mammal? I suggest that the ERV colonizations that triggered the VNO and OR changes were responsible for incapacitating the prior identity system (MHC-VNO-associated group identity) while superimposing a new system of group identity in place of the lost VNO. I suggest that the lost pheromone (odor) detection was largely replaced by another sensory modality which I will argue must have been a vision-based group identification that also used color vision. This evolutionary transition was retrovirus and ERV triggered.

Primate species vary significantly with regard to their use of VNO and color vision. For example, many New World monkey species have retained the use of scent markings for social identity. These New World species have also often retained a prominent VNO organ, but similarly retain small brains relative to the great apes and most do not use the same color vision as the Old World primates. We have already mentioned bats as recently evolving species that show major diversity in VNO organs. Clearly, even a highly conserved system of social identity like VNO can be selected to change.

Brain Changes

Compared to rodents, humans have a much reduced main olfactory bulb (relative to total brain mass). This is where pheromones and MHC peptide-transmitted nerve signals are processed. Using a fluorescent virus to trace these nerve connections, it has been shown that these rodent neurons can transport only in a backward direction from hypothalamus across junctions. With the mouse hypothalamus one can therefore trace back to olfactory organs and show connections mainly into olfactory region including cortex and even main olfactory epithelium, although not much tracing to the VNO was reported. Primates clearly underwent major morphological changes with respect to olfaction. New World primates (and prosimians) mostly retain VNO tissues. Marsupials also have retained their VNO tissue, but here morphology is distinct. A partial split in primate evolution with regard to color vision is also apparent. Some, but not all, New World primate species do have distinct version of X-linked color vision. This suggests that color vision was introduced into primate lineage around the time of the New–Old World split, but this appears to have occurred before the widespread loss of great ape VNO, consistent with HERV colonization patterns.

Human olfactory bulb is 0.009% of the total brain mass. This is very small relative to most vertebrate brains, smaller even than that of the chimpanzee, which are small relative to the olfactory bulbs of most other mammals. For contrast, the desert hedgehog has 10% of its brain mass in the olfactory bulb. Yet primates have larger brains overall and much greater social intelligence. Simians have a two- to fivefold reduced olfactory bulb relative to prosimians. The sensory neurons involved in odor detection (OSNs) have some unique features relative to other sensory neurons. OSNs have only two processes, a single axon and single dendrites which has cilia at the end. OSNs are also unique in that they are the only sensory neurons that project its axon directly into the brain. In fact the sense of smell has been considered as the most basal of all senses as it can bypass the cortex and link directly to the limbic system. No other sensory neurons do this. Typical VNO neurons project into olfactory bulb, which in turn send neurons to amygdala and hypothalamus. Thus, odor-based group detection has the capacity to communicate directly with limbic system and control basal emotional memory and behaviors. It therefore makes sense why odor detection would be well applied to emotional-based social bonding such as between a mammalian mother and her offspring.

Hominid Behavior Changes

In spite of lost VNO olfaction, the great apes do not show any obvious loss of maternal bonding to offspring or decreases in other sexual/social interactions. In fact the great apes have a longer duration of caregiving by mothers to very needy offspring. Also, apes clearly support large and complex social interactions and structures. Thus, these social interactions are enhanced in apes and ape mothers remain well bonded to their offspring for years. However, the great apes do not typically form pair bonds. Yet, maternal–offspring imprinting endocrinology is similar between humans and primates, involving prolactin and oxytocin, as used by other vertebrates, including pair-bonded species. Thus we expect similar neurological CNS circuits (such as emotional memory and addiction) to be affected by ape social imprinting as it was in rodents. However, it is clear that ape maternal imprinting hormones cannot be linked to VNO- and MHC-linked odor sensory receptors as seen in pair-bonded voles. Nor does partner MHC recognition (odor) seem relevant to ape mother–fetus development. There is no hominid Bruce effect as described in rodents, no induction of fertility by male odors and no estrus-linked odors for ape breeding. However, similar to vole species that have adapted to use audio calls for mate and group recognition, it seems clear that apes depend very much on visual recognition for social group recognition. For this to evolve, we would expect that significant developments must have occurred in the ape visual cortex, as these brain structures must now be able to recognize, remember and distinguish individuals visually and its neurons must communicate this memory to set emotional and addictive memory (such as via PRL and oxytocin). The primate evolution of X-linked red color vision along with the loss of VNO tissues can now be understood in the context of a displaced odor identity system. In the great apes, visual imprinting between mothers and offspring is expected to provide the basal social bond that has also allowed the development and evolution of more complex group identity also predominantly based on visual recognition and imprinting. Thus the evolution of this more complex social recognition required the development of sophisticated visual pattern recognition and this enhanced cognitive capacity also promoted the development of enhanced social intelligence. All this was the product of this basic shift in group identity.

A New Visual/Social Brain

The proposal is thus as follows: the HERV-K and other primate-specific ERVs superimposed new consortial identities with their successful colonization and set off waves of other hyperparasites that together eliminated the VNO and other odor-based group recognition in the great apes. Apes were thus liberated from the very ancient biochemical (odor-based) constraints of group recognition and identity and free to develop visual-based identity in its place. However, such a liberation should have been highly destructive. Thus, it is not likely that the visual system that was adapted was entirely new. After all, as presented earlier, fish and birds for example can both depend on vision for group identity. Thus, it is most likely that some version-based group identity was already in use by simians (especially for mother–offspring bonding) at the time of the great HERV-K colonization. Evidence of this is found in New World primates which show some uneven but clear capacity to see in color. However, the demands that a strictly vision-based identity system placed on the ape brain were considerable. Sophisticated visual pattern recognition linked to emotional imprinting would have been required in order for this to be used by ape mothers for the basic mammalian social bond to offspring. In addition other demands on this new ape brain would be numerous, including having highly developed visual pattern recognition capacity, decreased odor-based group recognition, dissociation of odors from breeding and mating physiology, association of visual cues to mating, long-term stable visual memory and visually imprinted emotional bonding. This latter requirement could be considered as a visual-based process of emotional addiction. For visual-based group identity to attain long-term stability, there would also need to be limited developmental periods during which these visual (mainly facial) patterns were acquired (such as between mother and offspring). This would correspond to a window of identity transfer as described earlier for all group identity systems. In distinction with almost all prior examples of vision based group identity in animals, the one developed by the apes is an inherently adaptable system as it uses a general solution that adapts to specific sensory cues. However, this adaptability is likely to operate along with some inherent (hardwired) visual recognition capacity (such as face recognition described below). This visual adaptability also provided the neurological substrate for complex pattern recognition that has allowed the development of general social intelligence. How could all the needed brain transformations be quickly acquired during evolution? I suspect that the great HERV-K colonization was also directly relevant for Old World primate brain development. There is no overlap between brain size of monkey relative to great apes. Chimps are quantally larger. Similarly, there is no overlap between human brain size and those of great apes. Humans are quantally larger. This looks like a step function, the result of a transforming event, not gradual selection. HERV colonization could clearly provide such a step function. There are numerous but curious similarities between immune cells and brain cells as I have presented previously. Since the brain has evolved to provide crucial functions for group identity, the immune systems and the brain can be considered as two parallel systems that both provide group identity. Any newly imposed system of group identity might also need to address both the immune system and CNS. As presented above, it seems likely that the exogenous virus that was ancestral to HERV-K was an MLV-like virus. Thus, it is most interesting that in inducing leukemia, MLV integrates and transforms hemopoietic stem cells, altering their developmental program and inducing proliferation. Since these same stem cells can also differentiate into functional neurons, an HERV transformation could also result in new hypertrophic, invasive and controlling neuronal cell layer. These are all features of the primate-specific brain adaptations in the neocortex. Given that the human neocortex also specifically expresses transcripts of many of the newer human-specific HERVs, a viral involvement in the origin of this tissue seems quite plausible.

Color Vision in New and Old World Primates

Primates have a prolonged fetal life and newborns have greater dependence on their mothers. I have asserted above that this basic social dependence is visually mediated. Most mammals have dichromatic vision, apparently due to early adaptations to a nocturnal lifestyle. Primates' vision is distinguished not only by its red color detection, but also by their powerful stereotactic visual system and increased neocortical brain that processes visual information. New World monkeys occur in 16 genera of over 70 species but their color vision varies significantly, thus they have species-specific variation for unknown reasons. Regardless of this species variation, all New World males have poor color vision (with one exception). Curiously, 1/3 of the females of 'color vision species' also had poor color vision, due to genetic polymorphisms, but other females had color vision equal to that of OWM. In the New World primates, the color vision gene is on a single X-linked polymorphic locus that undergoes random X-inactivation, thus allowing the heterozygous females (but not males, or some homozygous females) to see in trichromic color. However, reasons for this situation occurring are obscure. Although it is often proposed that food (red fruit)-associated selection is relevant, this should not result in subset of female-specific color vision, especially since these social foraging females are not the main finders of fruit. Also, males and females of NWM with color vision make the same diet choice. Thus, New World color vision cannot be explained by selection for finding fruit, and its association with evolution of X chromosome remains unexplained. Furthermore, a possible role for color vision in mate attraction is also problematic since it is a female phenotype, absent in males. However, if it were to be involved in social recognition, including mother to offspring identification, its expression in females could be rationalized.

In the Old World primates, the red receptor evolved from gene duplication onto the X chromosome. Here, both males and females see in trichromic color, although mutational loss in males (X-associated red color blindness) is not uncommon. However, a link to brain evolution seems clear. In macaques, for example, color-sensitive retinal neurons project predominantly into the outer brain layer whereas color-insensitive neurons project into deeper layers. In addition, macaques have large expansion of visual neocortical areas. Thus, as argued, these new neocortical regions seem to have evolved by addition (lamination) to prior version of the brain. Given this and the above New World situation, it seems the best explanation for female-biased color vision is for the purpose of social and sexual identification. Along these lines, it is known that in general mammals do not show colorful sexual displays, as do birds. However, numerous primate species are known to have brightly colored sexual tissue during estrus. Some species also have brightly colored facial markings, such as red faces of rhesus macaque, including males. Such color displays would be consistent with an enhanced role of color vision in social, mate and reproductive identity. Infant coloration and any possible effect on maternal imprinting have not been evaluated. In terms of matching brain adaptations, it seems clear that a neocortex would need to be able to 'learn' the new color vision since there would have been no preadapted mammalian brain structure for this purpose. It is interesting that mice can be genetically engineered to express human red pigment in photoreceptors on their X chromosomes. Behavioral assays indicate that such mice can perceive and differentiate the new color. This suggests that the mammalian brain is able to 'learn' from new modes of sensory input.

Enhanced CNS-Based Facial–Social Recognition by Great Apes

Facial–Social Recognition

Clearly, an increased dependence on visual systems occurred early in primate evolution as even early simians show a decrease in the size of their nasal cavities and have moved their eyes together. However, the evidence that supports the use of vision in social (maternal) bonds in early simians is thin and not well explored. Yet it clear that monkeys do have other special brain adaptations concerning visual recognition. Recent fMRI studies with macaques indicate that the monkey cortex clearly responds selectively to faces. In special regions of their brains, an astounding 97% of neurons respond to faces. Clearly, facial recognition has been under strong biological selection in monkeys. Interestingly, symbolic clock faces can also elicit response in these same brain regions whereas other visual stimuli do not, suggesting an inherent capacity for symbolic face representation in these monkeys. Interest in face-like schematic patterns is also seen in infant macaques less than 1 month old, especially regarding faces with schematic eyes. A similar early preference for face schematics is seen in human infants.

Altered emotional expressions (e.g., of familiar faces expressing fear or aggression) also drew attention of infant macaques. Furthermore, female macaques show ovarian hormone-linked facial preference, preference to male and agonistic toward female faces near ovulation. Thus, primates have specialized (and possibly symbolic) cognitive face perception brain machinery, known as the fusiform area (bottom surface of the right hemisphere) that is developmentally and hormonally regulated. In addition, fusiform adjacent regions in the macaque that respond to and recognizes body positions or gestures have also been identified. Human brains have retained both the face recognition and gesture recognition capacities in similar regions (discussed below). One curiosity of these studies is that it establishes a left–right hemisphere asymmetry regarding primate brain function and right hemisphere visual recognition.

Face recognition thus seems a most basal primate-specific system of individual and social recognition that involved brain-specific adaptations, although such recognition was not initially linked to color vision. This capacity further developed in the great apes. In studies comparing chimpanzees to rhesus monkeys, chimpanzees were shown to be much quicker at recognizing individual faces and this recognition was especially mediated by eye recognition. Clearly, facial cognition underwent further development in the great apes. Eye recognition was further developed in the evolution of humans which are the only primate with white eyes that aid in gaze and expression detection. Thus, facial cues became highly important for primates, but took on added importance in the great apes, associated with emotional communication.

The existence of specialized brain regions for face recognition also raises another issue concerning primate and human brains. This situation supports the concept of a modular mind, in which some brain areas take on specialized sensory recognition and processing. In the case of face recognition, very subtle patterns can convey emotional content and thus associated brain regions would need to recognize such patterns of social information. Since symbolic faces can also be recognized to have emotional content, it seems these brain regions have an inherent capacity for sparse or 'abstract emotional' recognition. Such brain regions appear more capable of artistic (abstract/emotional) responses. However, in humans, there is also good evidence for a less-specialized (non-modular) function. The human neocortex seems crucial for sensory pattern detection, including vision, but blindness can lead these regions to assume other sensory processing activities, such as for hearing. This indicates that a substantial but general plasticity evolved regarding hominid-specific brain function. Why then do some sensory functions, such as face recognition, get their own dedicated patch of cortex? I suggest such dedicated (facial) regions represent older primate strategies aimed mainly at establishing group recognition. Later in human brain evolution, less-dedicated regions of neocortex became able to participate in sensory-based (cognitive) visual and auditory group pattern recognition. However, some basic modularity and brain asymmetry (such as speech and vision) remain and underlie the structure of the human mind.

There is some evidence that the great apes have special facial recognition ability regarding their offspring. Both chimpanzees and humans have been shown to have male offspring-biased visual kin recognition. Fathers tend to recognize the face of their own male offspring even compared to their own female offspring. This capacity is of interest when we consider the origin of paternal and family social bonding in humans. Chimpanzees have very elaborate facial recognition, but there is clearly a developmental window associated with this ability, and during this developmental period, visual recognition capacity can be enhanced. For example, chimps can recognize themselves in mirrors, but this ability is not present in infants and does not develop until about age 2.5 years. This is slightly older than a similar mirror recognition which occurs in humans. It is very interesting that in humans, mirror recognition is also associated with other higher cognitive skills. Mirror recognition is interesting in other regards. Chimpanzee infants will imitate human facial gestures and expressions, such as tongue protrusion. This occurs early (neonates, by 2 months of age), but after 2 months of age, chimps no longer do facial imitation. Such behavior is consistent with a role for facial recognition during a window of identity transfer involved in establishment of group and social identity. Consistent with this proposal, chimps can be trained to recognize chimp faces in photographs. However, these chimps had to be trained during early critical period to establish this recognition. In addition, chimps that have been trained as infants to use visual symbols are also better able to recognize individuals as portrayed by even more abstract line drawings (visual symbols). Such trained chimps can also better categorize symbols from novel line drawing representations. For example, chimps can recognize pictorial (cartoon-like and line drawing) images of objects (flower) and sort them. Learning history was thus crucial for symbolic (abstract) competence and for developing a relationship between pictorial and symbolic information. It is likely that development of their visual cortex and hippocampus is involved. Chimps and humans also both perceive complex and related geometric figures similarly. Thus facial (even abstract) recognition capacity seems innate to chimpanzees and must have been biologically selected. As this capacity is learned during early development, the longevity of chimps would seem relevant. Chimps live longer than most other primates, such as relatively short-lived monkeys, but not as long as humans. This allows enough time for maturation and additional social learning from older chimps, as discussed below.

Other Visual Social Recognition

Chimpanzees clearly depend largely on visual cues for social communication. In addition to facial recognition, chimpanzees will naturally use their hands to gesture and also visually convey social information by body language. Chimpanzees are well known for their use of gestures to communicate aggression, threats, submission, reconciliation, courtship, maternal care and food request. These are all elements of social interactions. Here too, there is some evidence that abstract representations of body gestures can be recognized. Chimps can also recognize when their gestures are being physically imitated by humans. Use of gestures for social communication is a phylogenetically conserved feature of all great ape social groups, indicating a likely biological basis.

In contrast, monkeys are much less prone to use such gestures than chimps or great apes. Although all primates do some movements of arms, head and body to communicate, monkeys have more simplified gestures and social learning mechanisms and show a more innate, stereotypical and species-specific social patterns. We can recall, for example, that New World monkeys retained olfaction (VNO) for scent marking and social identification.

Thus, visual social intelligence of the great apes seems to stem mostly from the need to visually convey social meaning and this placed major evolutionary demands on the brains of great apes. From this perspective, we can best correlate the significant increases in primate brain size during evolution. Released from odor-based social communication mechanisms (VNO dependence) by the great HERV colonization which also likely stimulated brain hypertrophy, the great apes were able to develop cognitive brain-based mechanisms (mostly visual) for social identity and communication. Some of the visual social learning that chimps have developed is cultural. For example, some chimp troupes have developed an overhead hand clasp as a group-specific gesture. In these groups it is used for courting females. Chimps can also learn to visually sign words, whereas baboons or gibbons cannot be taught this. The great apes are excellent social manipulators and good problem solvers. And their social intelligence and general intelligence appear directly linked. This is well demonstrated in chimpanzee group hunting in which multiple (social) intelligences must work together (in a mirror-like linking of minds). This intelligence is mostly visual in that coordinated hunters must envision possible actions of their prey. Chimps must visually imagine what has not yet happened and respond accordingly by evaluating escape routes and preposition members for an ambush. Monkey prey does not have such social or general intelligence and are unable to anticipate the ambush. As such hunting is taught, it is cultural and the product of learning. This then defines a transmissive social state, and the origin of primate culture. Thus, the social links between the minds of great apes became a basis of social cohesion and identity in these species. In the great apes, social complexity and learning also provided the basis of their enhanced general learning ability.

Audio (Vocal) Social Recognition of the Great Apes

Although the primacy of vision for primate social communication seems clear, primates also have considerable ability regarding their audio (vocal)-based communication. However, in contrast to humans, no other great ape is a vocal learner. Monkeys do have stereotyped calls found in many species, such as a leopard call. But this is a general alarm call which lacks specific information as would be needed for the basis of a primitive language. As discussed above, gibbons, lesser apes, do use species-specific vocal calls associated with pair bonding. However, for the most part, these vocal patterns are genetically determined and inherent. Indeed, most animals' calls have limited repertoire with innate and fixed structure. Yet there is some evidence of semantic vocal combinations in some primate calls. For example, the syntax of calls by putty-nosed monkeys seems to involve two basic calls in various sequences to construct various urgent messages. Gibbons call regarding predators might be similar. Thus, although rather limited, it does seem that primates have some degree of abstract or syntax-based audio communication. The great apes clearly use audio communication for social purposes. For example chimps clearly pant and grunt as a signal of subordination and other social communication. Chimps also may use audio communication for group identification. For example, chimp males from distinct populations will produce distinct coordinated calls, a chorusing behavior. This behavior can also be associated with aggression between groups. This suggests some degree of a culturally transmitted vocal-based social identity in chimps. We can recall that in some mammals, audio calls have strong, even biological identity, such as pygmy marmosets which use vocal calls for pair bonding and affect reproductive biology. Other intelligent non-primates also seem to depend on audio communication for social purposes, such as dolphins that appear to have distinct whistles associated with pair bonding. However, this high degree of vocal-mediated social bonding is not seen in the great apes. Still, chimpanzees do have considerable social identification via vocal input. For example, chimpanzees can match individual vocalizations to individual faces and can correctly link a call to a face. Female chimps were able to match pictures of chimps to recorded vocalizations (pants, hoots, grunts, screams), including correctly identifying the two participants of duets. Thus chimps have considerably more vocal recognition and resolution capacity than was previously thought. For example, in language-trained chimps (trained for symbolic recognition) there are hemispheric asymmetries for processing language symbols as used for communication, showing a left hemisphere advantage. Thus, like humans, processing language information in these chimps tends to reside in the left brain. It therefore seems likely that the common great ape–human ancestor must have also had considerable vocal social abilities. However, it is clear that humans underwent a significant evolution in their vocal learning capacities relative to the other great apes and are the only extant primate that has acquired the ability to learn recursive language (see next chapter).

Mind of the Ape Stemming from Vision

Primate evolution does not appear continuous as their brains increase in non-overlapping steps from monkeys to chimpanzees to humans with no intermediate-brained species. This pattern resembles a cumulative series of transforming events, resulting in new laminated brain layers of a larger brain (such as human neocortex). However, as we transition from chimpanzee to human, we can consider how these brain transitions affect the chimpanzee mind in the context of social function and how it differs from that of human. Various definitions of the mind have historically been considered, often from philosophical and religious foundations. Some of these definitions would not appear to relate to the issue of a chimpanzee mind. However, in scientific terms that are susceptible to measurements by neuroscience and psychology, there can be little doubt that chimpanzees indeed have measurable and collective mental states we can call a mind. This collective state can also be considered as the state of consciousness that involves perception, thought and emotion which can affect imagination and will. In the great apes, we have seen significant expansions of the visual capacity of the primate brain, allowing complex vision-based group or social identity to be established. This enhanced vision-based group identity must also link directly to emotional memory if it is to compel behavior and set social bonds. This constitutes a learned (cognitive/emotional) state of group identity. For persistence and stability, it must allow group members to be identified and resist non-members. Visual cognition that directly links to emotional systems and memory are thus proposed to provide the foundation for the mind of chimpanzees. Chimps must learn and remember visual images that elicit the proper emotional (behavioral/social) response. It is such a social purpose that will provide a main selective pressure for the evolution of visual information processing, emotion, memory and ultimately evolve the mind of a chimpanzee. As mentioned previously, this view contrasts sharply with the widely held current view that the evolution of primate visual capacity was mostly the product of selection for the purposes of individuals finding food (i.e., red fruit). It also contrasts with arguments asserting the enhanced survival of the individual (not a group or collective) as the main determinants of evolutionary success (such as game theory). Consider, for example, that chimpanzees find food and defend themselves in collective groups. A system of compulsive social bonding would provide a mechanism by which collective behavior is under positive selection. Learning and memory in primates (activities of their mind) have taken on a crucial evolutionary role as the essential and central function for compulsive social bonding as well as group and social identity.

Social Learning Mechanisms

The sensory neurons (i.e., visual) must provide the pattern of information that needs to become memory during crucial developmental windows. For that, central neurons must be able to reproduce the sensory pattern but also recall it when similarity is detected. From this, we can see that such neurons must become capable to two clear but linked states. One is real-time detection and the other is detecting a reproduced equivalent pattern from memory in which the memory must be triggered by similarity (mirrored or self-detection). This latter memory will also need to link to emotional memory if it is to be applied for stable group identity. As face recognition became a primary system for group identity in primates, this was selected for a hardwired system of dedicated neurons in primates. As more complex social recognition also required differentiating emotional content, these dedicated structures were also selected for abstract (simplified, sparse) facial pattern recognition. The new sensory input and memory must mirror one another at the level of neuron function. Thus, face detection was under strong selection for similarity detection in primates. And this must also link to emotional memory (for group membership/behavior purpose), as this capacity is needed to be able to recognize subtle facial cues that conveyed emotional content. Thus, facial recognition and social memory and learning provided the foundation for ape and ultimately human mental development. Overall, we can see that the great apes indeed differ in several clear ways from most other animals and primates consistent with this idea. They are indeed highly social and have long periods of biological dependency during which social learning and bonding develops. They learn and develop more extended social bonds. They also live long, allowing a slow learning process to be incorporated into their neurological makeup and their social fabric. Such learning makes them smart. This was made possible by their bigger social brains. But enlarged brain tissue does not equal an enlarged mind. What then is distinct about the chimpanzee mind? As I have argued, their new visual and social capacity has given them considerable social intelligence. Liberated from olfaction, this social recognition no longer requires mating, offspring, alarm or aggression pheromones. However, the inherently abstract nature and capacity of this visual recognition system has set the cognitive stage for the development and use of symbolic information and social identity. In addition, this enhanced social intelligence was dependent on the evolution of generalized cortical tissues and can also be applicable to other non-social learning. It is from this that we can see the likely origins of chimpanzee general intelligence.

Thoughts and Identity

Thus in the great apes, the learned brain (mind) now principally controls social actions. Newborns are born with essentially no social capacity (social identity) and it must be learned or developed as a social mind. With this mind, primates have also learned to subordinate their reflexes to associated, willful, thoughtful and sometimes intelligent actions. Sensory input does not kick off stereotyped behavior via invariant emotional response (a pheromone-like situation of social insects). Sensory input, however, is needed for comparison to associated memory for emotional response and action. This results in thoughtful, learned but still often emotional action, not automated, pheromone-directed response of most terrestrial vertebrates. Thus thought, to a much larger degree, now controls social interaction and provides social membership. However, thoughts are no longer human-specific abstractions beyond science as once accepted and limited to the realm of philosophy. They are now part of measurable science and can be directly measured in animals (rodents, monkeys) with microelectrode implants into hippocampus and indirectly measured in humans with fMRI technology. These measurements can now record specific visual stimuli (thoughts) and be used to identify specific sensory input, as has been done with rats whose maze location can be thus measured.

Thoughts, it is asserted, thus became a main mediator to social action in primates, and they are central to the maintenance of group membership. Thoughts provide the impetus and media to control the reaction to the specific sensory input. The sensory mode operates from the visual cortex to hippocampus (site of emotional memory). The thought mode is now measured to operate from the hippocampus to visual cortex (or neocortex in humans). It is interesting that in rats, dreams are similar to thoughts as the dream mode is also from hippocampus to cortex and can activate vision centers (observed via electrode firing) very similar to those experienced by the visual cortex while awake. Such rat dream visions are obviously not real, but the products of thought and memory and possibly emotion. The mind is now creating vision, especially during sleep, and such vision creation likely relates learning and stabilizing of experienced visual patterns. Along these lines, the need for sleep and dreams in memory formation has always been is curious. The 'false' visioning during sleep (REM sleep) is especially associated with brains of warm blooded mammals. But essentially all animals that use CNS neurons to form memory (or establish group membership) also have some form of sleep, even flies. Thus the need to suppress the ongoing sensory mode (sleep) for extended periods is as ancient as the evolution of a primitive CNS. I suggest that such a role for central memory in group membership might provide a better way to trace the evolution of these memory functions and the need for sleep. For example, flies also must learn group (sex, mate) membership from olfactory and visual sensory neurons (see Chapter 10). Memory creation requiring the pattern of sensory input must also be reproduced and stable in central memory neurons, yet also have some independence from sensory input. I suggest that the sensory system cannot be both temporally producing input from visual receptors as well as temporally simulating input from central memory for the purpose of generating or recognizing pattern similarity. For the memory simulation to be set, I suggest the visual input must be transiently prevented for the time needed for the simulation, and this suspension became what we understand to be sleep. Parallel but temporal visual patterns, one sensory, one from memory, must be independently set, but linked and recognized at a cellular (neuronal) level. Thus, incapacitating the sensory mode with sleep became essential to allow the non-real vision to be replayed (dreamed), strengthened and set.

In this scenario, real and imagined (remembered) visions along with real and imagined thoughts have an essential role in CNS-based pattern of group recognition. Memory must keep a temporally accurate but non-real version of sensory input to recognize its real-time reproduction. Memory thus could have evolved from a mirrored sensory system most fundamentally aimed at establishing group identity. Such an evolutionary history would also have set the stage for a feature that was to become crucial for human consciousness, that is, imagination. Imagination and dreams are similar, as both can be unreal temporal scenarios. Imagination, literally, is to image what is not. The difference is that dreaming stems from fragments of sensory (emotional) records, whereas imagination can be less constrained by any actual experience. One can imagine scenes that are not possible and have never been experienced, even abstractions that cannot be experienced. In a sense, this is a defective version of memory. It is clear that chimpanzees can also imagine scenes and events that they have not experienced. They clearly envision what is not or has not yet happened and use this in order to set ambushes and position groups as they prey on monkeys during communal hunts as mentioned above. Such imagination is applied to cooperative hunting and involves various roles (specializations), such as driver, blocker, chaser and ambusher. This is a striking social ability of their much enhanced intelligence which they learn from older males. Chimpanzees must learn to imagine if a cooperative hunt is to succeed. They must learn the skill of unreal visions and unreal thoughts, both regarding their social membership behavior and for a general response to their environment.

General Intelligence and Learning

Chimpanzees with their high social intelligence also have high general intelligence. Their social intelligence is essentially visual in nature due to enhanced CNS visual capacity. In terms of general intelligence, they are the most proficient and versatile tool users of all non-human species. This proficiency is also culturally maintained (learned), such as the cracking of oil-palm nuts, whose learning is population specific, therefore cultural. Japanese macaques have learned to wash potato and wheat in specific populations. The most studied of such traditions is that of chimpanzees using sticks to retrieve termites. Chimps acquire this particular tradition during a critical period of learning (between 3 and 5 years). Juveniles are inherently most likely to explore and learn. It is thus likely that such learning capacity was biologically determined as a window of CNS plasticity associated, I suggest, with the transfer and establishment of social identity. Chimps also seem to have some 'folk' knowledge, i.e., display an innate practical capacity for heuristic abstraction. Such innate psychological capacity can be seen regarding group composition (small numbers) relative to group behavior and aggression. In group hunting, for example, when three or more males encounter a lone male from another troupe they will act together to kill it. A ratio of greater than three to one is necessary for this type of attack. Numbers are also important for other chimpanzee social ecology and group interactions. Chimps engage in lethal fighting of larger groups when numbers are also appropriate and will vocalize aggression to alert other group members (see Fig. 8.8). If, however, the chimp group is small, they will stay silent instead. Besides some 'folk' knowledge for small numbers, chimps also show some abstract ability regarding visual patterns. Using colored feeding bowls, chimps can learn to make symbolic associations between objects and

Fig. 8.8 Chimpanzees patrolling and calling in the forest (reprinted with permission from: Marc Hauser (2005), Nature, Vol. 437)

individuals' artificial visual language, including ownership. Chimps will sort objects into patterns of similar physical classification. They can also be trained to have some capacity to learn Arabic numerals, but unlike humans, chimps fail to generalize number concepts. Untutored humans (children) can represent small numbers but they can also represent comparative ratios between numbers. Thus stemming from their social intelligence, chimpanzees have developed many characteristics we can call a proto-human reasoning capacity. The capacity for abstraction thus sets the stage for abstract, symbolic and reasoned thinking and learning. However, it remains likely that any reasoned thinking that might develop would have at its roots social, group and emotional elements.

Emotions and Ape Group Identity Systems

Emotional memory is an ancient development found in most animals, including invertebrate arthropods and insects. It is based on 'sub-cognitive' mechanism for behavioral control that has been maintained in most animals. Groups of animals are susceptible to various forms of emotional contagion such as the spread of fear in shoals, flocks and herd in groups via body language. In many animals, however, olfaction is an important component of emotional memory (i.e., alarm, fear, mating, aggression).

As I have argued, emotions provide general tendencies for behavioral responses that are both harmful and protective, thus can be used as the T/A elements of an addiction module to attain stable social bonds. In mammals, the mother–offspring is the basal social bond. Thus this social bond presents us with exemplars of the mammalian social systems from which most others have likely been adapted and evolved. We have already noted that this maternal–offspring bond involves prolactin and oxytocin and these neuropeptides affect addiction centers in the brain. Primates have maintained this basic circuit, although minus the major olfaction role. The dependency of a mammalian infant on maternal care varies considerably within species, from the scale of several weeks for some rodents to between 7 and 15 years in the great apes. Humans have the longest period of infant dependency. The very extended infant dependency of the great ape infant has required a more stable social bond which must be attained by their large social brain, not via olfaction. I would argue that such enhanced social bonds have required the development of additional emotional (addiction) systems to provide the extended stability. For all mammals, the emotions of obsession and anxiety are seen in mothers. However, the infant must also be tightly bonded to its mother by similar emotions and this too has been extended in the great apes. Young primates separated from their mothers soon after birth display pathological anxiety or depression-like symptoms. This psychological pathology can be treated with serotonin uptake inhibitors. Thus, social deprivation leads to pathological (toxic) emotions and behavior consistent with an emotional addiction module. This type of pathology is never seen in field studies, suggesting that such social bonding is an invariant requirement for all chimpanzees. Psychological dependency is thus essential for normal bonding but likely needs to operate for extended periods to accommodate long infant dependency of great apes. How then was the stability of this bond extended in the great apes? And was this extended infant–mother social bond also used as a foundation to evolve additional and extended social bonding and identity generally used by the great apes? Below, I suggest that sequential windows of brain development that were linked to psychological development (and dependency) were evolved for the purpose of supporting extended and sequential social bonds.

Field studies support the idea that great apes show much more social affiliation than their ancestors. For example, studies of activity budgets of primates confirm that OW apes spend considerably more time in affiliative interactions. And the cooperative behaviors are overwhelmingly greater than are aggressive ones in these studies. For example howler monkeys show only about 1% of their daily activity in affiliative interactions. The average for prosimians was about 3.7%. New World primates average about 5% of their activity budget. This compares to a whopping 25% of affiliative activity in chimpanzees. As expected, the bulk of this chimp activity was due to various female–offspring interactions. Thus in chimpanzees, cooperation is providing much more social cohesion than is aggression (which is more associated with larger, i.e., troupe, identities and male behaviors).

Emotional Pleasure for Social Bonding

It thus seems likely that this social affiliation is more likely maintained by pleasurable emotions than it is by avoidance of negative emotions. Social primates simply 'like' each other and will seek (need) to be together. Mothers 'like' infants and infants 'like' mothers and other females. Pleasurable emotional states are thus the norm. Such interaction is therefore not the product of selection for 'individual gain' as is often rationalized in a strict Darwinian cost–benefit context. It is instead the product of an extended (social) group identity that has developed and used positive emotions to set, extend and stabilize group bonding. It operates via psychological addiction modules that in primates are essentially cognitive in nature. However, such social pleasure was not new to the primates, it was simply adapted and extended by them. All mammals are distinguished from other vertebrates in that their young play and while playing are happy and unaggressive. Such happiness is contagious and also involves dopamine. Even carnivores will play and can show empathy to others. I suspect this tendency is derived from maternal–infant bonding and its requirement for establishing social pleasure. Playing also involves mutual social pleasure of the participants, in which players typically observe others and do the same (mirror behavior). Their emotions become in a sense linked together. Empathy is related to this state. Below, I present a possible relationship of playing to mirror neurons and how biological selection might have developed this system in primates and provided the underpinnings of hominid empathy.

Emotional Pain for Social Bonding

The flip side of an addiction module is that it also requires a toxic, harmful component to compel stability. In psychological terms, painful emotions would provide the toxic component of a social identity, such as anxiety as experienced by mothers and infants during separation or the aggression mother's display toward threats to offspring. Thus we can expect the need for potentially painful psychological responses to compel behavior and why it would be conserved during animal evolution. Psychological pain thus serves a basic T-function as a T/A module in social bonding. Without this function, psychological pain would otherwise seem to only reduce fitness of those experiencing it and provide no selective advantage. Intense psychological pain, such as a mother's bereavement with the death of an infant, has the essential characteristics of a cognitive addiction module. The painful reaction is quickly induced (unmasked) by either the observation or even the thought of death of its offspring. Thus both the prevention of pain and its induction are set and maintained by states of thought or belief. The potential for strong pain is thus ever present, kept in check by the knowledge (a protective thought) of the infant. This then defines a purely psychological or cognitive addiction module and I am asserting that such a circumstance provides the basal bonding mechanism for more extended social structures and identities of the great apes, including humans. Social and emotional pain is thus an essential and conserved element of the primates.

Mirror Neurons: An Extended Mind and Emotional Net

Visual memory is proposed to be central in primates and the great apes for the development of group identity. It also needs to be linked to emotional memory. It is now clear that primates indeed possess an uncommon mirror neuron system that can be activated by visual stimuli and also involves motor control regions and also links to emotions. A macaque seeing someone grasp an item will activate the similar regions within its own brain. In macaques this mirror region is the F5 region which is similar to and asymmetric like Broca's area in humans. This F5 will fire during goal-directed hand and mouth movements, but the mirror subset of neurons will be active when the monkey watches another perform the same activity. It is worth noting that the homologous Broca's area in humans is thought to be crucial for language (see next chapter). Clearly this region is not providing a linguistic function in monkeys. Instead it houses mirror neurons. Such capacity likely reflects a social function important to primate brains. As mentioned above, in primates, gestures (in addition to facial expression) convey much social and emotional content. Others have suggested that gestures may have provided some seed function for the origin of language. However, I suggest that the common function between the role that F5 provides in monkeys and Broca's area in humans is not sensory specific (visual versus audio), but social. Both provide a common social function essential for group identity. The difference is that in humans, voice and language have evolved into a major system of social identity (see next chapter), thus humans evolved language primarily for social purpose. In primates, this mirror system allows the matching of manual execution of mouth and hand movements with visual observations of such motions as done by others. It has been proposed that this feature is important for learning by imitation in monkeys. Yet, curiously monkeys simply do not imitate each other as do humans. However, macaques clearly know when they are being imitated by humans, so some visual self-awareness is present. Another idea is that mirror neurons are providing a direct social function. In a sense, it allows vision to provide an extended, linked social mind. The ability of any group identity system to differentiate self from non-self is basic. However, group identity systems must also distinguish group members from non-members and if that social identity system is visual, vision must link members. Mirror neurons could provide exactly such social bonding by linking visual recognition, gestures and emotions of others to that self (via brain stimulation). Accordingly, these mirror systems were selected to provide a visually linked extended social identity, closely linking the brains of all group members. In a real sense, the brains of such grouped primates are connected via vision to the brains of other primates in the same group. Thus, strong visual-based group bonding is further mediated by mirror neurons. Are the emotions also connected to mirror neurons as predicted by social bonding? This question is difficult to answer in non-human primates. However, as presented in the next chapter there is strong reason to think that in humans, mirror system clearly links to emotions and social groups and autism may be affecting exactly such social–emotional systems. In addition, it seems likely that this mirror system may provide the neurological substrate for empathy between members. Such a capacity would provide a basis for the protomorality of primates as described below.

A visual social mind is proposed to provided the foundations for extended primate group bonding and a capacity for empathy with emotional linking mechanisms that use mirror neurons. A visual social mind also provides the foundations for envisioning actions that have not yet occurred (i.e., dream like). Empathy is a higher emotion that supports an extended social cohesion. It is clear that gorillas and chimpanzees can show empathy to each other and even other species when they observe distress in others. The observations that a chimpanzee will attempt to help an injured bird fly again clearly indicates some capacity of empathy for other species. Chimpanzees can also clearly show empathy and care for injured and elderly of their own group and will display contact comforting when they observe separation anxiety of infants. Similar documented observations of gorilla helping injured human boy after falling into their pen at a zoo shows they too have capacity for empathy with other species. These tendencies are inherent, hence must be biologically supported. In primate groups, social cooperation is the norm and positive emotional connections are also normally in place. In such associations, chimps have also been shown to have a nescient and inherent tendency to be 'fair' to one another. This tendency can be demonstrated by differentially offering foods, such as cucumbers versus fruits to groups. Chimpanzees will readily eat the cucumber if it is perceived as fair but reject if it is unfair due to fruit being offered unfairly to others. Preferences thus depend on others, consistent with the operation of an extended social mind. It appears that being fair feels good and being unfair feels bad to chimpanzees (and most humans). Along with empathy noted above, this interaction appears to identify the existence of a protomoral stage of primate evolution that likely involves an extended, social mind. In contrast to many human beliefs, such empathic primate behavior was not dependent on or the product of any moral or religious code. It is inherent to social structures and supports positive emotional states. Chimpanzees do not need beliefs to lead contented social lives. Such a need is human specific and the related role of cognitive addiction modules is discussed in the next chapter.

Not all primate social interactions, however, involve positive emotions as aggression is clearly an important primate social behavior. Chimpanzees very much tend to form extended group associations that employ lethal aggression to other groups to maintain their identity. Clearly, visual identification of a lone chimp as a non-member can induce strong aggressive group emotions. Chimpanzees also use aggression by the dominant alpha male to subordinate males and females. In contrast, the bonobos social groups are more cohesive than that of chimpanzees, but bonobos social structures are matriarchal and inherently much less aggressive. Bonobo groups are usually composed of greater than 20 individuals. However, bonobos do display aggression to other group members, although it is not usually lethal. Thus, like chimpanzees, interactions between groups are hostile. Yet here too hostility is often displayed by males usually against other males. Chimp males that hunt together will also show considerable kin bonding, unlike bonobos.

Chimpanzee hunting can also involve extreme cruelty to prey monkeys as they will hunt and consume live prey monkeys with apparent glee. Such seemingly pleasurable displays are also seen when chimpanzees hunt and kill members of other troupes. In such circumstances, there seems to be little empathy or positive emotion at work, and if anything, some pleasure is expressed during such negative emotional interactions. Human social structures contrast with those of chimpanzees in that humans lack the chimp alpha male breeding organization. In this, humans more closely resemble less-ordered, less-aggressive serial breeding structures as found in bonobos.

Why then is it so important to communicate emotion in primate social groups? Why is there so much biologically based brain structure directed to recognizing facial and gesture-based emotional content? The simple answer is group identity. In summary, we can say that primate emotions have undergone an expansion relative to other animals, but remain essentially in sets that are both beneficial and harmful, thus they generally provide the basis of social bonding. Since the social bond is mostly cognitive (visual), visual feedback regarding social status and affiliation is essential and expanded emotions must be transmitted (via face and gestures) to provide this status. Thus, faces, for the most part, have replaced alarm pheromones or MHC peptides (as in rodents) for providing much group information. Since identity status is mainly cognitive (thought maintained), it is prone to very rapid and dynamic social changes, thus thoughts must be frequently communicated emotionally. The visual capacity to expand and discriminate the subtle variations in facial expressions has placed strong demands on the visual centers of the brain and required significant brain adaptations to keep pace with evolving social group identity. Dedicated brain regions were needed to initially provide such high level and rapid facial resolution that also provided the inherent capacity for symbolic visual representation. Although primates were predominantly visual regarding the brain evolution and group identity, they also maintain some vocalization capacity that can also communicate emotion and group identity. The emotional component of such communication appears to have stemmed from brain regions that were involved in gesture communication.

Fitness Consequence of Social Bonds and Cooperation

The fitness consequences of not belonging to a primate social group are severe. This is most apparent regarding a lone male chimpanzee that may have become an outcast from a social group. Such a chimp is in mortal danger and will be actively hunted and killed by other groups of chimps. Even in situations where group aggression is not lethal, the ability of a lone primate to survive (find food, shelter, mates and provide protection) would be severely reduced when alone. The evolution of primate sociality has previously been explained as somehow being the byproduct of selfish genes. Game theory has been proposed to explain how selective pressures on the survival of an individual would lead to such cooperative behaviors. In addition, mate selection is also invoked to explain other social characteristics (mating, color vision). However, phylogenetic analysis of primate social systems indicates that social organization is often strongly conserved in some primate lineages even when there is considerable ecological variability. This supports the existence of underlying biological determinants to the evolution of such social organization. Selfish gene concepts do not provide explanations of such biologically determined and well-conserved social structures. Nor can these be explained by game theory or rational choice scenarios. The existence of a biologically determined and extended (mirrored) social mind as found in primates confronts the predictions of such theories. Often, selection for enhanced primate visual or mental capacity has been rationalized from the perspective of increased individual efficiency. As mentioned, both intelligence and trichromic color vision are rationalized from the perspective of food finding efficiency and survival of fittest individuals, and accordingly social food finding is more efficient thus invoking game theory. But the enhanced primate brain architecture is mostly dedicated to an enhanced visual social capacity, not efficient food finding or other characteristics that benefit individuals. I suggest such 'efficiency' rationalizations are failures, just like mate selection theory failed to explain gibbon pair bonding above. During human (hominid) evolution, there have been an estimated 20 humanoid species that were successful at one time, but except for *H. sapiens*, all have become extinct. They all had expanded brains to varying degrees. Clearly, their larger brains did not result in more efficient individuals. Compare this extinction rate to other recently evolved but currently diverse species (such as voles). The explanations of primate social evolution based on individual, competitive survival thus seem contrived, in opposition to the inherent cooperative nature of primate social structures and devoid of biological basis or data. As discussed above, I have already outlined some problems regarding explanations based on efficiency. In contrast to these explanations, selection-based group identity can readily account for the inherently biological tendency for social cooperation, the nature of primate social structures and why their brains would have been adapted to such a specific degree. This theory also accounts for the massive genetic changes and ERV colonizations specific to the primate genomes. All of these corresponding genetic changes (including VNO loss) are fully ignored by the above theories.

Another prevailing selection concept that has some characteristics of group identity is mating selection. Sexual identity and physical dimorphism have often been explained by mating selection. The great apes have a greater sexual dimorphism relative to humans as males have twice the body mass of females. This ratio was reduced in humans. The great ape social and mating systems vary greatly and include polygamous, monogamous and promiscuous situations. In addition, harem structures can exist, as in chimpanzees, and the emotion of jealousy appears to motivate males' possessive herding of female. This is often explained as a product of mate selection. However, bonobo females show large colorful visual sexual swellings, usually taken as a sign of mate selection, but as these are on females which are highly promiscuous, the mate selection concept is clearly not applicable here. Like bonobos, chimpanzee females also have apparent estrus, with often colorful display, although their relationship to males differs considerably. Given the rarity of color display in mammals, this sexual display in primates and use of color in facial marking of other primates (e.g., rhesus, mandrills), it would seem that social/sexual identity may provide a better explanation for the origin of primate color vision and its use in sex. Thus it is interesting that human females do not display estrus but maintain visually apparent mammary glands when not nursing as part of their sexual dimorphism. This visual ambiguity regarding reproductive status seems to have fundamentally altered human relationships from that of primates. Within chimpanzee females, antagonistic components in group associations are much less apparent than within males. This enhanced aggression by males is often explained by differential cost/benefit regarding the bearing of live infants and rearing them. Since males invest little in this, males will display the phenotypical consequences of mate selection, such as aggression, competition (large body mass) for females. Yet in some primate social situations, females have capacities absent from males. Curiously, the use of greeting grunts is somewhat sexually dimorphic in bonobos, but it is mostly a female to male greeting, not female to female or male to female. A distinct system of group identity could more readily account for the evolution of the bonobo social structure without invoking the need for mate selection.

Primate Social Structures Are Mostly Associative

The preponderance of associative social interactions in primates adds to the failure of selfish genes concepts as a driving mechanism of primate evolution.

Overall, these relationships are characterized by social pleasure, not aggression or cost–benefit scenarios. The majority of these associative interactions stems from female–young interactions, which is a basal and long-lasting social bond. Females initially travel alone with offspring within larger groups for the first 10 years. At about 6 years, females also provide some instruction to their young, such as learning nut cracking. At about 10 years, males will undergo sexual maturation, along with developmental hormonally controlled brain changes and begin to associate with older males from which they learn hunting. This is not a 'family' unit social structure in a human sense (lacking mate pair bonding and shared care of young). The more extended breeding units travel as a group in which it is the older males that provide group defense, mostly against other chimpanzee males from other groups. In a sense, the overall group identity has segregated the benefits (associative) provided mostly by females from the harmful (aggressive) emotions provided by males into sexually dimorphic social sets. The females are providing the long-term associative element and males the aggressive element that together provide a social identity. Most chimp social associations are composed of about 150 members that are hostile to each other. Although monkeys can have similar-sized communities, their social dynamics are much simpler. Unlike monkeys, however, great ape group identity and dynamics are more adaptable, not so set. In chimps, these groups have dynamic, non-stereotypical compositions that include sub-groupings, such as bands composed of several clans. A troupe is an aggregation of several bands, typically 30–60 individuals, but up to as many as 750. Both chimpanzee and bonobo social organizations are dominated by a senior individual (a matriarch with bonobos and an alpha male in chimpanzee). Interesting that in contrast to human sexual relationships and at odds with various mate selection theories, in bonobo social structures, older females appear to be more sexually attractive to young males. However, it is especially interesting to consider the cognitive mechanism by which a senior individual is providing anchor for an extended social structure or identity. Chimps, unlike bonobos, do not seem very loyal to their senior alpha male and will quickly change allegiances when the male sustains defeat in aggressive combat. Thus the social role and status of the senior individual male are not very stable (emotionally addicted). However, bonobos maintain more cohesive and stable social structures than chimps via an older female and loyalty seems more stable. How does this individual female exert influence and control to maintain group cohesion? What emotions bind the group to these older females? Is this social binding related to or derived from the type of social bonds between mother and infant? Since this bonding is likely to be mediated via individual visual recognition, memory, thoughts and emotions must be involved. The involvement of thought and memory, however, also suggests that belief (stable memory) in an individual also contributes to such social bonding and stability.

We should be reminded that the chimpanzees, although highly visual in their social communication, do maintain some vocal communication as well and can recognize individual calls. Range groups that move as a unit will not only maintain visual communication but also some audio communication. It seems clear that chimpanzees have the beginnings of a family-based social structure. However, in neither chimpanzee nor bonobo social structures is male–female mate bonding strong nor do males contribute to rearing of young. The absence of a strong paternal role in a family social structure clearly distinguishes chimpanzee and bonobo social structures from those of humans. For such a family-based social structure to evolve in humans, males must form lasting social bonds to females and to their young.

Paternal Role in Some Primates

As we conclude our consideration of the great ape systems of group identity, we have come to focus on some inherent distinctions in social structures with humans; mostly human paternal bonding. A paternal role in care of young, although common in avian species, is not common in mammals. Mammals are estimated at 5,000 species of which only 10% provide paternal care and only 3–5% of mammals are socially monogamous. Like the avians, in those mammal species that have been studied, prolactin levels in males appear associated with such care (i.e., native California mice and wolves). Curiously, this situation is more common in rodents and in these rodents, there is a strong phylogenetic conservation of paternal care, much more conserved than are mating structures. Why would such a characteristic have such a strong biological basis and how does this evolve? Given the PRL role, it seems likely that differential regulation of PRL-mediated social bonding circuits must be involved. What then might be the selection for this? Was this the product of selection for a new group identity? Since prolactin is central to maternal–offspring bonding, can we propose that human males (fathers) now have a female-like PRL response to infants? None of these questions can currently be clearly answered in humans. However, as previously noted humans and bonobos do share the altered genetic structure (and likely regulation) of their PRL receptor promoter relative to chimpanzees. But humans are not the only primates that provide paternal care. Examples in both New and Old World are known, although uncommon. Among New World primates, the Callitrichidae (marmosets and tamarins) are well known for extensive paternal care. Callitrichids live in family groups with one breeding pair that have offspring (usually twins) of successive births. All family members help rear the young. The father will carry the offspring on their back and provide food for them. The PRL biology of these males has been studied in two species, the common marmoset (*Callithrix jacchus*), and cotton-top tamarins (*Saguinus oedipus*). Marmoset fathers have significantly higher plasma PRL levels than non-fathers and tamarin fathers showed elevated urine levels of PRL. In the tamarin fathers, PRL urine levels began to increase 2 weeks before the birth of offspring and reached a maximum in the 2 weeks after birth. Thus, this increased PRL level was not simply caused by stimuli from the infants, but more likely was due to olfactory signals from the pregnant mother. If PRL is indeed needed to induce paternal care in tamarins, this induction needs to occur before birth, since fathers carry their infants and provide care immediately after birth. Unlike Old World primates, tamarins indeed maintain a VNO tissue, although its role in paternal care has not yet been evaluated. In monkeys, there may be an example of visually mediated bonding of a father to its offspring. The titi monkeys live in small family groups with an adult breeding pair and one to three offspring. In contrast to other primates, the father is the primary caregiver. With titi monkeys, fathers are visually attracted to newborns and carry them after birth. It is unknown if any olfaction is also involved. We have discussed the prevalent pair bonding and parental care of the gibbons above. However, the neurochemistry of these social bonds have not been studied in these lesser apes.

In humans, a role for PRL in paternal bonding to infants is not well studied. Systemic PRL plasma levels have been reported to respond in fathers, but in a complex way, decreasing in new fathers and increasing with subsequent births, although brain-specific PRL responses would not be apparent from such systemic studies. Interview studies also suggest that a high emotional attachment of fathers to their newborn does indeed occur, a state called engrossment. And, these emotions are often surprisingly strong to new fathers when they hold their infants (curiously, usually on the left side), suggesting that human fathers indeed love their newborns. Love could thus provide a strong state of emotional bonding specific to humans. A brain-specific PRL role in love states is presented in the next chapter. However, it may also be that human fathers, unlike all primates, have developed an audio cue regarding their infants. Infant cries have been reported to induce high PRL levels in fathers, increasing their alertness and feelings of sympathy. Such responses were increased in experienced fathers. Given the increased importance of vocal learning in humans compared to all other primates, a possible link of vocal signaling to human paternal bonding is most interesting.

Summary

In this chapter I started by outlining the major morphological changes associated with primate and hominid evolution in which it was the objective to understand the origin of human group and social identity systems by tracing the origin of the human mind. At the base of all primate evolution was the major genetic colonization involving numerous ERVs and hyperparasites (LINES, alu's, SINES). The source of this genetic colonization was mostly from exogenous viruses. Retroviruses of the MLV family seem especially involved as ancestors to the HERV-K viruses that are so prominent in Old World primate evolution, including recent human changes. One of the consequences of this genetic invasion for the Old World primates was the displaced group identity system via the loss of olfaction-based recognition, a system that had been conserved since the evolution of teleost fish. This colonization also had major effects on the system of immunity (MHC I/II) present in primates. Thus, the VNO genes and tissues were lost from both their MHC locus and other genetic sites of Old World primates. Other OR genes were also reduced. The resulting Old World primates no longer depended on olfaction for group identity. However, due to ERV colonization, they now had an altered relationship with exogenous retroviruses, supporting species-specific foamy and SIV viruses in non-hominid African species. This altered and stimulated the evolution of their systems of immunity and identity and made ERVs the major source of immune system evolution in primates. As a consequences of these changes, primates (especially Old World primates) evolved to be dependent on visual (mostly facial) cues for group identity. This dependence forced the evolution of significantly enhanced visual capacity (color vision, visual acuity, vision-specific brain adaptations). This included the evolution of brain-specific regions for facial recognition. Face recognition, rather than VNO pheromones or MHC peptides, became the main mechanism for individual identity. Emotions (especially prolactin mediated) were used to provide addiction modules for social bonding in response to visual identity (especially those of mother–offspring). The gibbons, lesser apes, also developed species-specific vocalization as an element of pair bonding. However, these vocal patterns were mostly genetically determined, not learned. The face recognizing regions of primate brains have some inherent capacity for symbolic recognition (mostly for emotions). A mostly vision-based system for group identity was also made possible by the evolution of mirror neuron systems that allowed emotional links between individuals mediated by vision. Thus the primates evolved an 'extended' visual mind for the purposes of group (social) identity (not efficiency). Learning visual recognition along with visually expressing emotion became the main mechanism to attain group membership. Group identity and its fitness, not game theory or selfish genes, better explains the evolution of social learning and highly social and dynamic primate groups. The great apes, however, did not evolve a strong paternal bond and hence did not develop biparental family-based social structures. Nor did they develop extensive vocal communication or learning for group identity.

Recommended Readings

Primate brain

1. Bartel, D. P. (2004). MicroRNAs: genomics, biogenesis, mechanism, and function. *Cell* **116**(2), 281–97.
2. Cooper, D. L. (2006). Broca's arrow: evolution, prediction, and language in the brain. *Anat Rec B New Anat* **289**(1), 9–24.
3. Gaskill, P. J., Watry, D. D., Burdo, T. H., and Fox, H. S. (2005). Development and characterization of positively selected brain-adapted SIV. *Virol J* **2**, 44.

4. Liman, E. R., and Innan, H. (2003). Relaxed selective pressure on an essential component of pheromone transduction in primate evolution. *Proc Natl Acad Sci U S A* **100**(6), 3328–32.
5. Lund, J. S., Hendrickson, A. E., Ogren, M. P., and Tobin, E. A. (1981). Anatomical organization of primate visual cortex area VII. *J Comp Neurol* **202**(1), 19–45.
6. Nakamura, A., Okazaki, Y., Sugimoto, J., Oda, T., and Jinno, Y. (2003). Human endogenous retroviruses with transcriptional potential in the brain. *J Hum Genet* **48**(11), 575–81.
7. Pfaff, D. W. (2002). "Hormones, brain, and behavior." 5 vols. Academic Press, Amsterdam ; Boston.
8. Sherwood, C. C., Broadfield, D. C., Holloway, R. L., Gannon, P. J., and Hof, P. R. (2003). Variability of Broca's area homologue in African great apes: implications for language evolution. *Anat Rec A Discov Mol Cell Evol Biol* **271**(2), 276–85.

Primate HERVs and Mouse ERVS

1. Adamson, M. C., Silver, J., and Kozak, C. A. (1991). The mouse homolog of the Gibbon ape leukemia virus receptor: genetic mapping and a possible receptor function in rodents. *Virology* **183**(2), 778–81.
2. Barbulescu, M., Turner, G., Seaman, M. I., Deinard, A. S., Kidd, K. K., and Lenz, J. (1999). Many human endogenous retrovirus K (HERV-K) proviruses are unique to humans. *Curr Biol* **9**(16), 861–8.
3. Barbulescu, M., Turner, G., Su, M., Kim, R., Jensen-Seaman, M. I., Deinard, A. S., Kidd, K. K., and Lenz, J. (2001). A HERV-K provirus in chimpanzees, bonobos and gorillas, but not humans. *Curr Biol* **11**(10), 779–83.
4. Bromham, L. (2002). The human zoo: Endogenous retroviruses in the human genome. *Trends Ecol Evol* **17**(2), 91–97.
5. Cordonnier, A., Casella, J. F., and Heidmann, T. (1995). Isolation of novel human endogenous retrovirus-like elements with foamy virus-related pol sequence. *J Virol* **69**(9), 5890–7.
6. Dunn, C. A., van de Lagemaat, L. N., Baillie, G. J., and Mager, D. L. (2005). Endogenous retrovirus long terminal repeats as ready-to-use mobile promoters: the case of primate beta3GAL-T5. *Gene* **364**, 2–12.
7. Dupressoir, A., Marceau, G., Vernochet, C., Benit, L., Kanellopoulos, C., Sapin, V., and Heidmann, T. (2005). Syncytin-A and syncytin-B, two fusogenic placenta-specific murine envelope genes of retroviral origin conserved in Muridae. *Proc Natl Acad Sci U S A* **102**(3), 725–30.
8. Fiebig, U., Hartmann, M. G., Bannert, N., Kurth, R., and Denner, J. (2006). Transspecies transmission of the endogenous koala retrovirus. *J Virol* **80**(11), 5651–4.

9. Hanger, J. J., Bromham, L. D., McKee, J. J., O'Brien, T. M., and Robinson, W. F. (2000). The nucleotide sequence of koala (*Phascolarctos cinereus*) retrovirus: a novel type C endogenous virus related to Gibbon ape leukemia virus. *J Virol* **74**(9), 4264–72.
10. Kim, H. S., Wadekar, R. V., Takenaka, O., Hyun, B. H., and Crow, T. J. (1999). Phylogenetic analysis of a retroposon family in African great apes. *J Mol Evol* **49**(5), 699–702.
11. Kim, H. S., Yi, J. M., Hirai, H., Huh, J. W., Jeong, M. S., Jang, S. B., Kim, C. G., Saitou, N., Hyun, B. H., and Lee, W. H. (2006). Human Endogenous Retrovirus (HERV)-R family in primates: Chromosomal location, gene expression, and evolution. *Gene*. **370**, 34–42
12. Kim, T. M., Hong, S. J., and Rhyu, M. G. (2004). Periodic explosive expansion of human retroelements associated with the evolution of the hominoid primate. *J Korean Med Sci* **19**(2), 177–85.
13. Kriener, K., O'HUigin, C., and Klein, J. (2000). Alu elements support independent origin of prosimian, platyrrhine, and catarrhine Mhc-DRB genes. *Genome Res* **10**(5), 634–43.
14. Mangeney, M., de Parseval, N., Thomas, G., and Heidmann, T. (2001). The full-length envelope of an HERV-H human endogenous retrovirus has immunosuppressive properties. *J Gen Virol* **82**(Pt 10), 2515–8.
15. Sverdlov, E. D. (2000). Retroviruses and primate evolution. *Bioessays* **22**(2), 161–71.
16. Wetterbom, A., Sevov, M., Cavelier, L., and Bergstrom, T. F. (2006). Comparative genomic analysis of human and chimpanzee indicates a key role for indels in primate evolution. *J Mol Evol* **63**(5), 682–90.
17. Yi, J. M., Takenaka, O., and Kim, H. S. (2005). Molecular cloning and phylogeny of HERV-E family that is expressed in Japanese monkey (shape *Macaca fuscata*) tissues. *Arch Virol* **150**, 869–82
18. Zhao, Y., Jacobs, C. P., Wang, L., and Hardies, S. C. (1999). MuERVC: a new family of murine retrovirus-related repetitive sequences and its relationship to previously known families. *Mamm Genome* **10**(5), 477–81.

SIV, HIV, FV and Primate Retroviral Resistance

1. Apetrei, C., Robertson, D. L., and Marx, P. A. (2004). The history of SIVS and AIDS epidemiology, phylogeny and biology of isolates from naturally SIV infected non-human primates (NHP) in Africa. *Front Biosci* **9**, 225–54.
2. Bailes, E., Gao, F., Bibollet-Ruche, F., Courgnaud, V., Peeters, M., Marx, P. A., Hahn, B. H., and Sharp, P. M. (2003). Hybrid origin of SIV in chimpanzees. *Science* **300**(5626), 1713.
3. Beer, B. E., Bailes, E., Goeken, R., Dapolito, G., Coulibaly, C., Norley, S. G., Kurth, R., Gautier, J. P., Gautier-Hion, A., Vallet, D., Sharp, P. M., and Hirsch, V. M. (1999). Simian immunodeficiency virus (SIV) from sun-tailed monkeys (*Cercopithecus solatus*): evidence for

host-dependent evolution of SIV within the *C. lhoesti* superspecies. *J Virol* **73**(9), 7734–44.
4. Carrington, M., and Bontrop, R. E. (2002). Effects of MHC class I on HIV/SIV disease in primates. *Aids* **16 Suppl 4,** S105–14.
5. Conticello, S. G., Thomas, C. J., Petersen-Mahrt, S. K., and Neuberger, M. S. (2005). Evolution of the AID/APOBEC family of polynucleotide (deoxy)cytidine deaminases. *Mol Biol Evol* **22**(2), 367–77.
6. Courgnaud, V., Saurin, W., Villinger, F., and Sonigo, P. (1998). Different evolution of simian immunodeficiency virus in a natural host and a new host. *Virology* **247**(1), 41–50.
7. Esnault, C., Heidmann, O., Delebecque, F., Dewannieux, M., Ribet, D., Hance, A. J., Heidmann, T., and Schwartz, O. (2005). APOBEC3G cytidine deaminase inhibits retrotransposition of endogenous retroviruses. *Nature* **433**(7024), 430–3.
8. Harris, R. S., and Liddament, M. T. (2004). Retroviral restriction by APOBEC proteins. *Nat Rev Immunol* **4**(11), 868–77.
9. OhAinle, M., Kerns, J. A., Malik, H. S., and Emerman, M. (2006). Adaptive evolution and antiviral activity of the conserved mammalian cytidine deaminase APOBEC3H. *J Virol* **80**(8), 3853–62.
10. Salemi, M., Strimmer, K., Hall, W. W., Duffy, M., Delaporte, E., Mboup, S., Peeters, M., and Vandamme, A. M. (2001). Dating the common ancestor of SIVcpz and HIV-1 group M and the origin of HIV-1 subtypes using a new method to uncover clock-like molecular evolution. *Faseb J* **15**(2), 276–8.
11. Sawyer, S. L., Emerman, M., and Malik, H. S. (2004). Ancient adaptive evolution of the primate antiviral DNA-editing enzyme APOBEC3G. *PLoS Biol* **2**(9), E275.
12. Song, B., Javanbakht, H., Perron, M., Park, D. H., Stremlau, M., and Sodroski, J. (2005). Retrovirus restriction by TRIM5alpha variants from Old World and New World primates. *J Virol* **79**(7), 3930–7.
13. Switzer, W. M., Salemi, M., Shanmugam, V., Gao, F., Cong, M. E., Kuiken, C., Bhullar, V., Beer, B. E., Vallet, D., Gautier-Hion, A., Tooze, Z., Villinger, F., Holmes, E. C., and Heneine, W. (2005). Ancient co-speciation of simian foamy viruses and primates. *Nature* **434**(7031), 376–80.
14. Wang, B., Mikhail, M., Dyer, W. B., Zaunders, J. J., Kelleher, A. D., and Saksena, N. K. (2003). First demonstration of a lack of viral sequence evolution in a nonprogressor, defining replication-incompetent HIV-1 infection. *Virology* **312**(1), 135–50.

Primate MHC Evolution

1. Adams, E. J., and Parham, P. (2001). Species-specific evolution of MHC class I genes in the higher primates. *Immunol Rev* **183,** 41–64.
2. Andersson, G., Svensson, A. C., Setterblad, N., and Rask, L. (1998). Retroelements in the human MHC class II region. *Trends Genet* **14**(3), 109–14.

3. Bontrop, R. E. (2006). Comparative genetics of MHC polymorphisms in different primate species: duplications and deletions. *Hum Immunol* **67**(6), 388–97.
4. Carrington, M., and Bontrop, R. E. (2002). Effects of MHC class I on HIV/SIV disease in primates. *Aids* **16 Suppl 4,** S105–14.
5. de Groot, N. G., Otting, N., Doxiadis, G. G., Balla-Jhagjhoorsingh, S. S., Heeney, J. L., van Rood, J. J., Gagneux, P., and Bontrop, R. E. (2002). Evidence for an ancient selective sweep in the MHC class I gene repertoire of chimpanzees. *Proc Natl Acad Sci U S A* **99**(18), 11748–53
6. Fukami-Kobayashi, K., Shiina, T., Anzai, T., Sano, K., Yamazaki, M., Inoko, H., and Tateno, Y. (2005). Genomic evolution of MHC class I region in primates. *Proc Natl Acad Sci U S A* **102**(26), 9230–4.
7. Gaudieri, S., Kulski, J. K., Balmer, L., Giles, K. M., Inoko, H., and Dawkins, R. L. (1997). Retroelements and segmental duplications in the generation of diversity within the MHC. *DNA Seq* **8**(3), 137–41.
8. Kriener, K., O'HUigin, C., and Klein, J. (2001). Independent origin of functional MHC class II genes in humans and New World monkeys. *Hum Immunol* **62**(1), 1–14.
9. Kriener, K., O'HUigin, C., and Klein, J. (2000). Alu elements support independent origin of prosimian, platyrrhine, and catarrhine Mhc-DRB genes. *Genome Res* **10**(5), 634–43.
10. Kulski, J. K., Anzai, T., and Inoko, H. (2005). ERVK9, transposons and the evolution of MHC class I duplicons within the alpha-block of the human and chimpanzee. *Cytogenet Genome Res* **110**(1-4), 181–92.
11. Vogel, T. U., Evans, D. T., Urvater, J. A., O'Connor, D. H., Hughes, A. L., and Watkins, D. I. (1999). Major histocompatibility complex class I genes in primates: co-evolution with pathogens. *Immunol Rev* **167,** 327–37.
12. Watkins, D. I. (1995). The evolution of major histocompatibility class I genes in primates. *Crit Rev Immunol* **15**(1), 1–29.

Y chromosome

1. Flockerzi, A., Burkhardt, S., Schempp, W., Meese, E., and Mayer, J. (2005). Human endogenous retrovirus HERV-K14 families: status, variants, evolution, and mobilization of other cellular sequences. *J Virol* **79**(5), 2941–9.
2. Gu, Z., Wang, H., Nekrutenko, A., and Li, W. H. (2000). Densities, length proportions, and other distributional features of repetitive sequences in the human genome estimated from 430 megabases of genomic sequence. *Gene* **259**(1–2), 81–8.
3. Jurka, J., Krnjajic, M., Kapitonov, V. V., Stenger, J. E., and Kokhanyy, O. (2002). Active Alu elements are passed primarily through paternal germlines. *Theor Popul Biol* **61**(4), 519–30.

4. Kim, T. M., Hong, S. J., and Rhyu, M. G. (2004). Periodic explosive expansion of human retroelements associated with the evolution of the hominoid primate. *J Korean Med Sci* **19**(2), 177–85.
5. Spierings, E., Vermeulen, C. J., Vogt, M. H., Doerner, L. E., Falkenburg, J. H., Mutis, T., and Goulmy, E. (2003). Identification of HLA class II-restricted H-Y-specific T-helper epitope evoking CD4+ T-helper cells in H-Y-mismatched transplantation. *Lancet* **362**(9384), 610–5.

Primate Social Behavior

1. Byrne, R. W. (1995). "The thinking ape: evolutionary origins of intelligence." Oxford University Press, Oxford ; New York.
2. Parr, L. A., Waller, B. M., and Fugate, J. (2005). Emotional communication in primates: implications for neurobiology. *Curr Opin Neurobiol* **15**(6), 716–20.
3. Russon, A. E., and Begun, D. R. (2004). "The evolution of thought: evolutionary origins of great ape intelligence." Cambridge University Press, Cambridge, UK ; New York.
4. Waal, F. B. M. d. (2005). "Our inner ape: a leading primatologist explains why we are who we are." Riverhead Books, New York.
5. Wilson, E. O., and Holldobler, B. (2005). Eusociality: origin and consequences. *Proc Natl Acad Sci U S A* **102**(38), 13367–71.

Face and gesture recognition

1. Aylward, E. H., Park, J. E., Field, K. M., Parsons, A. C., Richards, T. L., Cramer, S. C., and Meltzoff, A. N. (2005). Brain activation during face perception: evidence of a developmental change. *J Cogn Neurosci* **17**(2), 308–19
2. Eifuku, S., De Souza, W. C., Tamura, R., Nishijo, H., and Ono, T. (2004). Neuronal correlates of face identification in the monkey anterior temporal cortical areas. *J Neurophysiol* **91**(1), 358–71.
3. Gliga, T., and Dehaene-Lambertz, G. (2005). Structural encoding of body and face in human infants and adults. *J Cogn Neurosci* **17**(8), 1328–40.
4. Loffler, G., Yourganov, G., Wilkinson, F., and Wilson, H. R. (2005). fMRI evidence for the neural representation of faces. *Nat Neurosci* **8**(10), 1386–90.
5. Moody, E. J., McIntosh, D. N., Mann, L. J., and Weisser, K. R. (2007). More than mere mimicry? The influence of emotion on rapid facial reactions to faces. *Emotion* **7**(2), 447–57.
6. Morris, R. D., and Hopkins, W. D. (1993). Perception of human chimeric faces by chimpanzees: evidence for a right hemisphere advantage. *Brain Cogn* **21**(1), 111–22.

7. Okada, T., Tanaka, S., Nakai, T., Nishizawa, S., Inui, T., Yonekura, Y., Konishi, J., and Sadato, N. (2003). Facial recognition reactivates the primary visual cortex: an functional magnetic resonance imaging study in humans. *Neurosci Lett* **350**(1), 21–4.
8. Parr, L. A. (2003). The discrimination of faces and their emotional content by chimpanzees (*Pan troglodytes*). *Ann N Y Acad Sci* **1000,** 56–78.
9. Parr, L. A., Winslow, J. T., Hopkins, W. D., and de Waal, F. B. (2000). Recognizing facial cues: individual discrimination by chimpanzees (*Pan troglodytes*) and rhesus monkeys (*Macaca mulatta*). *J Comp Psychol* **114**(1), 47–60.

Mirror Neurons and Vocal Learning

1. Aziz-Zadeh, L., Koski, L., Zaidel, E., Mazziotta, J., and Iacoboni, M. (2006). Lateralization of the human mirror neuron system. *J Neurosci* **26**(11), 2964–70.
2. Burling, R. (2005). "The talking ape: how language evolved." Studies in the evolution of language Oxford University Press, Oxford; New York.
3. Cooper, D. L. (2006). Broca's arrow: evolution, prediction, and language in the brain. *Anat Rec B New Anat* **289**(1), 9–24.
4. Dinstein, I., Hasson, U., Rubin, N., and Heeger, D. J. (2007). Brain areas selective for both observed and executed movements. *J Neurophysiol* **98:** 1415–27.
5. Iriki, A. (2006). The neural origins and implications of imitation, mirror neurons and tool use. *Curr Opin Neurobiol* **16**(6), 660–7.
6. Lyons, D. E., Santos, L. R., and Keil, F. C. (2006). Reflections of other minds: how primate social cognition can inform the function of mirror neurons. *Curr Opin Neurobiol* **16**(2), 230–4.
7. Parr, L. A., Waller, B. M., and Fugate, J. (2005). Emotional communication in primates: implications for neurobiology. *Curr Opin Neurobiol* **15**(6), 716–20.
8. Paukner, A., Anderson, J. R., Borelli, E., Visalberghi, E., and Ferrari, P. F. (2005). Macaques (*Macaca nemestrina*) recognize when they are being imitated. *Biol Lett* **1**(2), 219–22.
9. Rizzolatti, G., Fadiga, L., Gallese, V., and Fogassi, L. (1996). Premotor cortex and the recognition of motor actions. *Brain Res Cogn Brain Res* **3**(2), 131–41.
10. Rossi, E. L., and Rossi, K. L. (2006). The neuroscience of observing consciousness & mirror neurons in therapeutic hypnosis. *Am J Clin Hypn* **48**(4), 263–78.
11. Uddin, L. Q., Iacoboni, M., Lange, C., and Keenan, J. P. (2007). The self and social cognition: the role of cortical midline structures and mirror neurons. *Trends Cogn Sci* **11**(4), 153–7.
12. Wang, W. S. Y. (1991). "The Emergence of language: development and evolution readings from Scientific American magazine." W.H. Freeman, New York.

Chapter 9
Human Group Identity: Language and a Social Mind

As presented in the last chapter, primates showed a significant shift in the mechanisms by which they attain group identity. Following the great HERV genome colonization, primates were no longer dependent on olfaction and MHC composition (as are all other mammals) for group recognition, but developed a strong dependence on visual information, especially facial and gesture recognition. This required significant brain developments to process visual information for social purposes. In humans, the HERV colonization has continued and possibly accelerated relative to other primates. Humans retain the primate's heavy dependence of vision (facial/emotional) for social purposes, and have further adapted the ancient link between the olfactory lobe to the amygdala, for visual-based emotional memory (especially fear) which is stored in the hippocampus. A vision-based system of group identity required adaptations in systems of emotional memories. It also required the development of additional social and emotional addiction states for extended social bonding. Using this vision-based system, primates initially extended the duration of the mother–infant bonds and also extended other social (troupe) bonds and structures. The human mother–offspring bond retains the strong ancestral primate visual sensory character, but further extends the stability and duration of the mother–infant bond. The use of song and voice assisted the extension of the maternal bond. The facial and gesture emotional recognition which was attained by the evolution of dedicated and specific brain structures had an inherently rapid, sparse and symbolic capacity. These brain structures (including mirror neurons) not only convey emotional content but also provide visual-based emotional group links between individuals. The maternal bond provided the basis for this development but conserved the prolactin, oxytocin, vasopressin and the opioid system to control and bind empathic/aggressive emotions. Humans, however, have a major distinction withal other primate social structures, in that humans evolved a paternal role in offspring bonding and a serial bonding between mates. Thus, human fathers became involved from infancy with their offspring and also form strong emotional bonds to their mates. The mechanisms of such bonding appears similar to the maternal–offspring bond (i.e., visual and prolactin), resulting in various forms of love (emotional addiction). Also in contrast to all other primates, humans have developed an

L.P. Villarreal, *Origin of Group Identity*, DOI: 10.1007/978-0-387-77998-0_9,

evolutionary novel form of group (social) identity that involved recursive language as an audio-based social bond. Vocal learning also required a considerably expanded brain capacity, especially in the neocortex, as well as brain lateralization and a necessary link to emotional systems and memory. HERV colonization and involvement in this new system of group identity seems likely. This created the large social brain of humans, one that literally connects to other minds via emotions and uses both sight and sound (with mirror neurons). It is likely that the use of song in mother–infant bonding was the basal system of vocal emotional bonding and vocal control of infants. A more temporally variable sound pattern as in song and later the use of a recursive language, however, allowed human populations to rapidly evolve their own diverse versions of group identity based on learning and essentially liberated the evolution of human social structures from dependence on genetic colonization events. Learning of language, however, retains the essential features of a group identity system. The greater brain lateralization that was involved also allowed left/right specialization in major cognitive processing tasks associated with social membership (visual and sound based). However, language also provided a much more symbolic version of group identity and promoted the emergence of ideas and abstractions. This development marked a major shift in the evolution of social identity systems. No longer dependent on variable biological characteristics of individuals (i.e., MHC makeup), the symbolic meaning of language became a major system for defining group identity. Our large social brains (which promoted a social consciousness or social mind) essentially evolved to acquire (learn) language and became a host for 'language colonization' that also transformed brain architecture and further promoted abstract pattern recognition. Thus language, rather than genetic parasites, became the primary transmissible information system that colonized host (social brains) and provided the host with diverse and competing group identity. With this, the stability of language (its meaning) became of central importance to maintain group identity. 'Meaning stability' is the foundation of 'belief states', and this required specific neural substrates to retain belief memory. Thus the emergence of a language-based group identity inherently led to the emergence of 'belief states' as designators of social identity and corresponding evolution of the needed brain structures. Social identity remains mainly a language-mediated learned state. And beliefs have assumed a much expanded role in human social identity, as beliefs provide emotional bonds to populations. However, like all group identity systems, language and belief states still required emotionally powerful addiction modules to maintain identity. Since this identity is learned, it must also resist (sometimes violently) subsequent learning to prevent identity displacement. Thus 'learning resistance' or 'cognitive immunity' becomes a core feature of belief states. The emergence of belief as an identity system was much enhanced by the emergence of writing which greatly stabilized and extended religious belief and meaning. This also expanded the reach of social membership to allow broader populations (including nations) to link into large, connected social minds with common social bonds (via belief in cultural ideas, political and religious leaders). However, the emergence of phonetic writing also greatly promoted brain restructuring, the expansion of vocabulary and the emergence of an internal voice of self-conscious reflection. From this quickly emerged the foundations for the individual-based, critical mind and a modern form of consciousness. Such a consciousness tends to acquire belief by distinct 'asocial' criteria, often based on evidence. Individual self-consciousness thus is a recent state to emerge, but depends on formal schooling (such as reading) to promote more abstract thinking. Such a process of belief acquisition has historically often generated conflict with social belief states and elicited anti-social reactions. However, since belief defined all prior social structures, scientific thinking has often been confused as another 'belief-based' identity system, although it is clearly not a social identity system. Science education should seek to clarify this distinction and educate people away from ancestral belief-based reasoning when it is used a force for unreason or the promotion of social conflicts.

The Shift to a Human Social Mind from a Chimpanzee Mind

A main evolutionary development in humans relative to chimpanzees is the much larger human brain which provided enhanced social capacity (hence social membership). Recent evaluations of the mental abilities of a pre-literate (2.5-year-old) human child relative to that of a chimpanzee or orangutan clearly establish that the humans are not superior in most mental activities (spatial, causal, quantitative), but were clearly better at social learning (by observing actions of others). The humans were also superior at 'theory of mind' tests. Thus humans seem to be 'ultra-social' in that their social intelligence, not general intelligence, seems to define the major difference between these species. Human social intelligence involves both learning from others as well as being able to infer what others believe or desire (theory of mind). This function develops after the first few years of life and involves characteristics such as seeing the eye gaze of others to know what they know. This is likely associated with the evolution of human-specific white eyes which aid in gaze identity, suggesting intense selective pressures. Such a gaze facility appears absent from cooperating chimps, although competing chimps may employ it. Curiously, domesticated dogs (but not wolves) seem better at following human gaze than do chimpanzees, suggesting that dogs were subjected to intense selection for such human social interaction. Human and chimp minds thus seem to differ in self-perception and envisioning mental abstraction of others (a clear social function). In keeping with this, chimps are unable to learn intentions via sign language, something human infants are inherently prone to do. Chimps learn words that are mainly sensory in character, but appear unable to learn metaphors. In this we see what appears to be another major distinction between chimpanzee and human minds. Chimpanzees can represent what they perceive, but humans can represent what they imagine (from unreal sensory or cognitive source). As described below, metaphors also require abstraction and are major elements of all human languages. This crucial distinction thus appears fundamental to human social intelligence. What forces in evolution might have favored such a mind shift? Why would imagination, inferring others' thoughts and abstraction, be subjected to intense selection during the evolution of our large social brain? Imagination can be considered as a synthetic (non-sensory) or altered (misremembered) memory. Both humans and chimpanzees have excellent memory for faces, which also serves core social functions for both species. Imagination seems unrelated to this. Imagination can be used in group hunting, but, as described below, chimpanzees are good at this. Humans can inherently infer what others might be thinking and this will often have few visible cues, thus requiring imagination. But how could this be under intense selection for group identity? Humans also have excellent memory for voices and word sets of language. Such memory requires the capacity to store temporal streams of sensory input, so that it can be recalled later from non-sensory (internal visions of faces or voices) as sources for recognition. Single neuron recordings indicate face recognition and gestures are indeed encoded in highly specific brain structures. But these are found in both humans and primates. Therefore excellent visual memory provides an enhanced capacity for comparison and recognition of sensory visual input to internal sources but is present in most primates. The human mind also appears to have an inherent capacity for recalling synthetic voices and visions which can also be elicited by dreams, hypnosis or disease (schizophrenia). These can all be considered as imagined. However, what forces might select for a mind that could provide such altered or synthetic memories? One correlation is the emergence of human language which could be relevant. Humans are the only species that evolved vocal (auditory) communication which uses syntax (recursive language). In order for syntax to convey meaning, the word with in a word pattern (not just the word itself) determines meaning. Meaning itself becomes conditional on temporal sequence and the relative placement of a word. A recursive language therefore requires multiple meaning, or abstraction, hence imagination, for any specific sound (word) to have meaning. Imagination could thus be a byproduct of this development. Perhaps the proper question that needs to be considered is what selective pressure led to the emergence of recursive language in humans. As discussed below, the issues of meaning and group identity can be directly linked when verbal information is used for specifying group identification. The basal question might then be not about imagination but how the evolution of extended social structures can explain the evolution of an ultra-social brain and why abstract and recursive language was needed? What was the initial social purpose of syntax and why would any social structure need it? The answer I suggest is that early human social groups needed to differentiate themselves from each other, beyond the visual systems used by other primates. Direct human ancestors additionally used vocal-based differentiation (protolanguage/song) as an early human group adaptation (such as in maternal/paternal bonding). Learning, however, often occurs via mimicry, and mimicry still appears to be an inherent process for initial learning of language (i.e., infants fists ma-ma, pa-pa). A vocal stream in which identity was not syntax dependent, however, is susceptible to copying by mimicry and social parasites or competitors. But mimicry could not reproduce proper syntax in which meaning depends on source. However, the use of syntax for meaning would make audio language-based identity or communication immune to copy or transmission by only mimicry. Thus syntax could provide an audio-based social identity that was differentiated from simpler systems based on position-invariant word meaning. A mind capable of abstraction would be needed for such syntax-based social identity.

The adaptation of recursive language for social membership is the thus major distinguishing development of human evolution that relates directly to our large social brain and our ability to infer what others think. Understanding the origin and consequences of this social adaptation is the main focus of this chapter. For language to mediate group membership requires that it also engages emotional systems and emotional memory to set social bonds. Language indeed has symbolic and strong emotional content (as does vision). Charged words, such as metaphors, elicit many emotions, even abstract ones. Posture also clearly has emotional content. The clarity of human gestures and expressions appears to be a cultural universal as seen in dance. When we observe others move, even abstract movements, the observers (a social set) experience common emotions. Thus our social brain uses both vision and recursive language to socially connect emotions of individuals. However, defects in the social brain can result in aberrant, sometimes enhanced, mental capacities. For example, some autistic people can demonstrate an incredible retention of factual detail, relative to most people (popularized in the movie Rainman). These people will often otherwise lack some of the most basic social skills and fail understand metaphors obvious to many children.

The ability to imagine, or originate alternate meaning and communicate abstractions has had many deep consequences for human cognition and culture. Although a capacity for recursive language may have been needed to initially promote abstract thinking, this has led to a positive feedback loop in that abstract thinking has very much expanded language, restructured brains and further enhanced abstract thinking. A much expanded capacity for abstraction and imagination thus developed which can now provide scenarios and concepts that cannot be observed or experienced by sensory systems, only by our mind. Let us now trace the specific steps in evolution that led to this.

Large Social Brains and the Great Viral Invasion

Morphologically modern humans are only about 150,000 years old. If we consider the types of genetic evolutionary events that might have happened during such a relatively short period of evolution, we are struck by the difficulty to explain these dramatic changes based on gene makeup. There are relatively few genes that distinguish humans from the chimpanzees, for example. Indels (insertion/deletion) in non-coding regions clearly account for the majority of differences between these genomes. Comparisons between their sequenced DNA indicate that they have relatively few ORF differences but have about 400,000 indel differences, the bulk of which correspond to LTR element differences that average about 300 bp in length. As suggested in the prior chapters, such genetic perturbations could easily affect gene regulation of many genes, in possibly coordinated ways (including transcription and post-transcription via RNAi). Thus dramatic changes in human function could be the result of wide-scale regulatory adaptations mediated by these and other genetic parasites. At a cytogenetic level, we previously noted the distinct C-bands found on chimpanzee but not human chromosomes. We know these regions are composed of highly repeated sequence elements, including HERV W and HERV-FRD (needed for placental function). We also know that some HERVs found in the human genome are much more recent than the time since chimpanzee divergence, and date to about 150,000 ybp. Thus HERV colonizations match the recent changes during human evolution. Since such elements are also highly represented in the Y chromosomes, and Y chromosomes between human and chimpanzee are distinct, these may also serve as marker of the most recent genetic changes and evolution specific to humans. Y chromosomes are indeed especially colonized by ERVs. Human evolution is now accepted to have been out of Africa, into Asia, then possibly returning to Africa to initiate modern humans. This view can be suggested based on the Y chromosome phylogenetics, which roots to Africa. It is interesting that India has highest diversity of Y chromosome haplotypes (12 of 18 total), relative to Africa (2–4). Although it would be highly informative to know the Y chromosome composition in other hominid lineages, only data from *Homo sapiens* is known. Neanderthals are dated to have become extinct about 40,000 ybp. Neanderthals thus appear to have overlapped with *Homo sapiens* for about 28,000 years. mtDNA evolution can be followed by using a distinct 344 bp D-loop method. Using such analysis, it does not seem that human and Neanderthal genomes were sexually mixed. Thus the two lineages seem to represent distinct species. Although we know little about how ERV DNAs might differ between *Homo sapiens* and Neanderthals, some things can be expected. In all mammals, ERV makeup provides clear (diagnostic) differences between related species. So in Neanderthals we would also expect distinct and recent ERV composition. Neanderthals also had large brains (slightly bigger brain than humans), suggesting a large social brain involvement is their group interactions. As noted below, *Homo sapiens* children undergo much brain growth and development after birth, and some of this development is crucial for our more advanced social features (such as language development, self-awareness and theory of mind). At about age 4, humans develop theory of mind and start to reason about the causes of other peoples behavior. In this social brain development, it appears that Neanderthals were more like chimpanzees in that the majority of infant brain development was completed by age 4. The inference is that the brain-based social functions that develop later in *Homo sapiens* children (e.g., theory of mind) did not occur in Neanderthals (or chimpanzees). Although we do not currently know the genetic mechanisms that expanded the human brain (especially the neocortex), we can guess that indel-mediated regulatory changes, not gene acquisitions, would be more likely to account for the numerous alterations. The human neocortex expresses more complex RNA sequences than any other organ, so clearly it underwent large-scale regulatory changes. Neanderthal brain would likely be similar (with hypertrophic cortex invasive of basal brain structures) and their increased brain size also suggests they had a large social brain. Various HERVs are known to be expressed in human cortex, but not in brains of other primates, including HERV-E, HERV-F ERV-9 and various HERV K members, although there seems to be some individual-based variation in these expression patterns. In the last chapter, I emphasized the significant shifts that occurred in primate genomes regarding HERV Ks as well as significant shifts in relationships to exogenous retroviruses (SIV and foamy virus). Since HERV K can also conserve dUTPase genes (see Fig. 8.6) and since dUTPase can be considered as an antiviral activity, its conservation in the human genome is most interesting. As viruses can potentially provide rapid and wide-scale shifts in molecular genetic regulation, it is reasonable to hypothesize that ERV and associated hyperparasites (LINEs, alus) could promote the wide-scale regulatory changes needed for rapid human brain evolution. It is thus intriguing that the most recently acquired of these HERVs (HERV K10 (HML2)) has also been shown to be significantly overexpressed (including env sequence) in prefrontal cortex in both bipolar disorder and schizophrenia relative to samples from healthy brains. Similar ERV overexpression in this mental disease was not seen with HERV W, HERV-FRD envs. Interestingly, some antipsychotic medications (haloperidol and clozapine) inhibit retroviruses in vitro. Schizophrenia is a complex and poorly understood neuropsychiatric disorder (discussed below), but one is closely associated with evolution of human language acquisition and human social intelligence. Thus viral footprints associated with this disease are intriguing.

The evolution of a large social human brain presents many dilemmas for evolutionary biology if looked at from a perspective of organ efficiency. For one, our big social brains are metabolically very costly. The human brain is 2% of total body weight, but uses 20% of total body energy; 60–80% of this energy is consumed for communication between neurons that do not appear to be doing much to support physical activity (as they are active during sleep). Human infants, with their disproportionately larger heads and brains, are even costlier metabolically speaking, plus their large heads radiate much heat, making young brains especially costly organs to feed and maintain. This high metabolism is continuous regardless of mental activity, thus big brains make humans prone to starvation. Why is the large social brain so active and costly? The human brain also has relatively disproportionately high level of connections in the visual cortex. Thus our brain is never at rest, even during sleep, and activity does not depend much on perception. In addition to being metabolically costly, an infant human has a functionally undeveloped brain and is the most helpless of all mammals. Aside from defecating, human infants have no preset knowledge or physical capacity. They cannot walk, crawl or even lift their head. And relative to a chimpanzee, they are significantly more incapacitated in that human infants cannot grip their mother and travel with her in trees. Human infants are truly helpless and must learn all their basic movements, and even must learn sensory functions. It seems that expanded human neocortex needs to develop before it can provide humans with all these needed functions. The growth of the infant brain is charted in Fig. 9.1, relative to the acquisition of language. Also, although such early incapacity seems associated with continued brain development, the large social brain does not complete its development until young adulthood (over 20 years). A schematic of the human brain, with some relevant regions labeled, is shown in Fig. 9.2. Thus the cost and developmental duration of a large social brain seem highly inefficient. Such a large social brain is also associated with longer life span. This long and extended brain development appears to provide various opportunities for much social learning to occur (theory of mind language, beliefs, etc.). Clearly, social learning must be crucial for human survival given these costs and biological problems. But in what way can the benefits of social learning offset such major and extended inefficiencies of our large social brain. Large social brains are not common to other mammals (including most primates) so it would not seem to represent a general issue for selection and fitness. What was special about human evolution and their social characteristics that favored or required this costly brain?

Let us consider what a 'brain-mediated' enhanced social capacity might provide humans to offset the major evolutionary cost outlined above. What

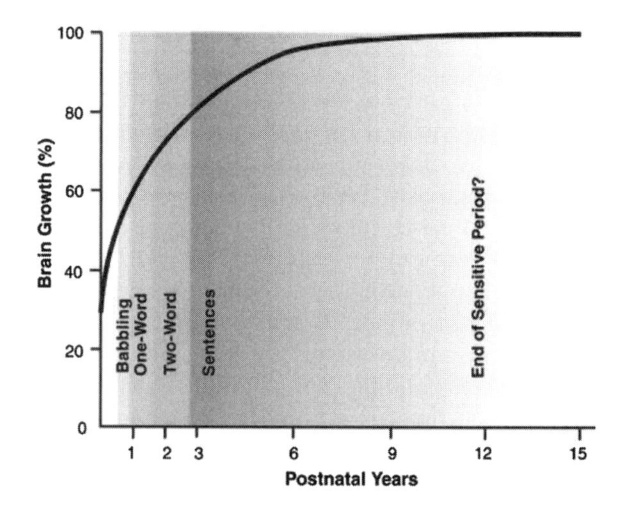

Fig. 9.1 Overall human brain growth and acquisition of language (reprinted with permission from: Sakai (2005), Science, vol. 310)

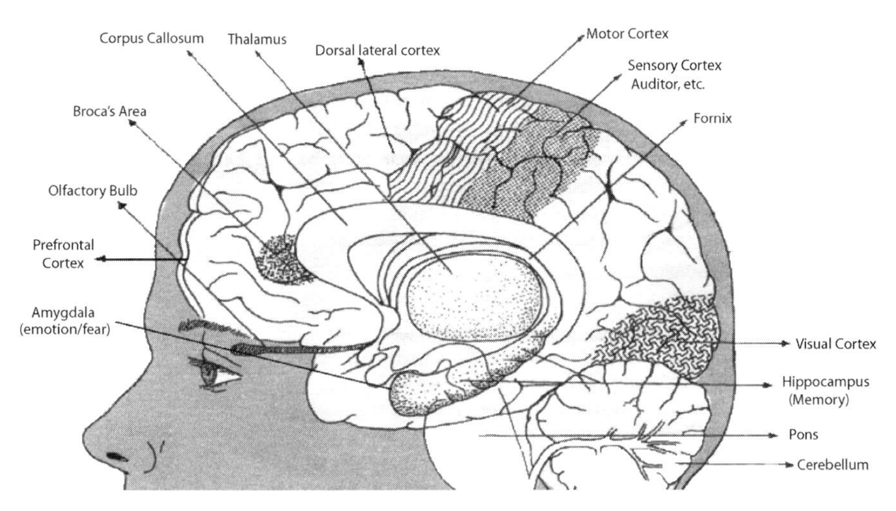

Fig. 9.2 Human brain schematic

about the extended maternal bond? In primate ancestors to humans, a visual-based group recognition system was mainly used for maternal bonding (see prior chapter). Many such species show extended maternal bonding, although not for the duration seen in the great apes. It seems likely that the initial selection for extended human social bonding would also use visual mechanisms, but the extended duration and strength of the human maternal–offspring bond is well beyond that of even the great apes (see below). This extended care promoted or allowed the development of an even more delayed social brain. Maternal obsession and compulsion are examples of emotions that would contribute to extend maternal bond. These could clearly be implemented as emotional addiction modules, but do not seem to require a bigger brain. We do not yet know how the human maternal bond was extended. However, besides an extended maternal bonding, and in contrast to all other apes (but seen is some New World primates), visually mediated parental social bonding appears to have been acquired by human fathers. Although not well evaluated, human fathers do show prolactin-associated changes in response to crying of their infants (discussed below). Clearly, emotions such as sympathy, love, anxiety must somehow be involved in such bonding. Thus in humans, both mothers and fathers were more socially bonded to offspring. Might this dual parental infant bonding account for some of our larger social brain? This seems unlikely. Given that much smaller-brained New World monkeys can also show visually mediated biparental bonding, a big brain would not seem needed. However, unlike the other great apes, both female and male human mates are also socially (romantically) bonded to each other. As discussed below, it is likely that the mechanisms bonding females to male mates and visa versa are similar, but not identical to each other. Clearly, humans do show extended social bonding, but similar bonds are seen in other vertebrates with much smaller brains. The distinction, however, may be that human bonding uses few biochemical signals, but is mostly cognitive, placing more demands on brains. There is, however, one human-specific social feature that places large demands on only the human the brain: language. As mentioned above, recursive language, if it is used for the purposes of group identity, might require a larger brain. But here, we can see the potentially large dividends in terms of group survival and adaptation and general intelligence. Language, besides possibly providing group identity, also transmits survival experiences and promotes social cooperation. Indeed, the human brain does appear to have specific structural adaptations associated with language acquisition, such as Broca's area as shown in Fig. 9.3. Thus the acquisition of language could offset the large and inefficient human social brain if such dividends are major. Furthermore, if language created the large social brain, the social mind that emerged from it becomes available as the substrate that can now be selected for an even more extended social membership. Accordingly, this emerged social mind required sensory, brain and emotional mechanisms that would compel greater social bonding and identity, placing even greater selection on a large social brain. These mechanisms involved would

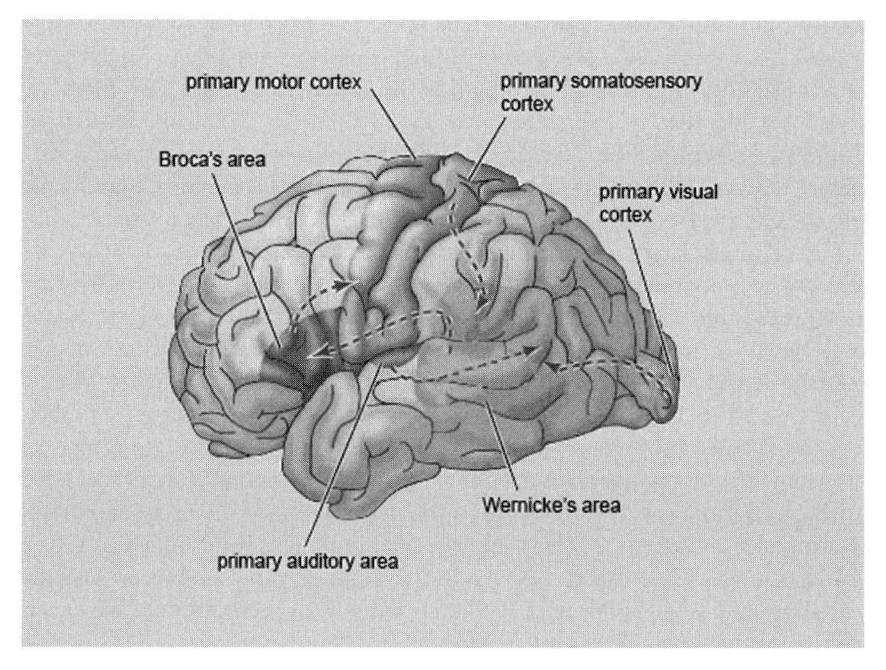

Fig. 9.3 Broca's area identified onto human brain (*See* Color Insert)

likely include social cognitive functions and social–emotional addiction modules. This would require the emergence of more complex social emotions that develop with socialization. Perhaps the human social emotions of pride, embarrassment, guilt that develop between 5 and 14 years of age are examples of such systems.

Human Social Brain: Addiction and Social Bonding

Humans, like primates, no longer depend much on olfaction, MHC or other biochemical identity markers as their social brains provide almost all of their group identity. By extending the basic mechanism of maternal and paternal social bonding, this social brain allowed the development of nuclear families, extended families, tribes, nations to large pan-national cultures, the most extensive group identity being that of religion that can cross all national, language, ethnic, racial and cultural barriers. However, all these group identities appear to have retained the basic group characteristics of addiction strategies and the mechanisms to define, stabilize and defend group identity. They are generally harmful to non-members and protective of members and are mostly acquired by learned information that colonizes a social mind. It is worth stressing again the fundamental importance of the concept of an addiction module. Although the application of this idea to stable phage colonization in bacteria was presented early in Chapter 1, it should be noted that the initial concept of an addiction module stems from the study of the human brain and drug addiction. Ten years prior to its application to phage and bacteria, R. Solomon (1980) proposed that human drug addiction could be understood by thinking of this state as having two components, one involved a rapid pleasurable process succeeded by slower but long-lasting toxic process (withdrawal) (see Koob and Le Moal). In terms of its application to social membership, strong emotions can provide both the pleasurable and toxic components (the T/A of a module). The internal brain structures associated with these emotions (such as maternal and romantic love presented below) are indeed similar to those involved in drug addiction. But human social group identity does not directly stem from the external action of opioids or endorphins. Our identity is learned during our development and the principle thing we learn is language. However, with the development of language as a transmissible group identifier, a transforming event happened in the human mind. Language led to the development of a cognitive, conscious and self-aware mind. This mind state, however, has not escaped the selective forces of group membership. Our mind has also been adapted for and used to evolve even more extensive group identity; a group identity based on cognitive content or belief of our mind will be emphasized below. This greatest of human identities thus has an ancient biological legacy.

Language, Speciation of a Group Identity System

As asserted, with the evolution of language, humans have been largely freed from many of the 'biological' constraints that drive the evolution of most other species, although we are still susceptible to large effects from viruses, for example. Our mind and culture (products of language) provide us with a very rapid behavioral adaptability to deal even with such threatening agents (like avian flu and SARS), well beyond the usual genetic adaptability of other species. For example in the 2005 outbreak of avian flu in China, 250 million domestic birds died whereas only about 250 humans died. This contrasts with an estimated 60 million human deaths from influenza during the 1918 pandemic. Social responses matter to our survival. Indeed, it appears that human evolution (genetic diversification) has accelerated in the last several thousand years and I would argue the chief reason is that we control infectious diseases and have learned to grow and protect our food, via social learning. No other species has even freed itself from similar biological constraints. Although language may have initially evolved as a group identity system, it now provides the basis of all our adaptability and social identity systems, such as culture and religion. Yet language still adheres to many of the basic principles of identity systems as we have considered in this book. Language is a transmissible information system that will superimpose an identity onto its host, much like a genetic parasite. Only the host of language is the large social human brain which is physically altered by the language that colonizes it and is also biologically adapted to be colonized by language. As presented below, languages retain many virus-like features: transmissive, colonizing, stable, preclusive, highly adaptive. But these features are no longer very dependent on our genetic content. Language thus promoted a major evolutionary transition in humans, well beyond even what chimpanzees are capable of. Yet languages, like persisting genetic parasites, still behave like species-specific entities. Language seems to be the DNA of culture and social identity. And like DNA, language can show some interesting patterns of speciation. In the DNA world, when we see situations where there are lots of related species in one particular habitat, we consider what external forces might be driving such high speciation. For example, in the relatively young Hawaiian Islands, there are 500 species of Drosophila found, yet in the combined total for the rest of the world, only 2000 Drosophila species are found worldwide. In such a circumstance, I have argued that we should look for the role of persisting genetic colonizers in speciation as they impose group identity onto their host. A language-based equivalent to this might be what existed in pre-Columbian California. When early Spanish explorers came to California, they were surprised to see the diversity of languages that existed there. An estimated 400 distinct languages, most of which were not understood by the other tribes are estimated to have been present. By contrast, it is estimated only about a dozen languages were present in India, a much older and larger human habitat. In both cases, it appears diversity is associated with more recent habitat introduction, and competition for persistence leads to reduced species. I suggest this is also a competition for group identity in the context of language.

The Human Mother–Offspring Bond: the Basal Social Link

Bonding with Infants: Nursing, Prolactin, Oxytocin and Vasopressin – the Face

As asserted above, the basal mechanism for extended primate and human maternal bonding is via visual cues, and primate brains are specifically adapted for facial recognition. Primates, including humans, have a very fast recognition of faces (200 ms) which occurs in the inferior temporal cortex. Even when presented at various viewing angles, faces are rapidly recognized, a task that is difficult for artificial intelligence systems. From this it seems that face recognition has an inherently abstract or sparse character to it. This may relate to why even primates are able to recognize clock faces as representing real faces. Interestingly, an even faster system of face recognition has been measured regarding the non-conscious recognition of the emotional content of a face (such as fear or disgust), as measured by facial responses. Thus both conscious (170–200 ms) and non-conscious (30 ms) face recognition exists that is able to read emotional content. The culturally universal ability to quickly recognize the ubiquitous smiley face may be due to similar built in pattern recognition capacity. The rapidity of this recognition appears to involve very few spikes per neuron, thus suggesting the existence of a sparse but dedicated neuronal system specific for faces. That this sparse system can also recognize the emotional valence of a face suggests the biological importance of social communication. By 3 months, human infants show clear preference (via eye tracking) to natural face recognition and will prefer attractive to unattractive faces. They also become able to better recognize, respond to and prefer faces of the same racial type, although exposure to faces of other races can reduce this preferential effect. It seems clear that the function of the fusiform area is involved in this face selectivity. Self-face recognition also seems to involve specific neurological domains involving frontoparietal structures that are part of the 'mirror neuron system'. This function would seen relevant for social cognition.

Faces, Fear and Racial Recognition

In humans, rapid facial detection is also linked to a rapid unconscious emotional response in the viewer, which likely involves mirror neurons. Within a second of viewing a facial expression, most people will match the emotional expression of the face. Thus emotions can be rapidly transmitted. It is also suspected that these emotional reactions most likely involve dopamine

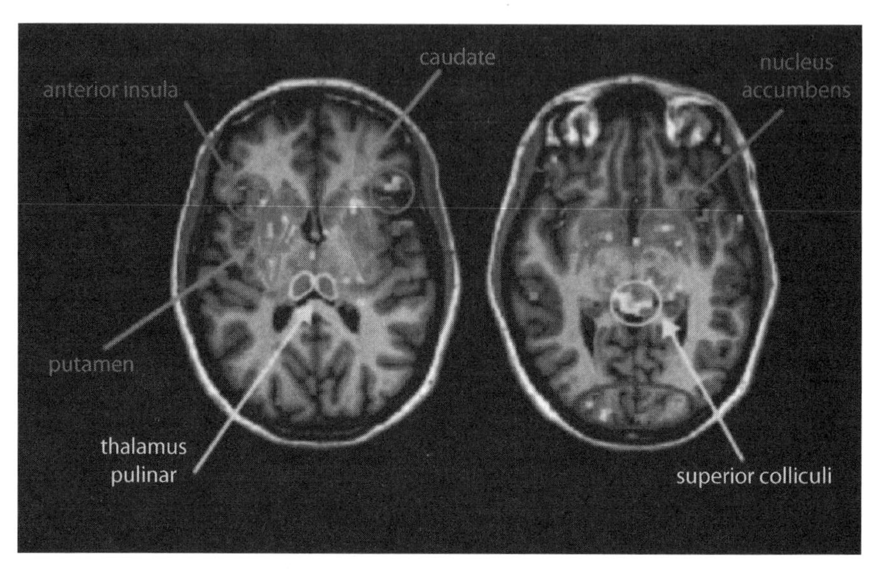

Fig. 9.4 fMRI study of brain response to fearful body positions (*See* Color Insert, reprinted with permission from: Gelder, Sgnder et al (2004), PNAS, Vol. 101, No. 47)

neurotransmitter. Such social contagion via facial recognition is also rapid. For example, facially expressed fear can be propagated as quickly as 150 ms and can be used for fear conditioning (association and learning). An fMRI study of face and body fear recognition is shown in Fig. 9.4. It now appears that there is an inherent tendency in human infants to associate other (non-mother) racial face types with fear and that the resulting fear association is persistent. Thus negative (aversive) attitude will be quickly expressed in response to facial-type recognition, which suggests that an implicit racial prejudice is inherent in the human face-recognition system. Such a reaction would be expected, if humans initially evolved to use facial appearance as a system of group recognition. Consistent with this idea, fear reactions (associations) like these can be measured even in human individuals that do not consciously endorse prejudice. However, it also appears that learning can counter such inherent association tendencies. Human infants repeatedly exposed to other racial face types in a non-fearful situation will lose their negative bias. It seems humans must learn not to respond negatively to other facial types as such a tendency is biologically favored.

Thus it seems clear that at least some of our big social human brain is associated with facial recognition, but this does not really explain the large increase in size relative to other primates. Although vision clearly serves a basal function with regards to group recognition, it too must be learned and developed after birth. Thus the cortical development of human visual capacity takes on special social interest to human evolution and why infants' brain development is so important. For example, if an infant's eyes are non-functional when they are young, subsequent cortical development is permanently affected. Any such individual will struggle with their pattern recognition capacity, especially the recognition of 2D and 3D patterns, if their vision is restored later in life. This indicates the existence of a developmental window during which their visual cortex has developed crucial pattern recognition capacity. Color vision is also interesting with respect to cortical development. For example, some types of cortical damage (extrastriate) can produce severe loss of color vision, but preserve non-color vision. Often this involves lingual and caudal fusiform gyri. This suggests that the basic function of vision was present prior to the primate's adaptations for color. Thus we have numerous reasons to think that higher primates in particular are highly visual creatures. For example, Chimpanzees, unlike humans, when they become separated from each other, do not vocally call out as do people. Instead, they search silently until they see one another then rush together. Vocalization, when it occurs, seems essentially reflexive in chimpanzees and is mostly associated with females greeting males. We might therefore expect that if chimpanzees had developed a language, it would have had to be visual (facial) basis. However, hand gestures could also provide the basis of a vision-based language but no chimpanzee (or other great apes) has developed any communication that resembles a gesture-based language (see Fig. 9.5). I suggest hand gestures would need to become important for ape group identity and be associated with their own dedicated brain regions to allow them to evolve into a language form.

Fig. 9.5 Chimpanzee gestures and expressions (Source: Photo researchers)

Even in visual primates, including human, it is likely that an even earlier more ancient system of mammalian maternal bonding involved the action of suckling and vasopressin and prolactin (described previously). It is known that longer breast feeding in humans leads to stronger maternal attachment, which is thought to be mediated by oxytocin. It is also known that vasopressin regulates prolactin and is associated with suckling-induced prolactin. We can recall that in mice-pup generated ultrasound stimulated the mother's prolactin production. Human babies do not generate ultrasound, but they do cry. Endo-opiates appear to be involved in mother's prolactin response. We might guess that in primates some significant adaptations might have occurred in the prolactin system. Indeed it is known that primates differ markedly from non-primates in their prolactin genes and that this difference occurs before New/Old World split. The human and chimp versions are very similar to each other (only 2 aa difference in the coding region), although human regulatory region has 64 bp satellite insert which is absent in chimp (but present in bonobos).

However, in the case of humans, facial vision does appear to contribute to the maternal bond. In terms of brain adaptations, forebrain neural circuits have been implicated in this maternal behavior, bypassing, perhaps, the older circuit from the olfactory bulb to the amygdala and the hippocampus (associated with long-term memories). fMRI examinations by Bartels and Zeki (2004) of brain response in mothers shown pictures of their infants is shown in Fig. 9.6. Here states of maternal love and romantic love are compared. We can see both states elicit partially overlapping activation of reward (and addiction) centers and regions expressing oxytocin and vasopressin receptors. With maternal love, the face-selective fusiform gyrus was activated whereas in romantic love this region was not activated. The deactivation response was fully overlapping between maternal and romantic love and involved brain regions associated with strong negative emotions (i.e., amygdala; fear and aggression). The implication is that these two states of strong social bonding (maternal and mate bonding) are related, can be visually elicited but are not identical. In other prior studies, infant vocal responses also elicit brain responses. Women, but not men, showed neural deactivation in anterior cortex in response to infant crying. However, the amygdala showed stronger activation from crying, independent of sex, but this is dependent on parental experience. Clearly these brain responses to infant vocalizations are able to affect emotions of both men and women, although there do seem to be differences between the sexes. There appears to have been additional brain-specific adaptations. For example, both oxytocin and vasopressin are strongly implicated to be generally involved in social memory and learning. In addition, humans seem to have a special version of neuropsin (associated with learning and memory) that is of recent origin in the hominoid species. I have already mentioned that oxytocin has also been associated with the social emotion of trust, apparently involved in social bonding. Trust allows one to develop the feeling they know someone else, such as their mother. The link between oxytocin and trust is not limited to maternal bonding and seems to be applied to other social interactions into adulthood. For example, in studies involving games of risk and trust, nasally administered oxytocin significantly increased tendency to trust the involved individual. Interestingly, this reaction was specific to a person (face) and was not applicable to a computer game partner. Thus oxytocin affected the human emotional response and learning to a facial interaction.

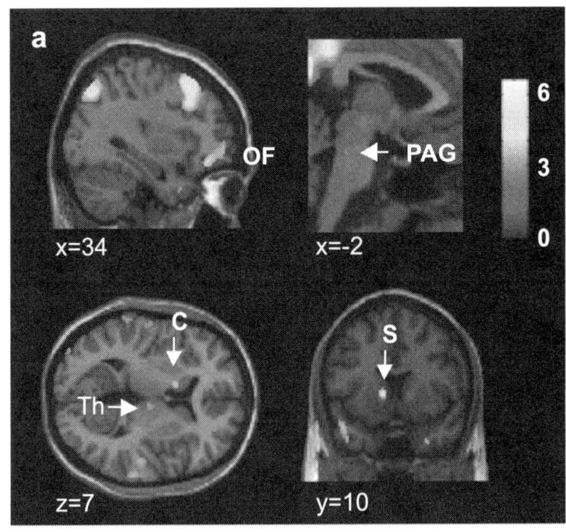

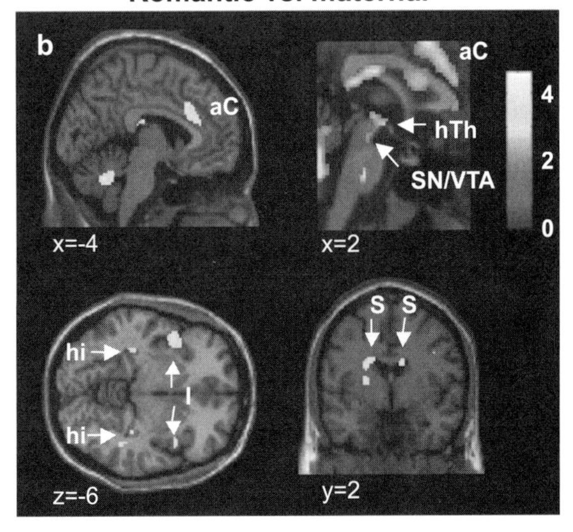

Fig. 9.6 fMRI brain study of maternal and romantic love (*See* Color Insert, reprinted with permission from: Bartel Andreas and Zeki Semir (2004), NeuroImage, Vol. 21, No. 3)

Brain Biology of Facial Emotions: Brain Damage

The emotional aspect of facial recognition appears to involve specific neural substrates, as implicated by brain damage. Bilateral damage to amygdala, for example, can impair the recognition of negative facial emotions (fear, anger), but leave the recognition of positive emotions (happiness) intact. This emotional reaction is also distinct from that of face recognition. For example, some stroke patients have difficulty recognizing a specific face, but can still recognize its emotional expression. Brain damage can thus result in poor facial recognition yet leave intact recognition of emotional expression. Thus there appear to be specific brain structure for personal face recognition and for facial emotional content. Since positive and negative emotions can be considered as distinct parts of an emotional T/A addiction module, such physical separation is not unexpected. In addition, there may be some sexual dimorphism with this facial emotional recognition. Female and male expressers can elicit different eyeblink startle responses, potentiated by happy faces but inhibited by angry faces. Normally the amygdala will activate before 170 ms when exposed to fearful faces. It seems right brain hemisphere is predominantly involved in control of these evoked emotional reactions. As presented below, schizophrenia is also associated with impaired facial recognition. Fear recognition and the amygdala is of particular interest here. As mentioned above, fear and emotional contagion is seen in infants, as young as 3 months of age. At around that age, infants become able to perceive facial emotional expressions from the mother. Indeed most studies of emotional communication have focused on experiments using facial expression, and such studies dominate the current literature. Since very specific neurons are involved with facial expression, clearly an intense biological selection for face recognition and emotional reaction occurred in primates. The amygdala is involved in communicating and assisting the remembering emotional events. In humans, it is massively connected to cerebellum and frontal lobe. Since the frontal lobe is the center for most advanced human cognition and more developed in humans than in other apes, its connection to this circuit is most interesting from the context of a large social brain. It seems that the most recent and advanced part of the human brain connects to the most primitive and uses emotional memory via amygdala. The clear implication is that our large social brain needs to link its most advanced cognitive functions to basal emotional reactions, which I suggest promotes social bonds and group identity.

Song, Human Emotions and T/A Sets

In contrast to all primates, which are indifferent to or avoid music in their habitat, in humans music communicates emotional content and provides a source of pleasure. Music can also communicate danger. Music seems to amplify the emotional content of visual stimuli and it is for this reason that music is so effective in setting the emotional valence of scenes in the movie industry. Besides music, fear contagion in the voice tone also exist between mother and infant (discussed below). But this contagion is also unconsciously modulated by facial expression. This face–voice emotional pairing also seems to be mediated by amygdala. Body position and gestures (lacking faces) can also transmit emotional and fear states. It seems clear that such emotional content of music is also the basis for use and appreciation of music and dance in all human cultures. There also appear to be other emotional links to sensory input. For example, there is a curious connection between emotion communication, movement and timing (tempo) in humans. Tempo (as used in music) can clearly add emotional content to visual stimuli, and this is also much exploited in movie soundtracks. In humans, the cerebellum (considered an older or primitive brain) is involved in maintaining tempo, but is also involved in sensing the emotion of music. It has been proposed that cerebellum is involved in modulation of emotional arousal. Monkeys with lesions in some regions of their cerebellum can show dramatic changes in arousal and can sometimes induce rage, perhaps the most potent of negative emotion. It is known that stimulation of a central region called vermis also leads to aggression in humans. Interestingly, lesions in other parts of cerebellum induce calmness (contentment) and have been used clinically to sooth schizophrenics. Yet other regions of the cerebellum can be stimulated to reduce anxiety and depression. Thus, various intense emotions (such as those induced by music) are associated with ventral striatum, the amygdala and the midbrain and are involved in reward, motivation and arousal. Ventral striatum includes the nucleus accumbus (NAc), the center of brains reward system and important for drug addiction and pleasure and also directly involved in transmission of opioids in the brain. This last region may be the end point of sensory stimulation of emotion when it must affect group identity. Why then is it so important for humans to communicate emotion? All of these emotional reactions can be considered as emotionally toxic and antitoxic sets and it thus seems likely that they constitute T/A pairs (addiction modules) that could define group identity if learned. Many of these same emotions are elicited during grief following death of close offspring (see below). As I have asserted, positive and negative emotions together would be needed to create extended social bonds, a core function of and crucial selection for the social human brain.

Biparental Social Bonding as a Base for Extended Social Bonds

It seems well established that evolution tends to build onto prior solutions and systems tend to develop more complexity. The mammalian maternal–offspring bond is in part mediated by the actions of prolactin. Also only in mammals, the placenta is a major site of prolactin expression and the placenta also has many placental-specific versions of prolactin. In humans, the placenta has three distinct types of placental trophoblasts and each of these cell types expresses its own version of prolactin. Thus it appears that the placenta itself was involved in the evolution of maternal–infant social bonding. This would clearly present a situation that applies only to the mother, not the father. In humans, however, it is clear that the brain is also affected by and involved in prolactin-mediated bonding. Evidence suggests that this also applies to human fathers. As presented above, visual input became prominent for social bonding in primates which allowed the extension of the maternal bond beyond the duration that could have been mediated by placental prolactin. The extended bond between mother and infant is maintained by visual and other stimuli and produces an emotional state we call motherly love. However, the involvement of the brain and vision (or voice) in maternal bonding would entail the use of a mechanism that could also be adapted by evolution to apply to fathers, who would otherwise lack bonding mediated by the placenta. All group identity systems need T/A sets. Thus any human paternal (and maternal) bond to infant must include negative or harmful reactions to non-members or those that threaten the bond. Thus toxic emotions, such as aggression and rage, are essential for social bonding. The positive emotions are the strong contentment and joy that is felt in the presence of the offspring. Negative emotions are strong fear, anxiety in the absence of offspring or aggression to threats. The sight of an infant is known to provide comfort to both mother and father. It seems likely that dopamine is involved in the pleasure part of this emotional bond. Both maternal and paternal bonds should increase aggression toward threats to offspring. For fathers to form such stable social bonds, visual (or sound) mediated sensory input needs to communicate and engage positive and negative emotional memory in a stable way. Thus the emergence of paternal bonding in humans was likely adapted from the visual and vocal emotional mechanisms that primates had already adapted for the maternal bond. However, the mechanisms employed for this paternal human bond promoted the evolution of even more extended social bonds. Since paternal and maternal bonding now involves all members of the species (male and female) in common social links, it becomes available for evolution to promote additional and extended group identity and social structures. As these bonds were mediated by learning and various social-CNS adaptations, social learning (especially language mediated) can now provide extended human group identity.

Grief as an Exemplar of Basal Social Bonding

If, as I have asserted, most human social bonds are derived from the maternal bond and maintained by sets of positive and negative emotions, then it is also expected the maternal bond will identify many basic features of general human social bonds. For example, breaking the maternal bond will disrupt emotional T/A sets in ways that are likely to be similar to other social bonds and we should expect that intense emotional discomfort and pain should result. The strongest bond in primates is the mother–offspring bond and it is clear that the great apes will experience intense grief with the death of their offspring. That humans also experience grief at the loss of a loved (bonded) one is a well established and one of the most intense emotional states they can experience that can incapacitate an individual. Bereavement and grief, however, is not usually a conscious choice. I suggest it results from the disruption of a cognitive-based (psychological/emotional) addiction module. Human grief indeed seems generalized in that it can also be induced by separation from other loved ones and is not restricted to the death of offspring. A romantic break up is a good example of this. Thus grief can provide the exemplar for how hominids form strong social bonds mediated by emotions and inform us about the inherent characteristics of such bonding. Humans experiencing romantic break ups will display changes in brain activity that can be seen by fMRI studies. Five stages of grief are currently recognized. The initial stage is disbelief. The affected person refuses to accept what can be compelling evidence. This basal reaction suggests that a belief mechanism is an early inherent part of the social bonding process. As developed below, belief will be presented as core component of extended social bonding. Next phase of grief is a yearning for the individual, the wanting of something positive that is now absent (reminiscent of yearnings for drugs). This is a less stable or transient phase. This phase is then followed by a period of anger. Anger would appear to represent the expression of the stable and toxic emotion of a T/A set. The anger phase is typically followed by a depression phase which can also result in a stable and toxic emotion that can last for an extended period and incapacitate an affected individual. It is interesting that depression seems to involve the same brain region as affected by romantic love. The final phase is that of acceptance in which the strong emotional reactions are over. This process normally takes about 6 months to complete for most people. This long duration is interesting and suggests some structural changes in the brain may be involved. It also seems possible that this long duration is associated with the time it takes to reset belief states. A grief response is considered pathological if it lasts longer than this period. When pathology occurs, it is associated with social withdrawal (group dissociations), thus it represents a clear disturbance to our social brain function and social activity. Extended pathology can result in post-traumatic stress syndrome (persistent negative thoughts), if the death of the loved one was witnessed. Also, children's reaction to parental separation will follow similar patterns. Related sets of grief responses can occur with the death of a leader (political, religious), thus this emotional dynamic clearly applies to larger and more extended social bonds.

Abstraction of Visual Group Recognition

Because our brains have inherent capacity to recognize faces via sparse and relatively abstract characteristics, we tend to rapidly and inherently recognize abstract or generalized facial difference (and other abstractions). Thus racial recognition, representing generalized variations in facial appearance, are easily recognized as representing other groups. However, aggression to non-group members also appears to be an inherent tendency. Yet, even when there is no facial or racial variation between similar groups (for example very similar but Paleolithic human island populations), there will still exist a tendency to create or adapt visual abstractions that will provide identity markers. Thus we know that face painting in Paleolithic cultures was common. Other abstract visual markers are also known, such as the tattoos found on the 'Iceman' that had been frozen into a glacier in the Alps. In New Guinea, distinct visual designs painted onto shields of otherwise identical warring tribes were common. Although such symbols have no meaning, they simply convey abstract patterns associated with the tribal identity. Thus, it seems that Paleolithic human cultures tend to use abstract pattern with no specific meaning that indicates group identity. Such a tendency can still be seen in national emblems and even in the hand signs of modern urban gangs.

Song: Early Vocal Amplification of Emotional Bonds

The human capacity for song and speech involved many evolutionary changes to brain and vocal architecture. These changes involve the human tongue, epiglottis, larynx and trachea, changes that for the most part make humans prone to chocking relative to apes, so their selection needs an explanation. These changes are estimated to have started about 2 million years ago and distinguish humans from the other great apes. What was the initial selective pressure for these vocal changes? As mentioned, humans clearly differ from the other primates in their preference regarding music. Monkeys and chimpanzees prefer silence to music. Tamarins and marmosets, for example, will spend the majority of their time in cages free of music including lullabies that are soothing to humans. In contrast, humans prefer a constant music environment (such elevator music, car radios or the ubiquitous iPods). As originally proposed by Darwin, music may have evolved prior to speech and was likely involved in maternal–offspring bonding. In chimpanzees, chorusing is most often done by females in response to males. Thus we expect human ancestors would likely have some vocal traditions in their females. In humans, this must have been adapted to extend maternal bonding. Maternal singing associated with bonding to babies is found in all cultures. Mothers sing to infants in all cultures. Typically, the mother's singing is relaxing to the baby and lullabies are used to help babies sleep (human babies have extended sleep, along with extended infant helplessness). Singing thus induce sleep via a process that likely involves endorphins. Other effects have also been documented, such as lower saliva cortisol levels and babies have a documented preference for song over spoken language. Thus, it is clear that singing provides an emotional link between mother and baby that could extend the duration of the bond. In addition, the recognition of the emotion valence in song is universally common; recognition of anger, sadness, threatening, happiness or joy of music is not culture specific. Why should music communicate such robust emotional content for all human cultures? If, as initially proposed by Darwin, music was part of a maternal–baby bonding system that promoted emotional links, then it would not be expected to have resulted from mate selection as proposed for bird songs. Indeed, mate selection has often been used to explain most social and sex behaviors in humans. As discussed previously in primates, mate selection cannot explain the social structures involving genetically unlinked females in infant care. A role for music in human mate choice is thus most unconvincing.

In terms of neurobiology, singing activates distinct brain regions from that of speech, although there is some overlap. In one fMRI study of music's effect on the brain, chord sequences were seen to activate regions that included Broca (left frontal cortex) and Wernicke's areas (left temporal lobe) (see Fig. 9.7). These regions had been previously thought to be domains specific for human language. Music is clearly emotional but it can be both pleasant or unpleasant;

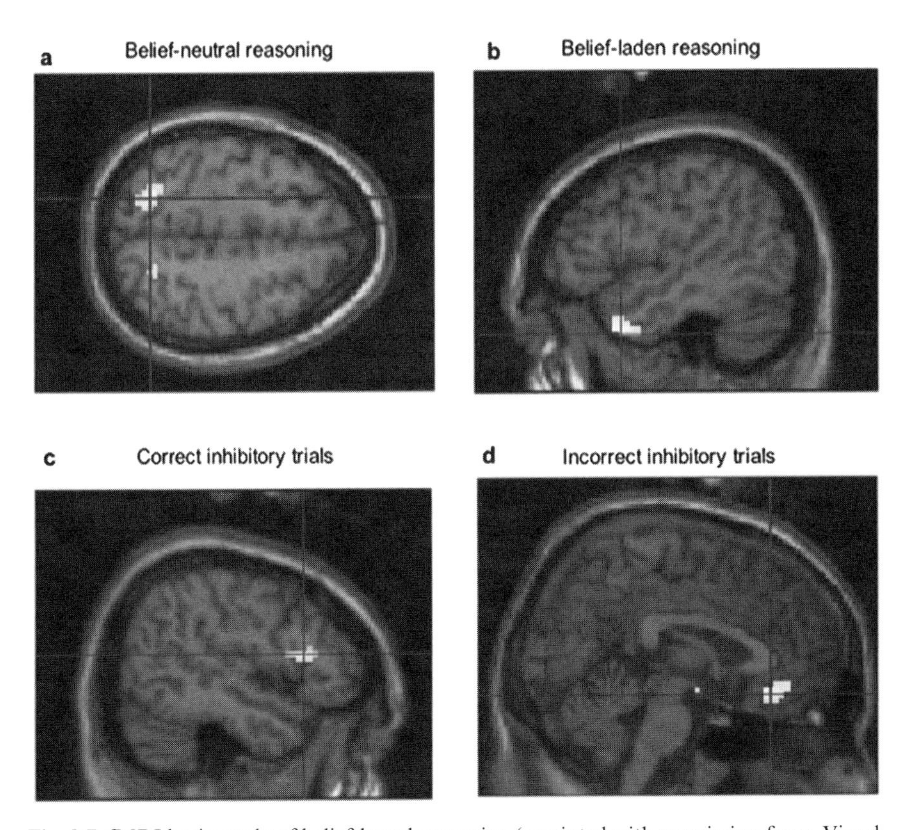

Fig. 9.7 fMRI brain study of belief-based reasoning (reprinted with permission from: Vinod, Raymond (2003), Cognition, Vol. 87, No. 1)

dissonance is unpleasant whereas consonance is pleasant. The cry of a baby for example, is unpleasant and emotionally links the distress of a baby to its mother and other adults. Baby crying also has a contagion character (like facial fear) and will induce other babies to cry as well. Thus this form of emotional audio communication exhibits 'mirror-like' affects on nearby individuals, clearly a feature of a social brain function. As mentioned above, music evokes emotions especially when applied to visual images and will activate various brain structures associated with emotion (amygdala). The amygdala also plays a role in recognition of negative, dangerous emotional social states produces when the amygdala is damage it can selectively impair recognition of scary and sad music, as well as facial fear and facial anger recognition. Congenital amusia is also known for some people; the existence of such non-musical humans clearly identifies the presence of a biological substrate for music perception. Some individuals with specific types of brain damage can also become indifferent to dissonance, yet can still differentiate the emotional valance of music (similar to face recognition). It is interesting that Broca's area and Wernicke's area are next to and linked with facial recognition and mirror motor system (found in two regions) which was not present in chimpanzee brain. These regions also overlap with regions involved in imitation, learning and theory of mind system of humans (all important human social functions). In homologous (but distinct) regions of macaque brain, these regions are activated by species-specific vocal calls. Thus singing is not the same as talking and likely represents an older, second mode of audio emotional communication, in which lateralization is different and uses distinct, although complementary (asymmetry), neural substrates. Along these lines, trained musicians will activate both left and right midfrontal brain regions when hearing music, an expanded pattern relative to non-musicians. Thus it appears that the human brain can be trained or developed into a more expansive and involved neurological substrate capable of identifying and producing more complex or sophisticated audio patterns. This capacity is the product of extensive sensory training and experience and is not simply inherent or genetically endowed. If such complex pattern recognition capacity can be used for group identity and membership, selection for it in a social brain would seem likely. However, since complex pattern recognition is also a basic feature of general intelligence (that includes sophisticated abstractions), this suggests that experience expands the social and general capacity of the brain. This concept will be of great significance when we consider the development of reading and the modern human social mind below.

Children will dance and sway to music well before they learn to talk, clearly indicating that they have inherently developed a sense of rhythmic sound–body connection prior to speech, However, we do not understand the neural substrates of rhythm. Yet it seems clear that music is not essential for emotional bonding as demonstrated by the deaf or people with amusia. Rather it appears that music and sound will greatly amplify the social emotional impact of visual information and can also affect larger groups. Music alters constitutively expressed opiate (mu-receptor) on mononuclear cells in listeners and has shown a robust ability to induce a sense of group calmness. This transmission and amplification of emotional response is the basis of its use in sound tracks of movies. The origins of using song this way date to the very earliest recorded human stories, such as songs used in story telling. Music not only remains a transmissible (social) amplifier of emotions, moving from one individual (mother) to another (baby) and exerting emotional content, but is also able to transmit and sooth emotions of adults in groups. As music is able to transmit pure enjoyment, it would seem to have no selective or adaptive value by traditional Darwinian reasoning. Its role in social bonding, however, explains a selective value in group identity. Musical instruments date back 35,000 years so music seems to have been present since the time of Neanderthals. The transmissive and often stable character of music is well established and music can also have a viral-like persisting character as anyone that has ever been colonized by a sound worm can attest (an internally reproduced song that cannot readily be forgotten). In addition to its likely role in maternal bonding, all cultures also use music for various social purposes, but mainly to set the emotional group ambience. When performed together, many voices singing in synchrony will acquire the same emotional valence, thus its use in choirs and chants to synchronize emotions. Endorphins appear to be released during such coordinated group singing, likely providing the soothing character to social music. Music can also diffuse social tension for example as used in social gatherings, parties and dancing. But it can also be used to stimulate or arouse emotions such as used for tribal music, fight songs, ritual bonding and religion. Clearly music provides a socially transmissible system for emotional communication and its role in group identity seems clear. Although it can be used to transmit cultural information, unlike language, it has remained active principally with emotional content, not information content and abstract meaning.

Paternal Bonding and Empathy

Why did humans need to amplify emotional communication by vocal means? In what way did this promote the evolution of much enlarged human social brain? Clearly, amplification of emotional communication could be used to extend social bonding and group identity. As previously noted, humans show two universal social features that are absent from the great apes: mate bonding and paternal role in upbringing of young offspring (both associated with nuclear families). Both of these social characteristics also appear to involve prolactin. In terms of mate pair bonding, we presented evidence that supports a role of prolactin and oxytocin in both romantic love and maternal bonding in which visual (facial) input was the primary stimulus. As outlined above, for human fathers to acquire this, their visual centers must engage emotional addiction systems that compel the father to bond with the young. One emotion that could provide this is empathy. The father must feel empathy for the helpless young. It has also been suggested that empathy is the main underlying emotion that mothers feel toward their young and thus is a main emotion for maternal bonding. We might now suggest a process by which this bonding might have also been developed in males via sex chromosome involvement, specific to human evolution. Empathy appears to be the central characteristic that underlies most positive human social and ethical behaviors. Its absence is strongly associated with social pathology (see below). Therefore, introduction of empathy as a social bonding system into males would represent a major development in human social evolution. It is therefore most interesting that recent fMRI studies suggest that humans performing selfless acts are actually tuned into the needs and emotions of others in ways that activate specific visual brain regions. The posterior superior temporal cortex (pSTC, near back of brain, involved in visual perception) becomes highly activated during altruistic acts. This is not part of the reward system, whose involvement had been hypothesized at the start of this study. Rather it appears to identify a more primitive system that may not simply define the emotion of empathy, but identifies a visually linked process that socially connects minds and emotions. The development of this system might also be related to the theory of mind as discussed below. In terms of inherent capacity for empathy, there is clearly some human variation. At a population level, females are generally stronger empathizers than males which clearly implies some role for the X/Y chromosome in this. Empathic ability also appears to be of relevance for autism, which affects males more often than females. Here, social sensitivity seems clearly affected and autistic individuals show clear impairment in empathizing (indicated by psychological profiles). Autistics also show an extreme bias in systemizing, a strong interest in narrow areas, insistence on sameness, repetitive behavior and obsessions with lawful systems (such as train timetables), including a strong interest in rule-based predictions. Autistics tend to show a form of a mind-blindness, or a delay in developing theory of mind during childhood. It is interesting that these same characteristics have also been proposed to be considered as an extreme version of the male brain, which is also more systemizing, more rule governed and less empathic as a population. As discussed below in the context of social minds, the conservative stance also appears to have many similarities to an extreme 'male' brain, but unlike autistics, a conservative or social mind shows characteristics of strong group identity.

Serial Mating Pairs, Bonding and Love

Although it can be argued that human pair bonding between mates is not fully monogamous and life long, relative to some pair-bonded voles (or birds), it is nonetheless clear that humans display much more pair bonding than do the other great apes. Accordingly, romantic love between mates is a universal characteristic of all human cultures. This bonding can be characterized as mostly being a form of serial monogamy. Examination of divorce rates in numerous and various cultures shows similar patterns. All these divorce rates peak around the fourth year of marriage, followed by a long-tailed decline afterward. Most of these divorces are followed by subsequent pair bondings. Thus, although absolute divorce rates vary with cultural setting, shape of their divorce rate curves is essentially the same. Thus it seems clear that human pair bonding has this inherent profile. Clearly, an early and strong emotion associated with initial pair bonding is that of romantic love. In romantic relationships, it is clear that visual input (facial pictures) can provoke a love reaction, and several fMRI studies of subjects deeply in love who were shown pictures of their partners suggest these subjects showed focused activity in medial insula and the anterior cingulated cortex plus other areas, all bilaterally. Deactivation was also seen in amygdala and was right lateralized. This pattern was distinct relative to fMRI studies of other visual (facial) stimulation. It has also been reported that happiness correlate with deactivation of right prefrontal and bilateral parietal and temporal cortices. We might consider this to be the A element of an emotional addiction module. In contrast, sadness and depression correlates with activation of some cortical regions (deactivated in this study). The deactivation of amygdaloid was also noted above in maternal bonding. Since it is associated with fear, sadness and aggression, it can represent the T element of an emotional addiction module. The activated foci seen here also overlap to a large degree with those activated by cocaine and mu-opioid agonist-induced euphoria. These two euphoric states (romantic love, opioid induction) overlap in brain structures and it has been proposed that romantic love resembles an addiction state.

A More Split Brain: Language and Belief

Brain lateralization especially with regard to language utilization appears to be a human-specific adaptation. However, some lesser lateralization is seen in primates. For example, most ape populations, such as chimpanzees, are right handed as are humans. The humans left brain controls right body. The left brain is also called the verbal half and is thought to be most involved in language. Language use is the most lateralized of human brain functions and as argued below represents the most recently evolved group identity system. Seizures in left brain account for large majority of language deficits. There seems to be some sexual dimorphism in language capacity as female populations generally perform better than do males. In addition to language, the left brain is also thought to be more involved in details, facts, order, patterns, names and math. For example, some left brain stroke patients can name objects, but not explain its use. The left brain is also considered to be more involved in rational thought. In savant individuals, the left brain is the usual site of damage or dysfunction, and it appears the right brain then takes over more functions.

In contrast, the right brain is considered 'non-verbal' half involved in explaining meaning and beliefs. The right brain is thought to process symbols and thus also appears to play some role in processing written language. Music and tonal stimuli are usually considered as right brain functions as are their emotional reactions. Thus symbols, emotions, imagination, religion, risk taking and beliefs are all more associated with the right brain. As will be presented below, belief states have a central role to play in the use of language as a system of group identity, so we might consider an overall lateralization has occurred in these two linked and interacting functions. The right brain which assumes a greater role in savants also seems to be more active in special perceptions of some autistic individuals (such as the fact-based detailed memory of the 'Rainman'). Thus the right brain is considered more important for emotional processing and appears involved in providing the 'big picture' for many individuals.

Sex and Brains

As alluded to above regarding overall language capacity, brain lateralization is associated with some sexual dimorphism (asymmetries, such as in the larger male amygdala). Sex and chromosome effect on lateralization have been reported. In this, it appears that the Y chromosome must have some activity since XX and XY chromosomes have distinct X dosage which affects various aspects of brain lateralization. XO individuals (Turner's syndrome) have non-dominant hemisphere deficits whereas individuals with an extra X (XXY, Klinefelter's and XXX syndromes) have dominant hemisphere, i.e., verbal deficits (mostly delays). Thus it appears that an extra X chromosome results in language deficits in individuals that are normal with regard to IQ and other aspects of mental performance. Since males (XY) do not show a single X effect (like Turner's), this indicates that the Y chromosome must provide some off-setting function preventing language deficit. It is thus very interesting that a recent genetic alteration in human evolution involves significant exchanges in the Y chromosome (involving recombination with sequences in the short arm of the X). This alteration in the Y makeup occurred after the separation of the chimpanzee and human lineage, and is also directly related to SRY (sex-determining factor on Y). We can recall from previous chapters that the Y chromosome has a distinct composition, composed of relatively few ORF but many ERVs, including recently acquired, human-specific HERVs, and other repeat and satellite elements. In the evolution of other vertebrates, similar Y-ERVs have been associated with group and sexual identity (see platyfish).

Applying the Split: Language, Acquisition and Bonding with Infants

First Words

In most mammals, early years of infant care and development are all almost exclusively female directed. This pattern was maintained in the great apes, such as chimpanzees in which females are responsible for the first 10 years of upbringing. With human infants, a mother's role continues to dominate, but a paternal role is now apparent. Both the maternal and paternal human social bonds are associated with the development of language. In essentially all cultures, the most basic words first learned by infants are usually a name for the mother, then their father (see Fig. 9.1). Thus these first words provide a social foundation absent in the great apes. These first words are typically produced by a reduplicated process, resulting in terms like ma-ma, da-da and pa-pa. The vocal reiteration appears to be important to allow setting or imprinting of the memory and association must provide meaning to the words. Visual facial learning occurs in association with audio learning. The words for parents clearly associate with emotional and facial states and are spoken at times of stress (a feature that can be maintained into young adulthood). Clearly, the presence of the parents (especially mother) provides comfort for the infant. Thus a basal maternal social bond has been mediated or enhanced by these first words. The basal importance of the maternal bond has long been recognized in all cultures, and was noted by Aristotle who viewed childbearing and sex as the basal social links for all human interactions.

First Grammar, First Insight

Language is thus used as an additional system to mediate the basal maternal bond. However, the character of language becomes rapidly and increasingly complex with infant development. Young humans are spectacularly good at language acquisition. It is an effortless, unstructured learning process that results in linguistic competence. But along with the acquisition of language, young humans also become inherently good at understanding what is going on in other peoples minds (theory of mind, presented below), thus its competence is directly associated with development of other social capacity. It appears that language acquisition may also be needed for these additional cognitive abilities to develop. Although simple language provides emotional links between parents and offspring, the human brains have clearly been adapted to learn complex language. As this language is recursive, its meaning depends on context, thus grammar or syntax becomes crucial for the specification of meaning (a skill that develops about 3 years). However, the nature of grammars that humans can learn appears to have biological constraints. Thus, it has been proposed that only certain universal grammars (UGs) exist and appear innate. The brain seems to have a grammar center (such as subregions of the left frontal cortex), much like it had a face recognition center as presented above. Consistent with this idea, not all grammar structures can be learned by children during primary language acquisition. In some cases, grammars can show competitive exclusion. In this, it appears that less-specific UGs can resist invasion by more specific UG, if learning is accurate. Accurate learning appears to stabilize UGs. It is interesting that some of the universal grammars also show evidence of competition and exclusion, although it is clear that other UG combinations can co-exist. Some neurological studies have suggested that language centers are linked with and adjacent to brain regions involved in gesture recognition, as present in primates. This has led some to propose that language may have evolved from sign language-like form of communication. However, there is little evidence that human vocal communication evolved from sign language. Still, there remains a curious link between gesture action and language. In one case study of viral-mediated brain damage, the patient lost all sense of touch below the neck, but continued to gesture automatically when speaking. Thus it seems that human hands are somehow linked to speech articulation. I suggest this link is due to the need for abstract (symbolic) pattern recognition in recursive language. When children are learning new words, they are guided by mouth gestures. This results in the McGurk effect, in that if one sees the mouth say 'ga' but it was recorded as saying 'ba', one will hear the sound 'ga'. It seems that language acquisition has somehow superimposed the system (Brocas) associated with visualization, manual motor control and mirror neurons. This could make sense if we consider both these to be core social systems involved in social communication with similar needs to link sensory input and access emotional memory. The ability to hear or perceive subtle verbal patterns, or nuances, is affected by the language learned as infants, who map language into the brain before they speak. It is interesting with English, the mapping of letters, words and sounds is more distributed than other languages. English has many inconsistencies due to the heterogeneous origins of its various elements. In this sense, English may not represent a fully natural grammar when compared to Latin language for example. PET-visualized activation patterns in English and Latin language speakers support this view. These characteristics of language are consistent with the idea that it may provide a vocal version of a group identity system.

Requirements for Language (and Emotion) to Provide Group Identity

An exemplar I introduced with mice was how colonization and persistence by extragenetic parasites (like mice with MHV) could differentiate otherwise identical mouse populations. Language can also have this viral-like feature to differentiate host (human) populations. If language initially evolved as a group identity system, language will also differentiate otherwise identical populations. Consider two neighboring tribes that speak distinct languages but are otherwise identical. Each has acquired their own social (group) identity systems via their respective language. Such a process of group identity should also generate diversity. For example, as mentioned, in pre-Columbian California, it is estimated that a high diversity (about 400) of distinct tribe-specific languages existed. In a sense, a population (tribe) that has been colonized by a specific language precludes the colonization by a second, adjacent tribal language. However, languages, unlike viruses, are not a genetic-based information system and do not evolve by the same principles. Yet it is an adaptive and dynamic information system that does have a transmissible character that stably colonized human brains (it host) and populations during crucial periods (childhood) of identity transfer. In so doing, it provides a common system of group identity and also provides a system that resists other identities. Because meanings are only fully maintained within one language group, the benefits of a common language are restricted to that language group. However, because language is not a genetic-based system, it is much more adaptable and can readily and rapidly be extended to very large social groups that have no genetic linkage. This could be accomplished within the duration of one generation in which offspring are now raised in the presence of the new language and their brains become colonized by that language. Thus, unlike facial (or racial) visual imprinting, language has a greatly enhanced social adaptability and promotes group communication and cooperation on larger and more dynamic scales. For this, language required a much more complex social brain to process and understand the large increase in audio complexity and abstract pattern recognition needed for a recursive language. It would also be expected that a language-based identity system must engage in some forms of addiction module involving emotions. Clearly language can convey a strong emotional content. And language can exert a potent transmissive control over the actions of others, clearly a social function. The command or urgent voice of a mother to her infant can exert considerable control over the independent actions of the infant. Fear and anger, for example, can readily be transmitted by such mother's commands. In terms of the basal maternal or pair bonds, it is clear that language can have a powerful role. Even mis-information transmitted by language can induce emotionally potent bereavement reactions, for example. In fact it can be argued that the majority of words in most languages are emotionally charged and in some instances they are highly charged. In English, for example (a language with various linguistic roots), most words are metaphors and expressions are rich with rhetorically charged emotional expressions, often associated with favorable or despised group identity.

Descriptive and Objective Language

It is also clear that language has a very functional and potentially objective or descriptive character to it. Language provides a media that can transmit information important for survival (concerning food or defense, or danger for example). Clearly, the benefit of language goes beyond any strictly social role in group identity, and in this feature language may have partially escaped from a strictly emotional basis for social or group bonding. By providing clear but functional information regarding survival, language is a crucial and emergent phenotype of our large social brain.

A Developmental Window

During language acquisition, many biologically based associations are seen. The most apparent is the neurobiological developmental window during which language acquisition is effortless. Numerous observations suggest that if language is not acquired by age 7, the person will never entirely catch up with training during childhood and will struggle with the higher syntax of language. In the case of feral children rescued as young adults, their speech level will typically attain that of only a 5 year old. Such feral children that have learned language late will also be unable to develop deep cognitive or thinking skills. Their brains appear to remain undeveloped and unable to develop more abstract or complex forms of pattern recognition. Thus a brain that has developed without a language colonization is a less capable brain. This clearly resembles a state of symbiosis between the brain and the language. The rate of normal language acquisition is impressive. During childhood development, children acquire about 8–10 words per day, a phase that has been called a vocabulary explosion phenomena. In order for such a high rate to be achieved, it appears that words must be acquired in a parallel, or distributed, process. The usual level of attainment for individuals after education at college level will be a vocabulary of about 60,000 words. This will have taken an estimated 75,000 hours during which much brain connection and remodeling has occurred. Thus during this developmental period, a new language identity has been acquired that has altered patterns of brain development, and transformed the cognitive capacity for abstract thought and learning.

As mentioned, a recursive language not only greatly expands the possible meanings of language but also required a large expansion in the ability of the brain to recognize very complex (perhaps infinite) vocal patterns. Such syntax-dependent meaning leads to the need for alternate meaning or alternate emotional reactions for the same specific word sound. From this, we can see how general intelligence was also greatly enhanced. The ability to think abstractly, including the ability to imagine alternate meaning and what has not or cannot be seen or heard as well as to imagine emotional reactions, was the product of recursive language. It is therefore likely that the much extended human brain development in human infants is associated with the demands of a recursive language on our social brain in order to promote effortless language acquisition. Thus not only does the development of recursive language place stringent requirements on the pattern recognition capacity of the brain, the acquisition of language itself transforms the cognitive and thinking capacity of the colonized brain by promoting more complex brain connections. It also provides a vocabulary that becomes used for internal dialog as used in conscious thinking (see below). In providing these features, language not only transforms human group identity, but also transforms human social intelligence and potential for general intelligence.

Social Learning, General Learning and the Social Brain

Our large social brains were thus especially enhanced as an evolutionary product of recursive language-based group identity. As a consequence of such enhanced social and abstract capacity, humans were indeed transformed in how they learn relative to the other great apes. In most mammals, mothers are most responsible in training offspring. One basal lesson they provide is in training what other species to fear. As a side note, mice inherently fear cats, but ablation of the mouse VNO olfaction yields mice that do not inherently fear cats, but can learn to do so. It is clear that in most mammal species, training can override fear of natural predators. Such fear is normally acquired observing the mother's emotional reaction. However, mothers can also contribute to non-fear-based training or learning. With chimpanzees, for example, mothers will teach offspring how to crack nuts which seems to be one of the more difficult skills young chimps are taught (around age 10). However, there are clear social differences in chimp mother–offspring training from that of humans. Although imitation seems crucial for both species, with chimps, young observe their mothers intently, but the mothers seldom observe their offspring when teaching. Chimp mothers seem unconcerned with the mental process that is occurring in their young, showing no eye contact with their pupils and not reacting to facial emotional states or reiterating parts of the lesson. Chimp young simply observe and attempt to copy their mother with no feedback from her. It has been proposed that chimps lack theory of mind and thus do not concern themselves very much with the mental states or processes of others. Thus when learning to crack nuts, (their most complex technology), infant learning is inefficient, taking years to acquire. This absence of the mental awareness of others may also relate to the different relative states of consciousness between chimpanzees and humans, which also requires self-awareness, as discussed below. Another major development in chimpanzee learning concerns young males that learn to hunt in groups. This occurs at the time of sexual maturation. But besides learning to hunt prey, they also form strong male in-group bonds and learn to hunt strange (non-group members) male chimpanzees, suggesting this learning might be also the development of a male-specific group identity process. The resulting competence of the trained young male is that not only do they participate in group hunting, but also participate in group attacks on lone males they do not recognize.

In contrast, humans are very much aware of the mental states of other humans that are learning. For example, in a human classroom where we can evaluate the gaze patterns of children, we see many children watching other children as well as watching their teacher: that is, child A watches B watches C watches the teacher. This is an inherently recursive behavior absent from chimps, in that children must be thinking about what others are thinking. The gaze pattern with chimps is one where animal A watches its mother as do animals B and C. In chimps, these non-recursive interactions are consistent with the absence of theory of mind. This learning pattern is social and identifies a crucial aspect of human social structures and group identity that differentiates human social structures from those of the great apes. Yet, chimpanzees are clearly better learners (by observing others) than are other apes. For example, in contrast to chimpanzees, baboons will not learn from watching the actions of others (like chimps) regarding learning to use sticks for termites. They simply wait until chimps have finished to go in and gather any remaining termites.

General human learning thus seems to be a byproduct of visual and vocal-based social learning that was much enhanced by acquisition recursive language and the resulting large social brain. In the field of human learning, general learning has often been thought of as a special creation, essentially from an a-biological perspective. The field of psychology has also identified many human behavioral tendencies and created many clinical evaluations that seem to provide some reproducibility of behavioral phenotypes. But such approaches have also essentially lacked attention to the underlying biological mechanisms, let alone how social learning behaviors were selected as evolutionary adaptations. With the concepts associated with group identity, we can now provide direct links between group behavioral characteristics and underlying biological mechanisms as well as associated evolutionary pressures.

Learning and Resisting Learning: a Stable Social Identity

Let us now summarize characteristics that support the assertion that language is a group identity system that has all the needed features for such a role. As mentioned, language differentiates otherwise identical human populations. It has an adaptive and transmissive character that is transmitted from old to young during a crucial period of development in young, where it is readily learned in a stable way. After brain colonization, language resists displacement by other language (identity) systems. As a result of colonization (stable learning), it marks population as distinct, adds both identity and capacity, inhibiting communication between populations that use other languages. It provides benefits to host (common communication, language-based general thinking, survival information). But it also identifies non-members as foreign. It is important to consider in greater detail the consequences of acquiring language by learning, since, as we will see, this will also identify an counter intuitive need to resist learning. Learning can be considered as the acquisition and acceptance of new information. If as asserted, language originated as an information system of group identity, the colonizing language must readily attain stability (be learned) during development, but once colonized must also resist displacement by subsequent languages (identities). Thus, the continued learning of subsequent languages becomes disfavored and the learning of more languages difficult. One direct implication of this assertion is that the resistance to learning (beyond a crucial developmental window) is fundamental to the use of language as a learned set for social identity. Both learning and resistance to learning are required. Indeed, resistance to learning is essential if learning is to serve a role in group identity. In the terms of the use of language, the resistance to learning a subsequent language appears to have an inherent biological basis associated with brain development in children. Learning becomes either closed or difficult afterward for most individuals after this window. However, the resistance to learning is not absolute, since language competence can continue to develop in older children. In addition, learning resistance need not theoretically be limited to developmental neurological windows. Such learning resistance can also be attained cognitively in a more developed brain (such as a mind). As presented below, such stable cognitive states for information are called 'belief states' which resist new information or learning (via closed minds). Thus the concept of 'belief' and its role in learning or resisting the learning of new information can serve a crucial role in group identity. We will now consider some of the biological observations that relate to this role of learning and belief.

Neural Substrates of Belief Indicated by Brain Damage

I have just asserted that belief, defined as the resistance of new information or new learning, is a needed feature of the use of learning in social identity. Previously, we have noted that a major region of the brain (the right hemisphere) is involved in belief attributions. We can now start to consider what role belief has in human social structures and if this has required special neurological adaptations. The development of the large human social brain also requires the development of an enhanced capacity for social learning. Any social information that is to be used for group identity will also need to be stable or attain belief states. Are there then any neurological substrates for such belief stability? The usual way we identify function- or domain-specific brain regions has been as the consequences of damage to specific brain structures. Accordingly, damage to right hemisphere can show specific belief deficits. In one example, an 85 year old woman developed a flaccid left limb paralysis following a stroke. Although her memory was normal, she lacked awareness (belief) of her deficit and denied its existence when questioned about it. She did not believe she had a problem and this belief state resisted learning and was not affected by demonstration of her paralysis or instruction about it. She was resistant to new reliable information. Instead of accepting the validity of such demonstrations, she would rationalize a confabulated explanation. A conclusion that was reached from such a clinical presentation was that the belief system of this patient had been damaged. Other similar cases are known. People with left temporal parietal junction brain also show it is needed for normal reasoning about beliefs. In some patients, it is clear that visual self-body part pointing can be affected by brain damaged. One such disease, autopagnosia, involves representation and pointing to body parts. Following brain damage, some patients are able to point to another persons body parts but not their own. Somehow, their self-representation has been damaged. The converse deficits are also seen. In other studies, the ventral medial prefrontal cortex was reported to be involved to override logical reasoning. An fMRI study of people undergoing reasoning that is either consistent with or inconsistent with beliefs shows that left temporal lobe is engaged during belief neutral reasoning. This parietal system is known to be involved in representing and processing special information, associated with mathematical reasoning, and numeric approximations. In contrast, ventral medial prefrontal cortex (VMPFC) is engaged when reasoning must overcome belief bias suggesting an influence of emotion on reason. VMPFC highlights its role in non-logical, belief-based responses. In general, it appears that the left temporal lobe is engaged during belief-based reasoning. Deductive reasoning and drawing valid conclusions should logically be a closed system. But clearly beliefs can compel people to logically invalid conclusions. During such reasoning, participants will rationalize or explain their beliefs (even falsely provided beliefs). These results taken together lead us to a very interesting idea. Normally, we consider the process of rationalizing some explanation as a simple attempt to explain what was true. However, one implication is that rationalization is often a confabulated mental activity intended to support or defend beliefs. Thus, we inherently tend to generate explanations that support our beliefs. Since our brain appears to have domain-specific reasoning functions regarding belief states, it seems that specific domains of the brain are involved in the contents of beliefs (belief attribution). There are also some developmental observations that suggest that a belief function appears to develop after perceptions and emotions functions are established. The existence of such brain structures in a large social brain also suggests they are also involved in social identity.

Clearly, belief, emotion and memory must be linked, but how they are linked is far from direct. Belief must involve stable memories in some way, but these may be distinct from short-term memories. One telling example of this idea was seen with amnesic Clive Wearing, a BBC music expert, that developed herpes encephalitis (March 1985). HSV-1 is a neuron-specific virus that persistently infects most humans. Clive Wearing developed an unusual encephalitis that involved his hippocampus (associated with long-term memory). In addition, there was some temporal (amygdala) and frontal lobe involvement. As a result of this infection, regions involved in transfers of memory were damaged, but in a particular way. Every day he appears to restart his consciousness when he awakes and he claims to have just woken up from a coma of many years, and disbelieves any evidence to the contrary. Yet, long-term memories and beliefs are intact. Thus he fails to accept any evidence that contradicts his false belief. For example, he remembers being in the active Navy (as a young man), although he was then over 60 years old; his early memory was more compelling (accepted as a belief) than the current logic he was presented with (i.e., he was too old to be in the Navy). This is like many patients with damaged belief reasoning that will confabulate denials of physical evidence (sometimes preposterous), including as proposing dead people coming back to life, in order to explain their memories or false beliefs. This tendency implies a biological basis for rationalizing, which need not involve causal reasoning. In the case of Clive Wearing, it was also clear that he still retained strong and recent emotional bonds (memory). He clearly still loved his second wife, Deborah, although they had been married only a year before his illness, which was clearly within the range of his other short-term memory that was lost. This retention of this recent emotional bonding clearly suggests emotional memory and bonds use distinct memory mechanisms. Also, he still remembered how to play the piano but would react with strong emotion upon stopping play as he seemed no longer able to control emotional memories elicited by the music. It thus appears his brain has distinct domains concerned with the emotional memory (romantic love and music) that were mostly undamaged. It is surprising that such mental specificity was brought on by a virus. A virus is usually thought of as a simple destroyer of tissue. Clearly its affect on Clive Wearing's consciousness was highly selective and reminds us of a computer virus that compels its host computer to continually reboot. This case also raises the question of what then is nature consciousness if no recent memories are needed? What constitutes self? Does this only require awake states plus stable memories including emotions? These questions are explored further below. Another relevant issue concerns this illogical belief that fully resists verifiable information. Similar resistance to logic, such as people believing in alien abductions, are also discussed below. Both these situations appear to identify specific alterations in consciousness associated with belief that resists logic.

We start to see a biologically based role for beliefs, or other stable forms of cognitive or emotional memory (including language), in the function of our social brain. Resistance to change in memory is essential for the stability of memory-based group identity. I have asserted stability is generally attained by the action of addiction modules. How then do beliefs and memory operate with regards to possible addiction modules? What, for example, might be toxic or antitoxic with regards to a stable memory? What emotional modules are engaged when stable belief is set that will resist evidence to the contrary? Language is the foundation from which higher cognitive capacity developed. We should now consider how language may have led to the developments of belief in group identity.

A Short History of Early Language

The topic of the origin of language should be approached with some caution. Extant humans are immersed in the complex fabric of intellectual life that has been made possible by modern language. The very consciousness of the modern mind is mediated, in part, by the internal voice provided by our language, thus language is hard to separate from our minds. As we now come to examine the basis for its development from the perspective of group identity, it needs to be acknowledged that the study of language evolution has a long, unproductive and checkered history. At one time, discussions on it were banned from some scientific societies due to their contentious and non-productive nature. Indeed, following the publication of Darwin's 'Origin of Species', after 1859, it was the topic of language evolution that was the source of strong anti-Darwin attacks. Chief amongst these attackers was Max Müller, who felt that Darwin's theories could not possibly account for something so complex as the evolution of language and thus used derogatory names (i.e., bow-wow theory) to attack Darwin's ideas. This compelled Darwin to consider this issue in his second book as he examined song as used in birds as a likely predecessor to language. Language dominates as an agent of human education and culture. How did we get to the point where the average college-educated person knows 60,000 words that represent essentially infinite set of ideas and took about 75,000 hours (about 9 continuous years) of learning to acquire. Recall that chimp's use of sign or symbolic words are sensory only. Chimpanzees are able to represent what they see or observe via their very limited vocabulary. But humans can represent what they imagine, or what cannot even be seen as mediated by language. Most of the words in human language are metaphors, not sensory or descriptive. Above, I argued that our large social brains were selected to evolve by the neurological demands of a recursive language used for social identity. However, the Neanderthals had brain sizes equal to or slightly larger than that of *Homo sapiens*. Does this suggest that they also used a recursive language? It seems likely that the large Neanderthal brain was also social. Neanderthals were in Eurasia for 200,000 years and were stable occupants of Europe for 50,000 years, until about 30,000 ybp when they became extinct. But Neanderthal child development was different from that of *Homo sapiens*. Although big brained, they developed their brains more like a chimpanzee in that adult sizes were reached by 4 years of age. Such rapid brain development would appear unable to provide the extended brain development that could be molded by the neurological demands of a highly recursive language. From this, it would seem their language capacity would need to be quickly attained and hence less developed or more biologically determined (such as universal grammar). We do not know if Neanderthals also underwent the late adolescent development in which the frontal cortex completes its development. However, it does seem likely though that the Neanderthals had developed language to the point where it was a group identifier. There is also good evidence Neanderthals held beliefs, since they buried their dead. Thus their language must have been able to communicate some degree of belief or acceptance and in this they resembled *Homo sapiens*. But they did not represent abstract symbols or paint symbolic faces so it would not appear that they had a strong tendency to associate abstractions with group membership (an indirect sign of recursive language capacity). It also seems clear that they were less imaginative than *Homo sapiens* since they showed little innovation in their rock tools for extended periods.

Homo sapiens are estimated to have emerged about 200,000 ybp. This aligns closely with estimates of the development of their physiology capacity for modern speech, which is also estimated to have developed about 300,000 ybp. Curiously, the FoxP2 gene, which is thought to be important for language is estimated to date from ~100,000 ybp. Full recursive human language appears to have been in use by *Homo sapiens* since about 50,000 ypb. Congruent with language development, we see that the emergence of humans of art (symbolic representation) and burying of their dead (belief attributes) becomes prominent. Thus, with the emergence of modern humans, we also see a sudden proliferation of various symbolic items, personal ornaments, abstract figurines and cave drawings. As mentioned above both musical instruments and art can be dated from 30,000 ybp and since both music and art convey and amplify abstract emotion content, such cultural forms can also apply to the formation and stability of social identity. This is consistent with the evolution of symbolism as an element of group identity. In terms of early forms of human language, it is interesting that 'click-speaking' as used in South Africa (Damin language) is thought to be ancestral to most other languages. In its current state, this language is used mainly during manhood initiation ceremonies. In this regard, it clearly appears to be used for tribal identity development.

A History of Early Group Conflict

Thus it is proposed that abstract and symbolic elements became used for social membership purposes. Symbolic stories also were adapted for use in group identity. About 30–40 kybp, there was a tremendous expansion in symbolic behavior of humans. In addition to the artifacts mentioned above, body painting became prominent, which can be considered as a visual symbolic group identifier. It is also at this time that not only bone flutes, but also spear throwers were introduced. The use of spears and throwers is generally associated with the development of more efficient hunting. However, I suggest that this technology would more likely initially have been invented for the purposes of human–human conflict between groups. In this regard, humans would be similar to chimpanzees that use group hunting for both prey and to attack other chimpanzees that are not group members. I suggest humans undertook a similar dual use for the development of spear technology, but that inter-group violence or group conflict was the initial selection for this technology. Chimpanzees did not develop weapons for group conflict purposes. But more imaginative humans did. Evidence that weapons technology was generally used for group conflict can be seen with the 'Iceman' who was found preserved after 5,000 years in a melting glacier in the Alps. It is now clear that his death was due to the action of other humans. The Iceman had a bow and arrow technology, generally considered as primary hunting weapons. But he died by bleeding to death from an arrow in the back. In addition, he had knife wounds on his hand indicating defensive wounds from attack by another human. It therefore seems clear that he was the victim of group conflict, much like a lone male chimpanzee might be a victim of troupe aggression. It also seems the Iceman had tattoos. Tattoos are symbolic emblems that could have also been used for group membership purposes, much as they are often used today in urban gangs. In humans, language, symbolic identity and beliefs, not just troupe association, became important for group identity. With the expansion of human social identity systems, we would also expect an expansion of conflict between otherwise similar human populations as we expect non-identity to likely be met with group hostility. Group identity remains a fundamental force in human social structures that drives innovation. Just like modern humans invented first fire arms then later the atom bomb for the purposes of group hostility in wars, I suggest that our ancient ancestors similarly invented spear throwers.

The Role of a Paleolithic Mind in Social Identity

A Social Mind Originating After Language

Humans of all cultures, including hunter gatherers, are story tellers in which cultural traditions and meaning are maintained by language. Clearly, social cohesion was promoted by language, allowing the transmission of not only information, but of oral traditions that are socially binding. And it is through such oral traditions that larger social identity can also be promoted. With language, the brain and the mind are the substrates for colonization (learned) by information-based identity. Thus with the emergence of the human mind we see the emergence of another entity available for social bonding; the mind. As our large social brain now supports an internal dynamic state we call the mind, this can also be selected for social membership. A mind can be considered to be the collective state of consciousness, memory and sensory input and that involves a sense of self. Evidence was presented above that mental states can also be socially linked between individuals (mirror neurons, emotional contagion, empathy systems). Thus our minds appear to have a considerable social interacting capacity as an inherent character. Given the strong biological basis of this, it is most likely that all these characteristics were also present in the minds of Paleolithic humans and contributed to their social structures. Early human cultures had memories, dreams and imagination that were available for social purposes. They held beliefs which must also serve social functions and like modern humans were also likely prone to hallucinations and religious experiences. As demonstrated above with Clive Wearing, a conscious mind needs sources of stable memory. But memory is the product of and can also be altered by learning. Thus, for use in social identity as described above, a mind would need to have both open and closed learning states in order to set its group identity but resist subsequent identities. Both language and belief acquisition have this open and closed character. A social mind would also necessarily require some form of social control over an individual (via addiction modules). If we consider the developing mind of an infant, it seems clear that the mother can initially exert considerable control over the will of the early infant. The urgent command, fearful voice and emotional facial expression of a mother can in some cases override the volition (conscious will) of her infant. Thus, a mother can to some degree command the mind of an infant as an extension of the mother's mind. However, with more development, the infant will become a toddler, able to (possibly needing to) resist the will of the mother as a sense of self develops. The mind can become closed to external agency. However, we can also see other circumstances in which human minds will remain accessible to external agency. The susceptibility of a mind to external (or alternate internal) directive or command appears to be an inherent feature in the developing mind. As presented below, the ability of most minds to undergo hypnosis appears related to the ability of vocal commands to engage in regulatory mechanisms that have been proposed to be integrated into the attachment or imprinting systems in individual behavior. Thus the volition control of individual behavior appears to have some inherent capacity to be overridden, most likely for social control of an individual mind (see also schizophrenia). In this case, sense of self (consciousness) is not always controlling behavior. How does the conscious mind participate in such unconscious social control? What is the relationship between sense of self (individual consciousness) and a social (possibly subconscious) mind?

Active Frontal Cortex Is Needed for a Conscious Mind and Sense of Self

The frontal cortex represents a newly expanded brain structure of the large social human brain. It is thought to be essential for the state we call consciousness. This state can be clearly recognized by its absence by neuroscientist, and involves being able to respond to sensory stimuli. Thus it appears consciousness requires a sensory stream or similar (memory) communication to be engaged or maintained. The cerebral cortex also contributes to consciousness, whereas cerebellum does not. Consciousness is the primary aspect of our lives. John Locke proposed that for the concept of a person to operate, one requires a mind with the capacity for conscious experience and its permanent loss is equivalent to the death of a person. The representation of self can be considered as a conscious construction of the brain. Although we are born with inherent tendency for maternal social links, it does not appear we are born with a sense of self. It develops along with greater consciousness during childhood. During such development, it might recapitulate the evolution of the mind and its association with recursive language. It is well established that the 'understanding other minds' also develops during childhood. At the age of 3–4 years children start to use whole sentences and represent mental states, thoughts and beliefs of others (theory of mind). Consciousness, memory and sleep are all linked. During slow wave sleep, for example, consciousness is much reduced but neural activity is as high as or higher than waking state. Thus consciousness seems to be a dynamic state that integrates a stream of information which is normally a non-static, temporal sensory stream (such as visual). Recent results with coma patients in minimally conscious states have added to our understanding of this state. Eye opening and visual sensation can be used to monitor consciousness. The thalamus (involved in motor control and sensory relay) sits between brain stem and the cerebral hemispheres and is considered a gateway needed to activate cortical networks. This gateway also allows verbalization. In one clinical study, deep electrical stimulation of the thalamus restored consciousness to a patient that had been minimally conscious for years. Thus it appears that consciousness requires dynamic consultation of information normally through the senses and memory. In the case of Clive Wearing mentioned above, it was clear that essentially no short-term memories were needed. However, alternative states of consciousness, such as during hypnosis (described below), appear to also exist, although such states may alter volition and relate to social control. Other studies also support the view that our thoughts (mind) are literally linked to our vision. Small eye movements known as 'microsaccades' can give away mental attention (thoughts) even when gaze is consciously directed differently. Such a linkage would provide a conduit needed for a social mind, and a skilled observer could literally see what others might be thinking. The ability to inhibit such a natural tendency (socially broadcast thoughts by eye and facial movements) may well underlie the skill needed for maintaining a poker face. Conversely, major politicians appear to be especially skilled at reading the gaze and facial expression of social groups. Indeed, it seems humans do have social minds. Yet a mind is an abstract dynamic state, not a thing that can be physically and statically defined. Understanding a mind required the abstract capacity of recursive language.

Another very recent study further informs us that consciousness is a dynamic construction of the brain that may have no firm biological residence. Sensory input seems generally important for the state of consciousness. However, such input can be manipulated to alter conscious perception. Recently, scientists in England and Switzerland used goggles that project video images and also used physical contact to communicate touch, creating sensory illusions in the subjects. The subjects were viewing the backs of their real body, while receiving unseen physical touch. This sensory illusion induced an 'out-of-body' experience in that subjects reported that their consciousness had drifted from their real bodies into the virtual ones provided by the video goggles. This suggest that this state of displaced consciousness was promoted by viewing of self (possibly activating the mirror sensory neurons) that promoted a mental association linking the virtual and real bodies. In this case, the 'other' vision was self and the self-perspective was a virtual self. This study supports the view that consciousness is a transient, dynamic construct of the brain created by multiple sensory sources and memory. I suggest such apparent transfers of conscious states were also made possible by the biologically based social human brain. Accordingly, it had been previously known that damage to specific brain regions could also induce such out-of-body conscious states. What then constitutes a conscious self if we can move it to virtual bodies? Our modern understanding of consciousness involves a sense of self which also involves a sense of agency. We strongly believe in human intentionality, as a culture. But the sense of agency, like self, also seems to be dynamic mental states that can be externally manipulated.

Hypnosis and Alternative States of a Social Mind

Hypnosis is a state (trance) in which one's mind becomes subjected to external agency. This is an altered mental state (consciousness) during which one becomes susceptible to the vocal suggestions of others. Since consciousness itself requires a dynamic sensory stream, it is most interesting that hypnosis can induce visual, audio and other sensory perceptions that are not actually happening (hallucinations). Hypnosis can also block or inhibit actual sensory streams, such as thermal pain or vision. Such hypnotic analgesia appears to be mediated by mechanisms involving release of opioid peptides in CNS, as it can be reversed by naloxone. These analgesic states involve activation of specific brain structures and deactivation of others. Thus it appears hypnosis may prevent nociceptive inputs from reaching higher cortical regions that perceive pain. One can also hypnotically induce pain, not just block it. The existence of hypnosis thus also identifies the existence of alternative, internally derived (dream-like) mental sensory streams, presumably derived from memory or imagination. The state of hypnosis is thus the product of external language instruction that leads to an altered state of consciousness involving a temporal flow of virtual internal senses. It appears that the external oral suggestions have displaced the inner voice of consciousness which also indicates the capacity of language (a social medium) to influence consciousness and control agency. Inducing a state of hypnosis is associated with mental relaxation and mental adsorption. It appears that hypnosis provides obstructive hallucinations that allow a hypnotic focus inward (e.g., a back flow sensory stream) as opposed to external sensory stimuli. During hypnosis, sensory cortical sources show decreased arousal. Much of the population is susceptible to various degrees of hypnosis. In men, 80% of subjects appear to be able to enter the first stage of hypnosis, and about 25% can enter second and third stages. Much of this susceptibility is heritable. Curiously, it appears individuals with strong belief-based reasoning are less susceptible to hypnosis. In contrast, individuals that do not hold such strong belief-based reasoning seem more susceptible, so some link to belief status seems present.

During hypnosis, a state of mental adsorption is reached, in which one shows a diminished tendency to judge or exert an independent will. One's own response seems to become automatic with a loss of agency. Thus internal volition becomes subjugated to external vocal stream. This sense of will may have a specific brain substrate. For example, some patients with lesions in anterior cingulate report developing an 'empty mind' state with no will to reply verbally. Seemingly, the normal internal voice has lost initiation control. Such patients show EEG patterns (theta activity in the cingulate cortex) that resemble people under hypnosis. Both hypnotic subjects and these patients lack desire to initiate activity, although in both the level of awareness is good.

Hypnosis and Sleep

A loss of volition and control also occurs during sleep. In addition, dreams involve internally generated sensory streams from memory thus resembling hypnosis in this as well. However, brain states during REM sleep and hypnosis clearly differ. The EEC patterns of hypnosis resemble those of an awake state, not REM. Therefore hypnosis does not seem to use a dream-based visual stream, and unlike dreams, it retains a open sensory stream. However, it is interesting that during REM sleep, this frontal cortex goes 'offline' allowing a dream stream to communicate. Clearly, dreams, hypnosis, memory and external and internal (memory) generated sensory streams, and memory (visual) shows some similarities. Hypnotically induced visions can also resemble other forms of hallucinations. Both types of visions require an unconscious synthesis. Thus hypnotic suggestion can override conscious will, and block external sensory streams and allow internal mechanisms for streaming sensory information. Although similar to dreaming or sleep walking, this in not a REM state.

Why should a capacity to be hypnotized by language exist in our large social brain? Was this the product of some evolutionary pressure regarding needs for social behavior or group membership? Did such control initially evolve for the purpose of maternal attachment and/or control of infant behavior? It has been proposed that indeed hypnosis is related to maternal attachment mechanisms in that hypnotic suggestions engage self-regulatory mechanisms that were integrated into attachment and imprinting instincts. In order for a mother's voice or her language to bond to or control her infant, volition control by the infant must be able to be interrupted by the mother's voice. Hypnosis suspends a self-regulatory mechanism (most likely involving the amygdala). This appears to be a direct product of evolution of attachment via voice, requiring control, and resembles visual attachment in birds. However, unlike avian visual imprinting, hypnosis requires learning the emotional content of the mother's voice for the infant to accept directives. For the mother's voice to communicate action, it must compel the emotions of the infant, thus her voice must have emotional (readily learned) meaning. Thus, it seems unlikely that such social attachment could be attained by a language that lacks emotional content. Clearly, explicit commands cannot be orally communicated to an infant, before meaning is learned. Only emotional tone might be communicated early in infant development. Human infants (especially their first year) do instinctively respond to parental communications (which requires a self-regulatory reaction). During this period, infants do not show emotional hostility or negativism, needed for social identity. Recursive language acquisition and theory of mind seem needed for this to develop.

Social Mind and Sense of Agency

Above, it was asserted that consciousness itself (the mind) can participate in human social identity structures. A social mind participation requires that social information (visual, facial inputs, spoken language) must also be able to exert some emotional control over actions. If so, social emotions must bind the individual minds, similar to the affect of a mother's voice on her infants. Thus the sense of self and agency (consciousness) must be open to some degree of social control. Accordingly, I would also expect such social bonding would need to use addiction strategies as do all other group identities. Clearly there exist both normal and diseased mental states in which the sense of self and agency are disrupted. The sense of self, or conscious self-awareness, develops later in infant mental development, but clearly needs stable memories. And hypnosis informs us that there clearly exists a capacity for alternative internal (virtual/vocal) sensory streams (an altered consciousness) that can control both sense of self and sense of agency. These observations together suggest that the mind has the characteristics needed for participation in social group identity. However, as a culture we hold strong belief in full human intentionality, which dismisses any significant effect of group or social consciousness. Human actions are considered as the results of individual consciousness, not under social control. It now seems clear that things are not so straightforward as we might wish to believe. Overall, human behavior must have a strong social component and instead resembles an assemblage of individual consciousness and social–emotional directives via suggestions, habits and urges from others we are bonded to in various degrees. Our actions only partially depend on volition, although education can alter this independence to some degree. The social circumstances during the mental developmental of a child can affect how 'social' their minds become, but so can genetic variables (see vole bonding genetic variation). A social mind thus resembles an imprinting-like situation that develops early. However, a social mind is not necessarily fully maintained during education. The continued development of 'self' and the resistance of self to social beliefs appears able to override strong social bonds.

Hypnosis is not the only process that can induce visual and audio (voice) sensory hallucinations. Psychoactive drugs, brain damage, mental disease and religious experiences can all induce various forms of sensory hallucinations. Interesting that many of these situations are also associated with a loss of a sense of agency or conscious control. Somehow, there seems to be a common relationship between a social mind, hallucinations and a sense of internal or external agency. Perhaps all these are epiphenomena of a recursive language-based social mind that must be open to external (social) agency. There are other circumstances in which self-generated actions can be attributed to external agents. Delusions of alien (external) control is one example that is also associated with schizophrenia (described below). Such states are associated with parietal cortex activation. Also common in schizophrenia are vocal hallucinations attributed to commands from God (perhaps the most potent and abstract of external social agents). Schizophrenia can clearly involve a loss of sense of agency, and induce alternative internally generated sensory streams (private internal voices) that can command free will. It is thus interesting that some schizophrenic patients make excellent hypnotic subjects and are able to hypnotically induce various psychiatric symptoms. It thus seems likely that schizophrenia may represent dysfunctional aspects of social mind.

Schizophrenia, Command Voice and Social Mind

The most diagnostic symptom of schizophrenia is hearing internal voices, that are often command voices (i.e., from God). It seems clear that language, meaning and memory must somehow be essential to this state and that language as a media of external social control has become a media for alternative internal control in schizophrenia. There are other links between schizophrenia and language. For example, some researchers have proposed a reversed language dominance in schizophrenia (with left hemisphere dysfunction). fMRI studies of normal people indicate a right-lateralized temporal lobe activation upon hearing human voices, but this seems impaired in schizophrenics. In addition, schizophrenics show verbal memory dysfunction as a most consistent cognitive problem. Thus the capacity to hear internal (command) voices appears to be a biological legacy from the development of a social brain (mind). Interesting, as proposed in 1976 by Julian Janes, early leaders of human social groups often heard voices that were then used to command social followers. Such capacity to hear internal command voices is still relevant to religious and other tribal social experience. That such internal dialogue may provide a source of new knowledge for many ancient cultures also seems common (discussed below). Clearly, the evolution of a brain able to understand the complex patterns of a recursive language is relevant to this disease.

Voice Memory in Schizophrenics

Humans are very good at recognizing voices, songs and emotional content with the minimal sets of sounds. Therefore our brains have potent audio pattern recognition that can work from sparse information. This suggests that an equally potent and sparse (i.e., abstract) memory systems must exist to recognize similarity. Memory and recognition must fundamentally involve a reversal of sensory stream to the cortex. As the main symptom is hearing voices (usually the same voice), schizophrenia clearly involves voice and language memory systems. Since these voices can be commanding, it also has emotional potential for social control. These verbal hallucinatory aspects of schizophrenia can often involve the reproduction (memory) of the same verbal content. In this memory, it seems auditory cortex (Herschl's gyrus) shows aberrant activity, which is mostly associated with emotional stress. However, hearing the same words expressed with similar emotional valence does not activate this region in schizophrenic patients in that such external sensory input did activate the orbitofrontal and medial prefrontal cortex, associated with cognitive control of emotional processing. Thus an external stimulus of similar content is not equivalent to the hallucinations from internal origin.

Aberrant Schizophrenic Social Biology

A schizophrenic state may thus be an aberrant version of normal social mind mediated by language. Since related mental states can be biologically manipulated and induced in normal subjects, this suggests that these voice and commands represent some inherent capacity of our social brain. For example, PCP (phencyclidine) and ketamine can induce both the positive and negative symptoms of schizophrenia in most normal people at adequate doses. During such induction, hyperactivity in cortex is observed. Interestingly, naloxone will inhibit unusual thought content of schizophrenics, but will not improve mood or other symptoms. Another reason to suspect some linkage to normal brain social biology is the curious linkage this disease has to the development of young adults and adolescents. Adolescent onset is a characteristic of schizophrenia. In addition, adolescent males are more affected in most populations. Since this is a period of considerable mental plasticity that is also associated with the last phase of cognitive development and the acquisition of male-like group (social) behaviors, and identity, including voice deepening, a derailing in the development of a social mind seems possible at this period. The deepening of voice is interesting from an evolutionary perspective. Recent estimations from vocal bone structures of Neanderthals suggest that Neanderthals had high-pitched, child-like voice. However, we are unable to answer the question regarding the occurrence of schizophrenia in Neanderthals, but with evidence of their reduced imagination, music and abstract art, it suggests schizophrenia may not have been a problem for Neanderthals.

Schizophrenia has often been proposed to be the byproduct of the evolution of a social brain and language in humans. About 80% of schizophrenia is estimated to be heritable (via twin studies). However, schizophrenia is not fully determined by genetics and also shows some epigenetic component. For example, dizygotic twin studies have shown that in some cases, one twin will develop schizophrenia, but not the other. Such a situation has been used to try to isolate the underlying biological differences between them. Schizophrenia involves frontotemporal and frontoparietal circuits. In one study, RNA expressed in the affected areas was subjected to an analysis which is used to isolate over-represented transcripts in affected brain regions (aka RDA). Interestingly, this resulted in the isolation of transcripts from an endogenous retrovirus, the SZRV-1 sequence. This ERV was similar to MSRV and ERV-9 described in the last chapter. Interestingly, related transcripts are also found in the placenta. As this is an endogenous virus, the mechanistic implications of this observation regarding schizophrenia remain obscure. However, this suggests some ancient but unknown process that was initially associated with maternal bonding (via ERVs) and has been adapted to language-based social bonding. Since ERVs are also under epigenetic control, they might also be relevant to the development of schizophrenia.

It is thus interesting that schizophrenic families tend to generally have intense interest in religion. Since religion and belief appear to be inherent characteristics of Paleolithic social structures, belief seems to represent a major and apparently new component of such early human social bonding. Acquisition of religious belief often involves transcendent states of rapture or euphoria. In this, belief can have an addictive (but cognitive) component of a T/A module needed for a social mind. If so, the related ability of psychoactive drugs to also induce states of rapture may be identifying the biological substrates involved. Language has a high emotional content. It is estimated that current languages are mostly composed of metaphors, words that are emotionally laden. Interesting that schizophrenics will sometimes speak only in metaphors, whereas autistic people (a different disease of a social mind) often fail to understand metaphors. Schizophrenics seem to have a disturbance in the emotional valency of their language usage. Conversely, some autistic savants, such as the Rain Man, although processing amazingly detailed memories, could not use or understand metaphors. In this case the emotional valency of language is diminished or lost entirely. Both these states seem related to brain lateralization in which right hemisphere damage is associated with singing (emotional) deficits and left hemisphere damage is associated with speech deficits. In savants, the left hemisphere is dysfunctional and right hemisphere takes over, thus seems to then provide largely accurate perception with little emotion. As discussed below, it is also interesting that religion, in particular, especially depends on metaphors for rationalization.

Beliefs and Their Learning as Basal to the Social Mind

One implication of the above discussion on schizophrenia and aberrant vocal hallucinations (including voices from God and beliefs) is that the process that sets language-based beliefs is part of the normal system that defines identity in a social mind. A mind can be defined mainly by its cognitive features: thoughts, emotions and memory expressed during periods of consciousness. We have already seen above that primate brains have neurological structures (mirror neurons) that allow vision to link the neurological and emotional reactions of individuals into social networks. Humans have an even more social mind and in humans, language allows the content of a mind to be transmitted between individuals. A 'social mind' needs to link the cognitive content of individual minds, thus thoughts, emotions and memories provide the basis of mind-based social identity. For these elements to be used for group identity, the cognitive content must link to stable toxic emotional components and work together with an unstable protective or beneficial emotional component. The combined T/A set would create a cognitive-based addiction or identity module that maintains the same or compatible cognitive content. A stable thought or memory also defines a belief state. Belief is the acceptance of cognitive information as stable or invariant (we call true), which is resistant to displacement. Its maintenance would be emotionally comforting and its loss would be emotionally painful. This state stems from the maternal–infant bond, we had presented above. We can now understand how cognitive communication (content) can induce the strong emotional pain of bereavement. For example, if a mother believes her infant to be dead, the basal social bond is cognitively broken (via belief), and intense emotional pain will rapidly follow. In this case, we can clearly see the basal role for belief in strong social bonding. Resistance to belief displacement is also a basic and robust feature of group identity (and language). Both beliefs and language are initially established (usually in young) by a process of learning (such as associated learning). This results in a pattern of thought or memory that resists other patterns or displacement by additional sensory information (subsequent learning). There are fMRI data which suggest that the human brain processes belief-based reasoning distinct from the regions that process information that does not have belief status. As shown in Fig. 9.7, clinical circumstances that examine belief neutral and belief laden-based reasoning show distinct patterns of brain activity. If such belief attribution originated as part of a social identity system, then the belief attribution system would also be part of a social mind. A social mind requires that learning be restricted once a social belief becomes set. In the transfer of group identity, there is generally only a developmentally limited period during which transfer is allowed (in this case learning). Thus, as mentioned above, open and closed learning become crucial aspects of identity transfer to social minds. The factual correctness of a belief need not matter, if it is not highly destructive to social survival. And if such social beliefs are factually incorrect, they will still resist factual correction. Human development seems consistent with this scenario. For example, the high tendency to acquire factually incorrect beliefs by children is considered an endearing feature of childhood (Santa Clause, boogy man, etc.). The learning needed for such early belief acquisition has various features. A source of authority (parents) is often involved. Also, coincident events that become generalized (not requiring an understanding) can set beliefs. The uncritical acceptance of such associations thus becomes the basis of superstition, a culturally invariant human feature. In contrast, the formal evaluation of coincident events for factual correctness usually requires formal education. Not all learning is conscious or uses declarative memory so consciousness itself would not have been needed for the early evolution of belief (or associative) learning. For example, associative learning such as habit memory is not conscious. Trial-and-error learning can also have a pre-conscious habit memory characteristic. Thus learning even complex pattern recognition that cannot necessarily be consciously explained can contribute to belief states (gut feelings). This scenario suggests the existence of distinct processes of belief acquisition. One is social, implicit, associative, pre-conscious (unschooled) and emotion laden (aka, gut feelings and religious beliefs). Another is less social (individual), formal (schooled), explicit, deductive and can be seen in some socially hampered individuals (savants). It is thus most interesting that researchers in learning have proposed that dual mechanisms exist for the theory of reasoning. These two mechanisms conform to the above characteristics.

Odd, Illogical Beliefs and Drug-Induced Mystical Experiences

It thus seems likely that a basal human tendency to associate with learning and a strong tendency for belief acquisition are due to our social mind and are involved in various forms of non-critical thinking. For example, jump-to-conclusion thinking (JTC) can be evaluated psychologically by various clinical measurements. JTC generally follows a process of data gathering and then some unconscious reasoning, but does not appear to involve critical probability judgment. It is very interesting that a JTC tendency is also found in people with delusional proneness (via Peters et al. Delusional Inventory; PDI). Thus JTC and delusional proneness appear linked. People with profound religious experience also show some pathology of the delusions via PDI, but in these cases they are not usually associated with other symptoms (such as those of schizophrenia). This suggests that irrational thought in the context of specific delusions is not a sign of abnormal pathology. Instead, delusions seem to be part of continuum that connects normal and pathological mental states. Believers in the paranormal can also be considered as delusional. They too display reasoning abnormalities. People that hold strong beliefs in the paranormal make many more errors in reasoning than skeptical individuals. There appears to be a dissociation between experience and beliefs. Possibly, these individuals have an exaggerated belief stability. Delusions can also be rather specific, not affecting other mental functions. Studies of people that hold alien abduction beliefs indicate that aside from these very unconventional beliefs, they appear relatively normal by most psychological criteria. Interesting that in many such individuals, the experiences associated with acquisition of the alien abduction belief mostly occurred during sleep–wake transitions, when sleep paralysis is still operating and frontal streams are likely internalized. From these results, it seems clear that belief attribution can be partitioned into specific domains of the mind, which allows cognitive dissonance and dissociates experience and logic from belief stability. In such states, individuals fail to adopt principles of scientific thinking and learning appears precluded by beliefs. Curiously, these odd beliefs can often provide emotional comfort. Even alien abduction beliefs appear to provide spiritual meaning to those that hold such beliefs, suggesting the engagement of strong cognitive addiction modules. Along these lines, the spiritual experiences induced by some psychoactive agents also appear to be relevant. Most relevant are psychoactive agents that induce altered conscious states and include transient loss of self-identity. That these agents can also induce intense mystical experience is especially interesting. Psilocybin is known to be able to induce intense mystical-type experiences and such experience will provide sustained personal meaning and spiritual significance to exposed individuals. In one study, 61% of participants reported a complete mystical experience, which was indicated as one of their most significant life experiences. Also, 79% reported they had increased their long-term sense of life satisfaction. However, not all experiences induced by psilocybin were positive. Some participants instead experienced extreme fear and were highly emotionally distressed. Both of these positive and negative reactions are consistent with the existence of emotionally intense addiction modules that relate to belief states and can be induced by either belief, psychoactive agents or mental disease.

Autism and the Social Mind

Human studies demonstrate that witnessing the actions, sensations and emotions of others appears to activate brain areas normally involved in performing the same actions or feeling the same emotions. Thus the social brain has dedicated neural substrate that links individuals. Understanding other minds also seems to be due to specific domains of cognition that are distinct from those of reasoning. We have discussed hypnotic induction, out-of-body experiments, voice of God, alien control as pathologies of the social mind related to a sense of self and volition and/or loss of agency. In addition, most of these states can also be drug induced, consistent with an underlying biological basis. Diseases of the social brain are not limited to schizophrenia. Recently, it has been proposed that autism may be due to problem in mirror neurons that are also involved in empathy. High-functioning children with normal IQ but with autism were reported to show no mirror neuron activity in inferior frontal gyrus (pars opercularis). Mirror neurons link observation of others to emotions and intentionality. Although autistic children can recognize a range of facial emotions, they seem to have specific impairment in recognizing negative emotions, such as fear, in others. Such a state resembles patients that have sustained specific damage in the amygdala. It thus seems clear that our large social brain also host a social mind that normally links us to extended social structures.

Interestingly, it has been suggested by D. Dennet that the neurological system which connects our minds to those of others (via mirrors and theory of mind) also creates an overactive mind-agent detection device that will inherently assign agency (actions of another mind) to unexplained, mystical events, by association. In other words, our theory of mind promotes us to explain events by the intentions of others. As described below, this inherent tendency has also evolved into assigning mental agency to mystical events due to the mind of God. However, as described below, by applying the social mind and connecting individual human minds to an abstract (or non-existent) mind of God, we also see the underlying biological basis that promoted the formation of the most extensive of all group identities, religion. Religion constitutes a strictly cognitive group identity in which only belief states define identity.

Social Mind and Extended Social Groups

With the emergence of the human mind as a participant in social group identity, we can see the congruent emergence of extended social structures mediated by the contents of social minds. Beginning from the maternal–infant bond, to the nuclear family to tribes, much of these social bonding could still be mediated by biological parameters (kin relationship). However, a mind-based social membership now extends group membership well beyond any specific biological constraints and will instead be based on shared learning experiences (memories, language) and beliefs. If such membership can also compel cooperative group behavior (both supportive and aggressive via addiction modules), the combined capacity of the group will far exceed any biologically based group fitness. Below, I trace the development of some of these social structures.

Adolescence and Male Social Groups

It seems clear that groups of adolescent males that have no kin relationship can form strong social bonds during the period of their common development post-puberty. Such social bonding between males is similar to that seen in chimpanzee adolescent males when they begin instruction from older males in group hunting and group defense and aggression. We might suggest that chimpanzee and human males express some XY-mediated phenotype along with puberty that promotes the formation of strong bonds that can even displace the maternal bond that was stable for 10 years (as seen in chimpanzees). However, in human adolescent males we see an additional development in their social brain that is not present in chimpanzees and involves the frontal cortex. The human frontal cortex is not fully developed until about 25 years old. This appears to offer a second round of brain development (post-language) during which group identity can be further developed. This is a period prior to adult maturity where young men are prone to risky impulsive decisions, physical risks or 'fraternity'-like behavior. This period is used by traditional tribal societies to indoctrinate males into hunting behavior and tribal warrior culture. Male group formation seems inherent. For example, spontaneous urban gang formation can occur in unsupervised male groups during this period requiring little external or adult promotion. From these features, it appears adolescent males have an inherent biological tendency to form social bonds and further develop group identity. This provides a window during which many young males can be developed to be aggressive to other social groups (such as in national armies). During this development, no kin relationship is required. Instead, group identity involves a shared learning experience and shared beliefs. Also, strong and stable religious indoctrination can occur during this period. Such stable states of mind, I suggest, underlie all other extended human social structures.

There appears to be some sexual dimorphism between men and women regarding group behavior. This can be seen on a population basis by measuring emotional reaction to unknown (non-group) faces. Men and women can differ with response to recognition of facial emotions. For example, arginine vasopressin (AVP) is known to influence behavior of humans and other animals in a gender-specific way. AVP can be administered by intranasal instillation, then subjects can be shown various faces of the same sex with different emotional expressions. The rapid facial responses of the observer can then be monitored to indicate their subconscious emotional reaction (via brow muscle movement). Such studies show that males react more with either anger or threat to emotionally neutral faces whereas women respond with an increased approachability. It is likely that this male characteristic develops after puberty. Males, however, do pay much attention to female faces and forms, and affects on the right amygdala can be seen in response to highly attractive female faces. Yet male mate choice is not simply the product of a facial emotional response. Evidence for this comes from a large cohort of twin studies involving 738 couples and examining 74 psychological variables. The results regarding mate choice indicate that the choice phenomenon is inherently random. The most common determinant seems often to be that of romantic infatuation and personality. Clearly, human mate selection does not follow any discernable biological marker. And it seems that cognitive features involving personality are important, but highly variable.

Emotions for Extended Social Binding

Extended social structures include gangs, armies, city states, cultures and nations, with religion being the most extended of all social structures. The common fabric that holds such structures together are similar learning experiences involving social pleasure and pain as noted above in adolescent males. Loyalty (adherence to group identity) is a core feature. The various social pleasures involved are amongst the strongest emotions. If we define such social pleasure as mental states of feeling of happiness, satisfaction and exultation, we can see that fame, adulation, praise and pride are also all strong social pleasures that can have addictive character. The rock star, sports figure, Hollywood lifestyle all involve similar pleasures. Such feelings are known to be associated with dopaminergic signaling and it is likely that endogenous morphinogenic mechanisms (endorphins) are involved. Clearly, these appear to involve addiction centers and it appears that various drugs can artificially tap into such states of social pleasure.

Negative emotions are clearly part of social bonding. We have already considered the intense emotions of bereavement expressed by a mother and father upon death of offspring. Similar types of emotional toxicity, involving depression and disassociation, are seen in other broken but more extended social bonds. Males bonded during adolescence, for example (such as warrior groups) can also display intense and similar bereavement at the death of a bonded group member during group conflict. In terms of larger extended social structures, there will typically be an individual leader for such extended groups. Although any individual member of extended group may not actually ever meet or know or even see their leader, here too bereavement can be intense upon the death of the leader. It will be interesting to consider how such distant but intense social bonding can occur onto individual minds (discussed below).

The Invention of Writing: Emergence of a Modern Individual Mind from an Ancient Social Mind

In the Paleolithic mind we see evidence for a large social brain that was promoted by the development of recursive language. This mind was predisposed to use abstraction and symbolism (both images and vocal) as identifiers of subtle (sparse) emotional content of faces and sounds associated with group membership. This social brain allowed the emergence of greatly extended social structures (culture). Such an enhanced brain and a recursive language had an inherent capacity for complex and abstract pattern recognition that has also promoted a large expansion of the general intelligence and the learning potential of humans. However, as presented above, since learning mostly evolved for social functions, social learning also had developmentally determined limitations (seen with language). Mostly oral and ritual traditions that supported group identity were maintained and provided by language. The learning of language was a natural competency for this large social brain (especially due to the left hemisphere). This native learning of language, however, was not highly demanding of cultural resources to maintain. Children, simply being raised by speaking parents, will become language proficient. This inherent symbolic capacity of our social brain, however, also promoted the development of the visual but symbolic recognition of words. Symbolic (abstract, artistic) representations of things and words takes advantage of and uses the two main sensory systems that were evolved for social (emotional) recognition: faces (visual) and voices (language). Consistent with this, the human brain has a patch of cortex next to fusiform area dedicated to face recognition that also responds very selectively to visually presented words. Clearly, such recognition capacity was present prior to the evolution of written language and words, so it seems likely to have existed for abstract pattern recognition, but also promoted the development of symbolic visual pattern recognition, pictographic words.

There is much evidence that early modern humans used abstract representation in ways distinct from Neanderthals. For example, in the Blombos cave of S. Africa, which is considered as the earliest to house modern humans, engraved geometric plaques on the cave walls were most likely for symbolic cultural purposes. Some arithmetic capacity and abstraction was also likely present in early culture. Early modern humans would likely have some degree of 'folk sense' of knowledge. Folk sense can be defined as knowing something in the absence of education or other cultural experience and can be found in great apes. Although humans seem least endowed with instinctual knowledge, relative to most mammals, humans, like great apes mentioned above, have an inherent sense of numbers. Numbers are important for chimpanzees' social behavior, such as knowing when to attack or when to be quiet regarding other chimpanzee groups and chimps will engage in lethal fighting when numbers are appropriate. (i.e., three or more males encountering a lone male from another troupe). Chimps also have some capacity to learn Arabic numerals, but unlike humans fail to generalize (abstract) number concepts. Untutored human children can represent small numbers but they can also represent comparative ratios between numbers. In contrast to chimps, children will quickly generalize numeric concepts and are able to represent a large, infinite list of numbers, even though not all human cultures can express large numbers precisely. This human ability with abstraction was likely present in Paleolithic cultures and may have evolved from recursive language that requires the ability to deal with infinite recursive word combinations. Thus humans seem inherently able to 'symbolize' and quantify large numbers precisely and to use symbolic representation of this.

Historic reconstruction suggests that the earliest forms of writing were symbolic in a pictographic sense, symbolizing things directly, not symbolizing phonetic content (i.e., ideographic writing of 3–5,000 ybp, Mesopotamia, Egypt). Logographic writing (cuneiforms, hieroglyphics) could be found in India and China also at around this time. It seems clear that much of this early writing was for the purposes of some form of folk math or for accounting purposes, such as for the fair trading of agricultural products and animals or taxation. Thus initially writing was mostly for arithmetic accounting but it slowly evolved to become more symbolic with additional images and definitions. With the Phoenician language (3,100–3,500 ybp), and Greek language we have an early example of languages that became fully symbolic, in a phonetic sense, representing the sounds of words and thoughts. This alphabet was also able to symbolize many emotions. However, such a visually symbolic but phonetic language poses some significant learning hurdles for our human brains. Symbolic visual patterns (letters and words) are mostly visualized via systems in the right hemisphere of our brain that appear adjacent to or related to regions for visual face recognition. However, phonetic learning and recognition of language is mostly a left hemisphere function. In order to learn a written language, our brain must link the capacity to recognize visually symbolic information to our capacity to recognize and express recursive language. Thus, such capacity requires the integration of the symbolic pattern recognition capacity of the full visual right and language left brain. This is not a type of learning that is innate or biologically promoted by the human brain. In sharp contrast to learning a spoken language, learning to read is inherently difficult, time-consuming process requiring extended effort, with much repetition and much instruction by adults. Starting just after age 5 learning to read extends into pre-puberty to master. During this extended period of learning, clear and major alterations have occurred in architecture of the human brain. Thus reading forces the combined use of our two major pattern recognition capacities inherent in our large social brain but this required a restructuring of nerve connections.

What about emotions and reading? Above I have stressed the central role emotions play in facial and vocal pattern recognition regarding our social brain. It is known that the emotional content of faces, for example, is recognized in distinct brain regions from emotional content of speech. The emotional content of written language must somehow reach this speech–emotion reaction from a symbolic visual input. For example, an emotional narrative activates amygdala (site of some negative emotions) as well as temporal pole (belief attribution). Thus, a significant reorganization of the brain (cortex) has occurred regarding pathways to emotional memory. This cross-connection I would argue has also resulted in a much enhanced capacity for abstract deep thinking and cognition in general. Highly abstract concepts are otherwise essentially impossible to learn without the type of immersed and extended learning that reading has required. In terms of evolution, since reading has a clear requirement for symbolic visual recognition, an inherent capacity in modern humans, as well as a period of extended brain development after age 4, it seems most unlikely that the Neanderthal brain with its limited symbolic recognition and limited development after age 4 would have struggled to learn to read a recursive language. The difficulty and duration of learning to read, however, placed major restrictions on human group functions and social structures. Teaching of and learning to read were much more demanding of cultural resources than other forms of instruction and remain so today. The product of this extended instruction was a much enhanced general mental capacity. The mind that emerged was equipped with the foundations that led to our modern mind and consciousness. This modern mind, with its large symbolic vocabulary, is much more capable of introspection and internal voices of consciousness. It is also more capable of learning independently of social identity. In this, reading promoted the development of a critical thinking by an individual mind as described below. However, this emergence was not without some significant and continued resistance from the prevailing social minds, basal to human cultures.

Writing: the Importance of Stable Ideas for the Social Mind

Clearly, extended human cultures were developed prior to the development of a written language, let alone any phonetic written language. In such cultures, we see many of the same social structures, characteristics and hierarchies. These include the prevalence of a leading class (chiefs, kings, queens, emperors), the presence of a religious leadership (priests, shaman, often the kings and emperors also become recognized as deity) and the frequent presence of a warrior class, all would not appear to have been dependent on writing. In these early societies, however, the cultural acquisition of new information (overall learning) came from various uncoordinated sources. This included the use of internal voices and visions, especially from leaders. Visions and voices remained an accepted and significant source of new information up until the time of the Greeks and their oracles. In general, early societies had no formal or schooled system, such as science, dedicated to a critical process of learning or objectively evaluating new things. Most new knowledge was acquired by a trial and error, associative and passive observation-based process. All such early cultures also held mystical beliefs that were mostly specific to their culture as well as an array of ritual behavior, often involving chants and dance (consistent with a social mind). Strong emotional links to leaders were also typical. I have made the assertion above that the forces that create these extended social bonds and structures are similar to and evolved from the basal maternal and paternal bonds. Social bonds now form with non-parental cultural leaders. But we might extrapolate how this evolved by considering the development of male bonding in post-adolescent great apes. Up until 10 years of age, male chimpanzees are tightly bonded to their mother. At puberty, this bond loses its tight grip and young males become associated with other males then form tight and stable social bonds with them. Similar transitions occur in humans. How do these post-maternal social bonds get established? In humans, the mind becomes a social element and belief is involved. The development of high social cognitive capacity in humans led to a mind that itself became available as a substrate for the formation of social bonds. Here, however, it is ideas and their emotions that link the group members and leaders to each other. Biology (kinship) is much less needed and not a necessary criterion. Humans thus have the capacity to become bonded to the idea of something, such as the idea of a king, belief or religion, for example. Such bonding can be very strong, even able to displace the strong and basal maternal bond in some cases. For example, a young adult male joining a warrior group will become much less dependent on his prior maternal bond. In this regard, the invention of writing was a powerful stimulus for the use of ideas (information) in forming much extended social structures. Writing stabilized ideas and their relevant metaphors (emotional rationalizations). Writing both expanded and promoted the use of ideas of leaders, religion and culture to create vast group identities. Indeed religious books have always been and remain the most numerous books ever produced. However, in contrast to kin-based parental or familial bonds, these more extended social bonds do not require any close visual, vocal or physical contact. These are established and maintained by ideas, cognitive mechanisms that operate on the mind via cognitive and emotional addiction modules. They are learned at crucial periods of development. Yet, to be able to maintain a social identity. These learned ideas must be stable and must resist alternative or subsequent sources of information. A belief state thus required for the stability of learned ideas. Such states are fundamental to being able to use cognitive information and the mind to create group identity. In this case, group membership becomes equivalent to believing the information, and that information can be written. Here we can clearly see strong promotional effect that the development of writing had on the evolution of idea-based group identity. Writing served as a powerful stabilizing and amplifying media better able to retain and transmit cognitive information for the establishment of even greater and more extended social structures: cultures, nations and religions.

Social beliefs are most responsible for defining current human social group membership (religion, nationality). Belief in cultural, political and religious identity and their leaders is core to such group membership. These are basal aspects of a social mind and belief stability (faith) is inherent in such states. And although social beliefs clearly prevailed prior to the development of writing, these were much smaller and less stable social structures. They depended principally on oral traditions and were limited by language. Nor were oral traditions as elaborately rationalized or communicated and they lacked stable written directives. Thus the invention of writing had a profound effect on the human mind, promoting the development of a much more extended belief states on a larger social scale.

However, there was a major unintended consequence that was to quickly emerge following the invention of more abstract phonetic writing. Learning to read indeed promoted belief-based social cohesion. But it also provided the vocabulary, internal voice, enhanced mental capacity and facility to develop and understand abstract concepts. This was to promote the emergence of an introspective mind, one that developed a sense of self, distinct from a social mind. This was to eventually lead to the emergence of the modern mind. We can define a modern mind with these very features; one capable of deep abstract introspection with a clear sense of self and a capacity for objective analysis not bounded by social beliefs. It is apparent that the Paleolithic mind lacked these characteristics. The difficulty of learning to read required the development of structured learning as a cultural resource. But the minds that were thus produced became able to examine the observeable consistency of their own, unquestioned social beliefs (described below).

Initially, learning to read was relatively restricted experience, used mostly for the ruling and religious classes since these individuals would be most responsible with the maintenance of cultural identity and the transmission of beliefs. Reading, however, likely transformed the brain architecture of these leaders. Recent studies suggest that the acquisition of literacy appears to affect various other cognitive skills, especially the ability to visualize two- and three-dimensional representations, including representations in pictures. Interestingly, when some members of isolated and illiterate tribal groups were first shown photographs of people, they had great difficulty in seeing what the photo represented. Curiously, chimpanzees that have been taught some limited symbolic reading are also much better at recognizing pictures. Learning to read has measurable consequences to brain organization as established by fMRI studies which show alterations in auditory–verbal language system of readers. It thus seems that the functional architecture of the brain is modified by literacy. Yet the brain does not need to be able to learn to read in order to establish high general intelligence. Humans with otherwise normal and sometimes high intelligence and schooling can have great difficulty in learning to read, for example due to dyslexia. This is a reading disability due to a left temporal region problem involved in converting written to phonological units (merging left and right brain functions). Dyslexia, although uncommon, is seen with all languages in all human populations. However, it is not likely that difficulty in learning to read could have been a problem for Paleolithic human cultures, thus the reason our brain mostly has a capacity to learn to read but can be biologically inhibited is not clear since this skill seems not to affect social or general intelligence.

Thus we can expect that writing also promoted the development of the cognitive capacity, especially in the ruling and religious classes. It aided abstract and deep thinking by this literate group that likely led to more elaborate and emotional rationalizations to defend and promote group identity. Writing also promoted the emergence of more ideas and words. One of the ideas that may have emerged from this would include the concept of a monotheistic God, an idea to which group members could form a personal relationship via a strong emotional social bond (i.e., to become one with God). Thus, the capacity of literate leaders to influence and convince their populations was greatly enhanced by the emergence of writing. As language was the main media for both group identity and its defense, writing became more diverse and expansive in its vocabulary for such purposes. Much of this word expansion, however, involved terms that carried an emotional charge (metaphors) and were often used to define both good and bad group membership. Perceptual, descriptive and objective words also increased, but remained a minority of all languages. The intense use of metaphors in all religious writings remains prevalent. This character of language has been culturally maintained by the Humanities, which has promoted and developed the use of rhetoric (emotionally charged terms) in argument. The purpose of this is to promote social dynamics. The emotional content and communication of such words can convince and bind social minds. Such rhetorical practice of language prevails to this day in social situations, especially in political debates and other social discussions. Thus, the ability to defend one's ideas (and group membership) by applying rhetorical methods and rationalizations is an essential political skill needed to sway most social groups. Rationalizing is a basic human tendency, a cognitive and language-based mechanism used for group defense that need not use or depend on objective evidence. It represents an inherent (almost instinctual) defense feature of the ultra-social human mind. Its importance to social identity, however, suggests that it likely has a biological basis. The use of rationalization for belief defense thus relates to the patients noted above with various forms of brain damage that dissociated sensory information (such as observing their paralysis) from their belief state (believing they were not paralyzed). They become biologically compelled to strongly rationalize their clearly false beliefs. Belief becomes a core and biological feature of a social mind and also requires the evolution of the corresponding biological substrate to maintain it. The state we call 'closed mind', as demonstrated by these same brain-damaged individuals above, is crucial for belief to function in group maintenance. Accordingly, 'belief-based' group identity also requires the establishment of a state I now call 'cognitive immunity'. This is a stable state of belief that can resist other beliefs, including believing sensory and objective information. Group identity then becomes defined by belief. However, like other group identities, this form of immunity not only resists displacement, but also needs mechanisms to defend itself. Our inherent instinct to rationalize based on emotionally charged terms is a defensive component of cognitive immunity. Since group identity can also be expressed physically, it can be expected that such cognitive defense reactions will also sometimes engage powerful and ancient emotional systems (such as group anger and denial of empathy) that can provoke lethal group violence against non-members. The emergence of written language thus promoted a much extended belief-based group identity.

The Unanticipated Emergence of the Modern, Critical Mind

In an evolutionary sense, writing was a basal novelty that promoted the emergence of new and greater forms of complexity and group identity, in this case, a mind-based social group identity. However, with the development of written phonetic and abstract language (such as Greek), we see an almost simultaneous emergence of an alternative mind state: what was referred to above as the modern mind. The emergence of this state appears to have been an epiphenomenon of culture and education of a social mind. As asserted by various authors, a modern mind is introspective, has a strong sense of self and is analytical in that it is able to accept sensory information that can displace belief states. Thus, a modern mind, can dissociate itself, to some degree, from the common belief state needed for binding a social mind. In a sense, such a mind partially transcends the ancestral, normal biological state. A modern mind is mostly the product of intense education or training, and unlike the learning of language or other mental abilities, it does not have a specific or dedicated biological substrate in the brain for its development and maintenance. Thus it has an inherently dynamic (fragile) character to it. It is a bit unnatural in that it is developed in an individual, thus can be somewhat asocial. Because of this, the evolution of this mind has been slowed and often opposed by prevailing culture (social minds) due to conflicts in belief states. It can even be internally opposed in individual's mind due to dissonance with their own underlying social mind (i.e., religious beliefs, see below). As noted above, written language, when used for extended group membership required much extended and structured education. But this restructured the brain and also developed deeper cognitive potential, including highly abstract pattern recognition. The emergence of this modern mind promoted the disassociation from the social belief states, based on objective sensory experiences (especially visual). Such sensory experience thus becomes a major source for the acceptance of new information. This state of information acceptance (learning) is similar to but distinct from a social 'belief state' (discussed below). This concept of the modern and social mind, as I have outlined it, resembles in several respects, an early idea put forward in 1976 by Julian Janes, the bicameral mind. The Janes idea of a bicameral mind also addresses the importance of writing for the emergence of a modern mind but this proposal has been mostly ignored or dismissed by neuroscientist. However, Janes did offer explanations for states such as voice hallucinations in early societies and hypnosis, which remain unexplained by neuroscience. Although many features overlap between the idea of Janes and the social mind (the importance of written language, left–right hemisphere integration, importance of internal voices, a commanded early social consciousness, etc.) the significant distinction between the two ideas that I am proposing is that a social mind is the biological predecessor needed for human social identity. In the context of a social mind, inner voices, hypnosis, schizophrenia and religious experience can all be accounted for. Furthermore, although this mind is biologically based and existed in all early human culture, it also required abstractions (symbolism) and stable belief states that are supported and maintained by selected brain systems. Thus all earlier human social group evolution was dominated by voices, visions and emotion as a mediator of social bonding. The capacity for reasoning was often used for the support of group belief. In contrast a modern individual mind is not looking to emotion or to hear internal command voices to attain new information or belief states. In an individual mind, external sensory information becomes the primary source of new information and is manipulated by thought, abstraction and imagination which provide the basis of new learning. The modern mind did not so much displace this more ancient social mind but rather it became able to superimpose itself and sometimes override these innate mental and emotional tendencies. Thus, with writing, we see a parallel development of an expanded social identity along with the emergence of a much more individual-based modern mind. Below, I will return to consider the details of the development of this modern mind. However, it is important to first consider the expansion of human mind-based social structures.

Modern Aspects of the Social Mind, Asocial Beliefs

The establishment of more extended social structures was directly aided by the development of language and its writing in the transmission of beliefs. Beliefs were used in support of extended social structures, such as beliefs in kingdoms and gods, which were clearly prevalent prior to the development of writing. Both a king and a God, as a socially bonded entity, appear to have evolved their bonding systems from the basal paternal and maternal bonds. In many cases there is little that emotionally distinguishes social bonds to a living king from the bonds to the concept of God, thus these two social states clearly overlap (including the power over life and death of individual social members). By including the idea of establishing a social bond to a God, belief-based social membership, however, now provides what may be the most stable of all social bonds: religion. Such resulting social bonds are sometimes so strong they can even displace the most basic and ancient of all mammalian bonds, mother–offspring bond. Social bonds of this nature are also clearly able to subjugate the individual, even induce self-sacrifice of group members, especially during group conflict (a 'god-like' control by leader over individual survival, resembling apoptosis-like death of the individual). We have seen that the death of a leader, even one that is never seen or met, elicits similar forms of grief to the death of a child or parent. It thus seems clear that belief-based group membership (religion) holds tremendous power over an individual and in some cases this power can be absolute. In such cases, no objective evidence may be able to displace such beliefs, and unending rationalization can be presented as a defensive cognitive response. The 'word of God', loyalty to emperor or king, for example, can often elicit an unwavering and rationalized cognitive defense, excusing all failings or foibles of the belief or leader. Such states are reminiscent of rationalizations made by stroke victims and may engage similar mechanisms. Other beliefs, such as persisting occult explanations are similarly defended unwaveringly. In such circumstances, the purpose to reason is clearly not for analytical purposes. In this situation reason is employed in the service of and defense of belief states. Thus, rationalization appears to originally function in a social mode that supports social bonds and counters the perception of cognitive dissonance that would otherwise occur between sensory experience and social bonds.

New Emotional Requirements for Highly Extended Social Structures

We might imagine that if bees had cognitive instead of olfactory-based group recognition, the queen would likely be their god as well and all the drones fully commanded by the will of and belief in the queen. Human social structures have historically approached this level of control by the social leader. If this leader is the 'idea of God', however, all humans become potential group members. Religion does not depend on biological, kin, family, language or national group identity for its membership. Religion uses strictly cognitive identifiers in defining membership. The individual must believe in the religious dictates or deity and submit their individual will to those of the group beliefs. The question is then raised if such extended membership would need to employ emotional addiction strategies beyond those we have so far considered. We do know that in some cases, non-belief can be the basis of out-group identification and violent attack, so clearly the strong emotion of aggression can be engaged. Examples from history of an inherent linkage between belief in social leader (kings, emperors) and beleifs that social leaders are also gods are too numerous to mention but include most Pharos, emperors, kings and queens. Does such a state engage distinct emotional modules? This seems possible. Even murderous leaders that may have had tortured or executed primary family members (including parents) of subjects, such as Stalin, were still revered and bereaved when they died by these same abused subjects. All truly was forgiven (rationalized) in this submission of will to a distant and sometimes cruel leader or cult. And even the most basal maternal bond can be violated by beliefs in emperors and gods (as witnessed by Japanese soldiers killing their own parents in service to the Emperor during World War II in Okinawa). Clearly very strong bonds indeed can be involved. The submission of will inherently demands the merger of and subservience of the individual mind and this would seem to be at odds with a fully independent self-conscious mind. There do indeed seem to be numerous emotions that are group associated and could provide the basis of emotional addiction modules which range from loneliness to being 'one' with a leader or deity. Negative or toxic social emotions include feelings of group hate, fear, loathing, disgust, shame, guilt or embarrassment. Positive social emotions include feelings of group superiority, purity pride, loyalty, joy, adulation and fairness. Euphoria and rapture can also be a potent group emotional experience, especially in religion. Such euphoria is interesting in that it is almost always accompanied by analgesia. Thus, strong emotional group reactions clearly exist. We know, for example, that during ritualistic dancing and choral chanting common to most tribal cultures, the release of endorphins is seen which may lead to global activation of opiate receptors. Other powerful extended social emotions are also clear, such as the adulation and popularity of entertainers and professional sports figures and teams, which must naturally tap into this reserve of group-based emotions. In addition, it seems some drugs can amplify such group emotional states, such as ketamine as used in extended contemporary social dancing.

Although we appear able to identify the existence of powerful social emotions, we do not currently understand the mechanisms of such extended emotions or how they mediated bonding. They seem to involve many of the same neurological and endorphin-based emotional systems as do other social bonds. That drugs can also induce powerful social emotions, such as mystic experiences that provide meaning to life, is consistent with the existence of neurological substrate for such emotional systems. Perhaps this is why these agents are so threatening to extant social structures, as they may undermine many cognitive-dependent (belief) and emotional states of social bonding and hence are often prohibited from all forms of study (such as psilocybin, LSD). Although it seems that social-based emotions evolved from maternal and paternal bonding systems, they have clearly developed beyond those parameters. Thus the development of a social mind appears to have introduced new sets of social emotions that were not likely present in our primate ancestors. However, it is not clear that such social emotions are present in feral humans that have grown up without learning a language or culture. For example, more socially complex emotions such as embarrassment are absent from these individuals and thus appear to require learning. Another strong social emotion can be called purpose. Most human social structures express a need for purpose in which purpose fulfills an emotional, not intellectual need thus it defines an emotional state. Human will often claim they need purpose to justify life. This too, it seems, is a byproduct of culture and education of the social mind as it is also absent from feral children. It seems learning language also provides the capacity for learning more complex social emotions and that purpose is one of these, which is otherwise not relevant to the lives of feral hominids. Nor is this relevant to the lives of our great ape relatives which share many of our less extended social bonds. Such emotions are relevant to our ultra-social mind. However, in having an emotional need for purpose, a social mind expresses dependence on beliefs in leaders and abstract deities which in turn provides the very 'purpose' and belief they seek. This situation has the clear hallmarks of a T/A emotion module that would help create, stabilize and extend group identity system. We now start to see how ultra-social emotional addiction modules can link abstract leaders and deities to individual minds, and why this would be supported by biological substrate that is susceptible to drug manipulation. Clearly, belief in religion is an extension and abstraction of belief in leaders. But religion trumps leaders in one crucial feature. Religion can be even more extended and more emotionally bonding than belief in a mortal king or actual leader. Religion provides social and emotional bonds that extend even beyond the death of loved ones, including dead social leaders. In this feature, only religion provides some emotional comfort (antitoxin) during the trauma that follows the death of a socially bonded person. Thus religion provides a meta-social (or trans-social) group identity, crossing the even boundaries of life and death to maintain an emotional social linkage to those that no longer exist. Religion appeals to and employs our most basic social emotional needs since the death of loved ones and the breaking of such social bonds is the most toxic of all emotional experience. Any social system that can offset this intense emotional toxicity can also potentially provide the most powerful T/A module. Religion and other related mystical beliefs provide such powerful emotional and social identity.

Much has been learned about mechanisms of social bonding from brain damage studies. Some brain researchers, such as Paul MacLean, have long been interested in the evolutionary basis of human social behavior as demonstrated by such brain lesions. Consistent with the above thesis, he and others have come to think that the parenting social bond is a basal bond from which other more extended social behaviors appear to have evolved and that this is reflected in observed evolutionary changes, such as human audio-vocal communication as used to bond parents to offspring. But such physical lesions (brain damage) do not inform us well regarding the evolutionary pressures or molecular mechanisms that created these brain and/or social-mind structures. For example, explaining an ultra-social emotions such as 'purpose' is not clarified by phenotypes that follow brain lesions. But here we can return to the overall virus-first perspective of this book for guidance. In explaining the origin of our large social brain, what has been consistently absent from the thinking of evolutionary biology was to consider the major consequences of genetic parasites to this evolution and how colonization by them have affected (and displaced) prior social identity systems. Human biological evolution has continued to be affected by genetic colonization, but in ways that remain mysterious (such as HERV Ks). For example, GLUS2 is restricted to hominids and is a brain- and testis-specific glutamate dehydrogenase (GDH) important for brain energetics that originated by retroposition to the X chromosome under positive selection for unknown reasons. Clearly, this has the hallmarks of a genetic event that affected hominid identity and deserves to be evaluated from this context. However, human social evolution has continued since the emergence of our social brain, resulting in a social mind that no longer appears to depend on genetic alterations for its development. The emergence of self-awareness and introspection is a recent and fragile state of mind, appears to depend on social education for it to develop, not genetic colonization. Biological or genetic alterations do not determine our expression of self-awareness. The emergence of such a mind seems to be a recent event in human history, thus it seems unlikely that any biological adaptations were involved. Yet the social behavior that has resulted has clearly affected the pattern and the rate at which genetic parasites affect humans. Our culture and technology allow us to survive the onslaught of genetic parasites like no other species in the history of life. Intentional vaccination is the product of culture. Without this culture and this technology, the product of our social minds, diseases like smallpox, influenza and HIV would have very different outcomes and very much limit the growth, diversity and evolution of our human population. Our minds thus provide us with the most potent antiviral immunity. But our minds have become host of different systems of information. Other type of 'brain colonization' events that involve language, ideas, beliefs and other cognitive addiction modules now provide group identity. But the resulting states of social addiction are also states of interdependency. Thus the origin of human social cooperativity stems directly from such states. Empathy involves visualizing and experiencing the emotions of others, likely mediated by emotional responses linked to the mirror neuron system. Diseases of the mind that affect social capacities, such as autism and schizophrenia, can also affect empathic abilities. A strong role of empathy in social bonding appears well established. Empathy and cooperation also seem closely connected and available for members of the same social group. Yet empathy is not necessarily applied to non-group members. It can clearly be withheld. Empathy can be denied to those that have different cognitive contents (beliefs). Cognition, not genetics, now mediates human group membership.

Cooperation, Empathy and Group Membership

The strong cooperative aspects and altruistic tendencies of human social structures have long been a puzzle for classic evolutionary biology. Applying kin selection or other conventional evolutionary models, such as game theory, has never worked well for explaining such extended social behaviors. To field researchers, such theories have often seemed contrived, overly specific or restricted by specialized calculations. They do not provide robust, generalized solutions for cooperative human social behavior. We know, for example, that humans clearly defy rational choice theory by cooperating in simple dilemma games. In this, how they think in groups is distinct from how they think alone or with non-living computers (a social mind set). In large human social structures, such as extended tribes, city states, nations, kinship is highly diluted, yet altruism remains (sometimes called indiscriminate altruism). Thus human cooperation remains an evolutionary puzzle. In this book, I have proposed the concept that addiction strategies can establish group membership, including extended group membership. This can explain and compel cooperative and altruistic behaviors. As mentioned, sympathy and empathy are core emotional elements that bind large social groups. Empathy also provides the foundations for most moral behavior, yet social groups will deny empathy in some conditions. Non-members can be perceived as having few emotional links to a group, in which case both empathy and morality can become compromised relative to them. The capacity for empathy is hard wired into our social brains, via mirror neurons, and is seen in our primate relatives. Thus an inherent moral tendency is an outcome of a social, cooperative brain, and need not be the product of any particular religion, in spite of the strong beliefs on this issue of most religious people. Clearly empathy exists free of religion. And clearly, damage to the human prefrontal cortex can leave intelligence intact, but massively affect moral reasoning. Our brain has a neurological substrate involved in empathic behavior. So when then is empathy is withheld? Under what conditions does one human decide that another human is a member of a positive group worthy of empathy? To what species, family, tribe, culture, nation, religion or belief must they belong? This is a core issue affecting moral behavior.

Semi-rational Decisions and Rationalization

Humans appear to develop beliefs by various processes and have been characterized as quasi-rational decision makers. Although economic theory assumes that humans operate by rational-based decisions, clearly this is often not correct. Group behavior has a significant impact. Instead humans apply heuristic-like processes that seem to underlie decisions and some of these processes can clearly be driven by emotional states in which rationalization can be applied after the decision is made. Others decisions are the result of reiterative adaptations based on experience. However, once a belief is developed, it tends to be defended. If the belief is odd, its defense can clearly defy rational thinking and confabulating of rationalized explanation is experimentally established. Psychological studies, for example, have provided strong evidence that people will often confabulate explanations to justify even false beliefs that were introduced as part of the experiment. In such cases, reasoning is clearly not applied for critical evaluation of evidence but simply to defend the belief. This is a very common mental tendency which can even be found in scientists (see below). Rationality in decision making is thus at best an uneven practice.

Charged Language, the Word 'Believe' and Group Identity

Our large social brain is particularly adapted to use language. And language itself has taken on a strong role in group membership. It is used to define and defend membership. Within language, there is an inherent tendency to specify out-groups in negative ways and in-groups in positive ways. The strong use of metaphor in language, as discussed above, demonstrates this tendency to attach an emotional charge to words. Our language readily allows essentially any group or circumstance to be tagged with a charged term and assigned to a bad (or good) group membership. For example, if one is knowledgeable about something that others do not know, one can be called a smarty-pants, arrogant, know it all. If one does not know what others know, one can be called a naive, simpleminded, know-nothing. Either way, a person with different knowledge on a topic is subjected to being tagged with language as a bad out-group (or a good in-group) member regardless of any objective information. And such rhetorical tags are emotionally effective. Thus, the ancient social tradition that promotes the use of charged terms is still used in teaching rhetorical argument and to develop debating skills. The objective is to influence beliefs and objective evidence is not crucial for such debates. Since beliefs define group identity and all groups defend their beliefs, this promotes some curious perceptions; beliefs are inherently equal. That is all systems of belief (group membership) have attained a perception of equal footing and are thus equally entitled to hold and promote their own views. This itself is a common belief. However, as outlined below, systems of belief are far from equivalent. Objective and reproducible criteria can be required for information before we come to accept (believe) it as correct, as in science. In contrast, accepted information based on faith, authority or other sources (voices, associations) need not adhere to any criteria and it can be simply a historic or cultural identifier with no basis in reality. Yet, these two systems are not in this same realm regarding reality and cannot be equally compared. The tendency to assign group membership to systems of thought is so ingrained, it is even applied to scientific thinking. What is often called 'Western thinking' or 'scientism' by some includes the formal and highly structured system of thought and analysis we call science. But as described below, science has clear criteria by which it comes to accept information and no one source of authority can provide or assure the validity of such information. It must retain consistency with reality (observation, experiment) and logic if it is to be retained. Its stability is not otherwise assured. If it should come to lose this consistency with future observations, it is discarded and does not attain a belief status. What is fundamentally confusing and promotes fallacious arguments is that we use the same terms to describe the mental states of accepting this scientific information as we use to accept any other belief. When we believe something, regardless of how we came to this state, it feels the same to us and involves related if not identical neurological and emotional systems for stable information. The word 'belief' itself is thus problematic because it applies equally to specific and reproducible criteria in science, as well as no criteria in the culture at large. Anyone can believe anything for any reason. These two states should no longer be considered equivalent, otherwise we continue to promote rhetorical debate between these systems as if they are on equal footing (e.g., science of evolution versus intelligent design). I strongly recommend that science abandon the use of word 'believe', just like science once had to abandon the use of the word miasma when germ theory was developed. As presented below, I recommend the word 'convince(ed)' be specifically defined and adapted for scientific use. Science is not a group identifier and should not allow itself to be so viewed. The science of Japan, India, Mexico or the USA can be written in different languages from different cultures but are not distinct, or 'Western'. They all, however, adhere to the same criteria for accepting new information.

The Scientific Mind

Early History

We now come to the last topic of this book, tracing the events that led to the development of formal scientific thinking and how science has slowly illuminated mechanism of human group identity. Scientific thinking must also stem from our social mind and was much influenced by language and the invention of writing. Scientific thinking closely associates with the emergence of the modern individual mind. The biological foundations of a social mind, however, with its dependence on belief states for religious and cultural identity are the necessary back drop for the emergence and development of scientific thinking. Because of this, the evolution of a scientific mind took much more time to attain than many might guess as it was (and is) often hindered by the social mind and group culture. The foundations of modern formal experimental science appear to be mostly monophyletic in human history, originating in Greece. However, the formal and high structure of science writing has only recently attained its current structure (i.e., formal separation of observation from interpretation), mostly being formalized during the last 200 years. Although elements of science are native to most human cultures, and even most children express scientific reasoning tendencies, in only this one place can we trace the direct development of science into the experimental, formal and written structure we now recognize. It was in Greece where we see these foundations. And the dependence on objective observation later separated from interpretation (beliefs) led to modern science. These descriptive and objective foundations can be found in the ancient writings of Aristotle (for example, his description of fish from the island of Lesbos). Written or objective descriptions such as these would not inherently appear to present a conflict with social or religious beliefs. In fact, the naturalism and natural philosophers of the sixteenth and seventeenth centuries, who sought to observe and classify all life, were mostly interested in collecting and writing observations that would support religious beliefs concerning the role of God in origin of the cosmos and life. However, it was eventually through such observations and their classification into patterns that the interpretation of these patterns as theory would often confront religious beliefs. This approach would transform the thinking abilities of humans and lead to the emergence of the critical, but individual mind. This would also eventually result in the emergence of the concept of 'scientist' (or a 'scientific mind') in the nineteenth century.

Observation and the Source of New Knowledge

Aristotle was a student of Plato, but differed from his mentor regarding sources of new knowledge. Plato had proposed the existence of psychic link to 'formopolis' as source of knowledge. This term is difficult to define with current concepts, but appears to be some type of subconscious or emotional knowledge, resembling an ether. It seems likely this represents an innate ability for complex pattern recognition that may not be conscious or the product of cognitive reflection or analysis. Such a view of new knowledge would be congruent with 'inner voice' or 'visions' as also contributing sources of knowledge thus consistent with the proposed Paleolithic mind as described above. Aristotle thought this idea was wrongheaded and instead introduced empirical bases, such as observation, as the starting point for new knowledge (hence the first fish descriptions from Lesbos). Aristotle became focused on the use of observation as a basis to categorize and understand life. Other ancient Greeks had similar views. These early systemic observations and categorizations by Aristotle would later be referred to by Darwin in his own writings. The importance of objective observation was to remain a core feature for the evolution of science and was often referred to by those that contributed to the development of science. For example, around 1,500 Leonardo da Vinci developed the 'conviction' that all science must stem from visual observations, hence his attention to visual records in notebooks. By 1,600, Galileo Galilei had further developed scientific thinking and has become considered by many to represent the first modern mind, judgment based on critical evaluation of experimental observation. He applied the approach urged earlier by Roger Bacon (13th century Franciscan friar), but unlike Bacon, confronted religious beliefs. Galileo is an important practitioner and founder of experimental method in science in which observations are no longer casual, but interventional, focused, formal and aimed at resolving ideas that were imagined to explain observations (theory). He rejected the authority (belief) from the ancient thinkers that did not adhere to experimental evaluation and clarified the application of critical and imaginative thinking from scientific observations. Imagination in linkage to experiment was to prove a lasting and productive process for science. Galileo imagined no friction in his thinking. Later, based on Galileo's observations, Newton would imagine no gravity whereas Einstein would come to imagine traveling at the speed of light. This process was so successful, scientists themselves came to 'believe' that reason and evidence were the core mental processes by which educated people all come to accept new information. However, this is not the inherent process by which our social mind establishes its identity and this 'belief in reasoning' by many scientists remains a problem. Interestingly, most of these scientists, including Charles Darwin, initially held religious beliefs that were often used to motivate and rationalize their initial investigation in which science itself was a manifestation of God's design.

Science Criteria for 'Belief' Status

The discussion above presents the argument that the word 'belief' is inherently problematic when used by science and I have suggested adopting the term 'convinced' to describe tentative acceptance of such objective information (as most or likely to be correct). With this distinct term and criteria, the acceptance of information in science can avoid the confusion of being considered as another 'belief' system or an element of cognitive-based group identity. I have presented various studies and observations that suggest that native thinking habits of most people that lack formal training can often allow a belief status to be attained by illogical or inconsistent reasoning processes. Deductive reasoning as process of drawing conclusions from a given set of premises can be considered as a closed system, from a logical view. Yet, clearly beliefs can offset such rational thinking. And it even appears that our brains inherently compartmentalize these two processes. Formal and deliberate reasoning, for example, is associated with the right brain function whereas religious beliefs, implicit or preconscious and emotional thinking is mostly associated with a left brain. Clearly, formal reasoning can override emotional thinking in some people, but the converse also occurs, as observed in stoke victims or highly religious scientist. One complication in reasoning is that humans can clearly learn to recognize very complex and rapid visual and audio patterns, which remain subconscious and elicit an emotional response. This is the gut-feeling phenomena familiar to most people and used by them to develop beliefs. These patterns can be correct and provide much insight. However, they can also be erroneous and many people accept or believe their 'gut-instincts' without subjecting them to formal or objective evaluation. Formal reasoning is necessary to sort this rapid pattern recognition from what should be permitted to attain a belief status. In addition, formal reasoning is essential to evaluate more abstract or counter-intuitive explanations. This is especially the case with regard to the natural tendency associated with the human reaction to other group members (race, class, sex, religion, etc.) as these tendencies have biological foundations. Group definition reactions require a rational and ethical examination before they should become accepted information. It is through the application of formal and ethical reasoning that we can provide a more coherent moral foundation. Religion has often been proposed (by its proponents) to have provided the foundations of most moral behavior. In my judgment, historical records do not support this assertion and such records are rather clear on this topic. Indeed, secular governments have been much more successful on this issue. Consider the example of the history woman suffrage and the attainment of the equal rights for half our human population. This issue did not benefit from religious dictates and remains a problem in many religious countries. Religious thinking also fails to explain some of the clear biological basis for empathic tendencies, noted above. It has been secular law and its protection of individual freedom from religion that has most promoted and protected basic human rights and overall advanced ethical judgment. The proposed role for religion in such development instead appears to be a rationalization offered in defense of 'belief' or group identity that does not depend on much objective data. As a scientist, I am thus not convinced by the historical evidence that religion has been anywhere near as positive regarding evolution of ethical behavior as its proponents maintain, although it has clearly provided a strong (but sometimes violent) sense of community and purpose.

Scientists' 'Belief' in Reason

The major early success of science was the strong impact it had on the ruling and religious classes when it first emerged in the seventeenth century. This period corresponded to the age of reason and it was predicted by many scholars that the decline of superstition and religious-based explanations would soon follow.

Science seemed to offer explanations for most of life's mysteries. However, this anticipated decline in religion did not happen then and is not happening now. The seventeenth century also saw the introduction of the printing press and the distribution of affordable reading materials to a broad populace, greatly expanding modern literate minds. In the USA, public literacy and education were introduced as a social right, a process that has since become global. Interesting that this major expansion in the number of people that learned to read did not significantly diminish interest in religion. Given that the Bible was far and away the most published book, perhaps this should not be too surprising. With regard to science and beliefs, the seventeenth century in particular saw development of natural history as a way to affirm the existence of God. Most naturalists thus initially approached their studies from a religious perspective. For some, however, scientific thinking clearly led them to accept reason over religious belief. In 1796, Von Humbolt wrote Kosmos, a compendium of scientific thinking that some thought of as anti-Bible for the age of enlightenment. However, such an anti-Bible never became popular and certainly did not displace the Bible or any religion. Such scholarly views at the time were based on the notion that belief was mostly the product of reason and that a strong treatise on reason should displace most belief-based reasoning. In a sense, it seems that scientist started to 'believe' in reason, rather than subjecting this hypothesis to scientific evaluation. As it now seems clear, beliefs are not inherently stemming from reason, but were evolved to support group identity, thus they can be expected to resist displacement. Instead a curious inversion of reasoning will often happen in that reason has been used to defend (confabulated) religious beliefs. This process is still prevalent (i.e., intelligent design). Thus, for a brief period in the seventeenth century it appeared that belief-based knowledge was 'believed' by many of the ruling class and promoted the development of some rational but secular political documents, such as the American Constitution. In current culture, however, many would seek to superimpose religious beliefs onto such political documents, such as 'Christian or Muslim law' claiming and rationalizing that such beliefs were the foundation of secular government. Many scientists still 'believe' in reason and consider that the ability to do causal reasoning is thought to be particular human strength. Such beliefs provide unending frustration for many scientists in their interactions with non-scientists.

The Individual Mind; 'Sense of Self' in Conflict with the Social Mind

We live in our brains, where our consciousness and sense of self reside. Consciousness can be considered as a collective and dynamic aspect of intellect involving thought, perception, emotion and memory. From a religious perspective, such a sense of self has been considered to be a God-given feature that cannot be studied or measured, a soul. Our sense of self, however, is a fragile feature of our mind that can be readily disrupted by clinical and pharmacological actions. Our concept of a 'person' requires the capacity for conscious experience, which needs an active frontal cortex. Science can indeed measure aspects of consciousness, such as cortical response to sensory perception. Recall the situation of Clive Weaver in that his consciousness was 'rebooted' every morning after awaking from sleep to new sensory perceptions. Our sense of self and person is also often associated with a sense of free will. In fact, it can be argued that society generally 'believes' in the existence of conscious free will. What then is free will with respect to consciousness? It seems these two issues can sometimes be separated. Consider that hypnotic techniques can induce a state of consciousness where a degree of volition, free will or sense of self becomes externally controlled and even sensory perceptions can be thus controlled. Meditation and psychoactive drugs can similarly suppress the sense of self and will. Nor does it appear that the sense of self or consciousness has a fixed biological residence. Sensory perceptions can be experimentally manipulated to create altered states of consciousness (such as out-of-body experiences) in which consciousness appears to reside in virtual bodies. Why would such a fluid and fragile state of self exist? Why is consciousness not firmly affixed to a specialized region of the brain, like face or language recognition? The sense of self can also be regarded as the perception of existence of an individual (asocial) mind; a mind that has independent will unassociated with mental states of others. One implication of this fragility is that sense of self is not a biologically ingrained feature. It is a dynamic state, the product of learning and language, hence a recent emergent mental phenotype from our social brain. If so, what is the relationship of sense of self to sense of other or social/external control? Do chimpanzees have a sense of self and self-awareness? Self-awareness requires mental (not sensory) visualization of the mind and would inherently appear to require an awareness of other minds, thus it may have emerged after the development of theory of mind (a social, empathic feature). This would infer that sense of self developed along with language following the development of a social brain. Until the recent introduction of fMRI and other neurological techniques, the study of the mind was mostly approached by amassing hundreds of case studies or anecdotal observations regarding mental states. This is a chaotic process which is also much like how education tends to be studied. With the application of CNS-based measurements, however, we start to see neurological correlates. Thus we realize that much of our recently evolved brain capacity seems to be dedicated to social functions, such as language or visual face/emotion recognition. The ability to rapidly communicate emotional information appears to be built into domains of our brain and can clearly be subconscious (rapid facial emotions, vocal fear). But the purpose of such emotion communication must be mainly social. Will is individual action from thought which requires ongoing communication to the frontal cortex and this must be where rational mind and personhood are generated. Yet the nature of action (will) is highly socially influenced. Clear examples of this include epidemic fear or anger which can control actions of individual in groups. Clearly, groups can respond in common to emotional signals and they inherently tend to perceive out-groups (both good and bad). This social will is a collective, group response that is biologically mediated and involves a social mind. Since this is biologically ingrained, it is also robust and not a recent development, as has the sense of self.

It seems clear that a sense of self and the reasoning abilities of an individual mind were crucial prerequisites for the development of deep and rational thinking. Conscious, formal thinking is an acquired, learned skill that is an essential aspect of scientific reasoning and the modern mind. Thus it seems both the sense of self and the capacity for conscious and independent scientific thinking emerged after the social brain had evolved. However, the emergence of individual-based critical analysis creates an inherent conflict between the individual mind and social minds. Such individual minds can confront the beliefs which define social minds (and their group identity) and thus can encounter group hostility.

Spontaneous Social Identity and the Unaffiliated Individual Minds

There is a biological tendency for humans to spontaneously classify other individual as a group member or non-members based on common beliefs. This tendency, however, also promote spontaneous group generation. We have noted spontaneous urban gang formation in young adult males of otherwise identical social, ethnic or cultural background (primate-like male group formation) that can sometimes engage in lethal intergroup conflict. However, in such cases there may be few beliefs (or other features) that distinguish such groups, aside from a belief in loyalty to group membership. Yet these groups still must share a social bond. An even more basic tendency to group membership assignment may differentiate between individual (independent) and socially bound minds. An individual that fails to establish social bonds, fails to believe in or adhere to group loyalty will likely be perceived as a non-group member. In this way, the modern individual mind, which has emerged with the acquisition of abstract language and reading, and become capable of much more self-reflection, can also create a mind that is perceived as a non-member. A mind with critical and scientific thinking skills can question social beliefs that are not consistent with objective observation or that conflict with formal reasoning. This will in turn weaken the hold that a social belief has in group identity. However, since humans have an inherent tendency to assign group membership to belief-based mental states as held by others, individual thinkers will also tend to be assigned into 'non-member' status and be perceived as belonging to another version of a belief-based group. How this tendency affects popular perception of the scientific mind is presented below. Yet, the situation as just outlined above is likely to be oversimplified. That is because the concept of an 'individual critical mind' appears to be more of an idealization than a reality. Most people will retain many elements of their underlying social mind which can often exert major (subconscious) influence on our reasoning.

Truly individual independent critical thinking is very difficult to dissociate from this underlying (often subconscious) social mind. Yet, as outlined above, there is some biological variation in the 'social-mind' phenotype of humans, such as autism. Indeed, scientists have some tendencies to think, as individuals may have some biological basis as well. Let us consider autism and various other mental disorders known to specifically affect social or asocial tendencies. Above, we discussed the apparent link of autism to mirror neurons and social empathy. However, some autistic individuals (savants) can have highly enhanced and specific cognitive skills, such as in the realm of math, memory or vocabulary. They can also have highly outstanding observational skills, accurately remembering details of scenes and landscapes well beyond that of normal individuals. Such savants can also show obsessions or compulsion toward creating order. Other autistics can also have an ability to focus their interest into narrow topics. The most basic tendency, however, is to be socially disconnected and not to perceive fear properly. It is thus most interesting that scientists, as a population, have many of these same cognitive and social tendencies. Scientists need to have an 'independent' individual mind that is not overly influenced by widely held views in order to explore and develop new domains of thinking. A mind that can resist the social consensus and stubbornly pursue narrow thinking and observation is well suited for science. Interestingly, scientists are often stereotyped to have a general reputation for being socially disconnected (socially labeled as nerds). Although I know of no population or scientific-based study on this inference, it could provide a most interesting topic of study. In this way, scientists would also tend to be perceived as non-members to many social groups.

However, scientists are not the only members of society that have developed critical thinking skills. Various other endeavors also come to depend on evidence as a primary criteria for establishing belief. However, scientists are probably an extreme example of this formal skill and mind set. Indeed critical thinking is a general objective of higher education and can be demonstrated in substantial sections of the population where education levels are high. Yet, it remains that regardless of occupation, the skill of critical thinking still emphasizes the abilities inherent in an individual mind needing a strong sense of self and it also remains that such a thinking style tends to place such people in conflict with 'belief-based' social reasoning. A spontaneous group assignment is thus expected to be applied to critical thinkers. They will appear to constitute a distinct 'system of belief', hence be assigned a common social membership. Thus critical thinkers can be considered as members of a social group that holds distinct 'critical thinking beliefs' (a clear oxymoron, but rhetorically effective). In this, they represent just another belief system. It is precisely such a general trend in perception that may underlie the prevalent concepts of 'conservative' and 'liberal' group identities. Although it seems clear that some degree of ideology can be applied to both sets, it is also clear that educated people that may otherwise hold no ideological positions or strong religious beliefs in common are also generally labeled as belonging to 'liberal' groups. For example, centers of higher education, regardless of the political foundations or histories, are almost always considered from such a perspective, and entire states have also been labeled liberal ('blue') states in the USA. The common denominator for these states appears to be the average levels of formal education that exists. There are some additional features, characteristic of 'group identity' labels, that appear to correlate with the conservative and liberal assignment. One measurable characteristic is relevant to issues of the sense of right and wrong social judgments. Psychological profiles can be used to distinguish overall group behavioral characteristics of conservative and liberal populations. Some of these profiles have evaluated the relative importance of harm, fairness, in-group membership, authority and group purity found within these populations. Interestingly, such evaluations have resulted in statistically clear results. Conservative populations tend toward counting purity, authority and in-group membership (including loyalty and obedience) as much more relevant than do liberal groups (which emphasize fairness and harm concepts). Moderates are intermediate in placing the importance of these characteristics. These characteristics are clearly relevant to social group membership. Conservatives display that overall features we would associate with more social minds, tending toward group membership and belief-based reasoning whereas liberal populations (the products of more education) show characteristics of individual critical minds, with weaker group affinities. These overall tendencies toward group membership also show some clearly associated tendencies in moral thinking and in right/wrong judgments. People with a strong sense of belief and group membership tend to judge non-membership harshly. Since such a social mind set seems more ancient, it was likely this tendency has resulted as some consequence of biological selection. In this light, the appropriate emotional communication in social settings can be very important (hence autism is under some negative selection). An autistic person would tend to communicate the wrong emotional valence regarding fear in a social setting. In early human groups this could easily have been a lethal phenotype. Consider, for example, a lone 'autistic' chimpanzee that mistakes anger for smiles in other male chimpanzees. This could easily mistakenly incite lethal attack and thus be a lethal social phenotype. Partially asocial human (scientist and autistic), with some highly specialized or developed thinking skills, would also not likely fare well in such early social group settings.

Conflicts involving belief-based group identities are still very much a part of current human culture. National and cultural identities continue to provide overall group reactions and retain the potential for promoting group conflict. To this day, such group (belief based) conflicts can escalate to social pathology and attempts at ethnic or religious cleansing have yet to be eliminated in various parts of the world. We as a species evolved with some inherent social tendencies to subconsciously recognize other groups (races, languages, cultures, beliefs) and react to them as out-groups, typically in a negative way. The sources of so many 'groupism' are thus ingrained into structures in our social brain, and these include cognitive or belief states. Unlike all other life on Earth, we have become less dependent on biologically (genetic) based identity (such as olfaction) and evolved to use the mind and what it has learned for the purpose of group identity. What our minds believe most define our group membership. Still, these group identities retain the same strategies (addiction modules) as were used to define essentially all ancestral group identities in all other life. Given the biological basis of this situation, it might seem depressing to contemplate the prospect of preventing human group conflicts. Yet, our social mind must learn and it is clear that education which does not promote virulent group identity can, to a large degree, offset these inherent group tendencies by developing a social mind that is controlled by morally sound individual and critical reasoning. Education can promote fair and ethical group behavior. We are endowed with an inherent tendency toward empathy which can further be developed to underlie much good behavior. We must exploit this tendency and inhibit another feature of the social mind, the development of virulent anti-group responses (denial of group empathy which can also be learned). Appropriate learning and education becomes the main key to offset human conflicts. This thus deserves the most serious cultural investment.

A Science Mind Is Not a Group Belief: the Difficulty of Recognizing and Unlearning Belief

A significant problem with beliefs is seen in how they affect science instruction and learning. Beliefs generally elicit subconscious cognitive defense (rationalizations), thus they can promote an often insidious capacity to resist learning and new knowledge. Individuals will often not realize they are defending their beliefs unless or until objective evidence is specifically consulted and formally evaluated relevant to a topic of new knowledge. Most people (even scientists) normally hold subconscious beliefs on various topics which are unapparent. For example, I have been involved with and experimenting with science education for several decades. Many of my colleagues similarly involved in science education have often expressed unquestioned consensus views regarding what is important for learning science. Clarity (simplicity) of material presented and developing motivation of the student to be interested in the topic are generally accepted as major teaching objectives that must be addressed. These views seem intuitively obvious. Yet when I ask for any relevant data on these specific topics, specifically what evidence supports these views, I will usually hear a rationalized answer as to why this should be the case, with no reference to specific evidence; not even anecdotal evidence or extended personal experience is usually offered. When they do consult their own personal experience, they will often realize their own development did not occur this way. These scientists readily become unscientific when defending such beliefs but will usually see the problem when I point this out to them. We are all prone to similar insidious and unapparent belief states and their defense. In the context of education, this is even a greater problem. Students believe in many things, especially regarding what constitutes a good educational experience, but almost never know of any relevant evidence that supports their beliefs. In spite of 30 years of experimenting on student education, and observing how students learn to think critically, it is not at all uncommon for a new science student to unhesitatingly dismiss my experience and defend their own views regarding education. In every instance I have experienced this, students have rationalized their views, with no acknowledgement to any objective evidence for any of their assertions. They are defending beliefs which are considered equal to all others (including mine) and they do not even realize that they hold beliefs on this. To formally train a scientist we must undo such mental states and the training generally begins in earnest as a graduate student. The student must learn to 'think' like a graduate student, that is with some independence and knowledge of relevant evidence and theory. They must learn to question their own rationalizations. But this strikes me as a highly delayed and wasteful process. After 16 years of formal education, why is this process just starting with college students? Why will it normally take over 20 years of formal education to train a scientist? Contrast this with the rapidity of learning to speak a language (also a very demanding but natural learning process). My conclusion is that it is basically 'unnatural' to reason by this objective and individual-based way. Such reasoning depends much less on belief-based reasoning and one must learn to inhibit (suspend judgment or belief) one's own associated instinct to defend beliefs with rationalizations. Thus this formal training of a critical 'unemotional' mind, in a real sense, goes against the evolutionary foundations of our social–emotional mind. We must dislink our thinking from our charged, emotional subconscious, reactions and not be overly influenced by group consensus and beliefs. This while keeping in mind that this subconscious-emotional system is capable of amazing pattern recognition, which can provide valuable insight. The mental discipline required for this takes years to develop and is probably never complete, even in scientists.

Learning to write was difficult relative to learning to talk. But learning a scientific and critical mind is much harder and less natural. An independent mind is not supposed to be influenced by what others think and should be free of social influences and dependent on evidence and logical support. This is much more difficult to do than is generally appreciated and even the most disciplined scientist will often struggle as described below. Since we all learn with our large primate social brain, we have many ingrained social learning tendencies that set beliefs and inherently limit how and what we learn. We very much tend to believe consensus, for example. Learning is not an inherently open state. If anything, I would suggest that there is an inherent resistance to learning. We must always be vigilant to guard against defending unconsciously learned beliefs based on rationalizations instead of evidence and logic. However, such defensive reactions are not too difficult to recognize when we chose to do so. Rationalizations and group defense will often use rhetorical methods, charged terms, negative group identification and name calling or invoke some convoluted logic for which direct evidence is generally absent.

Thus we come to understand why a consensus strongly affects individual learning. This tendency is from the foundations of our social brain. We want (perhaps need) to believe as others do. Beliefs are not simply a cultural attempt to explain the mysteries of the world, but they are a biological legacy of our social mind and cognitive group identity that elicits a cognitive immune reaction (defense). That all humans hold beliefs, regardless of educational status or knowledge of supporting evidence, should be acknowledged as a biological characteristic of our species. And since beliefs underlie identity, we will inherently defend them as valid, sometimes even in the face of convincing evidence to the contrary, similar to that seen in some stroke victims. We have come to accept the political reality that indeed beliefs define group identity and recognize in most societies that freedom of belief does limit group strife. Strong and even violent defense can be provoked if belief is questioned, regardless of the relevant evidence. However, it is one thing to realize this characteristic of belief and avoid confronting them in order not to inflame a violent defense, but it is quite another thing to accept such views as inherently equal especially those that stem from critical thinking (objective observation and logical consistency). It is because of such a tendency to treat all beliefs as equal that science finds itself being considered as merely another belief system and as such can also be considered as another (equal) group identity system. Thus, most religious people will react to scientific reasoning accordingly. It is ungodly, therefore an amoral and defines an out-group with bad characteristics. Even the 'most' developed and 'Western' countries (which ironically includes Japan) have been surveyed regarding the social significance of holding beliefs and rank 'atheist/scientist' as less desirable than any other social, sexual or ethnic grouping. It therefore seems clear that the *absence* of belief is used by most cultural groups as a way to assign a negative and false group identity to science (hence the use of the term 'Scienceism' by some religious communities).

Science is not a belief system or an identity system, nor is it a Western culture. Scientists do not adhere to any particular belief. And as I have suggested, science should drop the use of the term 'belief' to describe the acceptance of valid scientific information to avoid this confusion. Data and reasoning can 'convince' a scientist, but 'beliefs' should apply only to cultural membership. Science has developed slowly during history, often in strong opposition to prevailing local culture, sometimes with lethal outcome to it proponents. Unlike other writing forms, formal science writing was also long and slow to evolve and it took most of 2000 years to develop it into current formal (and emotionally dry) format (much to the dismay and criticism of the humanities). Science was not the product of an easy or natural birth, and developing a scientific mind remains a long and arduous process.

Beliefs of Scientists that Oppose Science

Since scientist themselves are the product of evolution, they too are prone to develop social belief states which are easily confused with scientific thinking.

Thus science like all other systems of human thought tends to operate via a group social process prone to consensus and beliefs. From this tendency, we can understand why new scientific ideas are often strongly opposed. Opposition to new scientific ideas is considered by many as a normal, positive and rational feature of science. However, such opposition can even occur in the presence of strong logical argument or supporting initial evidence. Such opposition, when it co-occurs, has the characteristics of a rationalized defense of belief. In some cases, it indeed seems clear that belief defense is being invoked and scientist can clearly become invested in consensus ideas and will defend them, even with emotional and rhetorical methods, including name calling. Thus essentially every major paradigm shift in scientific thinking has encountered such a group reaction. For example after the British astronomical team confirmed Einstein's prediction of light travel with respect to the Sun concerning his theory of relativity, a group of rightist physicist protested the positive public attention given to the theory of relativity and labeled it as the product of a liberal, Jewish and pacifist author. Many German physicists at the time resisted the theory on these group identity grounds. In the history of science, such reactions are not uncommon. Essentially all fields of science have similar stories. We previously mentioned some reactions to Darwin's theory of human evolution with respect to the evolution of human language (labeled the 'bow-wow' theory). Relevant to the thesis of this book (genetic parasites and viruses in symbiosis), however, there is also a relevant story of belief. One early report in the study of phage (bacterial viruses) concluded that they were part of or symbiotic with their cellular host (Bail and Der Colistamm). This report, however, was met with strong opposition. In the 1920s, D'Herelle and many others sharply disagreed with such a concept (that bacteria themselves had produced a lytic agent). He and others refused to 'believe' that phage could live symbiotically with their host, as their own experiments with lytic phage did not support such a view. This belief was also strongly held by J. Bordet. (The Theories of Bacteriophage. Proc. Roy. Soc. London B83, 398 (1931)). These researchers had come to believe that phage were always lytic of host. In this, Bordet and others attained a state belief which resisted displacement. Accordingly, they did not attempt to directly repeat the Bail and Der Colistamm results. They simply defended their belief based on their own experiments. Later, other prominent scientists, such as Max Delbrück and colleagues, similarly came to think that phage were always lytic based on their own experiments that also did not seek to reproduce the other results. They had observed that T-even phage always lysed with their host in a series of well-controlled experiments. Later, some famous scientists (A. Lwoff) even dispensed with any reference to observation and simply defended their beliefs by definition, saying things like 'a virus is a virus'. Thus according to its name, a phage as a cellular predator (eat) lyse cells and that infection necessarily results in replication which results in lysis. Thus we see that observations that did not conform to this view were simply rejected. This belief in strictly lytic virus remains widely held today, including most evolutionary biologists. Explaining these inconsistent results became a rationalization.

A prophage, if it indeed existed, was rationalized as simply being a seed, not virus, able to make lytic virus in the correct (germinating) circumstance. Yet the original observations that some bacteria can harbor a stable capacity for phage production remained valid, and is valid to this day as an observation that can easily be performed on the appropriate bacterial strains. And even more relevant, some prophage, like lambda, alter group identity making *E. coli* immune to T4,.for example. Thus even these very early observations showed that a virus was capable of providing group identity. Indeed, both sets of observations (lysis, genomic) could be reproduced. This situation raises a curious but important issue that has regularly occurred in science. Evidence that does not adhere to an accepted belief at the time it is presented is often rejected, usually up front by argument and usually without an attempt to evaluate the specific claim or observation. Such a response clearly adheres to that of a belief defense, but ironically, such beliefs were themselves the product of experimental scientific observation. In this we see a truly insidious characteristic of our social-based belief mechanism in that they tend to apply even to scientific conclusions. Scientists can also rapidly establish states of 'cognitive immunity' and reject clear evidence. This would seem illogical for scientists to be swayed by beliefs, given the objective character of the scientific thinking process and its fundamental dependence on observation. In this regard a belief held by scientists can at times be as vigorously defended by argument as any other belief of non-scientists. However, in the long run, experimental analysis will generally prevail to convince most scientists of the validity of an observation or theory. In this case, those early symbiotic phage experiments were not fully accepted until the late 1950s when molecular evidence was finally overwhelming. Both the genetic symbiotic virus (like lambda) and the strictly lytic virus (like T4) exist in *E. coli* and most other prokaryotes.

Another example that relates to the strong influence of consensus beliefs in science is the theory of symbiosis in the evolution of eukaryotes. In the 1960s Lyn Margulis submitted a paper for scientific publication in which she proposed the symbiotic origin of mitochondria from a bacterial cell. This paper was rejected about 15 times before being published in 'Theoretical Biology'. And the initial response of the scientific community was to heavily reject the concept, which seemed to defy the current neo-Darwinian thinking. However, the observations and scientific literature on which this paper was based did not change during this attempted publication or subsequent years. The symbiogenesis theory of eukaryotes is now orthodoxy and molecular genetic data now clearly support this theory. What changed mostly, I suspect, was the social beliefs of the community involved. This example identifies an inherent cognitive tendency in humans, including scientists, that favors the defense of beliefs, especially when they attain a social consensus. There is strong evidence that even scientists are often swayed by beliefs, as are all humans. A current consensus (and belief) in evolutionary biology is that genetic parasites are junk and viruses are not highly relevant to, and have little significant consequence to the evolution of, host complexity. Viruses are not considered as symbiotic with their host. This book has assembled a large number of observations and presents many logical arguments that counter such a view. Viruses and their hyperparasites indeed matter greatly to all life on Earth. However, I can expect that such evidence will most probably be rejected or even more likely ignored (since relevant observations are experimentally strong). Thus the views and ideas presented here are not likely to quickly displace those widely held views that have attained a belief status. Viruses and other genetic parasites have helped us define the basic mechanistic character of an addiction module and how genetic complexity can evolve from their accumulated action. In doing so, the concept of addiction modules has helped us to understand and define group identity (immunity) and also how it has evolved. And the concept of group identity has helped us to understand the origins of the large social brain and mind of primates and how olfaction-based group identity came to be displaced. It is from this foundation that we can now understand how the human social mind emerged and how learning and beliefs came to define our group identity. Our human group identity has come to be mostly learned and cognitive.

Beliefs and group identity operate from biological origins but are currently subjected to commercial and political manipulation, often not for purposes that benefit society. Such manipulation has been used with good effectiveness to counter rational and evidence-based reasoning. We can, for example, label global warming as a belief (a 'religion') and thus dismiss all the relevant scientific information that relates to this issue, as has been done by politicians and politically motivated news announcers. The public does not protest such manipulation since it is mostly prone to belief-based reasoning. Unfortunately, such manipulation provokes a dialogue that is based on group beliefs, not rational argument or evidence. And in such a setting, group beliefs often prevail in importance. Thus we witnessed on national television that a Republican Senator, who as a presidential candidate in the USA, openly rejected evolutionary theory as theology posing as science. He feels secure in making such statements because belief of the audience is on 'his side' and to them science is simply a minority belief system that can be dismissed as an out-group. Such a circumstance identifies a major flaw of the 'natural' social mind of humans and helps define a major mission of science, to displace belief-based reasoning via education in critical thinking.

The Mission of Science Education

The biology of our large social brain suggests that every newborn human that is not educated in critical thinking (or does not have an atypical social brain function) will most likely default and learn by associative and belief-based reasoning process. This will provide them with a social mind that promotes various forms of group membership, especially those that are determined by belief status. As science educators, this is the typical mental ground state from which we must work. Our objective is to promote the development of an individual mind that dissociates beliefs and inherently employs objective observation, evidence and logic to attain a convinced status for new information. However, we must also seek to promote empathy as the fundamental emotional link that provides sound moral judgment. Empathy must not readily be denied to our group membership. Since empathy also operates from a biological basis, it should be possible to develop educational paradigms that maintain this capacity. However, similar to scientists, science educators are also prone to hold various beliefs concerning education that may not be well supported by evidence. In addition, other professionals that work with people and transmit new information, such as physicians, will also be prone to a belief bias in their reasoning. Both physicians and scientists have an inherent tendency to anchor their thought process to initial associations or beliefs. For example, consider the point made above, in that we tend to adhere to the belief that if science is presented clearly, with strong supporting evidence, it will be understood and believed by students. But evidence for this 'belief' in science education is not compelling. Science educators in particular are often puzzled by the resistance that certain individuals can hold against science. Recently, it has been reported that such resistant attitudes are acquired as young children that can persist into adulthood. For example, a Pew Trust pole (2005) indicated that 42% of respondents said they believed that humans and other animals existed in their present form since the beginning of time, dismissing all scientific evidence that supports evolution. Other non-scientific beliefs are also prevalent. For example, belief in the efficacy of unproven medical treatments is very common, as is belief in ESP, astrology and other unsupported views. Much of the general population holds such unscientific beliefs. Given the prevalence and long history of most of these types of beliefs and that their prevalence has not been much diminished by public education, it needs to be acknowledged that science education is working poorly for large parts of the population. Why are they resistant? The answer by now should be clear, resistance to new beliefs is a basal state inherent to the human social mind. During the early development of these people, they learned (were colonized by) some version of a native belief and belief-based reasoning systems that left them resistant to formal reasoning. Thus it has been proposed that some resistance to scientific ideas is a human universal that stems from early childhood experiences which result mainly from associated learning mechanisms that are not declarative (conscious or formal). During such development, trustworthiness of source seems to matter a great. Recall that the trustworthiness of an authority figure can likely be mediated by oxytocin, and that this likely defines a critical period of group identity development and social bonding to group leaders that can be molded by education.

Most science educators essentially 'disbelieve' that their student body resists science education or holds many unscientific beliefs, in spite of clear and abundant evidence to the contrary. Clearly, science education themselves could benefit by adopting a more scientific approach. Our goal is already clear, education must provide the cognitive foundations to override the negative tendencies of belief-based reasoning. Also, our tendency to develop group membership, associate out-groups with negative stereotypes and a tendency to rationalization or confabulate explanations must all be offset by education. The historic purpose of education was to develop a social mind, which normally stemmed from religious education. But such education is really a version of social indoctrination and tends to develop belief-based reasoning, not individual critical thinking. We should recognize that such belief-based reasoning is inherently flawed and susceptible to being manipulated by social pressure and that such manipulation, if malevolent, can also result in virulent social identities. Science can inadvertently promote such reasoning when it avoids the examination of the mechanisms at work during religious and other social experiences. Also, science education has come to focus too much on passive lectures, in which students are asked to 'believe' in the authority of their teachers. This too promotes belief-based reasoning. Belief-based reasoning needs to be examined at all levels, from biological to social. For example, when people believe they act for the will of God, king or cult, political or gang leaders, they become capable of inhumane loss of empathy toward other groups. Our primate relatives retain both social empathy and its loss regarding social out-groups. Rational, ethical and empathic thinking can override such negative tendencies, but this is not easily learned. We need to now focus the full analytical power of science to better understand how to attain this rational–ethical learned state and how it relates to the biology of our large social brain. We also need to acknowledge that belief-based reasoning is not simply a flawed thinking process, but it is at core of our social mind that provides a social function, promotes the creation of cooperative communities and provides cognitive (mystical) satisfaction. Such deep satisfaction does not stem from a critical mind. The community of science, although clearly an international community of a similar mind, does not provide a cognitive addiction module nor does it provide deep emotional bonds. We do not love, kill and die for our respective scientific societies, even ones that are very fun and satisfying. Thus critical reasoning does not seem to engage these deeper emotional social bonds, although they do seem to be pharmacologically accessible. This could imply that a role of science education in transforming human social identity may be a highly unrealistic objective. After all, our group identities are and always have been fiercely defended thus it may be wholly unrealistic to think we can educate the population away from such biological tendencies. But we cannot predict the successes of science so this should not inhibit our search. Perhaps studies of empathy-inducing psychoactive agents, such as ketamine, could be very informative (if only they were not outlawed). It took 2000 years to develop modern science to its current state, so in another 2000 years we cannot yet imagine what insights might be attained and applied to advancing human development. We must initiate that path.

This book started by examining the possible and symbiotic role of viruses and related genetic parasites in the evolution of cooperation and host group identity. In tracing this evolution, we have ultimately been led to the biological origin of human group identity. Symbiosis has been a consistent theme of this book. It means living together, and cooperative living together is basic characteristic of humans and their social brains. Our large social brains may have been initially promoted to evolve by the action of symbiotic viruses, but successive colonization of this brain by other forms of information, language, then ideas has led to the creation of the social mind from which emerged the modern mind that transcends genetic identity. The mind, with its ideas and beliefs, is the new substrate for the continuing evolution of our group identity. And here, Science has a large and promising role to fill.

Recommended Reading

Genomes and HERVs

1. Barbulescu, M., Turner, G., Seaman, M. I., Deinard, A. S., Kidd, K. K., and Lenz, J. (1999). Many human endogenous retrovirus K (HERV-K) proviruses are unique to humans. *Curr Biol* **9**(16), 861–868.
2. Barbulescu, M., Turner, G., Su, M., Kim, R., Jensen-Seaman, M. I., Deinard, A. S., Kidd, K. K., and Lenz, J. (2001). A HERV-K provirus in chimpanzees, bonobos and gorillas, but not humans. *Curr Biol* **11**(10), 779–783.
3. Bromham, L. (2002). The human zoo: Endogenous retroviruses in the human genome. *Trends Ecol Evol* **17**(2), 91–97.
4. Flockerzi, A., Burkhardt, S., Schempp, W., Meese, E., and Mayer, J. (2005). Human endogenous retrovirus HERV-K14 families: status, variants, evolution, and mobilization of other cellular sequences. *J Virol* **79**(5), 2941–2949.
5. Fukami-Kobayashi, K., Shiina, T., Anzai, T., Sano, K., Yamazaki, M., Inoko, H., and Tateno, Y. (2005). Genomic evolution of MHC class I region in primates. *Proc Natl Acad Sci USA* **102**(26), 9230–9234.
6. Jern, P., Sperber, G. O., and Blomberg, J. (2005). Use of endogenous retroviral sequences (ERVs) and structural markers for retroviral phylogenetic inference and taxonomy. *Retrovirology* **2**, 50.
7. Kim, H.-S., Wadekar Rekha, V., Takenaka, O., Winstanley, C., Mitsunaga, F., Kageyama, T., Hyun, B.-H., and Crow Timothy, J. (1999a). SINE-R C2 (a *Homo sapiens* specific retroposon) is homologous to cDNA from postmortem brain in schizophrenia and to two loci in the Xq21 3/Yp block linked to handedness and psychosis. *Am J Med Genet* **88**(5), 560–566.
8. Kim, H. S., Takenaka, O., and Crow, T. J. (1999). Isolation and phylogeny of endogenous retrovirus sequences belonging to the HERV-W family in primates. *J Gen Virol* **80(Pt 10)**, 2613–2619.
9. Kim, H. S., Wadekar, R. V., Takenaka, O., Hyun, B. H., and Crow, T. J. (1999b). Phylogenetic analysis of a retroposon family in African great apes. *J Mol Evol* **49**(5), 699–702.
10. Lavie, L., Medstrand, P., Schempp, W., Meese, E., and Mayer, J. (2004). Human endogenous retrovirus family HERV-K(HML-5): status, evolution, and reconstruction of an ancient betaretrovirus in the human genome. *J Virol* **78**(16), 8788–8798.
11. Sverdlov, E. D. (2000). Retroviruses and primate evolution. *Bioessays* **22**(2), 161–171.

Primate Face, Gesture Recognition and Mirror Neurons

1. Aziz-Zadeh, L., Koski, L., Zaidel, E., Mazziotta, J., and Iacoboni, M. (2006). Lateralization of the human mirror neuron system. *J Neurosci* **26**(11), 2964–2970.
2. Cooper, D. L. (2006). Broca's arrow: evolution, prediction, and language in the brain. *Anat Rec B New Anat* **289**(1), 9–24.
3. Fischer, H., Sandblom, J., Gavazzeni, J., Fransson, P., Wright, C. I., and Backman, L. (2005). Age-differential patterns of brain activation during perception of angry faces. *Neurosci Lett* **386**(2), 99–104.
4. Gliga, T., and Dehaene-Lambertz, G. (2005). Structural encoding of body and face in human infants and adults. *J Cogn Neurosci* **17**(8), 1328–1340.
5. Gridley, M. C., and Hoff, R. (2006). Do mirror neurons explain misattribution of emotions in music? *Percept Mot Skills* **102**(2), 600–602.
6. Holt, D. J., Kunkel, L., Weiss, A. P., Goff, D. C., Wright, C. I., Shin, L. M., Rauch, S. L., Hootnick, J., and Heckers, S. (2006). Increased medial temporal lobe activation during the passive viewing of emotional and neutral facial expressions in schizophrenia. *Schizophr Res* **82**(2–3), 153–162.
7. Iacoboni, M., and Dapretto, M. (2006). The mirror neuron system and the consequences of its dysfunction. *Nat Rev Neurosci* **7**(12), 942–951.
8. Iriki, A. (2006). The neural origins and implications of imitation, mirror neurons and tool use. *Curr Opin Neurobiol* **16**(6), 660–667.
9. Itakura, S. (1994). Recognition of line-drawing representations by a chimpanzee (Pan troglodytes). *J Gen Psychol* **121**(3), 189–197.
10. Lametti, D. R., and Mattar, A. A. (2006). Mirror neurons and the lateralization of human language. *J Neurosci* **26**(25), 6666–6667.
11. Lyons, D. E., Santos, L. R., and Keil, F. C. (2006). Reflections of other minds: how primate social cognition can inform the function of mirror neurons. *Curr Opin Neurobiol* **16**(2), 230–234.
12. Moody, E. J., McIntosh, D. N., Mann, L. J., and Weisser, K. R. (2007). More than mere mimicry? The influence of emotion on rapid facial reactions to faces. *Emotion* **7**(2), 447–457.
13. Morris, R. D., and Hopkins, W. D. (1993). Perception of human chimeric faces by chimpanzees: evidence for a right hemisphere advantage. *Brain Cogn* **21**(1), 111–122.
14. Parr, L. A. (2003). The discrimination of faces and their emotional content by chimpanzees (Pan troglodytes). *Ann NY Acad Sci* **1000**, 56–78.
15. Parr, L. A., Waller, B. M., and Fugate, J. (2005). Emotional communication in primates: implications for neurobiology. *Curr Opin Neurobiol* **15**(6), 716–720.
16. Rossi, E. L., and Rossi, K. L. (2006). The neuroscience of observing consciousness & mirror neurons in therapeutic hypnosis. *Am J Clin Hypn* **48**(4), 263–278.
17. Schulte-Ruther, M., Markowitsch, H. J., Fink, G. R., and Piefke, M. (2007). Mirror neuron and theory of mind mechanisms involved in face-to-face interactions: a functional magnetic resonance imaging approach to empathy. *J Cogn Neurosci* **19**(8), 1354–1372.
18. Shmuelof, L., and Zohary, E. (2007). Watching others' actions: mirror representations in the parietal cortex. *Neuroscientist* **13**(6), 667–672.
19. Tsukiura, T., Namiki, M., Fujii, T., and Iijima, T. (2003). Time-dependent neural activations related to recognition of people's names in emotional and neutral face-name associative learning: an fMRI study. *Neuroimage* **20**(2), 784–794.

Autism and Schizophrenia and Relationship to Language

1. Burns, J. (2006). The social brain hypothesis of schizophrenia. *World Psychiatry* **5**(2), 77–81.

2. Berlim, M. T., Mattevi, B. S., Belmonte-de-Abreu, P., and Crow, T. J. (2003). The etiology of schizophrenia and the origin of language: overview of a theory. *Compr Psychiatry* **44**(1), 7–14.
3. Burling, R. (2005). "The talking ape: how language evolved." Studies in the evolution of language Oxford University Press, Oxford, New York.
4. Crow Timothy, J. (1993). Origins of psychosis and the evolution of human language and communication. *International Academy for Biomedical & Drug Research. Brunello, N., Mendlewicz, J., Racagni, G., (Eds.). International Academy for Biomedical and Drug Research; New generation of antipsychotic drugs: Novel mechanisms of action* **4**, 39–61.
5. Crow, T. J. (1997). Aetiology of schizophrenia: an echo of the speciation event. *Int Rev Psychiatry* **9**(4), 321–330.
6. Crow, T. J. (1999). Commentary on Annett, Yeo et al., Klar, Saugstad and Orr: cerebral asymmetry, language and psychosis–the case for a *Homo sapiens*-specific sex-linked gene for brain growth. *Schizophr Res* **39**(3), 219–231.
7. Crow, T. J. (2000). Schizophrenia as the price that *Homo sapiens* pays for language: a resolution of the central paradox in the origin of the species. *Brain Res Rev* **31**(2–3), 118–129.
8. Dapretto, M., Davies, M. S., Pfeifer, J. H., Scott, A. A., Sigman, M., Bookheimer, S. Y., and Iacoboni, M. (2006). Understanding emotions in others: mirror neuron dysfunction in children with autism spectrum disorders. *Nat Neurosci* **9**(1), 28–30.
9. Fecteau, S., Lepage, J. F., and Theoret, H. (2006). Autism spectrum disorder: seeing is not understanding. *Curr Biol* **16**(4), R131–R133.
10. Hadjikhani, N., Joseph, R. M., Snyder, J., and Tager-Flusberg, H. (2006). Anatomical differences in the mirror neuron system and social cognition network in autism. *Cereb Cortex* **16**(9), 1276–1282.
11. Highley, J. R., McDonald, B., Walker, M. A., Esiri, M. M., and Crow, T. J. (1999). Schizophrenia and temporal lobe asymmetry. A post-mortem stereological study of tissue volume. *Br J Psychiatry* **175**, 127–134.
12. Johnston, P. J., Stojanov, W., Devir, H., and Schall, U. (2005). Functional MRI of facial emotion recognition deficits in schizophrenia and their electrophysiological correlates. *Eur J Neurosci* **22**(5), 1221–1232.
13. Martelli, M., Majaj, N. J., and Pelli, D. G. (2005). Are faces processed like words? A diagnostic test for recognition by parts. *J Vis* **5**(1), 58–70.
14. Parr, L. A., Waller, B. M., and Fugate, J. (2005). Emotional communication in primates: implications for neurobiology. *Curr Opin Neurobiol* **15**(6), 716–720.
15. Quintana, J., Wong, T., Ortiz-Portillo, E., Marder, S. R., and Mazziotta, J. C. (2003). Right lateral fusiform gyrus dysfunction during facial information processing in schizophrenia. *Biol Psychiatry* **53**(12), 1099–1112.

Social Learning; Music and Language

1. Burling, R. (2005). "The talking ape: how language evolved." Studies in the evolution of language Oxford University Press, Oxford, New York.
2. Coch, D., Fischer, K. W., and Dawson, G. (2007). "Human behavior, learning, and the developing brain. Typical development." Guilford Press, New York.
3. Collins, J. W. (2007). The neuroscience of learning. *J Neurosci Nurs* **39**(5), 305–310.
4. Keysers, C., and Gazzola, V. (2006). Towards a unifying neural theory of social cognition. *Prog Brain Res* **156**, 379–401.
5. Levitin, D. J. (2006). "This is your brain on music: the science of a human obsession." Dutton, New York.
6. Mithen, S. J. (2006). "The singing Neanderthals: the origins of music, language, mind, and body." Harvard University Press, Cambridge, Mass.

7. Pedersen, C. A. (2004). Biological aspects of social bonding and the roots of human violence. *Ann NY Acad Sci* **1036**, 106–127.
8. Sacks, O. W. (2007). "Musicophilia: tales of music and the brain." 1st ed. Alfred A. Knopf, New York.
9. Sherwood, C. C., Broadfield, D. C., Holloway, R. L., Gannon, P. J., and Hof, P. R. (2003). Variability of Broca's area homologue in African great apes: implications for language evolution. *Anat Rec A Discov Mol Cell Evol Biol* **271**(2), 276–285.
10. Tham, W. W., Rickard Liow, S. J., Rajapakse, J. C., Choong Leong, T., Ng, S. E., Lim, W. E., and Ho, L. G. (2005). Phonological processing in Chinese-English bilingual biscriptals: an fMRI study. *Neuroimage* **28**(3), 579–587.
11. Wang, W. S. Y. (1991). "The Emergence of language: development and evolution: readings from Scientific American magazine." W.H. Freeman, New York.
12. Xue, G., Chen, C., Jin, Z., and Dong, Q. (2006). Language experience shapes fusiform activation when processing a logographic artificial language: an fMRI training study. *Neuroimage* **31**(3), 1315–1326.

Belief Acquisition

1. Goel, V., and Dolan, R. J. (2003). Explaining modulation of reasoning by belief. *Cognition* **87**(1), B11–B22.
2. Grezes, J., Frith, C. D., and Passingham, R. E. (2004). Inferring false beliefs from the actions of oneself and others: an fMRI study. *Neuroimage* **21**(2), 744–750.
3. Trimble, M. R. (2007). "The soul in the brain: the cerebral basis of language, art, and belief." Johns Hopkins University Press, Baltimore.

Neuroscience of Emotional States

1. Aron, A., Fisher, H., Mashek, D. J., Strong, G., Li, H., and Brown, L. L. (2005). Reward, motivation, and emotion systems associated with early-stage intense romantic love. *J Neurophysiol* **94**(1), 327–337.
2. Bartels, A., and Zeki, S. (2000). The neural basis of romantic love. *Neuroreport* **11**(17), 3829–3834.
3. Bartels, A., and Zeki, S. (2004). The neural correlates of maternal and romantic love. *Neuroimage* **21**(3), 1155–1166.
4. Delahunty, K. M., McKay, D. W., Noseworthy, D. E., and Storey, A. E. (2007). Prolactin responses to infant cues in men and women: effects of parental experience and recent infant contact. *Horm Behav* **51**(2), 213–220.
5. Eisenstein, M. (2004). Is it love...or addiction? *Lab Anim (NY)* **33**(3), 10–11.
6. Fleming, A. S., Corter, C., Stallings, J., and Steiner, M. (2002). Testosterone and prolactin are associated with emotional responses to infant cries in new fathers. *Horm Behav* **42**(4), 399–413.
7. Fisher, H., Aron, A., and Brown, L. L. (2005). Romantic love: an fMRI study of a neural mechanism for mate choice. *J Comp Neurol* **493**(1), 58–62.
8. Gazzola, V., Aziz-Zadeh, L., and Keysers, C. (2006). Empathy and the somatotopic auditory mirror system in humans. *Curr Biol* **16**(18), 1824–1829.
9. Hagstrom, C. (2001). "The passionate ape: bad sex, strong love, and human evolution." RiverForest Press, Bainbridge Island, Washington.
10. Insel, T. R., and Young, L. J. (2001). The neurobiology of attachment. *Nat Rev Neurosci* **2**(2), 129–136.

11. Koob, G. F., and Le Moal, M. (2006). "Neurobiology of addiction." Elsevier/Academic Press, Amsterdam, Boston.
12. Najib, A., Lorberbaum, J. P., Kose, S., Bohning, D. E., and George, M. S. (2004). Regional brain activity in women grieving a romantic relationship breakup. *Am J Psychiatry* **161**(12), 2245–2256.

Other Broad Overviews

1. Berlinski, D. (2008). "The Devil's delusion: atheism and its scientific pretensions." 1st ed. Crown Forum, New York.
2. Byrne, R. W. (1995). "The thinking ape: evolutionary origins of intelligence." Oxford University Press, Oxford, New York.
3. de Waal, F. B. M. (2005). "Our inner ape: a leading primatologist explains why we are who we are." Riverhead Books, New York.
4. Feist, G. J. (2006). "The psychology of science and the origins of the scientific mind." Yale University Press, New Haven.
5. Gladwell, M. (2005). "Blink: the power of thinking without thinking." 1st ed. Little, Brown and Co., New York.
6. Hawkins, J., and Blakeslee, S. (2004). "On intelligence." 1st ed. Times Books, New York.
7. Margulis, L., and Punset, E. (2007). "Mind, life, and universe: conversations with great scientists of our time." Chelsea Green Pub., White River Junction, Vt.
8. Ramachandran, V. S. (2004). "A brief tour of human consciousness: from imposter poodles to purple numbers." Pi Press, New York.
9. Russon, A. E., and Begun, D. R. (2004). "The evolution of thought: evolutionary origins of great ape intelligence." Cambridge University Press, Cambridge, New York, UK.
10. Sussman, R. W., and Chapman, A. R. (2004). "The origins and nature of sociality." Aldine de Gruyter, New York.

Index

Printed in the United States of America